Grzimek's ANIMAL LIFE ENCYCLOPEDIA

Volume 1

LOWER ANIMALS

Volume 2

INSECTS

Volume 3

MOLLUSKS AND ECHINODERMS

Volume 4

FISHES I

Volume 5

FISHES II AND AMPHIBIA

Volume 6

REPTILES

Volume 7

BIRDS I

Volume 8

BIRDS II

Volume 9

BIRDS III

Volume 10

MAMMALS I

Volume 11

MAMMALS II

Volume 12

MAMMALS III

Volume 13

MAMMALS IV

Grzimek's
ANIMAL LIFE ENCYCLOPEDIA

Editor-in-Chief

Dr. Dr. h.c. Bernhard Grzimek

Professor, Justus Liebig University of Giessen
Director, Frankfurt Zoological Garden, Germany
Trustee, Tanzania National Parks, Tanzania

VAN NOSTRAND REINHOLD COMPANY

New York Cincinnati Toronto London Melbourne

First published in paperback in 1984

Library of Congress Catalog Card Number 79-183178

ISBN 0-442-23045-1

Printed in Federal Republic of Germany

Van Nostrand Reinhold Company Inc.
135 West 50th Street
New York, New York 10020

Van Nostrand Reinhold Company Limited
Molly Millars Lane
Wokingham, Berkshire RG11 2PY, England

Van Nostrand Reinhold
480 Latrobe Street
Melbourne, Victoria 3000, Australia

Macmillan of Canada
Division of Gage Publishing Limited
164 Commander Boulevard
Agincourt, Ontario M1S 3C7 Canada

16 15 14 13 12 11 10 9 8 7 6 5 4 3 2 1

EDITORS AND CONTRIBUTORS

Editor-in-Chief
DR. DR. H.C. BERNHARD GRZIMEK
Professor, Justus Liebig University of Giessen, Germany
Director, Frankfurt Zoological Garden, Germany
Trustee, Tanzanian National Parks, Tanzania

DR. MICHAEL ABS Curator, Ruhr University	BOCHUM, GERMANY
DR. SALIM ALI Bombay Natural History Society	BOMBAY, INDIA
DR. RUDOLF ALTEVOGT Professor, Zoological Institute, University of Münster	MÜNSTER, GERMANY
DR. RENATE ANGERMANN Curator, Institute of Zoology, Humboldt University	BERLIN, GERMANY
EDWARD A. ARMSTRONG Cambridge University	CAMBRIDGE, ENGLAND
DR. PETER AX Professor, Second Zoological Institute and Museum, University of Göttingen	GÖTTINGEN, GERMANY
DR. FRANZ BACHMAIER Zoological Collection of the State of Bavaria	MUNICH, GERMANY
DR. PEDRU BANARESCU Academy of the Roumanian Socialist Republic, Trajan Savulescu Institute of Biology	BUCHAREST, RUMANIA
DR. A. G. BANNIKOW Professor, Institute of Veterinary Medicine	MOSCOW, U.S.S.R.
DR. HILDE BAUMGÄRTNER Zoological Collection of the State of Bavaria	MUNICH, GERMANY
C. W. BENSON Department of Zoology, Cambridge University	CAMBRIDGE, ENGLAND
DR. ANDREW BERGER Chairman, Department of Zoology, University of Hawaii	HONOLULU, HAWAII, U.S.A.
DR. J. BERLIOZ National Museum of Natural History	PARIS, FRANCE
DR. RUDOLF BERNDT Director, Institute for Population Ecology, Heligoland Ornithological Station	BRAUNSCHWEIG, GERMANY
DIETER BLUME Instructor of Biology, Freiherr-vom-Stein School	GLADENBACH, GERMANY
DR. MAXIMILIAN BOECKER Zoological Research Institute and A. Koenig Museum	BONN, GERMANY
DR. CARL-HEINZ BRANDES Curator and Director, The Aquarium, Overseas Museum	BREMEN, GERMANY
DR. DONALD G. BROADLEY Curator, Umtali Museum	UMTALI, RHODESIA
DR. HEINZ BRÜLL Director; Game, Forest, and Fields Research Station	HARTENHOLM, GERMANY
DR. HERBERT BRUNS Director, Institute of Zoology and the Protection of Life	SCHLANGENBAD, GERMANY
HANS BUB Heligoland Ornithological Station	WILHELMSHAVEN, GERMANY
A. H. CHRISHOLM	SYDNEY, AUSTRALIA
HERBERT THOMAS CONDON Curator of Birds, South Australian Museum	ADELAIDE, AUSTRALIA
DR. EBERHARD CURIO Director, Laboratory of Ethology, Ruhr University	BOCHUM, GERMANY

DR. SERGE DAAN
Laboratory of Animal Physiology, University of Amsterdam — AMSTERDAM, THE NETHERLANDS

DR. HEINRICH DATHE
Professor and Director, Animal Park and Zoological Research Station, German Academy of Sciences — BERLIN, GERMANY

DR. WOLFGANG DIERL
Zoological Collection of the State of Bavaria — MUNICH, GERMANY

DR. FRITZ DIETERLEN
Zoological Research Institute, A. Koenig Museum — BONN, GERMANY

DR. ROLF DIRCKSEN
Professor, Pedagogical Institute — BIELEFELD, GERMANY

JOSEF DONNER
Instructor of Biology — KATZELSDORF, AUSTRIA

DR. JEAN DORST
Professor, National Museum of Natural History — PARIS, FRANCE

DR. GERTI DÜCKER
Professor and Chief Curator, Zoological Institute, University of Münster — MÜNSTER, GERMANY

DR. MICHAEL DZWILLO
Zoological Institute and Museum, University of Hamburg — HAMBURG, GERMANY

DR. IRENÄUS EIBL-EIBESFELDT
Professor and Director, Institute of Human Ethology, Max Planck Institute for Behavioral Physiology — PERCHA/STARNBERG, GERMANY

DR. MARTIN EISENTRAUT
Professor and Director, Zoological Research Institute and A. Koenig Museum — BONN, GERMANY

DR. EBERHARD ERNST
Swiss Tropical Institute — BASEL, SWITZERLAND

R. D. ETCHECOPAR
Director, National Museum of Natural History — PARIS, FRANCE

DR. R. A. FALLA
Director, Dominion Museum — WELLINGTON, NEW ZEALAND

DR. HUBERT FECHTER
Curator, Lower Animals, Zoological Collection of the State of Bavaria — MUNICH, GERMANY

DR. WALTER FIEDLER
Docent, University of Vienna, and Director, Schönbrunn Zoo — VIENNA, AUSTRIA

WOLFGANG FISCHER
Inspector of Animals, Animal Park — BERLIN, GERMANY

DR. C. A. FLEMING
Geological Survey Department of Scientific and Industrial Research — LOWER HUTT, NEW ZEALAND

DR. HANS FRÄDRICH
Zoological Garden — BERLIN, GERMANY

DR. HANS-ALBRECHT FREYE
Professor and Director, Biological Institute of the Medical School — HALLE A.D.S., GERMANY

GÜNTHER E. FREYTAG
Former Director, Reptile and Amphibian Collection, Museum of Cultural History in Magdeburg — BERLIN, GERMANY

DR. HERBERT FRIEDMANN
Director, Los Angeles County Museum of Natural History — LOS ANGELES, CALIFORNIA, U.S.A.

DR. H. FRIEDRICH
Professor, Overseas Museum — BREMEN, GERMANY

DR. JAN FRIJLINK
Zoological Laboratory, University of Amsterdam — AMSTERDAM, THE NETHERLANDS

DR. DR. H.C. KARL VON FRISCH
Professor Emeritus and former Director, Zoological Institute, University of Munich — MUNICH, GERMANY

DR. H. J. FRITH
C.S.I.R.O. Research Institute — CANBERRA, AUSTRALIA

DR. ION E. FUHN
Academy of the Roumanian Socialist Republic, Trajan Savulescu Institute of Biology — BUCHAREST, RUMANIA

DR. CARL GANS
Professor, Department of Biology, State University of New York at Buffalo — BUFFALO, NEW YORK, U.S.A.

DR. RUDOLF GEIGY
Professor and Director, Swiss Tropical Institute — BASEL, SWITZERLAND

DR. JACQUES GERY — ST. GENIES, FRANCE

DR. WOLFGANG GEWALT
Director, Animal Park — DUISBURG, GERMANY

DR. DR. H.C. DR. H.C. VIKTOR GOERTTLER
Professor Emeritus, University of Jena — JENA, GERMANY

DR. FRIEDRICH GOETHE
Director, Institute of Ornithology, Heligoland Ornithological Station — WILHELMSHAVEN, GERMANY

DR. ULRICH F. GRUBER
Herpetological Section, Zoological Research Institute and A. Koenig Museum — BONN, GERMANY

DR. H. R. HAEFELFINGER
Museum of Natural History — BASEL, SWITZERLAND

DR. THEODOR HALTENORTH
Director, Mammalology, Zoological Collection of the State of Bavaria — MUNICH, GERMANY

BARBARA HARRISSON
Sarawak Museum, Kuching, Borneo — ITHACA, NEW YORK, U.S.A.

DR. FRANCOIS HAVERSCHMIDT
President, High Court (retired) — PARAMARIBO, SURINAM

DR. HEINZ HECK
Director, Catskill Game Farm — CATSKILL, NEW YORK, U.S.A.

DR. LUTZ HECK
Professor (retired), and Director, Zoological Garden, Berlin — WIESBADEN, GERMANY

DR. DR. H.C. HEINI HEDIGER
Director, Zoological Garden — ZURICH, SWITZERLAND

DR. DIETRICH HEINEMANN
Director, Zoological Garden, Münster — DÖRNIGHEIM, GERMANY

DR. HELMUT HEMMER
Institute for Physiological Zoology, University of Mainz — MAINZ, GERMANY

DR. W. G. HEPTNER
Professor, Zoological Museum, University of Moscow — MOSCOW, U.S.S.R.

DR. KONRAD HERTER
Professor Emeritus and Director (retired), Zoological Institute, Free University of Berlin — BERLIN, GERMANY

DR. HANS RUDOLF HEUSSER
Zoological Museum, University of Zurich — ZURICH, SWITZERLAND

DR. EMIL OTTO HÖHN
Associate Professor of Physiology, University of Alberta — EDMONTON, CANADA

DR. W. HOHORST
Professor and Director, Parasitological Institute, Farbwerke Hoechst A.G. — FRANKFURT-HÖCHST, GERMANY

DR. FOLKHART HÜCKINGHAUS
Director, Senckenbergische Anatomy, University of Frankfurt a.M. — FRANKFURT A.M., GERMANY

FRANCOIS HÜE
National Museum of Natural History — PARIS, FRANCE

DR. K. IMMELMANN
Professor, Zoological Institute, Technical University of Braunschweig — BRAUNSCHWEIG, GERMANY

DR. JUNICHIRO ITANI
Kyoto University — KYOTO, JAPAN

DR. RICHARD F. JOHNSTON
Professor of Zoology, University of Kansas — LAWRENCE, KANSAS, U.S.A.

OTTO JOST
Oberstudienrat, Freiherr-vom-Stein Gymnasium — FULDA, GERMANY

DR. PAUL KÄHSBAUER
Curator, Fishes, Museum of Natural History — VIENNA, AUSTRIA

DR. LUDWIG KARBE
Zoological State Institute and Museum — HAMBURG, GERMANY

DR. N. N. KARTASCHEW
Docent, Department of Biology, Lomonossow State University — MOSCOW, U.S.S.R.

DR. WERNER KÄSTLE
Oberstudienrat, Gisela Gymnasium — MUNICH, GERMANY

DR. REINHARD KAUFMANN
Field Station of the Tropical Institute, Justus Liebig University, Giessen, Germany — SANTA MARTA, COLOMBIA

DR. MASAO KAWAI Primate Research Institute, Kyoto University	KYOTO, JAPAN
DR. ERNST F. KILIAN Professor, Giessen University and Catedratico Universidad Austral, Valdivia-Chile	GIESSEN, GERMANY
DR. RAGNAR KINZELBACH Institute for General Zoology, University of Mainz	MAINZ, GERMANY
DR. HEINRICH KIRCHNER Landwirtschaftsrat (retired)	BAD OLDESLOE, GERMANY
DR. ROSL KIRCHSHOFER Zoological Garden, University of Frankfurt a.M.	FRANKFURT A.M., GERMANY
DR. WOLFGANG KLAUSEWITZ Curator, Senckenberg Nature Museum and Research Institute	FRANKFURT A.M., GERMANY
DR. KONRAD KLEMMER Curator, Senckenberg Nature Museum and Research Institute	FRANKFURT A.M., GERMANY
DR. ERICH KLINGHAMMER Laboratory of Ethology, Purdue University	LAFAYETTE, INDIANA, U.S.A.
DR. HEINZ-GEORG KLÖS Professor and Director, Zoological Garden	BERLIN, GERMANY
URSULA KLÖS Zoological Garden	BERLIN, GERMANY
DR. OTTO KOEHLER Professor Emeritus, Zoological Institute, University of Freiburg	FREIBURG I. BR., GERMANY
DR. KURT KOLAR Institute of Ethology, Austrian Academy of Sciences	VIENNA, AUSTRIA
DR. CLAUS KÖNIG State Ornithological Station of Baden-Württemberg	LUDWIGSBURG, GERMANY
DR. ADRIAAN KORTLANDT Zoological Laboratory, University of Amsterdam	AMSTERDAM, THE NETHERLANDS
DR. HELMUT KRAFT Professor and Scientific Councillor, Medical Animal Clinic, University of Munich	MUNICH, GERMANY
DR. HELMUT KRAMER Zoological Research Institute and A. Koenig Museum	BONN, GERMANY
DR. FRANZ KRAPP Zoological Institute, University of Freiburg	FREIBURG, SWITZERLAND
DR. OTTO KRAUS Professor, University of Hamburg, and Director, Zoological Institute and Museum	HAMBURG, GERMANY
DR. DR. HANS KRIEG Professor and First Director (retired), Scientific Collections of the State of Bavaria	MUNICH, GERMANY
DR. HEINRICH KÜHL Federal Research Institute for Fisheries, Cuxhaven Laboratory	CUXHAVEN, GERMANY
DR. OSKAR KUHN Professor, formerly University Halle/Saale	MUNICH, GERMANY
DR. HANS KUMERLOEVE First Director (retired), State Scientific Museum, Vienna	MUNICH, GERMANY
DR. NAGAMICHI KURODA Yamashina Ornithological Institute, Shibuya-Ku	TOKYO, JAPAN
DR. FRED KURT Zoological Museum of Zurich University, Smithsonian Elephant Survey	COLOMBO, CEYLON
DR. WERNER LADIGES Professor and Chief Curator, Zoological Institute and Museum, University of Hamburg	HAMBURG, GERMANY
LESLIE LAIDLAW Department of Animal Sciences, Purdue University	LAFAYETTE, INDIANA, U.S.A.
DR. ERNST M. LANG Director, Zoological Garden	BASEL, SWITZERLAND
DR. ALFREDO LANGGUTH Department of Zoology, Faculty of Humanities and Sciences, University of the Republic	MONTEVIDEO, URUGUAY
LEO LEHTONEN Science Writer	HELSINKI, FINLAND
BERND LEISLER Second Zoological Institute, University of Vienna	VIENNA, AUSTRIA

DR. KURT LILLELUND
Professor and Director, Institute for Hydrobiology and Fishery Sciences, University of Hamburg — HAMBURG, GERMANY

R. LIVERSIDGE
Alexander MacGregor Memorial Museum — KIMBERLEY, SOUTH AFRICA

DR. DR. KONRAD LORENZ
Professor and Director, Max Planck Institute for Behavioral Physiology — SEEWIESEN/OBB., GERMANY

DR. DR. MARTIN LÜHMANN
Federal Research Institute for the Breeding of Small Animals — CELLE, GERMANY

DR. JOHANNES LÜTTSCHWAGER
Oberstudienrat (retired) — HEIDELBERG, GERMANY

DR. WOLFGANG MAKATSCH — BAUTZEN, GERMANY

DR. HUBERT MARKL
Professor and Director, Zoological Institute, Technical University of Darmstadt — DARMSTADT, GERMANY

BASIL J. MARLOW, B.SC. (HONS)
Curator, Australian Museum — SYDNEY, AUSTRALIA

DR. THEODOR MEBS
Instructor of Biology — WEISSENHAUS/OSTSEE, GERMANY

DR. GERLOF FOKKO MEES
Curator of Birds, Rijks Museum of Natural History — LEIDEN, THE NETHERLANDS

HERMANN MEINKEN
Director, Fish Identification Institute, V.D.A. — BREMEN, GERMANY

DR. WILHELM MEISE
Chief Curator, Zoological Institute and Museum, University of Hamburg — HAMBURG, GERMANY

DR. JOACHIM MESSTORFF
Field Station of the Federal Fisheries Research Institute — BREMERHAVEN, GERMANY

DR. MARIAN MLYNARSKI
Professor, Polish Academy of Sciences, Institute for Systematic and Experimental Zoology — CRACOW, POLAND

DR. WALBURGA MOELLER
Nature Museum — HAMBURG, GERMANY

DR. H.C. ERNA MOHR
Curator (retired), Zoological State Institute and Museum — HAMBURG, GERMANY

DR. KARL-HEINZ MOLL — WAREN/MÜRITZ, GERMANY

DR. DETLEV MÜLLER-USING
Professor, Institute for Game Management, University of Göttingen — HANNOVERSCH-MÜNDEN, GERMANY

WERNER MÜNSTER
Instructor of Biology — EBERSBACH, GERMANY

DR. JOACHIM MÜNZING
Altona Museum — HAMBURG, GERMANY

DR. WILBERT NEUGEBAUER
Wilhelma Zoo — STUTTGART-BAD CANNSTATT, GERMANY

DR. IAN NEWTON
Senior Scientific Officer, The Nature Conservancy — EDINBURGH, SCOTLAND

DR. JÜRGEN NICOLAI
Max Planck Institute for Behavioral Physiology — SEEWIESEN/OBB., GERMANY

DR. GÜNTHER NIETHAMMER
Professor, Zoological Research Institute and A. Koenig Museum — BONN, GERMANY

DR. BERNHARD NIEVERGELT
Zoological Museum, University of Zurich — ZURICH, SWITZERLAND

DR. C. C. OLROG
Institut Miguel Lillo San Miguel de Tucuman — TUCUMAN, ARGENTINA

ALWIN PEDERSEN
Mammal Research and Arctic Explorer — HOLTE, DENMARK

DR. DIETER STEFAN PETERS
Nature Museum and Senckenberg Research Institute — FRANKFURT A.M., GERMANY

DR. NICOLAUS PETERS
Scientific Councillor and Docent, Institute of Hydrobiology and Fisheries, University of Hamburg — HAMBURG, GERMANY

DR. HANS-GÜNTER PETZOLD
Assistant Director, Zoological Garden — BERLIN, GERMANY

DR. RUDOLF PIECHOCKI
Docent, Zoological Institute, University of Halle — HALLE A.D.S., GERMANY

DR. IVO POGLAYEN-NEUWALL
Director, Zoological Garden — LOUISVILLE, KENTUCKY, U.S.A.

DR. EGON POPP
Zoological Collection of the State of Bavaria — MUNICH, GERMANY

DR. DR. H.C. ADOLF PORTMANN
Professor Emeritus, Zoological Institute, University of Basel — BASEL, SWITZERLAND

HANS PSENNER
Professor and Director, Alpine Zoo — INNSBRUCK, AUSTRIA

DR. HEINZ-SIBURD RAETHEL
Oberveterinärrat — BERLIN, GERMANY

DR. URS H. RAHM
Professor, Museum of Natural History — BASEL, SWITZERLAND

DR. WERNER RATHMAYER
Biology Institute, University of Konstanz — KONSTANZ, GERMANY

WALTER REINHARD
Biologist — BADEN-BADEN, GERMANY

DR. H. H. REINSCH
Federal Fisheries Research Institute — BREMERHAVEN, GERMANY

DR. BERNHARD RENSCH
Professor Emeritus, Zoological Institute, University of Münster — MÜNSTER, GERMANY

DR. VERNON REYNOLDS
Docent, Department of Sociology, University of Bristol — BRISTOL, ENGLAND

DR. RUPERT RIEDL
Professor, Department of Zoology, University of North Carolina — CHAPEL HILL, NORTH CAROLINA, U.S.A.

DR. PETER RIETSCHEL
Professor (retired), Zoological Institute, University of Frankfurt a.M. — FRANKFURT A.M., GERMANY

DR. SIEGFRIED RIETSCHEL
Docent, University of Frankfurt; Curator, Nature Museum and Research Institute Senckenberg — FRANKFURT A.M., GERMANY

HERBERT RINGLEBEN
Institute of Ornithology, Heligoland Ornithological Station — WILHELMSHAVEN, GERMANY

DR. K. ROHDE
Institute for General Zoology, Ruhr University — BOCHUM, GERMANY

DR. PETER RÖBEN
Academic Councillor, Zoological Institute, Heidelberg University — HEIDELBERG, GERMANY

DR. ANTON E. M. DE ROO
Royal Museum of Central Africa — TERVUREN, SOUTH AFRICA

DR. HUBERT SAINT GIRONS
Research Director, Center for National Scientific Research — BRUNOY (ESSONNE), FRANCE

DR. LUITFRIED VON SALVINI-PLAWEN
First Zoological Institute, University of Vienna — VIENNA, AUSTRIA

DR. KURT SANFT
Oberstudienrat, Diesterweg-Gymnasium — BERLIN, GERMANY

DR. E. G. FRANZ SAUER
Professor, Zoological Research Institute and A. Koenig Museum, University of Bonn — BONN, GERMANY

DR. ELEONORE M. SAUER
Zoological Research Institute and A. Koenig Museum, University of Bonn — BONN, GERMANY

DR. ERNST SCHÄFER
Curator, State Museum of Lower Saxony — HANNOVER, GERMANY

DR. FRIEDRICH SCHALLER
Professor and Chairman, First Zoological Institute, University of Vienna — VIENNA, AUSTRIA

DR. GEORGE B. SCHALLER
Serengeti Research Institute, Michael Grzimek Laboratory — SERONERA, TANZANIA

DR. GEORG SCHEER
Chief Curator and Director, Zoological Institute, State Museum of Hesse — DARMSTADT, GERMANY

DR. CHRISTOPH SCHERPNER
Zoological Garden — FRANKFURT A.M., GERMANY

DR. HERBERT SCHIFTER Bird Collection, Museum of Natural History	VIENNA, AUSTRIA
DR. MARCO SCHNITTER Zoological Museum, Zurich University	ZURICH, SWITZERLAND
DR. KURT SCHUBERT Federal Fisheries Research Institute	HAMBURG, GERMANY
EUGEN SCHUHMACHER Director, Animals Films, I.U.C.N.	MUNICH, GERMANY
DR. THOMAS SCHULTZE-WESTRUM Zoological Institute, University of Munich	MUNICH, GERMANY
DR. ERNST SCHÜT Professor and Director (retired), State Museum of Natural History	STUTTGART, GERMANY
DR. LESTER L. SHORT, JR. Associate Curator, American Museum of Natural History	NEW YORK, NEW YORK, U.S.A.
DR. HELMUT SICK National Museum	RIO DE JANEIRO, BRAZIL
DR. ALEXANDER F. SKUTCH Professor of Ornithology, University of Costa Rica	SAN ISIDRO DEL GENERAL, COSTA RICA
DR. EVERHARD J. SLIJPER Professor, Zoological Laboratory, University of Amsterdam	AMSTERDAM, THE NETHERLANDS
BERTRAM E. SMYTHIES Curator (retired), Division of Forestry Management, Sarawak-Malaysia	ESTEPONA, SPAIN
DR. KENNETH E. STAGER Chief Curator, Los Angeles County Museum of Natural History	LOS ANGELES, CALIFORNIA, U.S.A.
DR. H.C. GEORG H. W. STEIN Professor, Curator of Mammals, Institute of Zoology and Zoological Museum, Humboldt University	BERLIN, GERMANY
DR. JOACHIM STEINBACHER Curator, Nature Museum and Senckenberg Research Institute	FRANKFURT A.M., GERMANY
DR. BERNARD STONEHOUSE Canterbury University	CHRISTCHURCH, NEW ZEALAND
DR. RICHARD ZUR STRASSEN Curator, Nature Museum and Senckenberg Research Institute	FRANKFURT A.M., GERMANY
DR. ADELHEID STUDER-THIERSCH Zoological Garden	BASEL, SWITZERLAND
DR. ERNST SUTTER Museum of Natural History	BASEL, SWITZERLAND
DR. FRITZ TEROFAL Director, Fish Collection, Zoological Collection of the State of Bavaria	MUNICH, GERMANY
DR. G. F. VAN TETS Wildlife Research	CANBERRA, AUSTRALIA
ELLEN THALER-KOTTEK Institute of Zoology, University of Innsbruck	INNSBRUCK, AUSTRIA
DR. ERICH THENIUS Professor and Director, Institute of Paleontology, University of Vienna	VIENNA, AUSTRIA
DR. NIKO TINBERGEN Professor of Animal Behavior, Department of Zoology, Oxford University	OXFORD, ENGLAND
ALEXANDER TSURIKOV Lecturer, University of Munich	MUNICH, GERMANY
DR. WOLFGANG VILLWOCK Zoological Institute and Museum, University of Hamburg	HAMBURG, GERMANY
ZDENEK VOGEL Director, Suchdol Herpetological Station	PRAGUE, CZECHOSLOVAKIA
DIETER VOGT	SCHORNDORF, GERMANY
DR. JIRI VOLF Zoological Garden	PRAGUE, CZECHOSLOVAKIA
OTTO WADEWITZ	LEIPZIG, GERMANY

DR. HELMUT O. WAGNER Director (retired), Overseas Museum, Bremen	MEXICO CITY, MEXICO
DR. FRITZ WALTHER Professor, Texas A & M University	COLLEGE STATION, TEXAS, U.S.A.
JOHN WARHAM Zoology Department, Canterbury University	CHRISTCHURCH, NEW ZEALAND
DR. SHERWOOD L. WASHBURN University of California at Berkeley	BERKELEY, CALIFORNIA, U.S.A.
EBERHARD WAWRA First Zoological Institute, University of Vienna	VIENNA, AUSTRIA
DR. INGRID WEIGEL Zoological Collection of the State of Bavaria	MUNICH, GERMANY
DR. B. WEISCHER Institute of Nematode Research, Federal Biological Institute	MÜNSTER/WESTFALEN, GERMANY
HERBERT WENDT Author, Natural History	BADEN-BADEN, GERMANY
DR. HEINZ WERMUTH Chief Curator, State Nature Museum, Stuttgart	LUDWIGSBURG, GERMANY
DR. WOLFGANG VON WESTERNHAGEN	PREETZ/HOLSTEIN, GERMANY
DR. ALEXANDER WETMORE United States National Museum, Smithsonian Institution	WASHINGTON, D.C., U.S.A.
DR. DIETRICH E. WILCKE	RÖTTGEN, GERMANY
DR. HELMUT WILKENS Professor and Director, Institute of Anatomy, School of Veterinary Medicine	HANNOVER, GERMANY
DR. MICHAEL L. WOLFE Utah State University	UTAH, U.S.A.
HANS EDMUND WOLTERS Zoological Research Institute and A. Koenig Museum	BONN, GERMANY
DR. ARNFRID WÜNSCHMANN Research Associate, Zoological Garden	BERLIN, GERMANY
DR. WALTER WÜST Instructor, Wilhelms Gymnasium	MUNICH, GERMANY
DR. HEINZ WUNDT Zoological Collection of the State of Bavaria	MUNICH, GERMANY
DR. CLAUS-DIETER ZANDER Zoological Institute and Museum, University of Hamburg	HAMBURG, GERMANY
DR. DR. FRITZ ZUMPT Director, Entomology and Parasitology, South African Institute for Medical Research	JOHANNESBURG, SOUTH AFRICA
DR. RICHARD L. ZUSI Curator of Birds, United States National Museum, Smithsonian Institution	WASHINGTON, D.C., U.S.A.

Volume 11

MAMMALS III

Edited by:

IRENÄUS EIBL-EIBESFELDT
MARTIN EISENTRAUT
HANS-ALBRECHT FREYE
BERNHARD GRZIMEK
HEINI HEDIGER
DIETRICH HEINEMANN
HELMUT HEMMER
ADRIAAN KORTLANDT
HANS KRIEG
H. C. ERNA MOHR
RUDOLF PIECHOCKI
URS RAHM
EVERARD J. SLIJPER
ERICH THENIUS

ENGLISH EDITION

GENERAL EDITOR:
George M. Narita

PRODUCTION DIRECTOR:
James V. Leone
ART DIRECTOR:
Lorraine K. Hohman

SCIENTIFIC EDITOR:
Erich Klinghammer
TRANSLATORS:
Sharon F. Rinkoff
David R. Martinez
Erich Klinghammer
SCIENTIFIC CONSULTANT:
Marvin L. Jones
ASSISTANT EDITORS:
Peter W. Mehren
Sharon F. Rinkoff
Jeanine Grau
EDITORIAL ASSISTANTS:
John B. Brown
Karen Boikess
Rachel Davison
INDEX:
Suzanne C. Klinghammer

CONTENTS

For a more complete listing of
animal names, see systematic classification and index.

1. **CHIMPANZEES** 19

by A. Kortlandt in collaboration with D. Heinemann 19
Editor: A. Kortlandt

2. **HUMAN ORIGINS** 49

The phylogeny of human-like forms by E. Thenius 49
Modern humans by H. Hemmer 55
The biological basis of human behavior by E. Klinghammer 58
Editors: E. Thenius and H. Hemmer

3. **COLUGOS OR FLYING LEMURS** 64

by T. Schultze-Westrum 64
Editor: D. Heinemann

4. **BATS** 67

Introduction by M. Eisentraut 67
Editor: M. Eisentraut

5. **OLD WORLD FRUIT BATS** 93

by M. Eisentraut 93
Editor: M. Eisentraut

6. **INSECTIVOROUS BATS** 111

by M. Eisentraut 111
Editor: M. Eisentraut

7. **EDENTATES** 149

Introduction by W. Moeller 149
Phylogeny by E. Thenius 153
Present-day edentates by W. Moeller 154
Editors: H. Krieg and E. Thenius

8. **PANGOLINS** 182

by U. Rahm 182
Fossil history by E. Thenius 188
Editor: U. Rahm

9. **RODENTS** 191

Introduction by H. Freye 191
Phylogeny by E. Thenius 198
Editors: H. Freye and E. Thenius

10. **SCIURID OR SQUIRREL-LIKE RODENTS** 201

Introduction of sciurid rodents by H. Freye 201
Alpine marmot by H. Freye and B. Grzimek 206
Other marmots by H. Freye 206
Woodchucks and prairie dogs by P. Goodman 222
Flying squirrels by H. Freye 263
Southern flying squirrels by B. Grzimek 265
Pocket gophers by H. Freye 268
Beavers by B. Grzimek and R. Piechocki 275
Gundis by H. Freye 286

Scaly-tailed squirrel-like rodents, springhaas by U. Rahm with contributions by E. Thenius 289
Editors: H. A. Freye and R. Piechocki

11. MOUSE-LIKE RODENTS 296

Introduction, cricetid rodents and mice by R. Piechocki 296
Bamboo rats, mole rats and murid rodents by F. Dieterlen 346
Dormice and jumping mice by R. Piechocki 388
Editor: R. Piechocki

12. OLD WORLD PORCUPINES, MOLE RATS AND AFRICAN CANE RATS 407

Old World porcupines by F. Dieterlen 407
Cane and rock rats by D. Heinemann 418
Editors: H. A. Freye and D. Heinemann

13. CAVIES 420

Phylogeny by E. Thenius 420
Octodonts by D. Heinemann 421
Hutias by H. Wendt 428
Chinchillas and cavies by D. Heinemann 433
Pacaranas by H. Wendt 450
New World porcupines by D. Heinemann and E. Thenius 453
Editors: E. Thenius, E. Mohr and D. Heinemann

14. WHALES 457

Introduction by E. J. Slijper and D. Heinemann 457
Phylogeny by E. Thenius 472
Editors: E. J. Slijper and D. Heinemann

15. BALEEN WHALES 477

by E. J. Slijper and D. Heinemann 477

16. TOOTHED WHALES 493

Introduction, sperm whales by E. J. Slijper and D. Heinemann 493
River Dolphins by H. Hediger 502
Narwhals, porpoises, long-snouted dolphins by E. J. Slijper and D. Heinemann 503
Recent studies of dolphins by H. Hediger 509
Overview of dolphins by E. J. Slijper and D. Heinemann 514
The orca or killer whale by D. R. Martinez and E. Klinghammer 519

Appendix: Systematic Classification 525

Animal Dictionary 547
English-German-French-Russian 547
German-English-French-Russian 560
French-German-English-Russian 587
Russian-German-English-French 599
Conversion Tables of Metric to US and British Systems 611
Supplementary Readings 617
Picture Credits 621
Index 623
Abbreviations and Symbols 635

1 Chimpanzees

Chimpanzees, by A. Kortlandt and D. Heinemann

Of all anthropoid apes, the chimpanzee is most common in zoological gardens and therefore best known. The genus *Pan* consists of two species: 1. The CHIMPANZEE (*Pan troglodytes*; HRL 70–92.5 cm; see Color plates, pp. 25–36). The standing height of the ♂ can reach 170 cm; in the ♀ it can reach 130 cm. The arms are longer than the legs; the hands are longer and more slender than those of the gorilla; the feet are long and slender with a strong big toe. The skin of young animals often is quite fair; however, it darkens with age, becoming almost black or mottled. The hair is thick, brown-black to black, and shining; the face, anal and genital zones, hand palms and soles, and the dorsal sides of the fingers and toes are hairless. There are black whiskers of medium length and short white goatees. Infants and very young animals have a white tuft of hair above the anal zone. Old animals are often gray, and white hairs may appear on the head and back. Reddish-brown specimens are found occasionally especially in western Africa. There are three or four often disputed subspecies.

Distinguishing characteristics

2. The PYGMY CHIMPANZEE OF BONOBO (⦶ *Pan paniscus*; see Color plates, pp. 30/31 and 35) is much smaller and more slender and graceful. It is brown-black with reddish flesh-colored lips and heavy whiskers.

The chimpanzee

The chimpanzees are the most humanlike anthropoid apes; they are less removed from the common ancestors of anthropoid apes and humans than are the other species. They are not as well adapted to jungle life as, for example, the orang-utan. Formerly the chimpanzee, the orang-utan, and the gibbon were thought to be the perfect brachiators and treetop climbers in the tropical rain forests; the many chimpanzees in the zoos seemingly affirmed this opinion by their climbing skills. Moreover, close observation of chimpanzees in nature is very difficult. Grzimek may be the first to have done so in 1951 in the Nimba Mountains in then-French Guinea. He reported that they have a system of regular trails in the jungle. During his observation period they climbed trees only in order to pick fruits, to build nests, or to sleep. They regularly used a fallen hollow tree trunk for drumming. Bernhard and Michael Grzimek ate the same fruits as the chimpanzees when they once lost their way in the jungle. As early as 1930, Henry W. Nissen was able to study chimpanzees in the savannas of Guinea, although he did not distinguish individual animals. Only in recent years were my co-workers and I, as well as the young zoologist Jane van Lawick-Goodall, the English couple Frankie and Vernon Reynolds, and a Japanese group of scientists able to study the behavior of chimpanzees extensively. It appeared that the chimpanzee lives far more on the ground than was formerly believed. The chimpanzee is less agile when moving in trees than when on solid ground.

Fig. 1-1. 1. Chimpanzee (*Pan troglodytes*), † indicating areas in which the chimpanzee has recently become extinct; 2. Pygmy chimpanzee or bonobo (*Pan paniscus*).

According to Vernon Reynolds: "One can judge whether an animal lives mainly on the ground or in the trees by the way it flees when frightened. When we stumbled on chimpanzees they often panicked and swung by 10-m jumps down through the branches, landing on the ground some distance away from us, and then ran off in the underbrush. In this way they resembled the baboons and red guenons, while most other apes flee through the treetops. Perhaps the chimpanzees feel more at home when on the ground—who knows?"

Reynolds ascertained that the amount of time chimpanzees spend in trees or on the ground depends on the hour of the day and availability of food: "In the Budongo Forest in Uganda the chimpanzees spend fifty to seventy percent of the daytime in trees. During the night they also sleep in the trees. They should therefore be considered as arboreal (tree-dwelling) animals in that habitat. In this respect they are in between the more ground-dwelling gorillas and the almost completely arboreal orang-utans. The fact that they find their food mainly in trees and spend the greater part of the day feeding, accounts sufficiently for the long hours they spend in trees. However, they also spend long hours on mutual grooming and keeping watch; when young they chase each other in the branches or amuse themselves with similar activities. Often, when the trees in which food is found are close together, they swing from tree to tree, following each other at a safe distance, making exactly the same movements, even on branches which look dangerously thin to us. More often, however, they pick a new tree while walking on the ground, which means they climb out of their old food tree and walk to the new feeding place by their own routes. The way they walk depends on the distance to be covered. They climb bare, branchless tree trunks astonishingly well and in a peculiar way. As with all anthropoid apes, their arms are longer than their legs; their hands are extremely long. We observed how they climb with their long arms; they put their fingers somewhat around the back of the tree trunk, and move their hands alternately above each other. At the same time they push their legs along, keeping their feet at both sides of the trunk, close to the body, showing the small toe on one side, and the big toe on the other."

Fig. 1-2. This is how the chimpanzee moves comfortably on the ground. The weight of the upper half of its body is supported on its knuckles. The soles of its feet are placed flat on the ground.

Fig. 1-3. Slightly accelerated movement.

This climbing method is obviously the real "specialty" of the chim-

Fig. 1-4. When going faster the chimpanzee is slightly raised. Both arms grab forward at the same time. The legs, however, are moved forward alternately, when running.

panzees. They use their long arms mainly to climb those trees of large diameter which are too thick to be tackled by the smaller monkeys and therefore not yet stripped of fruit. But obviously the separate groups in different regions have quite different behaviors. The chimpanzees I observed in the Congo region spent far more time on the ground than those Reynolds studied in the Budongo Forest. For instance, they always returned to the ground with the papayas they picked, and consumed them there. According to my observations, the chimpanzees were spending nearly as much time on the ground as, for example, baboons. They walk by choice either on four, three, or two legs. Strictly speaking, they only climb trees because the fruits are there and sleeping there is safer when leopards are around. They seldom swing by their arms as gibbons or orang-utans do, for they are too frightened of falling. However, chimpanzees adapted to particular environments manage the brachiating techniques quite well. Reynolds often saw the Budongo chimpanzees brachiating: "They seldom missed their target when they swung between branches which were 3–7 m apart. We saw only two falls."

Fig. 1-5. The fastest motor pattern is the gallop. The arms and then the legs are moved forward, more or less at the same time.

On the ground or on thick branches the chimpanzees normally walk or run on all fours, supporting the weight of their bodies on the knuckles of their hands. Their feet are planted flat on the ground or on a branch. The chimpanzees are most like humans when walking erect. However, they do not do this very often. The most important reason for walking erect is to carry food from open terrain to the safe bush. In such a way they often run 30 m with large bunches of bananas in their hands. In case of danger they are able to sprint away very quickly in this position.

A chimpanzee looks most unusual when jumping a wide brook. It stands half erect and makes a powerful jump with both legs, at the same time swinging both arms forward. On the other bank it lands on its feet first and then on its hands. If one tries this technique, one will find that the stance is not at all peculiar and certainly is not so for an animal that cannot make a running start before jumping. However, it is quite a feat when the brook is some 2 m wide; even a mother with a half-grown young on her back manages to jump this far.

Fig. 1-6. Chimpanzees often walk erect for short distances. The knees are not straightened as in humans; therefore the chimpanzee always looks smaller than it actually is.

Chimpanzees are mainly vegetarians, but they also consume insects and in certain savanna areas they will even eat mammals up to the size of a small bush pig. Actually, their feeding habits vary largely with their environment. Their main food is fruit of all kinds, also buds, blossoms, leaves, plant stems, bark, and the like. Captive chimpanzees like to eat the bark and leaves of freshly cut willow twigs. Even today, many zoos make the mistake of giving the anthropoid apes far too much nutritious food. As a result, they grow too fat and may die from intestinal diseases. Generally much of their natural food consists of stuffing and ballasting substances with low nutritive values. Their feeding habits change with the seasons. They have to adjust to those food vegetations which are available in abundance or are bearing fruit at the time. Chimpanzees living in their natural environment sometimes have periods of relative famine.

Reynolds estimates that chimpanzees spend six to eight hours per day on eating or searching for food. He reports: "Ninety percent of their main food is fruit, the balance being leaves, bark, pithy stems, and small amounts of insects, mainly ants, which they find on the tree bark. It was, of course, impossible for us to ascertain how many kilograms of fruits were eaten per day, but it had to have been a wholesome amount considering a grown male has an average weight of 50 kg, and the females weigh approximately 40 kg. Feeding on fruits seems to supply the needed fluids, because only once did we observe an animal drinking, and that was while it was sitting in a tree! It took me some time to find out what the animal was doing. It dipped its hand in a hole in the tree, pulled it out, and put it over its head. By using binoculars I could see that water dripped from its hand onto its protruded lips. It had discovered a natural water supply 30 m above the ground."

Fig. 1-7. When climbing, the chimpanzee embraces the tree with both arms, while its legs push it upward. The big toes, which can be widely spread from the other toes, like a thumb, are a great help.

The chimpanzees defecate and urinate whenever they need to. Reynolds observed, however, that on the ground they often look for a fallen tree trunk on which they squat to defecate: "To be sure, we never witnessed such a 'session,' but we found their droppings near fallen trees or branches so often that chance seems to be highly improbable." In zoos they often use elevated places to defecate, so that their droppings can fall down. To urinate they like to hang on the bars or on a cord. When it had to urinate, an approximately two-year-old pygmy chimpanzee in the Frankfurt Zoo vigorously attempted to leave the bed in which it was sleeping for a short time with its nurse; when cradled in her arms, it climbed away before 'letting go.' Reynolds continues: "They do not foul their nests in nature either. We often observed that they defecated beyond the edge of their nests. It is interesting to know that the gorilla, in contrast to the chimpanzee, always fouls its nest (see Vol. X). What is the reason for this difference? Later, near Kisoco, we saw a gorilla nest with the usual pile of droppings in it. However, we were quite surprised to observe that the original shape of the droppings had hardly changed although a male gorilla of some 150 to 200 kg had slept on it. Schaller reports that gorillas do not get dirty when lying on their droppings. Chimpanzees do. Their food contains less fibrous materials and their droppings are therefore soft and spongy, with no real shape. If chimpanzees slept in their droppings for the whole night, their hair would be smeared all over, causing disease. Perhaps, during the course of evolution, only those animals survived that behaved more or less hygienically."

Fig. 1-8. The chimpanzee can move through the branches by brachiating. The large hands with their well-developed thumbs are quite useful here.

The savanna chimpanzees, observed by Jane van Lawick-Goodall for some years in the Gombe River Reservation in Tanzania, not only caught termites and other insects, but also hunted larger animals. Once in a while they even ate monkeys. Once a young chimpanzee managed to catch a red colobus by surprise. Six adults joined in sharing the booty, without a fight. A chimpanzee was seen with a small, dead bush pig.

Fig. 1-9. The chimpanzee often moves upright in trees. However, it supports itself with its hands when doing so.

Fig. 1-10. The chimpanzee brachiates from tree to tree. The arms grab forward, and the body is swung across as soon as the hands find a firm grip.

Fig. 1-11. A chimpanzee uses its arms and legs for steering when jumping from great heights.

After chimpanzees had hunted down an animal, they ate the flesh slowly and after every bite took an additional serving of leaves as a supplement. Jane van Lawick-Goodall and some Japanese researchers have observed that chimpanzees, when eating honey or insects, especially termites, use a stick or a grass stalk as a probe which they insert into the hole: the termite guards sink their mandibles into the stalk, are pulled out by it, and then are licked off by the chimpanzees. When the stalk is damaged during the process, the chimpanzee simply breaks it off in order to use it again. Often instead of using a stalk, the animal takes a twig, first removing the leaves. The chimpanzees therefore not only use tools, but also make and repair them. At times they prepare their twigs in advance to take with them to the next termite hill, sometimes over a distance of nearly 1 km. They procure honey in the same way from beehives. The chimpanzees at Wolfgang Köhler's research station on Tenerife, one of the Canary Islands, poked about with sticks and stalks in cracks and holes. They wetted stalks and held them across the paths of ants. As soon as some ants had crawled on the stalk, the chimpanzees passed it through their mouths and ate them. The chimpanzees I observed in the jungle did not eat meat. I offered them various fruits and other food but they ate only what they knew. Fruit from the oil nut palm, a favorite food of west African chimpanzees, was not touched. Small, live animals including snails, chameleons, chickens, and a young mangabey which we tied down, were not eaten. They were obviously frightened by a small dead monkey and a dead antelope fawn which I offered them. Eggs, too, were rejected. Only two or three chimpanzees ate them readily, with additional bites of leaves. On the other hand, chimpanzees living in the savannas, whether in Guinea or Tanzania, are principal egg eaters. These typical differences in behavior between chimpanzee groups in different areas seem to indicate that the anthropoid apes are far less dependent on inherited behavior patterns than some other animals. When a woodpecker uses a cactus thorn to explore cracks in trees for insects, this use of a tool is just as innate as the weaverbird's nest-building technique or a fish's swimming.

Instinct and tradition

In higher animals this innate basis of behavior can often be further developed or supplemented by training. As a rule, however, each individual animal in each generation must have its own relevant experiences. Only rarely are new habits picked up by other chimpanzees to the extent that they are passed on to the next generation and so become a "tradition." Often such "traditions" are merely unimportant elaborations on behavior patterns already present and which have a hereditary basis. The use of tools by some chimpanzee groups and the different feeding habits of various chimpanzee populations seem, however, to indicate that chimpanzees are indeed able to pass on important habits and inventions as a real tradition to other members of their own troupe, and especially to their young, just as, in the same way, humans pass on experiences to our fellow man, especially to our children and grandchildren.

Explanations for the following color plates:

Plate I
On Rubondo Island, Tanzania, in Lake Victoria, zoo-born chimpanzees (*Pan troglodytes*) were resettled and given their freedom. The chimpanzee male on the left in the picture shows the beginning of threat display: the upright posture on two legs, the thrusting forward of the head, hunching of the shoulders, bending of the arms and upper back. These are motor patterns which also belong to the display repertoire of man.

Plate II
Chimpanzees live in trees as well as on the ground. Baron Hugo van Lawick took this picture of a chimpanzee community in Tanzania.

Plate III
A chimpanzee mother with her child. The games chimpanzee mothers play with their small children are quite similar to those human mothers play with their babies, even though this female chimpanzee chose an airy branch as a resting place.

Plate IV
This large male chimpanzee in the Frankfurt Zoo, named Toto, is quite nice to his offspring, as are most chimpanzee fathers. Here he carries his small daughter, named Elisabeth, piggyback.

Plate V
The adult chimpanzee female named Uschi, in the Frankfurt Zoo, was already bald when young. It was previously believed that such bald-headed chimpanzees were representatives of a separate species. We now know that chimpanzees can look quite different even within the same group. It is therefore very difficult to draw an exact line between subspecies.

Plate VI
An adolescent chimpanzee (*Pan troglodytes*, left) and a nearly full-grown pygmy chimpanzee or bonobo (*Pan paniscus*; right) in the Frankfurt Zoo. Chimpanzee children usually have light faces; later on the majority of them have dark faces. The pygmy chimpanzee has a dark face from childhood on.

Plates VII and VIII
When one gives chimpanzees the opportunity to handle a brush and paints or a pencil, many of them will paint with enthusiasm. The English zoologist Desmond Morris found out that certain basic patterns can be found, similar to those in children's drawings.

Plate IX
The British zoologist Jane van Lawick-Goodall observed chimpanzees (*Pan troglodytes*) on the east African savannas using grass stems or small sticks for getting termites out of their holes.

Plate X
Pygmy chimpanzees or bonobos (*Pan paniscus*) are small jungle chimpanzees from the area south of the Congo River.

Achievements under human care

Chimpanzees can achieve surprising feats of performance when cared for by humans. At the research station of Robert Yerkes, the great American anthropoid ape researcher, these animals learned to handle vending machines (see Vol. X) and even managed to differentiate between the "buying value" of different colored coins, which they could trade in for bigger or smaller quantities of food. Having satisfied their appetite with the food from the vending machines, some of them really hoarded their "money," as John B. Wolfe has observed. Ula, one of the chimpanzees that grew up in the Grzimek family, imitated sewing with needle and thread and the window cleaning and scrubbing done by her human foster mother; she also played with dolls and kissed chimpanzee pictures on the mouth.

A chimpanzee raised by the Hayeses, a husband and wife team of psychologists, learned to speak the words "mama," "papa," and "cup," and used them correctly most of the time. Parrots, jackdaws, mynahs, and other speech-talented birds are indeed able to repeat mechanically all kinds of sounds, but do not comprehend the meaning of human language. The chimpanzee child of the Hayeses however, occasionally understood the meaning of already known words used in a new sentence. Cooperative efforts to reach a common goal also indicate a very high intelligence. Although chimpanzees can not talk with each other beyond a few simple sounds, they can make themselves understood. They often indicate to each other by signs that help is needed. Crawford put a banana in a chest that the chimpanzees could pull to their cell with a rope. However, he made the chest so heavy that one animal alone could not move it. Both chimpanzees began by pulling the rope individually. Then one of the animals beckoned to the other; the second also gripped the rope and both pulled in the chest.

◁ The facial expressions of chimpanzees are the same in many instances as those of humans. Here a large male chimpanzee "pouts" in the same way as human children do when disappointed. Whether the "pouting" of chimpanzees has the same meaning, however, needs to be further investigated.

How do the chimpanzees acquire these abilities which they do not use in nature or which they use only to a limited extent? One could suggest that the ancestors of the chimpanzees probably made more use of their intelligence than their present offspring. Some of the oldest known primitive men (Australopithecines) and their younger relatives had, to some extent, brains only slightly larger than those of the present anthropoid apes. However, they used and made tools from the very beginning, as is true for human civilizations. Possibly the chimpanzee ancestors living at the same time in western Africa had the same abilities.

Community living

Chimpanzees live together in far looser communities than other ape species and form no permanent groups. But there is a certain hierarchy in each existing group. The high-ranking males show their high position by an excessive display, tearing down small trees and brandishing sticks or twigs. The low-ranking animals then often are frightened, dispersing and scrambling for the bushes. When a member of higher rank displays with less intensity, the lower-ranking one reacts frequently with a "scared expression." They bare their teeth and curl their lips inside out, often screaming shrilly. It is interesting to note that the lower-ranking animals show this scared expression, without screaming, when the threat postures of the higher-ranking animals have not yet begun, but are expected to do so for some reason or other—mainly when a low-ranking member seems to have "something in mind" that could evoke a threatening posture of a high-ranking animal. I once saw a low-ranking animal going after bananas, which I had put up as a lure. A high-ranking animal followed it a short distance. The closer the lower-ranking animal came to the bananas, the more it showed the "scared expression," although the high-ranking chimpanzee acted absolutely indifferent and later passed by the bananas. That a higher rank can also lead to "dignified" behavior was shown by further experiments with food-vending machines at the Yerkes Primate Center: A high-ranking chimpanzee, named Bimba, and the low-ranking Bula were housed in adjoining cells. At first, the food-vending machine was placed with Bula, but Bimba received the operating coin. Following this, Bula begged for the coin at the cage bars and actually, received the coin. She operated the machine and ate the fruits. When the machine was placed with Bimba, and Bula got the coin, nothing happened. The higher-ranking Bimba did not beg. When, however, Bimba got the machine as well as the coin, she ate the fruits, but gave the peels (the less valuable parts) to the lower-ranking Bula. Both the "dignified" non-begging, as well as the hesitant giving of the less valuable parts, have human parallels which we see often enough.

Fig. 1-12. When jumping on the ground, the feet are brought forward between the arms and touch the ground ahead of them.

Displays

High-ranking chimpanzees can be recognized easily by how often and how vigorously they show off their strength. When displaying, the chimpanzee thrusts his head slightly forward, screams a few "ooh" sounds, stands erect, and bends his elbows slightly forward. He raises the long hairs on his shoulders and upper arms, so that the muscular body looks even larger and more ferocious, and he dances around uttering longer, louder, and higher-pitched "Hoo-hoo-hoo" shouts. He ends with a shrill screaming and raging, stamping and hitting everything that makes noise and booms. He often brandishes and throws sticks or other objects. However, his comrades and the females and young keep a respectful distance.

Other characteristics of the display of strong chimpanzee males include the far-traveling drumming sounds on tree trunks and the booming of the thin, boardlike roots of so many tropical trees. The muffled beating is so typical for these anthropoid apes that the inhabitants of the Budongo forest region call the chimpanzee *Ki-Tera,* which means "it beats." We humans also make rhythmical noises when we are excited; we drum with our fists or stimulate ourselves for attacks with the beating of drums.

Only the high-ranking animals are allowed to show a bad temper

Only high-ranking chimpanzees are allowed to be bad-tempered from time to time and to scare and chase any animal that comes in their way. Muscular power alone is, however, not sufficient to insure high rank. The old, white-headed chimpanzee, called Grandpa, that I observed regularly in 1960, and which looked older at that time than any of the 200 other chimpanzees I knew, was perhaps forty years old, and crippled. His back was somewhat bent, and he avoided unnecessary climbing and did not join the display dancing. This notwithstanding, he held the highest rank in the troupe, and a choleric and moody "gentleman" he was at that. He was the only chimpanzee male who was often mean to the females, including mothers. He intimidated the other adult males; once he took a papaya from an adult male, but no one dared to oppose him. This is quite remarkable, because with other primates (for example, the baboons and macaques), the highest in rank is dethroned as soon as its strength diminishes even slightly. (In the Frankfurt Zoo, a graying male hamadryas baboon was clearly dethroned, but was still accepted and cared for as usual by the females, who were not in heat, until his death.)

Three years later, Grandpa had been slightly lowered in rank and he was no longer so moody. A choleric temperament may be an important prerequisite for attaining high rank. Grandpa, however, still takes the bananas from other chimpanzees without any complaints from them. He is, notwithstanding his physical weakness, still not that much lower in rank. I presume that his earlier rather high rank is the main reason for the trust he receives as a "worldly-wise old man."

The rain dance of male chimpanzees

On several occasions, Jane van Lawick-Goodall observed a rather peculiar display of chimpanzee males; the "rain dance," often initiated by rain showers, and sometimes by other things, can last half an hour, in contrast with the normal displays, which are of shorter duration. The males leave their nests, race down the slope, and brachiate, swinging with alternate arms, in circles from tree to tree. They climb a tree and jump down. While jumping, they tear off branches and brandish them wildly. The chimpanzee females and young sit in the nearby trees and watch.

At other times, however, the displays of high-ranking males are triggered by fleeting moods and often last less than one minute. Afterwards, the animals are friendly again, and the high-ranking animals are quite nice to each other. They like to sit together and communicate by looks and gestures. During displays there is no biting or beating, nor aimed throwing of clubs or hitting. One could better describe this display as a kind of "war dance," by means of which social pressures are relieved as through a safety valve. Once I saw strong males wrestling and chasing each other. One bit the other playfully on the back, but that happened, so to speak, in a friendly way. The only victims are the papaya trees, which are torn down. Only one chimpanzee I knew, a female, had a scar on her back that might indicate that she had previously been bitten seriously.

Mothers are nice to each other

The mothers live together peacefully; often they leave their young in the care of others. Once, three big youngsters walked with one mother who also cradled a baby on her stomach. At the most, only one of her companions could have been her own, as chimpanzee mothers never nurse more than two children at one time. At that moment there was no other mother nearby. In another case, it took quite a long time before I found out to which of two mothers a certain young belonged. In short, baby-sitting belongs to the normal routine in the life of a chimpanzee. I also observed how a mother touched the shoulder of her young, which seemed to mean: "I am going, come quickly, ride on my back!" Children ask their mothers for food by stretching out their open hand, in a begging gesture.

Sex life

Although many manifestations of the chimpanzees' social life, and especially their facial expressions, seem quite human, their sex life seems entirely different. Chimpanzees show no courting, flirting, tenderness, intimacy, or any of those things which form the essence of the expression of human love, and with which there are numerous parallels in most birds. However, in the Yerkes Primate Research Station one could observe a kind of courting by chimpanzee males; whereas normally a chimpanzee male serves himself first, when offered food he leaves some for the female during the mating period, even when only a small amount is offered.

The chimpanzee females in heat can be identified by their large, swollen, brilliant rosy red genital and anal region. They are not able to sit normally on a branch. According to Reynolds, this swelling could be interpreted as a sexual release, one that can be seen from afar, because chimpanzees often roam, and separate from their companions. Jane van Lawick-Goodall observed that a female chimpanzee mated with seven males one after the other. The initiative for mating comes from either the male or female. The chimpanzees I observed mated without special signs of emotion; it appeared to be more a matter of rape than love and surrender. The female offers herself obtrustively to the male at first, but then at the moment of mating, she shows the "fear expression" on her face and screams. Afterwards she runs away. Many male chimpanzees, however, roughly and irritably fend off the female at first; then without further ado, they proceed with the mating. There are chimpanzee males that brandish and throw clubs as an overture for mating, or give the female a thrashing. However, other chimpanzees have a very different behavior in this respect. Some show neither rage nor fear; they prefer a certain partner, engage in "love-play," and huddle together after mating.

Fig. 1-13. Chimpanzees show very rapid changes in their expressions, depending upon their moods, as in these pictures of a smiling and laughing animal.

Mother groups and sexual groups

The social life of chimpanzees is far less organized than that of the gorillas. The individual groups are not that stable, their composition often changing daily. Often when two groups meet, the members start screaming and become excited, without one animal harming another seriously; then the groups either remain together or form new groups. The individual groups of a chimpanzee population in a certain area are united not only in families, but also by common interests. I could distinguish two kinds of such groups with common interests: mother groups and sexual groups. Both groups can, however, mix temporarily. In the mother groups, females with small young set the tone, although now and then a nearly grown male or a female in heat joins them. The sexual groups consist, however, mainly of adult males and sometimes females without young, which are therefore able to conceive. Chimpanzees reproduce very slowly, as is the case with all anthropoid apes, because the mothers are barren during the nursing period. The young are normally nursed for two to three years. Only after the end of the lactation period does ovulation again occur and make another pregnancy possible. As a result, if the previous one stays alive, the next young is born only after three to four years. The pregnancy lasts 216 to 261 days, or about eight months. Experienced and inexperienced chimpanzee females are their own obstetricians. They open the amnion with their fingers and pull the infant from the vagina. As is the case with all female mammals, they show labor pains. After the birth they cut the umbilical cord themselves. They eat the amniotic fluid and placenta (according to Plors-Bartels, there are parallels here with primitive tribes). Inexperienced mothers often act very clumsily. One can assume that the young animals in nature learn baby care by watching their elders.

Birth and "obstetrics"

The "examination" of the baby looks very much like that done by human mothers. Chimpanzee and pygmy chimpanzee mothers are not greatly concerned with their young at first; they lay it against their body, and only then do they begin to observe it carefully. They hold the baby in front of them, look it over scrupulously, finger and clean it, as if they only just then accepted it as "theirs."

Infancy and youth

The newborn young weighs an average of 1800 to 2200 g. Chimpanzees can also have twins. Newborn chimpanzees are absolutely defenseless. They can only drink, suck their thumbs, and hold tight to their mother's hair and climb around in it. The female chimpanzee infant named Akati, born in the Frankfurt Zoo, weighed 2460 g when taken away from her mother at five weeks of age to be raised on the formula bottle. When Akati was four months old she weighed 4200 g, at eight months, 6100 g, and at one year, 9000 g. Only after four months does the baby dare to venture a short distance from its mother. The young start to play with each other when they are one year old, and when two years old, they constantly play very actively. When living in the wild during this time, they can pick up a substantial part of their food; however, they still get a great deal of nourishment from their mothers. Even when they are five years old, they still occasionally ride piggyback on their mothers. They usually remain with their mothers until puberty. They reach the reproductive age at seven to nine years, and they are part of the social hierarchy of the adults by their twelfth year.

Observations by Jane van Lawick-Goodall

Whereas I only could eavesdrop on my jungle chimpanzees from a well-camouflaged observation post in a tree, Jane Goodall became quite familiar with the savanna chimpanzees of the Gombe River, Tanzania, after years of constant association. In the beginning they fled from her, but after many months they only stepped aside and looked at her curiously.

David, a male chimpanzee in the prime of his life, visited her tent one day and helped himself to bananas. He became tame and greeted her with a "hoo." He took bananas out of her hand and let her pet him. At first he always ate the bananas with the peel. One day, however, he started to peel them. David later brought Goliath and William with him, who both became extraordinarily tame. They stole blankets, clothes, and towels, which they sucked and hung in the trees. David remained very calm most of the time, but Goliath was often rather excited. When he did not get a banana one day, he took an ax and swung it over the head of the camp cook's wife. After her marriage to the animal photographer, Baron Hugo van Lawick, Jane van Lawick-Goodall went back to the Gombe River. The female chimpanzee, Flo, who had previously been rather friendly, often visiting the camp with her daughter Fifi, who was over two years old, and her five-year old-son Figan, now had a newborn baby. The van Lawick family moved to a new camp, fifteen minutes up the mountains in the forest, and there the chimpanzees were even less shy of people than previously. Soon forty-five chimpanzees moved around in the camp; they took whatever they could lay their hands on and chewed anything that would reasonably fit in their mouths. Flo, her newborn baby notwithstanding, remained as intimate as ever, and when her baby was ten months old, she even permitted it to investigate the hand of Mrs. van Lawick-Goodall.

The construction of the sleeping nest

Many sources report that chimpanzees build a new sleeping place in trees every night. If this were true, there should have been hundreds of nests in my study area in the Congo region, because an unused nest can remain intact for weeks or months. I counted, however, only 150 sleeping nests, of which twenty were on the ground. Obviously some of the chimpanzees sleep in old nests, while the rest build new ones. Perhaps this depends on whether the apes are migrating or have a fixed sleeping site, as was the case in my area.

Why do chimpanzees seem so human?

There are many reasons why the behavior of chimpanzees seems so humanlike. During conflict situations, they constantly show all kinds of scruples and uncertainties, the same as we do; one automatically takes part in each of their changes of mood. They hesitate, trying to decide whether or not they should move into the bushes; would it be safe over there, should they go to the left, or maybe it would be better to go to the right? Should they pick this tree, or maybe the other? Does their papaya fruit taste better, or not...? The abnormally large erraticism and vascillation of their behavior also seems so human. They walk alternately on two, three, or four feet; they hold papaya fruits differently whether it be in their mouths, with their hands, or under their arms. Sometimes they pick and eat fruits one at a time; other times they first pick a small stack of fruits and then eat them; still other times they climb a tree with a half-eaten papaya in their mouth, pick a fresh one, descend, leave the half-eaten one and start on the fresh one. It also happens that upon arriving in the top of the tree, their appetite seems to be gone, and they climb down without a papaya. They constantly use different trails, coming and going at different hours, appearing in different group assemblies. In short, their behavior is a constant surprise for the observer. Compared with chimpanzees, all other free-living animals seem to be far more dependent on rigid, inherited behavior patterns.

Interest in uncommon objects

The chimpanzee's interest in new objects is also very humanlike. When a male named Opa and some of his friends saw my telephoto lens for the first time, he and a graybearded male sat down together to study this strange-looking big eye comfortably and without any hurry. In a similar vein, another chimpanzee twice studied a length of barbed wire, previously unknown to them in this area, that I had strung across a buffalo trail.

Once a chimpanzee appeared at the forest edge, looked up at the sky for a quarter of an hour (it was indeed beautiful that particular evening), and disappeared again into the bush. It forgot to take its papaya with it. A mother put a vine around her neck and shoulders like a shawl, and walked around with this adornment for some time. A two-to-three-year-old chimpanzee pulled a corncob, that I had intentionally placed nearby, like a toy, just as human children do with their toy carts. The zoo chimpanzees, set loose by Grzimek on Rubondo Island, Tanzania, in Lake Victoria, took a pair of binoculars from one of the scientists and busied themselves with them for at least an hour, one after the other. One of the young chimpanzees observed by Mrs. van Lawick-Goodall pulled a dead rat after it by the tail. Half-grown apes also like to play with sticks. They manipulate them only for fun and never used them as weapons against dead or living snakes, tortoises, birds, mammals, or other frightening objects which I had placed or tied on chimpanzee trails. Dead or apparently dead (sleeping) mammals produced a fear reaction. Live mammals, however, and birds, dead or alive, produced a mixture of fear and curiosity. It seems, therefore, that chimpanzees have, however limited, a knowledge of life and death.

Fear

The fear of ever-present danger governs the whole life of the chimpanzees. To a great extent this danger comes from human beings. When the large males assemble in the vegetation for the display dance, and run around screaming, stamping and, roaring, they do not look at all frightened. However, these animals are cautious and listen for many minutes before they venture out into an open space; and as soon as a twig cracks or a papaya drops, they huddle into a frightened group.

Defense against enemies

The most important enemy next to man is the leopard. Some time ago I had already observed that zoo chimpanzees often "fought" their "enemies" (for example, the veterinarian, the zoo manager, or a keeper who was disliked) by throwing objects in their direction or even by striking out with sticks. It was interesting to note that a well-aimed throw or striking out occurred only when the apes were kept in large, free reservations. This made me think that free-living chimpanzees in open spaces defend themselves against large beasts by using simple weapons, because in such areas there is not always a tree available which they can climb for safety.

Experiments with a stuffed leopard

To check this idea I used a stuffed leopard, with an electrically operated movable head and tail. The experiment was made six times with free-living chimpanzees in the eastern jungle of the Congo region. During four experiments, the predator had a chimpanzee doll in its fangs; during the other two, only the stuffed leopard was present. The experiments were organized in such a way that mother groups and mixed groups were each involved three times. Recently my associates, van Orshoven, van Zon, and Pfeyffers were able to make the same experiment, once with savanna chimpanzees and three times with jungle chimpanzees, all living in Guinea, in western Africa. Because of this we are in a position to establish separately the influence of subspecies relationships, group composition, terrain, prey in the fangs of the leopard, and the repetition of the test with the same chimpanzee groups as separate causes affecting the behavior of the test animals.

When I made the test for the first time, the suspense for me and my camera man was, of course, tremendous. As far as we knew, no human eye had ever seen what happens when a chimpanzee group encounters its enemy. We waited daily for weeks in our hideout, as we wanted to test a large group of chimpanzees. At last on the morning of 3 August 1963, a large group appeared. I counted: "...fifteen, eighteen, twenty!" Gradually they arrived on the meadow. We watched and waited. The movie cameras and tape recorders started. Then a strong pull on the steel wire, and there, in full splendour, right in the middle of the meadow, was our leopard, wagging its tail majestically. In front of its paws lay a seemingly dead chimpanzee child on its back with arms outstretched—a toy animal with its glass eyes open.

Attack on the enemy

For thirty seconds there was dead silence. Then an uproar broke loose. With awful screams, individual chimpanzees attacked and charged the dummy predator, running mostly on two legs. They pulled and shook trees, jumped, and climbed on large papaya trees, rattling them frantically. They broke off smaller trees and waved small tree trunks in their hands, charging the leopard with them and often throwing their clubs more or less in the direction of the predator. In the meantime, the other chimpanzees stood or sat in a wide circle watching, or climbed trees, from where they had a better view. At the same time they scratched themselves all over their bodies vigorously. People, too, often scratch their heads when trying to solve problems, but chimpanzees scratch all over their bodies—for example, in the laboratory during intelligence tests, or in nature at a

Fig. 1-14. The leopard is the predator most greatly feared by chimpanzees. Kortlandt and his co-worker tested the behavior of a chimpanzee group against a stuffed leopard which had an ape doll in front of it: A highly excited chimpanzee mother, carrying her child on her back, grabs a large club and charges the enemy . . .

. . . and hits the leopard with great force.

forest edge, when they are wondering whether a plantation will be safe. Obviously, the leopard was a very big problem for them.

The care of their young apparently resulted in conflict for the mothers. Larger young were usually left behind when the mothers made a new attack; but often they turned back midway, ran to their young, and pressed them to their breasts. Smaller young were carried with them during an attack, holding on to their mother's chest and stomach.

There were short rests between charges. The chimpanzees calmly ate a few bananas and leisurely watched the dramatic scene in front of them, as if only the beautiful sunset mattered. Thirty-five minutes after the beginning of the spectacle the whole group retreated into the bush. Twelve bent or broken papaya trees remained on the battlefield. During both repetitions of the test in the following weeks, the attacks became weaker.

The next three tests were made approximately one year later at the same place and with some of the same chimpanzees; however, the monkey doll was not put between the leopard's claws until after the first of these new tests. As a result, the attacks against the leopard during the second test were the most violent. It appeared as if the doll was believed to be a real victim of the predator.

The most remarkable event happened the next morning. During the night we had returned the leopard to its hiding place; the doll, however, remained with its face turned upwards. When the chimpanzees again passed the site, the whole group gathered carefully and in silence at a considerable distance around the victim. Then a mother approached carefully, sniffed at the doll, turned to the others, and shook her head two or three times in a slow and negative way. They all left silently. On that morning I did not hear one single chimpanzee noise in the jungle, an unprecedented occurrence with such a large mixed chimpanzee group.

We do not know, of course, what the chimpanzee female expressed by shaking her head; maybe it was only an accidental gesture. As she had just sniffed the victim, it is probable that her head shaking meant: "Not one of us," or "Nothing wrong," and not "Dead, alas." The dead silence on this morning in the jungle, however, gave rise to the feeling that something had really died. The previous morning the doll obviously had seemed like a real child to the chimpanzees, and the tests with mammals showed that they do have a knowledge of life and death. Unfortunately, one can only make such tests on a limited scale, because the apes become very frightened and temporarily shun the observation site.

In none of the six tests made at the edge of the Congolese jungle was the leopard hit by a thrown object or club. These objects were used, therefore, more as a show to frighten and chase the enemy away, rather than as actual weapons. My associates got exactly the same results with the jungle chimpanzees in Guinea.

The results with the savanna chimpanzees, however, were quite different. They grabbed the largest of the available clubs, which was 2.10 m long, and they tore down small trees of about the same length; they slashed viciously at the leopard with these. With the aid of the film we made, we could measure impact velocities of approximately 90km/h, which would have been sufficient to break the back of a live leopard. In addition, there was teamwork in evidence during these attacks, again in contrast with what we observed in the jungle chimpanzees. During the final attack the dummy was encircled by five chimpanzees, while two others stood in readiness at some distance, in case they should be needed. Then the leader grabbed the tail of the leopard and ran away, tossing the predator so that the head flew from the body.

With that, the enemy was considered "dead." The apes showed no more fear of it, and the youngsters were allowed to touch it. The attacks on its head, however, continued during the whole day.

A side effect of the experiment was the observation that the savanna chimpanzees more often walk erect than do the jungle chimpanzees. Mrs. van Lawick-Goodall and my associates, van Orshoven, van Zon, and Pfeyffers noticed that on the savanna, most of the chimpanzees eat eggs, whereas only two or three of the jungle chimpanzees in the Congo region were egg eaters. The eggs are mostly eaten with a "leaf salad." All of this indicates that a savanna environment "humanizes" the free, natural behavior of these apes, making it more humanlike. A jungle environment, on the other hand, "dehumanizes" their behavior, making it less like human behavior. This result is understandable, as the human species can be called a savanna animal. The most remarkable fact may be that there are such large differences between chimpanzees of one species, which are not governed by hereditary factors such as subspecies formation, local races, etc., but which have to be considered as real cultural traditions. When chimpanzees temporarily move from the jungle to the clearings, as was the case in the Congo region, their behavior remains adjusted to the jungle.

Chimpanzees can live in a wide variety of environments and are therefore the most widespread of all anthropoid apes, although they are now extinct in nearly all savanna regions. In the Ruwenzori Mountains they live as high as 3000 m. Since they do not swim, the large rivers form an insurmountable barrier. Consequently, the chimpanzee populations on each side of certain large rivers look somewhat different. Admittedly, there are great differences within a single population; the animals may be large or small, fat or skinny, dark brown or fair-faced, bearded or beardless, some with long faces, others with short, wide faces, along with sex and age differences. A certain type, quite common in one area, can be rare or nearly nonexistent in another. Thus today one distinguishes three or four subspecies depending upon the prevalence of particular types. All of these chimpanzee subspecies live north of the Congo River. Small, dark-skinned chimpanzees, distinguished from those

The pigmy chimpanzee or bonobo

in the north by many characteristics, live in the southern part of the range. These PYGMY CHIMPANZEES OR BONOBOS (♁ *Pan paniscus*) are not only smaller and more slender than their large cousins, but also have a different behavior and voice. Other physical differences include the reddish-colored lips and the female genital organs, which are placed more toward the stomach. There are also behavioral differences in mating. According to the observations of R. Kirchshofer in the Frankfurt Zoo and of Pournelle in the San Diego Zoo, the mating in pygmy chimpanzees takes place nearly always in a lying or squatting position. In most cases the female invites her partner to mate by lying down on her back, at the same time embracing him with one or both arms. The copulation is done silently; this mating technique is also used sometimes by the large chimpanzees. The Frankfurt Zoo succeeded for the first time in breeding the pygmy chimpanzees, whose living habits are largely unknown. On 22 January 1962, the first, on 23 December 1963, the second, and on 17 June 1968, the third baby was born. The mother of the three children named Magrit, was nine to ten years old when the first was born; the father, named Camillo, was also ten years old at that time. The first two deliveries were close together because the first child died of pneumonia after one month. The second and third young animals, a male and a female, developed well. Even during Magrit's first parturition, she did everything correctly, as R. Kirchshofer reports. The whole delivery procedure was the same as in chimpanzees. The birth lasted twenty-five minutes. Two minutes elapsed between the entrance of the head into the vaginal canal and the bursting of the amniotic sac, because the mother opened the amniotic sac with her fingertips and pulled the child out. Everything was done standing up. She placed the infant to her breast in the correct way, and it clung there with its hands and feet. The placenta was expelled after ninety minutes. The female clearly had pain before the placenta was expelled. She ate the placenta hungrily. A nervous tremor of the legs after the birth and great physical fatigue were both particularly noticeable.

Pigmy chimpanzee births in zoos

The newborn cried when the mother laid it down, and its cry was the pygmy chimpanzees' typically shrill cry of fear. After ninety minutes the young animal was able to look around in a coordinated manner. It showed the innate searching movements for the nipples forty-five minutes after birth; however, it did not take its first drink until twenty hours later.

Chimpanzees under human care

The chimpanzee's great adaptability makes it possible for them to have a wide distribution in many different African environments. This is also the reason that chimpanzees in zoos develop better than do other anthropoid apes. As a result, one could see chimpanzees in zoos everywhere at a time when these zoos were still mostly bleak and narrow jails. Today, modern zoos keep chimpanzee families or groups in roomy indoor and outdoor facilities or on premises without bars. (To keep an anthropoid ape alone is torture for the animal.) They generally breed well in zoos today, and often reach advanced years. However, the mortality rate of chimpanzees born in human care is still fearfully high, although once the animals have survived the nursing period, their chances of survival are good. A chimpanzee in the Yerkes Primate Laboratory, in the U.S.A., lived forty-seven years, the oldest age attained by a chimpanzee in captivity. Because chimpanzee females are still fertile when they reach the oldest recorded ages, we must assume that these animals in the wild may live to be over forty years old. The Old Grandpa in the Congo region was physically disabled and looked older than the oldest chimpanzee I have ever seen in captivity, but he was allowed to take food even from strong males, and he looked well fed.

It is a pity, however, that time and again this physical adaptability of the chimpanzees leads to keeping them in small, poor, and therefore insufficiently equipped zoos, or even in side shows where these highly intelligent and sensitive animals eventually perish. We know now that certain chimpanzees, as well as other anthropoid apes, when inappropriately cared for, develop perverted behavior similar to those known as "hospitalisms" in children raised without love or in poorly run orphanages. They smear and eat their feces, pluck their hair and eat it, drink their urine or spit it, repeating these movements and others again and again (behavior stereotypes). However, my co-workers observed that chimpanzee young in the wild beg for and eat feces.

Unscrupulous businessmen even get away with advertising chimpanzee babies in their pet shops as "pets" for people who have not the slightest idea of the conditions necessary for professional chimpanzee care. It is high time we stopped this form of cruelty to animals, and left the procurement, trade, and ownership to scientific zoos and professional supply firms.

Danger from medical experiments

A great danger for all anthropoid apes, especially the chimpanzees, lies in man's efforts to use these closest relatives of humans for medical and pharmaceutical research. In view of the large quantities of test animals necessary to attain statistical significance in experiments to make even one statement, such testing could wipe out the anthropoid apes. The possibility of using organs or tissues in large amounts as human transplants is even more of a threat to the chimpanzees, as even now we do not know enough about these anthropoid apes through research in the wild to replace those animals through successful breeding. The extinction of the chimpanzees would be an unimaginable calamity for all of medical, psychological, and anthropological research.

2 Mankind and Its Origin

Phylogeny of human-like forms, by E. Thenius

Even before Darwin, researchers and thinkers were concerned with the origin of humans. Humans' next of kin are the apes, and more particularly the anthropoid apes; this fact has been known by important persons in the fields of philosophy and natural history for some time—people like Linné, Kant, Schopenhauer and Lamarck. However, it was only after Darwin realized that animals and plants developed from the simpler, ancestral forms under the influence of natural selection (theory of evolution, see Volume I) that it became possible for science to approach and find a valid answer to the question of man's origin.

Man's descent from ape-like ancestors has been demonstrated by science

Today, in view of the overwhelming amount of evidence, no one can seriously deny the fact that man is descended from the ranks of the primates, more specifically from the Old World apes. "The person who doubts this fact today, puts himself in the same category as the person who doubts atomic theory," says anthropologist Gerhard Heberer. Many people have wondered how, when, and where specifically human separation from the apes occurred, and opinions have changed many times during the last decades. Even the apparently simple question of which characteristics and qualities are specifically human (hominid) and which are anthropoid ape-like (pongid) cannot be clearly answered. However, the fact that scientists encounter these difficulties when examining fossils is one of the most convincing proofs of the close phylogenetic relationship between the anthropoid apes (family Pongidae) and humans (family Hominidae). For this reason, other theories regarding the origin of man have proved untenable. F. Wood-Jones and many other researchers believed, for example, that man did not originate from tertiary anthropoid apes, but from the ancestors of the tarsiers. In the meantime, however, it has become apparent that the physical similarities between fossil tarsier relatives and the higher ape species are not the result of a close relationship, but rather the result of parallel development.

Three different hypotheses

The protocatarrhine hypotheses

Because humans have some unique physical characteristics which are no longer found in the anthropoid apes or in many other Old World apes, some well-known scientists such as M. Boule, J. Kälin, J. Piveteau and W. J. Straus believe that the human branch separated from the other primates when the Old World monkeys and apes (Catarrhinae) were still in the "protocatarrhine stage," and thus were not yet divided into the various later and present groups. This "protocatarrhine hypothesis" is not, however, absolutely binding; the aforementioned unique characteristics of humans and the correspondingly "more advanced" state of the anthropoid apes can be easily explained through the retardation of certain developmental processes in one group and an acceleration of development in the other group. On the other hand, many similarities between the anthropoid apes and man in physical characteristics, chromosome count (see Volume I), blood composition (see Volume X), and other behavioral characteristics can only be understood when one accepts their close relationship and the common evolutionary development. However, we cannot, as a result of these similarities, conclude that the present-day anthropoid apes were the ancestors of man. Humans and the anthropoid apes are related in that they can be traced to common ancestors, but they have each developed in totally divergent directions. This is the core of the pongid theory as we know it today.

The prebrachiator and brachiator hypotheses

How might these common ancestors of the present-day anthropoid apes and humans have looked? When was the time of their separation? Scientists' opinions on these questions differ. Some believe that the common evolutionary pattern of anthropoid apes and humans did not extend to or include the brachiating anthropoid ape (prebrachiator hypothesis). Other scientists, however, believe that the common ancestor of humans and the anthropoid apes was already a genuine brachiator (brachiator hypothesis). From this problem alone we can clearly see that the question of man's origin cannot be solved with theories, however clever they may be, but only through scrupulous examination of fossil evidence.

Prehistory of the super-family of human-like forms

The prehistory of human-like forms (Hominoidea) is however, only partially clarified by fossil discoveries. The majority of the primates were and are forest animals. Skeletal remains deteriorate in forest ground; hence the fossil record of primates is incomplete. Many of the present gibbon and anthropoid ape characteristics were not yet developed in the ancestors of these apes. The forms from the Miocene period (about ten to twenty-five million years ago) which, according to their teeth, were either gibbon or anthropoid ape-like, were certainly not brachiators as are the present anthropoid apes, particularly the orang-utan and the gibbons. This lengthening of the arms did not occur until the early Upper Tertiary period, and is therefore a recent adaptation.

Family: Gibbons

The gibbons (family Hylobatidae) are particularly adapted to brachiating; their other characteristics have remained in their original forms. They must have branched off quite early from the other human-like species. The oldest known SHORT-ARM GIBBONS (Pliopithecinae) or the closely related forms (like the *Aeolopithecus*) were present in the Oligocene (approximately forty to twenty-five million years ago) in Egypt; in the Upper Tertiary period there were short-arm gibbons in Africa (*Limnopithecus*) and in Europe and Asia (*Pliopithecus*). Although these species were not the immediate ancestors of the gibbons, they were closely related.

Family: Anthropoid apes

The anthropoid apes (family Pongidae) apparently developed in Africa; remains of the most ancient forms (*Propliopithecus*, *Aegyptopithecus*) were also found in the Oligocene strata of Egypt. We have found only dental and jaw remains from these old forms. However, in the Miocene strata of East Africa, scientists have found the remains of skull, back, and limb bones of an anthropoid ape, which they called *Proconsul*; it was in the initial stages of becoming a brachiator. Its arms showed none of the typical brachiator characteristics that are present in all the anthropoid apes, and its skull did not have the protruding brow. *Proconsul*-like anthropoid apes also lived in Eurasia during the Upper Tertiary period. Their tooth and jaw remains were described under various names, but at present they are classified together with *Proconsul* in one genus, *Dryopithecus*. Some of these forms have teeth that show a relationship with gorillas or orangutans, but without further knowledge about the complete skeleton no statements may be made regarding their possible relationship.

Fig. 2–1. This is how the anthropoid ape (*Proconsul africanus*) may have looked according to the British scientist Oakley and Wilson the artist; this ape is thought to be the common ancestor of man and apes.

A giant Ice Age form from South China, which was described as a GIANT ANTHROPOID APE (*Gigantopithecus*), attracted a great deal of special attention, because large, very human-like molar teeth were the first remains to be discovered. Weinert and other researchers thought that this giant animal might have been an extinct human giant far larger than a gorilla. Today, however, various lower jawbones with teeth have been found. These indicate an anthropoid ape with giant teeth and jaws. Although the teeth and jaws are larger than those of a gorilla, without examination of other bones it is impossible to say how big the extinct giant ape really was.

Family: Oreopithecides

The genus *Oreopithecus* from the Lower Pliocene (approximately ten million years ago) from Tuscany is even more important than the giant ape from China. Remains of *Oreopithecus bambolii* have been familiar to scientists for almost one hundred years. However, the Swiss paleontologist, J. Hürzeler, succeeded in salvaging an almost complete skeleton only a few years ago. For some time it was believed that *Oreopithecus* was an ancient hominid, and thus to a certain extent "a man before the development of man"; other researchers believed it was an intermediate stage between the baboons and the anthropoid apes. Recent discoveries, however, show that *Oreopithecus* belongs to neither group. Although it has many characteristics in common with man and others in common with the baboons, one must classify it as an anthropoid ape, with a long and distinct evolutionary history. Today, we classify it together with *Apidium* from the Egyptian Oligocene in a single family (Oreopithecidae), which, within the human-like forms, is further away from mankind than the real anthropoid apes (Pongidae).

Family: Man

Where does man's evolution begin?

Where does the evolution of humans begin? What makes humans different from all other animals lies in the development of his mind which, from a biological point of view, is closely related with the development of his brain; consequently, the brain was formerly considered as the beginning of human evolution. The English anthropoligist Arthur Keith (1866–1955) believed that a cranial volume of 750 to 800 cubic centimeters formed the divide between the anthropoid apes and humans. Today we know that human evolution did not begin with an increase in the size of the brain, but with the acquisition of an erect posture. The protohominids, (Australopithecinae), who are generally considered today as geologically the oldest hominids, had a cranial volume of only approximately 450 to 680 cubic centimeters, not more than a large gorilla. Their pelvis and limb bones prove, however, that the australopithecines walked erect, and in this respect already had crossed the divide leading to man. We do not know what the reasons were which lead to walking erect.

Erect stance and brain development

The erect posture and walk meant that the hands were free and could be developed into the versatile tools that they are for us today. Those parts of the brain involved in controlling the new activities of the hands developed along with the hand mobility. Aside from the characteristic development of the brain in humans, and more specifically, the frontal lobes, most of the other differences between humans and the anthropoid apes are not qualitative, but rather quantitative, or differences due to changes in size or degree. Thus, those characteristic patterns typical of humans only can be explained by evolutionary development. Many of our characteristics indicate clearly that we originated from those higher animals that lived in trees; the binocular vision with two eyes would have been unnecessary and inexplicable in an ancient ground-dwelling animal. Our feet were originally supportive prehensile feet and these, together with our bipedal mode of walking developed into a foot that allowed us to stand upright. So it is with our arm and back muscles, and other characteristics as well. Even our apparently primitive teeth may have developed from an earlier stage which resembles the more advanced teeth of the anthropoid apes. Perhaps we can explain new fossil discoveries this way. G. E. Lewis and L. S. B. Leakey discovered jaw bone remains in the Pliocene strata of India and East Africa, and the teeth of these jaws fit somewhere between those of the Dryopithecines and man. These remains have been described under several different names (*Dryopithecus punjabicus*, *Ramapithecus thorpei*, *Ramapithecus brevirostris*, and *Kenyapithecus wickeri*); the American paleonthologist E. L. Simons groups these forms under the name of *Ramapithecus punjabicus* and considers them to be hominids.

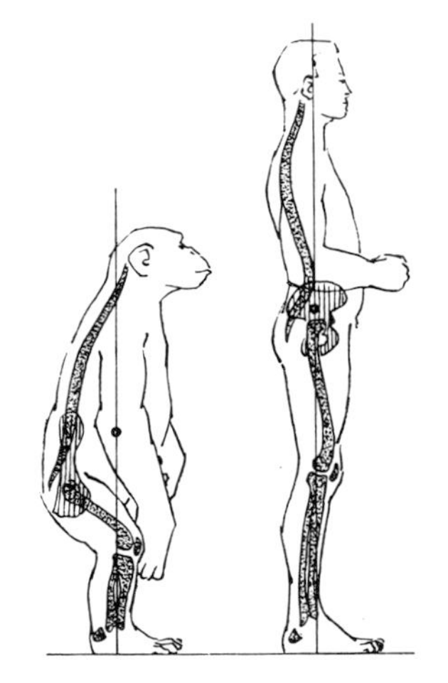

Fig. 2–2. The chimpanzee's erect posture is clearly different from that of man. In the chimpanzee (left), the spinal column remains evenly curved (kyphosis); the hip and knee joints are bent. In man (right), the spinal column is bent forward at the lumbar area (lordosis); the hip and knee joints are almost straight.

Future fossil discoveries will certainly increase and perhaps complete the picture of hominid origins. Most of the basic questions seem to have been solved. On the other hand, although the path of man's evolution during the Ice Age has been established by numerous fossil discoveries, there are still many unanswered questions before us.

Fig. 2–3. According to Oakley and Wilson, the ape-man *Australopithecus africanus transvaalensis* probably looked like this.

The oldest hominids, the protohominids (subfamily Australopithecinae), were found particularly in Africa in large numbers between the Tertiary and the Ice Age (one to two million years ago) and in the Lower Ice Age; for this reason, the assumption that hominids moved from Africa to other parts of the Old World cannot be overestimated.

Bones and tools from a wide variety of primitive men, ranging from the ape-man of the Lower Ice Age to the "*pithecanthropus*" forms of the Lower and Middle Ice Age, and including the Neanderthal man from the Upper Ice Age, and the more immediate ancestors of modern-day man from the Upper Ice Age and later, have been found in many different strata of the Ice Age in Europe, Asia, and Africa. This succession from older and younger forms of man seems to be a fascinating series of steps which proceed from the erect anthropoid ape-like forms, and from one level of cultural development to the next, to *Homo sapiens* of today.

The first discovery of an Australopithecine was made in 1925; however, this was the skull of a child, approximately six years old, from Taung in the Transvaal (South Africa), and its significance was only explained twenty years later when adults of this same form were discovered. Through the efforts of Robert Broom, several other discoveries have been made in South Africa since 1936; in Sterkfontein, Kromdraai, Swartkrans, and Makapansgat. Very important discoveries have been made by L. S. B. Leakey and his associates in East Africa (Olduvai Gorge in Tanzania) within the last few years. Countless remains of Australopithecines, which represent over one hundred different individuals at various stages, have been found in South and East Africa; these finds included not only teeth and jawbones, but skulls, limb bones and pelvic parts as well. These fossils proved that it was not a new species of anthropoid ape that was being discovered, as had been assumed; the remains, on the contrary, belonged to an erect, primitive type of ape-man, although one with a small anthropoid ape-like skull, and protruding jawbones, but with human teeth and human pelvic and leg bones. These observations notwithstanding, according to the zoological rules, these primitive men should be and still are classified as *Australopithecus africanus* (African southern ape), the name given to the skull of the child at Taung.

There have been other discoveries of ape-man remains in North Africa (Lake Chad area), in the Middle East (Jordan valley), in Java ("*Meganthropus*" *paleojavanicus*) and perhaps in China (? *Hemianthropus peii*). All of these remains have been found between the Lowest and the Lower pleistocene strata of the earth. Some of these fossils are 1.8 million years old, according to new physical chemistry tests (like the calcium-argon test, among others); other remains are considerably younger and are estimated to be half a million years old.

Subfamily: Ape-man

The protohominids (subfamily Australopithecinae) combine primitive and advanced characteristics. There is one genus with two species (previously considered as two genera): 1. *Australopithecus africanus*, (A-type;

Color plate, p. 78) is graceful and its skull does not have ridges. 2. *Australopithecus robustus* ("*Paranthropus*," P-type; Color plate, p. 78) is much stronger; the front teeth are small, but the molars are large. It also has powerful chewing muscles, which are attached to the ridge of the skull (they do not correspond to those of the anthropoid apes); they were probably a parallel evolutionary branch.

Whether the protohominids made tools and, in so doing, can be considered as a "hominid" could not be established at first from the South African discoveries. The South African anatomist R. A. Dart, the first to describe the child from Taung, believes that the bones, teeth, and horn remains, which were assumed to have been fashioned into tools of the protohominids, are probably only the remains of hyena kills. It is possible that "*Homo*" *habilis*, recently discovered in East Africa at the Olduvai Gorge, and a contemporary of "*Zinjanthropus boisei*" (*Australopithecus robustus*), made use of stone implements and tools. The remains of these forms were found in the lowest strata of the Olduvai Gorge, together with pebble tools (which indicate some sort of pebble culture).

Subfamily: Man, in a narrow sense

All of the other discoveries of early and primitive man are now considered to be of the genus *Homo*, to which we also belong. The most familiar form of those from the earliest evolutionary stages is "*Pithecanthropus*" (ape-man) or JAVA MAN (*Homo erectus erectus*; Color plate, p. 78) as he is now called, which was discovered by the Dutch physician Eugene Dubois in 1891–1892, near the Solo River at Trinil on Java. Related forms lived in China (*Homo erectus pekinensis*), North Africa (*Homo erectus mauritanicus*) and perhaps also in Central Europe, where a large ancient lower jaw was found, which may possibly belong to the species (*Homo erectus heidelbergensis*). Primitive man lived in the Lower and Middle Ice Age; his somewhat more modern-looking predecessors were still present in the Upper Ice Age, e.g. Ngandong man in Java and Rhodesian man of Broken Hill in South Africa.

Java man

Neanderthal man

The third group that appeared after the protohominids and early man is NEANDERTHAL MAN (*Homo sapiens neanderthalensis*; Color plate, p. 78). These ancient men (palaeanthropines) were given their name from the first discoveries which occurred in 1856, in Neanderthal near Dusseldorf (Germany). Many Neanderthal remains, some complete skeletons, were found later, especially in caves in France, Italy and the Soviet Union, but also in Spain, North Africa, the Near East and Central Asia. Neanderthal man is particularly familiar in his later "classical" form, from the latest European Würm Ice Age, some 50,000 years ago. These men were not a direct ancestor of *Homo sapiens sapiens*, because true *sapiens* humans were already contemporaries of Neanderthal man.

Cro-Magnon man

The first direct ancestor of modern-day man, CRO-MAGNON MAN (*Homo sapiens fossilis*), appeared at the same time as the late Neanderthal man, but with a much more developed culture. But where does this *Homo sapiens fossilis* come from? Attempts to trace him back to the so-called

PRE-NEANDERTHAL MAN from the last, warmer interglacial period between the Riss and Würm Ice Ages have not appeared to be very convincing in view of the separate evolution of Neanderthal man. In addition, the early man of the species *Homo erectus* probably belonged to an extinct branch of the human family tree and is not an ancestral species. The skull from Swanscombe (England) and those in Steinheim on the Murr River in Württemberg (Germany), from the great interim warmer interglacial (Mindel-Riss) Ice Age approximately 250,000 years ago, have shed some light on this question. These skulls are therefore, much older than those of the pre-Neanderthal man, although they have the same pronounced forehead, the high occiput, the flat sides, lack of the eyebrow ridges, and the small jawbones which are remarkably "modern," and *Homo sapiens*-like; for this reason they are designated the pre-sapiens type. Whether these pre-sapiens men, together with primitive man can be derived from the original A-type of primitive man, or whether these forms constitute an evolutionary branch like the heavy-jawed P-type, can only be decided by additional fossil discoveries.

We should not, in any case, consider the evolution of man as proceeding in a direct line; we should consider it as an evolutional "forest" with many branches, of which only one leads from the australopithecine-like forms through the pre-sapiens group, to Cro-Magnon man of the Upper Ice Age, and so to modern man.

Modern man, by H. Hemmer

There are many different opinions as to when the various modern human races originated. Some researchers believe that the roots of the major races go back even before the last Ice Age, perhaps even as far back as the primitive man. According to evaluations of skeletal and cultural remains from the final phase of the last Ice Age, the principal evolution might have been only in this later period. However, it is not yet possible to draw any conclusions on this question; we still need to wait for additional fossils.

In physical anthropology, which deals with the non-cultural, i.e. biological aspects of human evolution, the zoological concept of "subspecies" when referring to the subdivision of an animal species, is usually not used. Instead, anthropologists speak of "races" which in turn can be grouped into basic racial types. The latter can be compared to some extent with the true subspecies of an animal species. This means, for example, that someone belonging to the basic European racial type is, in principle, as close to a member of the Mongolian race as a South African leopard is to a member of an Asiatic subspecies. All modern humans are usually grouped together under the scientific name of *Homo sapiens sapiens* (as opposed to the fossilized early man, et al). There is only one subspecies, although the individual basic racial types might have been considered separate subspecies because of differences between them. The concept "race" is used for groups of somewhat uniform populations in certain areas, which are distinguished by the frequency of occurrence of various hereditary factors. One cannot speak of "pure races" in the sense of a racial ideology; the biological definition of race implies no value judgments.

Modern human races

The Human or Man (*Homo sapiens*; Color plate, p. 3, 75–78) is therefore, one species which can be subdivided into various basic racial types, each with a certain number of particular races:

Anthropological characteristics

1. The EUROPOID type has pronounced features and a small, high nose. His hair is sleek to curly, while the skin and hair color are normally fair. These men already existed in the final stages of the last Ice Age, and afterwards in Europe and western North Africa as CRO-MAGNON MAN. Modern races of today include, among others, the NORDIC race which is very light-skinned and blond, and is found mainly in Northern Europe; the MEDITERRANEAN race which is somewhat darker, and is found mainly in the Mediterranean area and as far east as Bengal; the ALPINE race, of which the DINARIC race has an especially short head; the ARMENIAN, ORIENTAL, and INDIAN races extend from eastern Mediterranean to Southeast Asia. In many non-European regions there are transitional races with Negroid and Mongolian races. During the last few hundred years, these races have moved from their original area and settled in all parts of the world, in the wake of colonization; they have partially displaced those races native to the areas they moved into, and in other cases they have mixed with those races.

Fig. 2–4. 1. Negro (and Khoisanic): original range; 2. Negro, further dispersion caused by slave trade; 3. Austral-Asian races.

2. The NEGROID type was originally found only in Africa south of the Sahara; its evolution is still largely unexplained. The jaws have a tendency to protrude (prognathism). The skin is very dark, the nose is flat, the lips are thick, the hair is curly, and there is little body hair. There are numerous specific races, including the SUDANESE with very distinct negroid characteristics; the ETHIOPIAN, which is a very tall transitional form between the Negro and the European, found mainly in Ethiopia and East Africa; the BAMBUTIC race, the PYGMIES of the Ituri forest, which are short (men are about 140 cm) and body and head proportions like those of children; the KHOISANIC race (Bushmen and Hottentot populations) which might be considered as a separate basic racial type; they were distributed all over the African continent after the last Ice Age (period of rains in Africa), but were later driven out to the more arid regions of Africa by the Negroes and Europeans. Today they are generally slightly built with fair skin, and a yellowish, flat face, and slanted eyes.

Fig. 2–5. 1. Europeans, original dispersion; 2. Europeans, through migrations since the Seventeenth Century have populated vast areas; they are also found in the areas in between.

3. The MONGOLOID type has a flat face with protruding cheekbones. The bridge of the nose is low, or may be hawk-nosed. The eye-opening is narrow and most have the Mongolian eye folds. The hair is straight and black; the skin is yellow. There is little body hair. According to Thoma, these people probably originated from forms related to the Neanderthals in Central Asia in the beginning of the last Ice Age. Some are especially adapted to a cold climate. There are many specific

Fig. 2–6. Mongolians, original dispersion.

races in Asia and America: the SIBIRID race and the TUNGIC race are found in Siberia, and include the Baikal types which closely resemble the ancient Mongolian; the SINOID race on the eastern Asiatic continent; the ARCHAIC MONGOLOID in the forest areas of Southeast Asia; the ESKIMOS from the North American Arctic area; and the INDIAN race, a grouping of mongolians in North and South America, including many such specific races one of which is made up of the Indians of North America.

4. The AUSTRALOID type (Australia), the VEDDIDS (India), and AINUID (ancient eastern Asiatic-Japanese population) races were previously considered partially as the ancient base of the Europeans race; however, they can also be combined as a part of a distinct very old super race (AUSTRAL-ASIATIC RACIAL GROUP). The MELANIDS (East India) and the MELANESIAN races (New Guinea and Melanesian Archipelago) also belong, in part, to this group while some other authors now place them with the Negroids. The POLYNESIAN (Pacific Archipelago) race probably came from a mixture of populations belonging to various super races. [Von Eickstedt's terminology has been simply translated, since in English classification systems of human races no comparable terms exist in many instances. Editor]

Aside from his erect posture, humans differ from the other higher animals among other characteristics in the development of his frontal lobes. Stephan's recent research showed that the ratio of cortical tissue and the rest of the brain between humans and chimpanzees is nearly three times as much as that between a simple hedgehog or shrew-related animal and a chimpanzee. This enlargement of the brain makes possible humans' large intellectual capacity. The results of this capacity, which have a wide range of effects, are for the most part, no longer only the concern of zoology, but also of philosophy, humanities, the arts, etc.

Man's speech

Human language ability is a relatively recent acquisition. It is a unique accomplishment in the world of living beings because of its specific symbolic character and the fact that language is learned, contrary to the vocalizations of most animals which constitute hereditary languages. Symbolic language makes it possible, as Schwidetzky says, to express events or ideas, independent of personal experiences from the past, present, or future, as well as those images that exist only in the mind. Through his language, man constructs for himself a view of life that extends beyond what is tangible, i.e. "physical," to the intangible, "the metaphysical."

Instinctive behavior and culture

Like the behavior of his relatives in the animal world, human behavior is partially dependent upon an innate basis. It is typical of humans that this innate basis is overlaid by culturally determined behavior and intellectual accomplishments; thus humans became to a large extent freed from innate determinants. In many areas, human actions are governed by learned elements, and not directly by hereditary factors. Even basic biological "functional systems" of behavior, such as propagation, finding food and feeding, are loosened up or changed in man by custom, tradition, through culturally determined social requirements (e.g. division of labor), or they include cultural components interspersed between the basic biological motivations and the final consummatory actions.

Thus humans have been able to separate themselves more and more from their natural environment, and even to master this environment in many instances, and often to destroy it. Like most animals, the Ice Age ancestors of modern humans lived in a certain kind of environment consisting of forest areas, broken up by shrubs and grassy patches in the immediate vicinity of water. With the progressive improvement of tool and garment making, the building of shelters and the use of fire, humans became so independent from their environment, that they could make their own. Consequently, they were able to settle in areas which were originally uninhabitable for them. This voluntary withdrawal by humans from the natural and the individual adjustment to a new, artificial environment is, to a certain extent, comparable to the domestication of animals, hence one speaks of the self-domestication of humans.

Contrary to all other living beings, humans are in a position to sidestep the laws of natural selection as they apply to themselves; laws which govern the evolution and transformation of species in other animals. Thus, humans can ultimately influence the evolution of their own species either for their own advantage or otherwise.

The biological basis of human behavior, by E. Klinghammer

Humans have always tried to understand themselves and find explanations for the motivations that underlie their behavior. Throughout human history groups of people, societies and nations have also tried to influence, shape and control behavior. Finally, people have always sought the good life, i.e. tried to improve their conditions in a material as well as spiritual sense. Since this encyclopedia is a biological work, we will focus on human behavior from an evolutionary perspective. Philosophers, theologians, physicians, historians, cultural anthropologists, sociologists, clinical psychologists and psychiatrists all have examined human behavior from their particular theoretical frame of reference. Each being concerned with particular aspects of human behavior, specific problems, special needs.

Since the environment in general and the social setting and the entire culture in which people grow up exert such a powerful and easily recognizable influence on behavior, it is not surprising that environmental factors as determinants of human behavior were so important in influencing the ideas of scholars over the centuries.

C. Darwin

With the rise of modern biology, especially with *Charles Darwin's* publication of the *Origin of Species*, humans were recognized as part of the natural world and as much the result of the evolutionary process and natural selection as all other animals. With his book *The Expressions of the Emotions in Man and Animals*, published in 1872, the study of human behavior was discussed from the point of view of phyletic descent, as was the behavior of all animals. In other words, it was not only structure that was considered subject to the forces of natural selection, but behavior as well.

C. O. Whitman and O. Heinroth

Along with the growth of biology in general, the emphasis on the biological basis of behavior in animals grew steadily. *C. O. Whitman* in the United States and *O. Heinroth* in Germany had studied pigeons, ducks and geese, respectively, on a comparative basis with emphasis on the evolution of behavior. While writing and lecturing on his bird studies, Heinroth suggested that humans could be studied from the same ethological and evolutionary perspective that had been so fruitful in studying the behavior of animals. He had laid the foundation for this methodological approach in his comparative studies of European birds, and especially through his technique of handraising birds in isolation from conspecifics and other normal environmental inputs which they would encounter in the natural situation.

The deprivation experiment

K. Lorenz

It was left to his student, *K. Lorenz*, a Nobel Laureate in 1973, to provide the impetus for looking at humans through the eyes of an evolutionist culminating in a controversial book, *On Aggression*, in 1962. Lorenz' personal synthesis of how he viewed human behavior through the eyes of an ethologist, while based on much solid animal data and his keen, albeit unsystematic observations on people, brought him not only praise but stirred up a storm of criticism as well. However, from this time on, the biological basis of human behavior was studied in earnest, and is very much a part of the science of humans today. Other popular books about human behavior from an ethological perspective appeared by D. Morris, R. Ardrey, and others. They raised more questions and made ethological concepts household words in a short time.

I. Eibl-Eibesfeldt, The ethological study of human behavior

At this time, *I. Eibl-Eibesfeldt*, then a research associate of K. Lorenz, gave up his animal studies and began to film sequences of human behavior. He travelled to Africa, the Maledive Islands off the coast of India, to Brazil, the Amazon Region, Japan and various places in Europe. Using a technique of filming people unawares, developed by his colleague *H. Hass*, by placing a prism before their regular camera lens which enabled them to film people at right angles from the direction in which the camera was pointing, they began to collect human behavioral data. Not since Darwin, and certainly not with modern equipment, had anyone looked systematically at simple and complex expressive behavior patterns in humans in different cultures. Demonstrating the existence of universals in human behaviors, i.e. those behavior patterns not modified by cultural influence, Eibl-Eibesfeldt's studies on the !Ko Bushmen in the Kalahari Desert and Waika (Yanoáma) Indians in the Amazon provided the beginning of a human ethogram, the behavioral repertoire which ethologists consider an indispensable prerequisite for any study of behavior in animals. His Human-Ethologisches Institut (Institute of Human Ethology) in Percha near Munich, Germany, was the first of its kind.

However, the time for these kinds of studies had apparently arrived. *P. Ekman*, a psychologist at the Institute of Non-Verbal Behavior in San Francisco, had studied large numbers of people in New Guinea, Japan and in the US in similar situations and recorded their facial expressions. He produced studies more quantitative than Eibl-Eibesfeldt's but he found the same universal expressions in cultures that had never had any contact with each other, and he was able to demonstrate how cultural expectations can modify the expressive behavior of people.

P. Ekman

Unknown to these investigators, an anthropologist, *E. Count*, had been quietly working over the years building an empirical model of the biological basis of human sociality which he called the *Biogram*. Trained in neurophysiology as well as in anthropology, he traced the beginnings of parental behavior, for example, from fishes through amphibians, reptiles, and mammals to the primates and humans. Pointing to the neural structures found in each group, which are correlated with advanced and increasingly more complex behavioral organization, he showed the presence of a basic vertebrate behavioral organization at all levels, with advances added at each higher level. Henceforth, the biogram of a particular species, including humans can no longer be ignored by anyone who is aware of the evolutionary history of animals.

E. Count, The biogram idea

E. D. Chappel, another anthropologist, added a new dimension to the understanding of the biological basis of human behavior by studying the effects of biological rhythms, e.g. circadian rhythms, on personality, emotional behavior, temperament and many other variables. Topics such as territorial behavior, the cultural dimensions of space behavior, and behavior in institutions, among many others, he showed, are firmly rooted within species-typical human constraints.

E. D. Chappel

E. T. Hall, an anthropologist, building on the studies of the Swiss Zoo director *H. Hediger* on wild animals in captivity, applied the latter's discoveries about individual distances to conspecifics and other animals to humans. The resulting research strategy called *proxemics*, focuses on when, how, and under what circumstances people maintain certain distinct distances between themselves and others in various social relationships to themselves. While these distances vary within sub-cultures, social classes and different societies, once acquired, these behaviors are extremely stereotyped and violations evoke various degrees of agonistic behaviors or at least feelings of discomfort. In other words, the behavior norms are learned, but the reactions to violations seem to be the same across cultures.

E. T. Hall and H. Hediger

A psychologist, *R. Sommer*, has studied the reactions of individuals toward physical space, and the effects of space on social interactions among people. Without dealing with the questions of a biological basis for humans' use of space, he showed how people are responsive to spatial characteristics in their environment of which most of us are not aware.

R. Sommer

A sociologist, *E. Goffman*, deals in an informal manner with observa-

E. Goffman

tions of people among themselves in various social settings. He provides no quantitative data or any other documentation, an omission for which he has been much criticized by some. However, in appealing to the personal experiences of his readers, he deals with human behavior in the everyday world. The rules of proper conduct in different cultures may vary, he says, but they show the same regularity and function as in other animals, especially social species, where coordination of activities and cooperation toward various ends are an important part of the adaptive behavior repertoire.

C. S. Coon

C. S. Coon, an anthropologist, in his book, *The Hunting Peoples*, gives an overview of the various adaptations of humans in different ecological conditions. From hunting strategies, to tools employed, to band and family and tribal organizations, he shows the wide ranges of human culturally adaptive strategies for survival, as well as common behavior patterns and needs shared by people living in different parts of the world.

A. Scheflen

In a clinical setting, *A. Scheflen* focused on non-verbal behavioral cues during psychiatric interviews and found, that, often more than the verbal responses of clients, they could reveal much about the person's behavior. He noted that postural cues preceded shifts in verbal messages, and that non-verbal cues emanating from the psychiatrist affected the client's behavior, although neither of them was consciously aware of them. Since then, people have become aware of body language in humans; behavioral cues long familiar to ethologists who were studying the postures and movement patterns of animals.

D. G. Freedman

D. G. Freedman, a psychologist, and his students have studied identical and fraternal twins and siblings from birth on, as well as children in different cultures in the first week of life. They found concordance of an extraordinary high degree among identical twins on the various measures such as smiling, level of activity, and crying to various stimuli. There was more variability on these measures in non-identical twins or siblings in the same home. In other words, the same environment did not bring forth identical responses in these children. Freedman also found that Oriental children differ from Caucasian children on several measures during the first few days of life. E.g. Caucasian children cry sooner and longer when put down alone than the Oriental children.

W. R. Charlesworth

W. R. Charlesworth, a child psychologist, is studying the greeting behavior of pre-school children, the development of tool-using in young children, and the role of intelligence in adaptive behavior. His approach is experimental and in a laboratory setting, but he focuses on behavior adaptive in the normal environment in which children are usually living, and from an ethological and evolutionary perspective.

R. Fantz

Responses to visual stimuli on the first day of life were investigated by *R. Fantz*, a psychologist. He found that some stimuli with face-like characteristics, e.g. eyes, shapes, etc., elicit orientation and focusing in these newborn children, whereas control stimuli did not elicit such responses.

E. H. Hess

E. H. Hess, an ethologically oriented psychologist, has done much to show the importance of ethology to the study of not only animal, but also human behavior. His work on imprinting, his development of pupillometrics—the measuring of pupillary responses in humans to various kinds of visual, auditory and olfactory stimuli, and his training of students who share his ethological, evolutionary attitude in behavioral research, mark him as an influential proponent of ethological research in human behavior.

N. Tinbergen

In Europe, *W. Wickler*, *P. Leyhausen*, *Blurton Jones* and *M. R. A. Chance* all have applied ethological methodology and thinking to the study of human behavior. *N. Tinbergen*, another pioneer ethologist and Nobel Laureate, and his wife, have applied ethological observational techniques to the study of autistic children with startling results in many instances. Recognizing motivational conflicts familiar to them from their animal studies, they hypothesized that the motivational conflict in autistic children was one between social or bonding tendencies and timidity or fear. They suspected that when the fear system is more sensitive to stimulation, the bonding system is starved to a degree that even normal parents may fail to establish a bond. Applying the analysis of non-verbal behavior to the assessment of the motivational conflict, in fact being able to detect it in the first place, it is possible to respond properly at the right time to the specific behavior patterns of the child and gradually reinforce behavior that leads to the establishment of bonds and a break-down of fear responses in these children.

A shared attitude

There are many more people who study humans from an ethological and evolutionary perspective. While many of these people come from various disciplines, they share a concern about human behavior as part of animal behavior, i.e. seeing humans as evolving in history long before cultures and societies as we know them today were present. They believe that these prehistoric adaptations are still with us and influence behavior as much as they ever did.

The environmentalist position

This is not the place to elaborate on various environmentalist theories which hold that humans at birth are a *tabula rasa* upon which experience engraves all that we know and do. For example, the concern with violence, war and aggression has led for a search on how these tendencies in humans can be modified. On the other hand, ethologists have called attention to the species-typical nature and adaptive function of agonistic behavior in animals and suggested that humans similarly possess a tendency for aggressive behavior, aside from any environmentally elicited aggression. They also make a distinction between intra-specific and inter-specific aggressive behavior, and they point out that, far from condoning violence and war, we must study humans as having both phylogenetic adaptations as well as culturally acquired behavior if we are to come up with realistic solutions.

These issues are often not discussed with the necessary detachment characteristic of scientific inquiry. Space limitations do not permit a comprehensive discussion of these questions. The supplementary readings at the end of this volume contain a list of the most important books on these topics to help the reader examine these problems further. Those authors who were not listed were not omitted for any special reason, other than that limitations of space and time prevented their inclusion.

3 The Colugos or Flying Lemurs

Orders: Flying lemurs, by T. Schultze-Westrum

The tropical rain forests of southeastern Asia are the homeland of various animal species that systematists cannot fit into one of the existing family groups. The ancient lanthanotid (*Lanthanotus*; see Vol. VI) is a reptilian representative of this particular group, while the mammals in this group are represented by the three shrews (Tupaidae; see Vol. X) and the colugos or flying lemurs. Some of the morphological characteristics of these animals are unique; other body features, however, show these animals to be close to more than one family group. Consequently, the COLUGOS or FLYING LEMURS (family Cynocephalidae) have characteristics of body structure which indicate relationships to bats, insect-eaters, and prosimians. However, they have other unique characteristics, so one cannot consider them as merely a transitional link between these groups. In 1886 W. Leche suggested that colugos should be considered an early evolutionary deviation; and today they are classified as an independent order (Dermoptera), in close proximity to the bats (family Chiroptera). Paleontologists have found *Planetetherium*, a very close relative of the present colugos, in the stony strata of the Paleozoic; these animals were unique even then.

Distinguishing characteristics

There are two living species of colugo, the TEMMINCK FLYING LEMUR (*Cynocephalus temminckii*), which is about the size of a cat, and the somewhat smaller PHILIPPINE FLYING LEMUR (*Cynocephalus volans*; see Color plates, pp. 83 and 84). Their color varies greatly. The basis color of the back is from brown to gray interspersed with yellowish-white spots. The underside is bright orange, brownish-red, or yellow. The hair is very soft and thick. This animal has a doglike head (in English Cynocephalus means dog's head). The neck is long and maneuverable, while the body is slender, with long extremities.

The most unique feature of the colugo is its "parachute-skin" which surrounds almost all of the body. During the gliding flight, the front flightskin (propatagium) is stretched between the sides of the neck and the arms; it is connected to the large side flightskin (plagiopatagium), which stretches between the front and back legs. The tail flightskin (uropatagium)

Fig. 3-1. Colugos: 1. Temminck flying lemur (*Cynocephalus temminckii*); 2. Philippine flying lemur (*Cynocephalus volans*).

stretches from the hind legs to the tail on both sides. The animal cannot actually fly with these gliding surfaces, as can bats, for example. Movement through the air is confined to gliding; steering is achieved by shifting the position of the legs and tail.

All other mammals which have developed flightskins for gliding—such as the pygmy flying phalagers (genus *Acrobates*), the honey gliders (genus *Petaurus*), and the greater gliding phalagers (genus *Schoinobates*) among the marsupials, and the scaly tailed squirrels (genera *Anomalurus* and *Idiurus*), and the flying squirrels (genera *Glaucomys* and *Petaurista*) among the rodents—have no tail flightskin, but, as a compensation, have a steering tail which is as long as their bodies. It is believed that colugos can glide from one tree to another over a distance of up to 70 m; of course, this is only when the take-off point is high enough, since the animal loses altitude during glides over long distances.

Colugos are strictly vegetarian. They eat only leaves, buds, and young seed pods. The intestinal tract has special features as a result of this specific foods preference; the caecum is large and the small intestine is shorter than the colon. The teeth have a special structure. The upper eyeteeth and the two outer upper incisors have two roots each, a characteristic that occurs among mammals only in a few insect-eaters (Insectivora); generally the eyeteeth and incisors of mammals have only one root. The lower incisors are notched, comblike, having about ten long tines on each tooth, for reasons yet unknown. H. Wharton surmises that the animals use these teeth like combs for their skin care. although he can offer no proof for this hypothesis. Colugos also have an involuted "under-tongue"; this feature is much more developed in the prosimians.

During the day, colugos, by nature unsociable animals, hang from the undersides of branches, gripping them with all four feet. We are familiar with this behavior in the South American sloths. In the dim light in the huge trees of the rain forest, these animals are well camouflaged by their gray-brown color. As night falls, they become active. They move along the undersides of branches with remarkable speed and they scramble up tree trunks with powerful jumps, thus reaching the necessary height for a glide to the next tree. According to Wharton's observations, these animals generally use fixed routes. The people of the Philippines take advantage of this by hunting colugos along their paths with bow and arrow or by placing traps in the branches the animals use.

The females give birth to only one young at a time. It is born in a semi-fetal state, almost as underdeveloped as a young marsupial. During the first weeks of life it fastens itself securely to one of its mother's two nipples, where it nurses. The mother carries her young with her, even during glider flights. Even the older young which have already grown their full hair are also carried along by the mother; they hold on to the hair on her breast.

The colugo's particular specialized life style calls for an additional range of special adaptations. J. L. Harrison, for example, noted that animals which hang head down or, as in the case of the colugos, parallel to and under branches, might be expected to soil their hair when defecating. Moreover, the tail flightskin forms a kind of sack underneath the anal opening when the animal hangs in a perpendicular position. However, colugos defecate infrequently, and when they do, they void large quantities of excrement at one time. In order to do so, they bring their body into a vertical position with their head upward. The tail and its flightskin is then raised and pressed high against the back in such a way that the anal opening is absolutely free. In this way colugos avoid dirtying themselves.

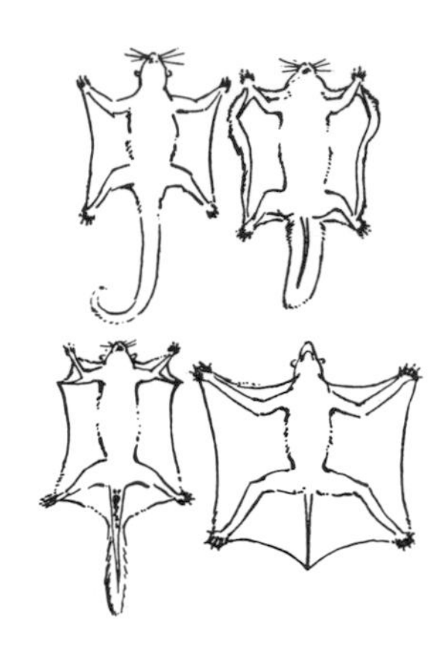

Fig. 3-2. Unconnected to each other, different animal groups have developed converging adaptations for gliding flight. The size of the flightskin differs, as does the widening of the tail. None of these aminals can really fly; they can only glide from elevated positions, although they may glide over very substantial distances. Top: left, honey glider (*genus Petaurus*; see Vol. X); right, flying squirrel of the genus *Pteromys* (see Chapter 10); Bottom: left, scaly tailed squirrel of the genus *Anomalurus* (see Chapter 10); right, colugo (genus *Cynocephalus*).

The colugo species are not directly threatened by man. The greatest danger for the survival of these interesting animals lies in the fact that their natural environment is gradually becoming smaller through deforestation. In 1948 Wharton reported that the Philippine flying lemur on the island of Bohol owes its survival to the superstition of the people there. The once-dense forests on that island were extensively harvested, but the trees on whose leaves colugos fed were spared. This, however, was not the result of the foresters' concern for the animals; the people believed that there were malicious ghosts living in these trees, and that if the trees were cut down, the ghosts would cause illness and other mishaps in revenge.

Colugos may indeed strike us as peculiar ghosts of the night. They are not afraid of human settlements. In the suburbs of metropolitan Singapore, surrounded by Malayan and Chinese homes and apartments, is the Bukit-Timah reservation with the last remnant of the island's original rain forest (primary forest). In 1965 I was able to observe colugos there, even through the noise of car horns and crying children in the nearest huts. I went to the reservation in the early evening with a powerful light. I had examined only a few trees when the light's beam struck a bright orange- red body; it was the underside of the lemur in flight. The animal landed noiselessly on a tree some 20 m from me. Before landing, the body axis of the animal, with the head up, was turned parallel to the tree trunk. At first the animal sat motionless with its head outstretched and its tail turned up against its back. The bright color of the underside of the tail flightskin contrasted sharply with the camouflage color of the topside. Possibly this bright color scares away enemies. In the excitement, the animal defecated. After a few minutes it climbed up the tree trunk with short strong jumps. Suddenly, without a noticeable push from the legs or a preflight orientation toward a target, it dropped from the tree. Again the animal passed me in its silent glide. This flight was directed, without a noticeable loss of altitude, to a tree some 30 m from the starting point. I was deeply impressed by the elegance and seeming weightlessness of the animal's body during the glide.

Unfortunately, no zoo has yet been able to keep a colugo alive for more than a few months. This is why our knowledge about the way of life of these remarkable animals is still so limited.

4 The Bats

Like the flying lemurs discussed in the previous chapter, there are other mammals which have the beginnings of a flightskin development. These mammals include the flying marsupials (see Vol. X), flying squirrels (see Chapter 10), and the scaly-tailed squirrels (see Chapter 10). These animals can spread their flightskins, which are attached to both sides of their bodies, by spreading their front and hind legs when jumping from tree to tree or branch to branch; the flightskins are extended much like parachutes. The animals are able to travel some distance through the air, although their flight is always downward, an inactive gliding, almost a jump flight. It is never an actual flight like that of the bat.

Order: Bats, by M. Eisentraut

The only mammals truly able to fly are the BATS (order Chiroptera). Their front limbs have developed into real wings. This development became possible when the upper and lower arm, particularly the metacarpal bones and fingers, with the exception of the thumb, all grew quite long. These bones serve as a support for the elastic wing membrane (patagium) which originates on both sides of the body. This membrane is divided as follows: The front or forward part of the membrane (propatagium) extends from the neck to the thumbs; from there, the finger membrane (dactylopatagium) extends between the fingertips and the tarsal bones (plagiopatagium); finally the tail membrane (uropatagium) stretches between the hind legs in such a way that the tail, when present, is totally or partially enclosed. Many species have a special bone buckle, the spur or calcar, which begins at the heel and follows the outer border of the wing membrane; in many species this calcar has special skin flaps. The end of the short thumb is not enclosed in the membrane; it has a sharply bent, pointed claw, as do the toes. The other fingers do not have claws; only the fruit bats, with a few exceptions, have a small, underdeveloped, and apparently functionless claw on their second finger.

The wing membrane is a continuation of the epidermis, the pigment layer, and the corium of both sides of the body. The wing membrane of some species begins high on the back. It is very elastic and interlaced with fine muscle fibers, nerves, and blood vessels. These blood vessels show independent rhythmical pulsations, which guarantee an equal supply of blood, even to the farthest regions of the membrane.

The special stress on the front limbs during flight is met by the strongly developed flight muscles and the very firm humeral ligament. The large pectoral muscle is attached to the ossified sternum, which has a flat rib in the middle, as in birds. The shoulder joint is a complicated ball-and-socket joint which allows the wings to turn in rowing movements. The elbow, hand, and finger joints are hinged joints, which, together with the corresponding muscles, give a firm support to the outstretched wing surface. During a rest the wings are folded and lie close to the body, while the flightskin shrinks and becomes smaller.

Distinguishing characteristics

The HRL ranges from 3 to 40 cm, the wingspan, between approximately 18 and 140 cm, and the weight, from 4 to 900 g. The front limbs have developed into wings. The back limbs are usually relatively weak and serve basically as body supports during walking and climbing. The knees, unlike those of other mammals which are oriented to the front and bottom, are directed out, towards the back, because the pelvic joints are turned to the side. The feet have toes with strong claws, used for gripping and hanging. The center of gravity is in the middle of the body because of the powerfully developed chest and weak development of the back legs and pelvic girdle. The body is held in a stable position during flight. While resting, bats position their bodies so that the head is down. They have between twenty and thirty-eight teeth (for tooth formulas, see individual families). The intestine is short, especially in insectivorous bats, and thus digestion is rapid. Females have two nipples (with the exception of the genus *Lasiurus*); they give birth to one, rarely two, young. There are two suborders: 1. FRUIT BATS Megachiroptera ; and 2. INSECTIVOROUS BATS (Microchiroptera); with nineteen families, approximately 200 genera, and approximately 800 species.

Bats' flight

In normal flight, bats do not move their wings from top to bottom and back again; instead they make a rotating motion so that the tips of the wings follow an elliptical course. The down stroke moves from high in back to low in front to high in back, with the narrow side of the wing facing forward. The idea that the flight of these mammals is clumsy fluttering compared to the elegant flight of many birds is incorrect; we are only concerned with flight and how it corresponds to all necessities of life for the separate species of animals which have the ability to fly. In this respect, we are familiar with all kinds of flight, ranging from flapping to hovering flight.

The number of wing strokes depends to a large extent on the size of the animal. In the large fruit bats, the rowing strokes are made much more slowly than in the smaller, insectivorous bats; the larger bats' flight can be compared to that of crows and ravens. The European LARGE MOUSE-EARED BAT (*Myotis myotis*) makes approximately ten to twelve wing strokes

Fig. 4-1. The flight patterns of specific species can be deduced from the shape of the wings: above, the long, narrow wings of the rapid-flying common noctule (*Nyctalus noctula*); below, the wider wings of the much slower greater horseshoe bat (*Rhinolophus ferrumequinum*).

per second during a flight of normal distance. In contrast, the much smaller LESSER HORSESHOE BAT (*Rhinolophus hipposideros*) makes approximately seventeen or eighteen wing strokes per second. The flight method also depends largely on the shape of the wings. As it is with birds, so it is among the bats: the fast fliers have long and narrow wings, while the species with broader wings fly more slowly. The LONG-WINGED BAT (*Miniopterus schreibersi*), with its noticeably long and narrow wings, is estimated to have a speed of 50 km/h; the NOCTULE BAT (*Nyctalus noctula*), another narrow-winged species, is also believed to be able to reach a speed of 50 km/h during normal flight. The broad-winged European little brown bat, on the contrary, travels only 15–16 km/h. Many smaller species distinguish themselves by the amazing maneuverability of their flight, something which no bird can match. One can make this discovery while trying to catch a COMMON PIPISTRELLE (*Pipistrellus pipistrellus*) with a butterfly net when it is flying around the ceiling of a small room. The animal evades each thrust like a flash and always finds a new escape route. The patterns of flight for the many and varied bat forms are all completely different depending on the way of life; each pattern is, without a doubt, the most perfect motor pattern for the specific species.

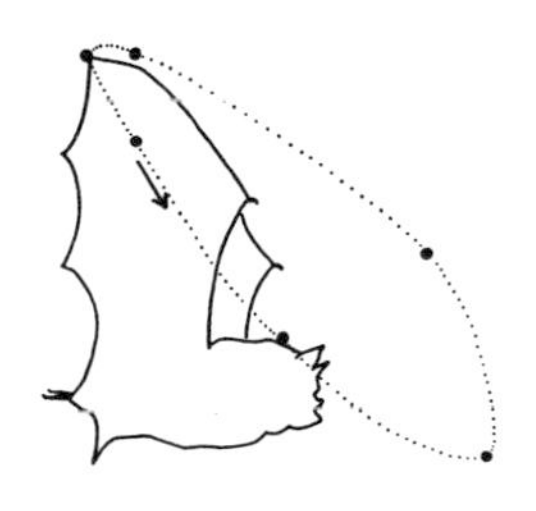

Fig. 4-2. The wing tips of an insectivorous bat follow an elliptical course during flight. During the upward stroke the flightskin is somewhat folded to reduce the air resistance. The bat shown is the lesser horseshoe bat (*Rhinolophus hipposideros*).

Many species, particularly the blossom-feeders, have a hovering flight during which the animal is able to suspend itself in mid-air for a long period of time, with its body erect. The rowing flight occasionally may also be broken up by gliding flights, when the animal uses the speed it has attained in its rowing flight to sail through the air without wing motions. Some larger fruit bats even seem to use upward (warm) air currents as a source of energy when flying long distances over land, and they glide along through the air with outstretched wings. Many bats build up the necessary initial velocity to be airborne by dropping from their resting place, stretching their wings, and flying away in rowing flight. Other species raise their bodies from the perpendicular resting position to a horizontal flight position with a few wing strokes; only then do they release their feet from their resting place. Most insectivorous bats are able to take off from the ground, however, by making a small jump. The flight speed is reduced on landing by a turning of the wings and the tail flightskin. When insectivorous bats wish to land on a wall or branch, they maneuver their bodies into a downward orientation with one final wing stroke during the last phase of the approach, and then immediately hold on to the resting place with the talons of their hind toes.

Climbing and running

Once the bat has folded its wing membrane, it becomes a real quadruped. When climbing or running, it draws itself up by the cushiony callous pads at its wrists and is supported by the soles of the hind feet. The feet are turned out, somewhat apart from the body, in straddle fashion. Bats do not move around awkwardly on the ground as is so often presumed; some species are even rather good runners. Bats can also climb excellently; the thumb and its claw play an important role in climbing.

The thumb is only partially enclosed by the flightskin, or it is not enclosed at all; therefore, it has retained a great deal of its maneuverability. When climbing a perpendicular wall or a tree, the bat grabs alternately with its front limbs, anchors itself with the thumb talons, and then heaves its body up. The insectivorous bats use a swinging, brachiating movement on ceilings and overhead trusses; in this case the claws are used as hooks from which all of the body weight is suspended. The fruit bats often brachiate along branches when in their resting trees.

Swimming

Although swimming is not one of the normal motor patterns of bats, an insectivorous bat is by no means helpless once it has fallen into water. It raises its wings, and by making fluttering movements, it manages to move on the surface of the water, more by hopping than swimming, usually reaching land rather quickly.

Evolution

Bats, without a doubt, must have developed from non-flying, running or climbing ancestors. The few fossil discoveries show that the development into a flying animal must have taken place very early. The oldest petrified remains of insectivorous bats date from the Lower Tertiary; the best-preserved finds are those from the Eocene layers at Messel, near Darmstadt, Germany. At that time, approximately 50 million years ago, there were fully developed insectivorous bats which differed from present-day forms by, at most, a somewhat more primitive body structure. Thus, the origin of bats must lie much further back in time.

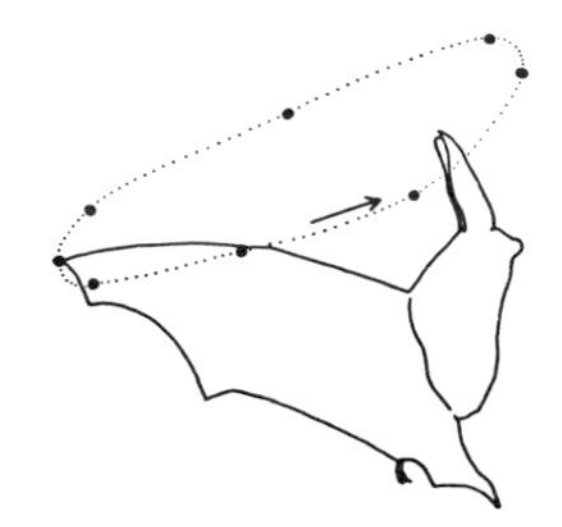

Fig. 4-3. Many insectivorous bats, like the long-eared bat (*Plecotus auritus*), shown above, are able to remain stationary in the air without forward motion; this is particularly important for the tropical species which visit blossoms.

Bats are very similar to other insect-eaters in their internal structure. The earlier forms from which all the higher mammals later developed at the beginning of the Tertiary period, came from the insect-eaters during the Upper Cretaceous period. We can assume with certainty that the bats descended from the primitive insect-eaters which lived in trees and moved about by jumping and climbing from branch to branch; they searched for their prey in cracks in the bark or in leaves and blossoms. During this development phase, the changes in the limbs of these ancestors must have ultimately led to the development of real wings. Perhaps skin folds first developed on both sides of the body, and a parachute-like, expandable flightskin may have appeared between the front and back feet. The flying marsupials (see Vol. X) had a similar development. The flightskin eventually enlarged and stretched between the neck and lower arm and between the back feet and the tail. At the same time, the fingers began to increase in length, opening the possibility for the development of actual flight. The front foot developed into a wing.

We do have indirect proof that the development proceeded as we have described it; it is a known phenomenon that evolutionary developmental stages often appear briefly during the embryonic stages of the individual animal or are suggested in some manner. Indeed, the early embryos of insectivorous bats initially have only the beginnings of skin creases on both sides of the body. These skin creases expand later on, due to a lengthening of the finger bones which continues even after birth. Only after some time do the various limbs acquire the final dimension ratios of the grown animals.

Evolutionary options

With the evolution of this new type of mammal, the bat, vast opportunities for further development became possible. As bats became able to fly, they moved into new areas of habitats and food niches. Their only feeding competitors, the birds, are diurnal animals, with only a few exceptions. Bats, however, adopted a nocturnal life from the beginning, in accordance with their origin from the insect-eaters. The night belongs almost completely to the bats. Thus, the primeval bats, of which we have not yet found any fossils, must have "settled down" in their new niche and begun to move toward increasing specialization and efficiency. The division into two suborders, fruit bats and insectivorous bats, must certainly have occurred very early in the development. Aside from the fossil finds of insectivorous bat remains in the Eocene, mentioned previously, only one fossilized fruit bat (*Archaeopteropus transiens*) has been found, from the Upper Oligocene in Italy.

Worldwide distribution

In the course of millions of years, bats became distributed all over the world. They have been able to settle on faraway ocean islands due to their ability to fly; on some of these islands, they are the only indigenous mammals. The polar regions and barren desert areas are the only places where bats cannot survive. The fruit bats are, of course, found only in the tropics and subtropics of the Old World. Some families of insectivorous bats are restricted to specific areas of the world, while other families are distributed worldwide. The wide variety of forms is indicated by the large number of species; next to the rodents, the bats have the largest number of species among the mammals. New species are still being discovered and described.

Orientation in space

The acquisition of the ability to fly resulted in a series of important biological adaptations. An extremely interesting capacity, which has been accurately researched only during the last decades, is the bat's echo orientation in space. Bats are active predominantly during the night, although many species appear in the late afternoon or have occasionally been observed leaving their sleeping quarters during full daylight. How do these animals find their direction during their nocturnal flight? How is it possible for them to avoid obstacles and to catch their food? Other nocturnal mammals, for example, the prosimians, have very large eyes, specially constructed for twilight vision. Such large eyes are also noticeable in fruit bats. The lowest light intensities are sufficient for them to find their way. However, when fruit bats are forced to fly in a totally darkened room, they land helplessly on the ground. Most insectivorous bats, however, have very small eyes; in many species they are barely the size of a pinhead. These eyes are sufficient to distinguish between light and dark, although they are not sufficient for pattern vision. Such underdeveloped visual organs are therefore hardly suitable for orientation in space. The insectivorous bats have developed another method: echolocation.

Explanations for the following color plates: Today mankind contains a large number of different races that we combine in a number of "super races." The following pages show a sample of the abundance of separate races which all belong to the one species of *Homo sapiens*.

PLATE I
Man (*Homo sapiens*): Aside from his erect walk man differs from the anthropoid apes and others by his relatively short arms. His posture is different, however, depending on age and sex. The proportions change during the course of development from child to adult. The head of the child, in relation to his body, is relatively large; this proportion decreases as the arms and legs grow longer. There are clear differences in general body structure between the sexes. The woman can be, on the average, up to ten percent smaller than the average man; the pelvis width is the only exception. The woman's longer torso and her relatively shorter arms and legs give her body proportions that approach more those of a child than of a man. The proportion of fat tissue is greater in women, but is lower for bones and muscles compared to men. Hence, the female form is softer and more rounded than the male forms which are harder, more angular, and more clearly determined by the configuration of the muscles.

PLATE II
A group of Chinese: In modern society man tends to live increasingly in anonymous groups. Where individuality of expression is minimal, this is often also reflected in similarity of dress. Hence, assimilation is facilitated by simple, unifying characteristics. Nevertheless, "mass man" remains an individual who has strong ties within smaller groups such as family, friends and other organizations.

PLATE III
Masai group: Originally man's life ran its course almost totally within a small group, where all members know one another. The personal bond appeases aggressive impulses; hence individuals are allowed to display ornaments and weapons.

PLATES IV and V
Table of some ethnic groups:

1. Nordic: man from the North Sea region
2. Alpine: man from Central Europe
3. Alpine: man from Central Europe
4. Dinaric: man from the eastern Alpine region
5. East European: woman from Czechoslovakia
6. Lapp: Swedish Lapp man
7. Lapp: Swedish Lapp woman
8. Mediterranean: man from Sardinia
9. Mediterranean: woman from Sardinia
10. Mediterranean: man from Tenerife
11. Berber: girl from Morocco
12. Armenoid: man from Turkey
13. Oriental: man from Afghanistan
14. Indic: woman from India
15. Nilotic: Karamojong man (Northeast Africa)
16. Nilotic: Turkana man (Northeast Africa)
17. Ethiopic: Masai man
18. Ethiopic: Masai woman
19. Sudanic: Fulani woman from Nigeria
20. Sudanic: man from the Upper Volta
21. Sudanic: woman from Nigeria
22. Kafroid: Zulu woman, South Africa
23. Bambutic: Efe pygmy
24. Bambutic: Efe pygmy
25. Khoisanic: Kung bushman
26. Tungic: Mongolian man from Ulan Bator
27. Tungic: Mongolian girl from the Baikal region
28. Sinoid: Chinese man
29. Sinoid: Chinese woman
30. Archaic Mongoloid: man from Bhutan
31. Sinoid-Archaic Mongoloid: woman from South Korea
32. Archaic Mongoloid-Malayoid: woman from Cambodia
33. Eskimoid: Eskimo from Alaska
34. Eskimoid: Eskimo from Canada
35. Woodland: Sioux Indian
36. Woodland: Iroquois Indian
37. Centraloid: Indian woman from Mexico, with child
38. Braziloid: Indian from Brazil
39. Pampoid-Braziloid: Indian from the Paraguay River
40. Polynesian: Maori from New Zealand
41. Polynesian: Polynesian woman
42. Polynesian: Polynesian woman
43. Ainoid: Japanese Ainu
44. Ainoid: Japanese Ainu
45. Melanesian: Papuan from New Guinea
46. Melanesian: Papuan from New Guinea
47. Australoid: Australian aboriginal man
48. Australoid: Australian aboriginal man
49. Australoid: Australian aboriginal woman
50. Australoid: Australian aboriginal woman

1
2
3
4
5
6
7
8
9
10
11
12
13
14
15
16
17
18
19
20
21
22
23
24
25

27 28 29 30

32 33 34 35

37 38 39 40

42 43 44 45

47 48 49 50

1
2 a
2 b
5e
3
4
5 a
5 b
5d
5 c

◁
Man: 1. Ape-man of the A-type (*Australophithecus africanus*); 2. Ape-man of the P-type (*Australopithecus robustus*), a) head, b) skull; 3. Java Man (*Homo erectus erectus*); 4. Neanderthal Man (*Homo sapiens neanderthalensis*); 5. Modern Man (*Homo sapiens sapiens*), a) Australian, b) North European, c) Vietnamese, d) West African, e) skull of a European.

A century ago the great Italian naturalist Lazarro Spallanzani suspended strings vertically in a room and found that free-flying insectivorous bats flew around without touching these obstacles. Insectivorous bats, even when their vision was blocked, adeptly avoided the strings. When their ears were plugged, however, they suddenly were hampered in their ability to orient. These first important insights were hardly noticed at that time, and were soon forgotten. On the other hand, under the influence of the important Frenchman Georges Cuvier the totally erroneous view that insectivorous bats orient themselves during flight with the help of a highly developed sense of touch became popular.

Echolocation

During the 1930s, two scientists, working independently of each other, approached the problem again. Using modern technological methods, a Dutchman, Dijkgraaf, and an American, Griffin, with his co-workers found that insectivorous bats produce sound waves of high frequency. These sounds cannot be perceived by the human ear. The animals perceive objects by emitting sound waves which are reflected by these objects and are then detected by the bats' ears. Bats can thus locate and avoid even the smallest obstacles during blind flight by means of this echolocation, which is like the mechanical sonar. Bats can detect strings of not more than 1 mm in thickness; they can find suitable resting places and insect prey with their echolocation mechanism. They are, therefore, able to orient precisely in total darkness. During further research on this important discovery, the German physiologist Möhres and his colleagues, among others, came up with the surprising conclusion that different insectivorous bats have developed very different orientation systems.

The true echolocating system

The vespertilionid bats (Vespertilionidae) produce very short sound impulses, each lasting only approximately one to two milliseconds; these have a frequency range of thirty to one hundred kilohertz (1 kilohertz equals 1000 cycles per second). These "ultrasonic bursts" are produced in the larynx and emitted through the slightly opened mouth. The ears intercept the echo of the reflected sound waves and thus a "sound picture" of the environment is communicated to the animal. As Möhres puts it: "The 'sound picture mechanism' of the insectivorous bats works presumably along genuine 'echo principles.' The animals detect the difference between the emission of a sound and its return as an echo. The time differential in the arrival of the echo into the different ears gives information about the direction of the reflecting object."

The "sound picture" of the horseshoe bats

A totally different echolocating system was discovered by Möhres in the horseshoe bats (Rhinolophidae). They have peculiar cutaneous growths on their noses (horseshoe-shaped nose flaps with the nostrils located at the base). The meaning of these growths was unknown until that time. It was also noticed that the live animal is able to make very active ear movements. Möhres found that horseshoe bats emit ultrasonic sounds of consistently high frequency (which in greater horseshoe bats is approximately 85 kilohertz, 110 kilohertz in the lesser horseshoe bat). Indeed, these sounds do not come from the mouth, but through the nose, and there is a close connection between the larynx and the nasal cavity. The horseshoe bats can produce their sound emissions with the mouth closed or while catching and eating insects. The horseshoe-shaped flaps around the nostrils are used as a sound cone, almost as a megaphone, during the emission of the ultrasonic sounds; and the curvature of these flaps can be changed by muscle contraction. Thus, the animals are in a position to change the width of the sound cone in accordance with the different distances of the sound-reflecting object. The ears receive the rebounding sound waves and operate independently of one another. The animal, therefore, does not need both ears for sounding objects as the vespertilionid bats do; one ear is sufficient for the echolocation.

Fig. 4-4. A bat is able to find its way even in absolute darkness. The ultrasonic sounds which it emits rebound from objects and the bat changes its direction accordingly.

Möhres, on summarizing this "picture hearing" of the horseshoe bats, characterizes it as follows: "These findings and the fact that one ear is sufficient to guarantee the full operation of 'picture hearing' show that the evaluation of time differences is not important for the sounding techniques of the horseshoe bats; the evaluation of differences in intensity in the rebounding signals is the most important factor in the echolocating techniques of these bats. The moveable ears, in this case, are used as directional receivers." The sounding system of horseshoe bats is far superior to that of the vespertilionid bats because of the cutaneous nose growth, since the directional emitter increases the distance of the acoustical sounding, while the ears, as revolving directional receivers, guarantee the possibility of a rapid and exact detection.

In some bat species, the ultrasonic sound bursts of the "vespertilionid bat type" and the ultrasonic cries of the "horseshoe bat type" are even combined (as in the genus *Asellia*). Other species with special nose growths also emit ultrasonic waves through the nose (for example, slit-faced bats and large-winged bats), although otherwise they use the echo sounding techniques of the vespertilionid bat type. Some fruit bats have evolved an ultrasonic sounding system independently from other bats. The cave-dwelling members of the genus *Rousettus* orient themselves with the help of their relatively large eyes when sufficient light is available like other fruit bats. In total or near total darkness, they can produce ultrasonic cries and use them as echo soundings as do the insectivorous bats. It is interesting to note that these sounds are not made in the larynx, but by a clicking of the tongue.

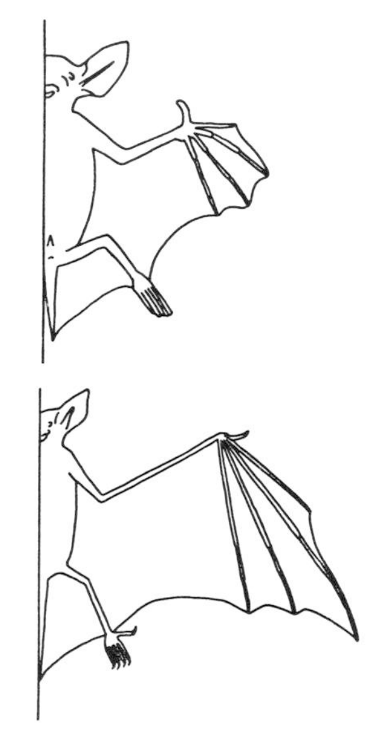

Fig. 4-5. The hand-and-arm skeletons of bats show great differences in ratios during adolescence (allometric growth); The bones of a young Natterer's bat (*Myotis nattereri*) which is only a few days old (above) are much shorter than those of the adult animal. (below).

Research on the echolocation in the insectivorous bats has shown that these animals live in a world of sound. Our old image of the noiselessly gliding night fliers is therefore incorrect. In reality these animals fill the air with their own vocalizations; human ears cannot hear these sounds because they are not able to detect such high frequency tones. It has been shown, however, that many moths hear these bat cries; they are even attuned to the frequencies of the ultrasonic waves emitted by insectivorous bats and react correspondingly. They do not have "ears," but hear with the so-called tympanic organ, located at the thorax or at the back of the body. When a moth enters the sound cone of an insectivorous bat, it immediately dodges and tries to escape; it may even let itself fall, "playing dead." This quick reaction saves many of them from becoming food in the jaws of the insectivorous bats. Here is a case of a surprisingly high adaptation of the prey animal to its predator.

Vocalizations within the range of human ears

Bats produce many other vocalizations which can be detected by the human ear. These sounds are generally produced when the animal is upset or, in the mating season, during the chase or to attract mates. Usually these sounds are very high and shrill, and are best described as chirpings, twitterings, whisperings, screechings, or even howlings. Sometimes one can hear a "rattle" in between the high frequency soundings.

Smell and taste

The senses of smell and taste are also well developed in the bat. Fruit bats are able to detect and discriminate between odors. Their sense of smell guides these fruit-eaters on their nocturnal flights to fruit-bearing trees. This procedure is probably very similar in the blossom-visiting bats. Both fruit bats and insectivorous bats have a variety of scent glands on their bodies which play a role in their social and sexual life, and which likewise are indicative of a well-developed sense of smell.

I tested the sense of taste in the long-eared insectivorous bats in the following experiment: On the first day, I offered the animals some sugar water, which they drank extensively. When I offered them some salt water on a following day, it was refused with lively shaking movements. The same thing happened when I offered water mixed with citric acid, which tasted quite sour. These animals' sense of smell seems to have a development similar to that of man.

Place memory

The memory of insectivorous bats for locations is surprising, especially when one considers that these animals fly around blind, and unlike an animal with eyes, they do not perceive a visual image, but rather a "sound image" of their surroundings. When we let an insectivorous bat fly into a room for the first time, it flew in apparently random circles. In reality, however, these movements were part of a strategy to familiarize the animal with the new area. It flies to each corner, even cupboard corners, and through ultrasonic soundings it "listens" to the walls and objects, building up a sound picture of the particular space. Finally it suspends itself in a suitable place. During subsequent flights, the animal regularly visits this resting place, since by this time it has oriented itself and knows the room exactly. It has a "sound-memory picture."

That an animal can acquire such an exact image of its environment purely by sound probing is an astounding achievement. The insectivorous bat must assemble the separately heard "mosaic pieces" into a complete image. It must, to a certain extent, construct and memorize a map of its habitat from very small clues; in addition, it must simultaneously memorize the movements it made during the orientation flight and record them on its map.

Numerous observations have proved how heavily an insectivorous bat relies on its memory images and how accurately it flies on its "memory map." A *Myotis emarginatus* which has been trained to come and fetch a mealworm, was, on various occasions in succession, fed from an outstretched hand. When the arm was lowered, the bat repeatedly flew to its old feeding place, where the hand would have been had the arm been raised, and searched in the empty space for the hand with the food. When insectivorous bats living free in a room are accustomed to sleeping in an open cage during the day, they will find the cage and its open door without difficulty, once they have completed their flight. When the cage is turned in such a way that the opening faces another direction, the animals will fly to their usual spot and look for the opening of the cage at the normal space. Sometimes, they will even crash into the wall of the cage. If the cage is removed, the animals will fly to the spot where it used to be and try to suspend themselves from the same place where they used to have their resting place in the cage. Once, a large quantity of insectivorous bats perished when, during their absence in the night, the door to the entrance of their sleeping cave was closed. The animals arrived in the morning with their usual speed and crashed with full force against the unknown obstacle, smashing their heads.

All of these observations indicate that insectivorous bats have an excellent memory for the special conditions of their environment, and that they rely heavily on this memory for orientation during flight. This reliance apparently goes so far that the animals seem to omit the constant echo sounding which they use in more confined living areas; otherwise any animal that can "hear" a thin string would not fatally crash into a door that is suddenly placed in its path.

▷ and ▷▷
The Philippine lemur (*Cynocephalus volans*), together with a second species of the same genus, represents a unique mammalian order.

▷▷▷
Most fruit bats are fruit eaters. This Indian short-nosed fruit bat (*Cynopterus sphinx*) feeds on a banana.

▷▷▷▷
Skeletal structure of the fruit bats: 1. Skull of a Tonga fruit bat (*Pteropus tonganus*); 2. Skull of an Australian giant false vampire bat (*Macroderma gigas*); 3. Large mouse-eared bat (*Myotis myotis*), a) Skull, b) Dentures; 4. Vampire bat (*Desmodus rotundus*), a) Skull, b) Dentures; 5. Wrinkled-faced bat (*Centurio senex*), a) Skull, b) Dentures; 6. Skull of the banana bat (*Musonycteris harrisoni*); 7. Skeleton of the Natterer's bat (*Myotis nattereri*).

Reproduction

The reproduction of the insectivorous bats is also singularly adapted to their mastery of the air. Most species have only one young at each birth. They do not prepare a nest or lair; instead, the newborn immediately fastens itself firmly onto its mother's fur, and later holds on independently with its hind feet. The wings at birth are still small and underdeveloped; the feet, which have claws, and the thumbs are large and out of proportion, because they are of vital importance to the newborn as clutching devices.

▷▷▷▷▷
Fruit bats: 1. Gray-headed flying fox (*Pteropus poliocephalus*)with fruit; 2. Sleeping colony of red-necked fruit bats (*Pteropus vampyrus*); 3. Head of the Indian short-nosed fruit bat (*Cynopterus sphinx*); 4. Head of the large tub - nosed fruit bat (*Nyctimene major*).

Birth

I was able to observe closely the birth procedure in a long-eared insectivorous bat. Around midday the female was greatly agitated in her cage. After climbing around for a long time, she fastened herself tightly to an upper corner of the gauge-wire cage with the claws on her toes and her thumbs. However, she did not hang her body upside down, as usual, but rather she was almost horizontal. The hind feet were spread widely apart, and the tail membrane was folded under the stomach, toward the front, forming a sort of bag. Then she repeatedly bowed her head in the direction of her stomach and made pressing movements. Apparently, the labor pains were now at full strength. On closer observations, I perceived a small foot emerging from the vagina. After another hour, the infant emerged with

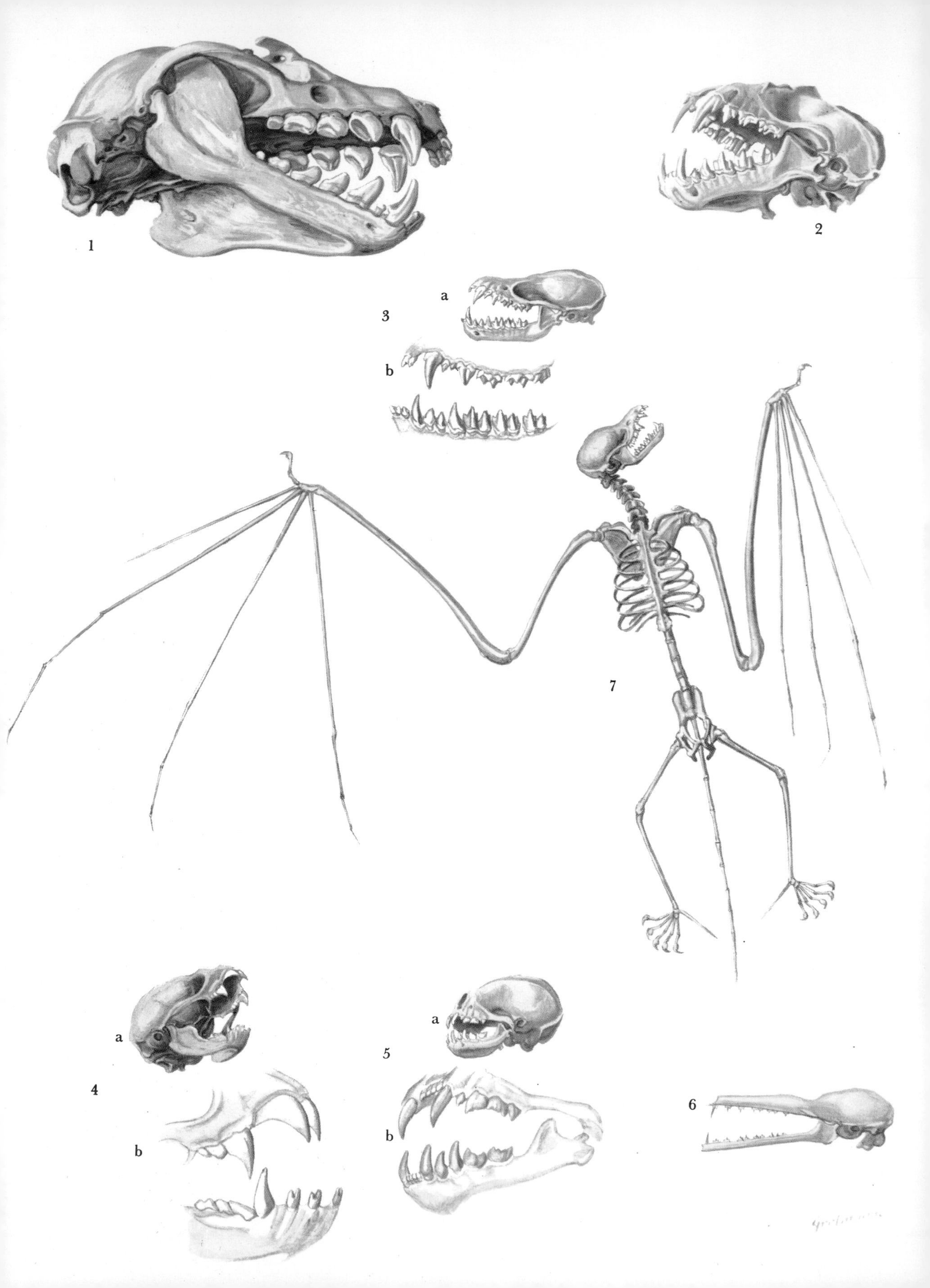
1
2
a
3
b
7
a
4
b
a
5
b
6

1
2
3
4
5
6
7
a
b ♂
8
♂
11
♀
9
10
12 ♂
Großmann

Continued from page 82.

5. Straw-colored bat (*Eidolon helvum*); 6. Colony of Egyptian fruit bats (*Rousettus aegyptiacus*), suspended in a cave and flying; 7. Hammerheaded fruit bat (*Hypsignathus monstrosus*), a) with fruit, b) head of a male; 8. Wahlberg's epauletted fruit bat (*Epomophorus wahlbergi haldemani*); 9. Head of a Zenker's fruit bat (*Scotonycteris zenkeri*); 10. Head of a dwarf epauletted fruit bat (*Micropteropus pusillus*); 11. African long-tongued fruit bat (*Megaloglossus woermanni*), male flying to a blossom, female on a blossom of the african sausage tree (*Kigelia africana*); 12. Head of a male Franquet's epauletted fruit bat (*Epomops franqueti*)

◁
Large mouse-eared bat (*Myotis myotis*) awakened from its daytime sleep in an attic (top) and sitting (bottom).

a final push and let forth a shrill "zick-zick-zick," its first cry. It rested for a while in the "bag" formed by the mother's wing membrane and was licked laboriously by its mother.

A short time later, the lively animal clutched the body of its mother and crawled toward her breast in search of the nipples. At this time it was still connected to the mother by its umbilical cord. Later the placenta was ejected and eaten by the mother. The next morning the insectivorous bat, with the baby on her breast, was again hanging in the normal sleeping position at the usual resting place. The entire birth in this case lasted for several hours (however, we must realize that captive animals are not confronted with the particular disturbances that can occur in the wild).

Recent observations have shown that there are many deviations from the procedure described above, both in the actual birth and in the position of the female giving birth. In the horseshoe bat and some related species, we find two so-called hind nipples, aside from the milk-giving nipples on the mother's breast; these are located at the lower end of the body, above the vagina. Apparently these nipples serve only as "clutching nipples," which the young suck, and to which they fasten themselves. In such cases the baby sits in a reverse position, with its head facing its mother's tail; the infant must turn around when it wants to reach the nipples which contain milk.

Development of young

The newborn young of the insectivorous bats are naked and blind. Initially, they are unable to keep their body temperature at a constant level, and they have to be warmed by the mother. During daytime rest, the mother usually wraps one wing around her young like a sheltering cloak. The eyes open after a few days, and a fine hair begins to grow, first on the back and then on the stomach. The close contact between mother and child gradually decreases, and the young hangs by its feet independently during the daytime rest period next to its mother. When the mother returns from her nocturnal hunting flight, she immediately looks for her baby, to nurse it.

During his observations of the little brown bat, Möhres determined the significance of the shrill noises of the newborn young, which are detectable by the human ear. These vocalizations can be considered as "contact calls," so to speak showing the mother where she can find her young.

Initially the young do not react to the mother's calls. "As the animals grow older, their contact vocalizations gradually changes over into orientation sounds typical of the vespertilionids," writes Möhres. "These diversify into a consistently larger number of impulses which become increasingly shorter and higher in frequency (and therefore undetectable by man)." This is how young insectivorous bats gradually develop the capacity to emit and perceive ultrasonic sounds, so that the mother and child can communicate. Tests showed that the mother can also find her own child among various young bats occupying the same space, because the sounds of each young are slightly different.

The young of many species of insectivorous bats, for instance the European vespertilionids, have milk teeth that are especially adapted for holding on. The front milk teeth are usually very pointed and directed backward to some extent. This makes it easier for the young to hold on to the nipple. The permanent teeth emerge in the third week, so that the large mouse-eared bat has a complete set of teeth, consisting of thirty-eight teeth, in between the thirtieth and thirty-fifth day. The milk teeth of the horseshoe bats do not actually emerge; they have already regressed before birth. These animals attach themselves to the nipple by sucking.

Flying is not a learned ability

Young insectivorous bats develop very quickly and become airborne without first having to learn how to fly. They soon follow their mothers on the nocturnal flights, and to some extent they are carried along in her "sonic slipstream." As Möhres ascertained, they use "ultrasonic impulses" emitted by the mother for their own directional orientation. Should the ultrasonic sounds of the mother cease for a few moments, the young then loses its connection and immediately emits "lost calls," whereupon the mother turns back and again takes her young in "tow." After six to eight weeks, young insectivorous bats are fully grown and independent; they are almost the size of the adults, and can be distinguished only by a slightly lesser weight and a different, generally grayer hair color.

Fruit bats develop somewhat differently. As Kulzer observed, the Egyptian fruit bats already have a thick hair on their back when born. The young remains with its mother for at least four months. Initially, the young develops rapidly, although it reaches its full size only after one year. I found that adolescent west African fruit bats are already airborne and join the adults on their nocturnal flights. Observations of captive animals and their young revealed that the young shared fruit with their mothers. They are, therefore, no longer dependent on milk, unlike the young of insectivorous bats.

As far as we know, most insectivorous bats in the temperate zones are not sexually mature within the first year; the young females can become pregnant, at the earliest, in their second year, and in the temperate zones they give birth only once a year. Apparently the females of some tropical insectivorous and fruit bats may be pregnant more than once a year; nursing females were observed to be pregnant again. The limited reproduction of bats is compensated for by a fairly high life expectancy, and a correspondingly long period of sexual fertility. We know from observations in zoos that fruit bats have reached an age of over twenty-three years.

High life expectancy

Since the beginning of the 1930s, marking and banding programs have been carried out with insectivorous bats. These have shown that these small mammals can become quite old. A maximum age of twelve to fifteen years has been ascertained in the European vespertilionid bats, while in the barbastelles, the maximum age is seventeen years; the maximum age in the whiskered bat and the greater horseshoe bat is over eighteen years. Such an old age is, of course, seldom reached. The average age of the insectivorous bat is significantly lower. Sluiter and his co-workers calculated the life expectancy for the whiskered bat at approximately five years.

Bats gather in large numbers at roosting sites

The bat's social instincts, which often result in large numbers of animals together in one place, are familiar to all. These mass assemblies are usually found at their daytime resting places. My first visit to a roosting site of STRAW-COLORED BATS (*Eidolon helvum*) on a small island near the Cameroon coast was enormously impressive. All my expectations were widely surpassed when a colossal swarm of straw-colored bats became airborne after a few shots were fired. The air was filled with fluttering and with the agitated shrieks of flying animals. New waves of animals kept flying off, until at last thousands were circling above the sleeping trees. The larger fruit bats of the genus *Pteropus* in the Indo-Australian area assemble in trees in giant sleeping colonies. Many species choose high, leafless trees as resting places, and they hang there like large fruit. We are familiar with "classic" fruit bat colonies which have used the same sleeping tree as far back as people can remember.

The sleeping colonies of insectivorous bats in caves are no less familiar; one of the most famous of these may be the Carlsbad Cavern in southeastern New Mexico, where unimaginable numbers of the MEXICAN FREE-TAILED BAT (*Tadarida brasiliensis mexicana*) used to live. As the Americans Bailey and Allison report, the swarm seemed like a cloud of smoke as it emerged from the cave entrance to begin the nocturnal flight; the diameter of the swarm was over 6 m, and it took eighteen uninterrupted minutes for all of the bats to come out of the cave. The complete mass could be seen over a distance of over 3 km. Allison calculated their numbers, by a sophisticated method, at nearly 9 million—an enormous number.

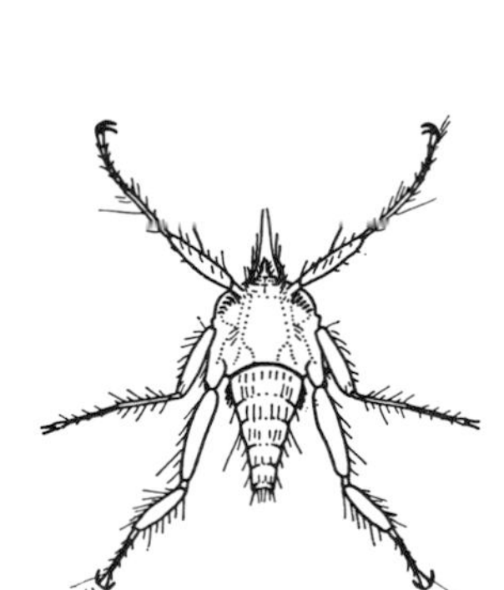

Fig. 4-6. Bats are often heavily infested with parasites. One of these is the bat fly *Nycteribia vexata*. Bat flies are bloodsuckers, and have lost their wings because of their parasitic way of life.

We are also familar with mass assemblies of specific insectivorous bat species in the temperate zones. Sometimes the animals assemble there in winter quarters to hibernate; sometimes there are summer quarters which are used as maternity wards for females giving birth and raising their young. The large mouse-eared bat, the most common native European species, prefers natural caves, artificial dugouts, and similar places for hibernation. As a "maternity ward," however, it prefers unoccupied attics. One can frequently find hundreds of insectivorous bats in these places, often huddled together in a small area.

There are also fruit bats and insectivorous bats that are unsociable from birth on. Often external factors affect these aggregations: a favorable site and the characteristics of the roosting place, better protection against enemies, the greater warmth obtained by huddling close together. Frequently, several species of insectivorous bats come together in the same place. Sometimes one can even observe how specific species in a cave, for instance, each show subtle differences in the selection of their resting place.

Thus, one species may prefer being near the cave entrance, while another may prefer sleeping farther back; one species may seek out higher humidity or temperature, while another may choose just the opposite for its resting place.

This tendency to aggregate notwithstanding, we cannot assume that individual animals in such a colony have a close personal relationship with each other. When species members fly to an insectivorous bat that is crying out in fear, they do not do so with any intention to help. Thus, it is not surprising to find that pair bonds between mates do not exist among these animals, or at least are not known to exist. Males and females mate at random, using a posture employed by most mammals. After the mating, the male takes no further notice of the female. The raising of the offspring is left solely up to the mother.

Enemies and parasites

Bats do not suffer much from natural enemies. Occasionally, an animal falls victim to a quadrupedal predator or is slain by an owl or by a bird of prey hunting late in the evening. The bat hawk (*Machaerhamphus alcinus*; see Vol. II), from the tropics of the Old World, is specialized on bats. Many snakes also hunt insectivorous bats in their daylight quarters. However, the resulting losses are moderate.

Bats are more severely bothered by parasites, particularly fleas, lice, ticks, and mites. One insect family, the wingless bat flies (*Nycteribiidae*) is parasitic and lives exclusively on bats. All these parasites have developed very remarkable adaptations to their host animals.

Endangered by human civilization

A serious and increasing danger for the native insectivorous bats has been created lately by man's progressive civilization, which destroys the bats' natural roosting places. The numbers of insectivorous bats have decreased significantly, thereby upsetting the balance of nature, in which these nocturnal flying "ghosts" play an enormously important role.

5 The Old World Fruit Bats

Suborder: Fruit bats, by M. Eisentraut

The OLD WORLD FRUIT BATS (suborder Megachiroptera) were often classified in contrast with the insectivorous bats. A classification of both suborders, according to their feeding habits is, however, not entirely appropriate. We are also familiar with fruit-eating, insectivorous bats. The German names for fruit bats and insectivorous bats is just as confusing, being respectively *Grossfledermäuse* and *Kleinfledermäuse*, meaning merely "large bats" and "small bats." There are many large fruit bats just as there are also smaller species, which are significantly surpassed in size by many insectivorous bat species.

Distinguishing characteristics

The HRL is 60–400 mm, the weight is from 15 to 900 g, and the wingspan measures 24–140 cm; the tail is short or nonexistent (with the exception of the genus *Notopteris*). The eyes range from normal-sized to very large, the ears are medium-sized, and there is no tragus. The external ear forms a fully enclosed circle at the base. The hair is of various thicknesses. The front parts of the head and the outer parts of the limbs and wing membranes are generally bald. The skin is usually brownish in color, although sometimes it can be lighter or almost black. The wing membrane usually is dark in color, often with a spotted pattern. Parts of the skin in the males of many species are conspicuously colored. The number of teeth varies; mostly, however, there are 34: $\frac{2 \cdot 1 \cdot 3 \cdot 2 \cdot}{2 \cdot 1 \cdot 3 \cdot 3}$, although sometimes there are 32, 30, or 28. The genera *Nyctimene* and *Paranyctimene* have only 24: $\frac{1 \cdot 1 \cdot 3 \cdot 1 \cdot}{0 \cdot 1 \cdot 3 \cdot 2}$. The skull is elongated with long, protruding jaws. The teeth often do not close completely. The back molars are flat and wide, and are used in crushing soft fruit. The facial structure of the short-nosed fruit bats is, as their name suggests, shortened, and the teeth close more fully. The tongue is proportionally long and agile. Most species are dependent solely on fruit.

Fig. 5-1. Fruit bats (suborder Megachiroptera)

There are three families: 1. FRUIT BATS (Pteropidae) with the subfamilies of FLYING FOXES (Pteropinae), EPAULETTED FRUIT BATS (Epomophorinae), SHORT-NOSED FRUIT BATS (Cynopterinae), and TUBE-NOSED FRUIT BATS (Nyctimeninae); 2. LONG-TONGUED FRUIT BATS (Macroglossidae); and 3. HARPY FRUIT BATS (Harpyionycteridae). Together there are thirty-nine genera with some 130 species. They are found from the tropics and subtropics of the Old World to the islands of the western Pacific.

Body structure

A typical characteristic of fruit bats, which distinguishes them from the insectivorous bats, is the development of a small claw on the second finger. This claw, reminiscent of the fruit bat's climbing ancestors, is not present in the genera *Dobsonia, Eonycteris, Nesonycteris*, and *Notopteris*. The second finger of fruit bats has the usual three joints; only members of the genus *Notopteris* have an underdeveloped last joint. The tail vertebrae show an extensive involution. Some genera do not have an outwardly visible tail. In others, the tail is only indicated. Only a few forms have a slightly more prominent tail. The *Notopteris* have a tail of particular length, which consists of ten vertebra and is reminiscent of the fully developed tail of the fossil fruit bat *Archaeopteropus*. The tail membrane is more or less underdeveloped and sometimes only a narrow skin strip along the inside of the lower tibia remains.

Hair

The male fruit bats of many species have hair tufts, shoulder brushes, or neck ruffles. The epauletted fruit bats get their name from the white or yellow hair tufts in the shoulder area. The hairs in this area apparently are connected to the pockets of the scent gland, and can be protruded or retracted. The most noticeable color difference between the sexes may well be shown by the graceful AFRICAN LONG-TONGUED FRUIT BAT (*Megaloglossus woermanni*). The males of this species have large yellow-white hair tufts on each shoulder area. These tufts meet in front of the breast, forming a kind of necklace. The females do not have this marking.

Most fruit bats are fruit-eaters

The exclusively fruit diet of the fruit bats has resulted in particular adaptations of the facial structure and teeth. The eyeteeth, which are well developed and relatively long, together with the small incisors, cut or open the skin of fruits. The flattened back molar teeth then crush the fruit pulp. When eating, the fruit bat often uses one foot to hold the fruit and bring it to its mouth. In most species, the mouth can not be opened wide. Thus, the back of the mouth cavity remains enclosed. This kind of cheek formation prevents the soft or juicy fruit pulps from slipping out of the mouth during chewing. There are many papillae on the upper surface of the long tongue, including the soft taste papillae and the horny papillae which are situated in the middle of the tongue, and which face the rear of the mouth cavity. These papillae apparently help in the breaking down of the fruit.

Many species do not eat the whole fruit. Instead, they crush it in order to release the fruit juice, which they drink. Research on the INDIAN FLYING FOX (*Pteropus giganteus*) has shown that this bat's stomach and intestines were filled with a white, milky mucus or a sticky fluid. Fruit bats do not eat the fibrous fruit pulp or kernels; they spit these out. One can find the waste of chewed fruits under fruit trees visited by fruit bats during the night. This may be the reason why natives in many areas believe that fruit bats do not have anal openings, but defecate through their mouths.

It is obvious that the large fruit bats in particular have to eat considerable quantities of fruit in order to become sated. As a result, they have a large, expandable stomach which increases on one side into a saclike form —in direct contrast with the long intestinal tract of the insectivorous bats. The droppings are relatively liquid.

Because of the abundant harvest of fruit trees in the tropical and subtropical areas, the diet of the fruit bats is well provided for during all seasons of the year. Fruit bats prefer the sweet fruits of the wild varieties of figs. They also like mangos, Mombin plums, papayas, avocados, pears, guavas, rose-hips, and bananas as well as the fruit of the Catappen, the pandanus, and the sapodilla trees. Fruit bats occasionally will eat oranges, grapefruits, dates, and the fruits of the Deleb palm. The fruit bats' taste is apparently much the same as ours, for they prefer sweet, aromatic, and juicy fruits.

It is, therefore, not surprising that the Australian fruit bats, for instance, raid fruit plantations. When they arrive in large numbers, they can cause substantial damage. Consequently, they are much disliked by the local settlers, and are relentlessly persecuted. Fruit bats sometimes carry fruits with them, in order to eat undisturbed in a quiet place. One result of such fruit transportation is the dispersion of fruit seeds and, thus, plants.

Blossom-visiting fruit bats

It has long been known that fruit bats also eat flowers. Initially, man assumed that this was done to get insects which were easily caught on the blossoms. Later research showed, however, that the bats were interested in the flowers themselves, which provide nourishment in the form of "meaty" petals, nectar, and pollen. This reciprocal relationship between plant and animal, similar to the relationships between plants and insects or birds (for example, hummingbirds), is called symbiosis, and developed during the course of evolution. Symbiosis occurs when the plant provides food for the animal, and the animal, in turn, pollinates the plant by carrying pollen from flower to flower in its search for food.

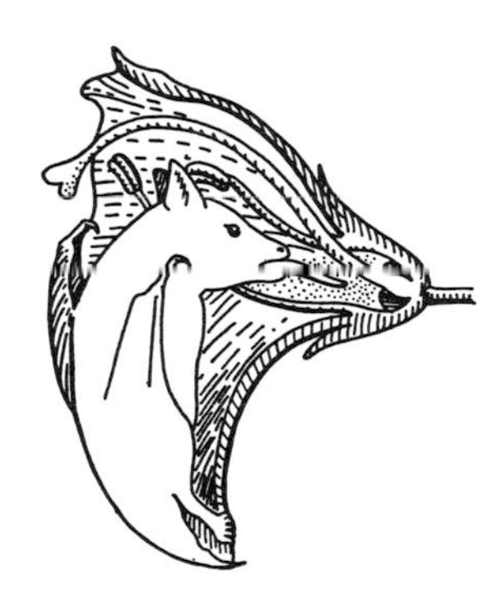

Fig. 5-2. The Indian short-nosed fruit bat (*Cynopterus sphinx*) clutches a *Kigelia* blossom with its thumb claws and feet in order to reach the nectar. For the most part, this species feeds mainly on fruits (compare with Color plate, p. 87).

Without a doubt, these particular fruit bats gradually changed from eating fruit to eating flowers. There are many reports of fruit-eating species searching for flowers in order to take the sweet honey and nutritious pollen. In the process, they occasionally eat the juicy petals. In these cases, only the animal is the beneficiary, and the plant itself is injured. Flying foxes especially damage flowers in this way.

The Frenchman de la Nux was probably the first scientist who observed fruit bats visiting a flower. He wrote in a letter to the zoologist Count Buffon, on 24 October 1772, that on Reunion Island he had observed the native fruit bat *Pteropus niger* visiting various umbel flowers. The next morning he found the ground covered with flowers that the animals had torn off during the night. On Ceylon, Tennent saw the Indian flying fox (*Pteropus giganteus*) eating the blossoms of the cotton tree. Büttikofer reports the same of the African STRAW-COLORED BAT (*Eidolon helvum*): "One night until midnight, near our station at Hokhie [Liberia], we watched swarms of these bats fly around a single *Bombax*, from which they ate flowers and young shoots." The Australian fruit bat *Pteropus alecto gouldi* eats the flowers of eucalyptus trees, among others. This causes its meat, which some local people eat, to have a very strong taste. According to Wroughton, the INDIAN SHORT-NOSED FRUIT BAT (*Cynopterus sphinx*) occasionally feeds on flowers. It seems to show a preference for the honey-rich blossoms of the banana, and therefore causes great damage to the gardens of India's Bandra coast.

According to many reports, the fruit bats mentioned above, which normally feed on fruit, occasionally pollinate flowers. McCann observed the regular visits of an Indian short-nosed fruit bat in Bombay, to the flowers of the African sausage tree. He has described how the flying animals hang on the flowers for a short time, put their head into the opening, and then fly away (see Color plate, p. 87, and Fig. 5-2). The fruit bats cling to the flower by grabbing the lower petals with their hind feet and the outside of the upper petals with their thumb claws. The upper part of the stem bends down with the flower under the animal's weight. As a result, the nectar flows to the lower petals, where the animal licks it up. During this procedure, the stamens which are situated at the upper part of the blossom discharge ripe pollen on the fruit bat's head. While visiting a second flower, the bat rubs its pollen-covered head on the protruding hilum of the flower and thus causes pollination.

Fruit bat and blossom adaptations

Fruit-eaters that only pollinate flowers occasionally do not have special adaptations. They are a transition, however, to the regular flower pollinators of the family of long-tongued fruit bats. These bats feed mainly or exclusively on nectar and pollen. They belong to the smallest group of fruit bats, and have a narrow, considerably elongated muzzle, and a slender protrusible tongue, which can be dipped far into the calyx. The tongue has fringelike papillae on its tip. These face the back of the mouth cavity and are used, like a paint brush, to take in the nectar and to brush the pollen from the stamens. Because these fruit bats do not have to break down their food, their dentures are largely underdeveloped. The molars barely protrude above the gums. The small needlelike holes later observed on the dropped petals are caused by the bat's claws, and are evidence of the fact that for a short time the fruit bats cling to the petals.

The flower, too, has become adapted to the bats. As ascertained by the botanist Porsch, these particular chiropterophile flowers—so named because they have become adapted to the bats—bloom during the night. Many plant species open their flowers only at night and by morning the blossoms have already withered away and dropped off. These flowers rarely have bright colors, because pollination takes place at night. However, they do have a strong, often pungent odor which attracts the bats. The nectar secretion is particularly copious and thus meets the needs of the bats. The animals can easily cling to the large, solid blossoms with their wide calyx openings. Most of these flowers are built so that a bat, upon putting its head into it, will wipe pollen on the hairs of its head and then transmit this pollen to the stamen of the next flower.

Porsch, and later Pijl, mention many plant families to which the chiropterophile (pollinated by bats) species belong. These include, among others, from the bignonia family, the genera *Kigelia*, *Markhamia*, and *Oryzalum*; from the bombax family, the genera *Ceiba* and *Duris* as well as the baobob tree (*Adansonia digitata*); and from the Sapote family *Illipae*. Proof of pollination of certain other plants by bats does not yet exist, although, because of their structure, these plants are possibly chiropterophile.

Fruit bats live only in the Old World

Fruit bats of this suborder are found only in the Old World. They have not migrated to the New World, although certain areas of the tropical zones of Central and South America would provide good habitats. Fruit bats live only in humid, forested areas of the warm zones of the Old World, and not in arid, treeless prairies and deserts. The genus *Rousettus* is found almost everywhere in the range of the fruit bats, except in Australia and the Polynesian archipelagos. The genus *Pteropus*, with a wide variety of species, is found mainly in the Indo-Malayan and Australian region; to the west it has extended toward Africa. Separate species are found in Madagascar, the Comores, the Seychelles, the Mascarenes, on the small island of Aldabra, and Pemba Island near the east African coast, although this genus is not present on the African continent. The reason for the absence of *Pteropus* species in Africa is a puzzle of animal geography. The distance between Pemba and the African coast is only slightly over 65 km, which could easily be flown by fruit bats.

Family: Fruit bats, in the narrower sense

The family of FRUIT BATS in the narrower sense (Pteropidae) includes the greatest number of species. This family is present throughout the entire range of the fruit bats. The first subfamily, the FLYING FOXES (Pteropinae), includes a number of the best-known species, some of which are to be found more or less regularly in zoos.

Subfamily: Flying foxes

The STRAW-COLORED BAT (*Eidolon helvum*; HRL over 20 cm; see Color plate, p. 87) belongs to the largest African flying foxes. The upper side of the body is brownish, yellowish, or reddish-gray; the middle of the back is darker. The underside is yellowish to brownish-gray. They also have a wide yellowish collar around the throat that can also be seen on the upper side. Males are more brightly colored than females. The pelage of younger animals is duller in color, and there are only indications of a collar band. There is a variety of subspecies including *Eidolon helvum helvum*, from the African mainland, *E. h. dubreanum*, from Madagascar, and the somewhat smaller form *E. h. sabaeum*, from southern Arabia.

Fig. 5-3. Genus *Eidolon*.

The straw-colored bat is a very sociable animal, forming large or small sleeping colonies in high deciduous trees. On Nicolls Island, off the coast of Cameroon, I found a giant daytime colony with an estimated 10,000 animals. The straw-colored bats also colonize in the middle of large cities.

without being bothered by the traffic. I found them, for instance, in the city of Santa Isabel on the island of Fernando Po, in Douala, Cameroon, and in Abidjan, Ivory Coast. I noticed that there was never complete stillness at these sleeping sites; there was always some movement. Neighbors fought each other and uttered angry noises. From time to time, one flew off to look for a more suitable place in the next tree. One wondered when these animals really slept.

When the twilight rapidly changes into night, the straw-colored bats gather in large or small groups and fly slowly toward the fruit trees, often traveling great distances in search of food. Once, when the fruit trees had ripened, I watched the straw-colored bats gather in an enormous mango tree which was at an elevation of 1000 m on the slope of Cameroon Mountain near Buea. The eyes of the animals, as they sat eating in the tree, were like glowing red points in the beam of the flashlight. The next morning, large quantities of partly eaten mango fruits lay on the ground. The fruit bats return to their sleeping trees at dawn. Occasionally, however, I saw some animals still circling their resting place at nine a.m.

Some females which were studied in early February in Cameroon were in an advanced stage of pregnancy, or carried their young on their body when they were startled from their sleeping tree. More research is needed to determine whether reproduction is limited to certain months of the year. The straw-colored bat is known to undertake long seasonal journeys in many areas. Presumably it does so in search of fruit-bearing trees.

The species of the genus of ROUSETTE BATS (*Rousettus*) are some of the few fruit bat species living in caves. They attain only an average body length and are found in the tropics of the Old World from Africa through southern Asia to the Solomon Islands. There are eleven species, of which only two will be described here: 1. EGYPTIAN FRUIT BAT (*Rousettus aegyptiacus*; HRL 15 cm, wingspan 60 cm; see Color plate, p. 87). The pelage is brownish-gray. It inhabits southern, western, and eastern Africa, Egypt, the Middle East, and Cyprus. There are three subspecies *R. a. aegyptiacus*, from North Africa, Cyprus, and the Middle East; *R. a. leachi*, from eastern and southern Africa; and *R. a. occidentalis*, from western Africa.

Fig. 5-4. Rousette bats (genus *Rousettus*).

The ANGOLAN FRUIT BAT (*Rousettus angolensis*; HRL 13 cm, wingspan 45–50 cm) is somewhat smaller. The pelage ranges from bright brown to a dark chestnut. The males have a sort of collar, especially on the lower side of the neck area, which consists of thicker, more rigid hair of a brighter brown. The females' neck area is only sparsely covered. The size of the three subspecies increases significantly from west to east: *R. a. smithi*, a small bat found from Sierra Leone to Nigeria; *R. a. angolensis*, somewhat larger, found from Cameroon to northern Angola; and *R. a. ruwenzorii*, the largest of the three, found in the Ruwenzori Mountains. It is especially long-haired and dark.

Egyptian fruit bats live in large colonies. They are found in caves, old graves, and temples in Egypt. Miss Bate observed them in dense trees during the summer on Cyprus. During winter, however, they were in old buildings and caves. I once found the datyime roost of the western African subspecies in an enormous mountain cave on the slope of a narrow erosion valley. When I visited the site with my companions late in the morning, a dense swarm of fruit bats fled from the cave entrance as we approached, shrieking loudly. An estimated 1000 adult and young animals of both sexes had been assembled there.

Egyptian fruit bats live on Cameroon Mountain up to altitudes of at least 1600 m. I observed the animals there as they gathered after dark in large numbers on fruit-bearing wild fig trees, and I caught them in nets which were stretched only a few meters above the ground. From October until March, I saw females with infants or with young. At the same time, however, I observed females which obviously were not in heat. This might mean that the Egyptian fruit bat does not have a limited breeding season.

The ANGOLAN FRUIT BAT is predominantly a cave dweller. I found them in Cameroon in deep mountain tunnels, where no light penetrated. These animals can, of course, orient in these tunnels by echolocation. Occasionally, I also found Angolan fruit bats during the daytime in the small confined recesses of brook ravines which had been eroded by water. On Cameroon Mountain, and on Fernando Po as well, I very often saw these species at altitudes of up to 2000 m.

Among the smaller species, I was able to observe so-called "maternity wards," which are common among the insectivorous bats but have rarely been observed in fruit bats. I observed all-female colonies in many caves. Sometimes, the females huddled close together on the ceiling of the cave or on its walls. All of the animals were either in some stage of pregnancy or had young with them. Presumably the mating season of these species is more narrowly defined than that of the Egyptian fruit bat. The young are born in the first months of the dry season (the European winter months). Final answers to these questions depend on further observations.

The COLLARED FRUIT BATS (genus *Myonycteris*) should be discussed at this point because they share many characteristics with the rousette bats (genus *Rousettus*) including the pattern of palate creases, the dental formula, and other morphological characteristics.

The collared fruit bats are the smallest representatives of their subfamily. They have a somewhat more pronounced shortening of the facial structure and were, therefore, previously classified as short-nosed fruit bats of the subfamily Cynopterinae. There are a few species from Africa, among them *Myonycteris torquata* (HRL 9 cm), from the Guinean-Congolese forest belt. Its males have special neck hairs.

Fig. 5-5. Flying foxes (genus *Pteropus*).

We know very little, as yet, about the life style of these small African fruit bats. In Cameroon, I trapped some animals with nets placed at forest edges and clearings. Some observations indicate that these fruit bats retreat into the forest during the daytime.

The FLYING FOXES (*Pteropus*; see Color plates, pp. 86 and 87) are the most varied genus of fruit bats. In 1912 Knud Andersen distinguished no fewer than eighty-two species, with over 100 subspecies. Since then, many more species and subspecies have been described. The habitat of many species is confined to small Pacific islands. However, this rigid division into species is no longer in keeping with our modern view. The number of classified species will certainly be drastically reduced as soon as a survey is made according to more modern systematic criteria. Many species will probably be reclassified as subspecies.

In this genus there is a transition from the largest known fruit bats, which have a wingspan of approximately 140 cm, to the medium-sized types with a wingspan of approximately 60 cm. As all of these bats have a very similar life style, only a few will be described here: 1. The INDIAN FLYING FOX (*Pteropus giganteus*; HRL 30 cm, wingspan 120 cm). The pelage varies from light to dark brown to black. The colors of the breast and stomach are faded. The shoulder and neck hairs are yellowish-brown and are consequently distinct from the rest of the body. The subspecies *P. g. giganteus* is found in the vast lowland areas of India and Ceylon, while the subspecies *P. g. leucocephalus*, with longer hairs, is found on the slopes of the Himalaya.

2. The RED-NECKED FRUIT BAT (*Pteropus vampyrus*; see Color plate, p. 87) has many subspecies which are characterized by differences in color and size. These subspecies are found in the Malay Peninsula, the Greater and Lesser Sunda Islands (Indonesia), and the Philippines. The Javanese subspecies, *P. v. vampyrus* (HRL 40 cm, wingspan 140 cm) is the largest. The hair on its back is blackish-brown, finely speckled with gray-white. The breast, flanks, and stomach are blackish, while the anal area usually is a reddish-brown. The hairs of the shoulder cloak vary from yellow-brown to orange-brown.

3. The GRAY-HEADED FLYING FOX (*Pteropus poliocephalus*; wingspan 1m; see Color plate, p. 87) has an olive-brown pelage and a tan cloak around its neck and throat. One of the four fruit bat species, the gray-headed flying fox, is found in large groups in the eastern coastal regions of the Australian mainland, sometimes traveling as far as Melbourne during its searches for food.

4. The RUFOUS FLYING FOX (*Pteropus rufus*) is the same size as the gray-headed flying fox. The upper sides of its body are dark brown, while its underside is yellowish-brown. Its cloak is a yellowish- to reddish-brown. The front of the head, throat, and neck are seal-brown, and are sharply distinguished from the back of the head, which is yellowish-brown. The rufous flying fox is found in the coastal areas of Madagascar.

The main diet of these "common" fruit bats consists of a wide variety of fruits; sometimes they eat blossoms as well. All these bats are mainly active at night, but reportedly, many of them occasionally move about during the daytime. Only one species, *Pteropus subniger*, from the islands

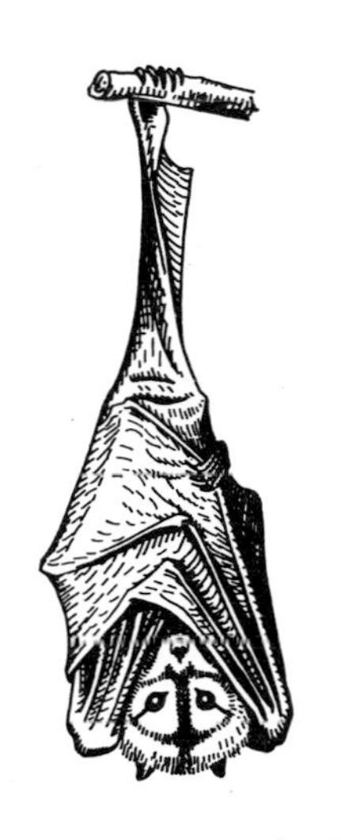

Fig. 5-6. During the day, fruit bats sleep, hanging upside down, wrapped up in their wing membranes. Pictured here is *Pteropus capistratus*. When the temperature rises, this "cloak" is opened up and used to fan the air.

of Reunion and Mauritius, was seen in tree holes or rock caves. Normally, the *Pteropus* species suspend themselves from trees during the daytime and when at rest. In this manner, they form large sleeping colonies. The animals are exposed to all sorts of weather as they hang from the barren branches—the sun, and rain and gales when the weather is foul. In bad weather, the bats' wing membranes, firmly wrapped around the body like a cloak, protect them against the cold and rain. On bright, sunny days, the bat airs out its wings, and then folds them tightly together again. In this manner, the air layer beneath the wings is replenished with fresh air.

Bats have been seen to stretch out one wing and fan it back and forth when the temperature rises. Aitken reports a surprising scene he saw when he visited a sleeping colony of Indian flying foxes in the noonday heat. All the animals hung from the branches, completely awake, making fanning movements with their wings, some with the right wing, some with the left. The wings were being used as "ventilators."

Kulzer checked the influence of intensive heat or cold on captive animals of the same type. When he lowered the room temperature a considerable amount, sometimes below 10° C, the fruit bat reduced its heat-radiating surface to a minimum, wrapped itself tightly in its wing membrane, and tucked its head and nose inside, so that it breathed the heated air between its body and wings. During this time it held onto its perch mostly with only one foot, keeping the other under its wing membrane, like many other fruit bats. At an average temperature of between 18 and 30°C, the animal folded its wings and laid them loosely against its body. When the room temperature was raised as high as the body-temperature range, the fruit bat spread its wings and fanned itself.

Kulzer reports on his tests: "When the temperature of the environment increases, the animal can only dissipate its body heat through evaporation of moisture. At an air temperature of approximately 37°C, the animals began to lick their breasts, stomach, and wing membranes extensively. The saliva secretion is raised to such an extent that the whole animal looks as if it had been dipped in water. At the same time, the fanning of the wings becomes more and more pronounced, thereby bringing fresh dry air to the body surface. The evaporation is thereby facilitated and the body is cooled." In this manner, the fruit bat makes use of the cooling process which in other animals, including man, is perspiration.

Fruit bats, being inhabitants of warmer climates, do not hibernate and do not fall into a lethargic daytime sleep, as do the insectivorous bats in temperate climates. They have a constant body temperature which only fluctuates within a slight range.

The Indian flying foxes usually prefer high trees, often in the middle of towns and villages, as resting places; here they come together by the hundreds. When twilight falls, they fly from their resting places to their food trees, which sometimes are a long distance away. Many times they first visit a watering place, touching the water surface briefly and drinking in flight. From this habit comes the incorrect assumption that fruit bats catch small fish. When the bats return to their resting trees the next morning, they often spend hours flying noisily over the best resting place, before finally falling asleep.

The Australian gray-headed fruit bat prefers dense, poorly accessible bush for its sleeping quarters. At night these animals also make long raids, and often visit fruit plantations—where they can ruin a whole crop in one night. The Australian farmers pursue them, therefore, wherever possible. The rufous flying fox of Madagascar prefers wild dates for its food and may become quite fat on the diet. Local people consider their meat a delicacy.

Subfamily: Epauletted fruit bats

The second subfamily of fruit bats in the narrower sense, the EPAULETTED FRUIT BATS (Epomophorinae), is confined to Africa. Its members range in size from large to very small. The front of the head is usually elongated. The teeth are fairly poorly developed for the eating of soft fruit, and are usually separated by large gaps. The males, generally larger than females, have shoulder pockets, from which sprout bright hair tufts, (epaulettes). However, this is not true of all species. These shoulder pockets may be either sucked in or thrust out, whereby the hair tufts are either moved back or are noticeably displayed. There are white hair tufts at the front and hind edges of the ears.

The epauletted fruit bats do not form large sleeping colonies like the flying foxes; they are seen only singly or in small colonies. Their daytime quarters are chosen in leafy trees or in the bush. They are rather quiet and difficult to detect, and consequently easily elude observation.

The members of the genus *Epomops* have long, soft hair. Included in this genus are: FRANQUET'S EPAULETTED FRUIT BAT (*Epomops franqueti*; see Color plate, p. 87), found in western African from Ghana to Angola, and east beyond the Congo basin; and BÜTTIKOFER'S EPAULETTED FRUIT BAT (*Epomops büettikoferi*), found in western upper Guinea; this is probably only a subspecies of *E. franqueti*. The upper side is pale brown to cream-colored; the sides are dark brown with large white stomach spots. The coloring may differ greatly among individual animals, and may vary with age. There is no externally visible tail. The tail wing membrane is present only as a narrow seam along the lower joints.

During the daylight, the epauletted fruit bats hang freely from the branches of low trees, with a special preference, it seems, for the banks of jungle brooks. At the fringe of forests in the Garamba National Park (Zaïre), Verschuren found them on branches reaching approximately 4–5 m over the water. I found them on similar sites at the northern foot of Cameroon Mountain. While wading through a stream bed, I often startled single animals. They let me approach until I was a few meters away; then they flew off suddenly and clung to another branch some distance away. I was unable to detect these animals in the foliage because of their perfect camouflage. At night, they visited fruit-bearing trees in large numbers and were noticeable by their loud nasal cries. Apparently only the males make these cries. Males and females that I caught in a net during the night emitted defensive squeaking noises. Some of the females observed in February and March were pregnant, and others were nursing. Therefore, this species may have a defined mating season.

The members of the genus *Epomophorus*, found from Senegal to eastern Africa and South Africa, are medium-sized. These epauletted fruit bats live more in open forest or in savannas. They are found in smaller colonies at their daytime quarters in trees and in the bush. Of the eight species (whose exact classification has yet to be determined), only one will be mentioned here:

WAHLBERG'S EPAULETTED FRUIT BAT (*Epomophorus wahlbergi*; HRL 14 cm). It is dark brown. In the female, the lower parts are lighter and the neck is whitish. The males have very distinct white epaulette hairs. Subspecies include *E. w. haldemani* (see Color plate, p. 87), found from western to eastern Africa; and *E. w. wahlbergi*, found more toward the south.

Kulzer reports his observations on the east African subspecies: "During the daytime, they hung in colonies of up to fifteen in tall *Grevillea* trees or in wild fig trees. They remained absolutely quiet. Only when twilight fell did they leave these sleeping places and fly to the nearby fruit plantations. Near Ol Donyo Sambu, at the southern foot of Mount Meru [Tanzania], they visited a fruit garden every night for several weeks. They flew around noiselessly between the trees and searched for ripe guavas. In order to chew these, they settled down on certain trees. They spit out the hard parts and fibrous material of the fruit. These fell down beneath the tree. By searching for these refuse dumps with a flashlight and, upon finding one, by directing the beam upward, one is almost sure to find the animals. They are not bothered by the light. Like all fruit bats, they cling to a branch with only one foot, pressing the fruit to their breast with the other foot, and biting off pieces. They often return to the same feeding places for days or weeks on end, just as they reuse their sleeping places. As a result, one can expect these animals at these feeding places at twilight."

The DWARF EPAULETTED FRUIT BAT (*Micropteropus pusillus*; HRL 9–9.5 cm; see Color plate, p. 87) also exists from western to eastern Africa. It is classified in a special genus. It has a lower-arm length of hardly more than 5 cm. The facial structure is wide and noticeably shortened. The teeth are close together. The males and females show very little difference in size. The hair is light brown; the underside even paler. There are distinct white tufts at the front and hind edges of the ears. The males have whitish shoulder tufts.

Little is known about the daytime roosts of this small fruit bat. Single animals have occasionally been found hanging in low bushes. I found only single specimens in Cameroon. Sometimes I caught them in a rat trap which was situated on the ground and baited with ripe bananas. This

F. Reimann

TWILIGHT NEAR A BROOK IN CENTRAL EUROPE
Mammals: □ Bats: 1. Greater horseshoe bat (*Rhinolophus ferrumequinum*); 2. Water bat (*Myotis daubentoni*). □ Rodents: 3. Vole rat (*Arvicola terrestris*). □ Predators (see Vol. XII): 4. Mink (*Mustela lutreola*), presumed extinct in Europe for many decades; 5. Ermine (*Mustela erminea*). □ Insect-eaters (see Vol. X): 6. European water shrew (*Neomys fodiens*), fishing.

Birds: Rails (see Vol. VII): 7. Water rail (*Rallus aquaticus*). □ Waders and gull-like birds (see Vol. VIII): 8. Woodcock (*Scolopax rusticola*). □ Owls (see Vol. VIII). 9. Short-eared owl (*Asio flammeus*). □ Sparrowlike birds (see Vol. IX): 10. Gray wagtail (*Motacilla cinerea*); 11. Wren (*Troglodytes troglodytes*); 12. Dipper (*Cinclus cinclus*).

Amphibians (see Vol. V); 13. Spotted salamander (*Salamandra salamandra*); 14. Frog (*Rana esculenta*); 15. Tree frog (*Hyla arborea*). □ Fishes (see Vols. IV and V): 16. Pike (*Esox lucius*); 17. Trout (*Salmo trutta*); 18. Minnow (*Phoxinus phoxinus*); 19. Merlin (*Noemacheilus barbatulus*). □ Insects (see Vol. II): 20. Mayflies (*Ephemera vulgata*): 21. The rare tiger moth (*Arctia casta*); 22. Moth (*pergesa porcellus*). □ Crayfish (see Vol. I): 23. Fresh-water crab (*Astacus astacus*).

indicated that these animals descended to the ground in search of their food.

The head of the HAMMERHEADED FRUIT BAT (*Hypsignathus monstrosus*; HRL 20 cm, wingspan up to 90 cm; see Color plate, p. 87), which is the largest epauletted fruit bat and also the largest African fruit bat, has a rather monstrous appearance. The muzzle is greatly raised, and widened at the front. Its sucking lips are shaped like flaps. Males, which are much larger than females, have an enormously enlarged larynx which reaches over the lungs to the diaphragm and is connected with the skin pockets. There are no shoulder hair tufts. It is found only in western and central African forest areas.

The enormous head of this fruit bat, which, in the words of Sanderson, looks like a "wild caricature of a horse head in profile," is used above all to suck the nutritious fruits. The animal surrounds the fruit with its flap-shaped lips and sucks in the juice in this manner. Earlier it was assumed that the enormous larynx was also a kind of suction pump for sucking the fruits. This idea, however, certainly is incorrect. The larynx, instead, may have something to do with the production of special sounds. At night during the reproductive period, the males utter loud, croaking noises at short intervals; these far-carrying calls are heard continuously. The first time I heard them, on the slope of Cameroon Mountain, I initially assumed that they were made by a frog.

On Fernando Po, where it is not so rare, I found the hammerheaded fruit bat grouped in small colonies during the daytime, hanging from the tops of palm trees, well protected from the bright sunlight. At twilight, the animals could occasionally be seen slowly winging their way between the trees. On Cameroon Mountain, I found them at an altitude of over 2000 m.

The shoulder hair tufts, which are so characteristic of the epauletted fruit bats, are not present in the hammerheaded fruit bat, nor are they present in both members of the genus *Scotonycteris*: the large SNAKE-TOOTHED BAT (*Scotonycteris ophiodon*), first described in 1943, and the small ZENKER'S FRUIT BAT (*Scotonycteris zenkeri*; HRL 8 cm, wingspan 30 cm; see Color plate, p. 87). Both species have a bright spot on the bridge of the nose and right behind the rear rim of the eyes, and off-white hairs on their lips. The hair on the upper side of the body is rust-brown; the hair on the underside is thinner, with a grayish-white central area on the chest and stomach.

The wing membranes of animals I collected had a particular saffron-yellow tint when the animal was alive. After skinning, however, this color largely disappeared. Both species are considered rare; some collections have only a few specimens from the west African forests. Only on Fernando Po, where the largest species does not exist, did I encounter the Zenker's fruit bat somewhat more frequently. The animals apparently live singly, hanging between thick bunches of leaves during the daytime. Little is known about their way of life.

Subfamily: Short-nosed fruit bats

The subfamily of SHORT-NOSED FRUIT BATS (*Cynopterinae*), which is classified into ten genera, is found in the Indo-Malayan area. The facial structure is short; the upper lip is split by a deep rut which starts above the nostrils and proceeds to the tip of the muzzle. Some species use caves for their daytime resting place. Others hang freely from trees. We mention here only the genus *Cynopterus*, which is divided into three species. The HRL is 9.5–12.5 cm, while the wingspan is 30–45 cm. The best-known species, the INDIAN SHORT-NOSED FRUIT BAT (*Cynopterus sphinx*; see Color plate, pp. 87 nd 88) has various subspecies living in India, Sri Lanka, southeastern Asia, and the Malay Archipelago. They are extremely common. They live in groups, prefering mountain caves, mine shafts, and occasionally tree hollows for their daytime quarters. These bats are also frequently found under the eaves of houses. They primarily eat fruits of various kinds, including guavas, bananas, mangos, and wild figs. When traveling in large groups, they occasionally damage plantations. They are often seen, however, visiting flowers, where they function as pollinators.

Subfamily: Tube-nosed fruit bats

The TUBE-NOSED FRUIT BATS are similar to the short-nosed fruit bats. They can be classified, however, as a special subfamily (*Nyctimeninae*). Their nostrils are greatly elongated and tubelike, giving them a very peculiar appearance (see Color plate, p. 87). The beginnings of such tube noses can be found among other species of fruit bats and even among insectivorous bats (genera *Murina* and *Harpiocephalus*). These bats are proportionally small (HRL up to 12.5 cm). The skin is brownish-gray and the wing membranes and ears have yellow spots. There are only twenty-four teeth: $\frac{1 \cdot 1 \cdot 3 \cdot 1 \cdot}{0 \cdot 1 \cdot 3 \cdot 2 \cdot}$. There are no lower incisors; they were replaced by the eyeteeth, which in turn gave way to a second set of strongly developed spaced teeth. The molars are particularly high and slightly flattened, with well-developed protuberances, as in the typical fruit eaters.

Their special tooth structure gave rise to the assumption that the tube-nosed bats probably fed on insects. S. Müller studied the stomachs of recently trapped animals of the local species, *Nyctimene cephalotes*, on the island of Amboina (Indonesia) and found the remains of bugs and flies. Felten, on the other hand, found animals of the species *Nyctimene robinsoni*, found primarily in northern Queensland (Australia), that took only fruits, preferably juicy pears, and rejected insects placed in their cage. Tube-nosed bats living in the wild were observed eating guava fruits. We distinguish two genera (*Nyctimene* and *Paranyctimene*), with a whole series of species. The daytime roosts of the tube-nosed bat may be between the leaves of trees.

Family: Long-tongued fruit bats

The LONG-TONGUED BATS should be mentioned here as a particular family (Macroglossidae). They are very small, and are mainly adapted to feeding on nectar. The facial structure is elongated and pointed. The tongue can be extended for some distance. They are found mainly in the Indo-Malayan-Australian area, although one species lives in the tropical forest areas of Africa.

The DOBSON'S LONG-TONGUED DAWN BAT (*Eonycteris spelaea*; HRL 12 cm) is the best-known of the four species of the genus *Eonycteris*. The upper arm length is from 6 to 8 cm. The upper side of the body is dark brown, while the underside is lighter. Males have a frill consisting of somewhat longer hairs on the sides of their neck. Females, on the contrary, have only a thin hair in that area. A small tail protrudes from the narrow tail wing membrane. This animal is found throughout Burma, Laos, Cambodia, Viet Nam, Thailand, the Malay Peninsula, and Indonesia.

As the scientific name indicates, this long-tongued fruit bat uses caves for its sleeping places. Bartels observed these animals on Java during their nocturnal visits to the agave flowers. They clung to the blossom for a short time, and were not frightened away by lamps in the area or by shots that were fired. The stomachs of captured animals were full of agave pollen. Heide reports that the same species visits the *Kigelia* and *Markhamia* flowers: "I noticed how, each time one of the bats visited a flowering bush, the twig supporting a flower was slightly bent down after the bat had left, and how, approximately one second later, it sprang up again. When a bat visited a flower and had landed on it, it folded its wings, pressed its thumb claws into the sides of the corolla, and put its head into the opening of the blossom. Each visit lasted not longer than one second, and in the course of approximately ten minutes I observed ten visits of bats to the flower-laden tree."

The long-tongued fruit bats are the smallest fruit bats of all; they are the representatives of the genus *Macroglossus*. The DWARF LONG-TONGUED FRUIT BAT (*Macroglossus minimus*; HRL 6.6–7 cm, wingspan 25 cm) also lives in southeastern Asia, including the archipelagoes. During the daytime the animals rest singly on the tree branches. Sometimes they have been found in rolled up leaves. At night they have been seen at the flowers of *Eriodendron*, *Eugenia*, and *Agave*; they have, however, also been observed eating the juices of soft fruits. *Macroglossus lagochilus* (HRL 6.3 cm, wingspan 24 cm), of Malaysia, Borneo, Celebes, the Philippines, and New Guinea, is still smaller.

The AFRICAN LONG-TONGUED FRUIT BAT (*Megaloglossus woermanni*; HRL 7–7.5 cm; see Color plate, p. 87) is the only species of this genus on the African continent. It lives in the upper and lower forest areas of Guinea, and reaches as far east as the Congo area. The most noticeable characteristic of the males is their off-white to cream-colored hair tufts in the shoulder area, coming together on the chest.

During travels in Cameroon and on Fernando Po, I watched this species very often, and was able to observe a flower visit for the first time. In addition to other plants, the animals flew to the flowers of the African sausage tree (*Kigelia africana*), which were specially adapted to visits of bats. I was also able to keep some of these long-tongued fruit bats in a cage and observe them. During the daytime, they hung quietly on a tree branch and folded their wings around their bodies. When darkness fell, they became animated, groomed themselves, and climbed around the cage. They soon found the small bowl filled with honeyed water, which they drank by quickly sticking out their tongues and pulling them in again.

Family: Harpy fruit bats

The third and last family of fruit bats, that of the HARPY FRUIT BATS (Harpionycteridae), is represented by only one species, the very rare WHITEHEAD'S HARPY FRUIT BAT (*Harpyionycteris whiteheadi*; HRL 14–15.3 cm). There is no externally visible tail. The upper side of the body is chocolate to dark brown; the lower side is lighter.

The harpy fruit bat differs from all other fruit bats in its dental structure. The molars have five or six distinctly developed points; the lower eyeteeth have three. Moreover, the upper incisors are directed noticeably forward, so that the eyeteeth are at right angles to each other when the jaw is closed. Nothing is known about the feeding habits of these animals. Only a few single animals have been found on the Philippines and Celebes, one of them in a cave.

Fig. 5-7. Whitehead's harpy fuit bat (*Harpyionycteris whiteheadi*).

6 The Insectivorous Bats

Suborder: Insectivorous bats, by M. Eisentraut

The variety of forms in the insectivorous bats (suborder Microchiroptera) is especially large. Until recently nearly all we knew about many species was what they looked like and how they originated. It is as if nature gave her "imagination" a free rein when it came to the insectivorous bats. The suborder, as a result, has an enormous assortment of shapes and sizes, as well as some quite unusual differences. This variety is most evident in the shape of the head, and more especially of the face, the features of which come very close to the grotesque.

Besides the many species that have a "normal" head shape, there are some that have an extended forehead and others with a shortened, "pug-like" head. The ears may reach gigantic dimensions in length as well as in width. The external ear, if it is present at all, is sometimes long and pointed, sometimes wide and round, and either smooth or notched. The noses have a so-called nose leaf, and there is a seemingly inexhaustible variety of nose shapes among these bats. The names of some families indicate as much: horseshoe-nosed, Old World leaf-nosed, American leaf-nosed, slit-faced, and bulldog bats. The lips may be smooth, serrated, wrinkled, or covered with papillae or warts. A look at the color plates gives one an idea of the unbelievably varied shapes of nose forms, whose special significance, for the most part, has not yet been determined. The research on echolocation has shown that the nose leaf definitely cannot be considered solely an ornament, but that it has a definite function. It was established that the nose leaf plays a part in the emission of the ultrasonic sounds, whereas the ears receive the rebounding echo. There is still much room for research. However, no other mammalian group can compete with the variety and peculiarity of the facial shapes of the insectivorous bats.

Distinguishing characteristics

The body length ranges from 3 to 16 cm, and the wingspan is from 18 to 70 cm. There is no claw on the second finger; there is only one finger joint (rarely two). The ears are not closed in a ring at the base, as they are with the fruit bats. The ear ligaments are usually attached at separate places on the head. Some families have a special shield in front of the ear, a tragus. The tail is mostly well developed and is more or less part of the wing membrane.

Only within the American leaf-nosed bats (family Phyllostomidae) do the tail and tail membrane show some signs of underdevelopment, apparently in adaptation to a particular way of life. The shape of the wings, whether they be wide or narrow, is important in the development of the capacity and manner of flight. Long, narrow wings enable the animals to be fast and elegant fliers. There are skin glands in a variety of body areas; facial glands secrete an oily substance for the oiling of wing membranes. Other skin glands on the head, throat, and near the genitals and the anal area secrete odor substances, which are often present only in males or are much more pronounced in the males. These odor substances may play a part in sexual behavior. The original teeth may be changed, in adaptation to different foods, through a decrease in number as well as changes in shape. The incisors are generally small, while the eyeteeth are long and strong. The molars of the insectivorous bats have knobby, pointed, and sharp-edged crowns or edges that help in crushing insect bodies. There are between twenty and thirty-eight teeth. The numbers of incisors, the premolars, and the molars have all decreased.

There are sixteen families, some of which are combined into superfamilies, with more than 150 genera in all, and some 650–700 species. More exact figures are not possible until the problems of systematic classification have been worked out.

Although insectivorous bats are much smaller than the large fruit bats there are rather extensive differences in size within the group. Among the larger types are the following, both of which are native to tropical America: LINNE'S FALSE VAMPIRE BAT (*Vampyrum spectrum*) and the AUSTRALIAN GIANT FALSE VAMPIRE BAT (*Macroderma gigas*; HRL 14–16 cm). The NAKED BAT (*Cheiromeles torquatus*) is not much smaller. It is found in the Malay Archipelago. On the other extreme, dwarf insectivorous bats of the genus *Pipistrellus* are particularly small and diminutive. The west African species, *Pipistrellus nanulus* (HRL 4 cm), is one of the smallest mammals known. The other insectivorous bats come in all sizes within this range.

In most species the body is generally covered with a thick hair which sometimes extends over part of the wing membrane. Usually, however, the wing membranes, ears, and nose leaves which appear in many species, are bare. Noticeably, the insectivorous bats do not have downy hair; only one type of hair develops. This hair, when viewed in individual strands, often is coarse rather than smooth. It is proportionally thin at the root; then it becomes thicker and forms peculiar scales, cones, chignons, or screw-shaped spirals at the outer cortex layers. In some species, the hair looks as if it had been divided into separate pieces and then put back together (see Fig. 6-1). The hair structure often is species-specific, and to a certain extent can be used as a distinguishing characteristic.

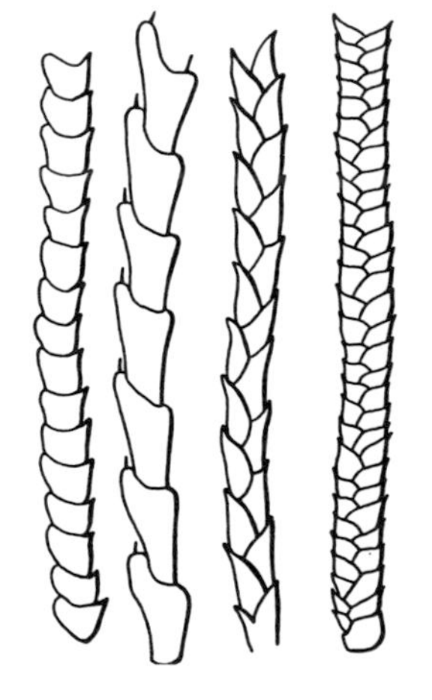

Fig. 6-1 The hairs of insectivorous bats have characteristic structures, which may help in classifying the species. From left to right: common pipistrelle (*Pipistrellus pipistrellus*), large mouse-eared bat (*Myotis myotis*), long-eared bat (*Plecotus auritus*), and barbastelle (*Barbastella barbastellus*).

Most species of insectivorous bats have a brownish or gray pelage, with variations from light to dark and in all intermediate shades, even including some blackish tints. Other colors include bright orange-red or reddish-brown. The underside of the body is generally lighter or paler than the back. There is a whole series of variations, including white. The skin and wing membranes of the GHOST BAT (*Diclidurus albus*), which lives in tropical America, are white to cream-colored. This is true of another Central American species *Ectophylla alba*. Nonpigmented, translucent, whitish wing membranes exist among other species, such as the small west African species *Eptesicus tenuipinnis*. Particular markings occasionally appear on the hair, including light spots or stripes down the back, as in the TWO-LINED SHEATH-TAILED BAT (*Saccopteryx bilineata*), the TENT-BUILDING BAT (*Uroderma bilobatum*), or the African genus *Glauconycteris*.

Sometimes we find great color differences between the hair and the wing membranes. The most beautiful example of this differentiation may well be the east African species *Myotis welwitschii*. Its color varies from straw-yellow to orange-yellow, and extends to the tail membrane as well as along the finger joints. The wing membrane, in contrast, is deep black. Color differences between the sexes are not too common. In the American RED BAT (*Lasiurus borealis*), the female has a dark reddish-brown color overall, while the male is a much brighter orange-red color. Similar color differences are found among the MEXICAN BULLDOG BAT (*Noctilio leporinus*). Many species, from various families, can have two color phases, one dark grayish-brown, the other a red-brown.

The diet of insectivorous bats is amazingly varied. We can see from this how nature "tries to fill up" all ecological niches. It was left to the insectivorous bats to occupy the heretofore unoccupied night skies and to exploit all available food resources there. Most insectivorous bats live on insects and have retained many of the feeding practices of their quadrupedal ancestors. The number of nocturnal insects is tremendously large. The insectivorous bats catch their prey in flight. Observers long believed that the extremely agile hunting flight of many species was little more than an erratic fluttering. We should be wary of such condescending judgements. However differently the flight capacities of the various species may have developed, they certainly represent the most perfect adaptation to the existing needs of each species.

The narrow-winged species, which like the nocturnal gliders have a fast and elegant flight, need a wide-open flight area, and hunt mainly at higher altitudes, above the treetops or open spaces. On the contrary, the broad-winged species are generally found at lower altitudes. They chase insects that are found in the shadows of trees or around bushes. Many species prefer to hunt above open waters, where insects are especially abundant.

The large amount of water needed by insectivorous bats also draws them to a wet environment. They drink as swallows do, in flight, flying close to the water's surface and dipping their mouths in for a brief moment. Many species noticeably search for water to quench their thirst before beginning their nocturnal hunts. If you want to keep insectivorous bats in captivity, it is important to keep a drinking bowl in the cage constantly. When they do not have water, their wing membranes will easily dry out in a dry atmosphere, which will cause the death of the animals. As a result of the changed conditions in a confined cage, insectivorous bats of course do not drink during flight but instead sip the water out of a bowl with their tongue, while sitting.

In captivity, the various species of insectivorous bats very quickly learn to eat the foods offered in the feeding bowl. This leads to the assumption that in nature they occasionally also catch insects that crawl on the ground or on tree leaves. There have been many observations supporting this idea; the European LONG-EARED BAT (*Plecotus auritus*) was observed taking caterpillars from leaves. Remains of non-flying beetles, crickets, and other insects were found at the feeding site of insectivorous bats. These must have been taken from the ground or from tree trunks. Felten has written about how a South African farmer found the feeding remains of a slit-nosed species, *Nycteris thebaica*, in the open hall of his house each morning. The remains consisted almost entirely of scorpion pieces. Obviously, in in this case, an insectivorous bat group had adapted itself to one certain food animal which was available in large numbers and easily caught. It is interesting to note that here a poisonous prey was involved, and conquering it required special abilities.

For some time, long lists of the food of European insectivorous bat species have been available, based on the study of food remains. These show that the nocturnal hunters are not overly fastidious. In recent times Kobb established from fecal analysis that the LARGE MOUSE-EARED BAT (*Myotis myotis*) often eats non-flying, ground-crawling insects, including walking beetles and dung beetles. Kobb was able to prove that the insectivorous bats find their prey while hunting on the ground, attracted by the sounds these insects make. Without a doubt, the flying hunters locate their flying prey initially by echo locations.

The insectivorous bats catch insects in their mouth, which has a wide opening that is especially useful to the flying hunters. Earlier beliefs that the tail membrane was used as a snaring net were based on the incorrect interpretation of what in itself was correct observation: When an insectivorous bat has not caught its prey in its mouth correctly, it turns its head backward, bows deeply, and uses its wing membrane as a buttress in order to gain a better hold on the insect. This behavior can readily be observed in captive insectivorous bats not only sitting, but also in flying animals (see Color plate, p. 134). The prey is either eaten in flight or brought to a resting place beneath which the remains are later found.

Food specialists

Beginning with the primitive type of "insect-eaters," particular kinds of food specializations have developed among the insectivorous bats; we

Fig. 6-2. This is how a Mexican bulldog bat (*Noctilio leporinus*) catches a fish.

can follow these changes and the accompanying changes in tooth structure. The insectivorous bats are directly related to the vertebrate-eaters. These species reach a considerable size and have particularly strong teeth. Similar specialists are to be found, for instance, in the family of large-winged bats (*Megadermatidae*). The INDIAN FALSE VAMPIRE (*Megaderma lyra*) was observed at night while catching small insectivorous bats of the genus *Pipistrellus*. Apart from insect remains, remains of small birds, among them white-eyes and nectar-feeders, geckos, frogs, and even fish have been found beneath their eating places. The related Malayan species, *Megaderma spasma*, and the Australian giant false vampire bat (*Macroderma gigas*), are assumed to have similar eating habits. The teeth of these insectivorous bats remind one of those of small predators. The SPEAR-NOSED BATS (*Phyllostomus hastatus*) of the New World are carnivorous. Dunn observed captive animals which preferred live food more than any other food. They easily overcame live mice, insectivorous bats, and small birds, eating them greedily. They took bananas and other fruits only when no meat was available. The adeptness with which these bats killed the captured animals suggests that the spear-nosed bat, even when living free, feeds largely on vertebrates.

Fish-eating insectivorous bats

A sole adaptation to eating fish is rare among insectivorous bats. Nevertheless, some species have become fish-eaters. It has long been known that the tropical American Mexican bulldog bat (*Noctilio leporinus*) catches fish; stomach examinations have revealed, apart from insects, a large amount of fish remains. Some observers report watching these insectivorous bats catch fish near Trinidad. Until recently it was not clear how this bat succeeded in catching fish in the water. It was supposed, however, that it gripped its prey in the water with its very large, extremely well-developed hind feet. Proof of this hypothesis was only recently provided by the American scientist, Bloedel. He photographed a bulldog bat dipping its feet in the water, gripping the fish in its strong claws, and lifting it out of the water. A second fish-eating species, the small *Pizonyx vivesi*, lives in northwestern Mexico on the Gulf of California. It also has extraordinarily large feet, which it uses presumably for catching fish. Its teeth have very high points, obviously an adaptation to feeding on fish. Other fish-eating mammals, like dolphins, have rows of pointed teeth with which they can easily hold the slippery prey.

Bloodsuckers

A totally new manner of feeding has evolved in the so-called "bloodsuckers" of the family of vampires (Desmodontidae). The vampires, which are very closely related to the American leaf-nosed bats (Phyllostomidae), are placed in three genera, with one species each. The best-known and most common of vampire bats is the COMMON VAMPIRE BAT (*Desmodus rotundus*). The totally different tooth structure immediately indicates a special feeding adaptation (see Color plate, p. 86). There are twenty teeth, all of which have razor-sharp edges, with which the vampires cut into the skin of their prey (nearly always warm-blooded animals). Using its tongue, the bat then licks up the blood which flows from this wound. The word "bloodsucker" is therefore not wholly appropriate. The "blood donors" of these vampires include cattle, horses, goats, dogs, pigs, and other domestic animals, even including poultry. Sleeping people are not exempted, as we know from many reports. We do not know which wild animals were the original "blood donors," as hardly any observations are available on this.

The vampires make their searching flights and approach their victims so carefully and quietly that sleeping people do not wake up. They do not notice the planting of the bat's feet or the quick bite of its teeth. Vampires normally land on their victim. They have small soft skinpads on their wrists and on the soles of their feet, and these help them to place their feet lightly and gently. Vampires are not difficult to keep in captivity and will eat blood, treated with an anti-coagulation agent, offered in a bowl. Normally they lick up so much blood during a meal that their bodies swell visibly and they can move only clumsily. Their stomach is a long intestinelike tube, which spirals several times. It is exceptionally elastic and is able to take in large quantities of blood when the bat eats. Similar (convergent) adaptations are also found among the bloodsuckers of totally different animal groups, such as leeches and mosquitoes. After drinking blood, the vampires retreat to their hiding places, usually in mountain caves, to digest their meal. They do not re-emerge for more food for a few days.

The bite of the vampire bats is harmless in itself, and the loss of blood in larger domestic animals is not too high. Afterward, however, infection can set in. Moreover, we know that dangerous infectious diseases can be transmitted by the bite. For instance, the horse disease called "murrina," which occurs in many Central and South American areas, is transmitted in this manner, as is rabies. The vampires are dangerous, then, not for their bloodletting, but as disease carriers. Various extermination programs against these insectivorous bats were carried out in many seriously endangered areas, with more or less successful results. Furthermore, it has recently been discovered that insectivorous bats carry rabies in North America and Europe. As a result, insectivorous bats were in danger of acquiring a bad reputation with public-health officials. These emotions may have quieted down by now, however, and it has been understood that such cases of disease transmission must be considered as rare, unavoidable events, which do not call for special preventative measures.

Insectivorous bats specializing in feeding on blood are found only in the tropical and subtropical zones of America. When the first reports about them arrived in Europe, exaggerated and fantastic images appeared, picturing vampires feeding on sleeping persons and putting them into a deep sleep by fanning the wings and sucking their blood. Other species which had rather grotesque features yet were absolutely harmless also came under similar suspicion. One representative of the American leaf-nosed bats, Linne's false vampire (*Vampyrum spectrum*), was treated in this manner.

It even got its scientific genus name, *Vampyrum*, as a result. Some Old World species, even including some European bats, were unjustifiably classified as "bloodsuckers."

In reality, Linne's false vampire is an insect-eater which also lives on fruits. This brings us, then, to those insectivorous bats that have totally or partially adapted to eating fruit. These are found exclusively in the large, diverse New World family of leaf-nosed bats. Some leaf-nosed bats are the American representatives of the fruit-eating bats of the Old World. These do not exist in large numbers, however. They live secretly and are therefore proportionally less studied. When discussing fruit-eating insectivorous bats, we should consider first of all those species from the subfamilies of the short-tailed leaf-nosed bats (Carolliinae), and the so-called "fruit vampires." (Stenoderminae). We can recognize the change from insect-eating to fruit-eating by the teeth. As in fruit bats, their molars are also wide, flattened, and knobby, and therefore suitable for crushing fruit pulp. There is no further evidence of insectivorous teeth.

Typical fruit-eating teeth are found, for instance, among the members of the genus *Artibeus*. Beneath the resting places of these cave-dwelling species one can find massive amounts of seeds, berries, half-eaten mango fruits, and pieces of aromatic fruit of the *Eugenia jambos*. The Seba's short-tailed bat was observed raiding the sapodilla tree (*Achras sapota*). Sometimes these species cause damage in fruit plantations. On the other hand, they also often carry the fruits away and thereby disperse the seeds of various plants.

The insectivorous bats have diverse ways of living, and it is not surprising that we can find species among them that feed on flowers. These bats are representatives of the leaf-nosed bat, specifically of the subfamily of the long-tongued bats (Glossophaginae). They are the New World representatives of the Old World flower-feeders from the family of long-tongued fruit bats, and they too have similarly adapted features as a result of their special food habits. On the evolutionary scale which leads from the exclusive fruit-eaters to the highly specialized flower pollinators, we find those bats which destroy the flowers when feeding upon them, and which do not show special adaptations for visiting blossoms. These bats may eat the juicy or pulpy petals, the nectar, or the pollen, and by doing so they destroy the flower. This is true of Linne's false vampire (*Vampyrum spectrum*), which otherwise eats fruits and insects. It was observed with other bats on *Caryocar nuciferum*, where they tore off blossoms and ate them.

A good adaptation, including pollination, was most often observed in the small LONG-TONGUED BAT (*Glossophaga soricina*). The bat initially visited the flowers of calabash trees (*Crescentia*), a plant species that is apparently exclusively adapted to pollination by insectivorous bats. The calabash flower shows special adaptations that we already know from plants pollinated by fruit bats; the blossoms open at night and secrete rich nectar, and they have large, sturdily built corollas. The long-tongued bats have a somewhat elongated mouth and a long tongue covered at the tip with long, thin hair papillae, with which they can reach deep down into the flower. The animals alight briefly on the rim of the blossom and cling to the petals with their claws. However, they can also hover in front of the flower like hummingbirds. In captivity, the insectivorous bats of these genera took insects and sometimes fruit juices or fruit pulp. Other representatives of the long-tongued bats, like the MEXICAN LONG-NOSED, BAT (*Choeronycteris mexicana*) and the Mexican BANANA BAT (*Musonycteris harrisoni*), are also highly adapted to visiting blossoms. Both have elongated beak-shaped jaws, and the teeth are largely underdeveloped. Only a few observations are available on the living habits of these species.

The suborder of insectivorous bats is found all over the world. Some species have even ventured into the Arctic. They are absent only in the polar regions, because there are no insects in these areas. The three families of the sac-winged bats, the vespertilionid bats, and the free-tailed bats are widely dispersed over the terrestrial parts of the globe. The families of the slit-faced bats, large-winged bats, horseshoe bats, and Old World leaf-nosed bats are among those species limited to the Old World. Meanwhile, the American leaf-nosed bats, the bulldog bats, the vampires, the funnel-eared bats, the smokey bats, and the disk-winged bats exist only in tropical Central and South America. The Madagascar sucker-footed bat (Myzopodidae) and the New Zealand short-tailed bat (Mystacinidae), each represented by only one genus and one species, are both confined to the small island areas noted with their names. These isolated cases testify to the great age of these islands.

Distribution and climate

The bats generally prefer higher temperatures. Therefore, we can find a great number of species in the tropical and subtropical areas, which are the centers of their evolution and dispersion. The numbers of species present decreases considerably as one enters the temperate zones. Many families, genera, and species are exclusively tropical. Europeans are familiar with only four families: the slit-face bats (solely on Corfu), the horseshoe bats, the vespertilionid bats, and the European free-tailed bats, of which only one species is still living in southern Europe. North of the Alps, in central Europe, only two species of the horseshoe bats are present; the northern edge of their range is in Germany. They are no longer found in the Baltic Sea area, and in northern Europe only a few representatives of the vespertilionid bats can be found.

Life in the temperate zones, with the regular alternation of warm and cold seasons, calls for special adaptations in the species that feed exclusively on the insects living there. They must survive during the cold season when food is scarce. Therefore they hibernate, like some other representatives of the insectivorous bats (Volume X).

Hibernation and lethargy

Insectivorous bats in the temperate zones not only hibernate during the cold season, but also exhibit a similarly "lethargic" condition during the summer. During the daytime, when the temperature drops beyond a certain point, their body temperature decreases until it is almost the same as the surrounding temperature. The animals fall into a hibernationlike sleep. Their metabolism slows down accordingly, and the need for food is thus lessened. Only in the evening, when again they become active, does their body temperature rise. This "daytime sleep" or "cold lethargy" is not detrimental to the animal at all; instead, it is a means of conserving energy. It presupposes a particular change in the animal's heat-regulation mechanism which according to observations, is also present in members of tropical insectivorous bat families. Thus it is a family characteristic and a prerequisite for insectivorous bats that migrate from tropical areas to colder zones.

Among the exclusively tropical species we find a far smaller variation in the range of body temperature. A slight supercooling occurs with various species; however, when stronger cold stimuli occur, a quickening of the metabolism follows, and their heat production is increased. Lengthy cold spells bring the metabolism of the tropical species to a halt, however, and death from cold is inevitable.

Most insectivorous bat species in the temperate zones look for especially sheltered winter quarters in which to hibernate. Many species choose mountain caves or similar places, while others hibernate in hollow trees; still other insectivorous bats prefer buildings. In most cases, the winter and summer quarters are not the same, and they are usually situated at different sites. For example, the European large mouse-eared bat prefers unused attics in buildings, preferably church attics, where the temperature is very high on warm days during the warm season. In winter, on the contrary, this bat normally looks for natural mountain caves, unused mine shafts, or quiet cellars, premises in which, during winter, the warmest temperatures prevail.

In the beginning of the 1930s, naturalists carried out a series of banding or tagging programs; these are still being continued in many countries. These procedures have shown that most insectivorous bat species are loyal to their home sites, to which they always return. Migrations can be related to the move from summer to winter quarters; species migrate over hundreds of kilometers. Some species even migrate occasionally during the day; migrating insectivorous bats, for example, noctule bats, were often seen on beautiful fall days, sometimes together with swallows.

Migrations

Only a few species of insectivorous bats migrate over long distances to warmer climates where they do not have to hibernate. This is especially true of some North American species; the red bat (*Lasiurus borealis*) migrates in the fall from its northern habitat to warmer areas in the south. However, according to more recent observations, not all members of this species migrate. Apparently some animals, particularly the males, remain in the northern areas and hibernate during the winter.

There is a great need for further research on the migration of the insectivorous bats. The question of how these animals orient themselves on their migratory flights has yet to be solved. Many species return consistently to their chosen sites. How is it possible for them to find the right route to their often-distant winter quarters when they generally travel at night during their fall migration, and how are they able to find their way back to their summer quarters in the spring? We will need more facts and observations, as well as the use of new research techniques, before we can answer this question.

Reproduction

The change of seasons in the temperate zones influences the reproductive habits of the local bats in a peculiar and unique way. Fetus development begins right after conception in those bats living in warmer climates, ending with birth. Former authorities ascertained that in European latitudes, particularly in the fall, female bats which have already mated can be found with their uteruses full of sperm. For many species, the new mating season begins very shortly after the raising of young. This takes place when the sexes are together in the common daytime roosts. Sometimes the males even visit the females in their "maternity wards."

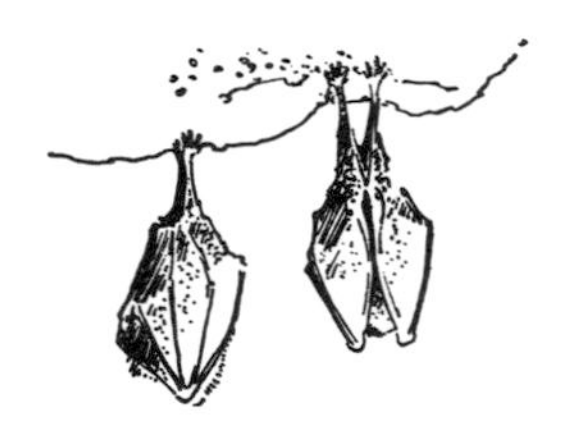

Fig. 6-3. The lesser horseshoe bat (*Rhinolophus hipposideros*) hangs from the ceilings of caves during its hibernation. The wing membrane is wrapped completely around the body.

During this time, there is no egg development in the females, and therefore no fetus development. The sperm remains in the female's fallopian tubes until spring, and only after the end of the hibernation period does ovum development take place. The developed egg then travels down the fallopian tubes, is fertilized by the sperm, and develops normally in the uterus. However, one can also find virgin females in the winter, and in such cases the sperm remains in the epididymis of the males from fall until spring. Spring matings have been observed, as well as mid-winter matings, when the animals sometimes wake up in their winter roosts. Consequently, we must assume that the males are ready for mating from fall until spring, and that they can copulate in the spring with those females that did not mate in the fall.

Fig. 6-4. Lesser horseshoe bat (*Rhinolophus hipposideros*) in Europe.

How can we explain these unusual phenomena in insectivorous bats of the temperate zones? I am inclined to assume that copulation originally took place in the fall. The retardation of the ovum maturation until spring must be considered as a later adaptation of those species that moved into the temperate zones; this then prevents the possibility of birth in the unfavorable winter months. A similar adaptation to the perilous winter months is found in deer and among some other native mammals. In deer, however, copulation and ovum maturation are not separated. The fetus development begins immediately after the mid-summer mating season; it then stops at an early stage and continues normally only after several months. The birth of the fawn occurs in the following spring, when rearing conditions are most favourable (see Vol. XIII). Therefore, the same goal, birth during optimum rearing conditions is attained in animals in very different ways.

Development of the fetus

According to available observations, fetus development in the European insectivorous bats begins soon after they awaken from hibernation,

Fig. 6-5. Greater horseshoe bat (*Rhinolophus ferrumequinum*) in Europe.

and birth normally takes place in June. However, an exact determination of the length of pregnancy is difficult to establish. First, for the reasons mentioned above, the exact time of conception is not known, and second, fetus development may be retarded by cold weather in spring. In the small WATER BAT (*Myotis daubentoni*) I calculated a gestation period of approximately fifty-four days. On the other hand, the basically larger COMMON NOCTULE (*Nyctalus noctula*) took some seventy-three days. Other observers came to similar conclusions concerning other native European species. If a female of one of these native species of insectivorous bats, which has mated in the fall, is brought from its winter quarters to a warm place, and and if it eats there normally, ovum maturation and fetus development will begin earlier. The animal gives birth much earlier than under normal circumstances in nature.

Birth and the rearing of the young

During the first half of the warm season, the females of those insectivorous bat species living in the temperate zones are fully occupied with reproduction. They gather in large or small groups, forming "maternity wards," in which the young are born and reared. The males generally live alone during this time, and are seldom seen together with females. The horseshoe males are an exception to this, as a sizable number can also be found in the "maternity wards." Some observations indicate that sometimes all-male colonies are found, as for example, among the PARTI-COLORED BAT (*Vespertilio murinus*) and the noctule bats.

Migrations in search of food

The bats' noctural flights in search of food may occur earlier in some species, later in others, and in many they occur at a specific time. The time is obviously related to the amount of light available. Thus the insectivorous bats fly later as the days lengthen in the spring, and in the fall they fly proportionally earlier as the daylight hours decrease. In the tropics, because of the very slight difference in the length of day and night, the nocturnal hunting flights of many species begin almost exactly to the minute at the same time every day.

The insectivorous bats have been classified into sixteen families, with over 150 genera. We will discuss only a limited number of the many, varied species. This limitation is quite justified, as we know hardly anything regarding the habits of many species. The insectivorous bats have escaped closer observation because of their nocturnal activities.

Family: Rat-tailed bats

The family of RAT-TAILED BATS (Rhinopomatidae) has only one genus (*Rhinopoma*), with three species. A typical characteristic of this family is the long, thin tail, which is almost the length of the body and is enclosed only at the base by the tail membrane, which forms a narrow skin seam. According to Mohre's observations, the animal uses its tail as a tactile organ when it retreats, crawling backward into a narrow crevice.

The best-known and largest species, *Rhinopoma microphyllum* (see Color plate, p. 123), lives in arid, treeless areas. It is particularly common in Egypt, and is also found from the Near East to India. It has large ears, interconnected by a skin strip in the middle, a well-developed ear flap, and a small, round nose flap at the front of the mouth. The hair color is brownish-gray. Large fat cushions are situated underneath the skin on the lower back and at the base of the tail; these serve as food reserves in unfavorable seasons. The rat-tailed bat lives sociably in natural and artificial caves, as well as in old buildings, such as the old Egyptian tombs. Underneath their sleeping places one can find mountains of guano, testifying to the fact that these daytime roosts have been used for hundreds of years. The somewhat smaller *R. hardwickei* is found in generally the same areas, as well as in Africa as far south as Lake Rudolf (Kenya) and in Asia as far as Indochina.

▻
Rat-tailed bats: 1. Rat-tailed bat (*Rhinopoma microphyllum*; from a photograph by Kulzer). Least sac-winged bats: 2. Sac-winged bat, head of *Balantiopteryx*; 3. Head of *Coleura*; 4. Head of *Emballonura*; 5. Head of a tomb bat (*Taphozous*); 6. Two-lined sheath-tailed bat (*Saccopteryx bilineata*).

Family: sac-winged bats

The family of SAC-WINGED BATS (Emballonuridae) is found mainly in the warmer areas of the Americas, although some genera and species live in the Old World, in Africa, Madagascar, southern Asia, Australia, and even on some South Pacific islands. The body length is between 3.7 and 10 cm, the tail length ranges from 0.6 to 3 cm, and the weight is from 5 to 30 g. The tail membrane is well developed and envelops only the lower tail vertebrae; the tail spine emerges free through the membrane, a specific characteristic of this family. Among some species, the nose is elongated and protrudes above the upper lip. There are thirteen genera (see Color plate, p. 123).

The nose length of the small PROBOSCIS BAT (*Rhynchonyteris naso*; HRL 3.7–4.3 cm) is especially pronounced. This bat's hair is a grayish-yellow blend, with designs on the lower back and underarms, which help to camouflage the animals at their resting places. This species is found in the New World tropics from southern Mexico to Peru and Brazil.

Proboscis bats, which live mainly near jungle waters, prefer to rest on the lower parts of branches and tree trunks that reach over and near the water's surface. There one can find them in small colonies of some ten to forty animals, generally one behind the other, as if connected in some way. If one approaches from the water in a boat, the whole sleeping colony suddenly flies off simultaneously, and lands in closed ranks at a similar place some distance away. Murie found a large number of such insectivorous bat groups at the upper Belize River in British Honduras (Belize). Sometimes up to thirty animals sat together on the lower side of the sloping trees near the water's edge. He counted no less than twenty such small colonies within a distance of 45 km, and it is probable that he may have overlooked many others. At night the proboscis bat hunts close to the water surface and avoids the open country.

▻▻
Bulldog bats: 1. Head of the Mexican bulldog bat (*Noctilio leporinus*). Large-winged bats: 2. Head of the Australian giant false vampire bat (*Macroderma gigas*); 3. Head of the African yellow-winged bat (*Lavia frons*); 4. Head of the Indian false vampire (*Megaderma lyra*). Slit-nosed bats: 5. Head of *Nycteris thebaica*. Horse-shoe-nosed bats: 6. Hildebrandt's horseshoe bat (*Rhinolophus hildebrandtii*; from a photograph by Kulzer).

▻▻▻
1. Greater horseshoe bat (*Rhinolophus ferrumequinum*), a) sleeping position, b) head; 2. Lesser horse-shoe bat (*Rhinolophus hipposideros*), a) sleeping position, b) head. Old World leaf-nosed bats: 3. Head of the South African lesser leaf-nosed bat (*Hipposideros caffer*); 4. Head of the Commerson's leaf-nosed bat (*Hipposideros commersoni*); 5. Head of *Hipposideros jonesi*; 6. Head of the trident leaf-nosed bat (*Asellia tridens*).

The small SHEATH-TAILED BATS (genus *Saccopteryx*), turn the next to the last joint on their third fingers up rather than down when resting. There are six species, including the TWO-LINED SHEATH-TAILED BAT (*Saccopteryx bilineata*; HRL 5–5.5 cm, TL 2 cm; see Color plate, p. 123), the largest species in the family. Its hair is a dark brownish-black, but lighter color tints also occur. Two whitish linear stripes which begin behind the shoulders and cross over the back down to the tail membrane are characteristic of this species. The arm pockets on the front wing membrane end

1
2
3
4
5
6

1
2
5
3
4
6

2b
2a
1a
3
6
4
5

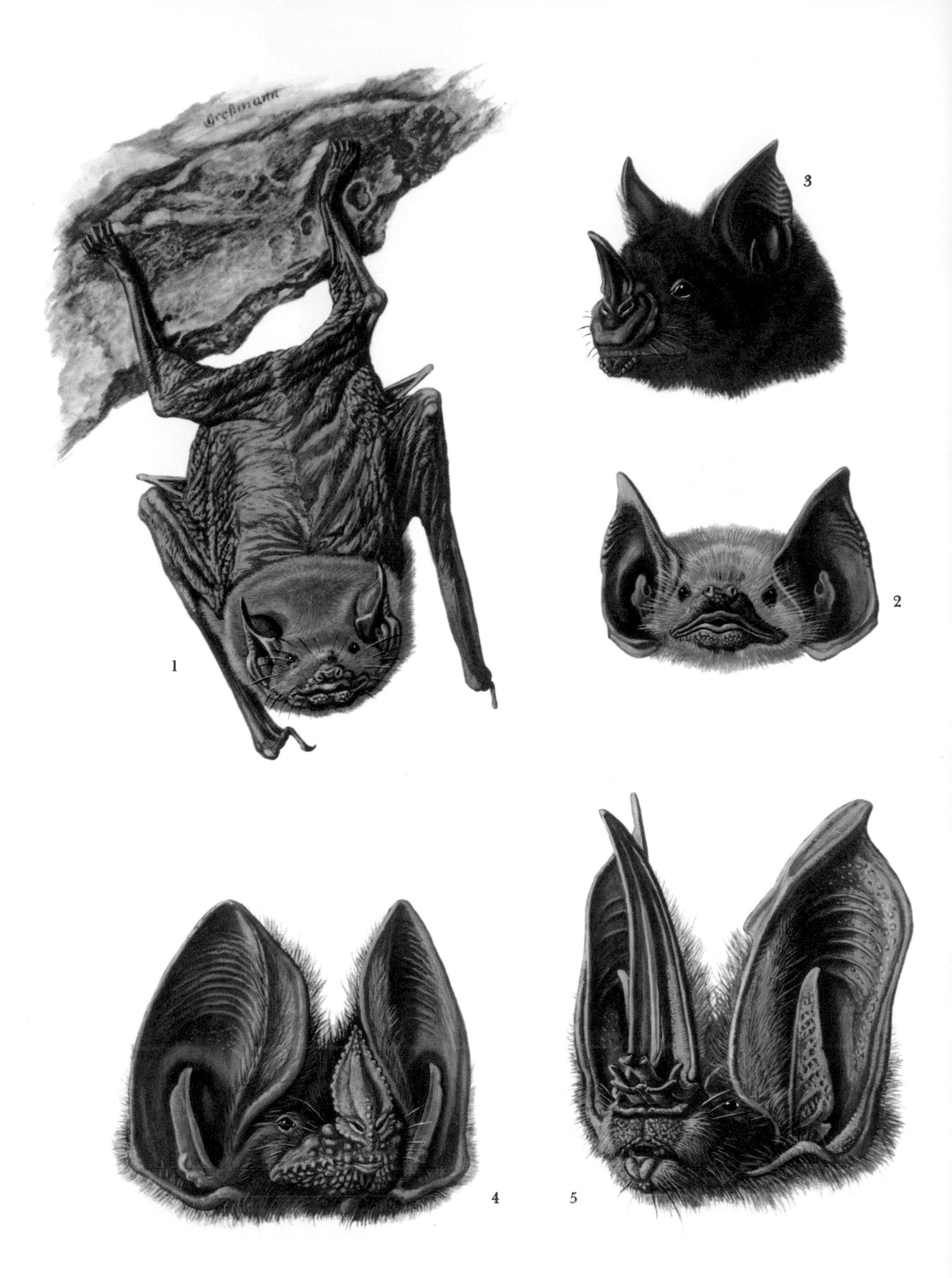

Greßmann
3
2
1
4
5

1
2
3
4
5

1
2
3
4
a
b

The sheath-tailed bats

American leaf-nosed bats: ◁◁◁ 1. Suapure naked-backed bat (*Pteronotus suapurensis*); 2. Head of *Chilonycteris personata*; 3. Head of the spear-nosed bat (*Phyllostomus hastatus*); 4. Head of *Trachops cirrhosus*; 5. Head of Tome's long-eared bat (*Lonchorhina aurita*).

◁◁ 1. Head of the banana bat (*Musonycteris harrisoni*); 2. Long-tongued bat (*Glossophaga soricina*) on a Cobaea flower; 3. Head of a Mexican long-nosed bat (*Choeronycteris*); 4. Head of of *Artibeus*; 5. Head of *Uroderma bilobatum*.

◁ 1. Head of the wrinkled-faced bat (*Centurio senex*); 2. Head of Seba's short-tailed bat (*Carollia perspicillata*). Vampires: 3. Head of the hairy-legged vampire bat (*Diphylla ecaudata*); 4. Common vampire bat (*Desmodus rotundus*), a) walking on the ground, b) head.

with a wide slit opening to the outside. In males these arm pockets, when viewed from the underside of the wings, are set off as pea-sized protuberances; they are also found in related genera. These bats are numerous in large parts of tropical America.

Starck proved that these arm pockets are free of sebaceous glands, and are not, as was formerly assumed, the sebaceous glands themselves. They are filled with a seemingly fatty mass, which apparently is secreted directly from the skin. However, the chemical composition of this mass has not been studied, although we can assume with reasonable certainty that it concerns aromatic substances which play a part in the sex life of these animals. These arm pockets are highly developed only in males. For its daytime roost, the El Salvador sheath-tailed bat prefers sites that are not totally devoid of daylight. In El Salvador, Felten found them in hollow trees and mountain caves, hanging in loose groups on the walls in such a way that they did not touch each other. They awaken and fly off at the slightest disturbance. There are remainders of small beetles and moths beneath their eating places.

Tomb bats

As Old World representatives of the sac-winged bats, we mention here the widely dispersed TOMB BATS (genus *Taphozous*; see Color plate, p. 123). There are three subgenera (*Taphozous, Saccolaimus*, and *Liponycteris*), with approximately a dozen species, ranging through Africa, Madagascar, southern Asia to the Philippines, the Solomons, and in Australia. These insectivorous bats are generally strong, splendid animals, distinguished by a powerful flight. Their body length is between 6.2 and 10 cm, the tail length is from 2 to 3.7 cm, and the weight ranges from 10 to 30 g.

The tomb bats were initially discovered in ancient Egyptian tombs by the scientists who accompanied Napoleon on his campaign. In Egypt we find two species, *Taphozous perforatus* and *T. nudiventris*. The species *T. mauritianus* (HRL 8–9 cm, TL 2.5 cm) is widely dispersed throughout sub-saharan Africa, including the islands of Madagascar, Mauritius, and Reunion. Its back is brownish-gray speckled with white, while the stomach is practically pure white. The males have a large throat pocket. The name "tomb bats" is basically unjustified, because these animals prefer open country. Their daytime roosts are usually in rather open spaces which are penetrated by daylight, such as underneath roofs or on tree trunks and walls. It has been reported that they occasionally fly around during the day and change their resting places.

Family: Bulldog bats

This brings us to the family of BULLDOG BATS (Noctilionidae); it contains only one genus (*Noctilio*) with two species: 1. The MEXICAN BULLDOG BAT (*Noctilio leporinus*; HRL 13 cm; see Color plate, p. 124 and Fig. 6-2). In males the hair on the back is a flaming orange-red; in females the color is gray or dark brown, and the underside is paler. The upper lip is divided by a vertical skin fold, forming a "harelip"; the lower lip has a smooth edge and a wart in the middle, under which the two skin folds meet. The ears are large and thin; the tragus is small with a serrated outer rim. The tail extends almost to the middle of the well-developed tail membrane, and emerges on the upper side, so that the short end of the tail protrudes freely. The species exists from Mexico to northern Argentina, and on the Antilles. 2. The SOUTHERN BULLDOG BAT (*Noctilio labialis*; HRL 7 cm) is found from Nicaragua south to Argentina.

Fig. 6-6. Mediterranean horseshoe bat (*Rhinolophus euryale*) in Europe.

We are already familiar with the Mexican bulldog bat as a catcher or fish, and have mentioned the special adaptations to this feeding habit, especially its large feet equipped with claws, with which it catches its slippery prey from the water. While the larger bats only sometimes hunt insects, we have learned that the southern bulldog bats feed exclusively on insects. During the day, groups of Mexican bulldog bats rest in mountain caves and hollow trees. These sleeping quarters can be identified from some distance by their sharp, penetrating odor. This scent originates not only from the piles of guano, but also from the animals themselves.

Family: Slit-faced bats

The family of SLIT-FACED BATS (Nycteridae) is an exclusively Old World family. Their nostrils are situated in a deep groove flanked by conspicuous nose flaps, which are used to focus the ultrasonic calls by which the animals orient themselves. These sounds consist of claps, like those of the vespertilionid bats; they are emitted through the nose, similar to the technique of the horseshoe bats. The ears are fused in the middle of the head, and they are very large and thin. The tail vertebrae are fully enclosed in the tail membrane and end in a V-shaped fork, a characteristic of this family. The hair of the slit-faced bats is remarkably long and loose. The coloring ranges from grayish-brown to reddish-brown. There is one genus with some ten species; the center of distribution is in Africa.

Fig. 6-7 1. Blasius horseshoe bat (*Rhinolophus blasii*) in Europe;
2. Mehely horseshoe bat (*R. mehelyi*) in Europe.

During the day, slit-faced bats can be found together in large or small colonies, usually in hollow trees, mountain caves, ruins, or roof structures of buildings. Many species occasionally fly into lighted rooms during their nocturnal hunting flights, lured by the numerous insects attracted by the light. The largest species, *Nycteris grandis* (HRL 9 cm, TL 6.5–7.5 cm), lives in western and central African forest areas. Another species, *Nycteris thebaica*; (see Color plate, p. 124), is found from Africa to Arabia and the Middle East; it was also observed on Corfu, and therefore belongs to the European animal world. The smaller *Nycteris hispida*, found from Egypt to Mozambique and from Gambia to Angola, is the only member of its genus which prefers to hang from the branches of bushes and trees when roosting during the day, rather than resting in caves. The JAVANESE SLIT-FACED BAT (*Nycteris javanica*) has a habitat that is far removed from the range of the other slit-faced bats; it lives in southeastern Asia from Tenasserim (Burma) to Malaysia and the greater Sunda Islands as far as Timor (Indonesia).

Family: Large-winged bats

Closely related to the slit-faced bats are the FALSE VAMPIRE BATS OR LARGE-WINGED BATS (Megadermatidae; see Color plate, p. 124); both families are therefore combined into one superfamily (Megadermatiodea). The nose flap, which has different shapes among the various species, gives

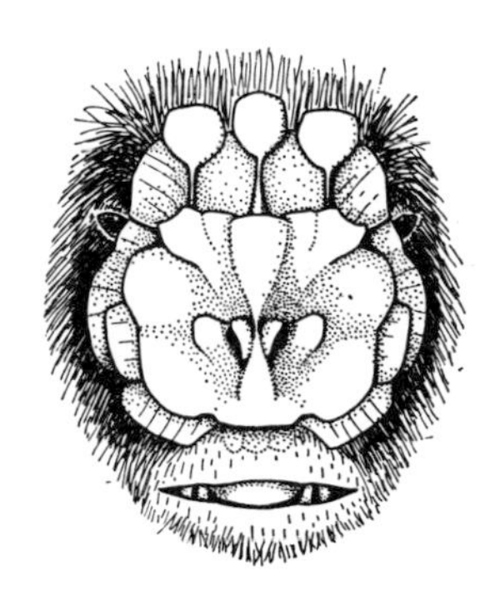

Fig. 6-8. The eyes almost disappear behind the strange nose "ornaments" of the flower-faced bat (*Anthops ornatus*).

these animals a unique appearance. The large ears are connected in the middle by a skin seam, and the ear flap is divided. A more or less well-developed tail membrane stretches between the hind feet; there are no tail vertebrae visible from the outside. We distinguish three genera with five species in Africa, southern Asia, the Malay Archipelago as far as the Philippines, and in Australia.

The only representative of the genus *Lavia* is the AFRICAN YELLOW-WINGED BAT (*Lavia frons*; HRL over 7 cm, ear L 4.5 cm; see Color plate, p. 124). These animals are pearly gray; the lower back is brownish, while the ears and wing membrane are brownish-yellow. It is found mainly in eastern Africa, and lives in open country with bushes and trees. When resting, it hangs from trees and bushes, but occasionally it will fly around during the day.

The FALSE VAMPIRES (genus *Megaderma*) are divided into three subgenera: a. *Cardioderma*, with the species HEART-NOSED FALSE VAMPIRE (*Megaderma cor*), cave-dwellers which are found from Ethiopia to Tanzania. b. *Lyroderma*, with the INDIAN FALSE VAMPIRE (*Megaderma lyra*; see Color plate, p. 124), from India and Indochina, where it is a cave-dweller, and also apparently largely carnivorous. c. *Megaderma*, with the species *Megaderma spasma*, a resident of south eastern Asia, where it is a cave-dweller and apparently carnivorous.

The AUSTRALIAN GIANT FALSE VAMPIRE BAT (*Macroderma gigas*; HRL 14 cm; see Color plate, p. 124), from western and northern Australia, is definitely carnivorous. It is the largest species of the family. The hair on the back is whitish with grayish-brown spots. The head, skin, and undersides are white. These animals are cave-dwellers.

Family: Horseshoe bats

The next two families are combined in one superfamily of HORSESHOE BAT RELATIVES (Rhinolophoidea). They are the HORSESHOE BATS (Rhinolophidae) and the OLD WORLD LEAF-NOSED BATS (Hipposideridae). Their nostrils are also surrounded by skin growths. In the horseshoe bats, these growths consist of three parts: the lower horseshoe-shaped nose flaps, with the nostrils at the base; a longitudinal ridge; and the higher, pointed projection, the "lancet." As we mentioned earlier in our description of echolocation, these growths help at least partially in directing and shaping echos. The large, pointed, projecting ears have no lobes. The female horseshoe-nosed bats have two nipple-shaped attachments on the lower backs of their bodies, near the vagina, so-called "back nipples," to which the young cling while they are carried around by their mother.

Fig. 6-9. 1. Suapure naked-backed bat (*Pteronotus suapurensis*) in Central America; 2. Naked-backed bat (*Pteronotus davyi*) in Central America.

The hair of the horseshoe bats is loose and soft, its color varying from blackish-gray and dark brown to a bright orangish-red tint. Most species are cave-dwellers and live together socially. While resting, they usually hang from the ceiling of the cave, with a certain distance between them, and wrap their wing membranes around their bodies like a cloak. Those species that live in the temperate zones hibernate during the cold season, although they interrupt their hibernation from time to time.

The genus *Rhinolophus* comprises some fifty species, found in all areas of the tropical and temperate zones of the Old World, east as far as Australia and New Guinea. We are familiar with large species, such as MACLAUD'S BAT (*Rhinolophus maclaudi*; HRL 7.5 cm), and dwarf species like the LESSER HORSESHOE BAT (*R. hipposideros*; HRL 4–4.5 cm; see Color plate, p. 125). This species is also found in Europe, and is the same approximate size as the DENT'S HORSESHOE BAT (*R. denti*) from southern and western Africa. We will mention two others of the many African species, which live in generally the same area in western Africa: the small *R. landeri* and the more stately *R. alcyone*. The males of both species have particular glandular hair tufts in the armpit area that are noticeable by their coloring. Otherwise, the only remarkable fact about these males is the presence of well-developed nipples, which apparently have some function; they do not produce milk; perhaps they secrete odor substances.

Two of the species have moved into the temperate zones; they are the GREATER HORSESHOE BAT (*R. ferrumequinum*; see Color plate, p. 125), and the aforementioned LESSER HORSESHOE BATS. The northern boundary of their distribution is just about at the edge of Germany's central highlands. The lesser horseshoe bat appears furthest to the north, although unfortunately this species has dwindled extensively during the last decades; other species are found on maps and on the Color plates on page 124.

Family: Old World leaf-nosed bats

The second family of the horseshoe bat relatives, the OLD WORLD LEAF-NOSED BATS (Hipposideridae), comprises nine genera with approximately forty species which are native in the warm areas of Africa, southern Asia to the Philippines and the Solomons and Australia. The shape of the nose flap is somewhat different than in the horseshoe bats. The horseshoe-shaped lower nose growth has a mostly flat, oval, multidivided flap attached to it. Among many species it is divided and widened in a rather grotesque manner. These cutaneous growths surrounding the nostrils serve in directing ultrasonic emissions for echolocation, as in the Old World leaf-nosed bats. Thus they can fly with their mouths closed, eating the insects caught in flight, without ceasing the emission of calls, just like horseshoe bats.

The COMMERSON'S LEAF-NOSED BAT (*Hipposideros commersoni*; HRL 11 cm; see Color plate, p. 125), with its west and east African subspecies *H. c. gigas*, is one of the largest species. They are found during the day in hollow trees and mountain caves. Loveridge visited such a large cave near the Mkulumusi River (Tanzania). As soon as he entered the cave, he heard a noise like the sound of rushing water; this sound was caused by the wing-flapping of thousands of insectivorous bats that were aroused by the beam of his flashlight. The roosting site of these animals was in another cave connected with the first one by a narrow shaft. Cries of all kinds greeted the visitors. The bats were so upset by this disturbance that even by the next day not one had returned there.

One of the most common leaf-nosed bats is the small *Hipposideros*

▷
Vespertilionids: Long-eared bat (*Plecotus auritus*).

▷▷
Funnel-eared bats: 1. Head of *Natalus stramineus*. Vespertilionids: 2. Large mouse-eared bat (*Myotis myotis*), a) hibernating colony, b) female with young; 3. Bechstein's bat (*Myotis bechsteini*) in flight, trapping prey by bending its tail membrane forward to form a trap; 4. Head of the Natterer's bat (*Myotis nattereri*); 5. Head of the water bat (*Myotis daubentoni*).

▷▷▷
1. Long-eared bat (*Plecotus auritus*) in hibernating position with folded ears and visibly protruding tragi; 2. Barbastelle (*Barbastella barbastellus*); 3. Common pipistrelle (*Pipistrellus pipistrellus*); 4. Serotine bat (*Eptesicus serotinus*).

▷▷▷▷
1. Common noctule (*Nyctalus noctula*), flying; 2. Red bat (*Lasiurus borealis*); 3. Head of the pallid bat (*Antrozous pallidus*).

Großmann
a
2
b
3
4
5
1

1
2
3
4

1
Großmann
2
3

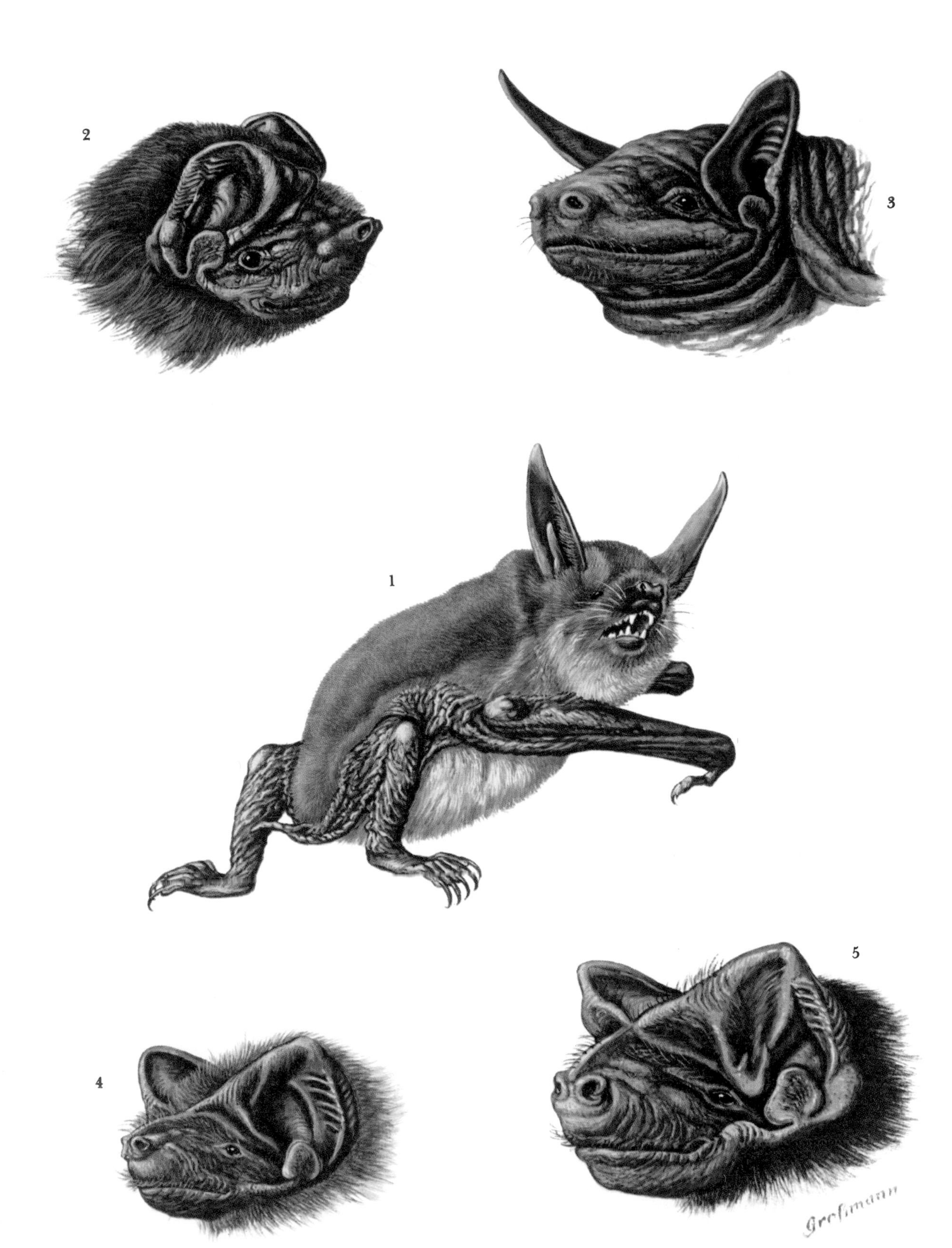
2
3
1
5
4
Greßmann

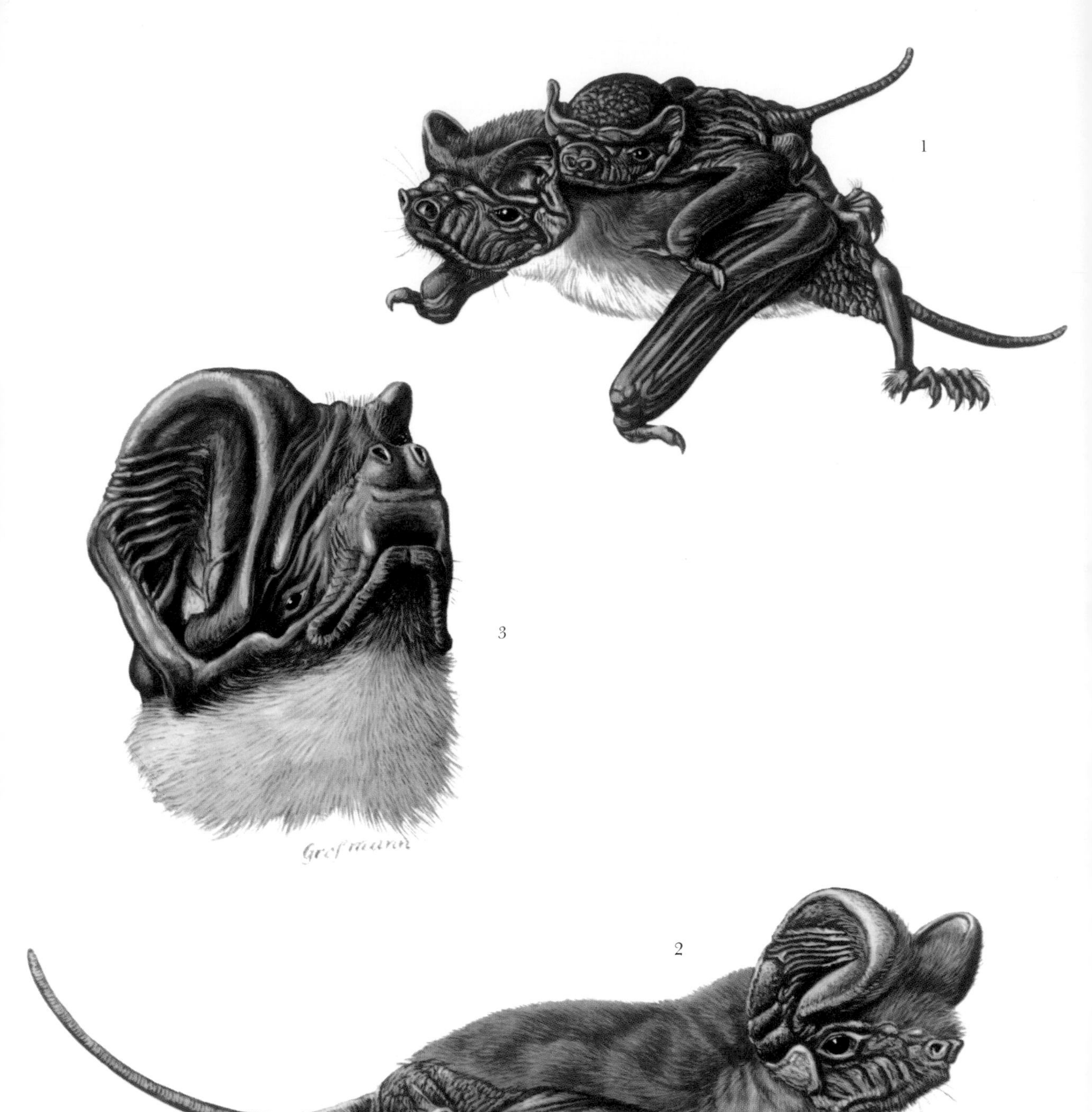
1
3
2

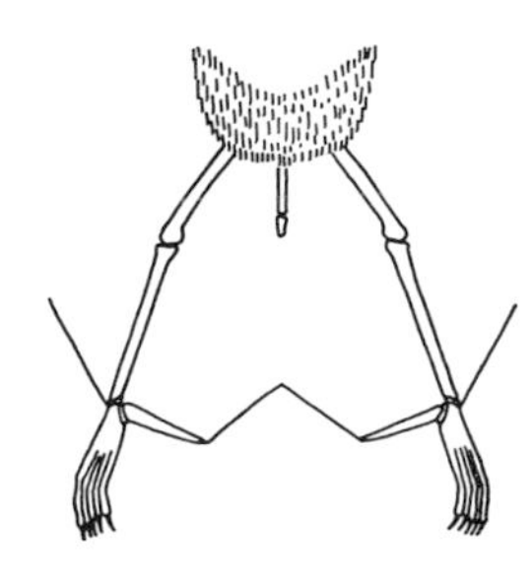

Fig. 6-10. The tail vertebrae of Seba's short-tailed bat (*Carollia perspicillata*) is so short that the tail membrane is attached between the long hind legs.

caffer (Color plate, p. 125). This species was present in almost all the mountain caves that I visited in Cameroon and on the island of Fernando Po. I found them in two color phases: a dark grayish-brown and a bright red-brown, although there may be a variety of colors in between. This species prefers to fly at night in places that are lighted and occupied. It has been reported that the species *H. armiger*, a native in India and China, is attracted by the shrill buzzing of cicadas. If the bat fails to catch the insect on its first attack, it patrols, waiting for the buzzing to begin again so that it can immediately try again for the catch.

We will mention the following species here as examples of the fantastic shapes the cutaneous nose growths can assume: the TRIDENT LEAF-NOSED BAT (*Asellia tridens*; see Color plate, p. 125), found from northern Africa to Somalia and Zanzibar, and in eastern to northwestern India; the FLOWER-FACED BAT (*Anthops ornatus*; see Fig. 6-8), from the Solomons; and the PERSIAN LEAF-NOSED BAT (*Triaenops persicus*), which is found from Iran to Egypt.

Family: Leaf-nosed bats

From the relatives of the horseshoe bat confined to the Old World, we will pass over to the family of LEAF-NOSED BATS (Phyllostomidae), which is divided into several subfamilies that live exclusively in the warm zones of the New World. The cutaneous nose growths are also important to these bats, although there are genera that have only underdeveloped or nonexistent nose growths. The size varies, body length ranging from 4 to 13.5 cm. The ears of some are very large, with earlobes. The tail vertebrae and tail membrane appear in all phases from fully developed to nonexistent. Theses animals are adapted to a wide variety of nutrition. There are fifty-one genera with some 140 species (see Color plates, pp. 126–128).

The representatives of the subfamily of MUSTACHE BATS (Chilonycterinae) have no nose growth; instead thay have skin flaps surrounding the chin and wart-shaped elevations at the edge of the lips (see Color plate, p. 126). We do not know the reason for these peculiar growths. The tail vertebrae protrude above the tail membrane as in the sac-winged bats. The flightskins do not begin directly at the sides of the body, but more toward the center of the back. The MUSTACHE BATS (genus *Chilonycteris*) show the first signs of this development. The strongest indications of this are in the NAKED-BACKED BATS (genus *Pteronotus*), whose wing membranes grow from the center of the back; thus the back looks as if it were naked. Actually, however, it has normal hair underneath the membrane.

◁◁
New Zealand short-tailed bats: 1. New Zealand short-tailed bat (*Mystacina tuberculata*). Free-tailed bats: 2. Head of *Molossops malagai*; 3. Head of the naked bat (*Cheiromeles torquatus*); 4. Head of *Molossus major*; 5. Head of *Molossus rufus*.

◁
1. *Tadarida condylura*, female with young on her back (from a photograph by Kulzer); 2. *Tadarida teniotis*; 3. Head of *Eumops perotis californicus*.

The genus *Pteronotus* comprises two species; the SUAPURE NAKED-BACKED BAT (*Pteronotus suapurensis*; HRL 6–6.5 cm; see Color plate, p. 126) and the NAKED-BACKED BAT (*P. davyi*). Both are found from Mexico to northern South America, and choose caves with particularly warm, humid, muggy air for their daytime roosting, in both small and large sleeping colonies. Felten has written: "Both *Pteronotus* species, because of their special need for humid and warm air, usually live in an atmosphere where it is impossible for humans to breathe. Temperatures of 38° C and air that is saturated with moisture and filled with unbearable odors from the immense guano heaps are the conditions of life for these animals and for the growth of their young."

Fig. 6-11. Mexican long-nosed bat (*Choeronycteris mexicana*).

All BIG-EARED LEAF-NOSED BATS (subfamily Phyllostominae) have a nose leaf, which in some species has developed into something like the point of a lance, with a tremendous length. This group includes among others, the TOME'S LONG-EARED BAT (*Lonchorrhina aurita*; HRL 6 cm; see Color plate, p. 126) and the LONG-LEGGED BAT (*Macrophyllum macrophyllum*; HRL 4.5 cm, TL 4.5 cm).

The nose leaf of the Tome's long-eared bat reaches a length of over 2 cm. The tail is nearly as long as the body, and is completely enveloped in the tail membrane. The ears are very large. The hair is reddish-brown. These animals are found from southern Mexico to northern South America, and have been observed on Trinidad. The long legged bat has a long nose growth which is very wide at the base. The limbs are noticeably long, and make the animal seem larger than it actually is.

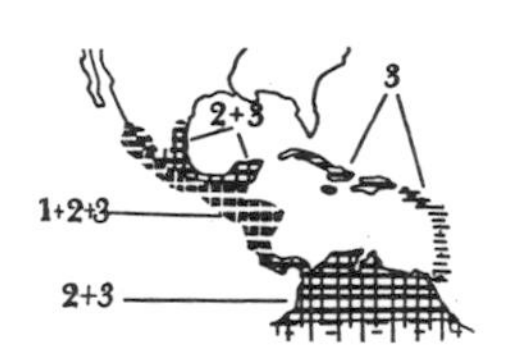

Fig. 6-12. 1. *Artibeus nanus*; 2. *A. lituratus* in Central America; 3. *A. jamaicensis* in Central America.

Tome's long-eared bats have been found both singly and in colonies of up to 500 animals. They were found in caves and subterranean shafts, frequently with other species. Pregnant females were found in these caves in February and March.

The SPEAR-NOSED BATS (genus *Phyllostomus*), which gave the subfamily its scientific name, are represented by four rather portly species. The largest of them is the SPEAR-NOSED BAT (*Phyllostomus hastatus*; HRL 10–13 cm, wingspan 45 cm; see Color plate, p. 126). These bats have strong teeth for eating insects and fruits and for dining extensively on small vertebrates. They are found from Honduras to Peru, Bolivia, and Brazil.

Fig. 6-13. Wrinkle-faced bat (*Centurio senex*).

Thousands of these spear-nosed bats were found in a cave in Panama. Felten was able to keep the related smaller species *P. discolor* in captivity for some time. Besides insects, the animals also ate fruits: "After the person in charge of feeding had left the cage, an animal appeared like a vanguard from the hiding place behind a truss of the wall; shortly thereafter, the others followed. They crowded halfway up the wall and then flew from there to the feeding bowl. They ate the food right there on the spot, or in particular eating places to which they carried the food in their mouths."

The last member of this subfamily is LINNE'S FALSE VAMPIRE BAT (*Vampyrum spectrum*; HRL 12.5–13.5 cm, weight 145–190 g, wingspan 70 cm). The ears are large and spaced far apart. The nose has a large, dagger-shaped process. Distribution is in the New World tropics from southern Mexico to Peru and Brazil, and on Trinidad.

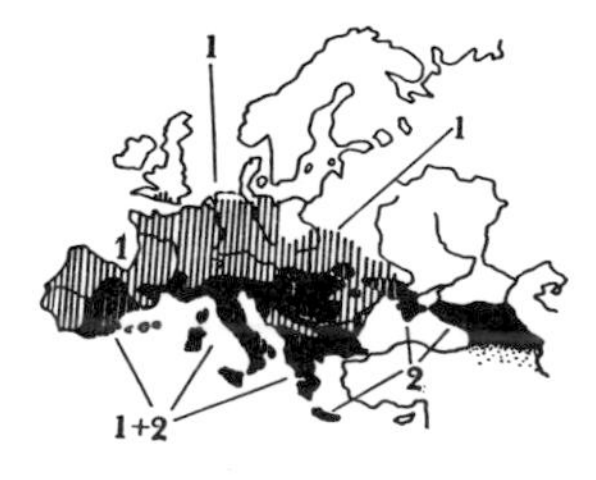

Fig. 6-14. 1. Large mouse-eared bat (*Myotis myotis*) in Europe; 2. Lesser mouse-eared bat (*M. oxygnathus*) in Europe.

Early authors erroneously described this species as being a bloodsucker, and its name is derived from this false characterization. It has been confirmed, however, that the species will feed on small vertebrates in addition to the regular diet of fruits and insects. Remains of feathers and mouse tails have been found in this species' resting sites in hollow trees. In captivity the bats have readily fed upon mice and chicks.

Fig. 6-15. Bechstein's bat (*Myotis bechsteini*).

Fig. 6-16. Pond bat (*Myotis dasycneme*) in Europe.

Fig. 6-17. Water bat (*Myotis daubentoni*) in Europe.

Fig. 6-18. *Myotis capaccinii* in Europe.

As mentioned earlier, the long-tongued bats (subfamily Glossophaginae) are to a large extent specialized feeders, concentrating on nectar and pollen. These bats are quite small, ranging in length between 4.5 and 8 cm. The snout is sometimes considerably elongated. The front of the long tongue has brushy papillae. There are several genera with numerous species, among which the best-known is the LONG-TONGUED BAT (*Glossophaga soricina*; see Color plate, p. 127). The ears and the tapering nose process are relatively small. The body is brown, with some different shades also present; the lower side of the body is somewhat lighter.

These long-tongued bats generally remain in groups of varying sizes in cliff or tree hollows during the day, but they also approach human settlements and are even found in homes. Felten, studying these bats in captivity, concluded that they feed primarily on insects, some caught in flight and others taken on the ground. He states, "Hydromel (a mixture of honey and water) and fruits were eaten throughout the year. At the beginning of the rainy season, hydromel consumption almost exceeded that of insects, while otherwise insects were consumed in far greater quantity. Intake of hydromel from freely hanging petri dishes occurred almost exclusively while in flight; the bats fluttered about the dish at first for a few seconds, then quickly dipped the tongue in, generally repeating this motion several times." Observations have also been made on the bats feeding on blossoming plants. The bats flutter before the blossom or often actually grab hold of the blossom while feeding.

Another great specialization to feeding on flowers may be inferred from other genera of this subfamily, which have a very unusual skull structure. This includes the Peruvian species *Platalina genovensium*, the MEXICAN LONG-NOSED BAT (*Choeronycteris mexicana*), distributed from southern Mexico to Guatemala, and the BANANA BAT (*Musonycteris harrisoni*), which was discovered in 1958 and apparently feeds only on banana blossoms.

Presumably the members of another subfamily of American leaf-nosed bats, Phyllonycterinae, also feed on fruit juices and blossoming plants. Like the long-tongued bats, they have a relatively long foreskull and a long, protrusible tongue. They are distributed on the Caribbean Islands, but little is known of their life.

The last three subfamilies of leaf-nosed bats are pure fruit-feeders. These are the SHORT-TAILED LEAF-NOSED BATS (Carolliinae), the YELLOW-SHOULDERED BATS (Sturnirinae), and the RED FRUIT-EATING BATS (Stenoderminae). All have particularly broadened and flattened cheek teeth, a feature specifically adapted to their feeding habits. As we have seen, the same structure is found in the fruit-eating bats of the Old World.

Of the short-tailed leaf-nosed bats, the SEBA'S SHORT-TAILED BAT (*Carollia perspicillata*; see Color plate, p. 128), found from Mexico to southern Brazil, has the largest distribution. The species has been observed feeding on guavas, bananas, and wild figs. The best-known yellow-shouldered bat is *Sturnira lilium*, distributed from northern Mexico to Argentina and Paraguay. The tail has completely disappeared in this species, and the tail membrane is present only as a narrow seam. Males have bushy growths of yellowish or reddish hair in the shoulder region, similar to the bushy hair on epauletted fruit bats. The molars have a deep longitudinal groove.

Fig. 6-19. Natterer's bat (*Myotis nattereri*) in Europe.

The subfamily of red fruit-eating bats has many genera and species, and the greatest degree of broadening of the molars is found in these. The eight species of the genus *Artibeus* are considered to be archetypal fruit-feeders. Their body length varies from 5 cm (*Artibeus nanus*) to 10 cm (*A. lituratus*). The body is plump, with a short, thick head and relatively small ears. The nasal process is lance-shaped, and there is no externally visible tail. The caudal skin is present only as a narrow seam along the thighs. The pelt is short and velvety, and may have dark brown, gray, or blackish coloration. Four brighter stripes run over the head. Distribution is from northern Mexico to northern Argentina.

Fig. 6-20. Whiskered bat (*Myotis mystacinus*) in Europe.

During the day, these fruit-feeding bats remain among the foliage in trees. Only *Artibeus jamaicensis* seeks hollows in cliffs and trees. Their food consists chiefly of soft, juicy fruits, whose remains can be found over the bats' resting sites. Since the bats drop the seeds and pits of the fruits, they assist in the distribution of these plants.

The TENT-BUILDING BAT (*Uroderma bilobatum*), distributed from Mexico to Peru, Bolivia and Brazil, and on Trinidad, has distinctive markings. There are four bright longitudinal stripes on the head and one bright medial line on the back. The body is stocky. The nasal process consists of a horseshoe-shaped nasal portion and the tapering leaf structure.

This species does not spend the day in hollows, but hangs under large leaves in small groups. Felten found them hanging on the central vein of a folded banana leaf. Other observers maintain that the bats form these tentlike structures themselves, by making a series of holes running across the middle of a large palm leaf. The bats then supposedly bend the outer half of the leaf around, so they can then rest inside this "tent."

Fig. 6-21. *Myotis emarginatus* in Europe.

I personally react to these observations very skeptically and doubt very much, in view of their tooth structure, that a bat is able to make such straight rows of holes in a leaf. On the basis of personal observations in tropical regions in Africa, I tend to believe instead that these holes were made by insect larvae while the leaves were still rolled up. A storm can then easily break the leaf along the line of holes and form the tent roof which is so convenient for the bats.

The last of the red fruit-eating bats cited here is the WRINKLE-FACED BAT (*Centurio senex*; see Color plate, p.128), a particularly grotesque-looking creature distributed from Mexico to Costa Rica and on Trinidad. The short, broad face is naked and has wrinkles and folds which give the bat its distinctive appearance.

Fig. 6-22. Common pipistrelle (*Pipistrellus pipistrellus*) in Europe.

The family of VAMPIRE BATS (Desmodontidae) is closely related to the

Fig. 6-23. The pipistrelle (*Pipistrellus nathusii*) in Europe.

leaf-nosed bats and undoubtedly has developed from them. Their specialization to feeding on blood has already been cited. The vampire bats inhabit arid regions and also hot, humid forests in the tropics and subtropics of the Americas. They are found primarily in natural and artificial rocky cavities, but also in hollow trees and inhabited buildings. They form sleeping colonies of various sizes. The bats do not leave these colonies until the onset of darkness, at which time they seek their "blood donors." Three genera have been identified, each containing one species.

The best known is the grayish-brown COMMON VAMPIRE (*Desmodus rotundus*; HRL 7.5–9cm; see Color plate, p. 128). The tail is absent. Caves have been found in which several thousand of these bats were together in a colony. Distribution extends to the South American subtropics. Two other vampire species, *Diaemus youngi* and *Diphylla ecaudata* (see Color plate, p. 128), are found in far smaller numbers and are restricted to purely tropical regions. The former is characterized by white wing markings. Little is known of their natural history.

Superfamily: vespertilionid bats

Six families are included in the VESPERTILIONID BATS (superfamily Vespertilionoidea). These species do not have prominent nasal processes or skin formations on the head. The five families of funnel-eared bats, smokey bats, disc-winged bats, sucker-footed bats, and New Zealand bats contain either few species or just a single species. But this group distributed worldwide, has the greatest diversity in form of any among the bats.

The FUNNEL-EARED BATS (Natalidae) are small, long-tailed bats with funnel- shaped ears and a short ear covering. They are distributed chiefly in tropical Central and South America and on the Antilles and a few coastal islands. For the resting site during the day they prefer caves and, less commonly, tree hollows. Their diet consists exclusively of insects. The family contains a single genus (*Natalus*), with three subgenera and about nine species. *Natalus stramineus* is pictured on page 134.

Fig. 6-24. 1. *Pipistrellus kuhli* in Europe; 2. *P. savii* in Europe.

SMOKEY BATS (Furipteridae), closely related to the funnel-eared bats, consists of two genera, each with one species, distributed in the tropics of South America. They are also characterized by large, funnel-shaped ears. The distinctive feature of these bats is the regression of the thumb, which is very small and is enclosed within the wing membrane so that only the small, functionless claw protrudes. The sharp, almost vertical band in the foreskull is a prominent feature on the skull. The species *Furipterus horreus* (HRL 4cm), distributed from Panama to Guayana and Brazil is among the smallest bats known.

Fig. 6-25. Common noctule (*Nyctalus noctula*) in Europe.

The few members of the next two families are characterized by the suction discs on their limbs. This is an unusual feature for mammals. The DISC-WINGED BATS (Thyropteridae) have suction discs on the wrist and the sole of the foot; they are circular and are on a small base. This enables them to maintain a firm hold on smooth branches and leaves and to crawl on them. A thumb is still present; it has a small claw. These bats prefer rolled-up leaves, for example, of bananas, for their sleeping site during the day; several animals can usually be seen sitting one behind the other, with their heads up. The family has just two species, SPIX'S DISC-WINGED BAT (*Thyroptera tricolor*) and the HONDURAN DISC-WINGED BAT (*T. discifera*).

In SUCKER-FOOTED BATS (Myzopodidae), the sucking discs are firmly attached. The thumb has almost completely regressed and is functionless. The single species, the GOLDEN BAT (*Myzopoda aurita*), has very large ears, and the ear covering terminates in a small round plate. The first beginnings of adhering discs are also found in a few vespertilionid bats, such as the FLAT-HEADED BAT (*Tylonycteris pachypus*) from India.

Fig. 6-26. Lesser noctule (*Nyctalus leisleri*) in Europe.

The many members of the VESPERTILIONID BATS (Vespertilionidae) are found in the Old and New World from the tropics into temperate zones, and some species are even distributed into the Arctic. Vespertilionids inhabit forests and open fields, both moist and dry regions, lowlands and mountains. There are dwarf forms (with an HRL hardly more than 3 cm) and magnificent specimens exceeding 10 cm in length. The tail is always well developed and is covered by the tail membrane, although in some species the tail protrudes slightly beyond the membrane. The ears have covers. Their diet consists almost exclusively of insects. There are forty genera with nearly 275 species (see Color plate, pp. 133–138).

Fig. 6-27. Serotine bat (*Eptesicus serotinus*) in Europe.

The MOUSE-EARED BATS (*Myotis*; see Color plate, p. 134) include some sixty species. They generally inhabit caves, or at least those in temperate zones do so during winter. There are seven species in central Europe. The biggest of these is the LARGE MOUSE-EARED (*Myotis myotis*, HRL up to 8 cm, wingspan 35 cm). The other European species are considerably smaller and are distinguished by the variation in size and shape of the ears, the ear covering, and the development of the tail membrane. Their habitats also vary. Some other species are: BECHSTEIN'S BAT (*M. bechsteini*), with relatively large ears; the POND BAT (*M. dasycneme*); the WATER BAT (*M. daubentoni*); NATTERER'S BAT (*M. nattereri*), with bent, brushlike processes on the rear tail seam; and the WHISKERED BAT (*M. mystacinus*), the smallest European species.

Fig. 6-28. Northern bat (*Eptesicus nilssoni*) in Europe.

After hibernation, which males and females spend together in natural or artificial cavities, the female mouse-eared bats gather in great collections at a particular location, where they bear their young and raise them. The birth of the single young normally occurs in the first half of June. In August, after the young are independent, the bats abandon these sites. Mouse-eared bats generally return to the same summer and winter roosts, which may be as much as 200 km away from each other. Another change of roosts is occasionally made in summer or in winter.

The smallest species are the PIPISTRELLES (*Pipistrellus*). This includes approximately forty species in North America, Europe, Africa, Madagascar, Asia, and on the Sunda and Philippine Islands, New Guinea, and Australia. There are often two young. A well-known European species is the COMMON PIPISTRELLE (*Pipistrellus pipistrellus*; HRL 3.5–4.5 cm, wingspan 20 cm).

Fig. 6-29. Big brown bat (*Eptesicus fuscus*).

Fig. 6-30. Particolored bat (*Vespertilio murinus*) in Europe.

Fig. 6-31. Red bat (*Lasiurus borealis*).

Fig. 6-32. Hoary bat (*Lasiurus cinereus*) in North and Central America; they are also found on Iceland and the Orkney Islands.

Fig. 6-33. Barbastelle (*Barbastella barbastellus*) in Europe.

During summer, pipistrelles prefer being on or inside buildings, and in winter they are found inside buildings or in caves and crevices in rocks. In some regions they undertake rather extensive seasonal migrations. Thus one animal marked in Russia was recovered in Bulgaria, at a distance of over 1600 km from the place of marking. The AFRICAN BANANA BAT (*Pipistrellus nanus*), distributed through much of Africa, invariably spends the day in young banana leaves which are still rolled up.

The NOCTULE BATS (*Nyctalus*), of which there are six species, are distributed from the Azores through Europe and Asia to the Philippines. The body length ranges from 5 to 10 cm, and the tail is 3.5–6.5 cm long. The COMMON NOCTULE (*Nyctalus noctula*: see Color plate, p. 136) is found in Europe, its smaller, much rarer relative, the LESSER NOCTULE (*N. leisleri*), is strictly a forest inhabitant. The largest noctule species, the GIANT NOCTULE (*N. lasiopterus*), is distributed from southern Europe to Asia.

The noctule is a typical tree-dwelling bat. It is usually found in hollow trees, particularly in summer and very often in winter; however, the species occasionally lives in or on buildings. Twins are often born in this group. This stately bat, with round, robust ears, small, rounded ear covers, and narrow wings, is capable of fast and skillful flight. The noctules leave their sleeping site early, and may begin the hunting flight in the afternoon, particularly in the fall.

The thirty species of BIG BROWN BATS (genus *Eptesicus*) are distributed worldwide. The SEROTINE BAT (*Eptesicus serotinus*; see Color plate, p. 135), which like many other species is divided into a series of subspecies, prefers living near human settlements. The North American representative of this group is the BIG BROWN BAT (*E. fuscus*). The European species is found during the summer in small groups among the rafters in ground-floor rooms. This bat is readily kept in captivity and has a calm temperament. Very small members of this genus inhabit Africa. In Cameroon I became acquainted with *E. tenuipinnis*, with its whitish, translucent wings; it is one of the smallest mammals on earth. One pregnant female which I kept gave birth to one young which was about the size of a bee. The NORTHERN BAT (*E. nilssoni*) has an extraordinary resistance to cold; its distribution extends into the Arctic.

Two members of genus *Lasiurus* penetrate into the northern parts of North America: the RED BAT (*Lasiurus borealis*; see Color plate, p. 136) and the HOARY BAT (*L. cinereus*). In winter they migrate, to some extent, to the southern U.S.A. During the day, they rest in branches or tree stumps, and their coloration so closely resembles that of a withered leaf or the tree itself that they are difficult to see. Interestingly, the females bear up to four young, and are the only bat species with four teats.

Two other vespertilionid bats are distributed over large parts of central Europe: the BARBASTELLE (*Barbastella barbastellus*; see Color plate, p. 135), with short, broad ears joined in the middle by a skin seam; and the LONG-EARED BAT (*Plecotus auritus*; see Color plates, pp. 133 and 135), with extraordinarily long ears, they can lay so far over the wing (done during the daily sleep and in hibernation) that only the long, pointed ear cover can be seen. Both species are found in caves in the winter. Comparative investigations have only recently shown that a "twin species" to the long-eared bat exists, the SOUTHERN LONG-EARED BAT (*P. austriacus*). This species differs from *P. auritus* in its grayer coloration and its body size. Long-eared bats are also found in the New World.

Fig. 6-34. Long-eared bats (genus *Plecotus*) in Europe.

The LONG-WINGED BAT (*Miniopterus schreibersi*) prefers warmer climates, as do the nine congenerics. The species is distributed chiefly along the Mediterranean in Europe and Africa, and is also found in southern Asia, New Guinea, and Australia. In Germany the species is found only in the Kaiserstuhl region on the upper Rhine, an area characterized by its mild climate. The long-winged bat is distinguished by its long, slender wings and the short ears, which are almost lost in the pelt.

Fig. 6-35. Long-eared bats (genus *Plecotus*) in America.

The last large subfamily of vespertilionid bats cited here, Nyctophilinae, contains the PALLID BAT (*Antrozous pallidus*; HRL 7.5 cm, wingspan 38 cm; see Color plate, p. 136), distributed from western North America to central Mexico. Robert T. Orr has devoted a rather extensive treatise to this species. This stately large-eared species has woolly hair, which is pale yellowish-brown on the upper side. Remains of food under feeding sites indicate that the pallid bat may take prey from the ground. Remains of bugs, crickets, and locusts have been found, and in some cases even scorpions, lizards, and geckos. The pallid bat sometimes feeds on vertebrates.

Fig. 6-36. Long-winged bat (*Miniopterus schreibersi*) in Europe.

The single species of the family Mystacinidae is restricted to New Zealand. The NEW ZEALAND SHORT-TAILED BAT (*Mystacina tuberculata*; see Color plate, p. 137) is the best-adapted bat species for running and climbing. The skin on the bottom of the foot is wrinkled and has longitudinal and diagonal furrows, giving the bat good traction when it runs along the ground. The sharp claws on the feet and the thumb are each equipped with a small secondary claw at their base, probably facilitating a firmer hold when the bat is climbing. Furthermore, the legs are short and firm, and the feet are oriented to the front along their sides, so that a forward motion may be readily executed. The thin wings can be set very closely together by a special folding mechanism. They disappear into a kind of sheath which is formed by the wing membrane running along the sides of the body. This bat's limbs are designed for running, and resemble those of other four-footed animals; from this we can infer that this species catches its prey primarily by running and climbing.

Fig. 6-37. Pallid bat (*Antrozous pallidus*).

The last bat family, the FREE-TAILED BATS (Molossidae), has ten genera and approximately eighty species; it is distributed across all temperate regions of the Old and New Worlds. The body length ranges from 4 to 13 cm, the tail length from 1.4 to 8 cm. The tail end of the body protrudes far beyond the rear edge of the wings. The tail membrane is extensible and can be moved forward and backward along the tail, altering the actual

Fig. 6-38. 1. *Tadarida brasiliensis* in North and Central America; 2. Mexican free-tailed bat (*T. b. mexicana*).

surface area of that membrane. The hind legs are short and powerful, and the feet are large and broad. There are well-developed calluses on the wrists. The outer teeth have a row of brushes which function in cleaning the hair. The head is thick and its snout portion is broad and flattened. In many species, the lips are folded and wrinkled, giving the face a highly unusual appearance. The ears have thick skin and are large and broad. The ear cover is generally short and rounded. Many species have a gland sack in the throat.

The powerful hind legs, the broad feet, and callosities on the wrists indicate that free-tailed bats are adapted to running on the ground; the wings can be folded close together and do not interfere with running. On the other hand, the long, narrow wings reflect the flight capability of this group, which can undertake long-distance flights. Only a few species are solitary; usually they form colonies of thousands or even millions in caves and tree hollows, behind trees, or in buildings, under roofs and similar cover. They generate a pungent musky odor. Sleeping sites in caves, which have been in use for thousands of years, have resulted in tremendous guano deposits. It took from 1901 to 1921 to dig out the valuable bat guano deposits in the Carlsbad Caverns (New Mexico), in order to process the guano into fertilizer. Tremendous colonies of MEXICAN FREE-TAILED BATS (*Tadarida brasiliensis mexicana*) are still found in these caverns. During the American Civil War, the forces of the southern states even produced salt-peter from the guano deposits in order to manufacture gunpowder.

Fig. 6-39. *Tadarida teniotis* in Europe.

The genus *Tadarida* is divided into four subgenera with approximately thirty-five species distributed through the tropical and subtropical areas of the world. They include small (HRL about 4.5 cm) as well as larger species (with an HRL extending to 8.5 cm). One species, *Tadarida teniotis* (see Color plate, p. 138), is distributed into southern Europe. Kulzer found one prevalent African species, *T. limbata*, during the day under house roofs in eastern Africa. One west African species allegedly is found with birds in tree hollows. The NAKED BAT (*Cheiromeles torquatus*; HRL 11.5–13.5 cm, TL 5–6.5 cm, weight up to 170 g, see Color plate, p. 137), distributed in the Malay Peninsula, Indonesia, and the Philippines, is one of the largest bat species known, together with its close relative, *Cheiromeles parvidens*, from Celebes. The body is naked, with barely perceivable hair on the tail membrane, the underside of the body, and on some places on the head. The neck has also black, brushy hairs. A large throat sack which extends over the entire front throat region exudes a pungent oily substance.

Fig. 6-40. *Eumops perotis californicus.*

The naked bat's lateral wing membrane originates near the central part of the back. A large pocket is formed on the sides of the body, in which the folded wings can be laid. Formerly, it was erroneously thought that these pockets were used to carry the young. Interestingly, the naked bat is subject to infestation by a parasite which belongs to the earwig group and which lives in the pockets. The parasite presumably feeds on substances excreted by the bat.

The last genus treated here is *Molossus*, which contains ten species of bats distributed from Mexico far into South America. As in other completely different families, this group is characterized by the appearance of two color phases: a bright reddish-brown phase and a dark brown to blackish phase. The lips in these free-tailed bats are smooth, not wrinkled. During the day, these bats crawl under roofs, often in large groups, especially under those covered with tin and others under which a high temperature can be created by the sun. The animals usually rest in a vertical position. They fly out quite early in the evening, in search of insects. It has been reported that they keep the prey in their cheeks, so that all can be ingested at the resting site. The widespread species *Molossus rufus*; HRL 8–8.5 cm; see Color plate, p. 137), one of the largest in the genus, has a short, rounded head with a blunt snout. The thick-skinned ears move in the middle of the head. A gland sack opens in the throat and is particularly well developed in males.

Fig. 6-41. *Molossops malagai.*

The bats are an order of mammals which contains not only a great number of species, but also a number of biological features which distinguish them from all other mammals. Bats are not readily accessible to observation, however, primarily because they are nocturnal organisms. Their nocturnal habits and unusual appearance are probably what have caused them to play a major role in the superstitions of many peoples. They are often connected with evil spirits, and the devil is often depicted with bats' wings, while angels have the wings of birds.

Even today many people are bent on destroying the bats as detestable creatures, a judgement which is completely undeserved, and they pursue and destroy any bats they locate. Very few species can inflict injury to man; those that can include the vampire bats, which can transmit diseases to people, and some fruit-eating bats which infect fruit upon which they feed. However, the vast majority are harmless creatures and even very useful animals which devour great quantities of insects that would otherwise have damaging effects on forests and farm crops. We should bear in mind that bats, too, have their functional place in nature, and that this gives them a right to their existence.

Unfortunately, environmental conditions have been altered to such an extent that there has been a great reduction in the number of bats. All bat species in Germany, for example, are under protection and have been so for a long time. However, legal measures alone are not enough to protect these animals. Everyone, wherever possible, should do what he can to enlighten others and bring about an understanding for these amazing creatures.

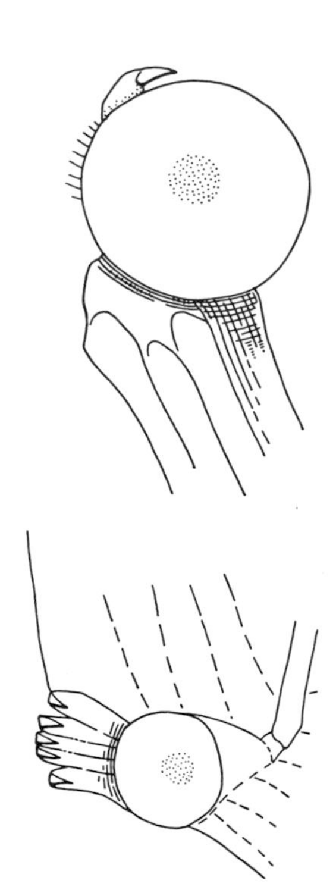

Fig. 6-42. Disc-winged bats (genus *Thyroptera*) have round sucking discs on the wrist (above) and sole of the foot (below).

7 Edentates

Order: Xenarthra

Suborder: Edentates, by W. Moeller

Armadillos, sloths, and anteaters belong to the most primitive and most distinctive mammals of the New World. They are the last survivors of the suborder Xenarthra, which flourished in the Tertiary period. They are grouped into the order Edentata; the Latin designation is erroneous. It stems from the Eighteenth Century, when the famed French zoologist Cuvier classified all edentates, as well as the aardvark (see Vol. XII), together with such true toothless creatures such as pangolins (see Chapter 8), the duck-billed platypus (see Vol. X), and echidnas (see Vol X). The only toothless edentate, however, is the anteater. The giant armadillo has as many as 100 teeth (one of the largest numbers of teeth in mammals). These small teeth, as well as those of related species, do, however, show clear indications of regression.

The three families of edentates living today are specialized in very different ways and thus differ widely from each other both in their external appearance and their structure. These are the armadillos (Dasypodidae), sloths (Bradypodidae), and anteaters (Myrmecophagidae).

Armadillos, sloths, and anteaters are distinguished by xenarthrales (extra articulations between the lumbar vertebrae); no other living mammals have them. These extra vertebral elements lend support particularly to the hips. Most armadillos utilize the reinforced vertebrae while digging for food. Sloths and tree-dwelling anteaters apparently do not make use of these peculiar structures; their method of movement does not require the xenarthrales. Edentates have a primitive brain with few convolutions, large olfactory lobes, and a well-developed olfactory region. The internal organs have primitive as well as more recently developed structural characteristics. The armadillos are the most primitive edentates: the neocortex is relatively small, while the olfactory lobe and the olfactory brain show the most extensive development within the suborder. Sloths have an intermediate position, while anteaters (on the basis of the external structure of the brain) have reached the highest level of development.

Even in earlier geological periods, the mammals with few teeth were

TUNDRA OF THE ARCTIC ZONE OF NORTH AMERICA
Mammals: Rodents: Greenland collared lemming (*Dicrostonyx groenlandicus*); 2. Arctic ground squirrel (*Citellus undulatus*). Predators (see Vol. XII): 3. Arctic fox (*Alopex lagopus*) in transitional coat. Hares (see Vol. XII): 4. Arctic hare (*Lepus timidus arcticus*). in transitional coat. Even-toed ungulates (see Vol. XIII): 5. Barren ground caribou (*Rangifer tarandus arcticus*); 6. Musk ox (*Ovibos moschatus*). Birds: Loons, divers (see Vol. VII): 7. Common loon (*Gavia immer*). Ducks and geese (see Vol. VII): 8. Canada goose (*Branta canadensis*); 9. Greater snow goose (*Anser caerulescens atlanticus*); 10. Emperor goose (*Anser canagicus*); 11. King elder (*Somateria spectabilis*); 12. Long-tailed duck (*Clangula hyemalis*). Fowllike birds (see Vol. VII): 13. Rock ptarmigan (*Lagopus rupestris*), in summer coat. Waders and gull-like birds (see Vol. VIII): 14. Arctic tern (*Sterna macrura*); 15. Long-tailed skua (*Stercorarius longicaudatus*); 16. Iceland gull (*Larus glaucoides*); 17. European knot (*Calidris canutus*); 18. Dunlin (*Calidris alpina*); 19. Gray phalarope (*Phalaropus fulicarius*); 20. Gray plover (*Squatarola squatarola*). Owls (see Vol. VIII): 21. Snowy owl (*Nyctea scandiaca*), catching a lemming. Sparrows (see Vol. IX): 22. Snow bunting (*Plectrophenax nivalis*).

Phylogeny, by E. Thenius

limited to the New World. The early history of some lines has been established by the fossil record, and its basic development is understood. These finds enable scientists to make clear statements about the origin of the entire order. The fossil record shows that edentates were once present in greater diversity (that is, as many different species) than is the case today. This group has almost ten times as many fossil genera as living genera. During the Ice Age there were giant edentates in North and South America. Remains of their bones, teeth, fur, and feces have been found.

The history of edentates (see Color plate, p. 176) is related to the early separation of South America from the other continents. Throughout the entire Tertiary there was no land connection with Central America; what is now Panama was formed at the end of the Tertiary period, permitting the intermingling of species between North and South America. However, during the earliest part of the Tertiary it seems that there was indeed a land connection between the Americas, for the ancestors of the oldest South American mammals (edentates, notoungulates, and primitive horse-like ungulates) were definitely from North America. The North American ancestral group of edentates were the early Tertiary palaeanodonts (sub-order Palaeanodonta). Even though this is a more primitive group than edentates, its development indicates a trend toward the edentate structure. The extinct PALAEANODONTS were small, armadillolike animals, lacking the bony armor. The vertebral column did not yet possess the extra articulations, but dentition had regressed and consisted almost exclusively of teeth, which lacked enamel.

The earliest edentates stem from the Upper Paleocene period in South America (about 60 million years ago). These are typical members of the edentate group, with bony armor and teeth which almost completely lacked enamel. Their ancestors were palaeanodonts from the Lower Paleocene or the Upper Cretaceous period. These mammals diverged from insectivores in the Cretaceous period (100 million years ago). For millions of years, the edentates remained in South America, separated from the rest of the world, where they formed numerous groups. Not until the Upper Pliocene (a few million years ago) did a few members begin to invade North America. These were giant sloths, followed in the Ice Age by glyptodonts and armadillos. Anteaters and tree sloths came as far as Central America and proceeded no farther. Originally all edentates probably had bony armor. If this assumption is true, and it has not yet been fully confirmed, then sloths, giant sloths, and anteaters subsequently lost their armor.

The GIANT SLOTHS (+ Gravigrada) had already separated into several groups in the Oligocene (40 million years ago), into the families + Mylodontidae, + Megalonychidae, and + Megatheriidae. Extremely large forms were prevalent in the Ice Age, of which + *Megatherium* was the size of an elephant, while + *Mylodon* and + *Megalonyx* were somewhat smaller. These clumsy, plump herbivores inhabited the shrubby savanna during that period. Some species began dying off in the Lower Holocene, ten to twelve thousand years ago, and thus were contemporaries of man. One cave which had been walled up by man in Ultima Esperanza in South America contained bones, pieces of pelt, and remains of feces of giant sloths. It was once thought that some giant sloths were kept as domesticated animals. One of these was + *Mylodon domesticum*, the "domestic giant sloth."

Of the ARMORED EDENTATES (Cingulata), only the armadillos (Dasypodidae) have survived to the present. From an evolutionary viewpoint, this is the most primitive edentate group. A second family, the palaeopeltids (Palaeopeltidae), died out during the Lower Tertiary. However, some glyptodonts (+ Glyptodontidae) developed into giant forms with a stiff armored covering and tail processes which in some species had spines and apparently served a defensive function. The present-day giant armadillo (*Priodontes*) belongs to Dasypodidae, but is not more closely related to the fossil giant armadillos. The largest species belonged to the genus + *Glyptodon* and were nearly as large as the rhinoceros.

Edentates are an example of a mammal group which developed during the period in which South America was separated from the rest of the New World. These edentates were among the few mammals which were adapted to certain ecological niches and were able to survive the predators and hoofed animals that invaded the area during the Ice Age.

Present-day edentates, by E. Moeller

Family: Armadillos

Distinguishing characteristics

The ARMADILLOS (Dasypodidae) are a stocky group of animals. Their body length ranges between 12 and 100 cm, and their body weight varies between 90 g and 55 kg. The upper side of the body has an armored covering. The belly may be sparsely or densely covered with hair. The head varies from wedge-shaped to elongated and narrow. The ears are long and cone-shaped, and may be pointed or round; they may even have regressed considerably. The tail has an armored covering. The limbs are short and powerful, and have grabbing claws. The skeleton is particularly massive. The forelimbs and the pectoral girdle have skeletal elements serving as points of attachment for the muscles, which makes these limbs very effective at grabbing. The first rib is short and broad; the ribs and sternum lack cartilage. There are from nine to twleve thoracic vertebrae; in the neck and sacral area some vertebrae have become fused. The digestive organs have a simple structure. The appendix is short or entirely absent. The simple uterus leads to a sinus urogenitalis instead of a true vagina. Testes are not visible, and the penis is quite long, probably as an adaptation to the long curvature of the ventral armor.

The armor of armadillos

Armadillos owe their name to the presence of a highly distinctive structure in mammals: they have an armor covering similar to that of turtles; it is, however, not rigid. It is interrupted by several folds in the skin along the middle of the body, forming girdlelike bony rings that open toward the belly side. The word armadillo has been taken from Spanish, and means armed or armored. During the first weeks of life the body of the armadillo is completely covered with horny scales. Ossified structures begin developing in the dermis underneath this horny covering, and during the course of development these become rectangular bony plates. These various plates form a complex pattern covering the head, shoulder, and pelvic regions. Bony plates also develop on the front side of the limbs and on top of the tail; they are poorly developed on the belly.

Hair and skin glands

However, in adult species the skin often has a thick hair covering. In some cases hard, brushlike hairs penetrate the armored plates. They stand between individual horny plates formed from the epidermis, and they cover the plates in particular patterns. All armadillo species have a collection of skin glands in small pockets in the anal region. Similar glands are found on the back in the six-banded armadillo and the hairy armadillo; their presence is indicated by two to four holes in the pectoral armor, and they secrete a yellowish, odoriferous mass. This dermal bone covering of the armadillo is an extremely effective means of protection from enemies. While many turtles withdraw their head and legs when threatened, the

Skin muscles and claws

armadillos' skin folds and powerful muscles permit them to roll up very rapidly. Even a jaguar has difficulty opening a coiled-up armadillo. However, the armadillos have a still-more-effective method of escaping from predators: the shovellike claws present in some species permit them to dig into the ground extremely rapidly, so that they disappear before the eyes of the predator.

Dentition and feeding

No mammalian family has as great a variety in the number of teeth as have the armadillos. Even within a species this number can vary considerably. Armadillos are characterized by cylinder-shaped teeth resembling each other (homodont teeth), with open pulps; the teeth grow continuously. Recent investigations have shown that the teeth lack enamel and that the dentine is surrounded solely by a cement coat; this means that chewing wears the teeth down rather rapidly. The largest and most powerful armadillo teeth belong to the six-banded armadillos, which feed on insects, snails, worms, carrion, and plant roots. They require their teeth for mastication more than do other species. The importance of solid nourishment for armadillos was discovered in zoos. When the food was too soft, the teeth did not get worn down as they should, and their excessive growth caused injuries to the jaws and gums. Grains, roots, and fruits have to be mixed with the food in order to keep the teeth in top condition. Only dasypodinid armadillos, which are pure insectivores and feed on adult and larval insects, have been found to grow new teeth. Seven double-rooted teeth were renewed in specimens just before the adult stage was reached, while the last (usually eighth) tooth in each jaw is normally not renewed. The giant armadillo, which has the most teeth (but also the smallest), has a specialized diet consisting of ants and termites.

Tongue and salivary glands

The worm-shaped tongue fulfills an important function when grabbing these insects: its upper surface has many small warts and is covered with a secretion to which the insects adhere. This saliva is secreted by large salivary glands in the lower jaw; it is stored in special cavities for use when large amounts are needed. In contrast to most mammals, the armadillo tongue has few taste buds, and thus taste is probably poorly developed. However, the sense of smell, important in finding food in the ground, is very well developed. The external brain structure indicates that armadillos are olfactory specialists. The small eyes lack cones in the retina; in other animals these are responsible for color vision. Thus, it would appear that armadillos cannot distinguish colors. As a primarily nocturnal creature, the visual senses in general play a subordinate role.

Sensory capabilities

Oxygen utilization

In the U.S.A., the nine-banded armadillo has been the subject of numerous physiological studies. It was found that this armadillo requires considerably less oxygen than do cats and dogs of about the same size. Even when strenuously exerting itself, such as in rapid digging, the nine-banded armadillo can hold its breath for as long as six minutes. This decreases the chance of stirred-up dust getting into the respiratory passages. The bronchi and trachea are quite extensive and serve as reservoirs for air. Some American investigators report that a nine-banded armadillo filled its stomach and intestines with air before swimming for a long distance. Few other armadillo species can cross wide rivers as successfully as this one can. Its swimming motions resemble those of a dog, but it can even walk along the bottom for short distances.

Blood circulation and body warmth

The armadillo circulatory system is highlighted by finely branched vessels forming a thick network leading to the heart. They afford more efficient oxygen utilization during periods of exertion. In contrast, temperature regulation is not well-developed and resembles the situation in reptiles. That is, body temperature is to some extent dependent upon the ambient temperature. At an external temperature of 16–18°C, the body temperature in nine-banded and three-banded armadillos remains a fairly constant 32°C. However, when the ambient temperature is 11°, the armadillo's body temperature can fall three degrees within four hours. Nine-banded armadillos cannot withstand frost periods of long duration, even within a well-insulated den; this animal's penetration to the north is probably limited by climatic factors.

Reproduction

Reproduction has been studied only in the nine-banded armadillo. Mating generally takes place in July. After fertilization of the eggs, only the first phases of growth occur; a cessation of growth then sets in. (This also occurs in deer and some other mammals.) Implantation occurs after about fourteen weeks. Further development continues then as in other mammals, lasting about four months, so that the young are born in February or March. Brehm still considered as supersitition the belief of Brazilians that the *tatu*, as they call the nine-banded armadillo, always has four young of the same sex. Actually, this belief is based on rather exact observations: quadruplets always develop from only a single egg, and therefore they are always of the same sex. They are fully developed at birth and are nursed a few weeks. Sexual maturity is reached after six months. Only dasypodinid armadillos have four teats; all other species have two teats and give birth to one or two young.

Tribe: Dasypodini

Dasypodinid armadillos are small to medium-sized. Their armor coat is thin and pliable in living specimens. The pectoral and pelvic armor is greatly arched and has relatively small bony plates covered by smaller horny plates. Coloration is from dark brown or gray-brown to yellowish-beige. There are six to eleven bands. The tail is almost as long as the body and is covered with bony rings. The cone-shaped ears are long and are placed quite close to each other. The eyes are small. The narrow head has a tube-shaped snout. On the average, there are seven or eight teeth in each jaw. These armadillos are long-legged and good runners. The forelimbs generally have four fingers, while the rear ones have five toes. The first and fourth fingers are equally long and are shorter than the second and third fingers, whose claws are as much as 3.5 cm long. The toes are symmetrical, with the third one the longest and the first and fifth, and second and fourth, equally long. There are four species distributed from Texas to Buenos Aires:

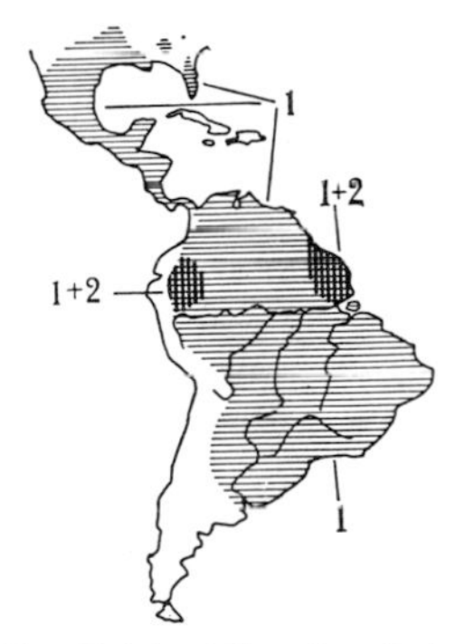

Fig. 7-1.1. Nine-banded armadillo (*Dasypus novemcinctus*), which has spread far to the north since the mid-19th Century; 2. Kappler's armadillo (*D. kappleri*).

1. KAPPLER'S ARMADILLO (*Dasypus kappleri*; HRL 53–56 cm, TL 40–43 cm, weight 8 kg), the largest species, is distributed in Surinam, eastern Ecuador, and Peru, but is rare throughout its range. It has seven or eight bands. The armor covering lacks hair. The forelimbs typically have five digits; there are generally eight teeth in each jaw; the largest teeth being in the middle of the row. There are four teats.

2. NINE-BANDED ARMADILLO (*Dasypus novemcinctus*; HRL 40–50 cm, TL 25–40 cm, weight 6 kg; see Color plates, pp. 173 and 189) is the most common and best-known species. They have eight to eleven bands (but usually nine). The armor lacks hair. There are several subspecies, only one of which (from Surinam) is distinctly recognizable, because of its skull structure. The number of teeth varies considerably. Of 192 skulls examined, eighty-nine had eight teeth in each jaw, while in thirty-one there were seven in the upper jaw and eight in the lower jaw; and in other skulls the number varied from six to nine teeth in each jaw. There are four teats. The developing embryos are carried for 120 days, and quadruplets from one egg are always born.

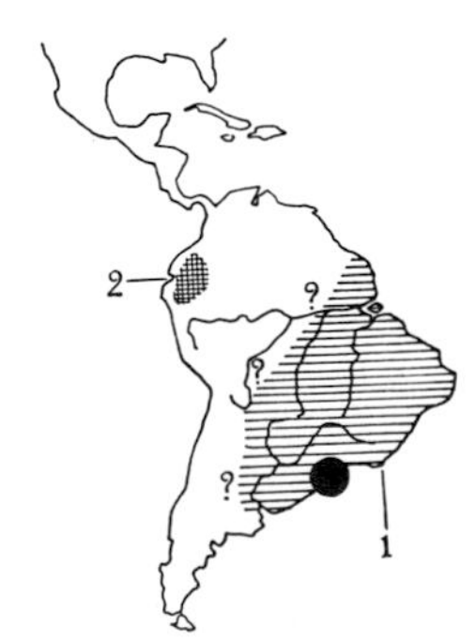

Fig. 7-2. Seven-banded armadillo (*Dasypus septemcinctus*); 2. *D. pilosus*. (sometimes called the hairy armadillo.)

3. The SEVEN-BANDED ARMADILLO (*Dasypus septemcinctus*) is not common in any part of its distribution. The species is smaller and has a shorter tail than the nine-banded armadillo. In other respects the two are very similar. Its weight is about 3 kg. It usually has six or seven, rarely eight, bands. The armor lacks hair. The number of teeth varies between six and eight in each jaw. There are four teats. The gestation period exceeds 120 days, and four, eight, or rarely twelve young are born.

4. *Dasypus pilosus* (HRL 37–40 cm, TL 25–27 cm), distributed in the highlands of Ecuador and Peru, is extremely rare. It has eleven bands, and the trunk is densely covered with rust-brown to dark gray brushy hairs which conceal the dermal armor. The snout is very long and narrow. The

teeth are smaller than in related species; there are five to seven in each jaw. It is sometimes popularly called the hairy armadillo, but this name is better applied to *Euphractus villosus*, described later.

The nine-banded armadillo

The NINE-BANDED ARMADILLO is a newcomer to the U.S.A. During the last eighty years the species has multiplied rapidly, spreading from its native Mexico and becoming well established in the U.S.A. Its popularity is high, due in part to the fact that it feeds on insects harmful to crops. Because of this, the species is often introduced into areas where it had previously not existed. Its distribution in Florida is a classic example of a successful introduction: during WWI a sailor brought a pair of nine-banded armadillos into Florida from Texas. He lost them near Hialeah, and during the 1920s, several nine-banded armadillos were seen in the vicinity of this city, the first reported in Florida. In 1924, a private zoo in Cocoa, on Florida's eastern coast, was destroyed by a storm, and a few nine-banded armadillos escaped from there and settled in the area. A third group escaped from a circus in 1936, near Titusville. The species then spread from the eastern coast toward the west. Today they inhabit one-third of the entire state of Florida, and, in addition to Florida and Texas, they have been found throughout Louisiana and Oklahoma. Scattered occurrences in Arkansas and Alabama are also probably due to human introduction. Factors limiting propagation include cold for long periods and a lack of suitable insects; thus the species will probably not spread toward the northeast. They prefer forested, slightly marshy regions as a habitat, the nature of the soil not playing any important role. They dig their dens particularly on stream and river banks, always in the vicinity of trees and shrubs. The armor provides excellent protection from thorns in the underbrush; only the ears get ripped frequently. In one particular area, more dens than armadillos were found. This probably means that some are only used temporarily for sleeping. The typical living den consists of a passageway 15–20 cm in diameter and up to 7 m long. At one end of the system, in a wider area, they make the den proper. There frequently are two or more entrances to the living area, although they only use one regularly. They pull leaves and grass into the nest, to use as bedding. Armadillos have been observed backing into the nest, pushing the nesting material in with the hind quarters. Nesting material is changed from time to time, particularly after severe rain showers, and decaying leaves can often be found in front of the nests.

The daily life of the nine-banded armadillo

The daily rhythm of the nine-banded armadillo is not only regulated by daylight, but also by the temperature. During hot summer months it leaves its nest only in the evening and at night; however, when the temperature is lower, it often appears in the sunshine while seeking food. As soon as it has left the nest, the armadillo begins to sniff, holding its snout just above the ground. It trots about, constantly changing directions, covering about 1 km in an hour, barely taking notice of its surroundings. With its keen olfactory sense, it can detect insects and worms as deep as

20 cm in the soil. If it is pursued, it changes from its clumsy gait to a fast gallop and may even escape capture by man over short distances. The most significant enemies of armadillos are no longer the large predators such as wolves, coyotes, and pumas, which have virtually disappeared from the southern states anyway, but man, together with his dogs and automobiles. In the early morning hours on the highways and country roads in Texas, one can see the bodies of numerous armadillos which have been run over. Every unexpected noise elicits a reflexive jump, so that instead of being run over by the wheels, the armadillos are hit in the air. Armadillos are once again being kept in the nocturnal-animal house of the Amsterdam Zoological Garden. Only a small artificial change is necessary to reverse the day-night rhythm of the nocturnal animals. At night they are provided with bright light, so that during the day, under weak lighting, they are active and can thus be viewed by the zoo visitors. I once had an opportunity to observe a nine-banded and a seven-banded armadillo together in one enclosure. At 10:00 a.m. the artificial dusk lighting was turned on and the armadillos awoke, winked a little, and moved their ears. Because of its long ears, the seven-banded armadillo is called *mulita* (the little mule) in its homeland. As they bore through the soft sand with their narrow snouts, leaving characteristic tracks with their tail pulling through the sand, they soon reappear with a morsel of food. They are fed chopped meat, raw eggs, and mealworms, for their teeth are too weak to eat solid food. We brought the animals into another room to be photographed. They scratched and moved around inside their transport box. As I lifted the nine-banded armadillo by its back, it stiffened at first, withdrew its head and legs, and then leaped out of my hands. The animal at once sought a corner and disappeared underneath a radiator. There it stayed with more strength than one would suspect such an animal would have. It stayed under the radiator for several minutes, and could only be brought out with a great deal of effort. Hensel reports a similar incident in the wild. Brazilians value the *tatu* for its tender white meat, and they hunt the species. "Even two strong men," writes Hensel, "cannot pull the *tatu* out of its hole, if it is narrow enough for the *tatu* to wedge itself in with its feet and back. One must remember that the tail tapers and is thus hard to grab. But when one man holds onto the *tatu's* tail and the other digs away the earth so he can grab the hind leg, the *tatu* ceases to struggle." Brazilians use the armor as a bowl or little basket. They tie the head to the tail end to form a handle. For tourists the inside is covered with bright silk. Often one can find musical instruments made of armadillo armor.

Tribe: Three-banded armadillos

Their ability to roll into a ball has been a feature of the THREE-BANDED ARMADILLOS. The Indian populations call them *mataco* or *apara*, while the Spanish-speaking people know them as *bolita* (the little ball), and *tatu naranja* (orange armadillo). Both medium-sized species form a single tribe: Tolypeutini. They are distributed from Guyana to central Argentina (the Rio Negro) in open grasslands; they prefer the plains. Their body length ranges from 35 to 45 cm, while the tail is from 7 to 9 cm long. The dermal armor is heavy and thick, with arched hemispherical pectoral and pelvic armor which protrudes over the sides of the body. The bony plates are generally hexagonal. Diamond-shaped horny plates cover the bony plates. The armor lacks hair. The coloration is dark to gray-brown beige, and there are two to four rings. The narrow head has very hard armor, and the short fleshy ears have a wide base. The species have long legs and are average runners and diggers. The fingers and toes are syndactyl; the third finger has a particularly long pointed claw. The first and fifth fingers as well as the first and fifth toes are set back. The second, third, and fourth toes have broadened, hooflike claws. Thirteen sacral vertebrae are fused together with the pelvis. The gestation period lasts five to six months, and births usually take place in November, consisting of a single young.

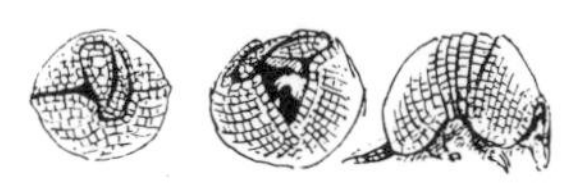

Fig. 7-3. A three-banded armadillo can coil into a ball which is so tight that it takes tremendous force to uncoil it.

1. The LA PLATA THREE-BANDED ARMADILLO (*Tolypeutes matacus*) is distributed in southern Bolivia, Paraguay, southwestern Brazil, and often in northern Argentina. There are two to four bands, four fingers and five toes. Each jaw typically contains nine cylindrical teeth. One form in eastern Argentina (*Tolypeutes muriei*) has been described as a separate species on the basis of the particular shape of the head armor. It is probably just a subspecies of *T. matacus*.

2. The THREE-BANDED ARMADILLO (*Tolypeutes tricinctus*; see Color plate, p. 189) generally has three bands, five fingers, and five toes. There are eight cylindrical teeth in each jaw.

Fig. 7-4. Three-banded armadillos: 1. *Tolypeutes matacus*; 2. *T. tricinctus.*

Hans Krieg has written the following on his experiences with three-banded armadillos: "Of all the armadillos, this one is decidedly the cutest. During the day, it runs through the dried grass with its back arched, sniffing about and scraping a bit, before running back in fright. If one comes too close to the armadillo, it first attempts to run away, but soon rolls up into a ball with a hissing breathing sound. But the more one repeats the process, the less tight the ball becomes, and eventually it loosens up to a point where the claws of the forefeet and the tortoiselike rear feet can be recognized. One can take the animal into one's hand and put it back down; when the armadillo opens itself up again, it turns around with a jerk and moves off somewhat stiffly, moving on the tips of its foreclaws. While coiling up is an unsuccessful tactic against man, it is extremely effective against all other enemies. Foxes and wolves can hardly open them up, at least on the basis of my own observations. All Indian tribes of the Gran Chaco hunt this armadillo. They roast it in its own shell, the usual practice with armadillos. Its meat and white fat are very tasty. During the dry season the Indians make camp fires in order to more readily find armadillos and rhea eggs." Elsewhere Krieg reported: "When riding through the open prairie, one can often see armadillos at a great distance, particularly at sites where camp fires had burned and left a black scorched area (where sometimes green grass was sprouting). The sun playing on their armor often reveals their presence. I observed *Tolypeutes* seeking food

Fig. 7-5. Giant armadillo (*Priodontes giganteus*)

Fig. 7-6. Eleven-banded armadillo (*Cabassous unicinctus*).

during the day more often than other species." On fright behavior, Krieg notes the following: "It first attempts to run away, but one can hold it without much difficulty unless it finds a hole immediately. It does not roll into a ball until the last moment. Reports on the digging ability of this armadillo are rather divergent. Indians claim that it uses only the holes built by other armadillos or viscachas [see Chapter 13]. I have never seen it dig itself in when pursued, as is the case in other species." Anton Göring, who kept an armadillo in South America, reports that it was unafraid from the beginning and would feed from his hand. In the presence of strangers it would eat peaches, pumpkin, and lettuce, but as soon as it was touched it would withdraw. Three-banded armadillos have lived over seven years in the Berlin zoo. In 1967 I noted that one of the residents of the Amsterdam Zoo's nocturnal-animal houses was a *bolita* (*T. matacus*).

Tribe: Giant armadillos

The largest species in the family, the GIANT ARMADILLO (*Priodontes giganteus*; HRL 90–100 cm, TL 50 cm, weight 50 kg; see Color plate, p. 189), belongs with the eleven-banded armadillos, to the tribe Priodontini: the armor is wide and flat and generally has smooth rectangular plates. The armor is sparsely covered with hair, if there is any at all. Coloration is yellowish-beige to dark brown. The medium-long, powerful tail has armor (in the giant armadillo) or a few irregularly arranged plates (in eleven-banded armadillos). The broad ears are rounded or cylindrical and are rolled up. The heavy, thick-walled skull has a broad snout as in Dasypodini and three-banded armadillos (Tolypeutini). There are five fingers and toes, and the third finger is particularly well developed, with a long powerful claw. The toes are somewhat syndactyl and have broadened claws. This species is an efficient digger. Thirteen sacral vertebrae have fused with each other and with the pelvis. Each jaw has from fifteen to twenty-eight small teeth.

The ELEVEN-BANDED ARMADILLO (*Cabassous unicinctus*, see Color plate, p. 189; distribution, Fig. 7-6) is somewhat larger than Kappler's armadillo, generally has thirty-four teeth, with nine on each side of the upper jaw and eight on each side of the lower jaw. Besides this species, three others within the genus have been described, their body lengths varying from 30 to 40 cm, their tail lengths, from 10 to 12 cm, and their weights, from 1.5 to 3 kg: the relatively long-eared *Cabassous lugubris*, from Colombia, Guyana, and northern Brazil; *Cabassous hispidus*, from southern Brazil; *Cabassous loricatus*, from northern Argentina and the Mato Grosso region. All three probably belong to one species, the SPINY ARMADILLO (*Cabassous hispidus*). The eleven-banded armadillo has also invaded Central America and is presently distributed from Honduras to northern Argentina, but nowhere in great numbers. The species inhabits open prairies as well as moist jungles. Large holes may be found at the foot of termite hills, leading as much as 5 m underground. This work has been done by the strong claws of the *rabo molle* (Spanish for "soft tail"). It destroys termite hills when seeking food.

Fig. 7-7. Spiny armadillo (*Cabassous hispidus*).

The giant armadillo also feeds chiefly on termites. Supported by the rear legs and the muscular tail, it strikes the hard termite hills and feeds on the termites with its sticky tongue. Often, whole termite hills may be found completely destroyed by the *tatu gigante*. Its traces are also found in regions with trees and shrubs. Hans Krieg writes: "The holes are wide enough for a man to crawl in. Often the ground has been quickly torn up, without regard for the roots of the quebracho and lapacho trees and the thorny shrubs. In some places I found entire woods that showed signs of the armadillo's destructive habits. The traces of their digging clearly may be seen for a long time afterward, and a single armadillo can leave a great deal of destruction in its wake." One encounters the giant armadillo only rarely; the local people report that they are nocturnal. It avoids regions which are settled and in which cattle are raised. If a giant armadillo occasionally stumbles into an area settled by man, the local people chase it, for the armadillo can cause tremendous destruction in the fields as it searches for insects, worms, and spiders. As with eleven-banded armadillos, its meat is not considered tasty, but the Indians use its tail as a megaphone. The distribution and biology of this primitive animal is poorly understood. It is reported fairly often in the Gran Chaco and the provinces of Misiones and Formosa, but its main distribution is probably north of Argentina, in the Brazilian jungles which still have not been extensively explored. Wolf Herre, writing on his experiences from expeditions in South America, states: "Unfortunately, there are many indications that this species is disappearing, and conservation-minded people now have a serious job in maintaining a place for this singular giant." Giant armadillos have rarely been kept in captivity. When a *tatu gigante* was brought to the New York Zoo in 1935, it fell while climbing about in its cage and fatally injured itself. The first specimen displayed in a European zoo was in the Berlin Zoo in 1930. Seven were exhibited before 1944, one living from 1934 to 1944. Agatha Gijzen described an interesting case of mistaken identity: the Antwerp Zoo received "a giant armadillo with young." The animals became accustomed to their new surroundings and presented no difficulties; yet people began to wonder because although they had been feeding regularly for months, the young did not grow any larger. Finally, it was found that at least one was an adult eleven-banded armadillo. The two genera resemble each other greatly, even in the way they walk. When running, only the claws of the forefeet touch the ground, while the entire soles of the hindfeet make contact.

The best diggers among armadillos are the SIX-BANDED ARMADILLOS and the pichiciago (formerly genus *Euphractini*; recently revised, placing it in the subtribe Chlamyphorina). A. SIX-BANDED ARMADILLOS (subtribe Euphractina): their armor is broader and flatter than in all other armadillo species and consists of multi-sided (four or more sides) bony plates and structured horny plates. The coloration is brownish-yellow to white; the bushy hairs vary in thickness. There are six to eight bands and short to medium-long ears spread far from each other. There is a powerful, medium-length-armored tail. The flat, broad skull has a protruding zygomatic arch. There are thirty-eight relatively large teeth with distinctive gable-shaped grinding edges. Few species deviate from the dental formula $\frac{9}{10}$. The short, powerful limbs have five fingers and toes, which correspond to the general armadillo type in structure, claw shape, and size. The gestation period is about two months, and there are two litters during the year, typically with two young. Distribution is from Brazil to the southern tip of Argentina, and the species are plentiful throughout the distribution. The three species are distinguished principally by size:

Tribe: Six-banded armadillos and pichiciagos

Six-banded armadillos

1. The SIX-BANDED ARMADILLO (*Euphractus sexcinctus*; HRL 40–50 cm, TL 20–25 cm, weight 3.5–4.5 kg; see Color plate, p. 189), has the northernmost distribution. The armor is sparsely covered with brushy hair. The gestation period lasts for seventy-four days. 2. The HAIRY ARMADILLO (*Euphractus villosus*; HRL 30–40 cm, TL 12–15 cm, weight 2.5–3 kg; see Color plates, p. 173 and 189) is found in the south. Two additional middle-sized species have also been described, but their distinguishing characteristics are transitional tending toward the hairy armadillo, so that it can be assumed that *E. nationi* and *E. vellerosus* are subspecies of *E. villosus*. The dentition in six-banded armadillos and hairy armadillos is characterized by two incisor teeth. 3. The PYGMY ARMADILLO (*Euphractus pichiy*; HRL 25–35 cm, TL 10–15 cm, weight 1–2 kg; see Color plate, p. 189), the smallest species of the group, often has teeth which have regressed. Its distribution is the most southerly of the different species. The armor is uniformly covered by 3–4-cm-long light- to dark-brown brushy hairs. There are several subspecies.

Fig. 7-8. Six-banded armadillo (*Euphractus sexcinctus*).

In Argentina the most prevalent armadillos are the hairy armadillos, called *peludos* (Spanish for "brushy", or "hairy"). Their dens are found in the dry savanna; they leave them even in bright daylight. The six-banded armadillos continually dig new passageways in search of food. They generally dig just 1–2 m into the earth, and then widen the underground area enough so that they can turn around. Defecation always takes place outside the den. When digging, the *peludos* do not throw the dirt to the side (as does the mole), but they scratch it up with their forefeet and then throw it behind themselves with their hindfeet. Except for the mothers, which nurse their young for several weeks, these armadillos are seen only infrequently. According to Rengger, the male and female meet accidentally at night and sniff each other for several minutes before copulation ensues. A larger number of six-banded armadillos can only be seen on the carcass of a dead animal. Krieg writes: "When night comes, the carcass is again visited. A gray, brushy armadillo (a *peludo*) comes sniffing under the horse's skull, where maggots swarm about. Soon the chest cavity moves and a large, yellow *tatu poyu*, a six-banded armadillo, becomes visible feeding on a few remaining pieces of meat." Even though armadillos do not live in groups, they are so numerous in some regions that they lay waste to the fields. According to Krieg, freshly plowed soil in a field of about two hectares in the province of Santa Fe attracted so many armadillos that control measures had to be taken. A bounty was set, and within a few days over 100 of them had been caught. Hans Krieg delivered the skulls to the Stuttgart Natural History Museum for scientific study. *Peludos* are also a thorn in the flesh of the gauchos, the South American "cowboys," because they cause dangerous accidents. Horses often break a leg while stumbling into a hole as they gallop across the pampas. Thus the armadillos are pursued even in non-agricultural areas. Creoles generally hunt armadillos at night with dogs, using clubs. The armadillos can save themselves only if they smell the dogs in time. But the dogs generally run faster than the armadillos can dig. "The fangs of the dog repeatedly slip off the back armor of the armadillo as it automatically attempts to flee or dig into the ground. When the dogs grab it on its belly, it is lost," writes Krieg. They hold it with their snout and paws until the gaucho comes and kills it with a blow to its head and a jab into its back with a knife. Those armadillos that get away from the dogs are forced out of their holes by pouring water in them. In dry, shrubby areas and on particularly hard ground, where they cannot dig into the earth, they can be seen pulling in their legs and pressing the edge of their armor against the ground so that the vulnerable belly is protected. This reaction also appears when the animals are molested by people, and apparently is a substitute for coiling up into a ball. Darwin observed this in the pygmy armadillo. In his *Voyage of the Beagle*, he wrote: "The *pichy* (as the native population calls the pygmy armadillo) prefers very dry ground and the coastal sand dunes where it can go without water for months. During a daily ride near Bahia Blanca I generally encounter several of them. As soon as an armadillo is spotted, one must dismount his horse immediately to catch the armadillo. If the soil is soft, the animal will have dug itself in before one can dismount. The *pichy* also frequently attempts to be inconspicuous by hugging the ground."

Fig. 7-9. Hairy armadillo (*Euphractus villosus*).

Fig. 7-10. Pygmy armadillo (*Euphractus pichiy*); 2. Lesser pichiciago (*Chlamyphorus truncatus*).

During the cold season the pygmy armadillo in Patagonia is said to hibernate. The Euphractina armadillos are among the more withdrawn species kept in small mammal houses in the zoo. Generally they cower in the corner and become aroused only at feeding time. At that time their protrusible tongue can be seen reaching at the bits of food; only the larger morsels are grabbed with the lips, and they lick up the last remains of their food noisily. In their banded armor they move back and forth within their enclosure, sniffing for more food. But soon they return to rest. Sometimes they lie on their backs and stretch all four legs in the air; their body twitches in a peculiar manner when they are in this position. At one time armadillos were displayed with apes in order to increase their spectator appeal. But the armadillos were no match against these more intelligent creatures. Baboons pulled them around by their tails or even rode on them. However, this play situation did not last long; soon the apes found that the armadillos were rather boring playmates. *Peludos* have never been seen climbing in nature. Apparently their climbing in zoo enclosures arise from attempts to escape. For example, two hairy armadillos in the old Hamburg Zoo, climbed out of their cage into the neighboring guinea pigs enclosure, and they chased one of the guinea pigs, apparently intending to eat it. The *Euphractina* armadillos have been successfully bred in captivity. The young begin to run about in their cage when they are about four weeks old. Until that time they are highly sensitive to any disturbance. This is probably the reason that until recently young were rarely raised in zoos. A hairy armadillo lived in the zoo in Zurich for seventeen years.

The PICHICIAGO (subtribe *Chlamyphorina*) is probably the most specialized and most recent branch of the armadillo family. There are two genera, each with one species:

The pichiciago

1. The LESSER PICHICIAGO (*Chlamyphorus truncatus*; HRL 12–15 cm, TL 2.5–3.5 cm, weight 90 g; see Color plate, p. 189), has armor which has regressed considerably and consists of thin, rectangular plates covered by thicker horny plates. Only the pelvic armor, which is fused with the pelvis and sacral vertebrae, has become hardened. There is no breast armor. There are twenty-three to twenty-five bands, separated by skin folds; they extend to the head armor. They are loosely situated, attached to the skin by a thin membrane along the vertebral column. A thick, silky, white-gray pelt is found under the dorsal armor. The tail, which is broadened at the end, is flattened and has an armor covering. The external ears have become tiny skin folds. The eyes are also very small. The skull is very thin, and head armor extends beyond the horny nasal plate. There are eight teeth in each jaw. The limbs are short, and the forelegs are more powerful than the rear ones. The five claws of the forefeet are also longer and broader than those of the rear feet. 2. The GREATER PICHICIAGO (*Burmeisteria retusa*; HRL 17–18 cm, TL 2–3 cm; see Color plate, p. 189) has armor which has not regressed as much as that of the lesser pichiciago; however, the breast plate is also absent. The twenty-four bands are all fused to the body. The armor elements form a unified structure only on the outer edge of the elliptical pelvic plate; toward the middle they are irregularly spaced in the skin. There are very small ear muscles and eyes. The forefeet have five grabbing claws, while the rear feet have five toes with short, blunt claws.

Structure and habits

The transformation of the pelvic region, already evident in other armadillos, reached its greatest modification in the pichiciago. Its pelvic plate is not separated from the skeleton, but has fused with the vertebral column and the pelvis. This portion serves to protect the armadillos in their underground passageways.

Because the pelvic plate stands almost vertically, the pichiciago's cylindrical body looks as if it has been cut off. The old Spanish name for the species, *Juan calado*, refers to the prominent white tuft of hair around its posterior end. Many South Americans admire these animals; if they manage to catch one, they keep it as long as possible, and finally they make mummies of them. Wolf Herre wrote: "If you ride through the province of Mendoza on the railway, you come to an isolated railroad station in the middle of a broad, flat pampa with legume bushes barely as tall as a man. The shrubbery is known as *pichi ciego*, the local name for the small pichiciago; this area is the center of its distribution. For hours we wandered under the bright sun, trying in vain to find tracks of the pichiciago. But in the vicinity, wooden railway ties were being replaced, and the *pichi ciego* was found in the rotting wood, probably attracted by the insect larvae inside." The distribution of the pichiciago extends from the center of Mendoza to the provinces of San Juan, San Luis, and La Pampa. Since the species prefers dry, sandy regions with thorny shrubs or cacti and high temperatures, the Andes in the west and the decreasing moisture in the east form natural barriers to the distribution. In the south, the barrier is the Rio Colorado and in the north, the excessively hard ground. The pichiciago infrequently leaves its underground nest, and then usually for short periods of time. When outside, it barely lifts its heavy grabbing claws from the ground; instead, it shoves them forward, leaving a characteristic track with its tail. But after a few meters it turns around several times, and shoves its horned snout into the sand repeatedly and begins to dig a cavity with its forefeet. The entrance is revealed by the presence of two piles of sand or earth. Friedrich Kühlhorn, who investigated the stomach contents of several specimens, found chiefly ants and spores of certain insects, as well as insect larvae. The greatest enemies of pichiciagos are the increasing development of the area and the accompanying cultivation of the pampas.

The greater pichiciago is believed to be distributed from Mendoza to southern Bolivia and in the western parts of the Argentine province of Formosa. This rare species was described by Burmeister over 100 years ago, but present knowledge is limited to its structural peculiarities. One was captured in Argentina in July 1967 and later sent to the Brookfield Zoo (Chicago), where it lived until December 1971.

Family: Sloths

The first family of ARMORLESS EDENTATES (partial order Pilosa) are the SLOTHS (Bradypodidae). There are two genera with about five species. Their body length is from 50 to 65 cm; the tail length (when a tail is present) is from 6 to 7 cm, and their weight ranges from 4 to 9 kg. The rounded head has a flattened face. The ears are round and the hair is thick and hard. Some species have distinctive facial and dorsal markings. The limbs are long and slender. Their toes are syndactyl with two or three sickle-shaped claws. The species inhabit trees.

They move hesitantly and deliberately through the trees of the tropical rain forest with movements that appear to be in slow motion. The belly is on top and the back is on bottom, and it is because of this striking pattern and the fact that they move so slowly, that we have come to call these animals sloths. One American zoologist has written: "Their sluggishness is one of the wonders of nature. The cellular fluids of a unicellular organ-

The adaptive value of slowness in moving

ism flow faster than a sloth fleeing from a boa." Sloths do not need to exert themselves when seeking food. With large, curved claws on their long arms and legs, they can anchor themselves firmly on tree branches and feed leisurely on the surrounding greens. They feed on leaves, young shoots, blossoms, and fruits. There is absolutely no necessity to move rapidly. On the contrary, their inconspicuous behavior is the best form of protection from their enemies. Nature has equipped them with a unique coat of fur: their hairs are coated with a thin layer of cells which form a series of diagonal ridges across the fur. Two species of blue-green algae (*Trichophilus* and *Cyanoderma*) grow in these ridges and thrive when the surrounding climate is moist and warm. They lend a greenish shimmer to the sloth's gray-brown hair and thus provide protective coloration. This is a true symbiotic condition, that is, a situation in which different organisms live together for their mutual benefit. Such is not clearly the case with reference to another organism associated with the sloth, the rare snout moth (see Vol. II). This moth lays its eggs in the sloth's fur, to which it can easily gain access, since the fur is not cleaned frequently. It is not known what advantage the sloth may derive from this association. The fur is interesting in still another respect: it is worn "reversed," so to speak. The hair is parted not along the vertebral column as in other mammals, but along the midline of the breast and belly. Because of this, rain and moisture from leaves do not stay in the fur on the belly, but run off down the sides.

Algae and moths in the fur

Sloths are found almost everywhere in the forests of Central and South America. Although many other mammals of comparable size have died out in these areas, sloths have been able to maintain their presence. Instead of derogatorily ascribing their survival to "success through laziness," we should see them as a splendid example of adaptation to the environment. What is lacking in alertness and reaction time is compensated for by advantages arising from their body structure and appearance. As Hans Krieg puts it, "Their inconspicuousness is their best protection."

Skeleton and limbs

The life style of a vertebrate is reflected in the structure of its skeleton, which permits rather precise judgements about some aspects of life, such as the means of movement, feeding mechanisms, and other characteristics. This is particularly true of sloths. The modifications of their limbs—which in some species are longer in front than in back—and especially the narrow hands and feet which terminate in long, sickle-shaped claws, enable sloths to move skillfully in the branches, while at the same time making it impossible for them to move on the ground. The fingers and toes, reduced in number to two or three, are greatly syndactyl. Sloths change trees only when the food supply is insufficient. But, if overhanging branches are not present, they hesitate a very long time before moving to the next tree, for their skill while in the trees is suddenly transformed to virtual helplessness when they are on the ground. To crawl they lie on the stomach and extend the arms and legs, seeking a hold for their claws. They then pull them-

Fig. 7-11. Sloths can crawl on the ground only with great difficulty.

selves very slowly for a few meters. Once they have reached the next tree, they must rest before beginning to feed. Sometimes they use just the claws of their hands to grab onto particularly lush branches. Lacking incisors, they tear vegetation off with their horny lips. The chewing surfaces of their teeth (which lack enamel) are rather hollowed out through masticating plant material. The open pulp cavities permit constant growth of teeth. The three-fingered (or three-toed) sloths have the most specialized diet of all species. They possess a very complex stomach, whose right and left halves are each divided into several subchambers. In the left half three incompletely separated compartments have been created by the development of thick, horny partitions. The largest of them, which is joined to the esophagus, contains approximately one-third of the stomach surface and contains only cardia glands. The ingested leaves are broken down by glandular secretions in this compartment. The right, horseshoe-shaped stomach half includes two chambers connected solely by a narrow opening. One of the two contains primarily peptic glands and is designated as the pepsin stomach; the other is apparently a masticating compartment (indicated by the thick walls covered by papillae), as is also found in anteaters. Both stomach halves can be closed off from each other. Precise studies of the exact pathway food takes have not yet been made. Food probably remains in the stomach for a long time, for the intestines are quite short. There is only a trace of an appendix.

Internal organs

Internal organs are also examples of adaptations which the sloth has made to its environment. The liver, which is a relatively heavy organ, has turned some 135° to the rear and is fully covered by the stomach so that it never touches the wall of the stomach cavity. The spleen and pancreas have also rotated in this way and are not on the left as in other mammals, but on the right in the vicinity of the stomach exit. The sloth's pancreas shows structural similarities to those in dogs and horses. Its prime function is as a blood reservoir. It is difficult to interpret this, because the blood reserves necessary in dogs and horses under great exertion are simply not needed by sloths. Another ambiguous feature consists of a network of blood vessels in the limbs. Respiration and circulation proceed at a leisurely pace, and all processes in general proceed slowly in sloths. They may go for a week without urinating. In this connection Krieg relates the following incident which took place in a park of a South American city: "The rain had stopped, and in spite of the humid heat I felt a perceptible coolness. This condition increases the peristalsis in the otherwise inactive intestines in sloths, whereby they feel the need to relieve themselves. I noticed that they all moved toward a small hill which was near me. Only when all six sloths came to the hill did I realize that it was formed entirely from sloth excretions. They held onto the trees which had guided their path and were not disturbed by people standing close by. They apparently did not notice us. Each of the animals left great quantities of feces. After more than half an hour the last one returned to his treetop." Hermann Tirler, who kept three-fingered (three-toed) sloths in Brazil for several years, often observed that they defecated on the ground. Krieg feels that this is derived from climbing down when the temperature greatly decreases during a storm. Sloths have a highly variable body temperature which is influenced by the external temperature. It varies from 24° to 33°C. Sloths spend an average of fifteen hours a day sleeping; they lay their head on their chest and bring their arms and legs in close to each other, thus protecting themselves against unnecessary heat loss. Often they look like a bundle of hay in the tree branches, although one leg is always attached to the branch for safety. Sloths do not drink in the wild, instead obtaining all necessary liquids from the leaves on which they feed and by occasionally licking up drops of dew. Hans Krieg found that the bladder is unusually large and can be distended almost to the diaphragm; it is surrounded by large blood vessels. This seems to indicate that water is stored in the bladder. Sloths are relatively insensitive to hunger and thirst, and are generally very hardy. They survive injuries which would kill most mammals in a short time. Unfortunately, sloths are heavily hunted in Brazil, for their meat has little fat and tastes somewhat like lamb. The strawlike pelt is used as a saddle covering. The curved claws are made into jewelry by the Indians.

Both sloth genera

Both genera of sloths are distinguished most clearly by the number of fingers. Since both possess three toes (i.e., three digits on the rear legs), it might be preferable to forego the old terminology of two-toed and three-toed sloths and instead call them two- and three-fingered sloths. This change has officially been made in the original German edition of this encyclopedia. In this edition we will use the new terms but retain the traditional terms in parentheses. Considerable differences between the genera also exist in terms of the number of neck and thoracic vertebrae The three-fingered (or three-toed) sloths (*Bradypus*) have nine neck vertebrae, the highest number known in any mammal. Traces of ribs are generally found on the eighth and ninth neck vertebrae. This indicates that during the course of evolution two of the sixteen thoracic vertebrae have taken over the function of neck vertebrae. This increase in neck vertebrae is probably closely related to the high mobility of the head in three-fingered sloths. Generally they hang by their claws, facing in the direction of their tail. Since they can turn their head 180°, it is possible for them to look forward without changing their body position; this can be done only by sloths and owls. The trachea is also a distinctive structure in that it divides into two large tubes before passing into the bronchi. This feature is found in no other mammal.

Two fingered (or two-toed) sloths

Distinguishing characteristics

In TWO-FINGERED (or TWO-TOED) SLOTHS (*Choloepus*), only one of the two species, HOFFMANN'S TWO-FINGERED (or TWO-TOED) SLOTH (*Choloepus hoffmanni*), with six neck vertebrae, deviates from the normal mammalian number. In the COMMON OR SOUTH AMERICAN TWO-FINGERED (or TWO-TOED) SLOTH (*Choloepus didactylus*, see Color plates, pp. 174 and 190), there are twenty-four or twenty-five thoracic vertebrae, the highest number among mammals. Two-fingered sloths are distributed in Nicaragua, Panama, Venezuela, Guyana, Ecuador, Bolivia, and northern Brazil. Their body length is from 60 to 64 cm, while the tail is barely visible. Their weight is approximately 9 kg. The hair on the upper side of the body is stiff, and its coloration is gray-brown to beige; the hair on the top of the back is up to 15 cm long, and is darker. The underside hair is thick, and the light, flattened face has a naked, protruding snout. The darkly edged, tiny eyes have brown irises. The small ears are generally hidden in the hair. The legs are slightly longer than the arms. The two syndactyl fingers (second and third) and the three syndactyl toes (second, third and fourth) have sickle-shaped claws up to 7.5 cm long. The soles of the hands and feet are naked and calloused. The teeth, which are regressed and lack enamel, grow continuously. There are five teeth in the upper jaw and four in the lower; all teeth are cylindrical and brown. The first tooth in each jaw, particularly in males, is elongated and pointed like a canine. There are two teats, and the gestation period is five to six months. One young is born at a time. The Hoffmann's two-fingered sloth is rarer than the common two-fingered sloth.

Fig. 7-12. Common two-fingered (or two-toed) sloth (*Choloepus didactylus*).

Sloths in the zoo

For many years the two-fingered sloths have been kept in zoological gardens. They are less specialized in their food than the three-fingered species and they readily become accustomed to a mixed diet of plant material and boiled eggs. Numerous breeding successes in the Prague Zoo and the National Zoological Park in Washington, D.C., have afforded a look at the birth and care of young in two-fingered sloths. One pair of common two-fingered sloths in the Prague Zoo had been living there fifteen years and both animals were very tame. At feeding time they climbed toward their keeper and took lettuce, germinated wheat, and fruit from his hand. The male and female are difficult to distinguish, for the scrotum is not externally visible. There is probably no fixed breeding season, for births have taken place in the Prague and National Zoos in all months except April, September, and November. When copulating, the pair hold onto a branch with just their arms and turn their faces toward each other. Zdenek Veselovsky reports that the gestation period is five months and twenty days. The birth also takes place in the tree. Often the female hangs by her arms, fully extended. The fully developed young is born head first, without the embryonic membranes. Breathing heavily, it assists in the birth until it can hold onto its mother with its claws. In the National Zoo it was observed that other common two-fingered sloths kept in the enclosure where a birth occurred pressed close to the mother, presumably to prevent the young from falling down. In nature, two-fingered sloths are solitary, as are all other sloth species. During the entire process of birth, which lasts fifteen to thirty minutes, the female licks its young as well as her own belly hair; the young climbs on this hair to reach the nipples. The mother grabs the legs of the young after biting off the umbilical cord.

Fig. 7-13. Hoffmann's two-fingered (or two-toed) sloth (*Choloepus hoffmanni*.)

Birth of a sloth

Development of the young

The newborn sloth is about twenty-five centimeters long and weighs 300 to 400 grams. Its woolly fur on the much darker back is 1.5 cm long, and the eyes are opened. The dentition has formed, and the teeth already have the characteristic brown coloration; they reach their full size within one year. During the first four weeks, the young stays hidden in the mother's hair and barely moving. After this time, its activity increases and the young sloth begins to reach for branches and sniff everything nearby. The mother carries the young about and continues her normal routine. However, if a man or another sloth disturbs the young, the young shrieks and the mother reacts immediately and will even bite if the cause of disturbance does not go away at once. The mother thoroughly removes feces and urine, which the young excretes into her hair at intervals of up to eight days. After ten weeks the young first begins eating the mother's food, still clinging firmly to her. It does not begin to move about independently until it is nine months old. Then every attempt to return to the mother is repelled. If the young still attempts to climb on her she will bite it. The adult size and weight are attained at an age of two and one-half years. Two-fingered sloths have a long life expectancy in captivity. A female common two-fingered sloth has been living in the Memphis, Tennessee, Zoo since at least February 1949, and was still in good health as of August 1974.

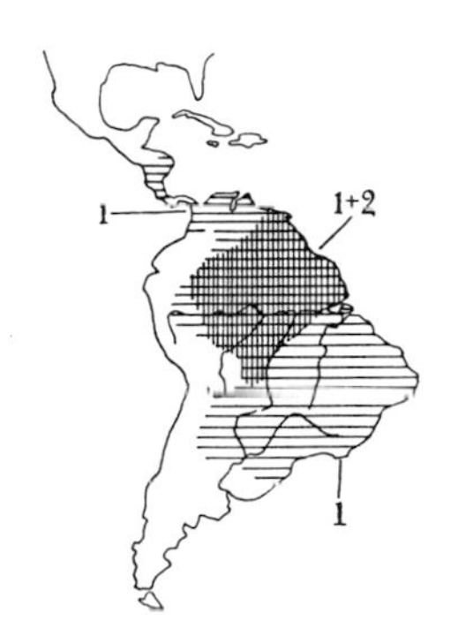

Fig. 7-14. Three-fingered (or three-toed) sloths: 1. *Bradypus tridactylus*; 2. *B. cuculliger*.

Sensory capabilities

The olfactory sense is not as highly developed in sloths as in armadillos and anteaters. Taste, however, is believed to be very sensitive. The eyes are nearsighted and the ears are apparently not very sensitive, for the animals seems to react only to certain noises.

Three-fingered (or three-toed) sloths

The THREE-FINGERED (or THREE-TOED) SLOTHS (*Bradypus*) are distributed from Honduras to northern Argentina. They are chiefly found on forest edges and river shores. These areas contain the ymbahuba tree (*Cecropia lyratiloba*), whose leaves, flowers, and fruits form the sloths chief diet. The body length of these sloths is 50–60 cm, the tail is 6.5–7 cm long, and the body weighs 4–5 kg. The hair is 5–6 cm long, lacks sheen, and is stiff. In the neck region the hairs are elongated and form a sort of mane; the hair on the lower body side is dense and soft. Coloration is gray-brown, often with bright spots. There is a distinct dorsal coloration which is less prominent in females and young than in males. The neck is relatively long, and the small round head has a blunt, hard-lipped snout and a small mouth opening. The round ears are hidden in the hair. The slender, muscular body has very long arms and slightly shorter legs. There are three syndactyl fingers (the second, third, and fourth) and toes (also the second, third, and fourth) with long hooked claws which are also used defensively. The soles of the hands and feet are covered with hair which extends up to the small, oval callosities on the inner side of the feet and hands. The brown, enamelless teeth grow continuously. There are five teeth in the upper jaw and four in the lower. The first tooth in the upper jaw is smaller, and in the young a smaller, frail tooth is present in front of the

Fig. 7-15. Necklace sloth (*Bradypus torquatus*).

remaining four in the lower jaw. There are two teats on the breast. The gestation period of about six months produces a single young at a time. There are three species with several subspecies: the most common and best-known is the SOUTH AMERICAN THREE-FINGERED (or THREE-TOED) SLOTH (*Bradypus tridactylus*; see Color plates, pp. 175 and 190). The sloth *B. cuculliger*, from Guyana and Bolivia, has long hair on the head and shoulders, which is parted in the middle. The NECKLACE SLOTH (*B. torquatus*) is distributed in northwestern Brazil and Peru. This species lacks the typical dorsal marking; its distinguishing mark is the black band which passes over the neck and shoulders.

Three-fingered sloths have rarely been kept alive in zoos for more than one month. They are highly specialized leaf-eaters, and as such are similar to Australian koalas. Hermann Tirler has shown, however, that it is easy to tame the docile three-fingered sloths in the Brazilian jungles. The *ais*, as they were called, came into his garden for several years. This local name, *ai*, is derived from the two-syllable call of the sloths; it is heard particularly during the breeding season in March and April. Other sounds besides the drawn-out a-iii are a snorting heard as a sign of displeasure and a guttural sound uttered especially by the young as they hang on their mother's belly. Tirler's ais bred successfully, giving him an opportunity to observe their family life. He found that the parents remain solitary. The male does not participate in any way in raising the young. Even the mother frequently appears to forget the young as she climbs about in the tree branches, because the young often cannot find a place to fit properly when the mother comes to a halt. Completely reliant upon itself, the young responds when any branch threatens to cut it off from the mother, and as Tirler has described, "the sloth climbs around the obstacle in order to jump back onto the 'train' passing through." For the young sloth, whose speed far exceeds that of its parents, the local name *Perico ligeiro*, is suitable: it means "fast Peter." Before the young sloth urinates or defecates in its mother's hair, it makes digging motions with its tail; the same motion is seen by adults on the ground.

The temperament of the sloth was aptly described by Tirler: "One evening we suddenly smelled something like a burning sloth, the odor coming from the neighboring room. Indeed, there it sat on the large electric bulb, half asleep. Its rear was already smoking, but it remained sitting. We brought it down, but it wanted to stay up there, grabbing on with its claws, making the "aiii" shrieks. The brown spot on the lower part of its back interests my scientifically minded visitors: 'Is it a new species?' 'No, this is only the brand from a 100-watt bulb'." Whenever Hans Krieg showed his film of this three-fingered sloth, the audience broke into laughter when close-up shots were shown. "Those who really know these animals find them fascinating examples of creatures which are well adapted to their environment and which are extremely tough."

Family: Anteaters

The ANTEATERS (Myrmecophagidae) are still more unusual than the

▻
The tamandua (*Tamandua tetradactyla*) is a smaller tree-and-ground-dwelling relative of the giant anteater (upper picture). The giant anteater (*Myrmecophaga tridactyla*) lives in open savannas and fields of South America (middle picture). The hairy armadillo (*Euphractus villosus*; lower left picture). The nine-banded armadillo (*Dasypus novemcinctus*; lower right picture).

▻ ▻
The common two-fingered (or two-toed) sloth (*Choloepus didactylus*, upper picture). Its peculiar face (lower left picture). Because sloths climb upside down, their hair pattern goes from the belly to the back (common two-fingered sloth; lower right picture).

▻ ▻ ▻
A three-fingered (or three-toed) sloth (*Bradypus tridactylus*) shown in its natural habitat (upper picture). When it is necessary, the three-fingered sloth can swim quite skillfully. It is also shown climbing a tree trunk.

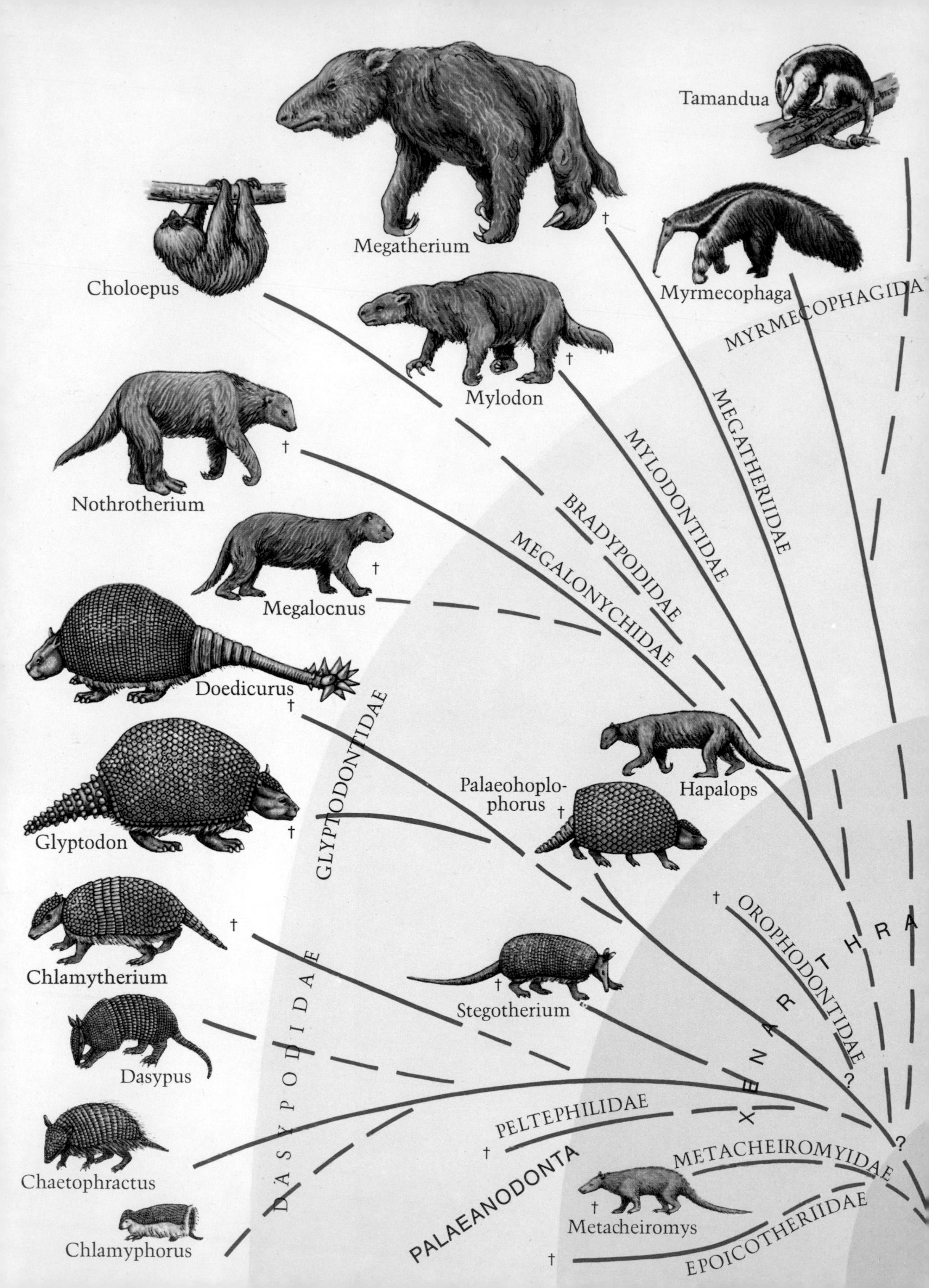

Tamandua
Megatherium
Choloepus
Myrmecophaga
MYRMECOPHAGIDA
Mylodon
MEGATHERIIDAE
MYLODONTIDAE
Nothrotherium
BRADYPODIDAE
MEGALONYCHIDAE
Megalocnus
Doedicurus
GLYPTODONTIDAE
Hapalops
Palaeohoplo-
phorus
Glyptodon
OROPHODONTIDAE
XENARTHRA
Chlamytherium
Stegotherium
DASYPODIDAE
Dasypus
PELTEPHILIDAE
Chaetophractus
PALAEANODONTA
METACHEIROMYIDAE
Metacheiromys
EPOICOTHERIIDAE
Chlamyphorus

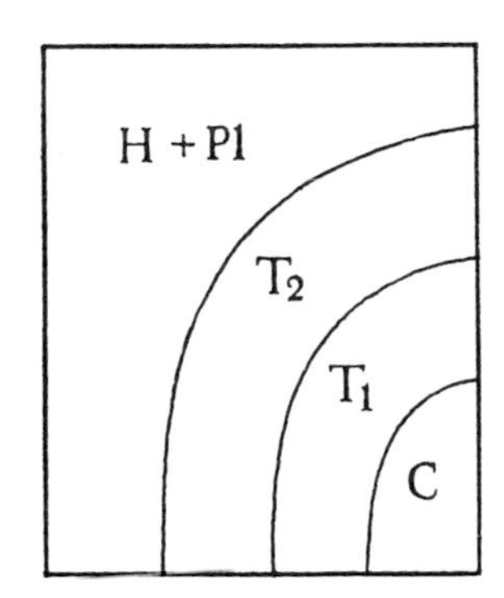

Evolution of the edentates. Geological periods: C = Cretaceous (65 million to 135 million years ago); Ti = Lower Tertiary (25 to 65 million years ago); T_2 = Upper Tertiary (2 to 25 million years ago); Pl = Pleistocene (10,000 to 2 million years ago); H = Recent (or Holocene; the period beginning 10,000 years ago).
A. Suborder + Palaeanodonta, which became extinct in the Upper Tertiary.
B. Suborder edentates (Xenarthra): 1. Armadillos (Dasypodidae), with living and extinct genera; 2. Sloths (Bradypodidae); 3. Anteaters (Myrmecophagidae). Extinct families: 4. + Peltephilidae; 5. + Glyptodontidae; 6, 7, and 8. Giant sloths (+ Megalonychidae, + Mylodontidae, and + Megatheriidae).

tree-dwelling sloths; their round head and long limbs almost look like those of apes. Their body length ranges between 16 and 120 cm, their tail length between 18 and 90 cm; their weight varies from 500 g to 35 kg. The elongated, sometimes tubular snout has a small mouth opening. The ears are short and round. The tail may have few hairs or it may be thickly covered with hairs. It is somewhat useful for holding things. The hair is short and thick or long and stiff. The short limbs have highly specialized toes and claws. Anteaters are arboreal or terrestrial. There are three genera, each with one species.

The bony covering of the armadillos is found only in the extinct ancestors of present-day sloths, but the giant anteater still has considerable scale development: the skin under the bushy tail is covered with scales. The giant anteater's striking external appearance deviates greatly from what we normally conceive of as mammalian. Its expressionless head looks like a tubular extension of the neck as it bends slightly downward. The head seems to have less significance than the hair-covered tail which is held erect and thus extends the mane along the back. The forelegs draw a great deal of attention principally because of their fur markings. The skull cap is barely visible, and the small eyes and naked ears are also inconspicuous. The long snout terminates in a narrow, tiny mouth opening. The jaws lack even vestigial teeth. Only its smaller relatives, the tamandua and the two-toed anteater, still have distinct vestigial teeth in their jaws, which indicates the presence of teeth in their ancestors. During the course of evolution the teeth were lost as the diet became more and more specialized. Thus, the anteaters are the only truly toothless species among edentates. They feed almost exclusively on ants and termites, as do the echidnas in Australia and pangolins in Africa and Asia. The anteaters have pointed claws like those in sloths for opening the anthills. All three species use these as powerful weapons when attacking if they are threatened. The anteater sits up on its rear legs and can seriously injure the attacker with blows by its forelegs.

The giant anteater

The GIANT ANTEATER (*Myrmecophaga tridactyla*; HRL 100–300 cm, TL 65–90 cm, weight 30–35 kg; see Color plates, pp. 173 and 190), is about the size of a German shepherd dog. Males are somewhat larger than females. The trunk is slender. The head is very long and tubular and the ear muscles, eyes, and mouth opening are small. The arms are more powerful than the legs, and the hands are greatly modified. The fifth finger is not visible externally, and the second and third fingers, with their long claws, are powerful; they are separated by a deep fold. The third claw, which reaches a length of 10 cm, is the strongest. The fourth finger is small and it has a pillowlike callosity at its base. This is separated from a similar callosity on the base of the hand by a softer skin. The first finger is very small and is raised. The foot has five toes; the first sits higher than the others; the second, third, and fourth are equal in length, while the fifth is slightly smaller. The hand rests on the outer edge of the last finger with the claws retracted, and the foot moves on its sole. The hair is straw-like and stiff; it is very short on the head and gets longer from the neck down. The hairs of the rear mane are up to 24 cm long, while those in the tail mane can be 40 cm long. There are two teats. The gestation period lasts six months, and only one young is born at a time.

The giant anteater is exclusively terrestrial. It is distributed in open lands as well as shrubby savanna from Costa Rica to the Gran Chaco in northern Argentina. Its enemies, the jaguar and puma, are very careful of the fore-limbs, because an embrace by an anteater can often prove fatal. Even dogs and man cannot escape this grasp. However, the *yurumi*, as the anteater is called in the Guarani language, is docile and never goes out to attack someone. It also has a short flight distance. "We could approach to within a few meters of it," wrote Hans Krieg, "without its making an earnest attempt to flee or even to take notice at all. It was very easy to move the animal about or to photograph it." Its visual sense is probably not well developed.

The female giant anteater stands while giving birth and lets the young independently crawl up on her back; she then licks the newborn anteater and lets it drink. According to observations in captivity, the breeding season takes place at particular times in the spring and fall. At birth the anteater weighs about 1700 g. It is carried about on its mother's back for a very long time, even though it is capable of running at a slow gallop when it is just four weeks old. Young anteaters often emit a short, shrill whistle. With its coloration, which corresponds to that of the mother down to the silver-white dorsal stripe, the young is very well camouflaged, and it presses so close to the mother that they seem to merge into each other. The young is fully grown when it is two years old, at which time it begins feeding independently. Giant anteaters are active during the day in desolate regions as well as those infrequently visited by man. They do not stay in one place, but move like nomads through the stiff grass of the savanna, holding their nose just above the ground until they encounter an anthill or termite nest.

The giant anteater rips a termite hill open with a single blow of its powerful hand; it sticks its long, narrow snout into the opening and digs in deeper and deeper until its worm-shaped tongue can penetrate to the center of the colony. The small insects are held fast by the sticky coating on the tongue. The insects are masticated by means of horny papillae on the roof of the mouth and sides of the cheeks. These are set into motion when the narrow lower jaw presses up on the tongue. The tongue-retracting muscles do not rest on the hyoid bone, but well beneath, on the sternum. The salivary glands have a capacity unknown in any other mammal; the large glands beneath the tongue extend to the sternum. They excrete saliva in the mucous membranes of the tongue. The anteaters' saliva supply is still larger than that of armadillos. It has been found that the 60-cm-long tongue can be pushed out and pulled back into the mouth opening up to

Fig. 7-16. Giant anteater (*Myrmecophaga tridactyla*).

Fig. 7-17. In the threat posture the giant anteater raises its powerfully equipped hands. When it is driven into a corner this animal is a very powerful adversary; it can deal strong blows with its arms.

160 times per minute. Tremendous amounts of saliva are necessary for this. This also indicates the extraordinary power of the muscles which move the tongue, for anteaters lack the hyoid mechanism found in some reptiles and birds. The insects, covered with saliva but still whole, pass into the stomach, where strong muscular walls and the horny stomach lining crush them. Sand and small stones have been found in the stomach, and apparently then help with this process. A giant anteater, which eats as many as 30,000 ants and termites in a single day, prefers those species lacking large jaws and barbs. Other food includes worms, larger insect larvae, and berries, which are sometimes taken up with the lips. Anteaters probably do not drink in the wild, but obtain their liquids by licking plants wet with rain and dew. "I saw the only tamandua drinking spontaneously from a puddle of water," wrote Krieg. "It lapped up small quantities with its tongue."

Giant anteaters sleep a great deal, whether in the zoological garden—where some have already lived for over twenty-five years, having become accustomed to a mixture of raw meat, milk, eggs, meal, and fruits—or when wandering about in the wild. They lie on their side and stick their head between their forelegs, which they then cover with their tail. Urs Rahm notes: "The giant anteater is apparently a light sleeper, because it awakens at the lightest sound and immediately extends its head and snout. It turns its head in all directions and sniffs the air. The olfactory sense is the most sensitive one in anteaters; it assists in escaping and in finding food." Hearing is also keen. The giant anteater can orient to and find a source of sound emanating from a point about 5 m away. Those who have observed it in the dry savanna will barely believe that it can skillfully cross rivers.

The giant anteaters in the Frankfurt Zoo regularly dig sleeping cavities in their outside enclosure. Rosl Kirchshofer found that they hollow out the base of an elm tree, and that they consistently prefer the corners of the enclosure. Thus, one can conclude that the giant anteater selects sleeping sites at those places where a tree trunk (or, in the zoo, a corner of the enclosure fencing) affords protection. In Frankfurt they dig these sleeping depressions so deep that when they are in them, the tops of their bodies are level with the surface. With this behavior and their protective coloration, anteaters are so well camouflaged that even attentive zoo visitors often fail to spot them.

The tamandua

The TAMANDUA OR COLLARED ANTEATER (*Tamandua tetradactyla*; HRL 54–58 cm, TL 54–56 cm, weight 3–5 kg; see Color plates, pp. 173, 190) is about half as large as the giant anteater and is not exclusively terrestrial. Its distribution is approximately comparable to that of its larger relative. The species prefers forest edges and tree savannas but is also found in treeless regions and in developed areas.

Its hair is thick, bristly, and generally short. The basic coloration is whitish-yellow to brownish with blackish red-brown markings, although some races are entirely yellowish-white with no dark markings. The hair covering of the powerful, scaly tail decreases from the upper third to the point. It is less useful for holding things than is the tail of the two-toed or silky anteater. The ears are spread far from each other and are relatively larger than those of the giant anteater; the snout is considerably shorter. The eyes and mouth opening are small. The arms and legs are powerful, and its hands and feet are the least specialized of any in the family. The third finger is the longest and strongest, with a claw up to 5 cm long. The second and fourth fingers are of equal length and have shorter claws. The first finger is insignificant, and the fifth is absent altogether. There are five toes, the first being appreciably smaller. The palm is more developed on its outer edge, while the foot soles are naked. When moving, the claws are retracted against the palm and are set down upon the entire outer palm edge; only half of the soles of the feet are set on the ground. There are two teats. The single young is typically born in spring. There are several subspecies.

Fig. 7-18. When the tamandua threatens, it sits on its hind legs supported by the tail, with its arms spread out.

The tamandua rarely feeds during the day. Its pace is still clumsier than that of the giant anteater and "almost appears like it is limping," as Krieg writes. One sees a "free step with a raised rump and stiffly held limbs." When the tamandua is frightened, it stands up like a bear, with outspread arms. But istead of embracing its enemies, it clamps anything or anyone between its strong hand claws and thus keeps the attacker away from its body. When greater danger threatens, it hisses and "drops onto its back from the upright threat position so that its legs are free to defend itself." The unpleasant odor which tamanduas emit when they are excited has given them their local name ("stinker of the forest"). Supported by its tail, it slowly climbs on the tree branches, breaks pieces off the tree termite nests, and, like its larger relative, licks up great quantities of ants and termites. The stomach of one young tamandua contained 500 g of ants and their larvae. The young tamanduas are carried about on the back or flanks of the mother for a long time. When she is feeding, the mother frequently sets her young down on a branch. In contrast to the case with the giant anteater, the fur coloration of the young tamandua is quite unlike that of the adult. Both yellowish-white and brownish-black young are found. The species is quite sensitive, making it difficult to keep one successfully in captivity. However, when an acceptable diet is provided it quickly becomes tame and may play with its keeper, whom it can differentiate from strangers. One lived in the National Zoo in Washington, D.C., for almost ten years.

Fig. 7-19. Tamandua (*Tamandua tetradactyla*).

The two-fingered (or two-toed) anteater

The third genus in the anteater family also contains but a single species, the TWO-TOED OR SILKY ANTEATER (*Cyclopes didactylus*; HRL 16–20 cm, TL 18–20 cm, weight 500 g; see Color plate 4, p. 190). It is exclusively arboreal, and, unlike the tamandua, never goes to the ground by choice. Its hair is soft, reddish-brown to golden yellow, and short. The muscular prehensile tail, important to the animal's movement, is naked at the end.

Fig. 7-20. Two-toed anteater (*Cyclopes didactylua*).

The eyes and ears are small, and the curved snout is only somewhat elongated. The mouth opening is large compared to those of the related species. The arms and legs are powerful, and the hand deviates from typical structure: the second and third fingers are syndactyl and almost equally large. The third, however, is more powerful and has a longer claw. There is no external first, fourth, or fifth finger. The palm is well developed and has grooves on its inner aspect which are indicative of the claws being retracted into the palm. There are two teats. One young is born at a time.

Distribution is in forested regions from Central America to Bolivia and Central Brazil. The two-toed anteater spends the day sleeping, rolled up in a tree branch or cavity, because it is difficult to see, the species is often considered to be rare. It moves through the branches slowly but with great skill. Like its larger relatives, it uses the curved claws of the forelimbs to open ant nests in trees and to ward off enemies. With the prehensile tail and feet firmly anchored, it pulls itself up and strikes with its arms. Raising the young is not the exclusive task of the female; both parents regurgitate semi-digested insects from their stomachs and feed the young, which is sometimes carried on the male's back. Two-toed anteaters utter a soft, whistling sound. They are the most difficult of the anteaters to keep in captivity, the longest period in captivity recorded to date (August, 1974), being that of one kept in Chicago's Lincoln Park Zoo for two years, four and one-half months.

8 The Pangolins

Order: Pangolins, by U. Rahm

The PANGOLINS are a peculiarity among present-day mammals because of their unique body cover and various other characteristics. They are, therefore, a separate order (Pholidota), with one family (Manidae). Previously they were classified with the edentates as they superficially resemble armadillos of the New World. However, the Old World pangolins are not related to the armadillos or to the edentates.

Distinguishing characteristics

They range from 75 to 150 cm long, with the tail amounting to from forty-five to sixty-five percent of the total length. The head is small, and the ears are underdeveloped. The body is covered with scales. In adults these scales, like those of fish, overlap one another like shingles; They do not overlap in young animals. There are no scales on the underside of the neck and stomach, nor are there any on the inner sides of the limbs or on the snout. The front feet have digging claws. There is no collar bone. There are from eleven to sixteen chest vertebrae, five or six loin vertebrae, two to four pelvis vertebrae, and from twenty-one to forty-seven tail vertebrae. The animals are insectivorous, preferring ants and termites. They have no teeth. Pangolins have a specialized stomach with a tough epithelium used to grind up food. The females have two nipples on the breast. The African species give birth to one young, while the Asian pangolins may have from one to three young. The duration of pregnancy is unknown. There is one genus, PANGOLINS (*Manis*), which has been subdivided by many researchers into various subgenera. There are seven species in all.

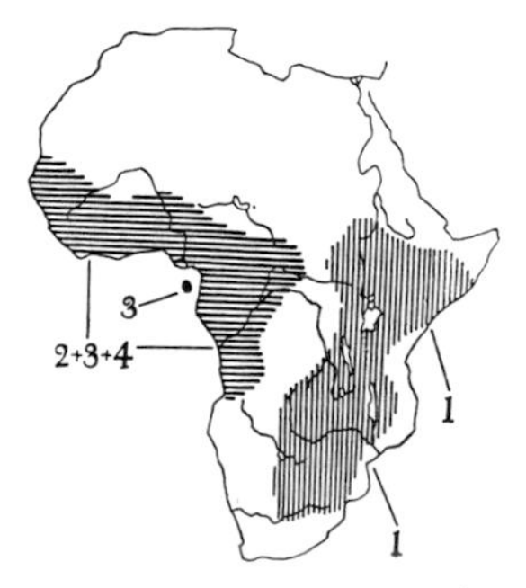

Fig. 8-1. 1. Cape pangolin (*Manis temmincki*), 2. Giant pangolin (*M. gigantea*), 3. White-bellied tree pangolin (*M. tricuspis*), 4. Long-tailed tree pangolin (*M. tetradactyla*).

The subgenus *Phataginus* includes the WHITE-BELLIED TREE PANGOLIN (*Manis tricuspis*; BL 35–45 cm, TL 40–50 cm; see Color plate, p. 199). The scales, which are relatively small, are brownish-gray to dark brown. This pangolin has particular back scales with three more or less pronounced points. In older animals these back scales are often worn. The skin and hair are white. This is a nocturnal, arboreal animal found in the tropical rain forests of Africa, from Sierra Leone to the Great Rift Valley. In some areas, such as Zambia and the southern part of Zaïre, it lives along the forest edges and the savannas.

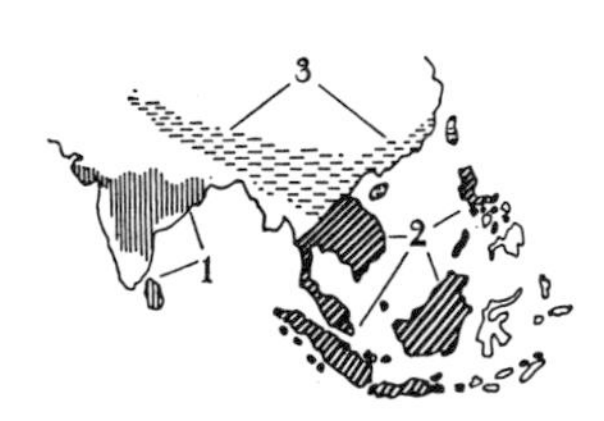

Fig. 8-2. 1. Indian pangolin (*Manis crassicaudata*), 2. Malayan pangolin (*M. javanica*), 3. Chinese pangolin (*M. pentadactyla*).

The subgenus *Uromanis* includes: the LONG-TAILED TREE PANGOLIN (*Manis tetradactyla*; BL 30–40 cm, TL 60–70 cm; see Color plate, p. 199), which has the longest tail of all pangolins (160–215 percent of the body length). Its scales are larger than those of the white bellied tree pangolin, but they appear in smaller quantities; they are dark brown, with a yellowish edge. The skin and hair are dark brown to black. This species lives in trees and is more active in the daytime than are the other species of tree pangolin. It inhabits the tropical rain forests of Africa, but not the forest edges.

The subgenus *Smutsia* includes two species: The GIANT PANGOLIN (*Manis gigantea*; BL 75–80 cm, TL 50–65 cm; see Color plate, p. 199), which is the largest species. The scales are large and grayish-brown. The hair and skin are whitish. This animal is nocturnal and a terrestrial hole digger. Its large digging claws enable it to rip open the hills made by ground termites. It is found in tropical Africa and in the savanna belt. 4. The CAPE PANGOLIN (*Manis temmincki*; BL 50 cm, TL 35 cm; see Color plate, p. 199), has large and grayish-brown to dark brown scales. The skin is whitish with dark hairs. This animal is also a nocturnal ground dweller, although it occasionally climbs bushes and trees. It is found in savannas and prairies of southern and eastern Africa. Reports on its habitat are scarce and incomplete.

The subgenus *Phatages* includes the INDIAN PANGOLIN (*Manis crassicaudata*; BL 60–65 cm, TL 45–50 cm; see Color plate, p. 199). The scales are large and light yellow-brown. The hair and skin are brownish. These animals are largely nocturnal; they climb with great agility and dig well, making tunnels that end in an enlarged cave. They live on the island of Ceylon and in much of India.

The subgenus *Manis* includes the CHINESE PANGOLIN (*Manis pentadactyla*; BL 50–60 cm, TL 30–40 cm; see Color plate, p. 199). The scales in young animals are purplish-brown; in the adults they are blackish-brown. The skin and hair are grayish-white. Its ears are better developed than are those of the other pangolins. It is found from southern China to Nepal and Indochina. The Chinese pangolin of the island of Hainan is classified as a special subspecies; *Manis pentadactyla pusilla*.

The subgenus *Paramanis* includes the MALAYAN PANGOLIN (*Manis javanica*; BL 50–60 cm, TL 40–80 cm; see Color plate, p. 199). Its scales are amber-brown to blackish-brown, and the skin is whitish with dark hairs. This pangolin is nocturnal; during the day it sleeps, curled up in a forked branch, in ferns, or in other plants growing on trees. It moves on the ground or in trees with equal agility. It is widely dispersed over Indonesia, the Malay Peninsula, Indochina, and Burma.

Scales as skin covering

The English popular name pangolin comes directly from French. The French name, which is now used in many other languages, stems from the Malayan word *pengolin*, which refers to the animal's ability to curl up into a ball. The scales are not "glued hairs," as was once believed; they are two-sided symmetrical elevations of the skin which are flattened from the back toward the stomach, and which face the back of the animal. The epidermis, as it grows into horn material, leads to scale formation. The horn scales, which are lost through wear, are constantly replaced by the skin base. Throughout its life, the pangolin always has the same number of scales. The pattern, quantity, shape, and size of the scales differ from species to species, and can also differ slightly within a species, depending on the part of the body they cover. The five arboreal species have a bald, cushion-like spot on the underside of the end of the tail. Whether this spot functions as a tactile organ in climbing has long been disputed. However, the tail ends of both terrestrial species are fully covered with scales. There are occasional differences in the scale coloring on the tails, especially in the long-tailed tree pangolin, which, incidentally, has the most tail vertebrae of all mammals, forty-six or forty-seven. Those parts of the body not covered by scales are sparsely covered with hair.

Claws and epidermal glands

Pangolins have short limbs with five toes. The front feet are equipped with large digging claws, the one on the middle digit being the largest. In the climbing species, the claw point reaches far over the finger and toe cushions. The eyes are small; thick lids protect them against ant and termite bites. There are no epidermal glands; the anal area, however, is surrounded by large bean-shaped gland sacs.

Internal body characteristics

Skeleton

The pangolins are distinctive not only externally, but also in their skeleton and in other internal structures. They show special adaptations to their particular way of life and nutrition. The skull has few protrusions, cheekbones, bone ridges or bone cams to which chewing muscles could be attached. The lower half of the jaws is underdeveloped and has regressed into narrow, toothless bone clasps. The upper jaw is also toothless. The snout is elongated and the mouth slit is narrowed.

The xiphisternum has gone through some adaptive changes in the pangolins. In the Asian species, it is rather long and ends in a disc which is widened at the sides, resembling a shovel. In the white-bellied tree pangolin, it forms two long strips of cartilage which combine toward the back and send out two more staffs of cartilage, which also fused toward the back. The total xiphisternum is so long that it reaches beyond the peritoneum along the lower abdominal wall to the pelvis, where it bends and leads along the upper abdominal wall to the kidneys. This adaptation is related to the extremely elongated tongue, which plays an important role in feeding. In the species of larger pangolin, the tongue is up to 40 cm long; in the arboreal African species, it measures 16–18 cm. In a resting position, the tongue is pulled back into a sheath which itself is moved back into the chest cavity. This requires a complicated muscular system attached to the xiphisternum. Extremely large salivary glands, which reach almost into the shoulder area, supply this heavily elongated tongue with a sticky mucus, to which the insects stick.

The stomach is also specially equipped. It serves almost as a "chewing

The stomach has "horny teeth"

apparatus." Because pangolins have no teeth, the ants and termites reach the stomach uncrushed, and they must be mechanically ground there. Therefore, the stomach has a horny laminated epithelium with "horny teeth" instead of an internal mucus membrane. A similar organ, also covered with "horny teeth" and moved by strong muscles, faces part of it.

What pangolins eat

Since pangolins feed only on ants and termites, they can, of course, live only in those areas where there are enough of these insects. Many species, like the long-tailed tree pangolin, have a preference for ants; others, like the white-bellied tree pangolin, prefer termites. Tree-dwelling pangolins can hunt and feed only in the nests of arboreal termites. Ground termite nests are too strong for them to open. If one has been "wrecked" by humans, however, arboreal pangolins will eat from it. The giant pangolin is able to rip open the strong nests of ground termites in the African rain forest. As true arboreal animals, the white-bellied tree pangolin and long-tailed tree pangolin especially prefer the nests of *Nasutitermes* and *Microcerotermes*, and they will reject the inhabitants of anthills (like those of the *Crematogaster* species). The giant pangolin also eats ants; up to eleven different species of ants were found in its stomach.

A grown white-bellied pangolin eats 150–200 g of termites nightly, and the stomach of the giant pangolin can hold 2 l of insects. The Cape pangolin feeds mostly on termites of the genera *Odontotermes*, *Ancistrotermes*, *Amitermis* and *Microtermes*. It also eats ants even including the feared army ants. The range of the Chinese pangolin corresponds approximately to those of the two most common termite species (*Coptotermes formosanus* and *Cyclotermes formosanus*). Many authorities have reported that the pangolins will ingest small pebbles and sand along with their insect food. These help crush the food. Arboreal pangolins apparently eat fragments of tree termites for the same reason. The dung of the pangolines is sausage shaped and very dry. It is black when the diet consists mainly of ants, but brown when the main food is termites. The Cape pangolin buries its feces in small depressions that it scrapes in the ground. The white-bellied tree pangolin, on the other hand, drops its feces anywhere; at least it does so in captivity. Pangolin urine has an extremely pungent odor. When dry, it leaves a whitish spot.

The pangolin special diet makes it very difficult to keep them in captivity. In their native countries, they can be fed ant and termite hills; obtaining these, however, requires a great deal of time and effort. Furthermore, pangolins do not always accept substitutes. Although pangolins can fast for from two to eight weeks, they eventually die when they have used up their fat reserves. Like the giant pangolin and the Cape pangolin, the Asian species seem to adapt best to food substitutes.

In the Antwerp (Belgium) zoo the giant pangolins successfully adapted to a replacement food mixture consisting of 25 g of cornflakes, 100 g of unsweetened condensed milk, 100 g of warm water, 150 g of sliced meat,

15 g of ant pupa, one spoon of wheat germ oil, and a drop of formic acid. In the Prague (Czechoslovakia) zoo, Indian and Chinese pangolins were kept on a mixture of raw and cooked chopped meat, minced boiled egg, carrots, curd, ant pupae and some salt. Occasionally they were also given moistened, honey-sweetened oatmeal. A Cape pangolin at the Basel (Switzerland) zoo ate chopped mealworms, boiled cockroaches, and chopped horseheart, which was partially mixed with fresh eggs and cooked oatmeal.

Movement

Pangolins are plantigrade animals. The small claws on the hind feet do not hinder the animals' locomotion; the digging claws of the front feet are folded upward so that the pangolins put down only the outsides of their soles. When running on the ground, all pangolins whether arboreal or terrestrial use their hind legs most of all. Ground dwelling pangolins are able to move greater distances in an erect position, without touching the ground with their front legs. In motion, they keep the body in a curved position; the tail is carried slightly above the ground and is used to maintain balance. The Cape pangolin can travel 5 km (3 miles) in an hour. The white-bellied tree pangolin needs one second to go slightly more than 1 m. When walking, the pangolins stop from time to time and raise themselves up like a kangaroo, supporting themselves on their hind legs and tail. In this position, with the front of the body erect, and the head protruding, they smell or look out for possible enemies.

When digging on termite hills, the terrestrial species use their hind legs and tail as a "tripod," digging with their front feet. With the feet held close to the left and right sides of the head, the claws tear open the insect nest. The snout gradually penetrates the structure. With its long sticky tongue, the animal catches the insects with quick, wriggling movements. The length of the tongue also enables it to reach and extract insects from the tunnels and shafts of the nests. The tree-dwelling pangolins anchor themselves to branches with their hind legs and tail when breaking up a termite or ant nest.

The white-bellied tree pangolin and the long-tailed tree pangolin are expert climbers. They climb vertical trees in the following manner: First, both front legs grip the tree simultaneously, then the body is curved. The hind feet loosen up and are anchored close behind the front feet. With the hind feet and the tail giving the body the necessary support, the front feet grip the bark further up. The jagged side scales of the tail function perfectly here; the tail can even wrap around thin trees. Climbing down is done head first, in the same way. In switching from one vertical branch to another, the animals have such a sure grip that they can loosen their front feet and keep their body at a right angle to the tree. When a tree pangolin is held by its tail, it climbs up "on itself," so to speak, placing its gripping claws between the side scales of its tail. All pangolins can more or less curl up into a ball.

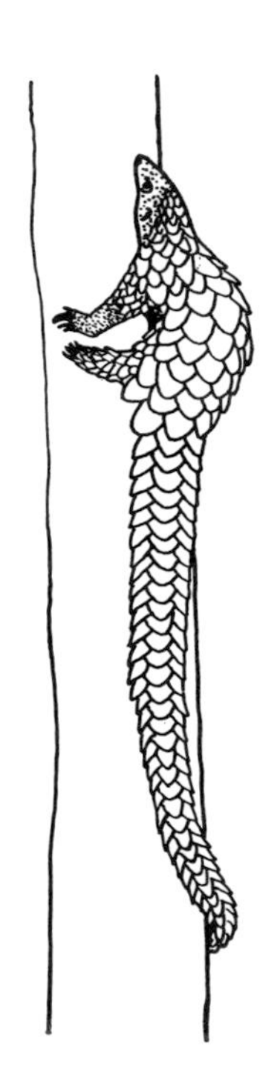

Fig. 8-3. A pangolin climbing a tree: its tail and front legs are rigidly anchored on the tree trunk; then both hind legs grip the trunk, and the arms move forward again.

Pangolins are nocturnal, with the exception of the long-tailed tree

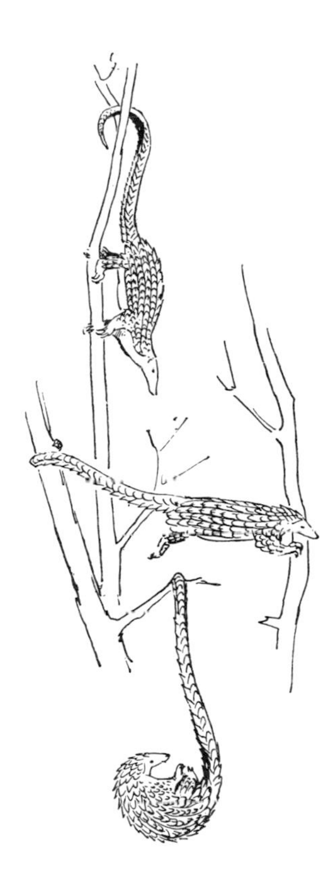

Fig. 8-4. The White-bellied tree pangolin (*Manis tricuspis*) is a very good climber, often using its long grasping tail.

pangolin, which is occasionally active during the day. The ground dwelling African species and the Indian pangolin dig dens in the ground in which they sleep during the day. The young are born in these dens. The tree-dwellers also sleep curled up, either in tree cavities or in the open between forked branches. The tree-dwellers are the better at curling up; they wrap themselves up with their tail so that they form a perfect ball. They take the same defensive position when they are attacked by a predator. Aside from man, leopards, tigers and other large animals prey on them. Smaller predators obviously can do nothing with the "scaly balls." It takes considerable force to unroll either of the African species. A human is barely able to unroll the other, larger pangolin species. When a smaller species is unrolled, it will, with surprising speed, spray its urine in defense. Whether the urine is mixed with the secretions of the anal glands is not yet clear.

Nothing is yet known about the lifespan of wild pangolins. Several pangolins have lived over four years in zoos. One Indian pangolin is still living (1974) in the Oklahoma City (U.S.A.) zoo after eight years in captivity. Another was still alive after five years in captivity. In captivity in their native countries, pangolins have sometimes given birth. This was the case in the Calcutta (India) zoo, and in the wildlife park of the University of Ife (Nigeria). In recent times other zoological gardens have also bred them successfully. A Chinese pangolin was born and successfully raised in the Ueno zoo in Tokyo. Two pangolins were born in the Evansville (U.S.A.) zoo, one of which lived two years.

Reproduction

These successes notwithstanding, we still know very little about the reproduction of these animals. We cannot calculate the gestation period from the few successful breedings thus far. Males and females may normally live together only in the mating season. A male and female and one young giant pangolin were found in a den. It has been reported that from January until March pairs of Indian pangolins live in caves with their one or two young. The African species give birth to only one young at a time. The Asiatic species may give birth to from one to three young. The giant pangolin from the Congo region apparently gives birth in November and December. The white-bellied tree pangolin normally gives birth to its young from November to January.

It is remarkable that all newborn pangolins, whether of the larger or smaller species, are the same length, approximately 20 to 30 cm. In the arboreal species, the newborn young cling to the base of their mothers' tail and are carried around riding on the tail. This particular method of transport was first observed in the white-bellied tree pangolin, long-tailed tree pangolin and the Malayan pangolin. Ground-dwellers like the Cape pangolin, the giant pangolin, and the Indian pangolin have a tail base that is too wide, and the new-born is not able to hold on to the tail. This may be the reason why the mother remains in the hole with her young for some time after their birth. When the female rolls herself up in order to sleep, she rolls the young up with her. The young lie against their mother's stomach and are protected by her tail. This safeguard is especially important for the tree-dwellers.

Although it is hard to imagine taming a pangolin or having a friendly relationship with one, because they seem so strange to us, some of these animals have adapted very well to captivity. The pangolin born in Tokyo was described by M. Masui as "very tame." An Indian pangolin in the Oklahoma City (U.S.A.) zoo seems to have grown accustomed to "being touched," according to the reports of P. W. Ogilvie and D. D. Bridgwater; "Previously it had always curled up into a tight ball. Now it is much more docile and lets itself be handled and treated for necessary examinations."

Fossil history, by E. Thenius

European pangolin remains have been found only from the Oligocene and Miocene periods, some 15 to 40 million years ago. Some giant forms which dated back to the Ice Age were found in southeastern Asia. These fossil finds gave us no information about the origin of this peculiar mammalian order, however, as the earliest forms must have lived much earlier. Certainly the pangolins developed into an independent branch at a very early time. They may have come from the same roots as the New World edentates, or they may be a direct descendant of the original insectivorous mammals.

▷
Armadillos: 1. Nine-banded armadillo (*Dasypus novemcinctus*); 2. Three-banded armadillo (*Tolypeutes tricinctus*); 3. Giant armadillo (*Priodontes giganteus*); 4. Eleven-banded armadillo (*Cabassous unicinctus*); 5. Pygmy armadillo (*Euphractus pichiy*); 6. Hairy armadillo (*Euphractus villosus*); 7. Six-banded armadillo (*Euphractus sexcinctus*); 8. Lesser pichiciago (*Chlamyphorus truncatus*); 9. Greater pichiciago (*Burmeisteria retusa*).

1
2
8
3
9
E. BIERLY
5
4
6
7

1
2
3
4
5
E. BIERLY

9 The Rodents

Order: Rodents, by H.-A. Freye

The rodents (Rodentia) are the largest order of mammals. More than half of the mammalian species alive today belong to the order of rodents. Furthermore, many rodent species surpass all other mammal groups in the total number of individual animals. It is then understandable that rodents have long played an important part in the life of humans. Rats and mice, together with fleas, lice, bugs, mosquitos, and flies—and, as a consequence, viruses, bacteria, and single-celled parasites—have played a very important part in the lives of people.

Distinguishing characteristics

The rodents vary in size from that of a pygmy mouse to that of a capybara, their length ranging from 12 cm to over 100 cm, their weight from 4 g to 50 kg. Many species have a long tail, often sparsely covered with hair. The body is cylindrical with short legs; the hind legs are often longer than the front legs. The neck is thick and the head in profile appears truncated, a result of the characteristic dentition (discussed below). There are many kinds of body hairs. Rodents have the remains of a scaly coat, mostly on the tail, but also on the limbs; this is especially true of the large scales on the flattened tail of the beaver, and on the round tail of the nutria. The toes have claws; in digging animals there are large scraping claws on the front paws. On the surfaces of the hands and feet there are bald areas ("sole pads"), which are slightly raised and can be species characteristics. There is a sebaceous gland at the hair follicles, and a larger local accumulation of various epidermal glands into so-called glandular bodies in the cheek area (for example in marmots), at the ear (lemmings), on the sides and in the groin area (cricetid mice), and in the anal zone (many porcupine-like rodents). In addition, there are "oil sacs" and preputial glands. The mammary glands, generally found in the chest and/or stomach area, occur high on the side of the body in the nutria, above the shoulder pits in New World porcupines, and on the thigh of the Cuban hutia. There are from two to eighteen nipples. The skull has a large zygomatic arch, a noticeable tympanic bulla and reciprocally moveable parts of the lower jaw (found in other mammals including kangaroos and

◁
Sloths: 1. Three-fingered (or three-toed) sloth (*Bradypus tridactylus*); 2. Common two-fingered (or two-toed) sloth (*Choloepus didactylus*). Anteaters: 3. Tamandua (*Tamandua tetradactyla*); 4. Two-toed anteater (*Cyclopes didactylus*); 5. Giant anteater (*Myrmecophaga tridactyla*) with partially extended tongue.

shrews). They are plantigrade or semiplantigrade animals. The thumb is usually nonexistent.

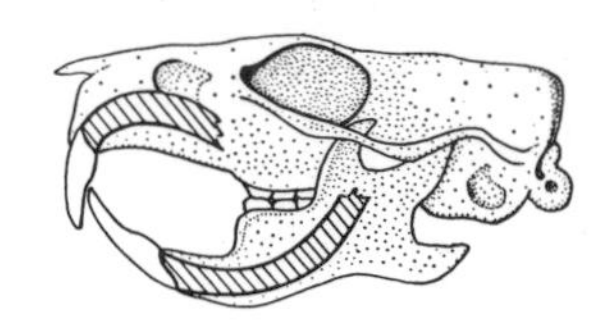

Fig. 9-1. Skull of a rodent with the rootless, continuous growing incisors, the large gap (diastema) between the incisors and molars, and the grinding teeth which consist of the pre-molars and the molars.

Rodents are either herbivorous or omnivorous. The entire order has a characteristic rodent dentition: two pairs of incisors (incisor teeth), one pair in the upper jaw and one in the lower. There is a large gap (diastema) because the canine teeth and pre-molars are missing. The incisors are always rootless, and therefore grow continuously. The mouth cavity is closed by the upper lip, which is often split ("harelip"), making the incisors visible.

Some species have cheek pouches. They have a simple stomach, shaped like a flask. The appendix is very large, and there is a long colon. The liver has five lobes; the gallbladder can be absent (as in mice). The lungs may be without lobes or they may have up to four lobes on the right, or three lobes on the left side. There is a larynx with vocal cords, although rodents make only simple vocalizations: squeals (marmots), barks (prairie dogs), grunts or squeaks (mice, rats).

The size of the eyes depends on their way of life. In digging species, the eyes are small, in palearctic mole rats, the eyes are rudimentary organs below the skin; in crepuscular and nocturnal animals the eyes are especially large; they protrude in squirrels, and when the squirrel is agitated, they protrude even more. Visual acuity is moderate; only a few diurnal species have color vision (golden hamster). Most rodents can see equally well backwards or forwards because their eyes are placed at the sides of the head. In water rodents, the eyes are narrow and are placed high on the head. The ear size varies; the external ear is quite short in subterranean or aquatic species. In northern ground squirrels ears are almost completely absent. Small rodents can also hear in the ultrasonic range (mice hear up to 100 kHz). In many species the sense of smell predominates. The sense of touch is sometimes very well developed being intensified by growths of tactile bristles on the head, the front limbs, the chest, and the stomach. The sense of touch often replaces vision when the animal is trying to find its direction. Many rodents have vibrissae which move rhythmically back and forth exploring space.

The brain is simple, usually with a smooth surface (however, distinctly furrowed brains are found in the beaver, marmot, capybara and other larger rodents). Many rodents (rats) have an excellent learning capability which generally increases with body size. Young animals also have a well-developed ability to play. Squirrels and beavers enjoy playing even when they are adults. Chattering with the teeth is a general behavior pattern found in perhaps all rodents. Orienting may consist of echolocation in mice, or it may be a sun-compass orientation as in the striped field mouse, vole, wood mouse and bank vole.

The penis often has a penis bone; the glans penis sometimes has thorns and spines (cavies, desert jerboa, paca); male beavers have a so-called uterus masculinus. Testes may be internal, but appear externally during the breeding season. The uterus has two horns and a disc-shaped placenta. Usually there are separate urogenital and anal openings (only the beaver has a cloaca). A plug forms in the vagina after copulation in many species (house mouse). Copulation follows a "mating fight" and is relatively short. Often a pair copulates several times in succession. Conception may occur immediately after birth in the smaller rodents. The beavers' copulation posture differs from the normal position used by mammals, because they mate in the water: stomach to stomach. Male flying squirrels envelop the females with their flightskins, which they use like a cloak. After copulation, mice have a "copulation rigidity," which is like a relaxation period. Gestation is from sixteen days (golden hamsters) to five and one half months (capybara). Young rodents are generally altricial although guinea pigs are precocial. In smaller rodents there is a quick succession of generations: the most recent young are still nursed when the mother is pregnant again. The rate of increase surpasses that of many insects. The paca normally has one young; the muskrat produces six to twelve; the golden hamster may have fourteen young, while the hamster may give birth to eighteen or more. The lifespan of smaller rodents is usually less than two years. Flying squirrels may live twelve or thirteen years, while the normal lifespan of the Old World porcupine is between twelve and eighteen years, with some living over twenty-seven years in captivity.

There are over 300 genera and nearly 3000 species of rodents. These can be classified into four suborders: 1. Sciuridor squirrel-like rodents (Sciuromorpha); 2. Myomorphs or mouse-like rodents (Myomorpha); 3. Porcupines (Hystricomorpha); and 4. Cavies (Caviomorpha). However, the classification of rodents is not yet complete.

Rodents as carriers of disease

Some rodents play a special role as carriers of disease. The dreaded plague bacillus is carried by brown rats, house rats, bandicoot rats, bobac marmots, northern ground squirrels, chipmunks, common voles and mice, as well as other representatives of various genera. This disease is transferred among the rodents and from the rodents to humans mainly by fleas on the animals. According to R. Müller, ship rats carried the plague to most of the ports on the earth and from there the local rodents became contaminated. The plague at first originates in an epidemic among the rodents; without an epidemic in the rodents, there can be no epidemic in humans. Although the death of the animals is not always noticeable, the Greek geographer Strabo (63 B.C.?–? A.D. 24), from Iberia, realized that the plague and similar diseases came from the local population of "mice."

Nobel prize winner Albert Camus, in his novel *The Plague*, accurately depicted the start of an epidemic of the plague among the rats of the town: "The quantity of rodents that had been collected grew continuously and the harvest was larger each morning. From the fourth day on, rats emerged in groups and died. They came from hideouts, basements, cellars and sewers in long, staggering lines; they fell down in the light, turned over and died near the people. At night one could clearly hear their faint dying cries in the lanes and narrow alleys. In the morning, one found them stretched out in the gutters of the suburbs, a little blood on their pointed nuzzles, one bloated, another rather putrid, and another with stiff vibrissae. In the town itself they were found in small heaps on the floor or in backyards. Often they died alone in the front rooms of official buildings, in the schools, or often on the terrace of a coffee shop. Our terrified fellow citizens discovered them all over town."

Rodents, however, are not only living carriers of the plague bacillus, but also for many other contagious diseases. Rats spread lepto-spirosis (through their urine), as well as rabies, toxoplasmosis, trichinosis, and hoof and mouth disease. Mice transfer tularemia, which causes swollen lymph nodes, and skin lesions; they also carry the bacteria of the Breslau enteritis, which often gives the impression of cholera or poisoning, as well as porrigo and tick encephalitis. The "lemming fever" has been well known in Scandinavia for centuries, in conjunction with lemming migrations.

Competition with man for food

Some rodents also destroy plants and vegetation. They thrive on cereals, seeds, vegetables, melons, and other useful plants. Since the end of the Second World War rodents have destroyed more grain in fields in the United States than America has sent abroad to hungry peoples. Even today millions of dollars' worth of food is consumed or spoiled by rodents in silos, warehouses and households. Wood products, paper, leather, linen and textiles, metal tubes, lead cable, electrical insulation, artifacts and many other things are not safe from their sharp teeth.

Humans had to defend themselves time and again from the damage and destruction of food by rodents. The well-known specialist on smaller mammals, K. Zimmermann, writes: "Man has been familiar with mouse plagues since agriculture began. These have been recorded in the Bible, in the ancient scriptures, and in the Middle Ages. As in the time of the pharaohs, so do we today continue to have years in which there are population explosions of mice in countries, when the number of rodents living in the fields is so great that a large quantity of the harvest is lost to them. Humans were noticeably late in studying this natural phenomenon." Charles Elton proved that these population explosions occur not only in agricultural areas, but also in uncultivated regions. These occurrences are not spontaneous but they take place in recurring cycles.

New results of behavioral research

The research of behavioral scientists has begun to give us basic knowledge about many of the biological characteristics of rodents. In 1958, Irenäus Eibl-Eibesfeldt published an extensive compilation of facts on rodent behavior. He discussed the structure of their environment, migration, motor patterns, nutrition, acquisition of food, grooming, social life, juvenile behavior and play, sexual behavior, the sensory faculties, activity, and learning and intelligence. The extensive, sometimes worldwide distribution of many rodent species, their often unbelievable adaptation to the most extreme living conditions; and their survival despite continuous persecution is made possible by characteristics (which are found in only a very small number of other mammals) that seem to be derived from a common, inherited behavioral organization. Many rodents are "specialized in the sense that they are unspecialized." Thus, in contrast to specialized animals, which can survive only in a limited habitat, rodent species can adapt to many living conditions. As very curious animals they actively investigate each new situation and thus learn their way about.

"Specialized in being unspecialized"

Rats and many other rodents can run as well as jump, climb and swim. The strong teeth serve not only as tools which are sharpened continuously through wear against each other (because the hard, resistant enamel is situated only at the front of the incisors, and not on the back), but also serve as weapons. The "harelip," the split upper lip of many rodents, is the result of the incomplete structure of the roof of the mouth cavity, whose development stops at an early stage. This harelip is an extremely useful adaptation for gnawing. Rodents can dig, hold food, cling to the top of cages, or build their nests with their hands. They can raise their young just as successfully in holes which they dig themselves as beneath the tiles of cold storage buildings, between the steel girders of a ship, or in the buildings of polar scientists. It is obvious that these animals are well equipped in the struggle for survival.

Social behavior

Rodents generally have an active and good social life, as can be observed in beavers. Only a few rodents, such as squirrels, hamsters, and dormice are genuine loners. Often the animals form a colony, which consists of several families grouped together, as in the marmots, or they may come together in large tribes or packs, as rats do. Mass migrations are the result of overpopulation or lack of food (which of course may be related problems). We know of the migrations of squirrels, lemmings and voles. On the other hand, when there is a surplus of food, various species like the housemouse, striped field mouse, and harvest mouse, give up their territorial habits by moving together and living as close neighbors with one another. Normally a family establishes its own specific area (territory) which is marked by scent (using urine or gland secretions) and is defended against any strangers from other families, even those of the same species.

Population density

Reproduction

The population density of many smaller rodents is subject to significant cyclic variations. An increase in the population is caused not only by increased reproduction, but also a capacity to live in more crowded conditions up to a point. This capacity can be recognized by changes in territorial behavior, decreased distances between animals, sometimes by the establishment of nest communities by several mothers, and an emigration of weaker males; consequently, the environment can be more efficiently utilized.

This population increase is counteracted by the short life expectancies of individual animals, by a drop in reproduction which occurs with a higher population density, by high winter mortality, damage to the hormonal adaptation mechanisms by stressful environmental influences such as rain, cold, food shortages and mutual disruption of normal behavior resulting from increased competition. According to Frank and Zimmermann, the higher the population density, the more sudden is the resulting population crash through deaths of individual animals. Some recent findings connect population cycles with changes in gene frequencies.

Historical significance in relation to humans

Because man has long been familiar with rodents, many species have acquired particular historical significance in human cultures. Mice, jerboas, rats, dormice, typical squirrels, Old World porcupines and beavers appeared in addition to domestic animals in the art and literature of the ancient Orient; for example, in clay and stone figures, grave artifacts, wall pictures and figures on seals. The North American Indians believed that beavers brought blessings and civilization; the Slavic peoples regarded voles and squirrels as symbols of the activity of forest ghosts; some North American spiritualists believed that mice were the animal forms of the human soul. Rodents were part of the religious beliefs of many peoples during the early classical period, and they were often represented in the fine arts. In Europe, rats and mice were believed to be helpers and allies of witches and wizards. The rodent's place in mythology and in superstition is reflected by the many fables, sagas and proverbs. One thinks of the Homeric story of the *Batrachomyomachia*, the "battle between frogs and mice," Aesop's fables, and the story of the Pied Piper of Hamelin.

Superstitions

Rodents were often used in medicine based on superstitions. The Roman wives often secretly smeared their husbands with mouse dung in order to stop their philandering. Hildegarde von Bingen (1098–1179) believed that mice were a means of curing epileptics. In the English *Pharmacopoeia* of 1667 one could find the following: "A dried and pulverized mouse, when eaten, helps those that cannot hold their urine or those with diabetes..." Even today in many areas of the Near East or in the Balkans, newborn rats or mice in olive oil are considered a universal folk medicine for sick people.

Rodents as fur-bearers

Modern man has found another positive side of the rodents—aside from the interesting studies of their lives and the many rodent species that are kept as pets by children and animal lovers. Muskrats, nutrias, chinchillas, beavers, typical squirrels, hamsters, ground squirrels, marmots and fat dormice are valued for their fur. For this reason, muskrats, nutrias, and beavers have been very successfully transplanted and introduced in the U.S.S.R. They are now distributed over vast areas there.

Finally, rodents today play a vital role as experimental animals in laboratories. As the experimental methods in biology and medicine advanced, so did the increasing need for laboratory animals. One can say without exaggeration, that an animal was involved in every great medical or biological discovery—and in ninety percent of these cases, the animal was a rodent. Initially the rodent, particularly the rat and the guinea pig, was used more as a "passive model" when scientists wished to study the

Rodents in research

life functions and compare them with the human organism. In the 19th century, however, rodents became experimental animals used by biologists, medical doctors and chemists. Rodents as "biological reagents" are a familiar organism, whose living conditions are easily standardized. Man has tested rodents' reactions to a wide variety of treatments: the injection of germs, the administration of medicines and vitamins, radiation, the growth of spontaneous tumors, and many more. Sufficiently large quantities of test animals are exposed to these treatments that scientists can obtain quantitively valid results and can compare these results with others and to evaluate them statistically.

The following figures may indicate the number of animals used for this purpose today. According to Weihe, 510,145 mice, 186,049 rats, and 59,085 guinea pigs were "used up" in Switzerland in 1957–58. In the National Institutes of Health in Bethseda, Maryland, one of the most important institutes of its kind in the U.S.A., some 800,000 mice, 300,000 rats, and 200,000 guinea pigs and hamsters were used there in 1965. Scientific research in microbiology, immunology, pharmacology, hormonology, and biological research on tuberculosis, cancer, nutrition, and radiation biology could not be carried out today without rodents.

Recently, because of the frightening rise in birth deformities among humans, people began to demand that medicine used during pregnancy be tested carefully to ascertain whether it may cause deformities in the fetus. In this case, scientists needed an "animal model" with a placenta that was as similar as possible to that of human mothers, where the fertilized egg was embedded in the mucous membrane in the same way as the human egg is embedded in the mucous membrane of the uterus. Here, too, rodents were used as experimental animals; e.g. guinea pigs, but also golden hamsters, mice and rats. Many tests are invalid because of infections in the experimental animals; more germ-free animals are raised today under rigorously sterile conditions. The majority of animals raised in this manner are rodents.

When an experimental animal is used constantly in large quantities for a certain purpose, it should also retain the characteristics of its breed. Therefore, it is necessary to know the animal's heredity and to maintain stable gene pools. The only way to maintain hereditarily similar animals is by inbreeding. Many such "pure breeds" of mice, rats and other smaller rodents are bred this way; these strains are not hereditarily pure with respect to all possible inherited characteristics; however, they are pure with regard to a few particular important characteristics. As a result of such work, our knowledge of heredity in laboratory rats, which is derived from the wild Norway rat, has increased to such an extent that laboratory rats, next to house mice, have become the most thoroughly studied mammals.

The house mouse was kept in laboratories as a basic experimental animal by geneticists long before researchers had acknowledged how important the use of hereditarily pure animals were for obtaining scientifically valid results of their experiments. Thus, in the 1930s, P. Hertwig, in Germany had already studied the effect of X-rays on male mice and as a result, had obtained mutations in their offspring. These allowed him to study many inherited characteristics. Facts and data essential to the study of genetics have been obtained through experiments with house mice, fruit flies (*Drosophila*), corn, snapdragons, peas, *Neurospora* mushrooms, viruses, and bacteria. The laws of inheritance apply to all creatures, including humans. Thus, the multifaceted picture of the rodents has both its light and dark sides. Some rodent species cause so much damage as carriers of disease, as raiders of fruit and food stores, and as destroyers of many important objects in our environment that we must stop their increase. However, as important and indispensable laboratory animals, rodents have made possible many medical and biological discoveries, without which we would not have risen to the levels of health and well-being enjoyed already by many people in this world. As a food source for many carnivores, and as modifiers of habitats, these animals play a very important role in the balance of nature. As small domestic pets they are often the first animals with which our children come into contact and through which they find joy in animals and living nature.

Evolution, by E. Thenius

Rodents have remained primitive mammals in many of their characteristics; only their skull and teeth are generally highly developed and specifically adapted. The earliest known rodents, the genus + *Paramys* (see Color plate, pp. 214/215) of the early Paleocene (some 60 million years ago) in North America, have the characteristic rodent dentition with the large incisor and molar teeth and the large gap (diastema) between the incisor and molar teeth. The eyetooth is missing. The time of their appearance makes plausible the idea that rodents descended directly from the insectivorous animals, which must be considered the ancestors of primates some 70 million years ago.

Scientists' opinions vary about the origin of the individual families of rodents. One of the essential unanswered questions is that of the origin of the South American rodents, and thus the classification of this related group in the zoological system. S. Schaub traces these animals back to an Old World rodent family that became extinct in the early Oligocene, the Theridomyds (family + Theridomyidae), considered by many scientists as the ancestral group of typical porcupines. A. E. Wood, however, traces them back to Eocene North American Paramyids as we show in our systematics at the back of this book. As a result, both Wood and Patterson classify the South American rodents as a separate suborder of the guinea pigs (Caviomorpha), thus distinguishing them from the typical porcupines (suborder Hystricomorpha).

▷
Pangolins: 1. Indian pangolin (*Manis crassicaudata*), 2. White-bellied tree pangolin (*M. tricuspis*), 3. Malayan pangolin (*M. javanica*), female with young on her tail, 4. Long-tailed tree pangolin (*M. tetradactyla*), 5. Chinese pangolin (*M. pentadactyla*); 6. Giant pangolin (*M. gigantea*), 7. Cape pangolin (*M. temmincki*).

1
2
3
4
5
6
7

10 Sciurids or Squirrel-like Rodents

Suborder: Sciurid rodents, by H.-A. Freye

Of all the families of rodents, the families of mountain beavers or sewellels (Aplodontidae) and typical squirrels (Sciuridae) have retained the greatest number of original characteristics; for this reason we place them at the beginning of this multifarious mammalian order. We include six other rodent families with them in the suborder of SCIURID OR SQUIRREL-LIKE RODENTS (Sciuromorpha). Some members of this group, such as the scaly-tailed squirrels, the springhass, and the gundis do not have a proven evolutionary relationship with the typical squirrels; thus their connection to this suborder is in a way a preliminary classification.

The sciurid rodents are divided into seven superfamilies: 1. Mountain beavers (Aplodontoidea), with the family of mountain beavers or sewellels (Apolodontidae); 2. Typical sciurids (Sciuroidea), with the family of typical squirrels (Sciuridae); 3. Pocket gophers (Geomyoidea), with the families of pocket gophers (Geomyidae) and heteromyid rodents or kangaroo mice (Heteromyidae); 4. Castorids or beaverlike rodents (Castoroidea), with the family of beavers (Castoridae); 5. Gundis (Ctenodactyloidea), with the family of gundis (Ctendactylidae); 6. Scaly-tailed sciurids (Anomaluroidea), with the family of scaly-tailed squirrels (Anomaluridae); 7. Springhaas (Pedetoidea), with the family of springhaas (Pedetidae).

◁
(Above:) When walking, the pangolins put the soles of their feet on the ground. The powerful digging claws are folded inward, the outside part of the paws is the only part put down. Pangolins frequently walk only on their hind feet, keeping their balance with their tail.
(Below): Pangolins can curl up into a ball (Cape pangolin).

The oldest original rodents which still are living, the MOUNTAIN BEAVERS (superfamily Aplodontoidea) have simple, rootless molars. There is only one family, the MOUNTAIN BEAVERS (Aplodontidae) with only one species, the MOUNTAIN BEAVER OR SEWELLEL (*Aplodontia rufa*; BL 33-46 cm, TL 2.5-3 cm, weight 800-1000 g; see Color plate, p. 229), which lives in North America from British Columbia (Canada) to San Francisco Bay (California, U.S.A.) The mountain beaver is the size of a housecat and has a sturdy, rodentlike posture. The hind feet are 4.5 cm long. It has a noticeably short and blunt head, with a convex profile. The eyes are small and the ears are short. The tail is indicated externally only by hair-tufts. The mountain beaver has numerous whiskers, a thick undercoat and long vibrissae. It has an apposable thumb, with a nail-like claw; all of its other claws are long and narrow. All four paws (its "hands" and "feet") are prehensile; the first toe of the hind foot is relatively large. The skull is thick-set and is widely supported in the back. There is no bony process on the back of the eye cavity (a primitive characteristic). The temporal muscles of the chewing musculature are strong and well developed. The mountain beaver is herbivorous. It has twenty-two teeth, arranged: $\frac{1 \cdot 0 \cdot 2 \cdot 3}{1 \cdot 0 \cdot 1 \cdot 3}$. The molars have a thin enamel cover, and are rootless. The ratio of brain to body weight is 1:62. The nose is the preferred sensory organ. As their small eyes suggest, their eyesight is poor. Mountain beavers make extensive structures in the moist earth near water. Females give birth to two to five young per litter.

The mountain beaver looks like a marmot or a beaver; hence its name, "mountain beaver," although it is not related to the real beavers and does not have their flat tail. Mountain beavers live in colonies. Their extensive underground structures are, at least partially, dug deep into the ground, and are almost always on the water's edge. These water-loving animals prefer a diet of juicy water plants; however, they will also eat twigs, leaves and leaf buds from deciduous trees, and conifer needles. These animals are crepuscular, emerging from their burrows only during the morning and evening twilight.

In captivity, the mountain beaver rejects bread, but will eat fruits, cake, and young pine, fir, and oak sprouts, as well as oak leaves and conifer needles. When eating, it often assumes a squirrel-like sitting position and holds the food with its hands. Like many other rodents, it does not drink. On the other hand, it likes to take baths. It bathes by sitting on its hind legs, dipping its hands into the water, and then washing its whole body from back to front. The long whiskers are also carefully washed. The droppings are oval, 10-12 cm long, and usually solid. The animals usually defecate at fixed dung sites.

When danger threatens, the mountain beaver quickly throws itself on its back. When defending itself, it lies motionlessly on its back and extends its feet, which are good defense weapons because of their long, sharp claws. The animals run very fast, keeping their bodies close to the ground. They are good swimmers, and can also climb into bushes and trees; they can even move by brachiating (swinging from branch to branch). They sleep on their stomachs or their backs during the day; their sleep is often so deep that it is possible to catch a slumbering mountain beaver by hand. In the past, natural enemies like the mink, prairie wolf, and lynx regulated the number of mountain beavers. However, the decrease in these predators due to human activities has allowed the mountain beaver to multiply to such an extent that today it has become a forest pest.

Very little is known about the behavior of these rare and ancient rodents. For a few days the brain anatomist G. Pilleri was able to keep and

Fig. 10-1. Mountain beavers or sewellels (*Aplodontia rufa*).

study a female weighing approximately 400 g. According to his reports, the animal was active mainly at night, although it was also lively for short periods during the day. In warm, humid weather, the animal generally slept on its back; more rarely, it slept on its stomach. The mountain beaver was very active when awake. Immediately after being freed from its box, it chose a dung corner, which it visited regularly; thereafter, the transport box was used only as a sleeping place. When living in the wild, these animals do not defecate within their dens. When defecating, the mountain beaver showed a unique behavior; it sat on its hind legs in front of the dung pile, bending as far forward as it could. After defecating, it took the single dung balls in its mouth, and "spit" them on the heap. Food and other objects were marked only during the the first few hours, by a brief touching with the anal area or by urinating on the particular object. The mountain beaver shows a marked preference for water; it bathes very often, especially after meals, and washes itself carefully. Pilleri's mountain beaver was quite sensitive to light, and rapidly became "drowsy" when in bright light. When walking, it detected obstacles in its path through smell, and did not avoid them until the last moment.

In females, the egg release takes place spontaneously; during the breeding period the Fallopian tube releases fertilizable eggs, independent of copulation. More embryos are always found in the left horn of the uterus than in the right one. The reproduction of the mountain beaver is reminiscent of that of typical squirrels.

Evolution of the mountain beavers, by E. Thenius

Mountain beavers are descended from the Paramyids of the Eocene (some 50 million years ago). They are also related to the extinct genus +*Paramys*. This rodent group once had numerous species; many lived in Europe and Asia during the Tertiary period. The Mylagaulides (+Mylagaulidae), which became extinct during the Upper Tertiary period, form another family which is closely related to the mountain beavers. One member of this family, the giant +Epigaulus (L 60 cm; see Color plate, pp. 214/215), from the Pliocene period (some 10 million years ago) in North America, had two bony horns on the back of its nose.

Superfamily: Typical squirrels

The largest number of squirrel-like rodents belong to the superfamily of SCIURID RODENTS OR TYPICAL SQUIRRELS (Sciuroidea) which has only one family, the TYPICAL SQUIRRELS, TREE SQUIRRELS, OR SCIURIDS (Sciuridae; BL 10-70 cm, TL 3-60 cm, weight up to 6 kg). These animals range in size from that of a mouse to that of a rabbit. Their tails may be either long or short. The body is generally cylindrical and slenderly built. The animals have large, protruding eyes. The hair can be soft or very soft, and it is often long. The tail is generally thick, and in many species it is covered with bushy hair. The muzzle is short and bald. The upper lip is split; the nostrils are divided by a furrow. The fore and hind legs of flying squirrels are connected by a skin fold (flightskin). The thumbs are usually underdeveloped; however, they often have an unusually

Family: Typical squirrels, by H.-A. Freye

Distinguishing characteristics

large claw (prairie dogs) or flat nail (marmots). All of the other toes have sharp claws. There are tactile bristles on the hands. The palms of the hands and the soles of the feet of these plantigrade animals have soft pads. The skull has a wide brow and a broad, bony palate. The lower jaw is relatively high. Members of this group have large, bony bullae whose development is important in classification. The cheek bones are long; the shoulder blades have a shorter, broader position. The shin bone and thigh bone are not fused. Arboreal red squirrels have a longer fourth finger and toe.

These animals are mainly herbivorous. They have twenty-two teeth, arranged: $\frac{1 \cdot 0 \cdot 2 \cdot 3}{1 \cdot 0 \cdot 1 \cdot 3}$. One upper front molar may be lacking. The incisors are chisel-shaped and are close together. The molars have cusps or abrasive chewing surfaces. They are square and have roots, as well as shallow enamel furrows which begin on the side of the teeth. The lower halves of the jaw are quite movable. Some genera have cheek pouches. The palate is covered with numerous sharply defined furrows. The tongue has three large papillae. The stomach has one cavity; the upper end of the colon is not connected with the appendix. The left lung does not have lobes. Many species, such as the prairie dogs, have a loud voice. The eyes are well developed, and the animals have particularly sharp vision. The retina in red squirrels, marmots, and prairie dogs has a resolution comparable to the visual acuity usually found only in the foveal region in other mammals; these animals, then, do not use the fixating movements so necessary to animals without visual acuity distributed throughout the retina. The penis and clitoris are split or asymmetrical. Males have a penis bone and a scrotum. These animals generally remain close to their nests. There are two subfamilies; ground and tree squirrels (Sciurinae), and flying squirrels (Pteromyinae), with forty-four genera, 380 species, and 1350 subspecies. Fossil remains have been found for only a few genera; however, not one of these is extinct. These animals are found in Europe, Asia, Africa, and North and South America; they are not present, however, in Madagascar, Australia, or Polynesia.

This large family takes in such varied forms as the marmot, the ground squirrel, the prairie dog, the lined ground squirrels and chipmunks, the American red squirrels, and the flying squirrel. According to Thenius and Hofer, these animals have "retained the greatest amount of their primitive characteristics, more so than other rodents extant today." The typical squirrels, then, are still relatively unspecialized, and are not very far removed from their original rodent forms. According to Irenäus Eibl-Eibesfeldt, all the typical squirrels which he tested (ground squirrels, marmots, prairie dogs, red squirrels, and southern flying squirrels) showed a common display, "the perpendicular twitching of the tail."

Subfamily: Ground and tree squirrels

The GROUND AND TREE SQUIRRELS (subfamily Sciurinae) are typical squirrels without flightskins between their fore and hind feet. Most of the family characteristics mentioned above essentially describe these animals. These squirrels, unlike the flying squirrels, are predominantly active during the day. The systematic classification of these animals is difficult; however, using Joseph Curtis Moore's information, we can classify them according to a series of species cranial characteristics, like the interior structure of the bony bulla tympani. Thus we distinguish eight groups of genera: the northern ground squirrels (Marmotini); the bristly ground squirrels (Xerini); the red squirrels (Tamiasciurini); the tree squirrels (Sciurini); the Asian squirrels (Callosciurini); the Indo-Malayan giant squirrels (Ratufini); and the oil palm squirrels (Protoxerini).

The northern ground squirrels

The range of the MARMOTS AND GROUND SQUIRREL (tribe or genus group Marmotini) includes the entire northern hemisphere, across the Arctic Circle, through the tropics to the southern ranges of the temperate zones. One may be surprised, perhaps, by the fact that zoologists include the prairie dog (which we find in our Indian legends), as well as the marmot and the ground squirrel, in this group. However, closer observation of these rodents reveals unmistakable similarities in body structure and behavior.

Marmots

Distinguishing characteristics

The MARMOTS (*Marmota*; BL 45-80 cm, TL 11-16 cm) have a plump, sturdy body which, in appearance, is not very similar to that of the slender, agile red squirrels. The head is broad and round, with clearly visible ears approximately 3-4 cm in length. The tail is short, with bushy hairs; the legs are also short. The tip of the thumb has a small nail, while the other digits (fingers and toes) have strong digging claws; the third finger is the longest. The hair is long and thick. The cheek pouches are underdeveloped. The upper molars have a somewhat more triangular circumference. Thirteen species and subspecies are found in mountain and prairie regions.

1. The ALPINE MARMOT (*Marmota marmota*; BL 53-73 cm, TL 13-16 cm, body height 18 cm, weight 4-8 kg; see Color plates, pp. 216 and 239). Its weight varies according to the season. The nasal bone is tapered toward the back; there is a circular hole in the back of the head. The forward, upper rims of the temporal bone diverge. The only existing part of the thumb skeleton is the metacarpal bone, which has an amorphous structure. The pelvic bone in adult animals consists of four vertebrae. The front incisors are brownish-yellow; in young animals, they are whitish. The first molar of the upper jaw is only half as large as the second one; the cross sections of the front molar and the molar of the lower jaw have a rhombic shape, increasing in size from front to back. The appendix is situated in the middle of the abdominal cavity.

2. The BOBAC MARMOT (*M. bobak*; see Color plate, p. 239) is heavier and larger, with a shorter tail (about one-fourth of the body length). The snout is also somewhat shorter. The skin color is brighter, and the front incisors are white.

3. The LONG-TAILED MARMOT (*M. caudata*) has a particularly long tail

Fig. 10-2. Alpine marmot (*Marmota marmota*).

(about forty-five percent as long as the body). It is the largest member of the genus. The skin is almost reddish; one-third of the tail is black. The AFGHAN MARMOT (*M. c. dichrous*), a subspecies found in Afghanistan, is light brown, with a lively reddish color on its lower side.

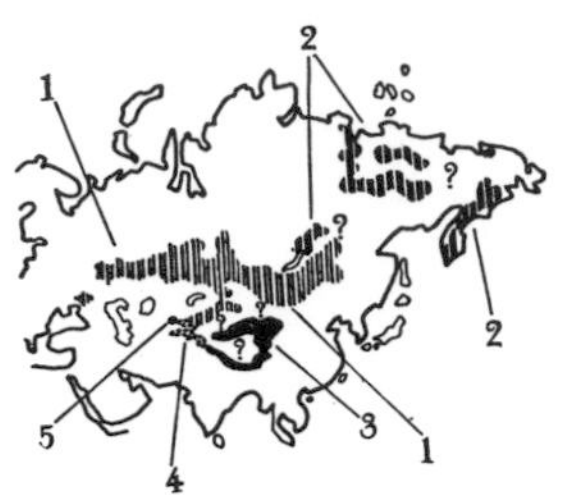

Fig. 10-3. 1. Bobac marmot (*Marmota bobak*); 2. Kamchatkan marmot (*M. caligata camtschatica*); 3. Himalayan marmot (*M. himalayana*); 4. Long-tailed marmot (*M. caudata*); 5. Tian Shan marmot (*M. menzbieri*).

4. The HOARY MARMOT (*M. caligata*; BL 41-53 cm, TL 18-30 cm; see Color plate, p. 216) has a bushy tail. Females have ten nipples. It is a mountain-dweller, preferring a habitat similar to that favored by the alpine marmot. There are a variety of subspecies, including the somewhat smaller, dark-brown *M. c. vancouverensis*, found on Vancouver Island (British Columbia, Canada) and the KAMCHATKAN MARMOT (*M. c. camtschatica*), found in the Verkhoyansk Mountains in northeastern Siberia.

5. The YELLOW-BELLIED MARMOT (*M. flaviventris*; BL 36-48 cm, TL 11-23 cm, weight 2.3-4.5 kg). The ears are relatively small; the head is dark. The lower snout is white; there is also a white area between the eyes. The back and sides are yellow-brown; the belly is yellowish. The sides of the neck have distinct, leather-yellow spots; the paws range in color from fawn to dark brown; they are never black.

6. The NORTH AMERICAN WOODCHUCK (*M. monax*; BL 40-51 cm, TL 10-18 cm, weight 2-4 kg; see Color plate, p. 239) has short feet. Its whiskers look as if they had been covered with hoar frost. The skull is broad and the nasal bone is long. A short thumb (8-11 mm) with a dull claw protrudes from the fore paw. The females have eight nipples.

The alpine marmot, by H.-A. Freye and B. Grzimek

The ALPINE MARMOT (*M. marmota*) has been known to man since ancient times. Hunters refer to the males as "bears" and the females as "cats." The Roman naturalist Pliny referred to it as the "alpine mouse" (*Mus alpinus*), because "it lives in holes and squeaks like a mouse." In 1909, Keller pointed out the fact that the French word *marmotte* actually means "mountain mouse." "Even in the 1800s," as J. Niethammer pointed out in his history of the "trained marmot," "the small marmotte was used to entertain in the annual trade fairs, along with camels, monkeys, and dancing bears. It traveled around with its master, usually a boy, from Savoy or Tyrol through Europe, dancing clownishly to the flute its master played." According to one of Pliny's stories, "the marmot remains in its hole, which it has filled in advance with food stores, for the entire winter. In order to fill the hole with food, it is said, one marmot lies on its back and is covered with hay. It holds onto the load, and lets itself be dragged into the hole by another marmot, which grips the loaded marmot tail with its teeth. The marmots then change positions, and that is why their backsides are so worn down." This imaginative explanation naturally is not supported by recent research; however, there are marmots whose back hair has been completely worn as a result of slipping in and out of their burrows.

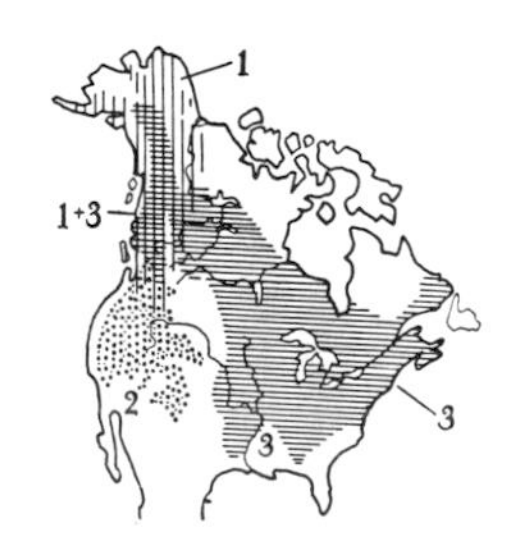

Fig. 10-4. 1. Hoary marmot (*Marmota caligata*), New World distribution; 2. Yellow-bellied marmot (*M. flaviventris*); 3. Woodchuck (*M. monax*).

In the Ziller Valley (Austria), as Tratz reports, melanistic (all-black colored) marmots (a then-unknown mutation), were sighted in the fall of 1963. Presumably, these animals were not particularly well adapted for survival, as they had a noticeable body twitch. All marmots have glands in the anal area; these give off a sharp secretion when the animal is agitated or when it is defending itself. The glands in the cheek bones are used for marking their habitat.

Marmot fat as medicine

Marmots may be rather fat when they go into hibernation, as a result of the large amounts of food they have consumed during the summer. Even today, one home remedy in alpine countries includes "Mankei fat" (as the marmot fat is called) as an important ingredient. Svolba found that the fat of a recently killed marmot differs noticeably, in appearance and consistency, from the fat of other free-living animals. When the fat is cooled to room temperature after being melted, it turns into a clear, transparent, non-viscous oil that soon separates into two layers; the one on the bottom is fairly rigid and fine-grained; the upper layer is clear and oily. Chemically, marmot fat consists of at least four different fatty acids, in addition to linolenic acid, a high iodine content, and vitamin D in more than trace amounts. A single liter of marmot oil costs 30 DM in West Germany.

People living in the Alps, as well as hunters in the mountains, are completely convinced of the healing powers of marmot fat. They use it especially as a remedy for chest and lung diseases. Some people came to the conclusion that marmot fat must be a good medicine for rheumatism because the marmot, in its damp hole, is not affected by arthritis even during winter. Such reasoning appeals particularly to uneducated people, and, as a result, marmot fat was much in demand as an expensive remedy for rheumatism. When a pharmacist from Schaffhausen (Switzerland) began to advertise marmot fat in all the newspapers, the marmot population began to diminish. The price of marmot fat rose, and in 1944, 16,000 animals were killed in Switzerland. However, it must be said in the pharmacist's defense that indignant naturalists later succeeded in convincing him of his mistake, and he terminated the venture which had raised such false hopes in so many sufferers from rheumatism and had caused them to squander their money needlessly.

Movement

The marmot has a waddling gait; its belly often touches the ground when it moves. However, it climbs and jumps with ease in its mountain habitat. When an animal wanted to climb up the rather steep side of a rock, reports H. Münch from observations in the Montblanc Alps, it gripped the irregularities of the rock with its fingers, and pushed its body up with the help of its feet. Its movement was not at all awkward, and it climbed crevices with an agility one would not have expected of a marmot. The ability to stretch not only its legs, but its whole body as well, was astonishing. However, the anatomy of its limbs explains the marmot's climbing skill. In addition to climbing, waddling, and hopping, marmots are also able to gallop. This is clearly indicated by their tracks in the snow. When galloping, the marmot often places its hind legs in front of its fore legs. Marmots can also sit erect because of their relatively short fore legs and their extremely agile hind legs.

Diet

The marmot's diet consists of roots, herbs, and grasses. From August on, the animals bite off grass and plant stalks in large quantities, and lay them out to dry; they "harvest" hay in this manner. Using their mouths, they line their nests with hay, which will be used as insulation and as winter food. The animals often sit on their hind legs when eating; they sit up and hold the plant pieces with their fore paws. They are able to find food even when there is snow on the ground; they scratch the snow away from a particular area and eat the plants beneath. They then enlarge the area, using their broad heads; in the process, they make large snow towers in front of their faces and brows. Because they shovel snow in all directions, the feeding area gradually becomes larger and larger. Their need for water is apparently satisfied solely from liquid in their food.

Vocalizations

Marmots are known for their shrill, squeaking vocalizations. When calling, they stand erect, letting their fore paws hang down. This squeaking is not only a warning call; the animal will squeak continuously when it has been disturbed. Marmots also squeak to communicate, and in the mating season their squeak is used as a "battle cry." The squeaking is a vocalization that comes from deep in the throat and is uttered with an open mouth. A tame marmot once squeaked continuously for thirty seconds when it caught its leg between the bars of its cage. When it heard the whirr of a vacuum cleaner for the first time, it squeaked for several minutes. According to the reports of Müller-Using, marmots produce, in addition to their squeaking, a wailing sound that is similar to that of young rabbits.

When a marmot utters the shrill warning squeak, all nearby marmots immediately dash into their holes. In isolated alpine areas they disappear into their holes when people approach within as much as 200 m. Hours pass before the animals cautiously reappear. Now, however, this practice has changed in many areas. Traveling south from Engadin (Switzerland), one can see an unusual triangular warning sign at the side of the road; it has a silhouette of a standing marmot, surrounded and enclosed by a red border. The marmots have a colony close to the side of the road; one can come as close as 2-5 m before the animals are alarmed. Only reluctantly, and almost unwillingly do they retreat into their burrows, where they turn around immediately and peer out. Like many other animals, they are not at all shy of cars, people, and other aspects of human civilization; however, they do not like to be hunted down and killed. Whenever one is friendly to them, they become trusting. The warning call of the marmot, incidentally, is also understood by the chamois, an antelope (see Vol. XIII).

Brain and sensory faculties

The brain of the marmot is of relatively simple structure, although its surface is convoluted. While, for example, a monkey of about the same weight (e.g., a rhesus) has a brain of about 6 g, the brain of a marmot weighs only about 3.6 g. Adolf Portmann has calculated the weight ratios of various brain parts, and has found that the marmot has a ratio of 4:3

between the cortex and the rest of the brain. Therefore, the marmot ranks far below bears and cats, and even beneath the hyraxes. However, it is above the armadillos. Naturally, such ratios are not an absolute measure of intelligence; they do indicate, however, something about the number of neurons in the higher parts of the brain, and thus, something about the importance of the activity centers of the brain, which control the behavior of an animal in relation to its environment. The eye is highly sensitive; marmots can view their surroundings without being betrayed by their eye movements. When out in the open in their territory, they rely predominantly on their sense of vision. Their sense of hearing is very well developed, although the ears are relatively short. However, the sense of smell seems to be rather poorly developed.

Marking with odor

Alpine marmots mark their territories with scents, using substances secreted by their cheek glands, and with sounds, by squeaking. Marmots trespassing on one another's territory are snapped at until they move away. Münch describes a marmot trespassing on another's territory as the result of being disturbed by people: "The disturbance had long since ceased; however, the marmot still showed great agitation. It did not use the well-trodden path which the marmots living in the area had made. Instead, it walked mostly at the side of the path, and made several cautious pauses. The whole demeanor of the animal showed that it was aware (by smell?) of the fact that it was trespassing. Soon it met two inhabitants of the colony. After a short face-off, the other animals made a thorough olfactory investigation of the intruder's anal zone. The intruder, in turn, adopted a submissive posture toward the two other animals. It put its tail between its legs, lowered its head, and bent down to a noticeable extent. Both inhabitants of the colony raised themselves into an upright sitting position and, with a gnashing of teeth, they began to make the marking calls typical of that region. The intruder retreated slowly in the direction from which it had come; the other marmots followed it with their eyes, while they remained behind, calling out loudly for several minutes more."

Behavior

Man has always been fond of the "murmelis," perhaps because they are able to sit up and, therefore, look like funny little people, or perhaps because they like to play so much. They play tag with one another, and roll down mountain slopes; they stand opposite each other, tilt their heads, and attack one another with their incisors, with such energy that one can hear the impact. While playing, they squeak "happily." They like "boxing"; one animal faces the other and presses its flat "hand" against the opponent's throat area or against its arms in such a way that the opponent can not catch its attacker. According to Müller-Using, these social animals live together in colonies ranging from two or three to fifty or more. We know relatively little about their social lfe, although their family life has often been mentioned in articles. In areas with a large population of marmots, they may form large colonies which live in an immense burrow system.

One expression of the marmot's social behavior is its very obvious need for body contact. Animals of several generations often lie closely together for hours in their favorite resting spots. Mothers and young are particularly tender toward each other. Münch observed "a marmot mother and her four young which were playing in front of a den. Suddenly one of the young (about four weeks old) moved away from the others and ran quickly to its mother, who sat in the upright position. The young cuddled close to its mother's body, pushing its own body up so that it, too, was sitting. Then the young animal tried to bring its face up to that of its mother. This became possible when she lowered her head. An ardent touching and hugging of the faces, more specifically of the snouts, followed." After a brief moment, the youngster stopped this hugging, ran away from its mother, and rejoined the game with its siblings.

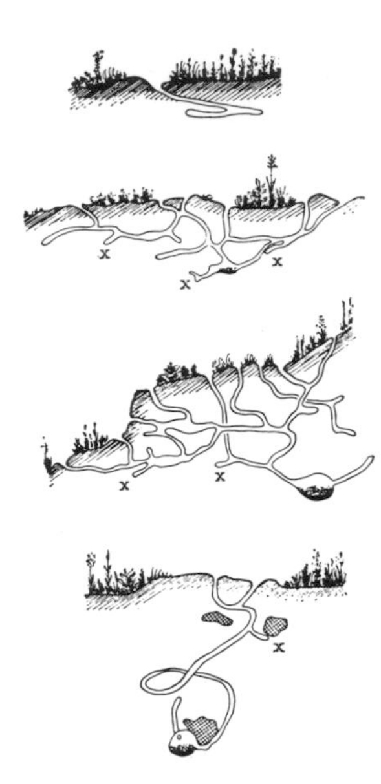

Fig. 10-5. The structure of an alpine marmot's burrow (from top to bottom): Escape tunnel or emergency shelter; summer quarters; permanent quarters (summer and winter structures); winter quarters. The Xs mark the dead-end tunnels which are used as place to deposit feces. The stones are put in the nest to help the drainage.

The true ALPINE MARMOT or EUROPEAN MARMOT (*M. m. marmota*) lives in the European Alps and the Carpathain Mountains. It has been restocked and resettled by man in areas where it had become extinct as a result of human persecution. In Germany, the original alpine marmot colonies are those near Berchtesgaden and in the western Allgäu. All the other colonies (as elsewhere in the Alps) have been restocked since 1880. Müller-Using believes that at least one-tenth of their present range in the Alps, as well as ten percent of the total marmot population (which he estimates at 50,000 to 100,000 animals), can be traced back to reintroduction efforts. Thus the colony near Hohenaschau in the Chiemgau, southeast of Munich, comes from eight animals which were brought from Berchtesgaden in 1887. All the alpine marmots in the Steiermark, Kärnten, lower Austria, Salzburg, Tyrol, and Vorarlberg areas are descended from animals that were resettled in these regions at the end of the 19th Century. Marmots have been successfully resettled in some cantons of Switzerland, in Yugoslavia, the lower Tatra (Czechoslovakia), the Pyrenees (France, Spain, and Andorra), and in the U.S.S.R. In 1954 and 1957, some animals were resettled on Feld Mountain in Germany's Black Forest; their population has increased substantially since then. In 1961, marmots were resettled in the Swabian Alps near Balingen (Germany). It has been found to be genuinely worthwhile to re-establish these animals; they settle down easily everywhere, even in the flat lands, and, so far, no one has been able to show that they cause damage.

According to Müller-Using, the habitat of the alpine marmot can be divided into three parts: valley settlements between the altitudes of 900 and 1200 m, where the species often appears in the lower areas of valley forests or grassy slopes; higher grassy slopes with wide plateaus, between the altitudes of 1200 and 1800 m, as well as in valleys of higher altitudes (their preferred habitats, along with the mountain valleys at even higher altitudes); and boulder regions at altitudes between 1800 and 2200 m, with mountainous ridges, boulder-glaciers, and the like. Marmots enjoy

Fig. 10-6. Before hibernation, the alpine marmot (*Marmota marmota*) brings large quantities of grass and other materials into its burrow, to use in lining and insulating its nest.

warm, sunny southern slopes with a wide view, close to the snow line. In summer the burrows are used by single pairs with their young; these dens are under roots, in the ground near the forest edge, or merely between boulders. The yearlings build their own burrows for the short summer season, which is just barely four months long.

In the fall, when all the alpine marmots have built up their body fat for the winter, they often migrate, according to Baumann, to lower altitudes, to the zone of higher alpine pastures, or even lower, below the upper tree zone. There they reoccupy their old winter burrows or dig new ones, which under favorable conditions are not far from their summer quarters. The entrance, which is generally the size of a human fist, leads into a narrow, enlarged "hallway," often over 1m long, which usually divides into two passages further back. On one side is a short passage where the earth has been dug away and used as camouflage for the entrance. On the other side is a long tunnel which may run continuously for 8-10 m before leading into a roomy rounded den, which is well lined with dried grass. All members of the family, which comes together again in the fall, help line the den. In relation to the entrance tunnel, the den is always located more in the direction of the mountains, and it may be as much as 1.5-3 m below the ground level, depending on the altitude.

Hibernation

Before they begin their hibernation, marmots plug up the entrance of their hole from the inside with hay, earth, and stones; these materials extend inward for about 1-2 m. The entire marmot family, which may often include as many as fifteen animals, remains in the den during the winter, huddled closely together in a deep sleep, with their noses between their hind legs. Hibernation lasts six months or more. The animals wake up every three to four weeks and defecate and urinate. Thus, the thinness of the animals as they leave their holes in spring is easily understandable.

In the spring, the family disperses again as the animals reoccupy their summer territories. Further observations are needed, however, before we can decide whether marmots actually do have both summer and winter burrows, or whether the summer colonies are merely gradually extended over the years, interrupted by the onset of winter, when the family draws together once again.

The duration of their hibernation depends on the altitude of the winter homes; it is shorter in valley settlements than in the higher locations. The temperature in these dens is rarely below freezing. The marmot is a true hibernator, which means that while it sleeps during the cold season, all of its metabolic functions are greatly reduced. When this happens, the oxygen use decreases from an average of 600 cc to 30.5 cc per kilogram of body weight. The breathing rate remains constant at two or three breaths per minute. The body temperature drops to between 4.6° and 7.6° C; the number of heartbeats decreases from between 88 and 140 per minute, to three or four. In the spring, when the animals awaken, they busy themselves immediately with spring cleaning; they throw out hay, nest-lining, and wastes. The mating season begins only a few days after the end of hibernation.

Mating season

The anal glands (mentioned earlier) also produce odor substances which stimulate sexual activity in both sexes during the mating season. These glands are located in three main pouches, connected with smaller pouches near the rectum. The openings of these glands can be extruded, like papillae, from the animal's anal opening. We do not know very much about the sexual and family life of these animals, because most of these activities take place underground. Mating occurs in May, while the animals are still in their winter burrows. The foreplay consists of mutual embraces and a meowing call; sometimes there are mock battles in which the sexual partners face each other, beat one another with their feet, and grind their incisors. They assume their "threat posture" and often attack each other with their teeth. Females that are not ready to mate throw themselves on their backs. Families consist of males and females, and the males are tolerant of one another even during the mating season. According to Müller-Using's observations, the males do not even compete for the favors of a female in heat. After mating, the males probably leave the females alone for some time.

▷ Red squirrel (*Sciurus vulgaris*, top pictures); Young red squirrel (bottom right); African ground squirrel (*Xerus rutilis*, bottom left).

▷▷ Evolution of the rodents:

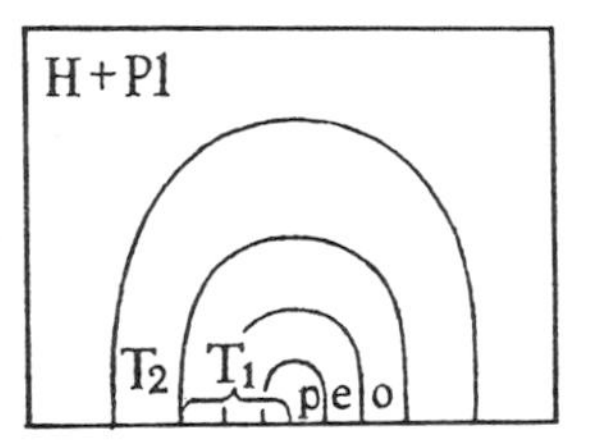

Earth periods: T_1-Ancient Tertiary (p-Paleocene, e-Eocene, o-Oligocene); T_2-Lower Tertiary; Pl-Ice Age (Pleistocene); H-Modern era (Holocene). Rodent families extant today: 1. Scaly-tailed squirrels (Anomaluridae); 2. Mountain beavers (Aplodontidae); 3. Springhaas (Pedetidae); 4. Typical squirrels (Sciuridae). (Continued on page 217.)

Pregnancy and birth

The gestation lasts approximately five weeks; the female gives birth to from two to seven blind, naked young, each weighing about 30 g. As Hans Psenner observed in his successful breeding groups in the Innsbruck Alpine Zoo, in Austria, the mother-to-be closes off her living quarters with hay several days before she gives birth. After birth, the young are always covered with hay whenever the mother leaves them for a long period of time. The young are already able to vocalize on their first day of life. After some time, the mother nurses them while lying on her back; the young bunch together around her belly. After the twentieth to twenty-fifth day of life, the eyes of the young marmots open gradually, and their incisors begin to appear. After approximately thirty-nine days, the young leave the burrow for the first time. They are fully grown at two years. Their lifespan is fifteen to eighteen years.

Raising of young

The young of one litter hibernate and live together in their parents' burrow for the next summer. The other members of the sleeping colony move out temporarily during this time.

Young marmots play a lot. They tumble about and roll down slopes (either on purpose or through clumsiness). Some members of the family always "stand guard" while the young play or while the young and adults eat and nap.

Predators

The marmot's chief predators are the golden eagle and the fox; young animals are often hunted by eagle owls and ravens, and, more rarely, by the pine marten. The marmot, however, is rather aggressive and is not afraid to fight with foxes or large dogs. A few years ago, a hunter observed a fox attempting to kill a marmot. The marmot defended itself so vigorously that at the end of the fight both animals were dead.

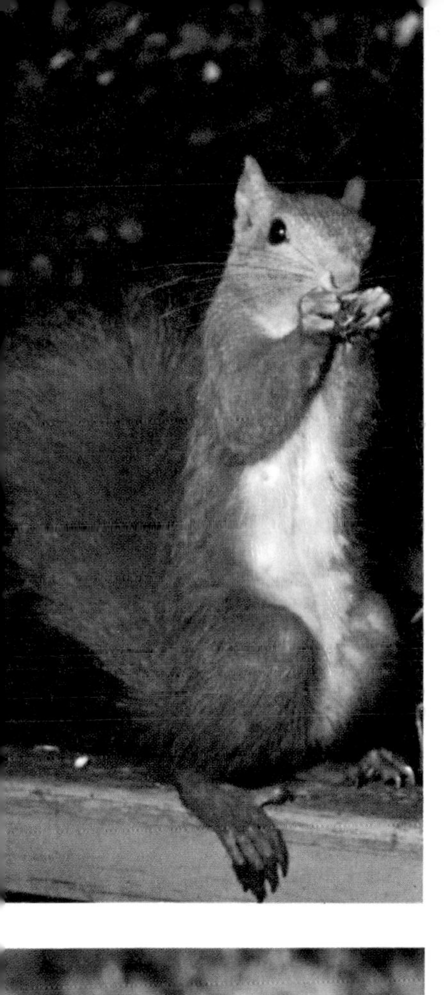

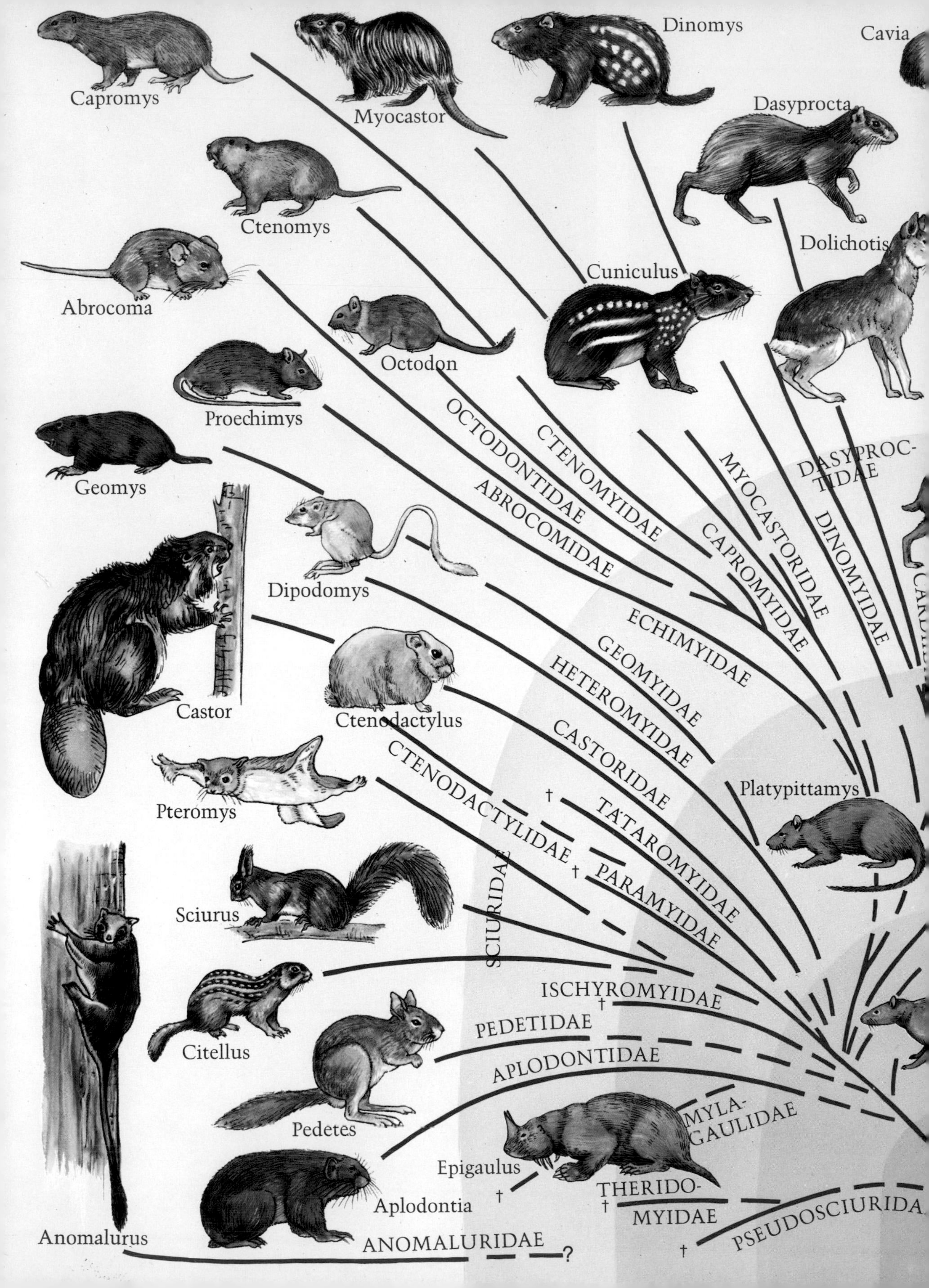

Capromys
Myocastor
Dinomys
Cavia
Dasyprocta
Ctenomys
Dolichotis
Abrocoma
Cuniculus
Octodon
Proechimys
Geomys
Dipodomys
Castor
Ctenodactylus
Pteromys
Platypittamys
Sciurus
Citellus
Pedetes
Epigaulus
Aplodontia
Anomalurus
OCTODONTIDAE
CTENOMYIDAE
ABROCOMIDAE
MYOCASTORIDAE
CAPROMYIDAE
DASYPROC-TIDAE
DINOMYIDAE
ECHIMYIDAE
GEOMYIDAE
HETEROMYIDAE
CASTORIDAE
† TATAROMYIDAE
CTENODACTYLIDAE
† PARAMYIDAE
SCIURIDAE
ISCHYROMYIDAE
†
PEDETIDAE
APLODONTIDAE
MYLA-GAULIDAE
†
† THERIDO-MYIDAE
† PSEUDOSCIURIDA
ANOMALURIDAE
?

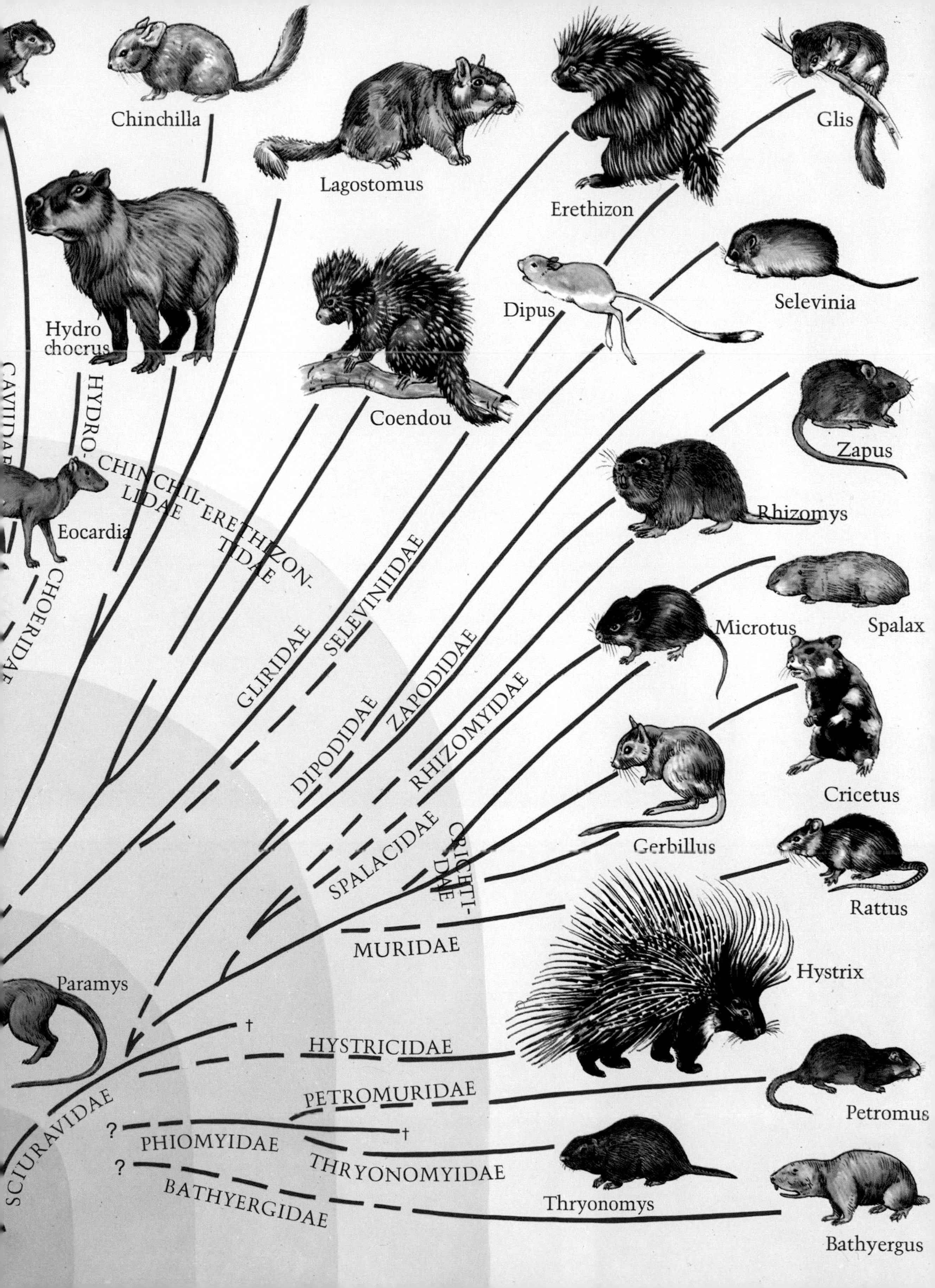

Chinchilla
Lagostomus
Erethizon
Glis
Hydro chocrus
Coendou
Dipus
Selevinia
Zapus
Rhizomys
Eocardia
Microtus
Spalax
Cricetus
Gerbillus
Rattus
Hystrix
Paramys
Petromus
Thryonomys
Bathyergus
HYDRO-
CHINCHIL-
LIDAE
ERETHIZON-
TIDAE
CHOERIDAE
GLIRIDAE
SELEVINIIDAE
DIPODIDAE
ZAPODIDAE
RHIZOMYIDAE
SPALACIDAE
CRICETI-
DAE
MURIDAE
HYSTRICIDAE
PETROMURIDAE
PHIOMYIDAE
THRYONOMYIDAE
BATHYERGIDAE
SCIURAVIDAE

Marmots in captivity

◁◁
(Continued from page 212.)
5. Gundis (Ctenodactylidae); 6. Beavers (Castoridae); 7. Heteromyid rodents or kangaroo mice (Heteromyidae); 8. Pocket gophers (Geomyidae); 9. Spiny rats (Echimyidae); 10. Chinchilla rats (Abrocomidae); 11. Octodont rodents (Octodontidae); 12. Tucotucos (Ctenomyidae); 13. Hutias (Capromyidae); 14. Coypus (Myocastoridae); 15. Pacaranas (Dinomyidae); 16. Agoutis (Dasyproctidae); 17. Cavies (Caviidae); 18. Capybaras (Hydrochoeridae); 19. Chinchillas (Chinchillidae); 20. New World porcupines (Erethizontidae); 21. Dormice (Gliridae); 22. Betpakdala dormice (Seleviniidae); 23. Jerboas (Dipodidae); 24. Jumping mice (Zapodidae); 25. Bamboo rats (Rhizomyidae); 26. Palearctic mole rats (Spalacidae); 27. Cricetid rodents (Cricetidae); 28. Murid rodents (Muridae); 29. Old World porcupines (Hystricidae); 30. Rock rats (Petromuridae); 31. African cane rats (Thryonomoyidae); 32. Mole rats (Bathyergidae). All other rodent families are extinct.

Because of their fascinating way of life, and their amusing behavior, marmots have long been kept in captivity. In Hellbrun, near Salzburg, Austria, there once was a "Mankei park" with about 100 animals that were probably kept to furnish oil for the pharmacy of the prince-bishop. In 1887, Girtanner bred marmots in his garden in St. Gallen (Switzerland). More recently, Hans Psenner dedicated himself to keeping and breeding marmots in the Innsbruck Alpine Zoo. In 1955 he was successful for the first time in breeding the animals. By 1970, five generations had been raised. His newly built marmot house, with outside quarters as well as indoor facilities, may well be the most modern structure ever built in any zoo for these fascinating rodents. The animals are protected by glass, through which one can watch them building their burrows. It is a singular opportunity to observe the subterranean life of the marmot; and, if one is lucky, one might even see a nursing mother with her young.

In zoos, marmots frequently live more or less free, much against the intentions of their keepers. All safety measures notwithstanding, marmots time and again will bite through the wires of their cage or they will dig tunnels under the walls of their outdoor premises. For many years, the marmots in the Frankfurt Zoo lived more outside their cages than in them. These animals dug tunnels underneath the walls surrounding their outside cage. The keepers and visitors to the zoo were not particularly annoyed by this; however, the marmots were a nuisance to dogs. When a dog passed, the male marmot rushed at it, jumped at its throat, and tried to hang on with its teeth. Zoo officials were worried about the possibility of a marmot digging one of its tunnels into a cage where the foxes were kept. Then either the foxes would kill the marmot and escape through the tunnel, or the marmot would kill the foxes. Once marmots have settled down, however, and have built their burrows, they no longer feel the need to run away or to make long reconnaissance trips. In the mountains a family of marmots "works" only 0.25 hectare (about 0.6 acre) of land. If one puts food in front of a marmot's hole, however, so that it does not have to cut grass, it becomes even less inclined to move about.

A marmot in a house

Otto Koenig, a behavior scientist, once chased a pine marten away from a small, thirty-five-day-old marmot in Vienna. He then took the marmot home with him. Initially, the animal walked freely around the house and was sociable to everyone it met. When it was seven months old, however, it gradually began to become hostile to unfamiliar people. Finally it recognized only the Koenigs and their large shepherd dog as true members of its group. It nuzzled its friends and invited them to play biting and snatching games. When a stranger approached, the marmot shook and beat its tail back and forth with the black hairs on end. It gnashed its teeth, stood erect, clung to the leg of the stranger, and tried to bite him.

◁
Alpine marmot (*Marmota marmota*, top and bottom right); Hoary marmot (*Marmota caligata*, bottom left).

The marmot became even more active the next spring. It climbed on all the furniture, nibbling at the sides; it bit buttons from cushions and clothes and turned over the coal basket daily. It collected socks, garments, and bed covers with which to line its living crate, and tore at bath towels hanging on the wall until they fell. In addition, it ate the house plants. It did not like closed doors, preferring an unobstructed pathway through the house. The marmot learned to open swinging doors by pushing with the help of its teeth and claws. When the job became difficult, it did not hesitate to gnaw at the door. As soon as someone closed a door, the marmot ran to it and opened it again. After it marked all of the sides of the furniture and the walls as part of its living quarters by smearing them with a secretion from its cheek glands, its owners were forced at last to move it to a marmot reservation, where it could live together with its fellow marmots. Later, when the Koenigs visited it and took it away from the reservation for a few hours, only to return it, its fellow marmots ran to meet it and nuzzled its snout. Obviously they wanted not only to reacquaint themselves with it; they also wanted to know where it had been and what it had eaten.

The reason people in so many alpine areas believe in the legendary *Tatzelwurm* (*Tatz*= paw print, *Wurm*= worm) can probably be traced back to the marmots. A marmot moves through the snow in such a way that one of its hind feet steps exactly into the print of a front foot. Thus its track consists of three rather than four footprints. This is particularly puzzling to a great many people, especially as all the footprints together look like a single large footprint.

The alpine marmot existed in much of Europe between the last two ice ages. It was driven into the central European areas, however, by the successive glaciers. Marmots were present in large areas of central and western Europe during the Ice Age; these were noticeably larger than the present-day animals. They have been classified as *M. m. primigenia*, a subspecies of the modern alpine marmot. Kratochvil believes that the relatively small marmot population which has survived in the high Tatra Mountains (Czechoslovakia) has been separated from the rest of the marmot population for over 25,000 years. As a result, he has classified these animals as a separate subspecies, *M. m. latirostris*. The marmots in the Siberian and central Asian mountains are related to the alpine marmots, according to Ellermann and Morrison-Scott. However, this group also shares some important characteristics with the related bobac marmot, so an exact systematic classification is rather difficult.

Other marmots, by H.-A. Freye

The bobac marmot

The BOBAC MARMOT (*M. bobak*; see Color plate, p. 239) lives in flatlands, prairies, and high-altitude grasslands. It has a much greater distribution than the alpine marmot; it is found from eastern Europe almost continuously to eastern Siberia. Moulting takes place only once a year, as in most hibernators. The animals' fur becomes considerably lighter in the spring, when the marmots come out of hibernation. The skin also becomes lighter; shedding begins first in the lumbar region, then continues gradually along the back, shoulders, and limbs. The market value of the skin increases as soon as all of it has turned lighter. The bobacs become increasingly fatter from July on. In the fall, before hibernation, a skinned adult bobac weighs 4–5 kg; 1.5 kg of this is fat. Mongolians value the melted fat not only as food, but also as a medication against tuberculosis.

Like its mountain-dwelling relatives, the bobac marmot digs an extensive underground tunnel system. According to the observations of the zoologist Ognew, these underground structures can reach a length of of about 20 m; the main den may be more than 3 m below the surface. Ognew describes a particular feature of the bobac's burrows: "The marmots build a series of high mounds around the wide entrance to their dens; these give the whole area a characteristically wavelike appearance. According to A. N. Formosov's calculations (1929), marmot colonies take up as much as twenty to fifty percent of the total area of the Mongolian prairies. If we assume an average size of some 3 km² for one marmot burrow system, and if we assume that a family den has five or six entrances, with at least one mound of earth in front of each one, then the earth which has been dug up around one single group of holes would cover 15–18 m²." The marmots digging activity may have a favorable influence on the dispersion of some prairie plants; and it certainly has an influence on the soil composition.

The social life of the bobac marmot is very similar to that of the alpine marmot. Formosov describes life in colonies of the MONGOLIAN MARMOTS (*M. b. sibirica*), which are called TARBAGANS in their homeland: "The marmots disappear into their holes whenever people appear. Those nearest the holes are the first to dive underground.Those that are somewhat further away squat in the entrances of their holes, leaving only the darker part of their head outside as they reconnoiter; they can hardly be seen against the black background of the entrance. The marmots that are furthest away from the danger zone form a "ring" of observers; they stand on their hind legs and whistle continuously. They are answered by those marmots that are sitting in front of their holes. Generally this "ring" changes position as the intruder moves about. Its diameter depends on the flight distance of the colony as a whole; it ranges from about 150 to 200 up to as many as 500 paces. The appearance of a dog or fox causes a totally different reaction among the marmots. These predators can come much closer, while each animal in the colony rapidly sits up and begins to whistle. The young stand about in groups; the older, more awkward animals are scattered and stand alone. All of the animals whistle; the adults make a strong, grunting, two-toned whistle, sounding rather like "kwi-kwitsch," while the young make a softer, more husky "fitsch-fitsch." The marmots produce the same whistle when a circling golden or white-tailed eagle appears above the colony (both birds of prey like to be in the vicinity of Mongolian marmot colonies). The animals back up to the entrances of their holes; however, most of them do not even stop eating. Apparently the marmots can usually hide from an eagle in time."

While summer ends anywhere from the middle to the end of September in high-altitude areas as well as the Altai and Changai Mountains (Asia), hibernation in the prairie areas does not begin until the middle of October, or even later. "Before hibernation, the marmots carry dry plants, old rags, pieces of felt, woolen knots, and even dried dung into their structures, as insulation," reports Dordshiin Eregden-Dagwa, an expert on Mongolian marmots. "Mongolian marmots of all generations help collect the dry plants. During the preparation period, before hibernation, the marmots stop eating, and they rarely appear outside their structures, particularly when the weather is dry and desolate. The contents of the stomach and intestine are emptied, and only two to seven hard bolusses remain in the anal portion."

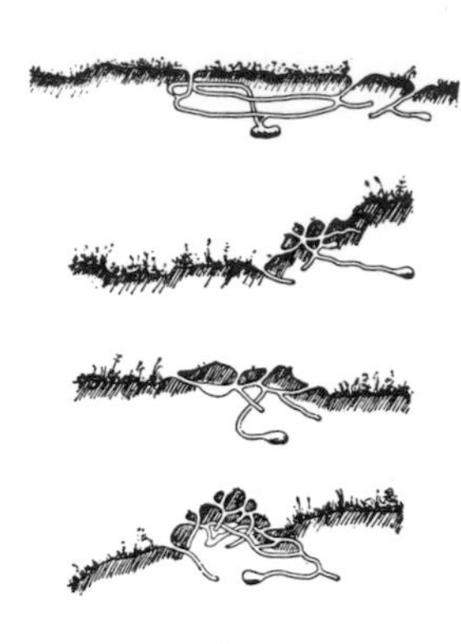

Fig. 10-7. The bobac marmot builds its burrow in a variety of shapes, according to the geography of the region (from top to bottom): in the dry prairies of Northern Kazakh (U.S.S.R.); in a pasture prairie on the mountain slopes of the Tian Shan (U.S.S.R. and China); in a pasture at a high altitude in the forest-prairie steppe area in Tian Shan; very wet mountain moor in the alpine belt of the Tian Shan.

When closing off the entrance to its burrow, the marmot begins with large stones and ends up using smaller ones; it mixes the stones with earth, dung, and old nest materials. Sometimes the materials may come from the side tunnels; sometimes a new tunnel is dug, and the earth and stones from it are used to close the entrance. The marmots tamp down the earth with their snouts; this sometimes makes loud sounds which can be heard far above on the surface. Only one animal, normally an adult family member, closes off the entrance. When a young animal hibernates alone, it closes the entrance and then covers itself with dry grass or burrows down into the grassy lining of the den. When a number of hibernating animals are together in one burrow, they do not cover themselves with grass at all. In prairie and forest-prairie areas, as well as isolated areas on southern and southwestern mountain slopes, Mongolian marmots normally wake up anywhere from the middle of March to the beginning of April; in the higher altitudes of mountainous areas, they do not awaken until mid-May. Often the last animal may not leave its winter den until the beginning of June, after the snow cover on the ground has melted.

The mating period begins as soon as hibernation is over. The males are sexually mature after the third year. In April, the adult male's testes weigh 4-5 g. However, at the beginning of hibernation, they weigh only 400 to 700 mg. The average weight of a female ovary during hibernation is approximately 30 mg; however, during the mating season, the average weight of the ovary is 70-80 mg. The gestation period in the bobac marmot lasts from forty to forty-five days; the female gives birth to an average of five or six, and in rare cases, sometimes ten or twelve, young per litter. Mongolians call the young *mundul*. The nursing periods lasts longer than one month; the animals appear outside their burrows for the first time in June. According to recent observations, the oldest recorded age of an animal is fifteen years.

Enemies, parasites, and diseases

The bobac marmot has many enemies, including wolves, dogs, foxes, bears, snow leopards, wolverines, and polecats, as well as prairie eagles, great bearded vultures, kites, and buzzards. When their holes are flooded by melting snow or heavy rainfall, the bobacs are forced up above ground, where they are easy prey for their enemies. Aside from the plague, bobac marmots are also susceptible to a disease of the gall bladder which Mongolians call *Som-Delu*; this disease can be quite infectious, and entire colonies may be wiped out. Ermines, weasels, ground squirrels, hares, and various reptiles and amphibians use the marmot burrows as escape routes. The temperature and moisture ratios, which remain constant in the burrows create a good environment for the many other animals which also live in them regularly, such as several species of flies, bugs, ants, grasshoppers, centipedes, spiders, mites, lice and fleas.

The fleas transmit the dreaded plague to the marmots, and also to people. The flea *Oropsylla silantievi*, in particular, carries the disease from the Mongolian marmot to people; the plague bacillus was detectable in them even after 385 days. As a constant source of infection, the bobac must be considered in any attempt to combat the plague. However "attempts to prohibit the hunting of marmots, and thereby limiting the spread of the disease, are unacceptable to Mongolians, as the hunt has an important economic significance for them," according to Dordshiin Eregden-Dagwa. "Furthermore, because Mongolians can recognize diseased marmots so well, they are able to safeguard themselves against contamination during the hunt."

Economic significance

The Mongolian marmot plays a major role in the Mongolian economy; it is the most important fur animal of that area. Its fur has a strong underside and an even, compact coat of hair; it is barely distinguishable from otter fur. Today, some 2 million furs are prepared annually in the Mongolian People's Republic. The marmots supply several hundred tons of fat as well. The local population uses large quantities of marmot meat as food, both for themselves and for their domestic animals.

Descriptions of bobac hunting written by Marco Polo, and later by Wilhelm von Rubruk (1210–1270), a Flemish traveler in eastern Asia, appeared in the 14th Century. The Mongolians have developed a wide variety of hunting methods since ancient times: they may tease the marmot's curiosity by waving small flags, white skins, or other objects, or they may hunt the animals with dogs; they may set snares and traps; they may dig up the marmot burrows; or they may use entirely different methods. Between 1947 and 1952, the Mongolians annually exported an average of 2.3 million bobac furs. However, the quantity of trapped furs was greater than the natural reproduction of the animals; this caused a reduction in the number of bobacs.

Distribution during the Ice Age

Fossil finds show that the bobac existed during the Ice Age in the Loess Prairie area of what is presently Thuringia, East Germany, and in areas somewhat more to the west. During the Ice Age their range was located north of the alpine marmot's. In ancient times, the range shifted more toward the east. The Ice Age forms of the bobac were smaller than those alive now, in contrast to those of the alpine marmot. A fossilized transitional form between the alpine marmot and the bobac was found at Thuringia. Today, this form is classified in the bobac group as the ALTAI MARMOT (*M. b. baibacina*), and it still exists in limited numbers in the Altai and Turkestan regions of Asia.

Long-tailed marmot

The LONG-TAILED MARMOT (*M. caudata*) was mentioned by the early Greeks. This animal's habit of making numerous hilly mounds in high-altitude grassy prairies may have reminded the ancient Greeks of ants. They had a fable which told of "gold-digging ants that were larger than foxes." The fur of this species was an important trade item in ancient times; they were even sent to India and China. Wild sheep and deer supposedly respond to the call of the long-tailed marmot.

Woodchuck, by Patricia A. Goodmann

The WOODCHUCK (*M. monax*) is a large fossorial (digging) marmot found throughout the eastern and midwestern U.S.A. from New York to the northern part of South Carolina and west as far as Kansas and Nebraska (*M. m. bunkeri*). *M. m. johnsoni* has been found on the Gaspé Penninsula (Quebec, Canada).

The name woodchuck is said to be derived from two Indian words: the Cree *otcheck* and the Chippewa *otchig*, which the Indians applied to the fisher (*Martes pennanti*). White trappers and traders in the Hudson Bay area are said to have corrupted these words into *wejack*, *woodshaw*, and *woodchook*, and misapplied them to the woodchuck.

Eight subspecies of *Marmota monax* have been described: *rufescens*, *preblorum*, *ignava*, *canadensisis*, *petrensis*, *ochracea*, *bunkeri*, and *johnsoni*.

Woodchucks are not highly vocal. Their most common vocalization is a sharp whistle given when the animal is excited or alarmed. It is followed by a series of hisses. The animal is double coated, having an outer coat of long, agouti-banded guard hairs and a soft wooly undercoat which may be entirely cinnamon or may be pale gray on the back and sides and reddish on the belly. Both albino (all-white) and melanistic (all-black) woodchucks have occasionally been captured, as well as the light grizzled-brown and the dark-brown forms. Individuals may have easily distinguishable markings. The vibrissae, which are well developed, are located on the upper lip, above the front part of the eye, and midway on the cheek between the eye and the angle of the mouth.

Woodchucks have three white, nipplelike scent glands located just inside the anus. When the animal is excited, they are everted and emit a distinct musky odor. The secretion is not sprayed in small droplets as is the skunk's; it is probably used for communication rather than for protection.

Records of woodchuck weights vary considerably, because the animal is able to consume as much as 453.6 g at one feeding. One of the largest specimens captured weighed approximately 6.9 kg. John Webb, of

Frankfort, Kentucky, had a captive adult woodchuck which weighed 16.7828 kg. Generally, the animals tend to be lightest in the spring, after hibernating, and heaviest in the fall.

The tracks, which resemble a racoon's, may be identified by the five toes on the fore feet and the four on the hind feet.

Reproduction

Though a few breed as yearlings, most woodchucks become sexually mature at two years. In Maryland the breeding season begins in late February; it probably starts later further north. Sexing of woodchucks may be done by examining the length of the peritoneum, which is about 2.5 mm long in the male and about 1.25 mm long in the female. The woodchuck normally has eight functional mammae: one pair is inguinal, one, post-abdominal, and two, pectoral. The gestation period is usually thirty-one to thirty-two days long. The young are born blind, toothless, and hairless; their hind legs are less well developed than the fore legs. The average litter size is 4.6 young, though as many as nine have been observed in one litter. The young woodchucks' eyes open at about four weeks of age; they are also well furred by this time. After their eyes are open, the young begin to venture from the den and to eat some adult food as well as milk.

Reports conflict as to how long male woodchucks consort with females. According to some, contacts are practically limited to copulation, and if the male stays much longer, the female drives him off before she gives birth. Others have occasionally seen males sharing a den with a lactating female. One captive female was kept caged with a male until twelve hours after she gave birth, when he was removed; at no time was she observed behaving aggressively toward him. Recent observations indicate that individual females may vary in their reactions to males outside the breeding season, some tolerating them, others not.

Diseases and parasites

Woodchucks do not seem to be subject to cyclical fluctuations in population. As a species, they seem to be remarkably free of disease and parasites, though individuals have been found infested with ticks, fleas, mites and roundworms, and bots. They can carry rabies, tularemia, and Rocky Mountain spotted fever. Woodchucks are subject to malocclusion, which may result in death from starvation as the condition of the teeth worsens. On the whole, predation rather than disease appears to keep their numbers in check. Man, foxes, and dogs are their most serious enemies, though badgers, snakes, and owls will prey on them occasionally. Like most animals, the woodchuck will flee from danger if given the chance; if cornered, however, it is a formidable fighter, able to hold its own against even a medium-sized terrier. A pack of dogs or individual dogs from the larger breeds will usually dispatch woodchucks they catch in the open. In the wild the woodchuck's lifespan is believed to be five to six years. The oldest woodchuck on record was ten years old when it died in the London zoo.

Dens

Woodchuck dens are nearly always located in well-drained soil, preferably sandy loam. The burrow, which may have from one to five entrances, may run directly from entrance to nest, or there may be a maze of tunnels leading to several nests, cul-de-sacs, and alternate entrances. These more elaborate dens may be the work of several generations of woodchucks who have enlarged the original burrow. The animals normally have two types of dens. The hibernation den is usually located in a well-drained woody area, often on high ground where water cannot flood it, and, when possible, under the roots of a tree or stump. It is often plugged off from the main passage. The summer dens are usually found in open areas near tilled fields. Woodchucks have shown preferences for fields of soybeans, clover, corn (other grains are generally avoided), and alfalfa. The animals move into their summer dens about one month after hibernation.

Woodchuck dens are often inhabited by other animals; if woodchucks are present there will be a mound of fresh earth outside the entrance, since woodchucks habitually clean out their burrows, depositing the soil outside the entrance several times a week. The dens have two types of entrances: a main entrance, with its mound of fresh soil, which leads downward at a moderate angle, and a plunge hole, 18 mm or less in diameter, with very little earth piled around it, which drops downward sharply. Below the main entrance the tunnel usually widens enough for the animal to turn around. From that point the tunnel may angle upward (which helps prevent the nest chamber from being flooded), branching out into several tunnels, one of which will end in the nest chamber. Cul-de-sacs are often used for the deposition of feces. The nest chamber is almost always higher than the lowest part of the tunnel; in one case it was within 45 mm of the ground surface.

Though normally fossorial, woodchucks have been seen climbing trees, which they do slowly but well.

Information on territoriality is conflicting. Grizell has seen only the immediate area of the den defended, and he believes that males show territoriality during the breeding season. Bronson has recorded a higher incidence of aggressive behavior in which threats are made by arching the back, rasing the tail and erecting its hair, and directing the head toward the opponent with the incisors bared. Avoidance of dominant animals by subordinates may consist of the subordinate crouching in a hole in the field, or mere avoidance of eye contact. These dominant-subordinate relationships, according to Bronson, are maintained irrespective of location; that is, being close to its own home den does not make a subordinate animal bolder or more aggressive. Young animals were usually lower ranking and tended to avoid their neighbors.

Hibernation

Hibernation appears to be progressive; older and fatter animals become dormant first, to be followed by yearlings and juveniles. They tend to emerge in the same order. Hibernation takes place roughly from mid-November (old animals may start sooner) to mid-February. The

Fig. 10-8. 1. Black-tailed prairie dog (*Cynomys ludovicianus*); 2. White-tailed prairie dog (*Cynomys gunnisoni*).

animal may lose from one-third to more than one-half of its body weight during this time. Hibernation is believed to be induced by cold, darkness, confined air, absence of external stimuli, season, absence of food, and physiological and hormonal factors. The normal, non-hibernating marmot has a basal metabolism lower than any other warm-blooded animal of comparable size.

Human civilization has tended to increase the woodchuck population. Farmlands are favored feeding sites for woodchucks, and gardeners have cause to complain about their activities. In turn, woodchucks have contributed to the survival of other species, particularly in areas where "clean-farming" practices have destroyed hedgerows, brush piles, and other areas which formerly provided cover for animals. The rabbit often uses woodchuck dens in the winter as a protection from cold; other animals found in woodchuck dens are skunks, foxes, chipmunks, ground squirrels, weasels, mice, opossums, racoons, and occasionally birds, especially quail, pheasants, and ruffed grouse.

The size of a woodchuck's home range depends on how close the den is to a suitable food supply. There appear to be three or four types of "migrations" in a year. In the spring the males may travel great distances in search of mates. The animals also move from hibernation dens to summer dens. Later in the summer the young disperse from the den in which they were born. Finally, in the fall the animals move back to their winter quarters.

Prairie dogs, by Patricia A. Goodmann

PRAIRIE DOGS (*Cynomys*; BL 28-35 cm, TL 3-10 cm, weight 700-1400 g) look like large, heavy ground squirrels. Their skeletons, including the skull and the number and location of the teeth, resemble those of marmots; the stomach is smaller, the colon shorter, and the cheek pouches less well developed. The guard hairs are agouti-banded, and the coat has a gray or reddish cast. The belly is buff, shading to light gray or white on the throat and the lower part of the face. The tail varies in length, and may be black-tipped (*Cynomys ludovicianus*) or white-tipped (*C. gunnisoni*). The BLACK-TAILED PRAIRIE DOG (*C. ludovicianus*) lives in the "buffalo grass" prairies, while the WHITE-TAILED PRAIRIE DOG (*C. gunnisoni*) lives in mountainous areas. Only these two species, of the seven described forms, can be clearly distinguished.

Prairie dog towns

These diurnal rodents live in social groups called coteries, whose members unite to defend their system of burrows from trespassing conspecifics (members of the same species) from other coteries. Large numbers of coteries live together in "towns." According to Koford's data, large towns may cover 65 hectares (160 acres). These interconnected burrows not only afford the prairie dogs a refuge and a place to rear their young; they also affect the animals' society: they add stability to the social organization. Burrows may be surrounded by the characteristic crater-shaped mound of earth, or by a dirt heap, or by no soil at all.

Prairie dogs breed once a year; under normal conditions this period is in late February and early March. Gestation lasts thirty to thirty-five days. The pups' eyes open at thirty-three to thirty-seven days. The mother nurses the pups for about seven weeks, during which time the young stay underground. When they emerge they seldom eat solid food for the first two or three days. The pups continue to grow until they are about fifteen months old; after that they usually cannot be distinguished from older animals by weight alone. Longevity has not been accurately calculated for free-living animals; King, however, found that mortality among animals older than three years is quite high. In captivity, prairie dogs have been known to live for eight and a half years.

Hibernation

Hibernation occurs in prairie dogs, but it appears to be less complete than in related species. Young, who observed several species of captive prairie dogs, noted that the black-tailed prairie dog hibernated throughout the winter, while the white-tailed prairie dog slept for intervals of no more than three to four days throughout the winter. In the wild, black-tailed prairie dogs have been observed to be active above ground during low temperatures, indicating that they do not have an aversion to cold. They do avoid rain and snow, by remaining underground.

As soon as fresh vegetation comes up in the spring, the prairie dogs feed on it, thus restoring lost weight. As long as food is abundant they continue to build up large fat reserves. Accumulations of fat appear to be correlated with a decrease in appetite; even when preferred foods are available, feeding declines noticeably by December and January.

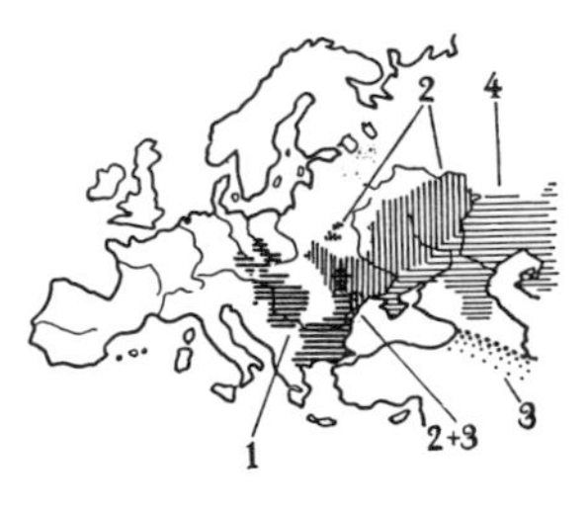

Fig. 10-9. 1. European souslik (*Citellus citellus*); 2. Spotted souslik (*C. suslicus*), European; 3. Little souslik (*C. pygmaeus*); European. 4. Long-tailed souslik (*C. eversmanni*), European.

The most common predators of prairie dogs are rattlesnakes, golden eagles, red-tailed hawks, bull snakes, coyotes, bobcats, and man. Prairie dogs show little fear of snakes, compared with their fear of predatory birds, which seems to indicate that the snakes perhaps prey on them for only a short time after coming out of hibernation. In contrast, prairie dogs herald the approach of a raptor with a distinctive series of short barks which sends all the prairie dogs that hear it scurrying for their burrows. Each will sit up at its burrow's entrance, allowing the bird to come quite close before diving below ground. They react similarly to humans and coyotes; they run to their burrows, where they sit up and bark until the coyote or human is quite close; then they go underground.

Social behavior

All animals in a coterie may use all parts of the coterie's burrow system. When coterie members meet, they frequently engage in contact by touching noses. One may touch its nose to the other's jaw, or both may turn their heads and open their mouths to permit tongue contact with each other. This contact may be a form of identification and possibly of greeting. We have observed pairs of coterie members passing single grains from mouth to mouth. One prairie dog will also allow another coterie member to take grass from its mouth. It may resist having its food taken by turning away from the other animal, by pulling away from the other animal, or by pushing the other animal away with its fore paws. Sometimes the second and third methods are used together.

Fig. 10-10. 1. Nelson's antelope squirrel (*Citellus nelsoni*); 2. Arctic ground squirrel (*C. undulatus*); 3. Thirteenlined ground squirrel (*C. tridecemlineatus*); 4. White-tailed antelope squirrel (*C. leucurus*).

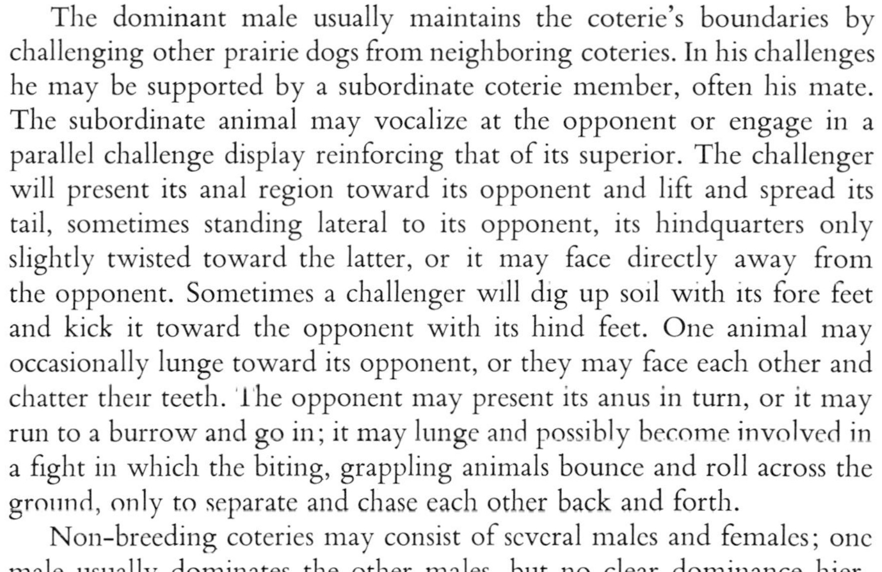

The dominant male usually maintains the coterie's boundaries by challenging other prairie dogs from neighboring coteries. In his challenges he may be supported by a subordinate coterie member, often his mate. The subordinate animal may vocalize at the opponent or engage in a parallel challenge display reinforcing that of its superior. The challenger will present its anal region toward its opponent and lift and spread its tail, sometimes standing lateral to its opponent, its hindquarters only slightly twisted toward the latter, or it may face directly away from the opponent. Sometimes a challenger will dig up soil with its fore feet and kick it toward the opponent with its hind feet. One animal may occasionally lunge toward its opponent, or they may face each other and chatter their teeth. The opponent may present its anus in turn, or it may run to a burrow and go in; it may lunge and possibly become involved in a fight in which the biting, grappling animals bounce and roll across the ground, only to separate and chase each other back and forth.

Non-breeding coteries may consist of several males and females; one male usually dominates the other males, but no clear dominance hierarchy is evident among the females. In breeding coteries, there are commonly four adult females and one adult male. Both sexes treat the young amicably during their first summer. The pups follow adults of both sexes about, climbing on them and over them and attempting to elicit play or grooming. An adult may thump a particularly importunate pup with its fore feet or dominate it occasionally, but no consistent hierarchy has been observed between age groups.

Relationships between coteries change with the seasons of the year. In summer, coterie boundaries are relaxed, allowing individuals from different coteries to become acquainted; "friendly" contacts with neighboring coteries are common then. By autumn, however, the coteries are once again becoming exclusive groups, and by December and January, according to the data of Smith, Oppenheimer, de Villa, and Ulmer, "the coteries were tense enclaves, and the male (or one of the males) of each was frequently squabbling with his neighbors, receiving vocal support from his mate and from some coterie associates." During the spring, boundaries are gradually relaxed, permitting cross-coterie contacts once again.

Fig. 10-11. 1. Franklin's ground squirrel (*Citellus franklinii*); 2. California ground squirrel (*C. beecheyi*); 3. Round-tailed ground squirrel (*C. tereticaudus*).

Prairie dogs have a variety of vocalizations. The most common is the sharp "bark," like a small dog's yip, from which the prairie dog gets its popular name. A prairie dog may bark when it sees an approaching predator or during a challenge display. They sometimes make a chittering noise; this is commonly heard as a female rebuffs her mate. A rasping noise seems to indicate that the prairie dog may attack. A clear, high-pitched scream may be heard when the animal is suddenly subjected to pain or is prevented from fleeing from danger. Prairie dogs produce a chattering noise by striking their incisors together; this usually seems to indicate belligerence, or perhaps frustration, on the part of the chatterer.

This tooth chatter is found among most rodents; it often indicates an intention to attack. The prairie dog also has a display called a "jump-yip," in which the animal throws its forequarters up so that it stretches vertically upward; occasionally it just tosses its head up. This jump is almost always accompanied by a loud yip. It is believed to indicate that the animal making the jump-yip will flee from an aggressive encounter, either precipitately or by a more gradual withdrawal. Thus the jump-yip may be used either by the retreating prairie dog after a challenge or by one who watches, but does not engage in, a challenge. These displays are sometimes made by an animal that has just successfully defended its boundaries. In captivity, jump-yips have been observed when the animals are very alert and nervous, and also in the winter, when territorial defense is most intense; jump-yips tend to decrease in number during the summer, when boundaries are relaxed.

▷ Mountain beavers: 1. Mountain beaver or sewellel (*Aplodontia rufa*). Typical squirrels: 2. Red squirrel (*Sciurus vulgaris*), a) winter coat in north-eastern central Europe, b) darker-colored species, c) normal color of animals in central Europe (*S. v. fuscoater*) in summer coat (notice the short ear plumes of the summer coat and the long ones of the winter coat), d) winter coat of the north-eastern European and Siberian subspecies; 3. Gray squirrel (*S. carolinensis*); 4. North American red squirrel (*Tamiasciurus hudsonicus*).

Among mammals, the young are usually expelled from the homesite by their parents, particularly if more young are produced during the same season. According to King's data, it is the adult prairie dogs who "emigrate," expanding the town's boundaries and leaving the old burrows for their offspring. Yearling prairie dogs have also been observed emigrating. King believes that there are different reasons for adult and yearling emigrations. The yearlings, he believes, especially the males, are subject to some aggression from the dominant males; young males show considerable aggression among themselves, too. Though he observed no overt aggression between adults and pups, King did find that pups constantly interrupt their elders' activities. Adults sometimes thump pups with their fore feet, but this has no lasting effect. Direct antagonism from adults to pups might stop these interruptions but would probably harm the structure of prairie dog society; King believes that the adults leave their territories to get away from their importunate young. This has survival value in that the older, more experienced animals leave and make new homes for themselves, leaving the inexperienced young in the familiar homes of their birth.

Parasites

Parasites found on prairie dogs include fleas, mites, and, sometimes, ticks. They seem to cause the prairie dogs considerable discomfort: the prairie dogs may bite, scratch, and roll in the dust to rid themselves of some parasite.

Prairie dog burrows are used as refuges by a variety of animals. Insects, reptiles (including snakes), and mammals have been found in them. Rabbits, red squirrels, various species of mice, porcupines, chipmunks, and burrowing owls may all use the burrows on occasion. Elk, deer, antelope, and bison all come to prairie dog towns to graze on the herbs which sprout after the prairie dogs have eaten the tall grasses. The elk and deer usually come at night and so have little direct contact with the prairie dogs; the antelope and the prairie dogs may feed fearlessly within a few meters of each other during the day. Bison are not feared

1
2 b
2 a
2 c
2 d
4
3

1
2
3
4
5
6
7
8
9
10
11
12

by the prairie dogs, but their greater size and numbers make their presence more disruptive. They frequently wallow on the craters of prairie dog burrows. When particular craters become favored wallows, the prairie dogs will often abandon them to the bison.

Prairie dogs are welcome and easily kept attractions in zoos because of their alertness during the day, their easy adaptability, and their lively mannerisms. Many zoos have built special open-air facilities for them, often modeled on small prairie dog towns. The prairie dog's lifespan is about ten to eleven years, depending on the condition of its teeth.

The ground squirrel and its relatives, by H.-A. Freye

Distinguishing characteristics

The SOUSLIKS AND GROUND SQUIRREL (*Citellus*; BL 11.5-38 cm, TL 3.5-25 cm, weight 144-830 g) are slender terrestrial animals which look like the red squirrel, but usually do not reach the same size. The Old World species are known as sousliks, while the New World species are called ground squirrels. Their ears, which are generally small, are almost hidden in the skin. The hair is gray-brown to sand-colored, and it is often somewhat irregularly waved. The spotted souslik has dense, light spots; the thirteenlined ground squirrel is marked with white transverse stripes and light specks. In some species, the tail is bushy; in most species, however, it is hairy on both sides and only somewhat bushy at the tip The third "finger" is the longest; the thumb, which may have a nail or claw, is underdeveloped. The hind foot is usually half as long as the tail; both outside toes are shorter than the middle ones. The underside of the fore foot has four cushions; the underside of the hind foot has five. The soles may be naked or hairy. The skull is like that of the marmot; the upper profile, however, is more curved, and has an egg-shaped circumference which narrows in the front. The nasal area gradually widens into the cheek bones. There are twenty-two teeth arranged: $\frac{1 \cdot 0 \cdot 2 \cdot 3}{1 \cdot 0 \cdot 1 \cdot 3}$. The first molar tooth of the upper jaw is about half as strong as the following ones, which are narrower toward the inside. The four lower molars, in cross section, have a rhomboidal shape. The incisors of the lower jaw are not so curved; they are directed toward the front, as in marmots (although they look more like those of the red squirrel). All species have cheek pouches which extend from the oral cavity to the hind end of the head; the right cheek pouch is used more than the left one. Females of the Old World species have four to seven pairs of nipples; females in the New World species have four to six pairs. The gestation period lasts from twenty-one to thirty days; usually there are from three to eight young, and in rare cases up to thirteen. The lifespan is approximately ten years.

◁ African Oriental giant squirrels and sun squirrels: 1. Oil palm squirrel (*Protoxerus stangeri*); 2. Ebien or African palm squirrel (*Epixerus ebii*); 3. The sun squirrel *Heliosciurus undulatus*. Callosciurini or "beautiful" squirrels: 4. Oriental pygmy squirrel (*Nannosciurus*); 5. Swinhoe's or Asiatic striped squirrel (*Callosciurus swinhoei*); 6. Perny's long-nosed squirrel (*Dremomys pernyi*); 7. Malayan black-striped squirrel (*Lariscus insignis*). African tree squirrels: 8. African pygmy squirrel (*Myosciurus pumilio*); 9. Congo striped squirrel (*Funisciurus congicus*); 10. *Paraxerus flavivittis*. Indo-Malayan giant squirrels: 11. Malayan giant squirrel (*Ratufa bicolor*). Bristly ground squirrels: 12. South African ground squirrel (*Xerus inauris*).

Of the seven Old World species found from eastern Europe to Mongolia, we will mention the following:

EUROPEAN SOUSLIK (*Citellus citellus*; see Color plate, p. 239). It either has no spots or barely visible spots. The eyes have elongated oval pupils. These animals live socially in colonies, but within these colonies they live singly. They are generally crepuscular (active in the morning and evening). This is the only souslik species found in Germany.

The SPOTTED SOUSLIK (*C. suslicus*; see Color plate, p. 239) is similar to the European souslik, but the fur on its back and flanks is spotted. The soles of its feet are hairy, and its tail is shorter than that of the European souslik. Its burrow system is especially complicated, with many branches. This species has been greatly reduced in numbers in recent years; it is even extinct in some places.

The LARGE-TOOTHED OR ARAL YELLOW SOUSLIK (*C. fulvus*; BL 38 cm) is the largest souslik species in Eurasia. Its cheek pouches are relatively poorly developed. It is found in deserts and semideserts in the lower Volga area, central Asia, northern Iran, and Afghanistan.

Fig. 10-12. 1. Golden-mantled ground squirrel (*Citellus lateralis*); 2. Mohave ground squirrel (*C. mohavensis*); 3. Ring-tailed ground squirrel (*C. annulatus*).

The LITTLE SOUSLIK (*C. pygmaeus*; BL 20-23 cm) looks like the European souslik, but the soles of its feet are bald. It is bright yellowish-brown to gray-brown, and more or less speckled. These animals are found in the U.S.S.R. from the Dnepr River to eastern Kazakhstan. Their preferred foods are spring wheat, corn, sunflowers, and clover. In dry years they also hibernate during the summer.

The LONG-TAILED SOUSLIK (*C. eversmanni*) has no hair on the soles of its feet. A subspecies, the AMUR SOUSLIK (*C. e. menzbieri*), along with the large-toothed souslik, belongs to the largest Eurasian sousliks. It weighs 1.9 kg in the fall. It is found in mountain pastures up to altitudes of 3000 m. It also lives in farmyards and barns. It is found in eastern Siberia, and is related to the North American forms.

There are fourteen species of ground squirrels in the New World, from Canada to northern Mexico. Many of these animals do not hibernate. We will mention the following New World ground squirrels, each of which represents its own subgenus:

Fig. 10-13. 1. Rock squirrel (*Citellus variegatus*); 2. Townsend's ground squirrel (*C. townsendi*); 3. Washington ground squirrel (*C. washingtoni*).

The ARCTIC GROUND SQUIRREL (*C. undulatus*; BL 21.5-35 cm, TL 7.6-15 cm; see Color plate, pp. 151/152). It is yellow to reddish-brown, with irregular off-white spots. It is found from Alaska to the Hudson Bay area (northeastern Canada).

The THIRTEENLINED GROUND SQUIRREL (*C. tridecemlineatus*; BL 11.5-16.5 cm, TL 6-13.5 cm, weight 150-240 g) is noticeably smaller. The basic coloring is light brown to dark brown. It has thirteen whitish stripes on the flanks and back, which are broken up into spots. It is a typical prairie inhabitant from Canada south into the central U.S.A.

FRANKLIN'S GROUND SQUIRREL (*C. franklinii*) is usually somewhat larger than the thirteenlined ground squirrel. Its tail is more than half as long as the body length. It has no stripes, but nevertheless, it looks very much like the thirteenlined ground squirrel. The back and abdomen are a light gray-brown. The teeth are like those of the red squirrel. Up to one-third of its diet consists of smaller animals; it prefers birds, and even successfully attacks ducks. It lives in eastern North America on damp, fertile prairies from southern Canada to Arkansas and western Indiana.

The CALIFORNIA GROUND SQUIRREL (*C. beecheyi*; BL 23-28 cm, TL 12-23 cm) is relatively large. Its head is granite-brown and the body is spotted with brown and white. The sides of the neck and shoulders are a dirty-white, with a dark triangle between the shoulders. The tail is strong and bushy; it is dark with lightly tipped hairs. This animal lives in semi-open land with low, green growth. It not only hibernates in the winter, but also "sleeps" (estivates) late in the summer. These animals regularly try to plunder the stores of woodpeckers, but they give way when these birds attack. They are found in the western U.S.A. in the hills and valleys of the Cascade Mountains and the Sierra Nevada.

Fig. 10-14. 1. Belding's ground squirrel (*Citellus beldingi*); 2. Columbia ground squirrel (*C. columbianus*); 3. Uinta ground squirrel (*C. armatus*); 4. Mexican ground squirrel (*C. mexicanus*).

The ROUND-TAILED GROUND SQUIRREL (*C. tereticaudus*; BL 13-16 cm), has a reddish-cinnamon color with a grayish overhue (camouflage coloring). The tail is not bushy, but rather like a brush. This animal sits on guard in the shadow of plants or rocks. It is a desert-dweller in the large area between the Sierra Nevada and the Colorado plateau.

The GOLDEN-MANTLED GROUND SQUIRREL (*C. lateralis*; BL 15-20 cm, TL 7-12.5 cm), unlike the other ground squirrels, has hair like that of the chipmunk. The head is copper-colored; the back is brownish-gray. The stripes on its flank are bordered with black. The abdomen is a light yellowish-brown; the tail has thick hair which is the same color as that on the back, mixed with black. This hibernating animal lives in mountain forests from southwestern Canada to the southern Rocky Mountains. It is a well-known and regular visitor to camp sites.

The WHITE-TAILED ANTELOPE SQUIRREL (*C. leucurus*) is about the same size as the round-tailed ground squirrel, and about the same color as the golden-mantled antelope squirrel, although the white stripes on its back do not have a black border. The white underside of its tail often flares up like a signal. It is found in open country in the southwestern U.S.A.

Fur

The fur of the east European and Russian sousliks is often sold under the trade name of *Susliki*. The skins are used as fur lining; the leather and hair are thin. These are considered second-grade furs in the U.S.S.R., and they are caught by the millions. The best souslik furs are those from animals caught in April and May. The fur is used in natural and artificial colors. Sousliks shed in the summer; during this time, the usual blackish gray undercoat falls out and a new summer coat develops. The molt starts with the head and proceeds backward over the neck and back, then to the flanks and abdomen. Ground squirrels and sousliks, like marmots, have three glands in the anal region; these, however, are relatively small. These animals become quite fat prior to hibernation.

Diet

The diet of ground squirrels and sousliks consists mainly of herbs, seeds, fruits, tubers, onions, and roots; they also search out and eat subterranean mushrooms. As inhabitants of steppes and prairies which are poor in vegetation, many ground squirrels and sousliks live mainly on prairie grass, mugwort, quaking grass (*Eragrostis*), and wild tulips (*Tulipa gesneriana*). In large numbers, they can cause heavy damage to agriculture. B. S. Vinogradov and S. J. Obolenskij have calculated that 600 million sousliks live in the U.S.S.R. alone, and that each year they eat approximately 2.4 million t of corn, sunflowers, legumes, potatoes, turnips, and similar food. Ground squirrels and sousliks also eat other small animals, such as insects, birds that hatch on the ground, birds' eggs, mice, and small rodents. In times of massive reproduction, they frequently become cannibalistic.

Many species store winter supplies; sometimes these stores are quite substantial. They are generally used only after hibernation, on those spring days when food is scarce. Erna Mohr has reported that her captive sousliks carried food containers and water bowls into their sleeping quarters, even when these made the sleeping boxes quite cramped. "One of them no longer had room for itself in its box, and it curled up in front of it. Its hoard consisted of fourteen glass bowls and approximately 1 kg of sunflower seeds." Sometimes the collection consists of colored pebbles, pieces of glass, white feathers, and other brightly colored objects.

Behavior

When the ground squirrel or souslik is threatened by danger or when it is afraid, it gives a loud, sharp whistle that is understood by all the other members of the colony. This warning signal is a single shriek when an enemy appears; when the animal is in trouble, it becomes a shrill screeching. When the animals fight among themselves, they make long or short growling sounds which are a combination of mumblings and a soft trilling, during which the animal nods its head. Other animals also understand these calls. According to the observations of Eibl-Eibesfeldt in Burgenland, hamsters run for cover when the sousliks whistle. We do not know however, whether this reaction is innate or learned.

The eyes are their most important sense organs. Ground squirrels and sousliks always look around very carefully before they leave their burrows. They watch over their young from a higher location, and they are particularly on the alert for the appearance of large birds of prey. "The head forms an obtruse angle above the eyes," writes Peus, "and the eyes lie directly under the vortex of this angle; when the head is held high in an elevated position, the eyes are at the top of the head. When a souslik that has hidden in its hole wants to come out again, it raises only the top of its head, so only its eyes, and nothing else, are out of the tunnel. As a second safety measure, the souslik moves only slightly forward so that its entire head extends from the burrow, thereby increasing its field of vision. Then the animal gradually advances out of the burrow and sits up in front of it. If it is still safe, the animal either walks away from its nest or begins to graze."

Fig. 10-15. 1. Richardson's ground squirrel (*Citellus richardsoni*); 2. Spotted ground squirrel (*C. spilosoma*); 3. Harris' antelope ground squirrel or antelope squirrel (*C. harrisi*).

As diurnal animals, ground squirrels and sousliks particularly enjoy the sun. They show a definite daily rhythm; apparently this develops gradually, as the newborn young are active at all hours of the day and night. The animals are particularly afraid of large birds of prey. Nelson's ground squirrel (*C. nelsoni*) react not only to the alarm whistle of their fellow colony members, but also to the alarm calls of larks and the yellow hammers. The enemies of the ground squirrel and souslik are mainly diurnal animals, including eagles, bustards, and especially polecats, which in many areas have specialized in killing ground squirrels or sousliks. The male polecats dig into the prey's burrows, and the smaller female animals enter the dens and kill their prey, often by breaking one or more vertebrae. After disposing of the victim the polecat moves into its den and falls into a deep sleep.

The fleas in the fur of the souslik have a great epidemiological significance because, as they do in many other rodents, they carry the plague bacillus. The souslik's role is similar to that of the marmot in spreading the disease. Since sousliks travel, they can spread the disease over an entire country. Destroying the animals does not help; on the contrary, the sousliks will then begin to move around or migrate, thus spreading the disease further. During the mating season, when the sousliks enter strange dens, they transfer the parasites. After the hibernation period, the souslik's fleas become "field fleas" as Tiflov-Potapov calls them, and they travel around in large masses, spreading the disease still further.

Social life

The social life of the members of the genus *Citellus* takes place in large, close-knit colonies. There are definite differences in behavior among the related species. According to Krumbiegel, the LITTLE SOUSLIK (*C. pygmaeus*) for example, lives alone, sharing a den with others only during the mating season. The adult EUROPEAN SOUSLIK (*C. citellus*) lives in its own den, which becomes more complex as the animal ages. Other ground squirrels and sousliks, on the contrary, live in large settlements, which are often adjacent to each other. Their massive tunnel system in the soil is often the reason why horses break through the surface into the burrows. Dams and other irrigation works are destroyed because of their activities.

Burrowing activities

The burrowing activity of ground squirrels and sousliks, as well as that of other rodents, exerts a considerable influence on the character of the soil. According to Ognev, "ground squirrels have built earth mounds in the cornfields of the steppe area around Sarepto. The moist, salty soil of the Sarepto-Otradinsko region is literally studded with the tunnels and dens of these animals. Each mound is built from a more or less loose mass, which is a mixture of earth from the upper and lower layers of the soil. Desert soil characteristically becomes richer in carbonates in its lower layers; salty soil contains more sulfates and chlorides in the lower layers. Sousliks, then, with their burrowing, can enrich the topsoil with salts. Water evaporates very slowly from the sousliks' mounds as a result of their looseness and porosity (especially in comparison to the soil that has not been disturbed by these animals). Consequently, the vegetation, especially in the spring, is particularly dense on the sousliks' mounds." This enrichment with salts and nitrogen is important for vegetation. In the area around Aktjubinsk, the little souslik creates a top soil that is too salty, and thus unfertile, because it digs into the saltier lower layers of the soil and brings this earth to the surface. When there are no deep salt layers, however, this uprooting can be very profitable.

Sometimes ground squirrels or sousliks dig particularly deep down into the soil. They remove the earth, like hamsters, rats, and other rodents do, by turning around and making alternate pushing movements with their fore legs so that they shove the soil ahead of them. According to Shaw, they can carry off the soil in their cheek pouches. Pregnant females do not pile the earth in front of their dens, but scatter it in the surrounding area. In the first year, they make only a simple summer shelter, nothing more than a shallow tunnel; in autumn they enlarge this with nest chambers for the winter. The animals close off the entrance and dig a perpendicular exit tunnel, which they do not open to the outside until the spring of their second year. Thus they make their dens more elaborate every year.

Muschketov found two types of structures in the Kalmjk prairies (U.S.S.R.): a permanent burrow with nest chambers in which the young are reared, and a temporary burrow which is used for only a short time. "The permanent burrows are constructed in firm ground; the temporary burrows are often in fields and soft, fallow land. The permanent structures may be as deep as 1–2 m, and they are either perpendicular or oblique to the ground surface. Most of the dens have only one entrance; others, however, have two or more entrances." The permanent dens are lined with hay. Before hibernation, ground squirrels and sousliks very carefully close off the entrance to their burrows. The North American ARCTIC GROUND SQUIRREL (*C. undulatus*) uses moist soil to make a plug, which has drainage holes; it then places this in the entrance.

Hibernation

The length of hibernation depends on the geographical region; it may be anywhere from five to seven months. The animal's body fat is a rich food supply; the animals are able to sleep when the outside temperature drops and when food is lacking. During the winter, this fat supply diminishes and gradually disappears. During hibernation, the SPOTTED SOUSLIK (*C. suslicus*) has a respiratory rate of one to fifteen breaths per minute. The heartbeat and body temperature also drop during hibernation. When the outside temperature is just above freezing, a ground squirrel or souslik has a body temperature of 1.8–2° C. The lowest body temperature in the spotted souslik occurs during the end of December, according to Kalabuchov; from January until the end of hibernation the body temperature rises gradually. Digestion ceases during hibernation; the intestines are empty and are often closed off. The ground squirrel or souslik has its lowest body weight toward the end of January. The little sousliks awaken from hibernation at different times: the adult males awaken first, followed by the females; the young animals awaken last.

Estivation

Some species, such as the large-toothed souslik, the arctic ground squirrel, the spotted souslik, and the little souslik not only hibernate during the winter; they also sleep (estivate) during part of the summer, whenever there is a drought and the ground is barren. The vegetation in the prairies of Turkestan withers in June, and while the water content of the vegetation is 68–77 % in April, by the end of May it has diminished to 4.5–18 %. For the large-toothed sousliks in Turkestan, estivation continues over into hibernation; these animals are awake and alert only from March to June. However, large-toothed sousliks in the territory along the Volga River, where the vegetation remains fresh until fall, do not estivate until August. Before estivation, the animals close off the entrances to their dens with grass and sand.

Migrations

As original prairie animals, ground squirrels and sousliks migrate all over the northern hemisphere, wherever forests give way to pasturelands. Most species generally remain in the same area; some, however, travel more than 100 km. Man's increasing cultivation of their environment has removed many of the conditions necessary for the sousliks' and ground squirrels' habitat, and they have been forced to move on. The further east one travels from central Europe, the more one encounters large numbers of these rodents. The souslik is protected in Germany; in the eastern countries, however, it is hunted and eaten, as the meat reportedly is especially delicious. In many areas, the souslik compensates for a decrease in its population by an increased birth rate during good years and by, as Krumbiegel puts it, "a wise utilization of the smallest pieces of uncultivated land." Sousliks have been resettled in some areas. The European souslik, for example, was relocated near Wilhelminenberg in Austria, and the large-toothed souslik was resettled on Barsa-Kelmes Island in Lake Aral, where its population increased thirty-five percent in three years. It is not known if the movement of the prairie polecat from southern Russia to Austria and Moravia has anything to do with the increasingly wide distribution of its principal prey, the souslik.

Reproduction

The males' testes become larger and the maturation of the spermatids begins during the first few weeks of hibernation. In February or March, toward the end of hibernation, the spermatids are transferred from the abdomen to the scrotal sac, where they reach their full growth. In females, the uterus is enlarged and becomes ready for reproduction. The mating period lasts approximately fourteen days; the foreplay is similar in most animals, and the mating is of short duration. During pregnancy, which lasts from twenty-one to thirty days, the females are very shy, generally remaining in or near their dens. They do not leave the den at all during the last few days before parturition. The young are born naked and blind. They open their eyes on the seventh or eighth day of life, and they make their first appearance above ground after twenty-two days. The mother nurses them for four to five weeks. Like many other rodents, the female ground squirrels and sousliks are very careful with their young; they will even drag the young from one den to another should they feel endangered. When the young are in the open, the mother watches over them from a high place, and warns them when danger is imminent. Young little sousliks are independent after thirty to thirty-two days, at which time they move out on their own. They are sexually mature in the next spring.

Distribution during the Ice Age

During the Ice Age, souslik species of the eastern steppes were found further west than they are today. The RED SOUSLIK (*C. rufescens*), which lives between the Volga and the Urals (U.S.S.R.), was also found in the Oberpfaltz (Germany) during the Ice Age. A series of souslik fossils from the Ice Age (which have been grouped together under the ancestoral species *C. primigenius* and which inhabited the prairies of Europe and Asia) indicates the original forms of the various species extant today.

Western chipmunks

The eastern chipmunks and western chipmunks are grouped together because both have cheek pouches, which extend as far as the back of the head and even to the shoulders. Unlike the prairie-dwelling sousliks and ground squirrels, the eastern and western chipmunks live in forests. They are all relatively small, with long and conspicuous black stripes down the back and sides of their brown to brownish-gray bodies. They also have stripes on the head, near the eyes.

The OLD WORLD AND WESTERN CHIPMUNKS (*Eutamias*; BL 8–16 cm, TL 8–10 cm, weight 25–125 g) live in the northern areas of the Old and New Worlds. There are sixteen species, including the BURUNDUK (*Eutamias sibiricus*; BL 13–15 cm, TL 8–10 cm). The tail is bushy; the back is gray with five wide blackish-brown lengthwise stripes. The sides of the body are yellowish-gray. The fur is rough with short, tightly packed hairs. This species is found in the northern forests of the Old World, from the White Sea to the Bering Sea, as well as in Mongolia, northern and central China, and the northern part of Japan. They prefer pine forests with underbrush, low bushes on the sides of fields and rivers, and river valleys with willows and bird cherry bushes.

The burunduk

Distinguishing characteristics

The limb structure of the western chipmunks, in contrast to that of the sousliks and ground squirrels, shows an obvious adaptation to arboreal life. As Ognev describes them, the ends of the toes are covered with distinct cushions (much more highly developed than those of the sousliks and ground squirrels). The balls of the feet are also more highly developed, and they have deeper clefts than do those of the sousliks and ground squirrels. The burunduk has shorter, sharper and more greatly curved claws than those of the souslik or ground squirrel. Like all other chipmunks, burunduks take care of their fur by removing dust and sand particles. They do so by lying on their backs or sides and rubbing the ground with quick twisting motions; then they shake the soil and dust from their fur.

The burunduk eats grains, berries, insects, amphibians, reptiles, and young birds. According to Klemm, who studied the life of Siberian chipmunks during travels through Burjato-Mongolia, one animal eats up to about 15 g of grain or about 94 g of berries daily. In eastern Siberia, burunduks plunder wheatfields and cornfields; they also eat linseeds, sunflower seeds, poppys, hemp, fruit, and vegetables. In Siberia their primary

▻
Northern ground squirrels: 1. Spotted souslik (*Citellus suslicus*); 2. European souslik (*C. citellus*); 3. Woodchuck (*Marmota monax*); 4. Bobac marmot (*M. bobak*); 5. Alpine marmot (*M. marmota*); 6. Black-tailed prairie dog (*Cynomys ludovicianus*).

1
3
2
4
5
6

1
2
3
3
4
5

◅
Scaly-tailed squirrels: 1. Fraser's scaly-tailed squirrel (*Anomalurus fraseri*); 2. Lang's small flying squirrel (*Idiurus langi*). Flying squirrels: 3. Southern flying squirrel (*Glaucomys volans*), in three successive seconds while landing on a tree trunk; 4. European flying squirrel (*Pteromys volans*); 5. Formosan giant flying squirrel (*Petaurista grandis*).

food is pine nuts; however, when there are not enough of these, they move to neighboring gardens and fields.

The burunduk can carry up to 9 g of grain in its pouches for over 1 km to its den. There the animal collects a winter food supply of seeds, buds, acorns, leaves, and mushrooms, each stored in different food compartments within the den. The total food supply is usually about 2 kg, although occasionally it may reach 6 kg. In the spring, according to Klemm, the burunduks that live near fields dig up the newly sown seeds from the soil and eat the seedlings. Farmers do not like these animals because they collect the ripened grains of wheat and oat. Sometimes, as Klemm writes, there are so many burunduks in the sheaves of the grainfields that "one beat of a stick on the sheaf makes fifteen animals jump out."

During hibernation, the chipmunks sleep in pairs under tree roots or under moldy tree stumps. The hibernation generally lasts from four to five months, from October to April. The animals wake up for brief intervals, to eat. Their burrows have one entrance; according to the type of soil, they may be as deep as 1.5 m. The burrow consists of storerooms, dead-end tunnels for wastes, and a nest chamber for the young. All of this is no longer than 1–2.5 m.

Burunduks are friendly, agile, and lively. Western chipmunks repeatedly stared curiously at Klemm while he was traveling around; they watched him from the forest sometimes from as close as 1 m. During their travels through Mongolia, Piechocki and Teichert observed burunduks licking the preserve cans and the other food remains that people left behind when they broke camp.

The burunduks' range has extended more toward the west during the last century. Long ago, these rodents lived only in Siberia; by 1850, however, they were seen in the Urals and in the pinewood areas in the former province of Kasan in the Tatar A. S. R. Some time after that, they crossed the Volga, and in 1935 they were seen in Temnikov and Sarov, where they lived in the bushes near millet fields. Burunduks have recently been seen in Finland. Their ability to move around, thereby increasing their range, shows that these animals can take care of themselves.

The mating period is in April. The female gives birth to from three to five young toward the end of May or the beginning of June, after a gestation period of thirty-five to forty days. The young are mature in July, after having been nursed for twenty-eight to thirty days. They are sexually mature after eleven months. The lifespan is about six to seven years. The burunduk's natural enemies include ermines, ferrets, martens, foxes, and especially the buzzard (*Buteo buteo*); up to thirty percent of this bird's diet consists of western chipmunks during the Russian-Siberian summers.

Eastern chipmunks

The CHIPMUNKS OR EASTERN CHIPMUNKS (*Tamias*) are limited to the northern areas of the New World; they are closely related to the Old World and New World western chipmunks. The subgenus *Tamias*, which has only two teeth in the upper jaw, has only one species, the

Distinguishing characteristics

EASTERN CHIPMUNK (*T. striatus*; BL 13–15 cm, TL 7–10 cm). These animals are terrestrial rather than arboreal. They can be clearly distinguished from other chipmunks because of the characteristic stripes on their faces. They are found from Hudson Bay to Mississippi, east to the Atlantic, and to the west as far as Minnesota and Iowa.

The subgenus *Neotamias*, which has four teeth in the upper jaw, consists of at least five species: The ALPINE CHIPMUNK (*Tamias alpinus*; BL 11–12 cm, TL 6.5–9 cm). The head and body are gray; the stripes on the side of the head and body are yellowish-brown. These animals are found on the edge of the tree zone, high (2500 m) in the Sierra Nevada (western U.S.A.) The LEAST CHIPMUNK (*T. minimus*; BL 9–11 cm, TL 7.5–11 cm, weight 30–50 g). The fur of adult animals ranges from yellow with reddish-yellow stripes to a grayish-yellow with black stripes that continue down the tail. The tail is held erect while the animal is running. This species is found all over North America.

Fig. 10-16. 1. Burunduk (*Eutamias sibiricus*); 2. Caucasian squirrel (*Sciurus anomalus*).

The YELLOW PINE CHIPMUNK (*T. amoenus*; BL 11–13 cm, TL 7.5–11 cm, weight 37–50 g). The upper side of its body is a bright reddish-gray; the flanks and the underside of the tail are reddish-yellow. The stripes may be black and white or gray. This species is found in the western pine forests of North America. The COLORADO CHIPMUNK (*T. quadrivittatus*; BL 13 cm, TL 8.5–11.5 cm, weight 50–65 g). The head and body are gray; the sides of the body are reddish-yellow. The reddish-yellow tail is bushy. The tip of the tail consists of black hair surrounded by white or pale-yellow hair. These animals are found in the fir forest belt and the mountain forests of the southern mountain ranges in the Great Basin and in the Sierra Nevada. TOWNSEND'S CHIPMUNK (*T. townsendi*; BL 13–15 cm, TL 11–13 cm, weight up to 120 g) is the largest chipmunk of the New World. The coat is dark with dull-yellow to light-gray stripes on the back and sides. The inside of the ears is reddish-yellow; their outsides are gray. These animals are found in forests on the western coast of North America, from British Columbia (Canada) to California.

The EASTERN CHIPMUNK is often seen sitting on a branch or on a fallen tree trunk, with its cheek pouches full of food. When building its underground nest in areas where it is likely to be disturbed, it carries the excess soil away from its nest in its cheek pouches, transporting it to some other place. These chipmunks live singly. When they are surprised or threatened, they rear up and then dash to the safety of their nests. The diet and habitat requirements of these chipmunks are similar to those of certain species of ground squirrels, so where the habitats of these two animals overlap, there is some stiff competition—for example, between the chipmunks and the California ground squirrel (*Citellus beecheyi*), as well as between the chipmunks and the thirteenlined ground squirrel (*C. tridecemlineatus*). With its "chuck-chuck-chuck," the eastern chipmunk is more often heard than seen. The chipmunk hibernates during cold winters only in the northern areas of its territory.

Fig. 10-17. 1. Eastern chipmunk (*Tamias striatus*); 2. Alpine chipmunk (*T. alpinus*); 3. Yellow pine chipmunk (*T. amoenus*).

The upper front molar is either absent or very small. These animals are distributed all over the world, with the exceptions of Australia, Madagascar, and southern South America.

Tree squirrels are able to live and make their nests in trees because of their excellent climbing and jumping abilities. These animals have powerful hind legs, which are longer than their front legs. They use these in climbing up and down tree trunks and branches, as well as in galloping and making jumping movements on the ground; they can also jump from tree to tree with ease. The tree squirrels' diet consists of seeds, fruits, buds, and insects. They prefer to sit on their hind legs, holding the food in their "hands." They do not hibernate regularly, but when the weather is cold they become sleepy and remain in their nests for many days.

Red squirrel

The genus of RED SQUIRRELS OR EUROPEAN AND AMERICAN TREE SQUIRRELS (*Sciurus*), with some 190 species and subspecies, is the most varied of the six genera in the family of typical squirrels. Only one species, the RED SQUIRREL (*Sciurus vulgaris*; BL 20 25 cm, TL 16.5–20 cm, weight 250–480 g; see Color plates, pp. 213 and 229), is found in central Europe. It is undoubtedly one of the best-known wild mammals in Europe. Its ears are 2.5–3.5 cm long. The hindlegs are between 5.5 and 6.5 cm in length. The southern subspecies are larger than those from the north. These animals have noticeable hair plumes on their ears. The tail has hair on both sides and is bushy; when the animal sits, the tail is usually laid up over the back. The ridge of the snout is bent. The hind legs are much longer and much more powerful than the fore legs. Young animals do not have ear plumes or a bushy tail. The back is foxy-red to dark blackish-brown; sometimes the sides are a yellowish hue, while the undersides are pure white. The tail has hairs that may be up to 10 cm in length; these are parted in the middle, as in dormice. The thumbs are involuted and have a single nail; the other "fingers" are longer, with high, round, sharp claws set closely together. The "toes" are all equally developed; the claws are like those of the fore feet, but more powerful. The stomach, which has only one compartment, is from 3 to 7.5 cm long; it is separated from the duodenum by a pyloric sphincter. The small intestine makes up eighty percent of the total intestinal length; the caecum is small, between 5 and 7.5 cm in length, while the rectum is from 15 to 22 cm long. These animals have a wide distribution in European forests and prairie areas, and in the mountains up to 2200 m.

Distinguishing characteristics

Fig. 10-20. Red squirrel (*Sciurus vulgaris*) in Europe.

Man has long been aware of the red squirrel and has regarded it with affection because of its liveliness and alertness during the day, its conspicuous coloring, and its comic mannerisms. Famous artists like Albrecht Dürer have painted the red squirrel; it has also been mentioned in folk stories, aphorisms, sagas, and fairy tales. Curiously enough, the Romans and Greeks made little mention of the red squirrel, although these animals were definitely known during those times. Roman women regarded squirrels as beloved play animals. Pictures of squirrels have been found on

Cultural historical significance

urns, on marble fountains in Athens, on a Roman sarcophagus, in two mosaic floors in Romanic Switzerland, and on gravestones in Attica. Grecian children also loved these animals. In a graveyard near Keszthely, on Lake Balaton (Hungary), Lipp found skeletons of squirrels on the shoulder bones of five children. The Grecian poet Oppian tells us that in the heat of summer, red squirrels would make a shadow for themselves with their tails. As a result, they were called "shadow tails." The scientific genus name *Sciurus* has come from this name. The German name, *Eichhorn*, originally had nothing to do with the horn and the oak (as one might think from the literal translation) as Keller realized in 1909: "In medieval Latin, the *Aichhorn* was called *spiriolus*, and this name has come down through the ages as *sciurulus*, 'the little animal with the beautifully twisted, spiral-formed tail'" The early Europeans called mound-builders were familiar with these attractive arboreal animals, and they probably also ate them; squirrel bones have been found in the remnants of villages built over many Swiss lakes. The Roman naturalist Pliny must have observed the red squirrel very carefully. He wrote that the squirrel could apparently foresee the weather, as it closed off the entrance to its nest when the wind was blowing toward the entrance, and then made another entrance on the side away from the wind. It used its tail as a blanket, and, using its fore legs as hands, it brought food to its mouth.

While the red squirrel did not play a significant role in Greek and Roman symbolism, mythology, or religion, it was particularly important in Indian and Germanic religions and myths. Its red color made it holy to the god Donar—this was also true of the red fox. The squirrel called Ratatoskr ran constantly up and down the huge ash tree named Ygdrasil. In Germany and in England during its Germanic era, squirrels were sacrificed at the feast of spring and at the feast of the winter solstice. In an Indian saga the squirrel suddenly dries up the ocean with its tail. In an Indian rendition of the age-old story of Amor and Psyche, the red squirrel plays the part of the gnome or goblin. Possibly the red of the dwarf's cap is related to the color of the red squirrel. Wolfgang Gewalt cites a quotation from the 18th Century which testifies to the fact that man has long observed the red squirrel very carefully: ". . . they diligently store acorns and nuts away in the holes of trees as reserves in times of need, putting man and his lazy ways to shame. Spectators find them very amusing, especially when they make large leaps or jumps." Only during the present century have various facts concerning the red squirrel's behavior patterns and reproduction become known, largely as a result of the studies made by Gewalt and Irenäus Eibl-Eibesfeldt.

Importance as a fur animal

The red squirrel has gray hairs in its winter coat; these mix with its normal reddish color, especially on the head and backsides, thereby producing a more subdued reddish color. The eastern and northeastern forms have a greater amount of gray in the fur. Gray is the dominant color in pure strains of some north European subspecies and particularly in the Siberian subspecies; these animals give us the expensive fur known in German as *feh*. "The more pronounced the gray is in a fur, the longer and fuller the hairs will be; some squirrel furs can compete with rabbit fur because their hairs are so long and full," writes Erna Mohr. According to P. Schöps, an expert on fur, the squirrels with the best fur are found in eastern Siberia: "The fur on the backs of the animals there is a beautiful pure gray color, while the belly is white." Furriers separate the back and the belly before doing further work with the fur. Quantitatively, the squirrel has a major role in the fur industry, and the U.S.S.R. sells the most squirrel furs. Man uses only the winter coat of the northern forms, such as those from the Kola Peninsula or from those animals in the Archangel region. The purity of the gray color, the thickness, and the softness (woolliness) of the fur makes these animals far more valuable than the southern subspecies. The best furs come the territory of the rolling taiga where the squirrels have a dark-gray wintercoat, and from western Siberia, home of the larger subspecies, the SIBERIAN RED SQUIRREL (*S. v. exalbidus*).

Color variations

The German subspecies, the CENTRAL EUROPEAN RED SQUIRREL (*S. v. fuscoater*) has a dominant reddish color in the lowlands; it is generally black in the mountains. In areas where animals of both fur colors live adjacent to each other, for example in central Switzerland, the number of animals with dark fur increases with the altitude. Almost all the squirrels in the high Alps have dark fur. This phenomenon has not yet been explained; it is possible that moisture and differences in temperatures as well as specific diets (such as oily pine nuts) have something to do with the predominance of one color. The black color of the Siberian tree squirrels may have a relationship with their diet of oily pine nuts. However, hereditary factors must also be important as both light and dark young can appear in a single litter. Perhaps these color changes are the result of interbreeding between two groups of different colors which gradually come together in their wanderings. Hereditary factors and territorial movements may combine to influence the color of the fur. There are, however, even more noticeable color deviations: In 1956, in Arolsen (in the Waldeck district of Hessen, Germany), Hornung reported that he saw a yellow squirrel with dark ears.

Seasonal fur changes

The long ear plumes appear only with the winter coat. Their appearance differs with each individual animal and according to age. The tail of the BRITISH RED SQUIRREL (*S. v. leucourus*), found in the British Isles, becomes much lighter after the new hair grows in during the fall, and in the spring the tail is a light cream color or almost white. The change of hair is related to the noticeable hairiness of the ear plumes and tail. The red squirrel sheds and renews its hair twice a year, in spring and in autumn. The spring molt begins on the head and continues over the neck, back, and hindquarters, as far as the base of the tail. The males change their hair first, and this change lasts until midsummer. The females grow their new hair more rapidly, and their change is finished before the young are born. The tail hair and the ear plumes change only once a year, and this change proceeds very slowly. Toward the end of winter, some of the bristly hair and the principal hair on the tail falls out; nearly all of the underwool falls out, leaving the tail with a barren appearance. The change of the tail hair begins in midsummer, starting in the center of the tail and moving in both directions as far as the base and tip of the tail. The ear plumes also begin to grow in at this time; they reach their greatest length in winter. During the autumnal molt, the tail hair and ear plumes do not change. The autumnal molt begins at the base of the tail and proceeds in two parallel lines along the spine to the head. Again, the males change first, then the young females; the pregnant females molt relatively late. The soles of the feet are covered with hair in the winter, but not in the summer.

Climbing and jumping

Almost all of the red squirrel's life takes place in trees. The tail is particularly useful in movement. It serves as a balance pole when the squirrel runs and climbs in the branches; it serves as a rudder when the animal jumps. The tail is used as a blanket to keep the squirrel warm while it sleeps, and lastly, although not the least important use, the tail is used in courtship, where it is held in a particular manner, moved, or spread out, and thereby serves as an optical signal. The four fingers have developed as "prehensile toes." The five powerful toes, with their sharp claws, are especially well adapted to clinging on tree trunks. When moving rapidly from branch to branch, squirrels are guided by sensory hairs located on the sides of their bodies and on the outer sides of their limbs. Thus, from a certain distance, the squirrel is able to sense the difference between a branch and a twig, and it can avoid one or the other with great dexterity.

Wolfgang Gewalt, an authority on squirrels, writes: "The animal goes up tree trunks quickly and surely, moving with a jerky motion as it characteristically puts the claws of both fore feet or both hind feet into the bark and then pushes. The squirrel balances easily on rough, swinging branches. When it wants to jump to the top of a neighboring tree, it gets as close to that tree as it can; this puts the squirrel out on the peripheral branches of the tree from which it will jump; there the animal crouches, a hesitant and unhappy figure. Because the squirrel cannot make a powerful spring from such a support, it loses much of the height of its jump, thus landing in a much lower location in the neighboring tree. The squirrel climbs down the trunk headfirst, sticking the claws of its hind feet (which are stretched as far to the rear as possible) into the bark of the tree, as supports; then the animal loosens this anchor and moves its fore legs forward to a new grip position, consequently moving down the trunk in a jerky 'hop-slide' motion."

On the ground, the red squirrel moves about in graceful jumps. In

Movement on the ground

between jumps, it often half raises its body and looks around for a moment in order to examine its surroundings. When possible, the squirrel avoids remaining on the ground for long periods of time. When it wants to cross an open area in the forest, it stops at every tree in its path and uses it for an observation post. It climbs a short distance up the tree and looks around; then it rapidly climbs down again and races on to the next one, where it makes another stop, and so on. Thus, the whole movement across a clear area is broken up many times before the animal reaches safety.

Flight and concealment

When it feels secure, the squirrel will climb all around a tree, visible to all. "If, however, it feels threatened by people," says Gewalt, "it is very careful to remain protected. It climbs the tree on the side away from the people; with hardly a sound, it climbs to the top of the tree, where it presses itself close to a branch, remaining there quietly. If it finds itself near a nest, it hides in that. Squirrels that are being hunted will hide themselves in a nest for hours." The tail is important in jumping; it is used partially as a parachute, but more significantly as a means of balance. Ognev noticed that a squirrel with an injured tail is much slower in its movements than is an animal with a normal tail. All in all, aside from the tree marten, the squirrel is the best climber of all the European mammals.

Swimming

Vianden (among others) observed, in 1952, that squirrels can swim and will go into the water willingly. In Vogtland (East Germany), 50 m from the mouth of the Trieb River, in the White Elster, he saw a squirrel swimming from one bank of the river to the other. "It was a little mountain stream, and the squirrel was swimming quietly. On the other side the animal shook itself and climbed into a high tree." The animal did not show any special abilities, however, and the reports of squirrels crossing large lakes and bays must certainly belong in the realm of fable.

Jumps on the ground

On the ground, squirrels not only run; they also hop quite skillfully. After a snowfall, one can easily see the tracks of hopping squirrels. When the squirrel jumps, it characteristically places its hind feet in front of its fore feet. The individual prints are a distance of about 55 cm apart; the length of one jump is between 30 and 90 cm.

Diet

The red squirrel prefers to eat buds, blossoms, fruit, seeds, young sprouts, and tree sap, quite an understandable diet for an arboreal animal. The squirrel prefers the seeds from the fir and yew trees (the latter are poisonous to man). In spring and autumn, trees provide the squirrels with a great deal of food, as Wolfgang Gewalt observed. Even a less observant person cannot fail to notice, when walking through a forest, the coniferous trees where the squirrels feed: on the ground and around the base of the tree trunk there are numerous remains of a meal, such as scales that have been torn off pine cones, and branches covered with pine cones that have been torn from the tree and thrown down or have accidentally fallen. If the squirrel responsible for the mess is still in the tree, one will hear the unmistakable crackling sound from some distance away, and perhaps one will even see a shower of leaf sheaves and stalks, which the squirrel has thrown away in its search for seeds.

The gathering and eating of pine cones proceeds very rapidly, according to Wolfgang Gewalt. The squirrel sits quietly on a branch, and, using its "hands," grabs all the pine cones within its reach, separating them from the branch with a quick bite. It keeps its hind legs in the same place as long as possible; when it can not reach any more pine cones by stretching its body and fore feet, it moves its hind feet. Before eating the cones, however, the squirrel returns to its old seat in the tree, as the thinner branches (where most of the pine cones are) are not large enough to give it any room to spread out and "work."

The squirrel carries the pine cones it has gathered in its mouth. When eating the cones, the squirrel takes them in its hands and begins to rip off the scales, beginning with the flat side of the cone. According to Eibl-Eibesfeldt's observations, the squirrel then licks the seeds out of the cone and, holding them in its crooked thumbs, eats them, using its other fingers to hold the cone. The little winged leaflets which are attached to the seeds are left over, and they are carried away by the wind. According to Holzmaier, one squirrel in Siberia cracked 190 pine cones a day.

Red squirrels have a particular preference for hazelnuts and green and ripe walnuts. Zoologists have repeatedly studied how the animals open these nuts. The squirrel grips the nut with its hands and places both of its upper incisors on a part of the shell as supports; then the animal gnaws a hole in the nutshell with its lower incisors. If this hole is large enough so that both of the lower incisors can fit in, these teeth break the nut. The mobility of the two halves of the lower jaw is of great importance in this activity, as are the individual lower incisors, which the animal uses like tweezers to help it pick out the pieces of the nuts. Eibl-Eibesfeldt has observed that squirrels initially work with a great deal of effort and take a long time at this activity. They begin by gnawing erratically; finally, however, they learn to combine their innate movements into an organized procedure. "Inexperienced squirrels gnaw anything that is the size of a walnut," he writes. "They gnaw 'glass nuts' and wooden bobbins. However, the animals rapidly learn to distinguish between those objects which are suitable and those objects which are not. Inexperienced squirrels open nuts with worms, and dried nuts; experienced squirrels merely sniff these nuts and throw them away."

Mushrooms, sap from birch, oak, and maple trees, berries, and fruits (some animals prefer pears) are also part of the red squirrel's diet. They will also eat other animals, including snails, ant pupa, insects and their larvae, and eggs and newborn birds, for which the squirrels take to robbing nests. Siberian hunters attract squirrels to their traps with dried fish.

When there is an abundance of food, the squirrel stores some away.

Stockpiling of provisions

It hides pine cones, nuts, seeds, and mushrooms in various places in the ground; it prefers to hide its store at the foot of a tree, in tree holes, or in old birds' nests. Eibl-Eibesfeldt studied and observed the squirrel's hoarding behavior, and he has concluded that, as a result of certain key stimuli, these actions are instinctive. "When a collecting animal finds a nut on a hazel bush, it picks the nut, throws away the green outer shell, and climbs down to the ground with its prize. There the animal scrapes away a certain area with its fore legs, and makes a little hole in the ground, where it deposits the nut. The nut is pressed into the ground quickly and powerfully as the squirrel uses its upper incisors to pack the nut down. Then the animal collects some soil and leaves with its fore legs, and puts these over the nut, finishing off its actions by packing down the ground with its feet. Not one of these actions is ever omitted. It becomes apparent that this procedure is innate when young squirrels, which have not had the opportunity to dig up the soil, hide their first nuts in their nestroom. After a long search, these animals scratch out a little area in the corner of the nestroom; using their mouths, they put the nut in the scratched out area, and then make scratching and packing motions in the air. We can see a relationship between these actions and the vain efforts of a young dog to bury a bone in a room."

Squirrels can smell pine cones that are as much as 30 cm under the snow. While searching for food in the winter, they find many of the stores they had collected during the summer; they then scratch these out of the snow. Naturally, a percentage of the buried seeds remain in the ground; the squirrel, therefore, does its part, however inadvertently, in increasing the distribution of trees and bushes, particularly the Douglas fir in North America.

Drinking

Unlike many other squirrels, the red squirrel must drink in order to survive. In dry years, it remains close to whatever water source is available in the forest; in winter, it quenches its thirst with snow. The vocalization of the red squirrel is a "tjuk-tjuk-tjuk" sound; however, these animals also coo and squeak, and, when in danger, they make sharp growling or snapping sounds.

Sensory organs and abilities

The eye is the principal sense organ. Squirrels have sharp eyesight and a wide field of vision, and they can easily fix their eyes on a distant object. Tame squirrels are able to distinguish individual people from some distance away. The squirrel is able to observe vertical objects particularly well because of the special structure of its retina; this is a particularly important ability for tree-dwellers. The squirrel evaluates distances by raising or lowering its head. It has a very weak ability to see color. The sense of touch is greatly augmented by brushlike hairs which are found not only on the animal's head (the well-known vibrissae), but also at the roots of all four feet, on the outer sides of the fore legs, and on the underside of the body, as well as the root of the tail. These hairs make it possible for the animals to keep the right distance away from objects, and, when they are in trees, these hairs enable them to keep contact with their supports.

Innate clasping reflexes to some extent protect the squirrel against falling. Rustling and crackling sounds are causative stimuli relating to security and flight. Squirrels escape from birds of prey by racing down the tree trunk in spirals; they attempt to escape from the pine marten by climbing into the highest branches of the tree and jumping from there to the ground, something the pine marten can not do. Rearing up when in danger is an innate behavior of the squirrel. If one places nesting materials that smell of skunk before a squirrel, it will become highly excited; squirrels even recoiled before a stuffed skunk.

Fig. 10-21. Red squirrel (*Sciurus vulgaris*) in a display stance. The animal attempts to appear larger than it really is.

Red squirrels may lose their fear of people after several positive experiences. In parks they may decrease their innate flight distance to such an extent that they will even take food from visitors. Squirrels show a very practical behavior when harvesting pine cones. Experienced animals do not attempt to tear the cones down; instead they draw the cones toward themselves, biting them off at the branch. Eibl-Eibesfeldt, when observing red squirrels being reared in captivity, felt that these animals were able to understand certain coordinate movements: "As soon as one went toward the door of the room, they rapidly ran and placed themselves in front of the prospective opening in order to escape the minute the door opened. When this technique was finally circumvented (after much effort on the part of the person involved), they discovered a new technique; they jumped on their guardian as soon as he made a move to leave the room."

Fig. 10-22. Red squirrel in defensive threat position

. . . and in accentuated aggressive threat posture.

Red squirrels assume the "typical squirrel stance" in order to groom their bodies. According to Bürger, the "head" and "hand work" is usually done together. Squirrels rarely scratch themselves with their hind legs, and it is even more rare for them to lick between their toes (a common practice with many other rodents). They like to scratch their rump and stomach against branches. They wipe their noses on tree bark after every meal. Red squirrels like to "bathe" on fresh moss; in captivity, they will also bathe on a wet cloth.

Red squirrels spend the night sleeping in their nests. They will not appear for days if the weather is wet and ugly, or if the dry winter frost sets in. Although they do not really hibernate, usually several animals—always including the young animals which remain together—will huddle together in one nest and sleep when it is cold outside, wrapping their tails around one another for warmth. However, one can also see red squirrels hunting for food when the temperature is very low and snow covers the ground.

Migrations

In eastern Europe and Siberia the movement of the red squirrels occasionally turns into a mass migration. According to Formosov, the animals travel alone, each out of its neighbor's field of vision, in extremely large fronts; they never travel in packs. Thus, in 1928, on the lower

Amur River (U.S.S.R.) there was a front line of squirrels that extended for 300 km. The squirrels rarely migrate in large numbers during the spring; in the fall, however, when the supply of pine nuts, other nuts, and other food plants is insufficient, these migrations are more common. Because red squirrels have a high birth rate in years when the food supply is adequate, the number of animals often far exceeds the available food supply during the next few years. The jays and crossbills generally migrate a year before the squirrels, as these birds can adapt more quickly to a poor harvest. In many cases, the squirrel migration in the fall is arrested only by the winter frost. Population densities and fluctuations are, in the final sense, "answers" to the fruit cycle of the fir trees. In central Europe, for example, the fir trees have "seed years" every three to five years; cedars of the western taiga are rich in fruit every three to five years, although this occurs only every ten years in the northern regions.

Nest of the red squirrel

Red squirrels build their round, roofed nests on forked branches in the older, higher trees of shady forests, bosky fields, and parks. The nests, made from moss and twigs, are built 5–15 m above the ground; they may be up to 50 cm in diameter, and some 30 cm high. The floor of the nest has a construction similar to that of a large bird nest. The main entrance leads down and to the side; it can be closed off during the brooding period. A tiny escape hole is made against the tree tunk. Squirrels like to rebuild large bird nests, especially crows' nests, strengthening the floors of these structures with soil and loam. They also build their nests in abandoned starlings' nests, holes in trees, and even in roof rafters. The young are raised in the more carefully constructed main nests. Squirrels usually have other nests which are used as hiding places or as sleeping chambers. The inner nest may be lined with grass, husks, lichen, moss, feathers, wool, and occasionally even paper. The outer edge of the nest is heavily covered with leaves. The squirrel carries nesting materials in its mouth, balancing equal portions on the right and left sides.

Mating behavior

The mating season lasts from spring through summer. The males become particularly aggressive during this time, challenging their opponents on the slightest provocation, and invading the females' territories. The females then attack the males, who, in turn, attempt to intimidate the females through threat behavior, imposing motions, and mock fighting. Then the males attempt to overcome the females' fear of contact. Eibl-Eibesfeldt writes: "The male expresses his readiness for mating through loud vocalizations which sound very similar to the sounds of the young. If the male catches up with the female, he stands before her so that she is sideways to him. The male wags his bushy tail a few times and then, with a particularly slow gesture, he lays it on his back. The female's flight, which initially is an earnest attempt to escape, ultimately becomes merely a symbolic running away. The female signals her readiness for mating by lifting her tail and by ejecting a small amount of urine (odor mark). After a courtship which may last anywhere from hours to days,

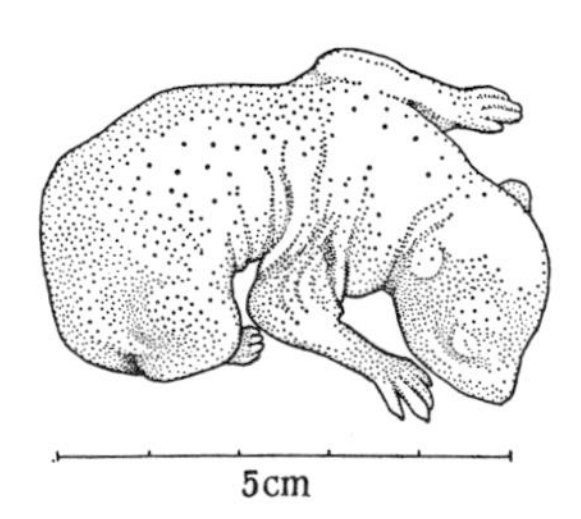

Fig. 10-23. Newborn red squirrel (*Sciurus vulgaris*) is blind and hairless at birth, a typical example of an insessorial animal.

the animals mate. The pair occupies the same nest for a short period of time. During copulation, the male mounts the female from behind, grasping her haunches. The female bends her back and lifts her tail. The copulation becomes possible only as a result of this active cooperation; thus rape is impossible."

Like all other squirrels, the red squirrel has a penis bone in the penis; this structure is one factor in the systematic classification of the squirrels today. It is shaped like a shovel, with a thick outer edge and a wide base. The female sexual organs in the area of the vagina become very horny during the mating season. The female has eight nipples. The pregnancy lasts for thirty-eight days; during this time the females again become irritable, and they chase the males from their hunting grounds. The litter size ranges from three to eight, usually five, young; they are pink and hairless at birth. The young also have a remarkable skin fold, extending from the elbows to the flanks, which is reminiscent of the flying squirrels' flightskins.

Pregnancy and birth

These tiny insessorial animals weigh 8–12 g at birth. They grow their first fluffs of hair after ten to thirteen days; by nineteen days they already are completely covered with hair. The eyes open after thirty to thirty-two days. The lower incisors appear after twenty-two days, while the upper incisors appear at about the thirty-fifth day. During this time the young animal develops its body movements and its cleaning, gnawing, and defense movements. The young animals leave the nest for the first time at about the forty-fifth day, and after eight weeks they become independent, after having been carefully nursed and cared for by the mother for this whole period. When nursing, the young use characteristic "pump sucking" motions, whereby their lips surround the base of the nipple, making it airtight. After some forty days, the animals take in their first solid food, tearing off pieces from the mother's food while she is eating.

Growth of the young

Fig. 10-24. 1. Eastern fox squirrel (*Sciurus niger*); 2. Western gray squirrel (*S. griseus*); 3. Arizona gray squirrel (*S. arizonensis*).

The squirrel mother is very careful about the cleanliness of her young; she holds them in her fore paws, turns them around, and licks them very thoroughly. When transporting her young or when, as a result of being disturbed, she carries them to another nest, the mother usually holds her young by the skin fold on their backs. She carries older young with her fore paws, thereby coaxing them into rolling up together; they hold tight to the mother's neck with their fore feet and tail. The squirrel does not have an active nipple transport. Young animals that have fallen or jumped out of the nest whistle loudly, alerting their mother with these "lost signals." When the family climbs through the treetops, the mother leads the young animals, keeping them together with a soft "duck-duck-duck" vocalization.

After they have been weaned, the young animals spend a great deal of time at play. Their skirmishing is clearly distinguishable from real fighting; social restraints protect the fighting partner from serious harm, and there is no introductory threat posture. Adult squirrels can also be rather play-

Playing and scuffling

Fig. 10-25. 1. Gray squirrel or Eastern gray squirrel (*Sciurus carolinensis*); 2. Kaibab squirrel (*S. kaibabensis*); 3. Abert's squirrel (*S. aberti*); 4. Apache squirrel (*S. apache*).

ful. A trusted human attendant is even allowed to play with them without being earnestly bitten, because the squirrels bite down very carefully. They invite people to play with them through certain gestures, even when they are no longer interested in playing with animals of their own species. Squirrels also like to play with pine cones and paper balls. Tame and wild squirrels alike turn somersaults, hold climbing practices, and play similar games of motion that are indicative of their strong activity drive.

In games of chase (escape), the important action is not the overtaking of the partner, but rather the chase itself. Eibl-Eibesfeldt has described it as follows: "The pursuer is evidently much less interested in the situation than is the pursued. When the squirrels play in this manner, each animal tries to hide out of sight of the other. If one squirrel sees its play partner coming, it rushes for cover on the other side of the tree trunk, waiting there to see if the partner will follow. The partner, however, 'thinks' that the movements of the other squirrel are directed against it, and so it too flees to the other side of the tree trunk; there the two squirrels meet, and each flees again. Thus, one can not distinguish in cases like this, which animal is pursuer and which animal is pursued."

Fig. 10-26. Gray squirrel (*Sciurus carolinensis*) in Europe, where it was introduced by man.

After they have become independent, young squirrels remain near the maternal nest for some months. They are threatened mainly by the hawk and the pine marten. Squirrels which do not fall victim to their enemies gradually spread themselves out over a wide area. They are sexually mature after eight to ten months. Only twenty to twenty-five percent of the young animals live to be one year old, although squirrels can live to be eleven or twelve years old. In years when the food supply is insufficient, up to forty percent of the young animals may die. In the northern regions of its distribution area, the European red squirrel has one or two litters per year; in the southern areas of its range, it has two or three litters a year. The female may become pregnant again even as she nurses her young (during the lactation period). In the Berlin zoo on 13 February 1957, while checking a titmouse nest which was 2.5 m above the ground and which had an enlarged entry hole, Wolfgang Fischer found two young squirrels which were approximately thirty days old. Births during the winter, when the weather is mild and food is plentiful, are certainly not infrequent.

The squirrel is a lively and beloved addition to our woods and parks, in spite of the damage it occasionally does to the forests. We want to be careful, then, to keep these animals alive for the future.

Gray squirrels and other species

We will mention a few of the other tree squirrel species: The GRAY SQUIRREL (*Sciurus carolinensis*; BL 20–25 cm, TL 19–20 cm, weight 340–680 g; see Color plate, p. 229) is larger and more powerful than the European red squirrel. The ears never have plumes. The backside of some individuals turns white in winter. The tail is bushy; in summer it has a white edge. Originally these squirrels were found only in the eastern U.S.A. The CAUCASIAN SQUIRREL (*S. anomalus*) is brownish with brown under-

Distinguishing characteristics

F. Reimann

MARSH AND LOWLAND FOREST SCENE IN CENTRAL EUROPE:
Mammals: □ Rodents: 1. Beaver (*Castor fiber*), extinct in central Europe except for a small population on the central Elbe, a) felling a tree, b) swimming, c) on a beaver mound; 2. Muskrat (*Ondatra zibethica*) imported from North America as a fur animal; 3. Vole rat (*Arvicola terrestris*); 4. Harvest mouse (*Micromys minutus*); its nest is usually hidden in the grass. □ Carnivores (see Vol. XII): 5. Ferret or fitchew (*Mustela putorius*); 6. Common otter (*Lutra lutra*), largely extinct in central Europe. □ Artiodactyla (see Vol. XIII): 7. European moose (*Alces alces*), almost totally extinct in central Europe.
Birds: □ Grebes (see Vol. VII): 8. Great crested grebe (*Podiceps cristatus*) in its nest. □ Grallatores (see Vol. VII): 9. Black stork (*Ciconia nigra*), appears only rarely in central Europe; 10. Gray heron (*Ardea cinerea*); 11. Male cinnamon bittern (*Ixobrychus minutus*). □ Ducks and geese (see Vol. VII): 12. Mute swan (*Cygnus olor*); 13. Greylag goose (*Anser anser*); 14. Mallard (*Anas platyrhynchos*); 15. Green-winged teal (*A. crecca*). □ Birds of prey (see Vol. VII): 16. Marsh harrier (*Circus aeruginosus*). □ Ralliformes (see Vol. VIII): 17. Coot (*Fulica atra*); 18. Common moor hen (*Gallinula chloropus*); 19. Spotted crake (*Porzana porzana*). Waders and gull-like birds (see Vol. VIII): 20. Black-headed gull (*Larus ridibundus*). □ Songbirds (see Vol. IX): 21. Great reed warbler (*Acrocephalus arundinaceus*). Coraciiformes (see Vol. IX): 22. Kingfisher (*Alcedo atthis*). □ Reptiles (see Vol. VI): 23. Ringed snake (*Natrix natrix*); 24. Swamp turtle (*Emys orbicularis*). □ Amphibians (see Vol. V): 25. Edible frog (*Rana esculenta*). □ Insects (see Vol. II): 26. *Calopteryx virgo*, male; 27. *Calopteryx splendens*, male.

sides. These animals do not have a special winter coat or ear plumes. There are six calluses on the soles of the foot; the tail is relatively short. The females have ten nipples. These animals are found in forests and mountains of beech trees and other nut trees in eastern Turkey, Syria, Iran, and Transcaucasia. The BRAZILIAN SQUIRREL (*S. aestuans*) is a representative of a single subgenus, *Guerlinguetus*. These animals are brown with a whitish-gray throat and a reddish-yellow breast and stomach. They are found in Brazil, Surinam, Guyana, and Venezuela. Two subspecies of the EASTERN FOX SQUIRREL (*S. niger*) are threatened with extinction, as is the KAIBAB SQUIRREL (⦶ *S. kaibabensis*), from Arizona.

The GRAY SQUIRREL was originally an inhabitant of North American oak woods, although it has long since settled in pine forests as well. In times of over-population as well as other times when the food supply is insufficient for even a normal population, large numbers of these squirrels migrate; in the past they even moved into fields. For this reason, they have been rigorously persecuted by man. In Pennsylvania, during the middle of the 18th Century, 640,000 animals were killed in one year when there was an excessive population. The gray squirrel competes for its food with the woodpecker *Balansphyra formicivora*. Usually several woodpeckers will defend a communal food tree together, chasing the squirrels away. The squirrels, in turn, avoid woodpeckers' trees, and, even when threatened, they will not climb these trees. The woodpeckers do not chase their competitors once they are on the ground, however. Gray squirrels are extremely quick, and over short distances they can reach a speed of 19 km/h (12 mph). It is interesting to note that these animals react not only to the alarm calls of their conspecifics (other animals in the same species), but also to the call of the junco (*Junco oregonus*), a New World bird.

Habits of the gray squirrel

The gray squirrel was introduced in England toward the end of the 19th Century. The first 350 animals were released in the county of Bedfordshire in 1889. Since that time, the animals have multiplied rapidly in their new home, driving out the native squirrels. The gray squirrels have spread out in forested regions throughout the British Isles. As in North America, however, these animals are also found in parks and city gardens, where they become very tame. Between 1945 and 1952, these animals spread out over an area of some 200 km^2, from which the native English squirrels disappeared. The number of gray squirrels is presently estimated to be over 1.5 million animals. Gray squirrels like to peel the bark from fir, larch, beech, birch, and ash trees, although they will also eat plant seedlings, sprouts, and buds. The trees from which the bark has been peeled die, causing the forestry industry in England to lose a great deal of potential income each year, sometimes several hundred thousand pounds sterling. The British hope to be able to control the gray squirrel, and with this in mind they have appointed official hunters.

Resettlement in England and South Africa

In South Africa, the well-known statesman Cecil Rhodes introduced gray squirrels in the newly planted forests around Cape Town at the turn of the century. The distribution of these animals has increased with the growth of oak and pine woods; today these animals even raid farmers' fruit gardens.

The CAUCASIAN SQUIRREL collects large winter supplies of beechnuts acorns, chestnuts, and other nuts, which it stores in holes in the ground or in trees, or under tree roots. The BRAZILIAN SQUIRREL lives in thick woods with dense underbrush, as well as in the shrubbery of abandoned plantations; it particularly prefers active cocoa plantations.

Dwarf tree squirrels

The DWARF TREE SQUIRRELS (*Microsciurus*) are small tree squirrels with short, straight, and narrow tails. They are found from southern Central America and as far south as Peru, Ecuador, and the Amazon area. There are twenty-one species and subspecies, which can only be distinguished by an expert. We will mention only the species *Microsciurus alfari*, from Panama and Costa Rica, where it has a position similar to that of the European red squirrel.

Groove-toothed squirrels

The GROOVE-TOOTHED SQUIRREL (*Rheitrosciurus macrotis*; BL over 60 cm), found on Borneo, is a large squirrel with remarkably long ear plumes. The squirrel's sides are striped, and it has a long tail which is particularly thick and bushy. These animals prefer to live on the ground, like ground squirrels.

Beautiful squirrels

Many of the ASIAN SQUIRRELS (tribe Callosciurini) are beautifully colored: red, brown, gold, yellow, with black and white markings; others, however, are quite drab and plain. In some species each individual hair is ringed with six different colors. This group, which has many genera and species, is found mostly in Indo-Malayan countries; its members are distinguished by a number of common skull characteristics. These animals do not collect supplies, and they do not hibernate.

Squirrels of the genus *Callosciurus* (BL 12.7–28 cm, TL 7.6–25.4 cm, weight up to nearly 500 g) are at the center of this tribe. There are twenty-five species, with 320 subspecies, all of which are divided into twelve groups, including SWINHOE'S or the ASIATIC STRIPED SQUIRREL (*Callosciurus swinhoei*; see Color plate, p. 230) and the SPOTTED SQUIRREL *C. notatus*, which is found on Java, Bali, Borneo, Sumatra, and many small Indonesian islands. These animals are divided into numerous island subspecies, some of which have an unusually small and even distribution. They prefer to live in coffee and cocoa plantations.

The ORIENTAL PYGMY SQUIRRELS (*Nannosciurus*; see Color plate, p. 230) are not much larger than a mouse. They are distinguishable by white ear plumes which are comparatively longer than those of other squirrels. These squirrels are found on the Philippines, Sumatra, Java, and Borneo. There are five species, including WHITEHEAD'S DWARF SQUIRREL (*Nannosciurus whiteheadi*; BL 8 cm), which is speckled with olive-gray. It lives in the thick treetops of the tropical rain forests in northern Borneo. The BROWN DWARF SQUIRREL (*N. melanotis*) lives in Java, only on the tree

Fig. 10-27. Southern flying squirrel's landing technique.

species *Pometia tomentosa*; two squirrels always live together on the same tree.

The LONG-NOSED SQUIRREL (*Rhinosciurus laticaudatus*; BL 20.3–24.2 cm, TL 9.5–12.7 cm) is the only representative of its genus. The nasal region of the skull is unusually elongated. Because of its brown coloration and the elongated snout area, this squirrel resembles the tree shrew (Tupaiidae), the ancient relative of the insectivorous animals, to such an extent that local peoples cannot tell the two animals apart, and so call them both by the name of *tupai*. These squirrels are found in Malaysia, Sumatra, Borneo, and the neighboring islands. They prefer to eat aromatic fruits and insects, particularly ants.

The LONG-SNOUTED SQUIRREL (*Hyosciurus heinrichi*) which lives in Celebes, has a proboscislike snout area. The tail is short. The claws are very strong, particularly on the fore feet, somewhat like those of the long-clawed ground squirrel. Its lifestyle is similar to that of the northern ground squirrel. Its teeth are very small; they are particularly well adapted to this squirrel's primarily insectivorous diet. Two other species of long-nosed squirrels are pictured on page 230, PERNY'S LONG-NOSED SQUIRREL (*Dremomys pernyi*) and the MALAYAN BLACK-STRIPED SQUIRREL (*Lariscus insignis*). Another related species, the BORNEAN BLACK-STRIPED SQUIRREL (Ø *L. hosei*), found in northern and northwestern Borneo, is presently in danger of extinction.

African tree squirrels

The Indian and African squirrel species grouped together in the tribe of AFRICAN TREE SQUIRRELS (Funambulini) prefer to eat palm tree nuts and fruit, although they will also feed on other fruit and on insects. The PALM SQUIRRELS (*Funambulus*) have an elongated snout area and pale stripes running down the length of their backs. The INDIAN PALM SQUIRREL (*Funambulus palmarum*; see Vol. XIII) occasionally looks for mangos or other cultivated fruits. Like other members of its tribe, the Indian palm squirrel has no definite mating period; the male's testes are active throughout the year.

The AFRICAN STRIPED SQUIRRELS (*Funisciurus*; see Color plate, p. 230) inhabit jungle areas. The WESTERN AFRICAN STRIPED SQUIRREL (*Funisciurus lemniscatus*) is found from Cameroon to Angola and parts of Namibia (formerly South West Africa); this squirrel is called "*osen*" by people in southern Cameroon. It is a tiny reddish-yellow animal with two pairs of black and white stripes on each side. These squirrels carefully and skillfully build spherical nests; when the mother leaves her young alone for awhile, she closes up the entrance.

The AFRICAN BUSH SQUIRRELS (*Paraxerus*), with eleven species and forty-four subspecies, are found in the forests of eastern and southern Africa. The MANTLED AFRICAN BUSH SQUIRREL (*Paraxerus palliatus*) is fiery-red on its tail (up to the tip), underside, legs, and ears; the back and tip of the tail, in contrast, are speckled with dark spots. These animals prefer corn (maize or Indian corn) and peanuts; they will rip the cotton from cotton bolls to get to the unripened seeds, which they eat. BOEHM'S AFRICAN BUSH SQUIRREL (*P. boehmi*) prefers to eat insects, particularly butterflies, large caterpillars, and even termites, which it catches in flight. The color plate on page 230 shows the species *P. flavivittis*.

The tiny AFRICAN PYGMY SQUIRREL (*Myosciurus pumilio*; BL up to 7.5 cm, TL 6 cm; see Color plate, p. 230), from Gabon, is the sole representative of its genus. Its tail is relatively thin; the skull has an unusually broad frontal bone.

Indo-Malayan giant squirrels

The tribe of INDO-MALAYAN GIANT SQUIRRELS (Ratufini; BL 25-50 cm), which adjoins that of the African tree squirrels, has only one genus, with four species and seventy-two subspecies. These squirrels are larger than pine martens. The upper and undersides are often colored quite differently; sometimes there are facial markings in the form of white or yellow stripes. The tail, which is thick and bushy, is almost as long as the body. The fore feet are unusually wide; they have inside sole cushions instead of thumbs. The hind feet are also wide; they have fully developed first toes. These animals are found from India to Indonesia.

The MALAYAN GIANT SQUIRREL (*Ratufa bicolor*; see Color plate, p. 230), found on Java, Bali, and Sumatra, does not have ear plumes and is thus easily distinguished from the Indian forms. It occasionally raids coffee, banana, and fig plantations. This species has been exhibited in several European and American zoos. They are black on the back, and light yellow on the undersides. They become quite tame in captivity, and they amuse visitors with their lively mannerisms. The life expectancy of this species may reach sixteen years; like most squirrels, however, they generally live about ten years.

The INDIAN GIANT SQUIRREL (*R. indica*), from India, has ear plumes. Its upper side is dark red; the shoulders, middle of the back, thighs, and tail are black. The undersides and throat are whitish-yellow. These animals have a pale stripe which runs in front of the ears, diagonally over the crown of the head. The Indian giant squirrel builds its large nest of leaves and twigs high in the tops of trees. It searches for food throughout the day, taking a break only during the hot noon hours. It can jump as far as 6 m from tree to tree.

Oil palm squirrels, ebien or African palm squirrels, and sun squirrels

The last tribe of ground and tree squirrels, Protoxerini, includes the OIL PALM SQUIRRELS (*Protoxerus*), the EBIEN SQUIRRELS or AFRICAN PALM SQUIRRELS (*Epixerus*), and the SUN SQUIRRELS (*Heliosciurus*). There are seventeen species in Africa, of which we will mention five: The OIL PALM SQUIRREL (*Protoxerus stangeri*; BL 30 cm, TL 40 cm; see Color plate, p. 230) is one of the largest tree squirrels. Its tail is short with white bands. The stomach is sparsely haired, often almost naked. There are numerous subspecies from Ghana to eastern Africa. It is always found on oil palms.

WILSON'S PALM SQUIRREL (*Epixerus wilsoni*; BL 25–30 cm, TL 28–30.5 cm) is found from Ghana to Cameroon and Gabon. Its stomach is also sparsely haired. The tail is longer than the body.

The EBIEN SQUIRREL OR AFRICAN PALM SQUIRREL (♁ *E. ebii*; see Color plate, p. 230). The WEST AFRICAN SUN SQUIRREL (*Heliosciurus gambianus*; BL 15.3–25.4 cm, TL 15.3–30.5 cm), the major representative of the sun squirrels, is found in fifty-two prairie and jungle forms over a large part of Africa, from Ethiopia to Rhodesia. None of the genera has a penis bone. The upper side is a mixture of grays; the underside is mouse-gray and reddish-gray to reddish-brown. The tail has black and white bands. The feet are a dirty-gray. When danger threatens, these squirrels press themselves flat against the side of the tree trunk away from the threat, with their legs outstretched; they remain motionless in this position. Another sun squirrel is *H. undulatus* (see Color plate, p. 230).

Subfamily: Flying squirrels, by H.-A. Freye

Like the flying marsupials and the flying lemurs, certain rodents are able to make gliding flights. These include the scaly-tailed squirrels and FLYING SQUIRRELS (subfamily Pteromyinae). Like the other gliders, these arboreal flying squirrels also possess a furry flightskin on the sides of their body, which works like a parachute whenever the animals jump. It extends from the front legs, where it is attached to a crescent-shaped bone or cartilage at the root of the fore paw, to the ankle joint. The flying squirrel uses its round, bushy tail as a rudder; the tail often has hair on both sides. With the help of its flightskin, the animal is able to glide through the air for relatively large distances, 30–50 m or more, and thus escape from its most dangerous enemy, the marten. It can change its direction in flight by changing the angle of its outstretched flightskin and the position of its tail. When landing, the flying squirrel raises its arms and tail, brings the longitudinal axis of its body into a more vertical position, and decreases the landing impact with the help of its flightskin.

Flying squirrels are active mainly in the evening and at night. The fore feet have four toes; the hind feet have five. All four feet have strong, curved claws, excellent climbing tools. The hair is thick, glossy, silky-smooth, and sometimes almost woolly. The head is generally rounded in the usual manner. The upper jaw has five molars on each side; the lower jaw has four molars on each side. All of the molars have relatively high crowns. There are no cheek pouches. The food is carried into special "food nests." The usual vocalization is a shrieking sound. The eyes are very large, and they protrude, indicating that these are nocturnal animals; the eyes and the ears are their most important sensory organs. There are thirteen genera with thirty-six species, distributed in the northern areas of the Old and New Worlds as well as in central, eastern, and southeastern Asia, and the Malay Archipelago.

All flying squirrels are genuine forest animals. There they eat buds, fruits, nuts, and insects. They are much slower and more awkward on the ground than in trees, as the flightskin is a hinderance to walking. As Ingo Krumbiegel has said, these squirrels are particularly fond of playing; they "glide back and forth in large groups, snatching at one another and competing with each other in speed and versatility." Their spacious nests or tree holes are very well lined. The northern species prefer to live in abandoned woodpeckers' holes, which they line with moss. The south Asian flying squirrels build their nests between palm fronds or in empty coconuts. The INDO-MALAYAN FLYING SQUIRREL (*Hylopetes lepidus*) hunts for nuts that other squirrels have already gnawed into; these nuts are consequently dried out. Flying squirrels sleep in their nests all day and during inclement weather; they also give birth to their young in these nests. The number of young varies between two and four; we have no idea of the number of young in some species. The sexual partners wrap their arms around each other and the male uses his flightskin like a cloak to surround the female.

Fig. 10-28. European flying squirrel (*Pteromys volans*).

Flying squirrels have various other enemies besides the marten; these include, among others, owls and many birds of prey. The bay owl (*Phodilus badius*) specializes in the smaller flying squirrels of Nepal, the island of Ceylon, Vietnam, Borneo, and Bali. It is believed that flying squirrels have a life expectancy of up to eleven years. The Asian forms are only rarely found in zoological collections, so the fact that we are presently unfamiliar with many forms is understandable.

Giant flying squirrels

The COMMON GIANT FLYING SQUIRREL (*Petaurista petaurista*; BL 60 cm, TL 60 cm) is almost a giant among the flying squirrels, with a wingspan of some 60 cm. The ears are short and wide, with a thick tuft of hair at the back. The tail is bushy and of an even thickness. The upper sides are grayish-black. The head, sides of the throat, upper side of the flightskin, and the legs are all chestnut-brown. The undersides are a dirty-gray; the tail is black. The underside of the flightskin is grayish-yellow with an ash gray border. These animals are found in eastern India and Burma, and on the island of Ceylon.

This giant flying squirrel prefers the highest and thickest areas of the trees, where it builds huge nests, up to 1 m in diameter. These squirrels can glide as far as 60 m on their nocturnal excursions. They are almost exclusively leaf-eaters. The FORMOSAN GIANT FLYING SQUIRREL (*Petaurista grandis*; Color plate, p. 240) is a closely related form.

Apparently even the ancient Greeks were familiar with flying squirrels. In an account of Alexander's campaign, a passage concerning "night foxes" says, "they rose up out of the sand and jumped as far as eight or ten yards [2.5–3 m]." Aristotle, too, mentioned a "fox with wings of skin." In as much as these ancient reports are not about fruit bats, we must assume that they were describing the EUROPEAN FLYING SQUIRREL (*Pteromys volans*; BL 16 cm, TL 10 cm, weight 140 g; see Color plate, p. 240) which is distributed, in nine subspecies, over northern and eastern Europe, Siberia, and the northern areas of eastern Asia. It is brown with white underparts; the flightskin and the outer sides of the legs are darker than the rest of the body. The tail is particularly furry on both sides; when the animal is young, this furry coat consists of very small hairs. During winter, the coat is silky-smooth and the tail and upper side of

European flying squirrel

the body are silver-gray.

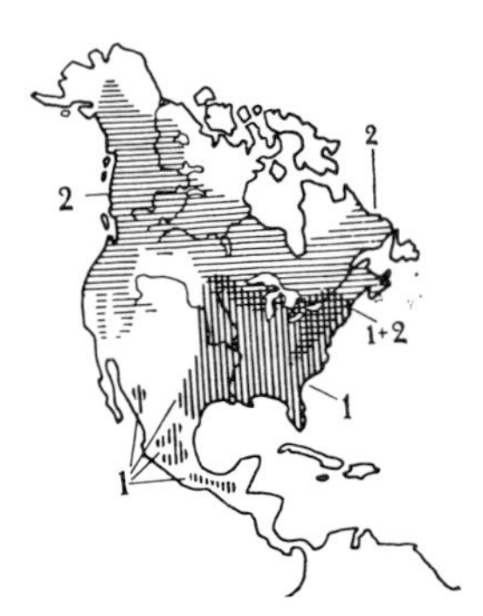

Fig. 10-29. 1. Southern flying squirrel (*Glaucomys volans*); 2. Northern flying squirrel (*G. sabrinus*).

Because they are such versatile climbers, these flying squirrels prefer birch, spruce, and Scotch pine forests; the northern limit of their range corresponds to that of the birch trees. They are able to glide as far as 35 m from tree to tree. They eat birch bark (which they peel from the tree in rings), birch leaves, sprouting buds from both coniferous and deciduous trees, pine seeds, alder catkins, berries, and mushrooms. In winter, the European flying squirrels in Yakut (U.S.S.R.) apparently eat larch buds almost exclusively. In virgin forests, they make their nests in the holes of old trees, where they give birth to between two and four young annually in spring. The young are blind for about their first fourteen days. In times of dry winter frost, the animals rest quietly in their nests for days, without hibernating. These animals have a remarkable "separation according to trees." Often there will be many animals on one tree, but, except during the mating period, all of these animals will be of the same sex.

The flying squirrel is gradually moving east out of Europe, as it flees from civilization. The reshaping of the land through development of the forestry industry, and, to a markedly lesser extent, activities of the squirrel's natural enemies, the marten and the forest owl, might be said to be responsible for this move. The flying squirrel is hunted in Siberia, its major distribution area, because of its fur, which is sold under the name of "Molenda" (erroneously as "flying dog"); it is usually dyed and used in fur borders. The European flying squirrel, like its related species, can occasionally be found in zoological collections; today it is generally kept with the nocturnal animals, becoming quite active in the artificial twilight. A giant flying squirrel lived in the London zoo for thirteen years and seven months. Occasionally flying squirrels in captivity have even reproduced.

The woolly flying squirrel

The WOOLLY FLYING SQUIRREL (*Eupetaurus cinereus*) is relatively large. Because it is a mountain animal, it has a long fur; even the soles of its feet are hairy. It lives on cliffs in northwestern Kashmir, where it springs or glides between the rocks with great versatility.

Southern flying squirrel, by B. Grzimek

Flying squirrels are also found in North America; they include the SOUTHERN FLYING SQUIRREL (*Glaucomys volans*; BL 13–15 cm, TL 9–11 cm, weight 50–70 g; see Color plate, p. 240), which prefers thick forests. It has thick, glossy hair. The NORTHERN FLYING SQUIRREL (*G. sabrinus*; BL over 15 cm, TL 15 cm, weight up to 125 g), found in forests with a mixture of pine and birch trees, is somewhat heavier. The stomach hair is white at the tip.

Distinguishing characteristics

Like all other flying squirrels, the southern flying squirrel has a narrow piece of cartilage on its wrist joint. When the animal spreads its arms for gliding, this cartilage is splayed and thus stretches the flight-skin back to the hind legs. When this tiny creature sits down to crack open a nut, in the same manner that squirrels typically do, it looks as if it were extremely fat, or as if it were wearing clothes much to large for it. This squirrel leads a nocturnal life that is so secretive that people rarely notice it. These animals live in gardens in the nesting boxes of birds, or even in house attics. Sometimes one can see them at a bird-feeding station.

These flying squirrels are fastidious and elegant. They can crack all kinds of nuts. They can climb any irregular surface, and they are also able to glide through the air from any high point for a distance of up to 45 m. The lifting or lowering of the arms serves to adjust the altitude of the glide. When these animals want to land on a tree trunk, they apparently use their tail as a brake; they hold it erect, which seems to move their bodies into a vertical position. At the same time, they extend their four long legs in order to cushion the impact and to grasp the trunk. As soon as they land on the tree, they race around to the other side of the trunk, probably as an innate cautionary measure to avoid any birds of prey that might have followed them in the air.

Southern flying squirrels in captivity

Southern flying squirrels sleep all day, often in groups in tree holes. One day I [Grzimek] put the first two animals sent to me by Ernest P. Walker into a box in my bedroom. They lay rolled up together in a tiny house in the box for the entire day. When one took them out of the box, they were still in a stupor and remained lying in one's hand for some time. It seemed to me that they had eaten an unbelievable amount of food the first day, particularly chestnuts; then I found out they had carried everything into their box, where they had stored it.

They were very lively at night; they romped around, gnawed, made noises, and quarreled until it seemed that the devil was loose in that corner of the darkened room. In the morning one could find ample evidence of their busy teeth on the wall of the box: there were serrations, scratchings, and scrapings. Twice I came home late at night and opened up the box to take a look at the little fellows, at which they jumped out and escaped. They led me on quite a chase as they raced about on the floor like mice; they crawled into incredibly tiny places, for example in between the dresser and the wall or between the bed springs. We clambered around on our knees and our stomachs for two hours before we were finally able to recapture the rascals.

Unfortunately, both animals were females, and at first I was unable to get a male from America. Finally, I was able to acquire a male from an animal dealer.

I didn't know whether this male flying squirrel would be compatible with one of my females. Consequently, one night I put the male's small cage inside the female's cage so that they could sniff each other through the wire and become acquainted. That night it was very noisy. There were squeals, rattles, and rasping sounds, as if there were a saw mill at work. That morning there was a hole in the male's cage and the two flying squirrels slept, united in their tiny house and probably dead-tired from the night's work. The male had been in his tiny cage for two or three days and had made no attempt to get out. That night, next to the female, it had taken him only two hours to find a way out.

Tame flying squirrels

The flying squirrels belonging to Ernest Walker in Washington, D.C., were much tamer than mine. He was able to acquire a tame pair that had been taken from their nest only a few days after birth. The animals had no fear of man at all, and they were as inquisitive and tame as a dog. They slept in the sleeping box of their cage during the day, but at night they began to awaken. They did not care for either daylight or electric lights, and rapidly escaped into the shadows; for this reason, a weak blue light was set up for them.

As soon as the cage was opened, they ran to their human companions, did a few gymnastics on them, and investigated the insides of jackets, shirts, and trouser legs. They particularly enjoyed crawling into pockets, where they liked to sleep. They treated people just like tree trunks.

They liked to glide down on a human companion from a higher piece of furniture; their owner could easily induce them to glide to him when he tapped his hand against his chest. Almost immediately they would land on his hand or on whatever place he had tapped. Thus he could indicate various landing points, and there they always landed, exactly on the spot, a good illustration that they understood and that they could aim their gliding flights. They glided with genuine delight, twelve times in succession. When they had to make an unusual flight, they first sat down and leaned to one side and then the other, raising themselves on their fore paws so that they could pinpoint the spot on which they wanted to land. They may have been estimating the distances in this manner.

Diet

The southern flying squirrels eat all sorts of nuts, including acorns, chestnuts, and peanuts; they occasionally eat berries, but never herbs. Instead, they, like so many other rodents, show a marked preference for mealworms, beetles, grasshoppers, and other insects. In the wild they will also kill baby birds in their tree holes. These little creatures never bit Walker. They showed their affection for him in many ways. Sometimes they stood on his shoulder and delicately nibbled on his ear, or they stuck their tiny noses in his ear and panted. They cleaned and looked through hair and clothes with their fore paws, teeth, and tongue. When they were still half asleep, they would allow Walker to pick them up and rub or scratch them. They would even hold out their heads in order to facilitate these caresses. Whenever any danger threatened—they were apt to be greatly afraid of rustling paper—they fled to as high a spot as they could find, or they sought refuge and security with people.

Reproduction

The female southern flying squirrel gives birth to two litters a year, the first in February or March, the second in midsummer; the gestation lasts for thirty days. The female is particularly conscientious in the care of her young, and, should any danger arise, she moves them to a more secure place. In order to do so, she rolls them up together like a ball, with her hands and mouth, and then carries them in her mouth. Fossils indicate that the genus *Glaucomys* had its evolution in North America during the Ice Age.

Superfamily: Pocket gophers, by H.-A. Freye

The POCKET GOPHERS (superfamily Geomyoidea), despite the fact that they may resemble rats or mice, are not related to the mouselike animals, but rather to the squirrels. Some are terrestrial and others lead subterranean lives. They range from small to medium-size. Their major distinguishing characteristics are their cheek pouches; these are covered with fur on the outside, and do not open into the mouth cavity, but instead open to the sides. They have twenty teeth, arranged $\frac{1 \cdot 0 \cdot 1 \cdot 3}{1 \cdot 0 \cdot 1 \cdot 3}$. The molars are almost rootless and, like the incisors, they are in a state of continuous growth; the enamel pattern is greatly simplified. The tibia and the fibula bones are fused. Some of these animals are delicate and beautiful; others are less "pleasing to the eye." We do not know very much about the way of life of these pocket gophers. They are much more highly developed than many other rodents. Their main area of distribution is in North America; some forms have moved south to Central America and northern South America. There are two families, the pocket gophers (Geomyidae) and the heteromyid rodents or kangaroo mice (Heteromyidae).

Distinguishing characteristics

Family: Pocket gophers

The POCKET GOPHERS (family Geomyidae; BL 11.5–23 cm, TL 4.5–12.5 cm) are adapted to a terrestrial and subterranean lifestyle; they live in extensive burrows. They are bulky and short-legged, about the size of a hamster, and similar to moles. The limbs used for digging are strong and have powerful claws. Each of the four feet has five digits; the "fingers" have long crescent-shaped claws, which are stronger than those on the "toes." The tail is of average length. The hair has stiff, rigid bristles which are particularly sensitive; the whole body is only sparsely covered with hair. The cheek pouches are large, extending back to the shoulders; they are covered with fur as far as the base. The upper lip is also covered with hair, and is not cleft. The color of the fur varies from white to yellowish-brown, chestnut-brown, and dark lead-gray; occasionally the color changes with the seasons. There is usually a dark stripe down the back, but this may not be present on the summer coat; piebald and albino animals also occur. The skull is rather flat and massive, with a strong zygomatic arch. The skin on the tail forms a sac in which the tail vertebrae can be pushed back and forth easily. These animals are herbivorous, and they lay in extensive food supplies. They have large, powerful incisors. The stomach has one compartment; the small intestine has a wide loop in the duodenum. The large intestine branches upward and has a noticeable loop (paracaecal loop). The eyes are very small. The external ear (pinna) is generally underdeveloped. These animals are found from British Columbia (Canada) south to Mexico and Panama, and in the southern U.S.A., east to Florida. There are nine genera extant today, all of which are only slightly differentiated; these

Distinguishing characteristics

include the EASTERN POCKET GOPHERS (*Geomys*), the WESTERN POCKET GOPHERS (*Thomomys*), LARGE POCKET GOPHERS or TUZAS (*Orthogeomys*), and the YELLOW POCKET GOPHERS (*Cratogeomys*), among others. There are 272 species and subspecies.

Adaptations to the subterranean life

The POCKET GOPHERS are subterranean root-eating animals that rarely appear above ground; consequently, they are not very well known. These thick-bodied animals show an excellent adaptation to their subterranean way of life in their appearance: they do not have a distinct neck, their eyes are very small, and their outer ears are barely visible. However, they do have powerful digging feet. The fore feet are particularly powerful, with strong, curved claws which are thick and dull in hard soil, but longer and more pointed when the animal lives in sandy soil. There are rows of bristly hair on the ends of the toes which may prevent dirt from getting in between them. The southern forms do not have hair on their tails, while the tails of the northern species are sparsely haired. Pocket gophers walk on hard ground with their claws drawn in to protect them; thus, they run on the outer rims of the soles of their feet. When the earth is soft, the animals walk normally, using their feet as shovels. They use their incisors like hoes to make holes in the earth's surface; the fore feet dig and shovel, while the hind feet push the loose earth that accumulates under the animals' bodies further to the rear and out of the way.

"If the pile of dirt grows too large," reports Ivan T. Sanderson, "the gopher turns around, places its hands with the palms together and pushes its hind feet against the pile, shoving it through the passage to the nearest opening. There, the earth is thrown out; the result looks similar to a molehill. The gophers often run through their burrows backward as fast as they do forward; the tail is used as a tactile organ whenever they do so." In many species a naked callus or swelling on the front part of the nose indicates that the head must also be used in the animal's digging activities.

Diet

The large incisors are also used for gnawing and biting off food. The pocket gophers prefer a diet of roots, tubers, onions, turnips, nuts, seeds, as well as corn (maize), wheat, barley, oats, rye, alfalfa, pumpkins, and melons. The animal's sparsely haired lips form a partition in the large gap between the incisors and the molars. As a result, the yellowish-brown incisors appear outside the mouth, and the gopher can gnaw at the ground or at woodchips or other objects which are too big for the mouth cavity, with its mouth closed. The animals collect large supplies of food in their cheek pouches, transferring these provisions to their subterranean storage rooms. For example, when a gopher bites off a large piece of root, it holds the root against its incisors with its fore feet and, almost like a lathe, turns the root and cuts it off. Then in a split-second move, one fore foot shoves the root out of the paw and into the cheek pouch. "Its cheek pouches almost bursting, the gopher then goes to its storage room," according to Sanderson's description, "where the cheek pouches, which are grotesquely filled, are emptied out as the animal presses them from back to front with its fore feet." The pocket gopher's diggings and burrowings cause damage to planted fields, pastures, dams, and banks; however, these burrowings are also beneficial to the soil, as they mix and aerate it.

Sensory organs and abilities

It is clear that any subterranean digging animal must have a very highly developed sense of touch. Pocket gophers have tactile hairs all over their bodies, and the tail has a pronounced tactile sense; this is especially necessary for the animal's very rapid backward motion. Although the pocket gophers live in continual darkness, they are generally crepuscular (active mainly early in the morning and in the evening). Sometimes, they will even leave their burrows for a short period at these times. Pocket gophers prefer to dig in rather moist, easily worked soil; they do not seem concerned with the degree of acidity. The numerous food tunnels, which the animals use when looking for roots, are usually fairly close to the surface of the ground; the living dens and the maternity wards, on the other hand, are located much deeper. There are several large food-storage rooms near the living dens, usually filled with extensive supplies. Thus, the spacious burrow with its many branching tunnels is, like the structures of moles, a combination food supply and living area.

Extent of the burrowing activity

Pocket gophers display a remarkable amount of burrowing activity, and for many hundreds of years they, like the beavers, have caused numerous changes in the landscape. When they burrow down into the ground, the sandy and gravelly subsoil which is brought to the surface can easily be distinguished from the rest of the landscape, as Ingo Krumbiegel (among others) has noticed. Initially, this diminishes the fertility of the soil, although the sand on the surface later disintegrates and is made fertile again through the actions of plant seeds; this is particularly true when these piles of sand lie in virgin soil. The pocket gophers spread roots, tubers, and onions as a result of their burrowing activities, and thus they contribute to the distribution of plants.

Fig. 10-30. 1. Plains or eastern pocket gopher (*Geomys bursarius*); 2. Texas pocket gopher (*G. personatus*); 3. Southeastern pocket gopher (*G. pinetis*).

The gophers use their cheek pouches not only for carrying food, but also to transport soft materials for their nests; they carry inflexible or rigid materials in their teeth. We do not know very much about their reproduction. In 1937, Mossmann discovered prominent glands in the peridium of the ovary; these must certainly release a hormone, a unique occurrence in the mammal world. During the mating period, in the spring, there are usually several copulations over a short period of time. Each litter usually consists of five or six young, with a maximum of nine.

The PLAINS OR EASTERN POCKET GOPHER (*Geomys bursarius*; BL 25 cm, TL 8 cm; see Color plate, p. 297) is found in the midwestern U.S.A. Its hair color varies from yellowish-brown in the west to brown or

Fig. 10-31. 1. Western pocket gopher (*Thomomys bottae*); 2. Southern pocket gopher (*T. umbrinus*); 3. Northern pocket gopher (*T. talpoides*); 4. Oregon pocket gopher (*T. mazama* 5. Sierra pocket gopher (*T. monticola*).

Fig. 10-32. Large pocket gopher or tuza (*Orthogeomys grandis*).

Fig. 10-33. Yellow pocket gopher (*Cratogeomys castanops*).

almost black in Illinois. The fur always has a noticeable silvery sheen. Each incisor has two lengthwise grooves. The eastern pocket gopher builds extensive tunnel systems in the deep soil of the plains and prairies. These animals will also approach human settlements, particularly during the colder seasons. Large numbers of them can ruin potato fields quite quickly; they can also destroy entire crops of fruit. One can find large quantities of roots, potatoes, nuts, and seeds in their storage chambers. The eastern pocket gophers have relatively small digging feet; they can run backward in their tunnels at amazing speeds. They also do this above ground, and they have been repeatedly observed as they fled backward into the darkness from the lights of an approaching car. Only an expert can distinguish the three related species of the genus.

The WESTERN POCKET GOPHER (*Thomomys bottae*; BL 12–18 cm, TL up to 8.5 cm, weight 70–235 g) is found in western North America from 43° north latitude south to Mexico. It is generally medium-sized, although the size, and the coloration, have a wide range of variations; the larger forms are found in the valleys, while the smaller animals live in the mountainous solitudes. The fur coloration along the Pacific coast is almost black; in the Imperial Valley (southern California) it is almost white. Otherwise, the normal coloration is a dull brown (there are probably three different hereditary factors for black, pale red, or albino).

The adult western pocket gophers always live together with their young in the extensive tunnels, although two adult males are never found together. The mountain forms are forced to make seasonal migrations because of ground water and melting snow. If they move into cultivated areas, they can cause a great deal of damage to gardens and planted fields. Some species of pocket gophers destroy coffee, orange, and sugar cane plantations in Mexico and Central America.

The native habitat of the LARGE POCKET GOPHER OR TUZA (*Orthogeomys grandis*; BL 27–35 cm, TL 9–14 cm, weight up to 1 kg) is Mexico's evergreen deciduous forests. One can find these animals by looking for the earthen piles they make, which are 40–60 cm high.

"Apparently, for these animals to be present, the soil must be neither too rocky and hard, nor too moist," according to Helmuth O. Wagner. "If the land is steep, the burrows will run along the slope; in wide, even plains areas, the tunnels are expanded to form a network. The animals do not like to construct their tunnels so that they run upward and downward. The burrow system may be only a few centimeters below the ground's surface in the deciduous forest and in newly planted coffee plantations where the humus layer is still loose. The large pocket gophers in plantations are more numerous than those in the neighboring forests. When the earth's surface becomes hard after many years of use, the animals disappear. They can cause a great deal of damage to coffee bushes by gnawing off the roots. They visit banana plantations on the Pacific coast in massive numbers, attracted by the readily available food; as a result, extensive damage to any harvest can be expected. During the growing season, people try to poison these animals with ricin (*Ricinus communis*). The Indians catch these *tussas* in their burrows with snares attached to flexible rods. Their meat does not taste bad, except for that of the older males, which is rather strong.

The YELLOW POCKET GOPHER (*Cratogeomys castanops*) can be distinguished from the eastern pocket gopher by the fact that its incisors have only one lengthwise groove in the front. Otherwise, these animals are the same size as the eastern pocket gophers, with a yellowish coloring and a longer, hairless tail. They prefer deep soil.

Family- Kangaroo mice

The HETEROMYID RODENTS or KANGAROO MICE (family Heteromyidae) look much different. They are slender, and range in size from that of a mouse to that of a rat. The hind feet are long and narrow, while the fore feet are relatively weak. These highly developed animals resemble jerboas. They are not adapted to walking on the ground, but rather to making large or small hopping movements. The head is round with a pointed nose; the tail, whose tip often looks like a paintbrush, is at least as long as the body. The hair is soft and woolly; it is heavily intermixed with stiff bristles and, at the back of the underside, with flat, grooved quills. The innermost digits are stunted, but they do have claws. The claws on the fore feet are not as large as those of the pocket gophers, but they are longer than the claws on the hind feet. Like the pocket gophers, heteromyid rodents have cheek pouches which are covered with short hair and which have openings at the sides of the mouth. The skull is delicate and somewhat arched; there is a prominent nasal bone and the zygomatic bone is involuted. These animals eat plants and seeds; they also build up large food stores. The incisors are not grooved, and they do not even approach the size of those of the pocket gopher. The eyes are large; the ears are rounded. Heteromyid rodents are found in western North America south to Mexico, Central America, and as far as Ecuador, Colombia, and Venezuela; the main area of distribution is in California. They live in grasslands, wastelands, semideserts, and deserts. They dig small burrows, consisting of supply rooms and nest rooms. There are five genera, grouped together into two or three subfamiles; these include the pocket mice (*Perognathus*), the kangaroo rats (*Dipodomys*), and the spiny pocket mice (*Heteromys* and *Liomys*). Fossils of these animals have been found from the Middle Oligocene period in North America.

Distinguishing characteristics

Fig. 10-34. 1. Bailey's pocket mouse (*Perognathus baileyi*); 2. Silky pocket mouse (*P. flavus*); 3. Plains pocket mouse (*P. flavescens*); 4. Wyoming pocket mouse (*P. fasciatus*).

Pocket mice

Originally the graceful heteromyid rodents were often combined with true mice or with jerboas because of their similarity to these animals. We now know that these similarities are due to parallel adaptations and that it is not unusual for very similar forms to come from two or three very different evolutionary roots. Without a doubt, the representatives of this family which seem to be most like mice are included in the POCKET MICE (*Perognathus*; BL 5–12 cm, TL 4.5–14.5 cm). The head tapers to a point, and the ears are usually small. The tail is generally longer than the

Fig. 10-35. 1. California pocket mouse (*Perognathus californicus*); 2. Hispid pocket mouse (*P. hispidus*); 3. Great Basin pocket mouse (*P. parvus*).

body, and is either hairless, sparsely haired, or bushy. The tail hair becomes especially thick and the white tufts on the end of the tail grow remarkably longer when the tail vertebrae are damaged. The hair is pale, and can be either silky-smooth or coarse and rough; in the latter case, it is intermixed with bristles and quills. There are twenty-five species, with over 130 subspecies, found from British Columbia (Canada) south to Mexico. In captivity, members of this genus have lived over eight years. We will mention three species: BAILEY'S POCKET MOUSE (*Perognathus baileyi*; BL 10–11 cm, TL 11.5–14 cm, weight 30 g), the largest species of the genus, is found in California, southern Arizona, and northwestern Mexico. The last third of the tail is bushy. The fur is usually gray intermixed with some yellowish hairs; the stomach and the underside of the tail are white. The SILKY POCKET MOUSE (*P. flavus*; BL 5–6 cm, TL 4.5–5.5 cm, weight 7–10 g) is one of the smallest forms. Its fur is silky-smooth and gray on the upper side, intermixed with some black; there is a yellow speck behind each ear. The underside is white. The tail is hairless and somewhat shorter than the body. These animals are found in sandy prairies. The CALIFORNIA POCKET MOUSE (*P. californicus*; see Color plate, p. 297).

Because they are definitely nocturnal animals, pocket mice spend the daytime in caverns, which they dig in the dry desert regions and the sandy steppes. They use their cheek pouches not only for transporting food, but also for carrying away dirt, as their small front feet do not seem to be particularly well adapted for shoveling. One can find extensive food storage areas in their nests; these are often filled with sunflower seeds or other seeds. The pocket mice can apparently go for an unlimited time without water.

The PALE KANGAROO MOUSE (*Microdipodops pallidus*; BL 7.5 cm, TL up to 10 cm, weight 10–16 g) is shaped very much like a kangaroo rat. The hind feet are long; the fore feet are relatively weakly developed. The hair ranges in color from white to a pale leathery color. The tail, which is almost completely hairless, seems to be somewhat swollen in the middle. These animals hop and jump like kangaroo rats; they are found in the sandy plateaus and prairies of Nevada.

Fig. 10-36. 1. Pale kangaroo mouse (*Microdipodops pallidus*); 2. Dark kangaroo mouse (*M. megacephalus*).

Perhaps the best-known KANGAROO RAT (*Dipodomys*) with the largest distribution is ORD'S KANGAROO RAT (*Dipodomys ordii*; BL 10–11 cm, TL 13–15 cm, weight 40–65 g). The upper sides are yellowish-brown, with a relatively distinct border between that and the white stomach. The tail is dark with white stripes; the last third is bushy. The hind feet have four or five digits. The lower incisors are not flattened off; rather, in cross section, they appear rounded. The BIG-EARED KANGAROO RAT (⊕ *D. elephantinus*) and the TEXAS KANGAROO RAT (⊕ *D. elator*) are both threatened with extinction. The DESERT KANGAROO RAT (*D. deserti*) and the PACIFIC KANGAROO RAT (*D. agilis*) are both pictured on page 297.

Kangaroo rats have become known the world over as a result of

Walt Disney's film, *The Living Desert.* Like kangaroos, they move about by making long powerful jumps; their unusually large hind feet are very important here, as is the long, pointed tail, which is used as a rudder. The forearms are so small that they disappear completely in the long, soft hair. The head is distinguished by large, protruding eyes and by a black or white (according to the species) pattern or marking on the face. There are also two white stripes running from the back of the shoulders, down the body on the left and right sides, to the root of the tail. Every now and then one can find an albino (all-white) animal.

Kangaroo rats are nocturnal animals; they inhabit the dry steppes and wastelands. They eat seed grains and the green parts of plants. They will also frequently visit cultivated fields, where they scratch up seeds and seedlings of corn (maize), sunflowers, melons, and similar fodder plants. These animals are also very much at home in vegetable gardens. They are able to get enough water for their needs from juicy plants. Some physiologists have established the fact that water taken from plants by the kangaroo rats is held not only in their kidneys, but also in the urinary bladder. The kangaroo rat's greatest natural enemies include the horned owl, the coyote, and the rattlesnake, but they are also threatened by cold and wet weather.

Fig. 10-37. 1. Ord's kangaroo rat (*Dipodomys ordi*); 2. Narrow-faced kangaroo rat (*D. venustus*); 3. Big-eared kangaroo rat (*D. elephantinus*); 4. Texas kangaroo rat (*D. elator*).

Ord's kangaroo rat is found in hard, packed soil. It often makes its nest, complete with the numerous escape tunnels and entrances, under some protective bush, thistle, or other plant. Occasionally one can find pocket mice (*Perognathus*) and small grasshopper mice (*Onychomys*; see Chapter 11) in the nests of the kangaroo rats. The large number of interconnecting tunnels that surround the nest form low piles of dirt which cave in very easily; as a result, mules and horses avoid these piles. The nest itself is well lined with grass and feathers. The neighboring storage chambers are filled, mostly with seeds. The Ord's kangaroo rat makes a sound similar to the soft chirpings of a bird.

The SPINY POCKET MICE (*Heteromys* and *Liomys*) are distinguished by soft quills and bristles which are, nevertheless, pointed at the end; these make the fur coarse and harsh. The tail, which is covered with scales, is sparsely haired, and is always longer than the body. The fore feet are short, and the hind feet, particularly the three middle toes, are long and narrow, although not as long as the kangaroo rats'. The head is pointed, like a rat's head; the rump appears to be fairly plump. The fur on the upperside is yellowish-red to reddish-brown; the under sides are always a brilliant white. The cheek pouches are relatively large. The stomach is small. The intestine is twice as long as the body, and any food that the animal eats is crushed to a particularly fine pulp. These animals are found in the desert plateaus and wastelands of Central America and northern South America. Recently, they have even moved into areas under cultivation.

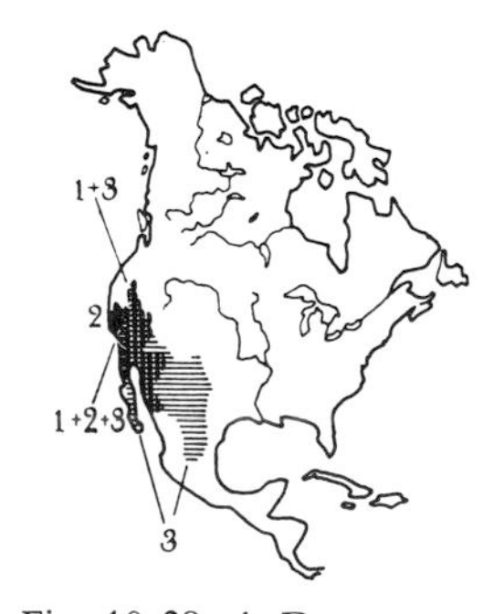

Fig. 10-38. 1. Desert kangaroo rat (*Dipodomys deserti*); 2. Pacific kangaroo rat (*D. agilis*); 3. Merriam kangaroo rat (*D. merriami*).

GOLDMAN'S SPINY POCKET MOUSE (*Heteromys goldmani*; BL 12-15 cm,

TL 17–20 cm) is the largest species of the Mexican spiny pocket mice. The center of its distribution is in the misty forests of the Sierra Madre, in the Mexican state of Chiapas. The closely related SOUTH AMERICAN FOREST MOUSE (*H. anomalous*), from Mexico, Venezuela, Ecuador, and Trinidad, is known as a crop destroyer; in Trinidad it even goes into grain elevators. The PAINTED SPINY POCKET MOUSE (*Liomys pictus*) gets its name from the faded brown and orange stripes on its sides and the white undersides. These animals are found in the dry savannas of Mexico along the Pacific coast as far as the Isthmus of Tehuantepec. The MEXICAN SPINY POCKET MOUSE (*L. irroratus*) is pictured on page 297.

Fig. 10-39. 1. Goldman's spiny pocket mouse (*Heteromys goldmani*); 2. South American forest mouse or pocket mouse (*H. anomalus*).

The spiny pocket mice are agile jumpers. Many species are also good climbers, able to go up bushes and trees; despite their stiltlike hind legs, they are able to move about in the branches with great agility. Occasionally they laboriously pull themselves up into the heights, using all four feet. Some animals build subterranean nests, others put them in grassy bushes or tufts, while still other animals build their nests in hollow tree stumps. Helmuth O. Wagner observed Goldman's spiny pocket mice in the Sierra Madre, where they build their burrows in the stone-free soil of mountain slopes and in the slopes of river banks; the animals then often lived in these nests for many years. Although several animals may rest together in the same burrow, only one pair of animals lives together during the mating season. Three or four litters, each with three or four young, are born between May and December. These animals gather their food exclusively at night; they eat only in their burrows. Because of their hoarding instinct these animals do not always fill their cheek pouches with edible items. Wagner found pieces of shell from a land crab, pebbles, red *Salvia* blossoms, a 20–30-cm-long piece of black agaric (gill fungi), seeds, berries, and coffee beans, all of which were in the animal's cheek pouches. Wagner saw 1–2 l of seeds, berries, and pieces of leaves, all piled up in a single storage room of one of these animals. Spiny pocket mice use their food inefficiently, but this does not seem to matter, because of their extensive food stores.

Fig. 10-40. 1. Painted spiny pocket mouse (*Liomys pictus*); 2. Mexican spiny pocket mouse (*L. irroratus*).

The painted spiny pocket mice also rest together, several animals in one den; pregnant females, however, isolate themselves. Like the house mouse, these females have a vaginal plug after copulation. They give birth several times a year, to between three and six young each time.

All of the kangaroo rats extant today are descended from the species of the genus + *Heliscomys*, from the Oligocene Period, which were very close to the common ancestral forms of the pocket gophers.

Superfamily: Beavers, by B. Grzimek and R. Piechocki

The next superfamily in the suborder of squirrel relatives is the BEAVERS (Castoroidea). This group contains only one family, the BEAVERS (Castoridae), with a single species, found in the northern forest belt of Eurasia and North America. In the past, the Old World beavers and the New World beavers were considered as two separate species. H.-A. Freye's basic research, published in 1960, proved the untenability of this division.

Distinguishing characteristics

The BEAVER (*Castor fiber*; BL up to 100 cm, TL up to over 30 cm, weight up to 30 kg; see Color plates, pp, 283 and 284) has a heavy build and is one of the largest rodents. The body becomes large as one looks at it from front to back. The hind feet have web membranes between the toes. The small fore feet are particularly good gripping tools. The feet all have five digits, all with strong nails. The second toes on the hind feet carry a weak double claw which is used for cleaning the fur. Beavers have strong, bristly hair, and their woolly hair is curled thickly. The tail is flat, 12–15 cm wide, and shaped like a ladle with scaly, leather-like skin; the beaver uses it as a rudder. The eyes and ears are noticeably small. The skull is sturdy. There are twenty rootless teeth arranged $\frac{1\cdot0\cdot1\cdot3}{1\cdot0\cdot1\cdot3}$. The incisors are like chisels, with an orange-red outer enamel. The pre-molars and the molars have folds in the enamel. The beaver looks awkward on land, but in the water it is an outstanding swimmer and diver. These animals are exclusively herbivorous. In summer they eat the juicy shrubs found on river banks, bulrushes, buds from softwood trees (especially poplars, aspens, and willows), as well as root pieces from water lilies. In winter, beavers will eat bark from shrubs and from trees they cut down. There are several subspecies, including, from Eurasia: the SCANDINAVIAN BEAVER (*Castor fiber fiber*); the ELBE BEAVER (⊕ *C. f. albicus*); the RHONE BEAVER (⊕ *C. f. galliae*); the POLISH OR BYELORUSSIAN BEAVER (*C. f. vistulanus*); the URAL BEAVER (*C. f. pohlei*); the MONGOLIAN BEAVER (*C. f. birulai*); and, from North America: the CANADIAN BEAVER (*C. f. canadensis*); the MICHIGAN BEAVER (*C. f. michiganensis*); the NEW-FOUNDLAND BEAVER (*C. f. caecator*); the RIO GRANDE BEAVER (*C. f. frondator*); the GOLDEN-BELLIED BEAVER (*C. f. subauratus*), as well as other subspecies.

Fig. 10-41. The Old World subspecies of beaver were originally distributed over much of Europe.

The beaver

Beavers live much of their life in water; they settle along the banks of streams, rivers, and lakes in the thick underbrush of lowland woods. When the beaver lives near a large river full of rushing water, or when it lives near a river which is too wide to be dammed up, then this "master builder," as it has been called in animal fables and by sportsmen, will build itself a simple house in the river bank, with at least two entrances, and sometimes four or five, all of which are underwater. The Elbe beaver often makes these simple living quarters in river banks. The entranceway rises on a diagonal from just below the water's surface to immediately below the surface of the ground. There the animals build a room which is some 1.2 m wide and 40–50 cm high, and very carefully smoothed out on the inside. When the surface of the water rises, the floor of the room must also be raised; in order to accomplish this, the beaver merely gnaws or scrapes dirt from the ceiling, and this dirt then falls on the floor. Usually the ceiling is so strong that several people can stand on it together without mishap. However, if the beaver is forced to make its house somewhat higher, then it will strengthen the ceiling with mounds of twigs which it lays outside over the ceiling; these can become regular mountains of twigs. As the water level rises, the height

Beaver mounds

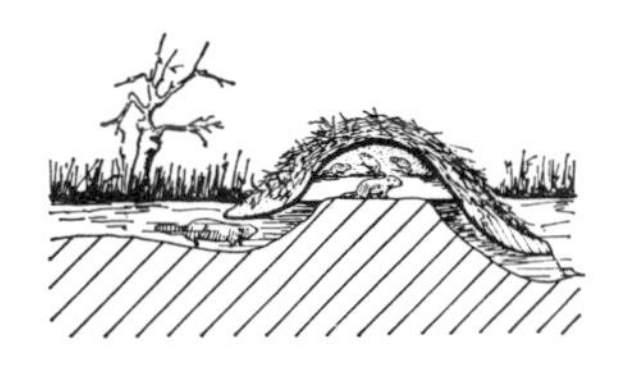

Fig. 10-42. The beaver mound may not always look the same, but the entrance to the den is always under water.

Fig. 10-43. Beavers are still found in North America wherever the environment is suited for their needs.

of the mound must be increased, and finally it becomes an island in the water.

This is probably how the beaver's water stronghold originated. Today beavers sometimes (although not often) also build these houses in the middle of their reservoirs, by making underwater mounds of mud and water-logged branches. The resulting mounds may rise 1–2 m above the water's surface. The insides of these structures are then hollowed out and supplied with burrows and entrances. Although these beaver mounds are carefully waterproofed with mud and loam, the beavers often will not pack down the center of the ceiling, so air can escape. However, the branches can still hold the weight of a man, and they cannot be easily broken into by predators. Should one observe a beaver mound in the winter, one can see water vapor rising from the top of the structure, like tiny wisps of clouds. Older structures have several stories, so the living area is always about 20 cm higher than the water's surface. These mounds are inaccessible to most predators, and even if a bear, with all its stregnth, manages to break into the top of the structure, the beavers will have long since escaped through their entrance holes and disappeared into the water.

Beaver dams

If a river or lake should dry up, the entrances to the beaver mound will be above the water. Beavers build dams for just this reason, to keep the water at a constant level and, even more, to create artificial reservoirs where they feel secure and in which they can easily transport their food. The animals build their dams by placing tree trunks and branches, which they have cut down and prepared, in the ground, perpendicular to the river bed, and by fastening and anchoring these uprights with stones, mud, reeds, and whatever else might be at hand. The beavers will often make a support out of a tree that has fallen across or floated down the river, so they can expand their construction. The more the water behind the dam rises, filling the creek or riverbed, the higher the beavers must make their dam, extending it on the sides to the higher river banks. The longest dam in the Voronezh region in the U.S.S.R. is near Marinika; it is 120 m long, 1 m high, and from 60 to 100 cm wide. The dams across the mountain streams at the foot of the Rocky Mountains in North America are up to 3 m high. The largest beaver dam of all is that on the Jefferson River, near Three Forks, Montana; one can follow it for 700 m. This dam has been maintained by successive beaver generations for decades, perhaps even centuries, as long as enough food for these animals can be found in the area. There is a dam in Colorado that has been there for seventy years; trees and bushes finally grew on top of it, and the dam itself keeps getting wider and stronger. A man could ride over it with ease. It takes a family of beavers about a week to build a dam that is 10 m long. When the river has a strong current, the beavers build several auxiliary dams facing the current so that the main dam will not be destroyed. If the water breaks through the dam at a single point, the beavers will rapidly repair the leak the next night. Our expression for feverish activity, "to work like a beaver," comes about as a result of the industriousness of these animals. The dams differ according to area and subspecies. The dams in the flat swamplands of Mississippi often reach gigantic dimensions, several hundred meters in length, and the area that has been inundated provides the beaver with a large habitat. The beaver is able to guarantee itself a sufficiently large body of water only through its dam-building. However, if the animal lives on a river with stony banks, such as those in the hills of northern Michigan or Wyoming, it builds dams that are shorter, wider, and generally higher.

The beaver's dam-building ability indicates a superior adaptability to changing ecological situations. I [Piechocki] was able to confirm this in the southwestern regions of Mongolia, in the semidesert of Dzungaria. There, on the Bulugun River alone, I found numerous dams which were no longer able to hold the water back in the lowland forest and thus were quite dry except around the bases. As a result, there was a chain of more or less separate water holes, interconnected only by beaver paths. The proximity of the desert exerted a special effect on the life conditions of the beavers living in this unusual area. During the short, hot summers the air temperature would rise to 50°C, while at night it would drop to freezing. As beavers suffer from overheating whenever the temperature rises above 20°C, it is possible for them to live in that area only because they are able to dip their naked tails into the rushing Bulugun and so keep their body temperature normal. According to Steen's research, the beavers are able to give up twenty percent of their body heat in this manner.

Fig. 10-44. The beaver has survived in modern Europe only in the areas designated above. The black sections indicate those areas where man has recently resettled the beaver population. 1. Rhone beaver (*Castor fiber galliae*); 2. Elbe beaver (*C. f. albicus*); 3. Scandinavian beaver (*C. f. fiber*); 4. Polish or Byelorussian beaver (*C. f. vistulanus*); 5. Polish beaver in Voronezh region; 6. Ural beaver (*C. f. pohlei*) in western Siberia; 7. Ural beaver in Tuva region; 8. Mongolian beaver (*C. f. birulai*); 9. Resettlements of Canadian subspecies; 10. Hybrid animals, cross between Canadian and European beavers.

At a point of influx on Lake Michigan, some beavers even used an abandoned railroad bridge as a support for their dam. They built it up to 2 m high, and were thus able to dam up the water. The beaver dams in Canada are often 4–6 m wide at the base, and 1–2 m wide on the top ridge. Sometimes, their reservoir will flood a road or a stretch of railroad tracks in a single night. If one is able to break through the dam again quickly, the stubborn struggle between man and beaver begins, and it will last until the animal is trapped and moved to another location.

Beavers are excellent landscape artists, particularly in the mountains. Turbulent brooks are checked, and the dams protect the fields and pastures below from becoming silted up with sand and gravel. They deliver the water to the valleys very slowly. The beavers' artificial reservoirs are soon filled with trout and other fishes as well as with waterbirds.

The U.S.A. once had 60 million beavers, and the enormous number of tiny reservoirs and dams that these animals built must have certainly greatly reduced the amount of flooding. This was also true in Germany 1000 to 2000 years ago. Now, man attempts to achieve the same results

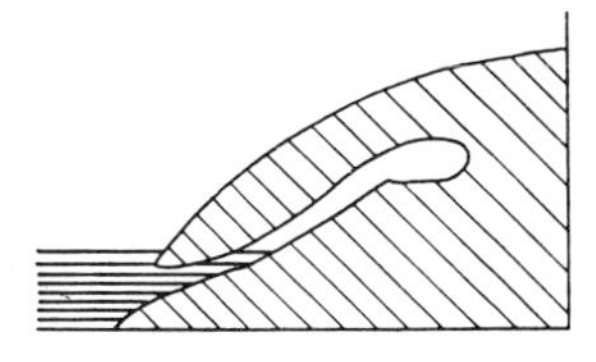

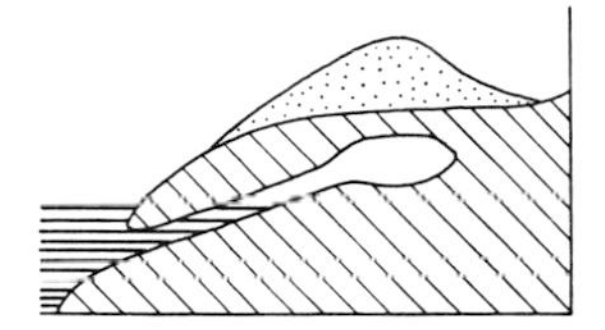

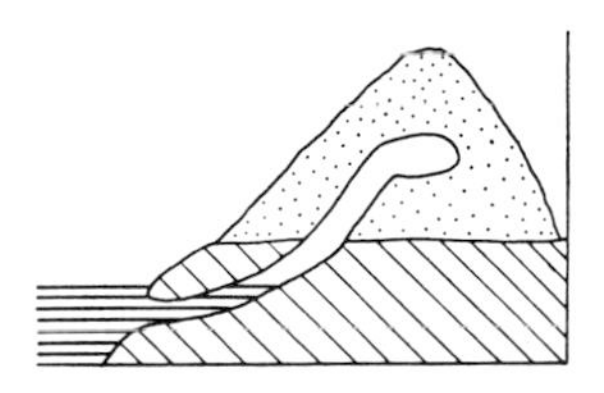

Fig. 10-45. Different kinds of beaver dens: above, earth structure; center, small mound; below, large mound. The type of den the beavers build depends on the terrain.

with dams he builds in the mountains. People now spend millions of dollars, not including what we spend on converting water power into electricity (which may soon be a thing of the past as atomic power develops), on maintaining the inland waterways and on preserving artificial reservoirs to meet human water needs; in other words, man has had to do, by himself, what nature and the beaver had done for eons, before man's senseless damage to the balance of nature through his persecution of the beaver.

Man has always praised the beaver because it can fell a tree so that the crown always lies toward the water and because these animals calculate the fall of the tree so exactly that no beaver is killed in the process. In addition, beavers seem to be able to judge the direction of the tree's fall so that the crown of the tree is not caught in nearby trees and so that the tree itself does not fall in an area where it would be of no use to the beaver. All of these beliefs are not quite true. Certainly trees growing very close to the water almost always fall with their crown in the water; this, however, is because the branches on the side of the tree facing the open water are more highly developed, consequently making the tree heavier on that side. Trees further from the water which are felled by beavers may fall in any direction. Beavers have also been killed by falling trees, although this happens very rarely. Once someone observed a tree whose trunk had been carefully gnawed through by a beaver. However, the tree stood on a slope, and when it fell it was supported by the neighboring trees, and thus only fell about 1 m. Once again the beaver gnawed through the trunk. When the tree remained standing, still supported by its neighbors, and only slid a little lower, the beaver gave up.

Beavers like to gnaw through tree trunks between 8 and 20 cm in diameter; they like aspens, willows, and poplars, and are somewhat less enthusiastic about birch and wild cherry trees. They almost completely ignore the coniferous trees, and thereby avoid harming our most important timber and hardwood trees. The softwoods do not play a very important role at all in our lives. Beavers can cut through a willow tree 8 cm in diameter in five minutes; they will work on a larger tree for several nights in succession. The beavers on the Bulugun River must work several months before they can fell a poplar with a diameter of 85 cm. Each beaver chews through 200–300 thin poplar trunks annually; 0.5 hectare (about 1 acre) of land will supply a beaver colony with food for one to one and a half years. Two beavers will often work together on a thick tree; usually one animal cuts while the other looks around. Of course, beavers prefer trees that grow close to the water. They drag thin tree trunks as well as branches they have cut off of larger trees to the water by the shortest possible route. They do not cut down trees that are more than 200 m away from the water. When they have cut down everything within the circumference of their colony, they like to move to a new area. They cut the trees they have felled into pieces—the

Fig. 10-46. The beaver's characteristic tooth marks are unmistakable.

thicker the tree trunk, the smaller the piece. They leave tree trunks of more than 10–12 cm in diameter wherever they fall. If the bark of the tree is edible, they peel it off and eat it.

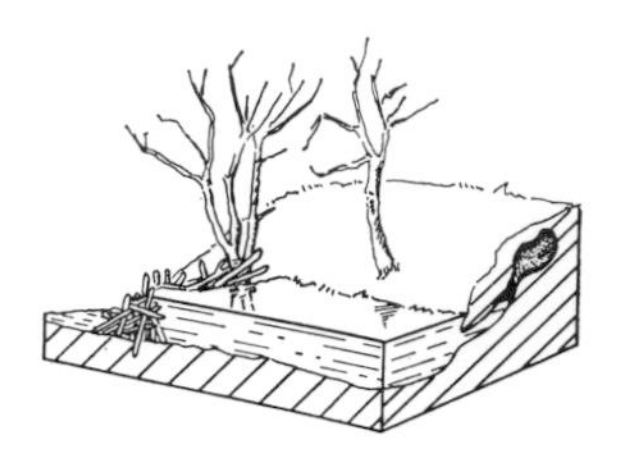

Fig. 10-47. The effects of a beaver dam.

The variety of these activities gives one a good idea of the productive accomplishments of an individual beaver. Consequently, it is not surprising that the beaver has an unusually highly developed brain for a rodent, according to brain anatomist Pilleri. Beavers are rather awkward on land; they cannot move as fast as a man, and they tire easily. For this reason, those animals living in flat, watery areas will often build canals leading to their grazing areas, a relatively long distance away. Perhaps these canals first came about as the result of regular paths that sank deeper and deeper into the soft earth; later, however, these were systematically widened and deepened until they were sometimes 50 cm deep, enabling the beavers to dive underwater and swim. The beavers could also transport branches of trees much more easily in these canals.

Beavers will place several of their harvested branches in the water around their mounds, making sure that the end that has been gnawed off is stuck firmly into the mud of the riverbed so the branch will not float away. These branches are provisions, particularly for the winter months, because in the summer these animals can also eat water plants, berries, swampwood roots, and, where possible, cultivated fruit. However, their main diet consists of fresh green bark and soft wood. Hard nutrition of this type is in keeping with the strength of these animals' teeth. A single beaver exerts a chewing force of 80 kg with its incisors; man exerts a force of only 40 kg. In addition, the beaver usually weighs 18–20 kg, 30 kg at the very highest. Every family of beavers in the Voronezh region of the U.S.S.R. had an underwater store that averaged 108 aspen trees, ranging from 8 to 35 cm in diameter. Every beaver in the Grafskaja research station receives 7 kg of aspen branches, with bark, 2 kg of birch branches, and 2.5 kg of willows daily. An adult beaver uses 4197 kg of wood, with bark, in addition to its green fodder, a total of 7.5 m³ of wood annually. The beavers living in the Voronezh region eat 148 of the 580 different species of plants growing in that area.

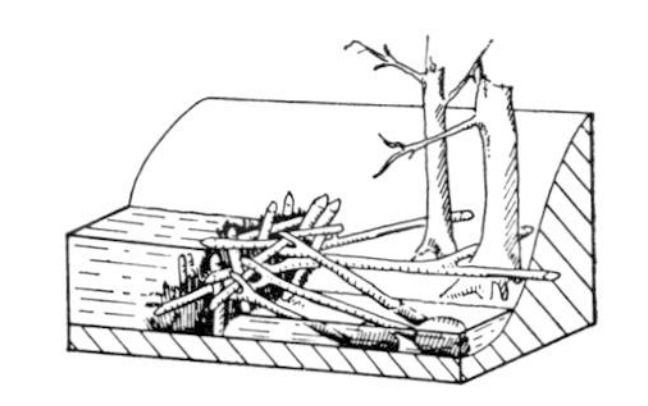

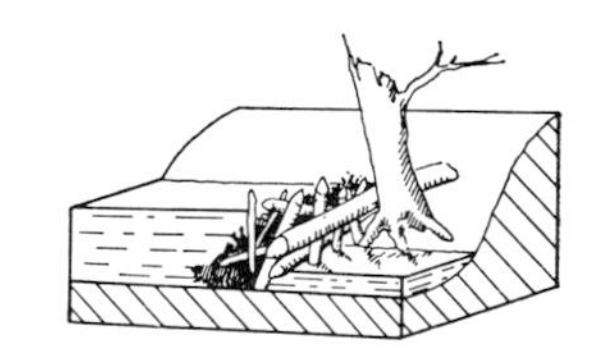

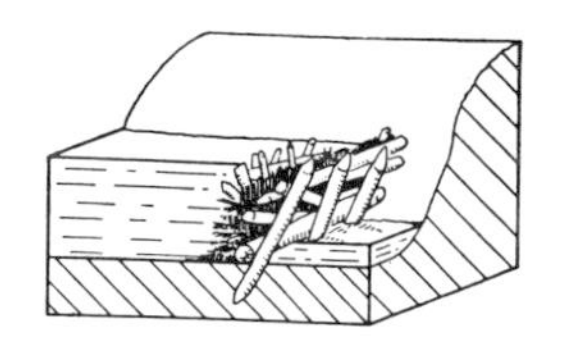

Fig. 10-48. Beavers are very skilled dam builders. Above are three different forms of dams built by the Rhone beaver.

Beavers do not hibernate, although they become much less active during the cold season. One may not see them for as long as a week. When it is −15°C outside, the air in the beaver's house will remain at about the freezing point. As soon as the river freezes, the beaver swims under the ice to its storage supply of branches. Often they do not even have to make air holes in the ice, because the ice in rivers and lakes begins to freeze first at the edges, near the banks; then, in the course of the winter, the water level sinks to some extent, especially in rivers. Thus the ice in the middle of the river sinks, while the ice near the banks of the river is held up by the ground and the roots of trees, leaving safe and secure air pockets for the beaver.

Beavers are also excellent divers. Older books on natural history report that these animals could remain underwater for two minutes, however, in February, 1939, Rudolf Berndt timed a beaver that stayed underwater for ten minutes, near Steckby on the Elbe River. In Grafskaja, someone timed a beaver that stayed underwater for fifteen minutes. The beavers usually defecate in the water, never in their houses. They make soft sleeping areas for their young out of finely grated wood.

Reproduction

Beavers are sexually mature after three to four years. Copulation takes place in the water between January and March; the male swims under the female, with his stomach facing upward. According to most observations, it appears that beavers form permanent pair bonds, although when the young are born in the beaver mound (after a gestation of 105–107 days), the male, and the young from the previous year, must move out of the structure for some time. Young beavers have hair on their bodies and are able to see at birth. Their eyes are half closed at birth and covered with a thick fluid. The incisors are already visible. Young beavers nurse for almost two months; they grow noticeably during this time. The beaver mother often places her tail under her stomach and lifts one leg so that the young can sit on her warm tail for support while they nurse.

Fig. 10-49. The beaver mother carries her young by supporting them in her arms.

Young beavers are also able to swim and dive very early. If they remain in the water, the mother will often forcibly transport them back to the house. Bolau, director of the old Hamburg zoo (which is no longer in existence), once wrote that when she travels on land, the beaver mother carries her young in her outstretched fore legs and walks on her two hind legs. This description was hard to believe, but, some fifty years later, this unbelievable method of carrying the young was observed and photographed in the Zurich zoo. The mother can also carry her young about by having them sit on her tail.

Usually three generations of beavers live together in one mound. The parents always drive the oldest offspring out of the den, by biting them, in order to make room for the new litter. Beavers can live to be ten to fifteen years old. One beaver in the Voronezh beaver reserve lived for twenty-three years.

These animals were formerly found in North America, Europe, and Asia as far as Siberia; they lived in rivers, streams, and ponds where only ash, willow, and birch trees grew. There never were any beavers in Ireland; they were exterminated in England in the 12th Century—England was the first European country where this happened. The last beaver in Switzerland was killed in 1705, in Birs, near Basel. In Germany, beavers were exterminated in the Rhineland and in Saxony in 1840, in Bavaria in 1850, in Würtemberg in 1854, in Lower Saxony in 1856, and in North Rhine-Westphalia in 1877. The names of villages and towns are the only testimony to the fact that these small, industrious rodents once were rather populous,—names like Biberach, Bevensen, Biberstein, Biebrich (as *Biber* is the German word for beaver), or villages and rivers called Bober, Böbersbach, Bobitsch, and Boberow (because the Russian,

Polish, and Czechoslovakian word for beaver is *bobr*). In Germany alone there are over 200 places whose names at first glance seem to be derived from that of the largest rodents in the northern hemisphere. In the New World as well, man's excessive hunting of the beaver has greatly damaged its populations; this has only been controlled by the passage of several very strong protective laws. Since the passage of these laws, the beaver has begun to resettle much of its early distribution area in the U.S.S.R. and North America, and for this reason a small amount of beaver hunting has once again been allowed in these areas. This is not true of the few remaining groups of beavers in central Europe; these animals are protected, and yet the population of the Elbe beaver is still diminishing. However, on a happier note, Müller-Using reports that the Rhone beaver population, which is distributed over 300 km along the Rhone River, in some areas very close to human habitations, has increased from 100 to over 1500 animals.

▻
Beavers will often carry rather large branches or pieces of tree trunks into the water, and then float these down to their beaver mounds (above). In Canada, beavers dam up their rivers in order to prevent changes in water level, thus keeping the entrance to their den below the surface of the water. One can see the head of a swimming beaver, center front (below).

Man, driven by greed, has long been interested in the beaver. In 1782, C. Gottwaldt wrote: "One should not wonder that I spend so much time and trouble on my observations and studies of an animal that so carefully avoids humans; this is because it knows how well it is hunted, and this hunting is not in vain, because the animal is very useful. Everything that this animal has can be used, and to good advantage." What is the reason for man's merciless persecution of the beaver? In the first place, the beaver's scent glands, which are pear-shaped and found in males and females at the rear of the body, contain the much praised "castoreum" which was used in the medicine of the Middle Ages. Lonicerus' *Book of Herbs* contained a wide range of prescriptive applications for this "wonder medicine," apparently to be used in fighting all sicknesses, from headache to dropsy. In 1685, Marius and Frank published a *Castorologia* in Augsburg, with over 200 different recipes. This was certainly very expensive medicine; its healing ability undoubtedly came from the salicylic acid which the beaver acquired by eating willow bark. In 1852, one paid 720 marks ($240) for a "castoreum"; and even today it is occasionally prescribed by homeopaths.

The beaver was also hunted because of its apparently scaly tail, which was considered a delicacy. In 1754, the Jesuit father Charlevoix wrote: "As far as its tail is concerned, it is completely fish, and has been legally declared as such by the Faculty of Medicine in Paris. Following this declaration, the Faculty of Theology decided that the flesh may be eaten on days of fast." During the Middle Ages, beaver flesh was very highly esteemed, with the result that the strong smell of a beaver roast frequently hung over the tables of the castles and cloisters. Otherwise, why would the beaver need such an official classification?

Demand for beaver fur

Mainly, though, man hunts the beaver because of its fur. Not only crowned heads, but rich merchants as well had these expensive furs made into caps and capes. The fur of the beaver was the most important article

1
1
2
2a

◅
1. East African springhaas (*Pedetes surdaster*), in a variety of movements drawn by Wolfgang Weber in eastern Africa.
2. Beaver (*Castor fiber*) with a young (a).

of trade in the golden age of the Hanseatic League. In the beginning of the modern era, continuous persecution had reduced the Old World beaver population to such an extent that the only fur available for sale came from eastern Europe and Asia. Fortunately, the dwindling number of beaver skins brought the end of the production of hats made of beaver fur, which required the largest number of beaver pelts in the 17th and 18th Centuries.

The North American figures indicate man's disgraceful persecution of these animals. In the 1600s, there were between 60 and 100 million beavers; these were hunted by the Indians, who respected these animals. In contrast to that relationship, the Europeans—against the will of the Indians—annually exported 50,000 beaver pelts until 1800. This export of beaver pelts reached its peak in 1875, with the sale of 270,903 skins by the Hudson's Bay Company. At that point, however, the beaver population rapidly began to diminish. Around 1900, the beaver was almost extinct in the U.S.A., and these animals had completely disappeared from large parts of Canada. Almost too late, the North Americans passed several protective laws, and they began to resettle the beaver in areas where it had once been common. The beaver population recovered, and during the First World War, the Canadians were again allowed to hunt beavers. In 1924, 169,172 beaver pelts were sold on the market. However, many beaver colonies were destroyed and, once again, the white man was forbidden to hunt beavers, although Indians still could. The beaver population recovered its strength rapidly in the almost unlimited regions of North America and in 1961 the ban on hunting was lifted slightly, so that 33,400 animals were trapped. Today the population in the U.S.A. and Canada is so high that some 175,000 beaver pelts are obtained annually.

Threat of artificial landscaping

Certainly the beaver has a rather difficult struggle for existence in the intensely cultivated, technically developed, and urbanized areas of Europe. After I [Piechocki] had studied the remains of over eighty Elbe beavers which had been found dead, I ascertained that the reasons for the unnatural decline of the beaver population were the incredibly high losses of young animals, the high average age of the population, and man's illegal killing of animals of breeding age. Researchers in the U.S.S.R. found that about half of the young beavers die within their first two years of life, without human influence. This is certainly much the same with the Elbe beaver. Many young beavers in the central Elbe region are killed every year by the floods that come almost regularly every spring and in the last months of summer. Small hills have been erected in the middle of the Elbe River in an attempt to rescue some of the animals caught in these floods. However, experience has shown that even the presence of

Efforts to protect the beaver

these rescue hills has not kept the loss of the beavers within reasonable limits. In addition, when the beavers fight and bite each other, especially in dirty or polluted water, the resultant wounds often become infected, and this infection is frequently fatal. Old beavers are especially susceptible to such injuries. Studies of beavers that had been found dead revealed that older animals particularly, but also individual animals between seven and twelve years old, suffered from injuries to their teeth, from a fusion of separate cervical vertebrae, and from extensive bone tumors (exostosis) on parts of the vertebral column. For the first time, this high average age of the Elbe beaver population was definitely ascertained and identified as an intrinsic cause of the decrease in the Elbe beaver population within its tiny, isolated, almost insular area. All of the causes mentioned above work together in such a way as to make a stable population of the Elbe beaver in its present living area impossible. Today there are only around 150 beavers on the Elbe. The society founded in order to protect the Elbe beaver is presently attempting to capture some of these animals, breed them in the Magdeburg zoo, and then resettle families of Elbe beavers in those areas where they formerly were common. Only in this way is it possible—or so we hope—to preserve this threatened subspecies for posterity, as successful operations of this sort in Scandinavia, Poland, and the U.S.S.R. have led us to believe.

Evolution of the beaver, by E. Thenius

There are many opinions about the evolutionary origin of the beaver. We trace them back to the Paramyids (just as we also traced the sciurid rodents back to the same evolutionary ancestors). The oldest forms are the + *Steneofiber* from the European Oligocene and the + *Agnotocastor* from the North American Oligocene (some 35–40 million years ago). There were giant beavers (+ *Trogontherium* and + *Castoroides*), as large as small bears, during the time of the Ice Age.

Superfamily: Gundis, by H.-A. Freye

For a long time, the GUNDIS (family Ctenodactylidae) were a completely isolated group among the rodents. They are small, sturdy, clumsy terrestrial rodents with blunt-nosed heads and long vibrissae. They have bristles above their claws, which work like brooms when the animal digs into the desert sand. Until recently, these animals were connected with the collective suborders of rodents; they were finally placed in their own unique suborder. Within the last few years, however, several different forms from the Tertiary period have been discovered and described; these seem to be related to the gundis, and they may also be related to the Paramyids from the Eocene, some 50 million years ago. Because the sciurid rodents may also be descended from the Paramyids, we might best classify the gundis as a separate superfamily (Ctenodactyloidea) within the suborder of sciurid rodents.

Family: Gundis

Distinguishing characteristics

The GUNDIS (Ctenodactylidae; BL 16–20 cm, TL 1–4 cm) range in size between that of a hamster and that of a guinea pig. The body is short, stout, and bulky; the head is thick, with a blunt nose. The ears are short and round. The vibrissae are very long and bristly. The limbs are powerful. Each of the four feet has only four short digits, with short, very pointed claws. The soles of the feet are hairless. The inside digits of the hind feet, and sometimes other toes as well, have a diagonal row of horny, pectinate tips above the claw; there are stiff white bristles above these tips. The tail is stubbed and has long bristles. The skull is oblate with wide frontal bones which have a wide sweep toward the back. The zygomatic bone is large and extends to the lachrymal bone; the mastoid process is inflated. The malleus and the anvil, both of which are auditory ossicles (bones), are fused. The hind feet are longer than the fore feet. The mandible is low, and the sympyhsis is flexible (although both halves are united, each can move independently). There are between twenty and twenty-four teeth arranged: $\frac{1 \cdot 0 \cdot 1-2 \cdot 3}{1 \cdot 0 \cdot 1-2 \cdot 3}$. The molars are almost rootless. The lower teeth increase progressively in size from the first to the last tooth; their outline is shaped like a figure eight. The premolars are small; often they are absent. The incisors are weak, but heavily bent. The tongue has two papillae. The stomach is round and has only one compartment; the large intestine has a spiral-shaped position. The rectum is unusually long. The left lung has three lobes, the right lung, four. The eyes and ears are well developed. There are two glands near the anal region. The underside of the penis tip has a skin fold that runs lengthwise. The animals may possibly be precocial. There are four genera: *Ctenodactylus*, *Pectinator*, *Massoutiera*, and *Felovia*. These animals are found in northern Africa, from Senegal and Morocco as far as Somalia. They live in rocky mountains and in dry steppe areas on the edge of the desert.

Brehms Tierleben described the unique bristles on the gundis' back feet: "Directly over the short, curved hind digits is a row of horny pectinate points; there is a second row of stiff bristles above these, and a third row of long, flexible bristles above that." This brush comb is used not only to clean the animal's fur; it is also important when the animals dig into the desert sands, working like a broom in brushing the sand away. Furthermore, this particular structure, together with the distinct balls of the toes, helps the gundi move rapidly across rocky surfaces.

The gundi

The major representatives of this family belong to the GUNDIS (*Ctenodactylus*). On each side of the upper and lower jaw, there is one premolar (often absent), as well as three molar teeth. There is only one species, the GUNDI (*Ctenodactylus gundi*; BL 17.5 cm, TL 1.5 cm; see Color plate, p. 426). Its fur is very soft, without stiff, straight hairs. The fur is similar to that of the South American chinchilla, and, like the fur of those animals, it offers the gundi no protection from rain. The scapula has a characteristic structure; the neck and shoulders are relatively broad. The collar bone is well developed. The sternum, which is in three sections, is unusually short and wide. There are only six real pairs of ribs. The undersides of all four feet are soft, with cushiony balls which are not clearly distinguishable. The masseter muscles are poorly developed, and these animals' gnawing ability is not very pronounced. Gundis live in crevices and natural cracks in rocks and mountains, as well as on the edge of large desert steppes.

When the weather is wet, the gundi's hair sticks together in tufts. In order to groom its extremely soft coat, the animal often uses the pectinate bristles on its hind feet, and it takes particular care in making sure that its fur remains loose. In order to groom itself, the gundi sits, slightly bowed, on its hind legs, and carefully curries its fur. Gundis can move very rapidly, even over rocks. Kock and Schomber observed gundis in the Tebessa Mountains (Tunisia), on the northern border of the Sahara Desert, and were able to take the first photographs of these animals in the wild. They reported: "The animals' bodies were barely lifted off the ground when the gundis moved slowly; only when the animals were in a hurry did they run with their legs straight. They rarely sat up. They looked around by sitting on their hind legs, leaning the front part of their bodies over their outstretched fore legs, and raising their heads slightly. When sunning themselves, they lie stretched flat on their stomach." These animals are also excellent climbers, effortlessly going up walls that are almost perpendicular. When climbing, gundis take advantage of even the slightest irregularities in the surface, and they always press their bodies close to the support. Their diet consists of grasses, herbs, the few fruits growing in that area, green grain, other plants, and seeds. They take the food directly from the ground with their mouths. They drink only a little water, although they do need water regularly. They defecate only in specific areas.

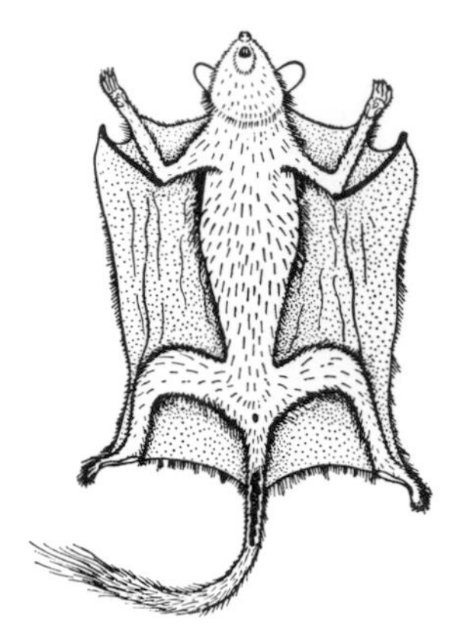

Fig. 10-50. Like the flightskin on the flying squirrel, the flightskins of the scaly-tailed squirrels (above, Jackson's scaly-tailed squirrel, [*Anomalurus jacksoni*]) and the African small flying squirrels (right, large-eared small flying squirrel, [*Idiurus macrotis*]) are also reinforced by a bar extending from the arm. However, the flightskin of the flying squirrels is connected to a bone strip at the wrist, while the scaly-tailed squirrels and African small flying squirrels have their flightskins connected to a bone strip at the elbow.

Kock and Schomber have written the following description of the gundi's daily habits: "Gundis become active after sunrise, about 5 a.m. One hears a peculiar birdlike whistling and peeping before these rodents appear outside their dens. After their sudden appearance they remain motionless either near or still half within their burrows, sniffing the air and watching the surrounding area." If they do not discover anything suspicious, they begin to eat. Next, they take long sun baths. "The animals are active until about 10:30 a.m., after which their energy decreases as the temperature rises. At about 5 p.m., they begin to be active once again. They spend the night in the shelter of their rock-bound crevices. Gundis appear to live predominantly in family groups, although one can also see individual animals roaming around. Neighboring groups do not seem to have distinct regional boundaries, as one never saw a roaming animal being driven away from a particular area, and no fight were ever observed. When the animals were frightened or surprised while sunning themselves, they gave a short whistle and then disappeared very rapidly. Their warning signal, which they made at any suspicious object, was a striking chirping whistle, which increased in intensity as the cause of the alarm persisted. When these animals are greatly disturbed several times in one place, they change their burrows."

Because gundis regularly feed in twilight, the local peoples have called this time of the day "the hour when the gundis appear." Gundis exhibit a particular immobile, trancelike state whenever they come in

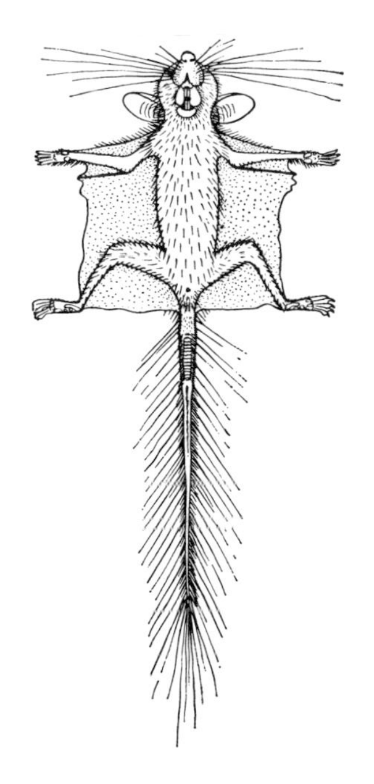

contact with their enemies (snakes, lizards, foxes, jackals, and cats), most of which hunt by sight. H. Roth has described this behavior: "If one grabs a gundi, it will not bite, but it will go into this trancelike state and become completely motionless. If one then lets the animal go, sometimes it will remain immobile for several more seconds. If the animal is greatly frightened, this state will last longer. The gundi lies on its side with its legs outstretched. The animal may stop breathing for as long as a minute. The mouth is half open; the eyes are wide open and slightly sunken. The gundi looks as if it were dead. The breathing begins again in successive gasps which gradually become deeper. The mouth, which had been slightly open, is opened wide with each successive gasp. After another two or three minutes the rigidity wanes, and the animal reassumes its normal stance, although it sits quietly for a few minutes before it flees."

We know very little about the reproduction of the gundis. Apparently the young are born with hair and with their eyes open, and are able to run at birth; thus they are precocial. Kock and Schomber were able to observe "how one of the adult animals, presumably the mother, carried four young to safety, holding them in her teeth by the skin of their necks, and transporting them some distance away to another crevice in the rocks. The second adult animal kept a lookout from a location near the old den. All of the young animals (each was 7–8 cm long) were carried directly out of the old den, with the exception of one young gundi which already sat at the entrance of the den." At no other time were the gundis observed away from cover.

Speke's pectinator

SPEKE'S PECTINATOR (*Pectinator spekei*; see Color plate p. 426) is a close relative of the gundi. It is about the size of a guinea pig; its tail is bushy and somewhat longer than that of the gundi. The characteristic bristles are found on the two inside digits of the hind feet. This species lives in the mountains and desert steppes of Somalia and Ethiopia.

Superfamily: Scaly-tailed squirrel-like rodents, by U. Rahm

The SCALY-TAILED SQUIRRELS (family Anomaluridae) also have an isolated position among the rodents; their systematic classification is still disputed. Some zoologists feel that these animals should be included with the porcupine; others feel that they should be included with the sciurid rodents. This is where we will classify them for the time being as a unique superfamily of SCALY-TAILED SQUIRREL-LIKE RODENTS (Anomaluroidea). The evolution of these animals is presently unknown. They are definitely not more closely related to the squirrels; they probably have a long and unique history of development.

Family: Scaly-tailed squirrels

There are two subfamilies, the SCALY-TAILED SQUIRRELS (Anomalurinae) and the FLIGHTLESS SCALY-TAILED SQUIRRELS (Zenkerellinae). In both subfamilies the body is between 7 and 45 cm long, the tail is from 7.5 to 45.6 cm long, and the weight ranges from 20 to 700 g. Two genera (*Anomalurus* and *Idiurus*) have parachutelike flightskins, and one genus (*Zenkerella*) does not.

Distinguishing characteristics

The skull of these animals is very different from those of the true squirrels. Scaly-tailed squirrels have four fingers and five toes. The body is covered with hair and at the base of the underside of the tail there are horny scales, a characteristic of these animals. The incisors are thick and well developed; the molars are small and usually have somewhat lower crowns. Females have one pair of nipples on the breast. They give birth to one or two young after a gestation of unknown duration. There are about eleven species, found in the rain forests and the wooded savannas of subsaharan Africa.

The name scaly-tailed squirrels comes from the unique horny scales found on the underside of the base of the tail. If one observes the scaly-tailed squirrels or the African small flying squirrels, one will immediately notice the parachutelike flightskin which is stretched along the sides of the body between the fore and hind legs and the first quarter of the tail. Only the flightless scaly-tailed squirrel (*Zenkerella insignis*) does not have such a structure. This parachutelike arrangement may remind one of the flying squirrels, although it is not connected to the wrist with a bone spur as is the case with the flying squirrels. Instead, the flightskin of the scaly-tailed squirrels is connected to the elbows with a bar of cartilage situated on a lamellar widening of the ulna. This cartilage bar may be between 8 and 8.5 cm long in the larger species, although in the African small flying squirrels it is only 1.8–2.2 cm long.

In some species the horny scales on the underside of the tail extend over as much as one-third of the tail's length. They are arranged in two rows, and are directed toward the back. Each scale is almost triangular. The unattached edge is sharp and often has a small point. Jackson's scaly-tailed squirrel (*Anomalurus jacksoni*) has between twelve and fourteen of these scales, of which the first and the last are curved. The scales of the African small flying squirrels take up only fifteen percent of the underside of the tail. These scales are small, shaped like ceiling tiles, and are arranged, on the average, in eighteen diagonal rows, which, in turn, form four rows running lengthwise. The flightless scaly-tailed squirrels have between thirteen and eighteen scales; these cover only 3.5 cm on the underside of the tail.

Scaly-tailed squirrels

Distinguishing characteristics

The SCALY-TAILED SQUIRRELS (*Anomalurus*; BL 20–45 cm, TL 14–45 cm, weight 500–700 g) are the largest representatives of the family. The fur is black, gray, or reddish on the upper side of the animal, and somewhat lighter on the underside. The tail ends in a beautiful tuft of long hairs. All of the digits have sharp, pointed, crescent-shaped claws. The soles of all four feet are hairless and calloused. The external ear is oblong and relatively large. The vibrissae are unusually long. The nose is rounded and protrudes slightly. There are nine species. PEL'S SCALY-TAILED SQUIRREL (*Anomalurus peli*; BL 45 cm, TL 35 cm). The fur is black. These animals are forest-dwellers. The PYGMY SCALY-TAILED SQUIRREL (*A. pusillus*; BL 25 cm, TL 15 cm). The upper side of the body is grayish-brown or dark gray; the underside is gray, mixed with yellow. JACKSON'S SCALY-TAILED SQUIRREL (*A. jacksoni*; BL 35 cm, TL 25 cm). Its body is a dark grayish-brown to black, while the underside is whitish. These animals live in rain forests. NEAVE'S SCALY-TAILED SQUIRREL (*A. neavei*) is light gray on its upper side and white on its underside. It lives in savanna areas. The GABON SCALY-TAILED SQUIRREL (*A. chrysophaenus*) is reddish-brown with whitish-gray undersides. It lives in rain forests. FRASER'S SCALY-TAILED SQUIRREL (*A. fraseri*; BL 25-35 cm, TL 20-30 cm; see Color plate, p. 240); its coloring varies with each subspecies. These animals live in rain forests and wooded savannas. BELDEN'S SCALY-TAILED SQUIRREL (*A. beldeni*) is an ash color; the middle of the back is bright red and the underside is whitish-yellow. BEECROFT'S SCALY-TAILED SQUIRREL (*A. beecrofti*) has a variety of colors and sizes, according to each individual subspecies. These animals live in rain forests. *A. erythronotus*.

African small flying squirrels

The AFRICAN SMALL FLYING SQUIRRELS (*Idiurus*; BL 7-10 cm, TL 9–13 cm) are much smaller. They have a thick, soft, woolly fur on the upper side of their bodies and on their flightskins; all four feet are also very hairy. The breast and stomach are sparsely haired, and there is very little hair on the underside of the flightskin. The coloring can be a creamy-brown, a grayish-brown, or reddish, depending on the species. The tail has a ridge of lashlike hairs on its edge and in the middle of its underside. The nasal area of the skull is small and compressed, and shows a noticeable protuberance; the nose is button-shaped. There are three species: ZENKER'S SMALL FLYING SQUIRREL (*Idiurus zenkrei*; BL 7–8 cm, TL 9–11 cm) is brownish-red or tobacco brown; the underside is red with a gray fold. These animals are found in very few localities. The LARGE-EARED SMALL FLYING SQUIRREL (*I. macrotis*; BL 8–10 cm, TL 12–13 cm) is gray mixed with light brown except on the undersides, which are beige and light gray. LANG'S SMALL FLYING SQUIRREL (*I. langi*; see Color plate, p. 240).

Distinguishing characteristics

Flightless scaly-tailed squirrels

The FLIGHTLESS SCALY-TAILED SQUIRREL (*Zenkerella insignis*; BL 18–22 cm, TL 16–18 cm), the only representative of its genus, does not have a flightskin. These animals are a uniform ash color; their undersides are somewhat paler and are mixed with silver. These animals are similar in shape to a large dormouse. The base of the tail has a border of short, soft hair; there is a remarkable tuft of long black hair on the tip of the tail. A hump-shaped gland area at the ankle joint is covered with black hair tufts. The flightless scaly-tailed squirrel is found in very limited areas of southern Cameroon, Gabon, and parts of the Central African Republic.

Distinguishing characteristics

The scaly-tailed squirrels, the African small flying squirrels, and the flightless scaly-tailed squirrels are all definitely arboreal, and they are excellently adapted to this lifestyle. Their pointed claws enable them to cling to tree trunks that appear to be relatively smooth. Their climbing motions on a perpendicular tree trunk remind one of a caterpillar's movements; the squirrels fasten both hands on the tree trunk, and then pull the rest of their bodies up to secure a firm anchor for their feet; during this activity, the scales on the tail provide another firm anchor. We watched a scaly-tailed squirrel climb a large, old tree one evening, using this technique. The scales on the tail are particularly useful as a firm support; the squirrels press their tails against the tree whenever they rest. Scaly-tailed squirrels are also very agile on thin branches. When a Jackson's scaly-tailed squirrel paused on a branch for a moment, it placed itself in a diagonal position relative to the branch, holding on with all four feet and maintaining its balance with the front of its body and its outstretched tail. The scaly-tailed squirrel hops around on thicker branches; these kangaroolike jumps are quite pronounced when the squirrel is on the ground. These rodents are unwilling to descend to the ground; once there, they become helpless and awkward. We were able to observe this after we had put our animal on a lawn. The squirrel was easily overtaken and recaptured. When one handles a scaly-tailed squirrel, it is advisable to wear gloves, as the claws can make severe scratches and the animal's bite is certainly not pleasant.

Fig. 10-51. 1. Pel's scaly-tailed squirrel (*Anomalurus peli*); 2. Pygmy scaly-tailed squirrel (*A. pusillus*); 3. Neave's scaly-tailed squirrel (*A. neavei*).

Scaly-tailed squirrels and African small flying squirrels climb trees in search of food, and also to be high enough to make a gliding flight to another tree. We had the rare opportunity of observing wild Jackson's scaly-tailed squirrels climbing and gliding in the forests of the Congo region. The animals took off from a horizontal branch or a fork in the branches, and glided in S-shaped paths to another tree, where they landed a little closer to the ground. When the squirrel glides over a longer distance, such as 20 m, it does not open its flightskin until a few meters after take off. Pel's scaly-tailed squirrel can make glides of up to 50 m, while the African small flying squirrels are able to glide as far as 100 m. These animals have rarely been observed in the wild, and we can only speculate on their lifestyle.

All the species of the family are definitely nocturnal animals, becoming active as dusk falls. During the day, these animals retreat into a hollow tree, or they merely hang onto the tree trunk with all four feet. The zoologist Johannes Büttikofer wrote of some Beecroft's scaly-tailed squirrels in Liberia: "I found these animals near the Du-Queah River, where it seemed as if they spent the day pressed against a tree trunk; they looked so much like a piece of old bark that only the natives with their sharp eyes were able to discover them. One of our specimens from the Du-Queah River, however, was found in a hollow tree trunk that was lying on the ground." According to Bates, another species, *Anomalurus erythronotus*, also sleeps during the day, clinging to a tree trunk; it rarely looks for hollow trees. According to our observations in Zaïre (formerly the Congo), both the Jackson's and the pygmy scaly-tailed squirrels cling to the inside wall of hollow tree trunks, where they sleep during the day. Two of the trees we studied were completely hollow and had no crown, so we were able to look down into the trunks and see

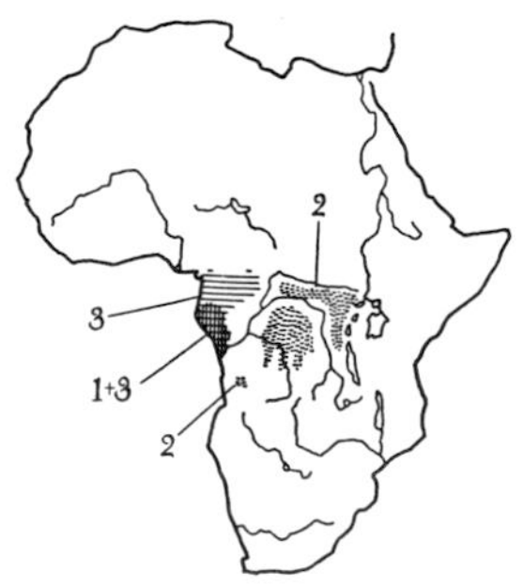

Fig. 10-52. 1. Gabon scaly-tailed squirrel (*Anomalurus chrysophaenus*); 2. Jackson's scaly-tailed squirrel (*A. jacksoni*); 3. Belden's scaly-tailed squirrel (*A. beldeni*).

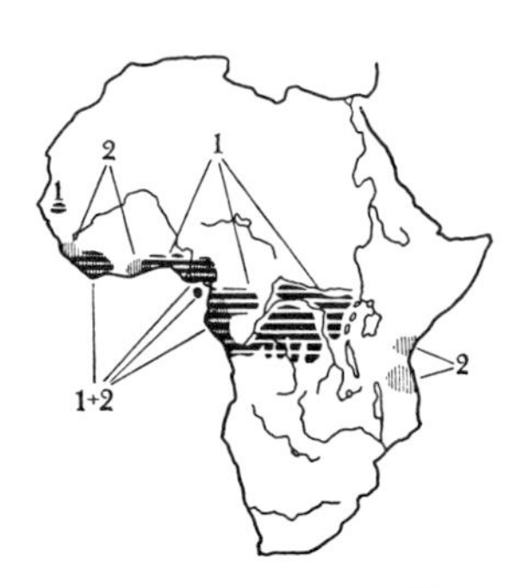

Fig. 10-53. 1. Beecroft's scaly-tailed squirrel (*Anomalurus beecrofti*); 2. Fraser's scaly-tailed squirrel (*A. fraseri*).

the animals. One Jackson's scaly-tailed squirrel, which we observed for two weeks, left its shelter every evening between 6:25 and 6:40, with the coming of twilight.

According to the local peoples, scaly-tailed squirrels often live together in pairs for quite some time. These animals return consistently to a particular area; one animal was found to have lived in the same tree trunk for at least two years. The local peoples believe that a scaly-tailed squirrel will occasionally move to a new sleeping area for one or two days, and then return to its old shelter. When the animals are repeatedly disturbed while in a certain tree trunk, they will definitely change their homes. One can easily wake a scaly-tailed squirrel and drive it away by beating a stick against the hollow tree.

The African small flying squirrels also use hollow tree trunks for shelter, but, in contrast to the scaly-tailed squirrels, these animals live together. According to our conclusions and those of another observer, these animals sleep together in large or small colonies, often together with the insectivorous bat, *Mops leonis*. Sometimes one finds the large-eared small flying squirrel and the Zenker's small flying squirrel together in the same tree cavity. A few African small flying squirrels, which we had in captivity for a short time, slept very close together, holding onto the wire of the cage, similar to a colony of insectivorous bats. Apparently the kind of tree makes no difference to either the scaly-tailed squirrel or the African small flying squirrels in their choice of living quarters; however, the tree must be hollow. Henry Walter Bates is one of the very few researchers who have been successful in observing the flightless scaly-tailed squirrel in the wild. According to Bates, this species is nocturnal, sleeping in hollow trees during the day.

Diet

We know very little about the diet of the scaly-tailed squirrels. According to Dekeyser, Beecroft's scaly-tailed squirrel feeds on palm nuts; other authorities believe that this species also eats tree pulp and leaves. We found several branches gnawed by Jackson's scaly-tailed squirrel. Apparently these animals ate both the bark and the pulp of the tree. The stomach contents of a pygmy scaly-tailed squirrel consisted exclusively of the remains of fruit from the *Musanga* tree. We do not have any definite knowledge of the diet of African small flying squirrels.

Vocalizations

Büttikofer has described the vocalizations of the scaly-tailed squirrels in Liberia: "While at my nocturnal post in the mountains, I heard, here and there in the treetops, a shrill chirping which I had formerly attributed to the galagos. This chirping apparently comes from the flying squirrels which eat the fruit of the tree and which glide from tree to tree under the cover of night." Recently Martin Eisentraut has added, "the linked succession of calls consists of a single, loud, trilling syllable which is constantly repeated with increasing speed and a rising pitch; then it is suddenly broken off." We also often heard such calls in forests in the Congo region, although we were not able to determine whether these vocalizations came from the scaly-tailed squirrels. We were never able to verify such vocalizations in Jackson's scaly-tailed squirrel.

We know almost nothing about the reproduction of this fascinating group of rodents. The gestation period is also unknown. Most of the pregnant scaly-tailed squirrels observed gave birth to only one young; Neave's scaly-tailed squirrel apparently may give birth to one or two young. We were able to establish that each one of five female large-eared small flying squirrels gave birth to one young; all of these births occurred between June and August. The life expectancy of the scaly-tailed squirrel and the African small flying squirrel in the wild is unknown.

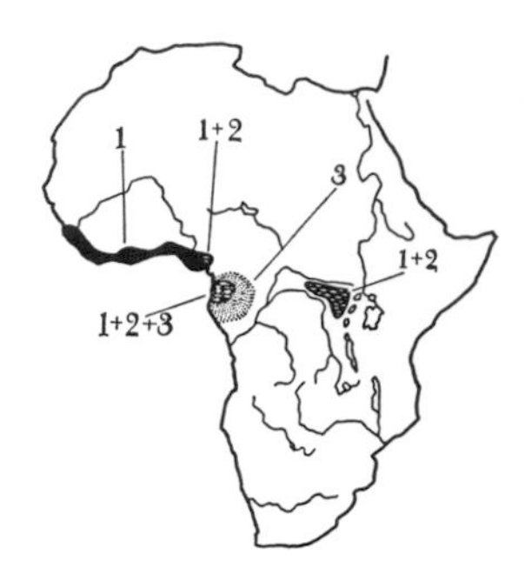

Fig. 10-54. 1. Large-eared small flying squirrel (*Idiurus macrotis*); 2. Zenker's small flying squirrel (*I. zenkeri*); 3. Flightless scaly-tailed squirrel (*Zenkerella insignis*).

Scaly-tailed squirrels have only rarely been held in captivity outside of their homeland. Two Beecroft's scaly-tailed squirrels survived for about a month in the Antwerp zoo; a Jackson's scaly-tailed squirrel that was brought to the Bronx Zoo in New York by a noted animal collector Charles Cordier, lived for only six weeks. We have kept a Jackson's scaly-tailed squirrel outside in Zaïre, in a very high and roomy enclosure for more than a year and a half. Its diet consists of a mixture of 50% cooked rice, 10% oatmeal, 10% cornmeal 10% milk powder, 5% yeast, 3% mineral salts, and chopped, cooked meat. In addition, we gave the animal bananas, papayas, grated turnips, and the leaves of sweet potatoes. Its feces resembled that of rats; the animal defecated during its rest period, and a tiny manure pile began to appear at the foot of the tree in which the squirrel slept during the day.

Superfamily: Springhaas, by U. Rahm

The SPRINGHAAS, SPRING HARE, or JUMPING HARE (family Pedetidae, single genus *Pedetes*), which superficially resemble the common hares or, to an even greater extent, tiny kangaroos, were formerly classified with the jerboas, then with the porcupines, and, finally, close to the scaly-tailed squirrels. According to our present knowledge, it seems probable that the springhaas may have descended from the ancient Paramyids; therefore, they are classified in the suborder of sciurid rodents as a particular superfamily, that of the SPRINGHAAS (Pedetoidea). The only fossil discoveries we are presently familiar with are from Africa (+ *Parapedetes* and + *Megapedetes* from the Miocene, and *Pedetes* from the Pleistocene).

Distinguishing characteristics

These animals are about the size of a common hare (BL 35–40 cm, TL 40–50 cm). The tail is long and bushy; the head is sturdy. The ears are relatively large (7–8 cm long) and are sparsely haired on the outside; the inside of the ear is hairless. There is a tragus at the base of the ear; the outer ear can be completely closed, as the edges of the pinna lie against one another; this prevents sand from getting into the ear when the animal is digging. There are five short digits on each fore paw, with curved and pointed claws. There are two sole pads on each fore paw; one is large, the other, small. The hind legs are 14–16 cm long; each hind foot has four digits, all of which have claws. Parts of the sole pads

Fig. 10-55. 1. Cape springhaas (*Pedetes cafer*); 2. East African springhaas (*P. surdaster*).

on the feet are completely hairless. The fur color may be brown, tan, leather-colored, or red, depending on the species and subspecies; the underside is white. The upper side of all four legs is also usually white. The tail is brownish above and white below; the end of the tail has a black tuft. The nasal bone and frontal bone of the skull are broad. There are twenty teeth arranged: $\frac{1\cdot0\cdot1\cdot3}{1\cdot0\cdot1\cdot3}$. The molars are subdivided by lateral folds on the outer face of the teeth in the upper jaw and on the inner face of the teeth in the lower jaw. The chewing surface has a simple relief pattern. There are two species: EAST AFRICAN SPRINGHAAS (*Pedetes surdaster*; see Color plate, p. 284), with three subspecies. These animals are found in relatively few areas; and the CAPE SPRINGHAAS (*P. cafer*), with nine subspecies, some of which are only very slightly differentiated.

Springhaas inhabit desertlike steppes and dry savannas. Their long hind legs enable these animals to make jumps that are 6–10 m long, although the individual jumps are usually only 2–3 m long. When these animals do not jump, they move about in the manner characteristic of most quadrapedal (four-footed) mammals. Springhaas prefer sandy areas or places with loose soil where they can easily dig their dens. The tunnel system is often deep in the soil and very complex. The loose earth thrown out of the system is piled in front of the tunnel that leads down to the den. The side tunnels end in small round holes at different places in the surface of the ground. There are no earthen mounds in front of these holes, as the animals use them for entering and exiting. According to some reports, the animals close off the main entrance, near the opening, with loose sand or soil. In many areas one can find several of these burrows together, indicating that the animals in these areas live together in colonies.

The springhaas are nocturnal animals whose diet consists exclusively of plants, chiefly the roots and tubers of various grasses, although they will also eat seeds and green shoots. The female has two pairs of nipples on her breast. She gives birth to her young in a room of the burrow, without preparing any sort of nest. The Cape springhaas apparently gives birth to between two and four young each litter. Until the advent of daylight-reversal lighting systems, these lively animals were only rarely exhibited in zoos; now they are being imported in increasing numbers. They have bred successfully on several occasions. Two East African springhaas have lived in the Bronx zoo for more than ten years.

11 The Cricetid Rodents

▻
New World mice (members of the cricetid rodents): 1. Deer mouse (*Peromyscus maniculatus*); 2. Piñon mouse (*P. truei*); 3. Bushy-tailed wood rat (*Neotoma cinerea*); 4. Rice rat (*Oryzomys palustris*); 5. Pygmy mouse (*Baiomys taylori*). Pocket gophers (members of the sciurid rodents): 6. Pacific kangaroo rat (*Dipodomys agilis*); 7. Desert kangaroo rat (*D. deserti*); 8. California pocket mouse (*Perognathus californicus*); 9. Mexican spiny pocket mouse (*Liomys irroratus*); 10. Eastern pocket gopher (*Geomys bursarius*).

▻▻
The golden hamster (*Mesocricetus auratus*) has become quite popular within the last decasde.

▻▻▻
Long-tailed field mouse (*Apodemus sylvaticus*).

We will group the next nine rodent families together in a suborder of CRICETID RODENTS, which we will divide in turn into three superfamilies. As has been the case with the sciurid rodents, however, we must remember that in many instances the classifications given here are open to debate. For this reason, we will dispense with general characteristics of the suborder and the superfamilies.

The suborder of MYOMORPHS (Myomorpha) consists of three superfamilies and nine families: A. Murid rodents (superfamily Muroidea): 1. Cricetid rodents (Cricetidae), 2. Palearctic mole rats (Spalacidae), 3. Bamboo rats (Rhizomyidae), 4. Murid rodents (Muridae); B. Dormice (superfamily Gliroidea): 5. True dormice (Gliridae), 6. Spiny dormice (Platacanthomyidae), 7. Desert dormice (Seleviniidae); C. Superfamily Dipodoidea: 8. Jumping mice (Zapodidae), 9. Jerboas (Dipodidae).

Within the superfamily of MURID RODENTS we find the families of cricetid rodents and murid rodents, both of which have a wide variety of forms and are easily distinguishable from one another, together with the families of Palearctic mice, and bamboo rats. Because the Palearctic mice and the bamboo rats are both so highly adapted to a subterranean lifestyle, the classification of these two groups is unusually difficult, and is by no means final. The superfamily of murid rodents offers us many interesting examples of parallel development and converging adaptation because many of the members of this group have similar lifestyles even though they may not be at all closely related to other members.

Family: Cricetid rodents, by R. Piechocki

Most species of CRICETID RODENTS (Cricetidae) eat plants and invertebrate animals. Hamsters show obvious hoarding tendencies as they store their food, especially grain, in their burrows. Only a few species of this family are basically carnivores. The grasshopper mice (*Onychomys*) feed mainly on invertebrates, although they will also eat small rodents when they can overpower them. The aquatic rats (*Ichthyomys*), which are among the New World mice, are completely adapted to a predatory life in the water; these rats prefer to feed on fish of up to 15 cm in length.

1
2
3
4
5
6
7
8
9
10

1
11
5
13
2
6
12
14
3
4
7
8
9
16
17
15
10

◅ Hamster: 1. Common hamster (*Cricetus cricetus*). Mole mice: 2. Transbaikal zokor (*Myospalax aspalax*). Malagasy rats: 3. Lamberton's Malagasy rat (*Nesomys lambertoni*); Maned rat: 4. Maned rat (*Lophiomys imhausi*).

Cricetid rodents multiply regularly in the warmer regions of their distribution range; reproduction occurs only during the summer months in the temperate regions of the distribution range. The young reach adulthood within a few weeks, becoming sexually mature at that time. In the wild, most cricetid rodents live less than two years. These animals have been divided into the following subfamilies: 1. Cricetinae; 2. Malagasy rats (Nesomyinae); 3. Maned rats (Lophiomyinae); 4. Microtine rodents (Microtinae); and 5. Gerbils (Gerbillinae). There are approximately 2000 species.

These rodents have a variety of body shapes. The length of their bodies ranges from barely 10 cm to about 60 cm. There are generally four digits on each of the front feet, although some species do have five. There are five digits on each of the back feet. The tail is usually covered with short hairs; in a few forms it is bushy. The New World mice and other forms have larger eyes and ears, as well as a longer tail. The microtine rodents and the lemmings have smaller eyes and ears, and usually a fairly long coat of hair. The shape of the skull varies. There are sixteen teeth, arranged: $\frac{1\cdot0\cdot0\cdot3}{1\cdot0\cdot0\cdot3}$. The gestation period lasts from sixteen to thirty-three days, after which from one to eighteen young are born.

Subfamily: Cricetinae

We include four generic groups in the subfamily Cricetinae. These are the New World mice (Hesperomyini), the hamsters (Cricetini), the white-tailed rats (Mystromyini), and the mole mice (Myospalacini).

New World mice

The NEW WORLD MICE (Hesperomyini) are found from Alaska to Patagonia. One of the best-known genera is that of the WHITE-FOOTED MICE (*Peromyscus*), named for their white feet. These animals are very similar, in both appearance and behavior, to the wood mice (*Apodemus*) from the Old World. The genus of white-footed mice consists of some fifty-five species and subspecies, including the PIÑON MOUSE (*Peromyscus truei*; see Color plate, p. 297), the GOLDEN MOUSE (*P. nuttalli*), the DEER MOUSE (*P. maniculatus*; see Color plate, p. 297) and the WHITE-FOOTED MOUSE (*P. leucopus*). Some species are smaller in size than a house mouse; others may become as large as a black rat. The body length ranges from 8 to 17 cm, the tail is from 4 to 20.5 cm long, and the weight is from 15 to 50 g. The color of the fur varies within each species. The lighter subspecies are found on the narrow coastal stretches, while the subspecies living inland in the rich undergrowth of vegetation have a darker coloration, which probably evolved as a protective camouflage through natural selection.

◅ Microtine rodents: 5. Bank vole (*Clethrionomys glareolis*); 6. Mole lemming (*Ellobius fuscocapillus*); 7. Common vole (*Microtus arvalis*); 8. Snow vole (*M. nivalis*); 9. Eastern meadow mouse (*M. pennsylvanicus*); 10. Norway lemming (*Lemmus lemmus*). Gerbils: 11. Tristam's jird (*Meriones tristrami*); 12. Large North African gerbil (*Gerbillus campestris*); 13. Great gerbil (*Rhombomys opimus*); 14. East African gerbil (*Tatera vicina*); 15. Bamboo rat (*Rhizomys*); 16. African mole rat (*Tachyoryctes daemon*); Palearctic mole rats: 17. Lesser mole rat (*Spalax leucodon*).

White-footed mice are nocturnal animals. They live in woods, on prairies, among rocks, and near buildings. Like the voles, most of these species are terrestrial. A few species, however, have adapted to an arboreal life, even building their nests in trees. This is particularly true of the golden mouse, which can be easily identified by the golden color of its back. Their nests have a diameter of 10–15 cm, and they are built at heights of up to 4.5 m. Each nest accommodates two or three young. The young open their eyes two weeks after birth. The golden mouse may even climb up into the crown of the tree if the footing is firm. Eating places used by these mice, which resemble the nest both in structure and appearance, have been found in trees at heights of up to 15 m.

The white-footed mouse, which follows a sedentary lifestyle in its preferred habitat of the deciduous forest, becomes more nomadic when it lives in the semi-desert areas of New Mexico. There these mice occupy a certain territory for a short time and then move on again. We do not yet know whether this unusual behavior is caused by the availability of food or by other factors. Unlike the cricetid mice, the white-footed mouse rarely causes much damage to farm crops; instead these animals are helpful to man, because they feed on insects which damage forests and croplands. Up to thirty percent of their diet consists of animal matter. One subspecies, *Peromyscus leucopus tornillo*, found in the remotest reaches of the Carlsbad Cavern (New Mexico), where they breed in total darkness, feeds almost exclusively on the cave cricket *Ceutophilus*, which is common in that area.

Fig. 11-1. 1. Piñon mouse (*Peromyscus truei*); 2. Golden mouse (*P. nuttalli*).

The PYGMY MOUSE (*Baiomys taylori*; see Color plate, p. 297), is smaller than the white-footed mouse. The GRASSHOPPER MICE (genus *Onychomys*; BL 9–13 cm, TL 3–6 cm, weight 40–60 g) are easily confused with the white-footed mice. One northern species, *Onychomys leucogaster*, and one southern species, *O. torridus*, have been differentiated.

These mice live in prairies and deserts. They are very beneficial to man, as ninety percent of their diet consists of insects, grasshoppers, cicadas, caterpillars, moths, and beetles, while only five percent consists of seeds from planted crops. Grasshopper mice store supplies of seed away for the winter, although these animals sleep during most of the cold season.

Fig. 11-2. 1. White-footed mouse (*Peromyscus leucopus*); 2. Deer mouse (*P. maniculatus*).

The AMERICAN HARVEST MICE (*Reithrodontomys*; BL 5–14.5 cm, TL 6.5–9.5 cm, weight 10–20 g) bear a great resemblance to the house mouse, both in size and in appearance. There are sixteen species, including the WESTERN HARVEST MOUSE (*Reithrodontomys megalotis*), found from southwestern Canada as far south as Colombia and Equador.

These mice live in habitats ranging from salt marshes to tropical forests. They prefer low grassy flatlands. They build an unusual conical-shaped nest of grass. There are usually four young per litter, each weighing about 1 g at birth. The young harvest mice leave their nests after about three weeks, and they reach their full size at five weeks. Some females in this species are capable of reproducing when they are only seventeen weeks old.

The RICE RATS (*Oryzomys*; BL 9.3–19.7 cm, TL 10–23.5 cm, weight 40–80 g) are among the larger representatives of the New World mice. There are about 100 species spread over South America, Central America, Mexico, and the southeastern U.S.A.

Fig. 11-3. 1. Northern grasshopper mouse (*Onychomys leucogaster*); 2. Southern grasshopper mouse (*O. torridus*).

Rice rats are also found on the Galapagos Islands. Unfortunately,

Fig. 11-4. 1. Pygmy mouse (*Baiomys taylori*); 2. Western harvest mouse (*Reithrodontomys megalotis*).

the rice rat populations on some islands have died out due to the domestic cats and the black rats brought intentionally and accidentally by man. Irenäus Eibl-Eibesfeldt gives us an interesting insight into the life of the rice rats on the barren slopes of the Galapagos:

"During the night the temperature sank to 17°C and it was unpleasantly wet. I awoke at 3:00 a.m. and two fat rice rats crawled out of a hole under a stone. They were somewhat smaller than the brown rat. Their fur was a grayish-brown color and they had large, membranous ears. Once in front of their living quarters, they sat up, one after the other, sniffed, and looked around with their dark eyes. Then they wriggled around me curiously, looked at me with their large, round eyes, and nibbled every now and then on a grass stalk. They sat down gracefully on their hind legs, took the stalk in their fore feet, and gnawed on the sprout. It was so quiet that I was able to hear the soft rasping sounds quite clearly. If I made any movement, they rapidly disappeared. However, after a short while they reappeared."

These rodents are also active during the day, and they can swim and dive rather well. They like to build large nests of grass near the water, and they make their feeding place close by. Rice rats feed mainly on grasses, seeds, fruits, fish, and invertebrate animals. In the Mississippi Delta, the RICE RAT (*Oryzomys palustris*; see Color plate, p. 297) breeds from February to December. The gestation period lasts twenty-five days, so that theoretically a female may give birth to up to nine litters a year, with an average of three or four young per litter.

Fig. 11-5. 1. Rice rats (genus *Oryzomys*) in North and Central America; 2. Rice rat (*O. palustris*).

As representatives of the numerous New World mice in South America, we will describe only the VESPER MICE (*Calomys*; BL 6–12.5 cm, TL 3–9 cm, weight 30–38 g). There are ten species, including the VESPER MOUSE (*Calomys musculinus*).

According to the observations made by F. Kühlhorn while he was on the South American expedition led by Hans Krieg in 1937–8, the vesper mouse lives in the shrubbery of the southern Mato Grosso (Brazil), in the thinner areas of woods, and the open edges of woods. These animals also prefer to live in windbreaks in the vinecovered and overgrown areas around fallen tree trunks. Vesper mice are apparently crepuscular. Females caught between April and May had no embryos, and there were no young in their nests. We do not know when the mating period begins. Many of the vesper mice observed had damaged tails. Often part of the skin on the tail, or even pieces of the tail itself, was missing. The new skin, which was always covered with white hairs, was not gray like the animals' original skin, but rather flesh-colored. Often a piebald coloring appeared as a result of the wounded tail. According to Kühlhorn, these injuries were probably caused by small predators that were only able to catch the tails of the fleeing mice. According to our observations, such skin wounds are more often caused by biting fights between animals of the same species.

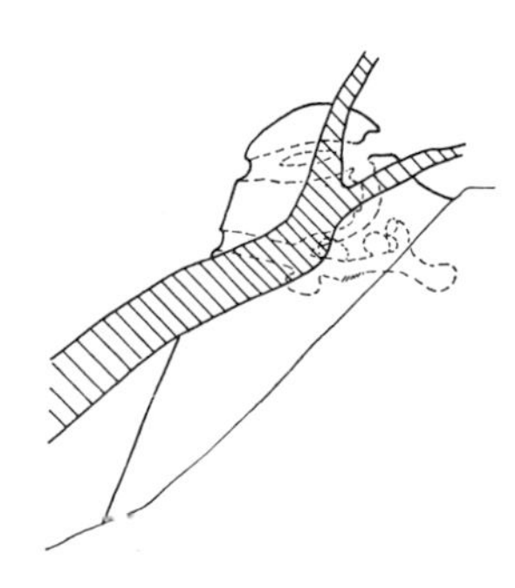

Fig. 11-6. Burrow of dusky-footed wood rat.

The WOOD RATS (*Neotoma*; BL 15–23 cm, TL 7.5–24 cm, weight 200 g to over 400 g) are able to adapt to a wide variety of environments, whether they be hot or cold, wet or dry. These rats are usually larger than brown rats. The tail is basically round, and in some species it is bushy (like the tails of dormice or northern ground squirrels). There are twenty-two species spread over western North America, east to South Dakota, Missouri, and Alabama, south to Nicaragua and Guatemala. Two of these species are the WHITE-THROATED WOOD RAT (*Neotoma albigula*) and the BUSHY-TAILED WOOD RAT (*N. cinerea*; see Color plate, p. 297).

Fig. 11-7. 1. White-throated wood rat (*Neotoma albigula*); 2. Bushy-tailed wood rat (*N. cinerea*); 3. Eastern wood rat (*N. floridana*); 4. Dusky-footed wood rat (*N. fuscipes*).

Wood rats usually live far away from people, although they occasionally move into crop lands, where they can cause extensive damage. Generally these animals are good climbers; consequently, they like to build their nests in trees or bushes, as high as 6 m above the ground. Other species build their large nests (made of various materials) in rock crevices or in the ground, regardless of whether the chosen site is close to water or far from it. Wood rats carry any available material into their nests, even when these materials are useless. They show a particular preference for shiny objects. Mouse traps, tin cans, pieces of glass, and silverware have all been found in a single nest. The white-throated wood rat is able to climb cacti without injuring itself, and it carries the spiny needles back to its nest to use as a defense against enemies. This defense notwithstanding, Reed and Carr watched a skunk work its way through the tight, spine-protected opening to the nest, while the white-throated wood rat rapidly escaped through another opening and jumped on a branch of a live cactus. After a few minutes the rat jumped down again, undamaged, and disappeared into its nest.

The COTTON RATS (*Sigmodon*; BL 12.5–20 cm, TL 7.5–12.5 cm, weight 70–200 g) live in North and Central America, as well as in the tropical areas of South America. These animals are among the most common rodents in all of their many areas of distribution. The upper side of the body is grayish-brown to blackish-brown and the under side is grayish. The ears are short. There are perhaps only two or three species, including the HISPID COTTON RAT (*Sigmodon hispidus*).

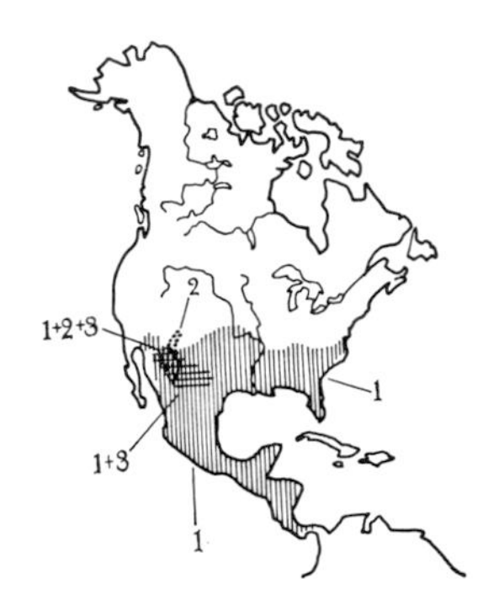

Fig. 11-8. Distribution of the cotton rats: 1. Hispid cotton rat (*Sigmodon hispidus*); 2. Least wood rat (*S. minimus*); 3. Yellow-nosed wood rat (*S. ochrognathus*).

Because cotton rats are able to reproduce all year long, these animals have regular periods of overpopulation during which, according to von Goertz, more females occur than males. More males occur, on the other hand, in less densely populated areas and during mild winters. During the warmer months of the year there are cotton rats of all sizes and ages. However, most of the youngest and oldest animals probably die during the cold season.

When conditions are favorable for reproduction, often millions of these omnivorous rodents will devour not only field cultivations like sugar cane, corn, or sweet potatoes, but also eggs and chicks. Lack of food eventually leads to cannibalism, as the stronger animals overpower and devour their weaker conspecifics. This behavior can be understood when we read what von Ihering wrote about cotton rats in South America: "Everything that was not made of iron, stone, or glass, whether it be furniture, clothing, hats, boots, or paper, everything carried the mark of these rodents' gnawing teeth. These rats may even gnaw horn and wooden knife handles. They may ruin leather, cloth, or linen, destroy books and paper, and even test the strength and sharpness of their teeth on cherry pits. We do not know why they gnaw and bite through pewter vessels, lamps, lead balls, and small shot. However, this is not all; cows have had their hooves gnawed even as they stood in their stalls, these rats have literally gnawed on fattened pigs, and they have even wounded sleeping people."

Hamsters

The next genera group of the cricetid rats and mice is that of the HAMSTERS (Cricetini). These animals are small to medium-sized. They have five digits on each of the front and back feet. The protuberances on the molar teeth always occur in two parallel longitudinal lines. Most of these animals have cheek pouches, which are used to store food. There are seven genera found in the Old World from central Europe to Asia.

The MOUSELIKE HAMSTER (*Calomyscus bailwardi*; BL 7–8.5 cm, TL 8.3–10.1 cm), the only member of its genus, provides a transitional bond between the white-footed mice of the New World and the hamsters of the Old World. Unlike the other hamsters, these animals have long tails. Superficially these hamsters have a surprising similarity to the wood mice. The fur is a dull sand color (ash-gray in the young) while the undersides and feet are always white. The tail is covered with rather thick, closely packed hairs which come together at the tip of the tail in a soft brush. Unlike the other hamsters, these animals have no cheek pouches.

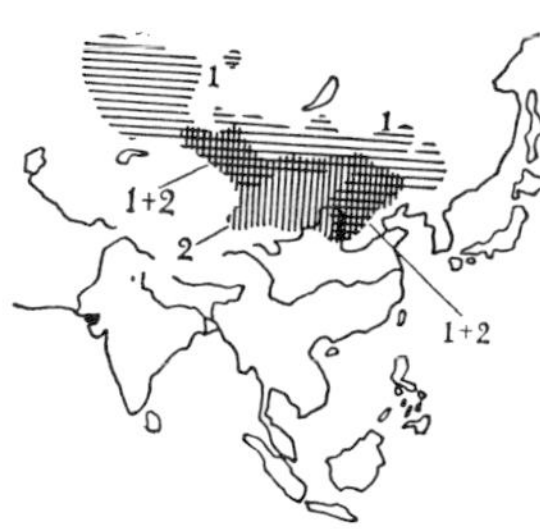

Fig. 11-9. 1. Striped hairy-footed hamster (*Phodopus sungorus*; 2. Roborovsky's dwarf hamster (*P. roborovskii*).

Heptner gives us the following report on the habitat of this species in southern Turkmenia: The mouselike hamster apparently does not appear in the prairies at all, but remains a true mountain animal. One cannot find these animals even in prairie areas that are only a few meters from the rocks. These animals have a rather large distribution in the mountains, and they can be found from the foot of these mountains up to altitudes of 4000 and 5000 m; they also appear in belts of evergreen trees. The similarity of the habitats preferred by the mouselike hamster is striking; the principle distinguishing characteristic of these areas is their desertlike appearance, so that observers often wonder just what these animals can find to eat. These hamsters avoid moist areas, areas with even a half-way healthy plant growth, and areas built up by man. When attempting to find these animals, one should look for their burrows on slightly stony slopes at middle altitudes, in mountain deserts with thorny plant growth and wormwood thickets. During the summer, the mouselike hamster is active only at night; in winter and fall, however, these animals are diurnal. Reproduction takes place from March until June. The young are hairless at birth, and they open their eyes after

Fig. 11-10. 1. Eversmann's dwarf hamster (*Allocricetulus eversmanni*); 2. Short dwarf hamster (*A. curtatus*); 3. Mouselike hamster (*Calomyscus bailwardi*).

thirteen days. The mouselike hamster's diet is made up mostly of plant material, although they will also feed on animal matter.

Fig. 11-11. 1. Migratory hamster (*Cricetulus migratorius*); 2. Striped hamster (*C. barabensis*).

The DWARF HAMSTERS (*Phodopus*; BL 5.3–10.2 cm, TL 0.6–1.8 cm) are tiny animals with a tail which does not protrude from under the fur (in living animals). The soles of the hind feet are densely covered with hair. There are two species: 1. The STRIPED HAIRY-FOOTED HAMSTER (*Phodopus sungorus*) has a gray-brown to ocher-brown coloration, which, on the upper side of the body, is crowded into three curves between the white-colored flanks. There is a black stripe down the length of the back. The undersides are white or gray-white. In the northern regions of their distribution range, these animals become partially or completely white when the cold weather begins. 2. ROBOROVSKY'S DWARF HAMSTER (*P. roborovskii*) is somewhat smaller. Its upper sides are a pale yellow, with a rust-colored shimmer; the fur has a grayish tinge during the change of hair. The undersides and feet are white. The ears are dark with a white border. There are white spots over each eye. These animals live in sandy deserts and in semi-deserts.

Fig. 11-12. 1. Long-tailed hamster (*Cricetulus longicaudatus*); 2. Tibetan hamster (*C. lama*); 3. Rat-like hamster (*Tscherskia triton*).

The genus *Allocricetulus* includes dwarf hamsters of medium size with a relatively short tail, which, nevertheless, is always longer than the feet. EVERSMANN'S DWARF HAMSTER (*Allocricetulus eversmanni*; BL 13.6–16 cm, TL 2–2.8 cm) is in this genus. The animals are a uniform dark-brown to rust-sand color; the underparts are white or gray-white. There is a rust-brown or gray spot on the chest, between the fore feet. The SHORT DWARF HAMSTER (*A. curtatus*) is somewhat smaller. These animals are more yellowish-gray in color (young animals are ash-gray). There is no dark spot on the breast.

The RATLIKE HAMSTERS (*Cricetulus*) are small animals with relatively long tails. The snout is more elongated than in other dwarf hamsters. In summer, the calluses on the naked soles of the feet are easily distinguishable. The STRIPED HAMSTER (*Cricetulus barabensis*; BL 8.2–12.7 cm, TL 2–3.3 cm) is in this group. The upper side of the body is dark brown, reddish-brown, or reddish-gray. There is a more or less distinct black stripe down the center of the back. The gray undersides are not readily distinguishable from the coloring on the animal's back. The ears are dark with a white border. The soles of the feet are almost naked in the summer; in the winter they are covered with hair. The CHINESE HAMSTER (*C. griseus*) has a lighter and more pale coloration; often there is no stripe down the back. Until recently this species was regarded as a subspecies of the striped hamster. It was recognized as a separate species as the result of chromosome studies. The LONG-TAILED HAMSTER (*C. longicaudatus*) is a pale yellowish-gray or ochre-gray color with white undersides. There is no stripe down the back. These animals have dark ears bordered with white. The MIGRATORY HAMSTER (*C. migratorius*; BL 9.6–13.4 cm, TL 2–3.7 cm) is usually dark-gray to ochre-sand colored on the upper side of the body, occasionally, however, being almost straw-yellow. The hair along the spine is usually somewhat darker. Dark patches divide the pale flanks into two or three parts. The underside of the body is white or grayish-white. These animals were first discovered in southeastern Bulgaria in 1959. In 1962, remains of these hamsters were found in fur balls left by owls in northeastern Romania. The TIBETAN HAMSTER (*C. lama*) may be a subspecies of the migratory hamster.

The largest of the dwarf hamsters is the RATLIKE HAMSTER (*Tscherskia triton*; BL 18–25 cm, TL 7.3–9.9 cm) which is the largest species of its genus. These animals are a uniform dark gray, with whitish-gray undersides. The coloring on the back gradually becomes lighter near the flanks. The ears are all one color or they may have an indistinct border of white. The tail is short and thinly covered with hair.

Toward the end of May 1956, I [Piechocki] caught a small animal with a soft, silky fur, above the timber line on Tai Shan (1550 m high) in the Shantung province of China. The animal had very tiny feet and a short, stubby tail; its body seemed to be only head and rump. As the animal was being dissected, I realized that it must have been a Chinese hamster because of the two rows of ordered protuberances on the molar teeth. This first experience with a representative of this genus was a memorable occurrence for me. We were careful to watch for further evidence of other dwarf hamster species. It was not until we were in northern China [formerly Manchuria], however, that we were able to catch striped hamsters, which are common in that area. These animals lived near Tsitsihar in and on the edges of huge millet fields. They lived near Hailar in the moist depressions of the sand dunes, in light pine growth, and in the steppe areas near Dalai-nor. The dwarf hamster was a close neighbor of this species in the steppes. We found the ratlike hamster only in the mountainous areas of northeastern China [Manchuria]. These animals, which are much like water voles in size and appearance, lived on valley slopes under a thick growth of oak and hazel as well as near meadow bogs. All the other species of dwarf hamster prefer a dry habitat; this species, however, likes a moist environment.

Our encounters with dwarf hamster species distributed in Mongolia were no less impressive. On 15 June, 1962 our expedition camp was located in the Echingol Oasis, in the middle of the Transaltai-Gobi desert. The *arates* (cattlemen) of the area soon found out that we were interested in any and all animals native to the region. When these people told us that they had some small rodents living in their *jurts* (tents), we promptly set mouse traps to catch these animals. The next morning we discovered that we had caught not only house mice in our traps, but also several male short dwarf hamsters. Another species, the dwarf hamster, lives in steppes and semi-deserts near human civilization. Roborovsky's dwarf hamster has a particularly good camouflage coloring which helps it blend in with the desert; we found these animals near Bulgan, in the governmental district of South Gobi-Aimak, in the middle of May. They live in loose sand dunes without a close plant cover. We were able to catch the striped hamster in 1964, in the western regions of the country on steppe slopes at an altitude of 1350 m, and in the light larch forests of Charchira Mountain at an altitude of about 2000 m. In the same area of Mongolia, in the middle of August, we were finally able to catch a male long-tailed hamster near the Galutyn River, as it crossed over the broad, almost barren quicksand area to the Caragana semi-desert.

The short dwarf hamster is a characteristic inhabitant of the semi-desert. We found this animal in the Sargyn-Gobi, in the Zavchan river-bed, and in the northernmost desert area of the Old World, Dacst Somon, which is not far from Lake Ubsa Nuur. Occasionally we even saw a fleeing dwarf hamster reflected in the headlights of our Land Rover as we drove over the semi-desert at night. We were able to stop the car, catch the animal, and put it in a bag of strong cloth; everything seemed to go perfectly. However, this tiny character was able to escape, and we did not discover our loss until we returned to camp. The dwarf hamster had gnawed a hole in the sack while we were driving across the desert, and thus regained its freedom. We noticed that the freshly caught hamster lay quietly in the hand of its captor and did not try to bite. Veselovsky and Grundova noticed the same behavior in larger groups of *Phodopus sungorus*. According to their report, these animals were not only friendly with people; they were also less aggressive among themselves than are the common hamster or the golden hamster.

Our observations indicated that, for the most part, dwarf hamsters live in relatively dry environments. All of the species with a large distribution area also live in steppe areas and wooded steppes, as well as in planted or cultivated fields like grain fields or even gardens. When large areas of the northern Kasachia steppes were plowed and planted with wheat, the population of dwarf hamsters in the area increased greatly, and these animals even began to move into storage barns and houses. Most wild dwarf hamsters generally feed on grass seed, shrubs, beetles, spiders, snails, and other invertebrates. These rodents eat fresh plant material, as a rule, only when there is a shortage of their regular food. We are especially knowledgeable about the diet of many species of dwarf hamsters because these animals have often been caught with their cheek pouches full of food. These cheek pouches, which are particularly elastic and are not always packed symmetrically, extend back to the animal's shoulders so that not only the head of the hamster seems "out of shape"; the breast area also has this appearance.

Unlike the common hamster, the dwarf hamster does not belong to the group of skilled rodent architects and builders. The dwarf hamster usually lives in burrows abandoned by other animals, or in natural hide-aways. Many times they will live as secret "subtenants" in the subterranean burrows of other rodents. The summer burrow is often nothing more than a tunnel with one chamber. The winter quarters are deeper in the ground, and they also have supply chambers, along with a well-padded nest room. The distinguishing characteristic of all dwarf hamster burrows is a vertical shaft. As we mentioned before, these burrows (whose other characteristics vary from species to species) are used not only as living quarters, but also for storing winter food supplies. The supplies are usually piled up in widened areas of the subterranean tunnels or in dead-end tunnels that have been widened into a sort of room.

The amount of food stored away is usually not very large; striped hamsters generally store about 200 g, and migratory hamsters store 400–500 g. These quantities amount to a food supply for six to nine months. Some species like the ratlike hamsters take in large amounts of winter supplies. Nikitin found up to 10 kg of potatoes, soy beans, and wheat in the burrows of these animals.

Often an animal will build several supply or storage structures separate from its major burrow; the hamsters visit these storage burrows regularly during the winter, as long as these burrows still contain food supplies. The storage burrows of the ratlike hamster are easy to find; all one has to do is to follow the animal's tracks through the snow. The dwarf hamsters are active during the morning and evening hours. We should mention that some dwarf hamsters are active all day long, especially in winter, when they do not hibernate.

The length of the mating season varies with each species. It may begin as early as the end of February, depending upon climatic conditions, and it may continue through September or October. As far as we know, the gestation period varies between seventeen or eighteen and twenty to twenty-two days, according to the species. The female gives birth to three or four litters a year, with an average of five or six, and occasionally even more than ten, young per litter. The size of the litter is dependent chiefly on the available food, but it is also related to weather conditions. The young grow to be about the size of their parents in two months.

The common hamster

The COMMON HAMSTER (*Cricetus cricetus*; BL 24–34 cm, TL 4–6 cm, weight up to over 500 g; see Color plates, pp. 300 and 321/322) is characterized by its coloration, which involves three different colors. The lighter areas of the fur are quite distinct from the dark stomach area, and consequently they play an important role in the animal's threat posture. Aside from hamsters with a normal pattern of coloring, we find a wide variation, from albinos with red eyes, to various kinds of lighter colors, to melanistic (completely black) animals. The tail is almost hairless. The head is pointed. The eyes are large and black. The ears are medium sized and membranous. The feet are unusually short, so that the hamster's body almost touches the ground when it runs.

The common hamster is one of the best-known wild European mammals because of its fur, which is used by furriers, and because of its particular lifestyle. Man has long hunted these tiny creatures; the hamster's lifestyle and behavior were discussed as early as 1679, in Adamo Lonicero's *Herb Book* (in the chapter about "Animals of the Earth").

The oldest known monograph on the hamster was written by Friedrich Gabriel Sulzer (1749–1830) and was published in 1774; it was entitled *Versuch einer Naturgeschichte des Hamsters* (The Natural History of the Hamster), and it discussed the hamster's anatomy, its living quarters, its lifestyle, and its hibernation. Sulzer's monograph presented so many important facts about these animals that it is still useful. In 1936 Hans Petzsch produced the next important description of these small rodents, concentrating on the hamster's reproduction, an area that was largely unexplained by Sulzer. After the Second World War, scientists again began to study the hamster. Finally a two-part film on hamster biology was produced with the cooperation of Irenäus Eibl-Eibesfeldt. The hamster is now one of the best studied of the smaller mammals.

Hamsters live in steppe areas and cultivated agricultural areas on the plains, up to an altitude of 400 m. In Württemberg, these rodents were found at an altitude of 625 m. The hamster is not always evenly distributed among the field cultivations in its major distribution area. According to Kramer's studies, these animals prefer established field crops of several years standing, especially red clover and their particular favorite, lucerne. Unlike red clover, lucerne usually grows in a dry habitat, which offers the best living conditions to the hamster. Experimental plots of lucerne had three times the number of hamster burrows as did neighboring fields of grain.

Why do hamsters prefer lucerne fields? Lucerne grows quite quickly: within eight to ten days after the last crop has been mowed, a new crop has already become thick, and the hamsters are thus able to find cover while the grain fields are filled with nothing but short stalks. The extensive root system of the lucerne plant favors the construction of burrows, and it protects the entrances and exits of these burrows from weather damage. Finally, the green plant material of lucerne is particularly rich in vitamins and minerals, and thus the hamsters can take advantage of a rich and tasty food during both the fall and the spring, when there is not much available food in the neighboring fields. The hamster, however, is not the only animal that lives off the lucerne. Many other smaller animals, including worms, anthropods, snails, and even a variety of tiny mammals, all of which provide a wide variety in the hamster's diet, share the benefits of the lucerne plant. Sulzer, Löns, Ludwig Heck, and Petzsch have all noted the hamster's preference for animal food of any kind.

Fig. 11-13. 1. Common hamster (*Cricetus cricetus*); 2. Golden hamster (*Mesocricetus auratus*); 3. White-tailed rats (genus *Mystromys*).

The hamster is a crepuscular animal. Especially during the harvest, one can often find these animals in the fields, not far from their burrows. While digging for hamsters as a child, I [Piechocki] discovered just how different each hamster burrow can be. After the wheat had been harvested in northeastern Halle (Germany), we used to take a spade out to the wheat

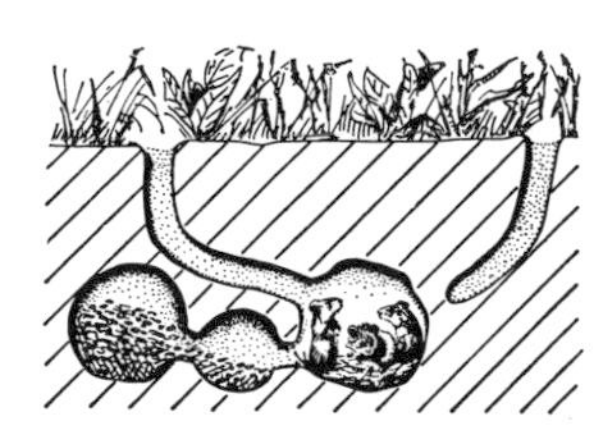

Fig. 11-14. Basic diagram of a common hamster burrow.

fields and dig up the hamster's winter stores of grain and corn. Often we were unsuccessful in our attempts, particularly when the burrow we were digging up was only a simple escape tunnel or when the tunnels were so deep that we boys could no longer throw the excess dirt out of the pit. We were often amazed when unemployed men (there were quite a few at that time) would bring a rucksack full of grain up out of the earth from a depth of over 2 m. Even today some people in Magdeburg (East Germany) are able to bring up almost 50 kg of corn for their chickens from a hamster burrow.

One fall, in Great Kyhna (Germany, not far from Halle), Martin Eisentraut dug up about fifty hamster burrows, and by so doing he was able to find and report on the burrows of the young hamster, the late summer and autumn burrows of the adult hamster, and the winter burrows. The burrows were each constructed so differently that not one looked like another. It was apparent, however, that the extensiveness of the burrow and the diameter of the tunnels were generally dependent on the hamster's age. The rooms and tunnels of the autumn burrow always lie in the same plane, usually at a depth of only 50 cm. The winter burrows are often placed at a depth of more than 2 m. Hibernating hamsters in central Europe dig their winter burrows in flat fields, keeping a large distance between each burrow. In the foothills of the Altai Mountains in Asia, these rodents always dig their burrows in groups on southeasterly slopes. The dense crowding of these burrows is not due to any tendency on these animals' part to live together in colonies; it is due, rather, to the limited number of possible sites suitable for burrow construction. The hamster "mother" builds her burrows in many shapes. Six to eight escape tunnels leading to the nest and supply rooms have been observed; there may also be several vertical shafts. Usually only the floor of the nest is lined; the material used for this purpose consists mainly of grass stalks. The hamster lines the entire nest in its winter burrow. In late autumn every hamster retreats into its nest alone, and stops up the entrance with earth. Hibernation begins as soon as the animal's body temperature sinks from over 32°C to about 4°C. The hibernation is interrupted every five to seven days by short periods of wakefulness during which the hamster takes advantage of its winter food supplies. The hamster is a particularly clean animal, and it regularly defecates in a certain dead end tunnel of the burrow.

Fig. 11-15. Winter burrow of an adult female common hamster as seen from the side: 1. Vertical tunnel; 2. Nest room; 3. Supply rooms; 4. Branching tunnel stopped up with earth; 5. Defecation area.

The mating period begins in the first part of April. Males remain sexually active throughout the summer. The solid, muscular males move into the living area of the females, where they mark the surroundings with scent secretions from their flank glands, by rubbing their flanks in a particular manner against plant stalks growing in the region. According to Eibl-Eibesfeldt's observations, the mating process occurs as follows: "Once the male has occupied the female's territory, he then approaches the female by assuming a special copulation behavior which allows him

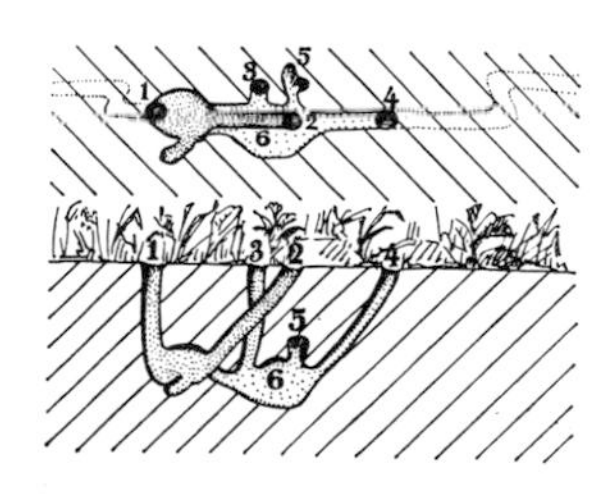

Fig. 11-16. Nest burrow used by a female common hamster two years in succession. Top, as seen from below; bottom, as seen from the side: 1. and 3. Vertical tunnels; 2. and 4. Normal, oblique tunnels; 5. Defecation area; 6. Living room.

to overcome two difficulties: the female's aggressive behavior and her flight behavior. These solitary animals have a very strong tendency for keeping away from one another, and this tendency must be gradually overcome before copulation can occur. When the male and female meet for the first time, they usually sniff rather superficially at each other's noses. The female hisses and many times she also resists the male by biting. Then she flees, often so rapidly that her partner is not able to follow. The female in heat, however, renews the contact she has just broken off, by running back to the male after a short time; the same behavior may then be repeated again. Finally the two animals sniff one another with more interest, first on the noses, then the flank glands, and lastly, in the anal region. Then they raise themselves up, each for the other. Both animals are very cautious and they attempt to avoid touching one another, a contactual shyness that can also be observed in other solitary rodents."

Eibl-Eibesfeldt continues: "As the mating foreplay progresses, the flight of the female becomes less and less serious in intent. Ultimately, the female runs away very slowly so that the male is always able to keep in contact. The male follows with his snout on her, and he emits a soft rhythmic panting (a characteristic sound of sexual activity) while he does so. This vocalization is always an indication of readiness for sexual contact, and it has a characteristic courting tone. This vocalization is used to suppress the female's flight response. Often one can hear sexually aroused males grind their teeth. The lower incisor is rapidly rubbed against the upper incisors and the resultant noise sounds much like the whirr of a sewing machine. This rubbing together of the incisor teeth also serves as a threat posture, and it has been observed in hamsters confronting their enemies as well. The male adopts this behavior during the mating process when the female reacts negatively to him by hissing and grinding her teeth. If the female shows none of these actions, then the male does not assume his threat behavior. This threat behavior may well be a way of subordinating the female to the rank order; on the other hand, it may also be a brief sign of an uncontrolled aggressive impulse."

Copulation takes place in the female's burrow, after a long foreplay. Then, after one or more days, the female becomes irritable and drives the male, who makes no attempt to defend himself, out of her burrow. The gestation period lasts eighteen to twenty days, and the female gives birth to four to twelve young; hamsters reproduce twice a year, on rare occasions giving birth to up to eighteen young a year. As the mother has only four nipples, the extra young are bitten at until they die, according to Petzsch, and then they are eaten either by the mother or by the remaining members of the litter. The newborn young are naked and blind; they weigh 7–8 g. When danger threatens, the mother puts her young into her cheek pouches, or she lays them diagonally across the toothless area of her mouth, between the incisors and the molars, for

Fig. 11-17. Hamsters are not very social animals, and when two hamsters meet they are apt to become quite aggressive. Here, two hamsters sniff each other very carefully.

safety. The older young which have already found their way out of the nest are carried back, one by one, by the mother. Those young being carefully carried in their mother's mouth go into a "carry paralysis" and thus they let themselves be transported without resistence. After a week, the young hamsters (whose eyes are still closed) begin to nibble on green food. They are able to eat green food at such an early age because their upper and lower incisors have already broken through the gum at birth. By the time the young open their eyes, after two weeks, they already have a thick coat of hair. The young are weaned after three weeks, at which time they leave their mother's burrow and become independent. The young animals reach their mother's weight after eight weeks.

In 1950, Rosl Kirschhofer was the first to breed a hamster pair that had been kept together in captivity. Since then, the Petzsches have repeated this experiment. The successful results of these two attempts have shown that hamsters are not always irritable solitary animals, as was long believed, but that these animals may also raise their young in pairs and that they may live together peacefully. This fact also explains why large, even massive, populations of hamsters often live in limited areas.

Although the hamster will eat almost anything, most of its diet consists of plant matter. They feed largely on farm products of all kinds. Of the animals it will eat, the hamster prefers earthworms, and field mice. The hamster annoys the farmer not only with its digging activities in the fields, as a result of which many young plants are covered over with dirt, but also by biting off pieces of roots in order to keep its 15-cm-wide runway open. When the weather is dry, these animals strip kernels of corn from the ears, carrying them to their burrows in their cheek pouches. Occasionally, when the weather is wet, the hamster will bite off the entire ear of corn and carry it back to the burrow. If the hamster is unable to bite through the shoots of a fruit-bearing plant, as is the case with sunflowers and corn, it agilely climbs up the stalk to the fruit. Corn (also called maize or Indian corn) is one of the last crops to be harvested and, as a result, herds of hamsters from neighboring fields will often ravage a corn crop, down to the last stalk.

A shortage of food will force the hamster to travel to other areas where the supply is not so scarce. In 1924 and 1930 thousands of members of the small subspecies of the hamster, *Cricetus cricetus nehringi*, traveled from the Ukraine to Bessarabia, and, in so doing, they had to swim across the Dnjestr-Liman. Generally only the older hamsters survived the crossing of the Liman, while the younger animals were carried far out to sea. Shortly thereafter, the ocean waves brought hundreds of dead hamsters up on the beach. Once the surviving animals reached the Romanian shore, they immediately hid themselves under the overturned fishermen's boats lying on the beach; forty to fifty animals were under each boat. When evening fell, the animals left their hiding places and moved into the village. There they chose houses, courtyards, cellars, and sheds as their living quarters. Onions were their favorite food. They remained hidden during the day. People in the city of Cetatea-Alba (Romania) saw them when they re-appeared after 10:00 p.m., climbing up the vines in the house gardens and the vineyards, in order to eat the grapes. Because the hamsters in this area also destroyed from fifteen to twenty percent of the corn harvest, each human inhabitant of the area was obliged to kill fifty animals.

Although the hamster has any number of natural enemies among the birds of prey and the smaller carnivores, man too must fight these small rodents whenever there is massive hamster reproduction. However, the hamster is not always destructive (in human terms); it can also be very valuable to farmers when it hunts field mice and other pests. The hamster's fur is used by furriers and "one must not forget," as H. Petzsch writes, "that there are many people who, even today, are quite fond of hamster meat, especially since the Saxon elector August de Starke included the hamster in his list of wild animals suitable for coursing (hunting with dogs that follow their prey by sight, not smell); this list appeared in a hunting decree in 1717." We must, however, be careful that these old and popular animals do not fade out of the European animal population.

Hamsters in captivity exhibit a remarkable learning capability. When these animals are raised in captivity from their early youth, they can become really tame. These small rodents can even be trained to certain tasks when one has as much patience as Elizabeth Naundorf, for example, who gives us the following report on her pets: "My second hamster, 'Nickels,' had a particular preference for inedible objects. He stuffed whatever he found into his cheek pouches, and carried whatever would fit in: rubber erasers, buttons, rags, and especially ribbons. He collected the latter wherever he could find them. He even pulled hair bows from the hair of small girls and tucked these bows into his cheek pouches. It was always very funny to watch the difficulties he had when the ribbon was too long for the cheek pouch and thus remained hanging out of his mouth; he was not satisfied with this arrangement, however, and he would constantly pull the ribbon out, rearrange it, and stuff it back in. I supported this preference because it looked so funny, and I began my hamster demonstration by letting him stuff a long ribbon into his cheek pouch while I held the other end. I put the hamster on a table and pulled the ribbon out, much to the hamster's annoyance, while he tried unsuccessfully to hold the ribbon in his cheek pouch. As soon as I let go of the ribbon, the hamster rapidly stuffed it back. For a second trick, I had hung small tidbits from the chain of a lamp, so that the hamster, after a barely noticeable hand movement on my part, immediately sat up and pulled the chain. Then came the 'play dead' game where he remained lying on his back until he saw a hand signal that allowed him to stand up.

Fig. 11-18. Common hamsters groom their bodies vigorously.

Fig. 11-19. Golden hamster emptying out its cheek pouches with its hands.

All of my hamsters learned this trick. Finally, when 'Nickels' cheek pouches were filled with rewards, and when I saw that he wanted to get away, I would sit down at the table and call him, whereby he immediately looked for me, made a large leap into my lap, and climbed onto my shoulder. He could pick me out from several other people seated around the table, and he never went to anyone else but me."

The four species of the genus *Mesocricetus* live in the prairie steppes and mountains of Asia Minor, the Caucasian Mountains, and the lower part of the Donau River Valley. Included in this genus is the GOLDEN HAMSTER (*Mesocricetus auratus*; BL 17–18 cm, TL 1.2 cm, weight up to 130 g; see Color plate, p. 298 and distribution, Fig. 11–13) which is particularly well known today as a laboratory animal and is popular as a pet. The gestation period is only sixteen days, the shortest gestation period of all mammals except the marsupials. Females may give birth to seven or eight litters a year, and each litter consists of between six and twelve young. The young hamsters are sexually mature after almost two and one-half months.

The golden hamster

When, in 1839, Waterhouse began his description of the golden hamster with the following words: "This species is smaller than the common hamster and it is particularly noticeable for its deep golden color," he could not have foreseen the enormous popularity these animals would achieve. After this tiny rodent was identified and given its name, it literally disappeared from sight for over 100 years. It was not seen again until 1930, when I. Aharoni, while on an expedition through northern Syria, near Aleppo, the same area from which the first discovered golden hamster had come, succeeded in digging out alive an adult female and twelve young (each of which measured about 3 cm) from their burrow, which was 2.5 m deep. According to the subsequent reports by Aharoni's daughter, the golden hamster is active mainly in the twilight and during the night. It lives in holes that it digs in richly cultivated corn fields. In addition, according to her reports, in captivity these animals will eat not only seeds, but also pickles, bread, and meat. One young female in the group of hamsters kept by Aharoni gave birth to six, then eight, and finally ten young. Within one year, the Aharoni couple had 150 hamster young as the result of breeding one hamster male and three females. Golden hamsters were first brought to England in 1931, and to the U.S.A. in 1938. As far as we know, all the hundreds of thousands—or even millions—of hamsters being bred today are descended from the breeding group caught by Aharoni. The incredible fertility of the golden hamster led to its role as an experimental animal in many research laboratories soon after its rediscovery. The fact that these hamsters have significantly assisted our scientific exploration becomes obvious when we realize that there were almost 3000 articles about the golden hamster in the *Bibliography of the Golden Hamster*, compiled by R. Kittel in 1966.

Fig. 11-20. Golden hamster carrying its young; the young hamster becomes rigid as soon as the mother grasps it.

White-tailed rats

The systematic classification of the WHITE-TAILED RATS (*Mystromys*;

BL 14–18.4 cm, TL 5–8.2 cm; see distribution, Fig. 11–13) has been the subject of much controversy. Voronzow clarifies the classification of these animals by placing them in their own genus group (Mystromyini) within the subfamily of hamsters. These animals are medium-sized and they have relatively short tails. The hair is grayish-brown and silky smooth. There is no division between the straight hair and the under hair. White-tailed rats do not have cheek pouches. The gestation period lasts twenty-seven days, and the female gives birth to four or five young per litter. Presently we are familiar with two species, the WHITE-TAILED RAT (*Mystromys albicaudatus*), found in southern and central Africa, and the LONG-TAILED RAT (*M. longicaudatus*). Both of these species are largely nocturnal. They live on the ground and in their burrows. There are three other fossil species.

Biologically, at least, the white-tailed rats are very close to the hamsters. They replace their European-Asian relatives in the African savannas, steppes, and semi-deserts, up to altitudes of 1500 and sometimes even 2000 m. However, the white-tailed rats do not have cheek pouches, because their diet consists mainly of green plant material. Reproduction occurs all year long. The only known natural enemy of the white-tailed rats is the suricate (*Suricata suricatta*), which belongs to the family of civets. Other carnivores avoid the white-tailed rat because of its particular smell. White-tailed rats can also damage cultivated crops, and, like so many other rodents, they can carry the plague.

Mole mice

The last group of the hamsterlike rodents is that of the MOLE MICE (Myospalacini), which has only one genus (*Myospalax*) with five species. These animals are very similar to moles, and they are highly adapted to a subterranean lifestyle. Their body is 15–27 cm long, and the tail length is 3–7 cm. Their weight ranges from 150 to 250 g. The hair is soft and silky, and is usually a light gray or yellowish-brown color; there are generally some white flecks or stripes on the animal's head. The fore foot is used as a digging tool, and it has a large claw on the third finger. This claw does the major part of the digging, in conjunction with the two neighboring claws. There are five species distributed over central Russia to northern China (including Manchuria), Mongolia, and eastern Siberia. There animals live either in soft black soil or in sandy soil up to an altitude of 2000 m. Included among the five species are the ALTAI ZOKOR (*Mysopalax myospalax*) and the MANCHURIAN ZOKOR (*M. psilurus*), and the TRANSBAIKAL ZOKOR (*M. aspalax*; see Color plate, p. 300).

We were completely convinced of the MANCHURIAN ZOKOR's fantastic digging ability when we watched an animal that had been placed on a hard-packed loam floor; within four minutes the animal had disappeared, and after twelve minutes it had dug a tunnel 70 cm long at a depth of 30 cm. When digging, the animal chopped off troublesome roots with its incisors, and it pushed the loose earth to the side with its wedge-shaped head. As soon as the zokor had dug itself completely into the ground, it turned around in the tunnel and pushed the loose earth out to the surface with its front feet and head so that a large mound of earth gradually developed in front of the entrance to the burrow. When the animal had dug a tunnel of 3–5 m, it broke through to the surface again in order to throw out the loose soil. We were always able to detect the presence of this species by the accumulated piles of dirt.

Fig. 11-21. Hand (fore foot) of an Altai zokor, with digging claws.

Mole mice are industriously hunted in the Chinese province of Honan because, next to the moles, these animals cause the most damage to crops. Fan Shou, an official hunter from the collective, showed us an ancient rock pile where he had caught forty animals in two months. According to him, the mole mice are also active during the winter. Litters, of four to six animals, are found only during March and April. Possibly the mole mouse has only one litter a year, like the mole. In any case, of six females caught on 22 May, only one was pregnant with two embryos. The nest room is built at a great depth in the ground. We gave up our search for a nest when after hours of working and digging up the soil while in a cramped space, we found another group of branching tunnels.

According to Fan Shou, the burrow consists of a nest chamber (located at a depth of about 2 m), a supply room next to that, and a special chamber for fecal deposits. Between one and four food tunnels lead from the nest room to the surface, depending on the food supply; these food tunnels may reach a length of 50–100 m. The tunnels are built at a depth of 20–30 cm below the food plants. Wheat stalks are carried down into the burrow one by one, while onions are gnawed off their stalks from below. The animals use dirt to block off their food tunnels near the surface after they have finished using them. Storage rooms are not only built close to the nest chamber; they are also constructed at various places along the food tunnels. We found a well-chewed wheat stalk about 5 cm long in one of these food tunnels; this particular tunnel had a vertical diameter of 9–10 cm and a horizontal diameter of 8 cm. According to the Chinese farmers, there will be good weather whenever the mole mouse leaves its tunnels open; whenever it closes off its tunnels, however, the weather will turn bad.

We found another species of mole mouse, the ALTAI ZOKOR (*Myospalax myospalax*), in northeastern Mongolia. The distribution of this species is related, more or less, to river areas, although these animals also appear in forest areas. We were able to catch Altai zokors in this region relatively easily by digging up the food tunnels which were situated at a depth of 20–40 cm. We placed a trap in front of each tunnel opening, and then we re-covered the whole tunnel. We found several verifications of the fact that these animals are active throughout the day. The trapped mole mice made a slight snarling sound, which was apparently some sort of defensive vocalization; when they were touched, they made light peeping sound.

Zoologically and geographically speaking, the island of Madagascar is the only remaining, isolated land area of what once was a large continent. Madagascar is not only the major distribution area of the prosimians it also is the home of the ancient insectivores and the fossas, as well as a rodent group that does not exist anywhere else on the earth. This is the subfamily of MALAGASY RATS (Nesomyinae). Very little is known about the lifestyle of these animals. We distinguished seven genera with a total of fifteen species, most of which are about the size of a rat. The VOTSOTSA (*Hypogeomys antimena*) is the only exception, reaching a body length of 30–35 cm.

Subfamily: Malagasy rats

The votsotsa lives only in dense virgin forests, digging long, deep holes in the ground. It apparently takes the place of the rabbit in Madagascar. It feeds largely on fallen fruit. Unfortunately, the population of these unique big-eared animals has declined sharply because their chosen habitat has become too small. Many trees have been cut down, and, as a result, much of the Madagascan landscape has been turned into a barren, treeless land, with dark, gloomy mountains and arid prairies.

The smallest species are those in the genus *Macrotarsomys*. These Malagasy rats are particularly noticeable because of their long ears and their tail, which is longer than the body. The species *Macrotarsomys bastardi* builds burrows which are up to 1.5 m long, under rocks or bushes. These animals live in breeding colonies. Usually there are two or three young per litter. The genus *Nesomys* includes three species, all of which live in the forests of eastern and northeastern Madagascar. One of these species, LAMBERTON'S MALAGASY RAT (*Nesomys lambertoni*; see Color plate, p. 300), may reach the size of a muskrat. The species *Brachytarsomys albicauda*, which lives on the neighboring islands, reaches almost the same size. These animals have a grayish-brown back and a sparsely haired tail. The root of the tail appears black, while the tip looks white. This species can be distinguished from the closely related members of the genus *Nesomys* by its shorter feet and smaller ears and eyes, as well as a different cranial structure. These animals do not dig holes in the ground; they live in cavities at the foot of trees.

The genus *Eliurus* includes five species. Their characteristic feature is their tail, which is longer than the body and is covered with scales at its root; the rest of the tail is covered with relatively long hair. These animals resemble dormice in that they have large eyes and generally hairless ears. Their weight fluctuates between 35 and 100 g. These rodents are usually arboreal. The species *Eliurus minor*, which lives on the eastern part of Madagascar, is the only species which digs holes in the earth. The two species of the genus *Brachyuromys* (neither of which is uncommon) live on the central plateau of Madagascar. These animals resemble vole rats in size, color, and body shape. They are diurnal and live in tunnels.

Subfamily: Maned rats

The subfamily of MANED RATS (Lophiomyinae) includes only one genus, with one species (*Lophiomys imhausi*; BL 25–36 cm, TL 14–17.5 cm;

see Color plate, p. 300), with several subspecies. The tail is bushy. The females are usually larger than the males. These animals have a series of black and white markings; the back appears silvery. Maned rats live together in pairs in holes in tree or in rock cavities. They have an apposable thumb. The digits (fingers) are mobile, as an adaptation to climbing life. The maned rats are found in dense forests at altitudes above 1200 m. They are distributed in eastern Africa, from Kenya across the Ethiopian mountains to the Sudanese border.

Albert Kull, an animal painter, was able to observe two maned rats rather closely, and some of his observations are presented here: "They have beautifully articulated, rounded heads, with black-pearl eyes, small rounded ears, and long vibrissae. They sit on their back legs and look about, just like squirrels; they clean their nose with their delicate paws. Their coloring, hair growth, and movements remind one of a badger. They are about the size of a hedgehog. Their body shape, however, is more delicate than bulky and the body itself looks larger and more shapeless than it actually is because of the approximately 5-cm-long hairs growing on the animal's back and sides. The distinct and very noticeable black and white coloring makes this animal really beautiful. However, it can also make its appearance quite interesting. When it is irritated, it raises the hair on its back to a vertical position, which makes it look like a porcupine. When this hair is raised, it forms a brushlike comb which has a width of some 4 cm and is sharply divided at the side. At the same time, the maned rat reveals two longitudinal stripes, each about 2 cm wide, which begin in back of the ears and run down the side of the body to the pelvis. These stripes are covered with short, dense yellowish-brown hairs. This change gives the maned rat a totally different appearance, and anyone who is unprepared for such a transformation will be highly astonished. It seems as if the erection of its hair provides this defenseless animal with a means of terrifying its enemies, of which the caracal is the most dangerous."

Subfamily: Microtine rodents

The subfamily of MICROTINE RODENTS (Microtinae) is distributed throughout both the Old and the New World. One definite characteristic of these animals is the enamel pattern of the molar teeth, which varies with each species. There are three generic groups: the LEMMINGS (Lemmini), with four genera; the VOLES (Microtini), with sixteen genera; and the MOLE-LEMMINGS (Ellobiini), with one genus. There are over 200 species in all.

Lemmings

The popularity of the lemmings, which is quite unusual for such small mammals, is doubtless due to the fact that many people never before believed that regularly recurring massive overpopulations of these rodents end in "suicidal" migrations to the sea. We will return to this question in our discussion of the Norway lemming.

The lemming is certainly the most common smaller mammal in the northern countries. The long, thick, waterproof fur of these animals is

F. Reimann

ON THE EDGE OF A FIELD IN CENTRAL EUROPE Mammals: Rodents: 1. Common hamster (*Cricetus cricetus*); 2. Yellow-necked field mouse (*Apodemus flavicollis*); 3. Common vole (*Microtus arvalis*). Carnivores (see Vol. XII): 4. European polecat (*Mustela putorius*); 5. Weasel (*Mustela nivalis*); 6. Beech marten (*Martes foina*); 7. European fox (*Vulpes vulpes*); 8. Domestic cat (*Felis lybica domestica*) with common vole in mouth. Lagomorphs (see Vol. XII): 9. European rabbit (*Oryctolagus cuniculus*) in front of their burrows; 10. European hare (*Lepus europaeus*). Birds: Fowl-like birds (see Vol. VII): 11. Gray partridge (*Perdix perdix*); 12. Common quail (*Coturnix coturnix*). (see Vol. IX): 13. Hoopoe (*Upupa epops*) with erected and relaxed crests. Song birds (see Vol. IX): 14. Skylark (*Alauda arvensis*); the male is singing in flight and the female is feeding the young in a nest on the ground; 15. Red-backed shrike (*Lanius collurio*); 16. Whitethroated warbler (*Sylvia communis*); 17. Magpie (*Pica pica*).
Insects (see Vol. II): 18. Field-cricket (*Gryllus campestris*) in front of its burrow entrance; 19. Mole-cricket (*Gryllotalpa gryllotalpa*) in front of its burrow; 20. (*Agriotes lineatus*); 21. Ladybird (*Coccinella septempunctata*); 22. Colorado beetles (*Leptinotarsa decimlineata*) with larvae on a potato plant; 23. Bumble bee (*Bombus terrestris*); 24. *Bombus lapidarius*; 25. *Chrysophanus virgaureae*; 26. Clouded yellow (*Colias croceus*); 27. Cuckoo wasp (family Chrysididae).

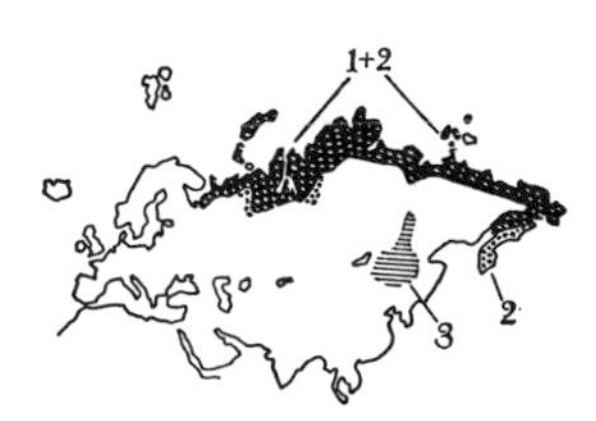

Fig. 11-22. 1. The arctic lemming (*Dicrostonyx torquatus*) occasionally travels as far as Spitzbergen on ice flows; 2. Siberian lemming (*Lemmus sibiricus*); 3. Amur lemming (*L. amurensis*).

Fig. 11-23. 1. Greenland collared lemming (*Dicrostonyx groenlandicus*); 2. Collared lemming (*D. hudsonicus*).

Fig. 11-24. 1. Southern bog lemming (*Synaptomys cooperi*); 2. Northern bog lemming (*S. borealis*).

a good indication of the fact that their principal distribution area is in the polar regions. Their bodies are rounded and bulky, and neither the short snout, the small ears, nor the short tail protrudes distinctly; all of these characteristics help the lemmings resist the cold. These animals, which have long claws on their front feet, are excellent diggers. During the winter months, the lemming lives almost exclusively under the snow cover, thus avoiding the harsh snow storms which begin around the end of October. Because the lemmings usually occur in large groups, and because they do not hibernate, they are the principle prey of the arctic and red foxes, the wolverine, the snowy owl, and other northern birds of prey. Lemmings may also be eaten by wild caribou, which have an occasional appetite for meat.

The COLLARED LEMMINGS (*Dicrostonyx*; BL 12–15 cm, TL 1–2 cm) are distributed over all of northern Asia and in the North American Arctic as far as Greenland. The summer coat ranges from light to dark grayish-brown; the undercoat is much lighter. The winter coat is pure white—very few of the murid rodents have such a complete change in color. The third and fourth digits of the front feet have double claws which grow in every autumn and then disappear every spring. This is a unique occurrence among the mammals, and probably is an adaptation to the more powerful digging that the lemming must do during the winter half of the year in order to get under the frozen earth, and to their need to grip the ice and snow firmly. The bones of the forearm of older animals have an unusual broadening of the shaft of the ulna and a nodular extension of the radius; this can also be attributed to heavy digging activities. The auditory canal does not have a concha; instead it ends in a hairless skin fold and can be closed off with a brush of stiff hairs, an adaptation to life in subterranean runways. There are five species, including the ARCTIC LEMMING (*Dicrostonyx torquatus*), which occasionally drifts as far as Spitsbergen (Norway) on ice floes, and the GREENLAND COLLARED LEMMING (*D. groenlandicus*; see Color plate, pp. 151/152).

The collared lemmings migrate only when they are forced to do so by a lack of food. Dubrowskij, for example, noted that collared lemmings left their living areas whenever large herds of molting bean geese (see Vol. VII) had already eaten all the available food in the district. Three or four young are born in early summer in nests of grass or moss. The gestation period lasts three weeks. The next litter follows during the warm summer months. The Eskimos use the magnificent winter coats of these animals as decorations for their walls; their children use lemming fur to make things for dolls.

The BOG LEMMINGS (*Synaptomys*; BL 10–13 cm) are found from eastern North America west to the Pacific Ocean and north to Alaska. These animals live in moist bogs or meadows. They are two species, the SOUTHERN BOG LEMMING (*Synaptomys cooperi*) and the NORTHERN BOG LEMMING (*S. borealis*).

The southern bog lemming lives mainly in dense bluegrass growth, probably because these plants make up the major part of its diet. They build their nests out of dried grass; the nests may be either on the ground or under it. The mating season begins in March and ends in October. There are between one and seven (on the average, three) young in each litter. Bog lemmings are active all day long during both the summer and the winter. Because they are so active, they often quite literally wear themselves out, and die. The bog lemmings are rather numerous in their home regions during some years; during other years, however, they may disappear almost completely.

Fig. 11-25. Wood lemming (*Myopus schisticolor*).

The WOOD LEMMING (*Myopus schisticolor*; BL 7.5–11 cm, weight 20–30 g) is the only species of its genus. These animals have a varying slate-gray color with rust-brown markings on their backs. The females give birth to two litters during the summer.

The wood lemming prefers altitudes of 600–2450 m, where it lives in marshy swamps during the summer and in pine woods rich in moss during the winter. These animals are very rarely visible in their natural habitat because each uses a system of runways, constructed, for the most part, underneath the moss cover. We were able to find this species in western Mongolia during the latter part of August 1964. These animals were living near some bank voles on a somewhat marshy northern hillside near the Iderün River. We also found the wood lemming in tamarack woods at an altitude of almost 2000 m, where the ground floor was covered with cranberries and myrtle berries. The wood lemming also makes massive migrations during the so-called "lemming years," but unlike the much livelier Norway lemming, it only travels or migrates locally.

Fig. 11-26. Norway lemming (*Lemmus lemmus*), normal distribution. This distribution is greatly increased during migration years.

The BROWN LEMMINGS (*Lemmus*; BL 10–13 cm, TL 1.8–2.6 cm, weight 40–112 g) are either famous or notorious for their migrations, depending on one's point of view. There are five species, most of which have a noticeable coloring. These animals are found in northern Europe, Asia, and North America (see Figs. 11-27 and 11-28). We will discuss the NORWAY LEMMING (*Lemmus lemmus*; see Color plate, p. 300) here. The fur coloring of these animals varies.

Kalela has reported that one can observe two migrational periods in the Norway lemming during years when there is a massive overpopulation. The first of these migrations occurs in the spring; the second comes in the late summer and lasts through autumn. During the summer, this species lives in the open swampy flatlands of the tundra; during the winter these animals live in the snow basins of the mountains. The Norway lemming lives in woody marshes only during the transitional period in the fall. The animals establish runways in this habitat that, with the exception of the entrances to the nest, extend far out into the marshy plain, passing under stones, bushes, or grassy hillocks along the way.

Fig. 11-27. 1. Brown lemming (*Lemmus trimucronatus*); 2. Black-footed lemming (*L. nigripes*).

The Norway lemming usually builds its summer burrow under tree stumps or fallen treetrunks in the treed tundra; in the treeless tundra these animals build their nests under rocks and especially in hillocks. The nest, which is only 15–18 cm long, often has an entrance way which is 30–40 cm long. The winter nests, on the other hand, are often built on the ground in the snow on slopes, on small elevations, or even in the branches of a tiny shrub. These nests are usually cone-shaped, with thick grass walls. The mating period lasts from spring through late autumn. The gestation period is sixteen to twenty-one days. The female gives birth to three litters a year, with between one and twelve young per litter. The young are nursed for fourteen to sixteen days. The life expectancy of the Norway lemming is only one and one-half or rarely two years.

The incredible reproductive ability of the Norway lemming is due to the fact that in good years not only the females from the first litter, but also some of the females from the second litter are able to reproduce, as long as they have reached a weight of 38 g or more. When Frank was breeding these animals, he found that a female which was only thirty-nine days old was already able to give birth; as the gestation period lasts twenty days, this female must have been impregnated in her nineteenth day of life, at the latest. According to Frank, the pregnant female has a noticeable change in body shape: "It becomes wider and wider, and just before the birth it bears a greater resemblance to a turtle or to a bedbug than to a small mammal. The proportion between length and width is about 10: 7 (usually for a moving animal it is 10: 4 at most). The lemming female reminds one of a turtle or a crawling dung beetle in its locomotion as well, because it has a very slow and clumsy walk, and no desire whatsoever for climbing."

Frank's excellent description of the raising of young Norway lemmings indicates that the female lemming shows an unusually firm initial bond with her young. "The young are tucked away in the voluminous nest material, where they remain extraordinarily warm for long periods of time." Usually the female nurses her young from the prone position, a position which is relatively easy for her to assume because of her broad body shape. The young lie on their backs to nurse; when they are able to walk, they will also nurse from the prone position, reaching their mother's teats from the side. According to Frank's reports, the lemming mother washes her young just as other female small mammals wash their young, by licking them repeatedly and thoroughly; this also removes fecal and urine remains. Healthy lemming young always look so clean that the developing fur acquires a silky shine. The mother lives in harmony with her young until she is about to give birth again.

Newborn lemmings have an average weight of almost 4 g. The regular initial growth rate decreases slightly by the tenth day of life; the rate of growth then accelerates after the thirteenth day as the animals begin to eat other food besides milk. This increased rate is greater than the initial growth rate, although it gradually begins to decline after the young animals have reached a weight of 40 g. The lower incisors usually break through the gum on the sixth day of life; the upper incisors generally break through a day later. The eyes open on about the eleventh day. The hair begins to grow on the second day of life, and the fur is already so complete by the twelfth day that it resembles that of an adult animal. The color change from nesting coat to juvenile coat begins at about the twentieth day, and it proceeds very rapidly. The insignificant juvenile coat changes into the first adult coat, which is distinguished by clear, strong colors, at around the thirty-fourth day of life.

Pleske gives us the following description of his encounter with a long line of lemmings on the Kola Peninsula: "One could hear the continual sound of lemmings quite some distance away; all of them were moving in the same direction, and they all crossed the path. The animals approached from the right, making soft whistling sounds, flitted rapidly over the foot path, and disappeared in the shrubs to my left, still whistling. One could hear the next group approaching even when the first group was still audible. Occasionally one could hear the lemmings as they met one another on their way. The whistling sound became louder and one could distinguish the soft barking and growling which indicated a small fight, after which the animals continued on their way in the same direction as before.

"When they came to a larger barrier, for example, a river or lake, they would all gather together in crowded herds on the bank, forgetting their individual differences in the general emergency. If the lemming is such a courageous, one might even say confident, creature which, like the hamster, defends itself against dogs and even from man rather than fleeing, we must certainly admire the foolhardiness with which these tiny animals overcome all obstacles in their path of migration. Thousands of these animals gather together on the bank of a wide and rushing mountain river; then they all plunge into the water and swim to the other side. Hundreds of animals die, wrestling bravely with the waves. Such losses, however, hardly diminish the herd, which continues unperturbed along its chosen route."

According to Frank, the Norway lemming, unlike the microtine rodents, swims with its back out of the water. This is certainly related to the large amount of air which their long, dense fur is able to hold; because of this ability, the lemming usually emerges from the water almost completely dry. The lemming is a better swimmer than the microtine rodent because it has a more favorable body position in the water, and its continuously alternating rudderlike feet are larger than those of the miniature cutter when it swims through the water. Just as these lemmings do not shy away from wide rivers and lakes, as well as other seemingly impossible obstacles, so too are they not afraid to throw themselves into the ocean once they have reached the coast. The ocean connotes the end of their migration and the animals plunge into the water exhausted and powerless, only to drown or to be swallowed up by the larger fishes and sea gulls.

To what can we attribute these migrations, which have often been described? In his extensive study of mammal migrations, Kalela has observed that the spring migration of the Norway lemming begins about the time of the spring thaw, and that this migration is of relatively short duration. During this period the lemmings travel quite rapidly because they eat very little food. Initially some seventy percent of the animals in the migrating herds are male. Later this number rises to as much as ninety percent because the females, many of which are already pregnant, settle down in suitable locations or habitats. Apparently the basic drive behind the spring migration is the lemmings' need to find a suitable nesting area. Large numbers of these animals also traveled to unsuitable environments, where they die.

The lemmings are on the move again from August (and sometimes even July), when they leave their summer homes, a good food supply notwithstanding, until the beginning of winter. This is the fall or autumn migration, and during this period the animals appear in dry, wooded areas where the moss is plentiful. These areas will be the future winter quarters. Occurring simultaneously with this resettling process are regular migrations into unfamiliar lemming-free areas. According to Kalela, these animals often cross bodies of water, and many are killed on farms by dogs. We have long known that migrating lemmings are particularly irritable and aggressive, even toward man. Lemmings which have already settled down in wooded areas also exhibit this behavior before they have completed their system of runways. The migration reaches its zenith about the middle of September, and it gradually fades away during October.

Kalela has noted that when the increase in the lemming population is average, most of the animals that left the summer habitat apparently remain in the winter quarters, and the migration, as a result, is rather short. Migrations of a longer duration occur when there is a massive population increase; these longer migrations are also apparently brought about when the moss ground cover in the winter quarters (which the animals occupy in the fall) has been largely destroyed by grazing lemmings during the previous winter. Most of the animals that participate in these longer migrations die. A few of the migrating lemmings, however, are able to settle down in habitats along the migration route. Often there will be a large population increase in these areas, while the lemming stock in the original habitat remains significantly weaker. Unlike the spring migration, the late summer-fall migration is interrupted by frequent stops for food. Nevertheless, the speed of travel is often considerable. In one instance, a speed of 15 km per twenty-four hours was estimated. The animals prefer to travel during the night, but when there is a particularly large number of migrating animals, they will also travel during the day. When there is only a small group of Norway lemmings, these animals will migrate only when forced to do so by environmental conditions. Individual animals, on the other hand, make obvious attempts to follow the trail of a lemming that has passed them by. Once Kalela was able to observe sixteen successive animals, each of which was attempting to follow the animal in front of it. The whole observation lasted an hour.

Kalela's observations tie in very nicely with the final conclusions of Frank, who raised several generations of lemmings and who studied the behavior of these animals in the laboratory. Frank concluded that the lemming phenomenon is definitely not the model for regularly repeated rodent migrations, as was formerly believed. Instead, Frank concluded, the lemming migration is much more of an exception, a complex modification of the massive rodent fluctuations we find in the temperate zones of the world. The lemming migration must then be considered as a special adaptation to the restrictions of the arctic climate.

Both of these scientists, through their careful studies of lemmings in the wild and in the laboratory, have proved just how fanciful the ancient stories of great herds of these animals descending from the heavens to consume all living plants are.

Voles

The branch of the order of rodents which contains the VOLES (tribe Microtini) includes more species than any other group within the order. These animals are sturdier than mice. They are exclusively terrestrial. The tail, which is sparsely haired and has distinct rings, ranges from very short to just barely body length. The head is broad, the snout is stubbed, the eyes are small, while the ears are short and do not usually protrude beyond the hair. The surface of the molar teeth is more or less characterized by triangular enamel patterns.

Because it would be impossible for us to discuss all the many forms of this generic group within this book, we will pay particular attention to those species which come in contact with man and those species which can damage cultivated fields, woods, gardens, and even water areas used by man.

The bank vole

During the winter an observant walker may often see elder trees (*Sambucus canadensis*) near the edges of woods or on valley slopes; a large amount of the bark on these trees is missing so that the lighter sapwood shows through. Upon closer observation, it is not hard to discover traces of a rodent. Anyone who wishes to discover the identity of the marauding rodent need only set up several spring traps at the base of the elder tree during the night. The next morning he will probably have caught some foxy-red mice whose tails are half as long as their bodies. If these characteristics prove correct, the animal in question is a BANK VOLE (*Clethrionomys glareolus*; BL 9–11 cm, TL 4–5.5 cm, weight 16–34 g; see Color plate, p. 300). Klein, a well-known expert on smaller mam-

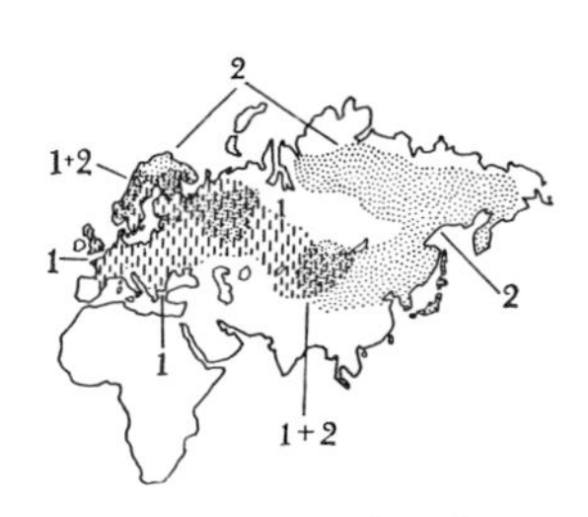

Fig. 11-28. 1. Bank vole (*Clethrionomys glareolus*); 2. Large-toothed red-backed vole (*C. rufocanus*).

mals, found that the bank vole eats the bark of such shrubs only in an emergency, when the local population is faced with starvation, and not, as was formerly believed, because of a general nutritional requirement. The youngest, lightest, and smallest bank voles are the first to die in such cases, because only the stronger animals are able to climb up into the bushes and gnaw off the bark which serves them as a substitute for their normal food.

The bank vole is a so-called "shade animal," as it prefers to live in wooded areas, in shrubs or banks, and in swamps. Pregnant or nursing females often weigh more than others. The gestation period is eighteen to twenty days. Three or more litters are born between spring and fall. The newborn bank voles are completely naked, and weigh barely 2 g. There are many geographical subspecies with a variety of colorings and body shapes. Bank voles from the Alps and the central highlands (Mittelgebirge) are larger and have longer tails than do the lowland forms.

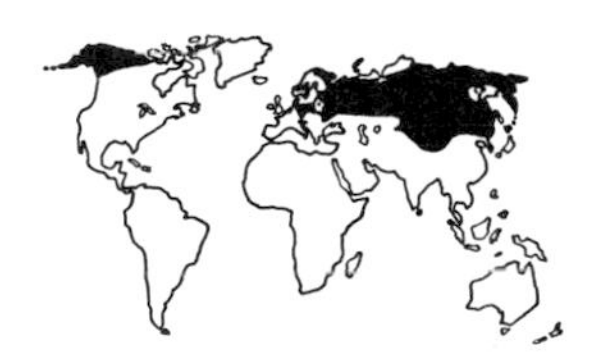

Fig. 11-29. Northern red-backed vole (*Clethrionomys rutilus*).

The female bank vole has four pairs of nipples. She gives birth to between three and five, rarely up to seven, young per litter. The nest is made of grass, moss, and leaves. One can find these nests either underground or slightly above the ground in rotten wood or under dead wood. If the mother is disturbed while nursing her young, and if she is able to escape, the young instinctively retain their hold on her nipples; as a result they can be carried for some distance. If one of the young becomes lost, then it probably announces its location with vocal signals which only the mother can hear. When the bank vole finds her young, she picks it up and carries it back to the litter with her head erect. Maternal care is continued until the mother is no longer able to keep her increasingly active brood together.

After eight days, the young bank voles have already acquired enough hair that the characteristic reddish coloring becomes apparent. Their eyes open after about twelve days. They are sexually mature after nine weeks. This rapid development and the rapid succession of litters often leads to massive overpopulation. When this occurs, these nocturnal and diurnal animals (which are better climbers than the other microtine rodents) move into nurseries and seed beds, where they can cause extensive damage. They strip the bark from trees up to a height of several meters, and they consume the needles from larch trees and other coniferous trees. The usual diet of the bank vole consists of seeds, fruits, grasses, and lower animals. The bank vole uses leaves to cover over food that is left lying out in the open as well as its winter storage supply. During harsh winters these animals will sometimes seek shelter in buildings near wooded areas.

Fig. 11-30. 1. Gapper's red-backed vole (*Clethrionomys gapperi*); 2. Western red-backed vole (*C. occidentalis*).

The bank vole has a territory of 900–1000 m^2. It is really remarkable how bank voles which have become lost are able to find their way back to the nest, over distances of up to 700 m. Schleidt, during his studies in the area around the Lunzer Obersee, found that bank voles could find their way back to the nest over a distance of 300 m, with a travel speed of 1.5 km/h.

We found the small NORTHERN RED-BACKED VOLE (*Clethrionomys rutilus*) together with the larger LARGE-TOOTHED RED-BACKED VOLE (*C. rufocanus*) in the mountains of northeastern China [formerly eastern Manchuria] as well as in Asia and Asia Minor. The large-toothed red-backed vole was prevalent in mixed woods of deciduous trees wherever there was a dense undergrowth. The northern red-backed vole, on the other hand, lived mainly in coniferous forests. We also found both species in the wooded areas of western Mongolia. The red-backed voles in that region are generally found in larch woods, partially in dry habitats and partially in humid, moss-covered rocky crags, where we were able to capture them. However, the animals were not so numerous in this area as they were in the woods of northeastern China, where they cause extensive damage by eating the relatively large, nutlike seeds of the Korean scotch pine tree. Several species of this genus are also distributed through the wooded areas of North America.

Père David's voles

Voles of the genus *Eothenomys* might be considered as a transition between the red-backed voles and the voles. These animals bear a close resemblance to the red-backed voles in coloring, size, and lifestyle. There are six species, found in China, the northern part of Indochina, Japan, and Formosa. Included among these species is the PERE DAVID'S VOLE (*Eothenomys melanogaster*) which lives in wooded and mountainous prairies at altitudes of 1800–4400 m.

Western voles

The WESTERN VOLES (*Alticola*) take the place of the European snow voles in central and eastern Asia. There are five species, including the HIGH-MOUNTAIN VOLE (*Alticola strelzowi*) and ROYLE'S HIGH-MOUNTAIN VOLE (*A. roylei*), which live at altitudes of 900–5800 m.

In Mongolia we found that large rocky crags in areas of poor vegetation are the typical habitat of the western vole. One sure indication as to the location of these rodents' colonies was always the large amount of droppings we found in between cracks in the rocks. Large populations of the high-mountain vole lived in the Charchira Mountains under piles of larchwood branches at an altitude of 2000 m. According to M. Stubbe, Royle's high-mountain vole begins to store away its winter supplies in June in rubbish piles and clefts; plants are dried before they are stored away. The fruit of the wild almond bush (*Amygdalus pedunculata*) is a favorite food of the western voles living near the steep banks of the Char-usnur River in northeastern Asia. The female voles give birth to, at most, two litters of young during the short time the wild almond bush is in bloom. The average number of young per litter is five or six. The environment of the western vole is characterized by the abrupt changes in climate; the animals are equipped with thick, soft hair which allows them to withstand such changes. The change of hair for the winter coat begins in the middle of August.

European water vole

The EUROPEAN WATER VOLE (*Arvicola terrestris*; BL 12–19 cm, TL 7.5–12.5 cm, weight 70–180 g; see Color plates, pp. 105/106 and 257/258) is the largest terrestrial microtine rodent in the Old World (except the muskrat, which was once restricted to the New World, but which has now been introduced to parts of the Old World as well). The European water vole is similar to the mole both in its size and lifestyle. The hair is thick and shiny. Particularly large animals that are completely black have been known to occur rather often in lowland areas. The earlier name for these animals, water rats, is insupportable both systematically and biologically as the European water vole is not really a rat at all, nor does it have any distinct adaptations to an aquatic lifestyle. Usually, these rodents may live far away from water just as often as they live near natural waters. Their preferred habitats are woods, meadows, croplands, and gardens. There are numerous subspecies. The IBERIAN WATER VOLE (*Arvicola sapidus*) has recently been declared a separate species.

The reproduction of the European water vole occurs throughout the summer. The males mark their territory with odoriferous secretions from their side glands. These side glands are situated in two horizontal rows, up to 2 cm long, located just in front of the back legs. Frank found that these glands always become more active whenever the animal is aroused, whether it be through an encounter with an unknown conspecific, a sexual partner, or an enemy. When the European water vole is aroused, the secretions from its side glands are transferred to the soles of its feet by rapid movements of the back legs, reminiscent of the animal's cleaning maneuvers; the animal then stamps its secretion into the earth by making alternating trampling motions with each of its back feet. These movements are accompanied by a rapid beating of the tail from side to side.

When the population is very dense, the males will often participate in fierce biting matches during which they make high, shrill squeaking cries (which are both threat and fear vocalizations). One June, I [Piechocki] caught a male in heat on Hiddensee Island; his skin was covered with several bite wounds which looked much like the injuries that might result from a round of birdshot. The nest of the water voles is built either in underground tunnels or, when the area is wet, on the ground, where it is generally constructed out of grass, depending upon the habitat. The gestation period lasts twenty-one days, after which, around the beginning of May at the earliest, two to eight young are born. There are three or four litters per year. Depending on the size of the litter, the weight of the newborn young ranges between 3.2 and 7.8 g. The young open their eyes after about nine days. They become independent at four weeks.

Fig. 11-31. 1. European water vole (*Arvicola terrestris*); 2. Iberian water vole (*A. sapidus*).

Although water voles often live in rather large densities, they do not form social groups. They build extensive tunnel systems which are generally immediately under the surface. These branching food tunnels are usually situated so close to the surface that one can easily follow their slightly elevated roofs. The animals toss the excess earth out of their tunnels just like moles, although the resultant earth piles are always flatter than those moles make; in addition, the piles built up by the water voles contain uprooted plant growth. The European water vole regularly uses the tunnel system of moles; thus it is able to reach the roots of plants, which constitute its basic food in environments which are rather removed from the water, without too great an expenditure of labor. This dietary preference on the part of the European water vole can be quite damaging to man, especially when these animals inhabit tree nurseries or orchards. Healthy fruit trees may have their roots so badly chewed up that they ultimately wither and fall over. These animals also show a preference for the roots of rose bushes. In addition, the European water vole will eat farm produce, vegetables, and fruit found on the ground. In the Netherlands, these animals feed mainly on tulip bulbs during the winter.

Fig. 11-32. 1. Muskrat (*Ondatra zibethica*); 2. Newfoundland muskrat (*O. obscura*); 3. Round-tailed muskrat (*Neofiber alleni*).

The European water vole is found all over Europe, although there are great differences in the numbers of animals found in particular areas. These animals are particularly numerous in the foothills of the Bavarian Alps, according to Ballow, especially in the vicinity of the Chiemsee (Lake Southeast of Munich), but also in the Bavarian part of Swabia. The European water vole is just as numerous in neighboring Württemberg (Southwest Germany) along the banks of reed-filled ditches, rivers and lakes. Ballow's studies indicate the enormous numbers of European water voles which occur. In North Württemberg, in a good vole year, 403,000 animals were caught. This impressive number appears small, however, if one considers that in one small village in the County of Wolfratshausen in Upper Bavaria 43 people killed in one year more than 50,000 voles in a meadow no larger than 330 hectares (~ 840 US acres). Near Tomsk (Siberia) over 4 million water voles were caught in a single October. When we realize that only a part of the water vole population can be caught at a given time, we can understand what great numbers of water voles live on this earth.

Fig. 11-33. The muskrat's (*Ondatra zibethica*) approximate distribution in Europe. The muskrat population in Europe can be traced back to a few imported animals.

Water voles carry and transmit the virus for tularemia, a disease of rodents to which man is also susceptible. Thus, of the 300 hunters of water voles who were active near Tomsk, about eighty-five were stricken with tularemia. Whoever hunts water voles, whether it be in croplands or in cellars, must be very careful not to be bitten by any of these animals when they are alive in the traps. Another precautionary measure which is, or should be, taken to avoid the risk of further infection is to bury dead water voles at a depth of 50 cm; then one should wash one's hands carefully.

The muskrat

The MUSKRATS (*Ondatra*) can reach the size of a wild rabbit. There are two species, of which *Ondatra obscura* is found only on Newfoundland (Canada), while the well-known MUSKRAT (*O. zibethica*; BL 30–

36 cm, TL 20–25 cm, weight 600–1500 g; see Color plate, pp. 257/258) has spread from its North American habitat to large parts of the Old World. The fur is shiny and dark-brown to chestnut-brown. The underside (the "belly" to furriers) is a lighter brownish-gray. The body is relatively heavy and thickset. The head is short and thick, and is directly connected with the body. These animals have a flat, rudderlike tail. There is a fringe of bristles on the edges of the toes of the back feet, commonly known as the swim fringe. The inner ear can be closed off by a skin fold (an adaptation to aquatic life; particularly to underwater swimming).

The somewhat smaller ROUND-TAILED MUSKRAT (*Neofiber alleni*) is a close relative of the muskrat. These rodents are more terrestrial than the muskrat. They are able to reproduce all year long in the coastal swamps and bogs from Florida to Georgia, wherever the local climatic conditions are favorable. The two young are born in each litter.

In the spring of 1905, Prince Colloredo-Mannsfeld brought three pairs of muskrats with him to Europe after a hunting trip to Alaska; one of the male muskrats died enroute. The prince put the remaining animals in two natural ponds on the estate of Dobrisch Castle, 40 km southwest of Prague. All the muskrats in Europe today are probably descended from these five animals. Initially the muskrats were fed with corn, carrots, and potatoes; shortly thereafter, however, they escaped, and their descendants spread themselves out through Germany and Austria, to Switzerland and Yugoslavia. The muskrat became particularly prevalent in Germany, where it occasionally must be kept under control by specially designated muskrat hunters, in order to prevent its undermining of dams and dikes. Such a prevention program is especially important in the Netherlands, where the muskrat is man's most important natural enemy; the Dutch hope to eliminate the muskrat population completely in their country.

Muskrats were stocked in Finland in 1922, because of their valuable fur. Five years later muskrats were introduced into Scotland, England, and Ireland; however, they vanished from these areas within ten years because they were hunted and killed to excess. In France, some muskrats escaped from farms in 1938, and presently there are still wild muskrats in northern France. Muskrats were introduced into the U.S.S.R. in 1929–30. According to Ognev, these animals are widely distributed there today, from the Kola Peninsula to the Sea of Okhotsk and the Bering Sea, and to the Kazachstan Desert. Seven years after the muskrat was introduced in these areas, man began to use muskrat fur in various industrial processes. (The use of muskrat fur has been greatly increased to date). In some areas, the muskrat is the most important animal for the fur industry. These rodents are also valuable fur animals in their natural homeland of North America, where there are several subspecies. Initially man was particularly active in increasing the distribution of these animals. However, as greater and greater areas of North America were given over to cultivation, the muskrat became a pest, just as it had become in Europe, and as a result, man was forbidden by law from introducing new settlements. The increasing range of these highly adaptable rodents is related to the decline of their natural enemies, especially the otter and the white-tailed eagle.

Because the muskrat is so well equipped for aquatic life, it is an excellent swimmer and diver. On land, however, it moves very awkwardly. The powerful incisors, which protrude from the oral cavity, are generally used underwater. Muskrats are crepuscular and nocturnal animals. They feed largely on the roots, sprouts, blossoms, and the fruit of plants growing near the water. In addition, they will also eat willow buds, farm produce, and garden fruits, as well as small amounts of mussels and water snails.

One unmistakeable indication of a muskrat settlement in a particular body of water is the presence of large pieces of plants, which have been gnawed off and left to float on the water surface in the channels which these animals construct through the marginal plant growth near the edge of the water. Another sign of these rodents is the runways they build on the bank, near which one can often find definite areas where piles of dark-green feces of 2–3 cm in length are deposited. In addition, the muskrat digs its burrow in the steep banks of a lake or river; the entrances are tunnels which begin underwater, and the burrow itself is equipped with small ventilation shafts. When the land around the water is flat, the muskrat builds a cone-shaped mound of collected plant material. The nest is inside the center of this mound. The winter mounds are much larger, reaching a height of 1 m; these mounds contain two or three dens and a supply room. The muskrat does not hibernate and is not particularly adversely affected by the mild European winters. According to Bujakovi's reports, these animals have also adapted well to the stringent conditions necessary for survival in the extreme northern areas of the U.S.S.R. They spend the winter in deep lakes where the water near the bottom remains unfrozen. The muskrats make "breathing holes" in the ice layers; these holes are 30–50 m from the muskrats' burrow, and they are located at distances of 20–30 m from each other, parallel to the lake shore. The muskrats stop up these holes with plugs of frozen plants which they change once or twice a day. During the dark nights, the muskrats leave their ice holes and look for food under the snow. The muskrat population in this area must be quite large, as 500,000 muskrats are killed annually in Yakutsk.

Fig. 11-34. A large muskrat mound built in shallow water.

The mating season reaches its peak in April and May, after a preliminary heat period in January and February. Breeding in central Europe and in the northern areas of the New World occurs only from the beginning of spring until fall. Muskrats living in the southern part of North America are able to breed all year long. In the breeding season males mark their territory with odoriferous secretions from the glands in the perineal area. Bitter fights over the possession of the females are often bound up with the mating games; these games take place both in the water and out. We are familiar with two different vocalizations in the muskrat, a squeaking sound like the squeak of a hare, made by the female and probably meant for the male, and a noisy, crunching sound which is emitted by excited animals. The gestation period lasts twenty-eight days. There are three or four litters, or more, a year. One litter contains seven or eight, and in some cases fourteen, young. Newborn young weigh about 20 g; they weigh two times this amount after one week, and forty times as much after ten weeks. Their eyes open at eleven days. When the young muskrats are three weeks old and have reached the size of a bank vole, they make their first dive. The young of the first litter are sexually mature within the same calendar year. Wild muskrats live a maximum of four years.

Fig. 11-35. Southern European pine vole (*Pitymys savii*).

In addition to muskrat fur, man also uses the scent glands and the meat of these animals. The scent glands are used in the American perfume industry. Muskrat meat is much like that of the best domestic fowls. In order to avoid any possible prejudice against "rat meat," muskrat meat is sold in the U.S.A. under the trade name "marsh rabbit" or under the Indian name, *Musquash*.

Fig. 11-36. 1. European pine vole (*Pitymys subterraneus*); 2. North Italian pine vole (*P. multiplex*); 3. Mediterranean pine vole (*P. duocedimcostatus*) in Europe.

Eleven species of microtine rodents are classified together as PINE VOLES (*Pitymys*). The EUROPEAN PINE VOLE (*Pitymys subterraneus*; BL 8.5–10 cm, TL 2.5–3 cm, weight 13–23 g) is a native European animal. The AMERICAN PINE VOLE (*P. pinetorum*), found in North America, has become a pest because of the damage it does to the roots of fruit trees.

According to Kratochvil, pine voles live in woods in both lowland and highland areas. Many species can also be found high above the timber line in subalpine or alpine meadows. From wooded areas, these animals may move out into some of the more humid areas of wooded steppes. The European pine vole was long one of the lesser known small mammals. However, in recent years, several studies of this species have appeared and we now know more about its lifestyle. The European pine vole is quite similar both in shape and in coloring to a half-grown common vole. When observing these animals, one should pay particular attention to the small black eyes and the soft, dense fur. These animals live in shallow tunnel systems, and they build runways on the ground, which they cover with fallen leaves and grasses. According to Grummt, these animals like to use the large subterranean tunnels of the water vole. In winter, when the ground is completely covered with snow, the European pine vole makes branching systems of runways in the snow as well as under it. Langenstein-Issel has noticed that European pine voles are hardly ever found on grassy pasturelands, because grazing cattle usually trample down their tunnels and burrows.

Fig. 11-37. American pine vole (*Pitymys pinetorum*).

The decreasing distribution of this species may be due to the fact that the common vole and the European pine vole are mutually exclusive. Certainly the common vole, which is larger and has a higher rate of reproduction, is able to supplant the European pine vole, which gives birth to only two or three young per litter. These small rodents are herbivorous and afraid of light; consequently, they rapidly carry their food to their nests, where they consume it. One adaptation to a lifetime spent in the small tunnels is shown by the fact that European pine voles eat with all four feet on the ground; even in the open these animals rarely eat with their hands, unlike the technique used by most other mice.

The VOLES (*Microtus*) include almost fifty species, of which we are only able to discuss a few. The COMMON VOLE (*Microtus arvalis*; BL 8.5–12 cm, TL 3.5–4.5 cm, weight 30–50 g; see Color plate, p. 300) is one of the most common European mammals. Males are usually somewhat larger and heavier than females. The FIELD VOLE (*M. agrestis*) is similar to the common vole. Its hair is coarser and somewhat darker. The TUNDRA VOLE (*M. oeconomus*), which is very similar to both of these species, is distinguished by its particularly long tail. The EASTERN MEADOW MOUSE (*M. pennsylvanicus*; see Color plate, p. 300) is found only in the New World. BRANDT'S VOLE (*M. brandti*) lives in the high steppes areas of Mongolia. The SNOW VOLE (*M. nivalis*; see Color plate, p. 300) is found in the Alps and other high mountain ranges above an altitude of 1000 m. Distribution ranges for more species are given in Figures 11–39 to 11–44.

The common vole

Man has had to come to terms with the common vole ever since we took up agriculture. During the Middle Ages people believed that these mice could be driven away with the help of the clergy or through special repentance days. The Bavarian duke Albrecht V (1550–1579) even ordered that special prayers be offered up in all the churches, prayers that would protect the crops from voles. At one point, in November 1648, the Bavarians instituted a law prohibiting the hunting and killing of foxes over a period of three years whenever the common vole became too numerous and began to do serious damage to the croplands. This law seems modern by today's standards, as it attempted to maintain a biological control over these pests through their natural enemies.

That fact that man was still unable to control the common vole, in spite of all his remarkable efforts in that direction, is clearly indicated by the following disposition in the will of Baron Max von Spek-Sternburg, from Sachsen (Germany), in the 19th Century. A prize of 150 talers was to be provided to whoever was able to answer the question as to which method of combating the common vole had in general proved the most reliable, cheapest, and most adequate for the agricultural industry as a whole. Jäckel mentioned the following as possible measures to be considered in this effort: setting mouse traps, constructing bore-holes, burying jars, trampling down the vole holes, flooding the vole holes with water or liquid manure, driving the voles away with sharp and unpleasant odors, moving pigs into the woods and fields, using sheep

Fig. 11-38. 1. Common vole (*Microtus arvalis*); 2. Field vole (*M. agrestis*).

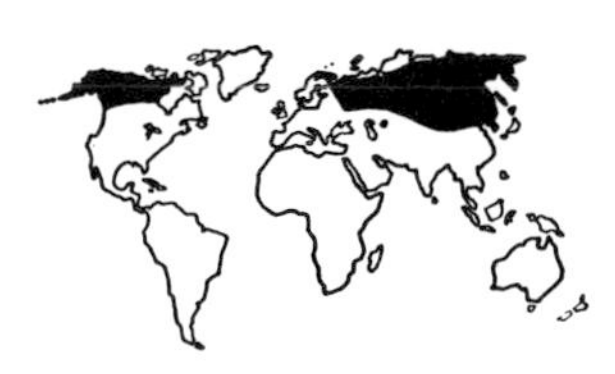

Fig. 11-39. Tundra vole (*Microtus oeconomus*).

to protect crops, smoking the voles out, and using poison. His main recommendation however, concerned the protection of the common vole's natural enemies. Jäckel ultimately concluded "that control of the destructive activities of these mice will not be man's doing; a higher power must set the limits and goal."

Since that time, zoologists have repeatedly concerned themselves with these animals, and, according to Frank, the common vole is one of the most thoroughly researched of all wild mammals, at least regarding its population biology. In spite of all the research, however, we must agree with Jäckel's fatalistic pronouncement of more than 100 years ago, that man is unable to control the massive populations of these rodents. This is because the common vole has an unusually strong reproductive pattern. This small, insignificant microtine rodent species completely overshadows all of what we presently know about the reproductive abilities of the higher animals, as Frank was able to ascertain when he carried out a large laboratory breeding program that included over 4000 animals. In addition, 1150 common voles in the wild were each marked so that the validity of the results could be supported. The results of this program were surprising—one might say almost amazing.

Fig. 11-40. 1. Snow vole (*Microtus nivalis*); 2. Gunther's vole (*M. guentheri*) in Europe.

The high rate of reproduction is equaled, under natural circumstances, by an equally high mortality rate. Wherever man plants field and garden crops, thereby offering his animal competitors a rich supply of food, he has unknowingly decreased the mortality rate of these animals, and, as a result, they change from annoying but harmless competitors to dangerous "pests." The destructiveness of these species increases as man's methods of cultivation become more intense and as man drives away the natural enemies of these animals. Certainly some of the circumstances which turn these animals into pests arise as natural by-products of an ever-increasing human population. However, on the other hand, many of these problems could be avoided if man would modify his actions on the basis of a greater consideration of biological relationships.

Fig. 11-41. 1. Eastern meadow mouse (*Microtus pennsylvanicus*); 2. California vole (*M. californicus*); 3. Mexican vole (*M. mexicanus*).

The common vole is able to reproduce rapidly and prolifically because young voles become sexually mature incredibly early in life. Frank found that female sucklings were already sexually mature. Such common vole young, which were eight to twelve days old and weighed only 5 g, already had an expanded uterus which was supplied with blood by a vascular system visible to the naked eye. Beginning with the thirteenth day, the young females, which weigh only 7–9 g, mate with older males, many of which weigh over 40 g. Females in the wild, which themselves weighed only 10 g, already carried embryos that were 1–2 mm in size; females weighing 12 g carried embryos of 3–4 mm. Thus a female common vole may give birth to her first litter when she is only five weeks old. This early sexual maturation undoubtedly produces the highest efficiency rate of all the mammals. It is due to an extraordinarily rapid growth rate which allows animals that are only forty days old to attain a body weight of more than 30 g. There are, on the average, four to seven young in each litter, although the highest possible number may be twelve. To date, one breeding female holds the record with thirty-three litters and a total of 127 young.

Fig. 11-42. 1. Prairie vole (*Microtus ochrogaster*); 2. Rock vole (*M. chrotorrhinus*; 3. Yellow-cheeked vole (*M. xanthognathus*); 4. Townsend's vole (*M. townsendi*); 5. Creeping vole (*M. oregoni*).

Newborn common voles weigh almost 2 g. They open their eyes eight to ten days after birth. During the summer it is possible for the female to have a new litter every three weeks. Occasionally, but rarely, young voles may also be born during the winter, when the weather is mild or when the food supply is particularly plentiful, as for example, in storage shacks or corn bins. Frank was able to detect signs of winter litters in north-western Germany in the particularly harsh winter of 1962–3, especially in hollows where the snow shut out the cold or even in dense ground cover which had been blown together into a pile of more than average height on some deserted meadow.

The common vole is definitely herbivorous; it lives on a schedule of activity and rest periods, each of which lasts two to three hours before giving way to the other. These animals settle in colonies. They build systems of branching tunnels which are usually situated fairly close to the surface and which are equipped with nest rooms and supply chambers. There are several exit tunnels leading to the surface; these are connected with each other aboveground by runways. The voles also use these runways to get to their feeding areas, which may be in croplands or pastures or even in light pine woods. If one observes the life and behavior of a colony of common voles aboveground, one will soon notice that these animals always move about on their paths because, when they flee back to their burrows, they can reach a greater speed on the runways. The common vole is a particularly adaptable species, and, as such, it lives in a wide variety of habitats. One can often find these animals on farms or in piles of straw or grain. In barns which have been completely filled up, one may find that the plants which have been stored away have all been gnawed at and have holes in them, as well as being coated with layers of droppings, so that they no longer have any economic value. The economic damage that the overwhelming numbers of common voles cause in croplands (our man-made dreamlands for mice) amounts to millions of dollars every year.

Fig. 11-43. 1. Long-tailed vole (*Microtus longicaudus*); 2. Singing vole (*M. miurus*); 3. Richardson's or American vole (*M. richardsoni*).

Georg H. W. Stein, in his excellent monograph on the common vole, stresses the fact that, according to our present knowledge, the prevention of massive populations of these rodents is much more important than the fight against individual animals. Among the preventative measures he suggests for this purpose are: elimination of barren (waste) lands, slopes, and other winter retreats of the vole; careful and rapid recovery of the corn harvest; plowing the fields immediately after the harvest; harrowing the winter furrows at the right time so that the burrows and nests of the common vole are destroyed and all of the harvest remains, which the animals use both for food and for cover, are buried;

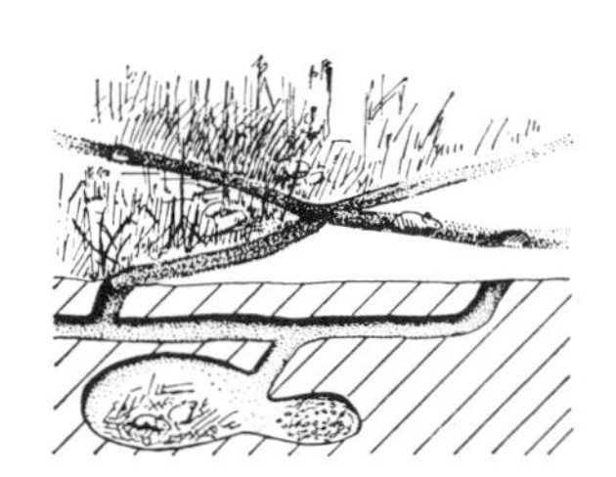

Fig. 11-44. Burrow of a common vole.

and finally, plowing up all the intermediate crops like sweet lupine, at the end of the year. If these measures were carefully applied, we would probably find that large-scale vole populations were a thing of the past.

The field vole

Many of the areas that were deforested as a result of the Second World War have now been covered over with high grass. Consequently the populations of FIELD VOLE (*Microtus agrestris*), in Germany at least, have increased substantially. Unlike the common vole, which prefers a warm environment, the field vole prefers a cold, moist habitat, and of all the small mammals, the field vole is found most often in the high moors. In the Alps one can find these rodents up to an altitude of 1800 m. The reproduction and fluctuations in population of these animals are similar to those of the common vole. Foresters persecute these animals vigorously because they cause extensive damage, especially in reforested areas.

The tundra vole

The TUNDRA VOLE (*Microtus oeconomus*) is quite common in some areas of Europe as well as in Alaska. Separate populations of this basically east Asian species have probably been in existence since the end of the Ice Age; there are tundra vole stocks in the Netherlands, Scandanavia, Czechoslovakia, Austria, and Hungary. These rodents prefer a moist to wet environment with dense plant growth. Certain Mongolian subspecies live near the Gobi lakes and even in the basins of the larger lakes. Frank and Zimmermann have characterized these rodents as quiet animals with a preference for water. They reproduce more slowly than the common vole, and they are less inclined to create large families. The tundra vole is much like the common vole and the field vole in its body development, its sensory faculties, and its behavior. The snow vole, on the other hand, which belongs to the same genus, exhibits rather strong differences in the development of the young.

Brandt's vole

BRANDT'S VOLE (*Microtus brandti*) is found all over the Mongolian high steppe areas. These rodents live together in large colonies. They "sit up" much like the ground squirrel. We found colonies of Brandt's voles on the Tuula Plain, near Ulan-Bator (Mongolia), in a state of hectic activity in the middle of May; voles were running around everywhere and whistling. These animals build complicated burrows under the plant cover; the burrows present a very real danger to grazing stock. Plants in this region have a very short growing season, and, as a result, the voles must store hay away in their storage rooms, which may be up to 90 cm long. Dawaa found that the fluctuations in the population of Brandt's vole have a twelve-year cycle; the last major increases in population occurred in 1928–9, 1942–3, and 1956–7. Because plant and animal life in Mongolia is still governed largely by natural conditions, the increase in the vole population is accompanied by an equal increase in the populations of the natural enemies of these animals. Several times we watched as the first young animals became easy victims of the numerous brown shrikes (*Lanius cristatus*). The upland buzzard (*Buteo hemilasius*) and the steppe eagle (*Aquila rapax*) are particularly common in areas where the vole is found. The population of the small fox (*Vulpes corsac*) increases threefold whenever the vole population reaches massive proportions.

The snow vole

Just as the chamois is considered to be the larger mammal most characteristic of the Alps, so too, according to Zimmermann, is the SNOW VOLE (*Microtus nivalis*), the most characteristic small mammal of the Alps. These animals were found in 1844, in southern France, near Nimes, at an altitude of only 180 m. Recently Heim de Balsac found these animals in the Rhone Valley, near Valence, at an altitude of 125 m. The snow vole can also survive in altitudes of up to 4000 m when the ground is not too moist. They are not dependent on snow, as their name implies, but rather, on rocks. We can surmise as much by the fact that snow voles have long vibrissae, a necessary characteristic of all rodents living in rock clefts; these vibrissae allow the animals to feel their way in the darkness.

"Anyone who travels through the Alps with his eyes open," says Zimmermann, "will be able to see snow voles, because they remain out of their burrows throughout the day. The silvery-gray coloring on the upper side of their bodies makes them easy to distinguish from other rodents; another characteristic of these animals is the fact that they run with their tail hanging loose. On older animals, the tail turns white on the upper side. Like all of its relatives, the snow vole is almost exclusively herbivorous. During the summer the whole range of Alp flora is at the disposal of these small creatures. They show a preference for pink (clove) root and rhododendron. During the winter the food selection is more limited, but the voles are able to reach rhododendron branches, roots, and hay under the snow. They do not hibernate. Haystacks and similar shelters make the winter months more bearable, but they are not absolutely necessary for the snow vole's survival; the snow cover also offers protection from frost and storm. Küsthardt observed a particular behavior when the snow suddenly began to melt in the spring: overnight these animals built earth walls 8 cm high in front of each entrance to the burrow to protect it from flood waters.

The female snow vole gives birth to one or two litters a year after a gestation period of three weeks; there are between two and seven young in each litter. Although the young are relatively heavy at birth, weighing about 4 g, their development is rather slow. They open their eyes for the first time at thirteen days. The nursing period lasts for three weeks. It is very possible that the slow, one might even say careful, raising of the young and the long bond with the mother are related to the demands of life in the higher altitudes. The growing season for the plants is so short that the young must be in good condition for the winter months. Frank has observed an additional characteristic which might be regarded as an adapatation to this environment: "Instead of rapidly crawling away from a disturbance, as other microtine rodent nestlings do, snow vole nestlings exhibit an innate fear of falling, even though the ability to climb develops very early in these animals."

Steppe voles

According to their systematic classification, the STEPPE LEMMINGS (*Lagurus*) belong more to the microtine rodents than to the lemmings. There is one species in western North America, the SAGEBRUSH OR RABBIT-TAILED MEADOW VOLE, (*Lagurus curtatus*) and two in the Old World, from the Ukraine to Mongolia. The STEPPE LEMMING (*Lagurus lagurus*) is one of the Old World species, both of which inhabit semi-arid wormwood steppe areas. Theseanimals are the size of a common vole. The upper areas of their bodies are gray and the sides are an ocher; the undersides are yellowish. There are narrow black stripes from the nape of the neck to the root of the tail. The tail is very short.

Fig. 11-45. 1. Steppe lemming (*Lagurus lagurus*) 2. Long-clawed vole (*Prometheomys schaposchnikowi*).

According to Heptner, the diet of the steppe lemmings consists of herbs, of which wormwoodlike plants are the favorite. These animals also show a preference for tulip bulbs, roots and other underground parts of plants, young corn, and seeds. The steppe lemming generally reproduces during the warmer seasons. These animals do not hibernate. They are active during the day. They prefer to build their subterranean nests and burrows in black soil or in other areas where digging is not difficult. The gestation period lasts an average of twenty days. The female gives birth to four or five litters a year, with twenty-five to thirty days between each one. The average litter contains six young. Young females give birth for the first time when they are about two months old, and this pronounced reproductive capability rapidly leads to the development of a large population. After periodic population increases, the steppe lemming appears even in areas where it has not been seen for years or where it was rarely encountered before.

Runways made by steppe lemmings are much wider than those made by common voles. Hall noticed two grassy green hillocks on such a runway, which the steppe lemming had to climb over every time it passed by. Common voles dispose of such obstacles immediately by biting them off, eating them, or at least by pushing them out of the way.

The long-clawed vole

The LONG-CLAWED MOLE VOLE (*Prometheomys schaposchrikowi*; BL 13 cm, weight 70 g), the only member of its genus, has a cylindrical body which gives it a superficial resemblance to the mole. The feet and tail are short. The eyes are small. The fore feet are equipped with long, sickle-shaped claws, which identify these animals as diggers.

The long-clawed mole vole lives in the central and western Caucasus as well as in the Lesser Caucasus, in moist meadows in alpine and subalpine zones. These voles also live in open areas in the woods, on the banks of mountain streams, and in small rocky fields and pastures at altitudes of 1500–2800 m. In 1962, H. Steiner found long-clawed mole voles in the Yalnizcam Mountains (Turkey); this was the furthest south these animals have been found, and, at the same time, it was the first population of long-clawed mole voles found outside the U.S.S.R. Like the other earth-dwellers, these voles build earth mounds. Their digging activities occur mainly in the morning hours and in the evening until around 10 p.m. The burrows have open exits which may be sheltered by a bush or which may be completely unprotected. Several older males live together with the females in a single burrow. The size of the litter ranges from two to six young, even though the female is able to nurse an even larger number, as she has eight nipples. The long-clawed mole vole seems to prefer the poisonous buttercup *Ranunculus elegans* as its main nutrition.

Mole lemmings

The MOLE LEMMINGS (Ellobiini), with only one genus (*Ellobius*) and two species, is the last group of microtine rodents. These animals have a subterranean life style. Their bodies are cylindrical. The body is from 8 to 15 cm long. The head is blunt. The incisors are large and protrude foreward; these teeth do most of the work when these animals dig out their burrows. The edges of the feet have stiff side hairs which help in pushing away the earth. These animals do not have an external ear. Their eyes are very tiny, as is the case with many specialized earth-dwellers. The brain is smooth, and the sense of vision has regressed. The hair is velvety, brownish-yellow, and has almost no grain. As a result, the shorter hairs may lie facing either the front or the back; this allows the animal to crawl in either direction without difficulty. The two species are the MOLE LEMMING (*Ellobius talpinus*) and the AFGHAN MOLE LEMMING (*E. fuscocapillus*; see Color plate, p. 300).

The mole lemming lives mainly in steppe and semi-desert areas, only rarely moving into wooded steppes. According to Serebrennikov, the mole lemmings in the northern region of the southern Transurals settle largely on the edges of birch and aspen groves because they prefer barren land to that which has been brought under cultivation. These animals avoid salty areas and cattle ranges, where the ground is hard. In Mongolia we found that the mole lemmings preferred to live in deep, moist soil on the banks of rivers, lakes, or streams. We also found the earth piles characteristic of this species in the flat, arid areas of the Dzungarian Gobi. Mole lemmings have been found at altitudes of over 2500 m in the Gobi-Altai.

The mole lemming burrow is an interbranching tunnel system which is built at a depth of 20–30 cm, and which includes storage rooms for food supplies. Only the nest chamber is built at a depth of at least 50 cm. Mole lemmings feed on underground parts of plants; tulip bulbs and tubers which are particularly pulpy are favorite foods. In the southern Ukraine, Subko and Ostrjakov found ninety-day-old mole lemmings which were already sexually mature. There are six or seven litters a year, with between three and five young per litter. The gestation period lasts twenty-six days. The young remain in their parents' nest for two months before becoming independent.

Fig. 11-46. 1. Mole lemming (*Ellobius talpinus*); 2. Afghan mole lemming (*E. fuscocapillus*).

Mole lemmings are not at all easy to catch, and they are not represented in most zoological collections. In order to catch these animals on our Mongolian expedition, we broke into freshly used burrows by digging rectangular trenches with widths of 30–40 cm. Then we put spring traps on the flattened floors of the trenches, in front of the open section of the burrow, and covered over the trench, taking care that no light filtered through into the burrow. We caught ten animals with this method. As the controls to our cases showed, the mole lemmings' activity is not governed by definite times of the day. These animals do not hibernate, and they suffer severe losses whenever the ground frost is of a particularly long duration. Their greatest natural enemy in Mongolia is the eagle owl, in whose excrement we were able to find the remains of mole lemming bones.

Subfamily: Gerbils

The last subfamily of the cricetid rodents is that of the GERBILS (Gerbillinae). These rodents are very similar to rats in size and appearance, in spite of their elongated hind feet. There are more than ten genera with over 100 species:

Distinguishing characteristics

1. The GERBILS (*Gerbillus*; BL 8–13 cm, TL 7.5–12 cm) are small, long-eared animals, pale-sand in color, with relatively large eyes. The tail is very long and sparsely haired; it has a small brush at the tip. There are fifty-four species, including the FIELD GERBIL (*Gerbillus nanus garamantis*) and the LARGE NORTH AFRICAN GERBIL (*G. campestris*; see Color plate, p. 300).

2. The JIRDS (*Meriones*; BL 11.4–13 cm, TL 9–11 cm, weight up to 62 g; see Color plate, p. 300) include twelve species.

3. The GREAT GERBILS (*Rhombomys*) are diurnal animals with a massive appearance. The one species, the GREAT GERBIL (*Rhombomys*; BL 15–20 cm, TL 13–16 cm, weight up to 200 g; see Color plate, p. 300), is found in Turkestan as well as in the deserts and semi-deserts of China and Mongolia.

4. The FAT-TAILED MICE (*Pachyuromys*), unlike all the other gerbil-like animals, can rely on fat stored in their tails for nutrition when the normal foods are scarce. There is one species, the FAT-TAILED MOUSE (*Pachyuromys duprasi*; BL 10.5–13.5 cm, TL 4.5–6 cm), found in northern Africa from the Algerian Sahara to south-western Egypt.

5. The LARGE GERBILS (*Tatera*; BL 9–19 cm, TL 12–24.5 cm, weight 30–100 g) have longitudinal grooves in the incisors in the upper jaw. The soles of the front and back feet are hairless. The head and body are narrow. The tail is thin. There are ten species, including the EAST AFRICAN GERBIL (*Tatera vicina*; see Color plate, p. 300), found over most of the African continent, in savannas, wooded areas, and in croplands, and the INDIAN or LARGE INDIAN GERBIL (*T. indica*), found from India to Arabia and on the island of Ceylon.

Fig. 11-47. Gerbils (subfamily Gerbillinae).

6. The SMALL NAKED-SOLED GERBILS (*Taterillus*) include only one species, the SMALL NAKED-SOLED GERBIL (*Taterillus emini*; BL 10–14 cm, TL 14–17 cm), which prefers areas without trees, or areas overgrown

with thorny bushes in Africa. Like the other gerbils, these animals live in tunnel systems which have living rooms.

All of the gerbils and jirds, with the exception of the great gerbils, are basically nocturnal animals which live harmlessly in open desert or steppelike areas. They drink only rarely or not at all, because the plant matter and to some extent the animal matter they eat supplies their bodies with practically all of its water requirements. When eating, gerbils sit on their hind legs and use their fore legs to bring the food to their mouths. When they are disturbed, they run away, often using their hind legs to make large kangaroolike jumps. These animals always remain within the boundaries of their territories, which are marked by both young and adult animals with secretions from glands on the stomach.

To date, there have been very few observations of these animals in the wild. Rosl Kirchshofer was able to study the FIELD GERBIL in Algeria, south of the Biskra Oasis, and she was also able to breed these gerbils in captivity. According to her observations, the gerbil burrows which are dug in sandy steppes have round entrance tunnels which are 3–4 cm in diameter. These entrances are closed off during the day. The tunnels in a shallow burrow, which may extend over several square meters, descend to between 5 and 10 cm below the surface. The tunnels run in a relatively straight line; they have offshoots here and there to other tunnels, and thus complete tunnel systems develop. Just in back of the entrance hole, one can regularly find a supply room which is always filled with flowers and leaves of a mallow species. In contrast to these relatively simply constructed male burrows, there are complicated female burrows in which the tunnels and storage rooms are not only arranged horizontally, but vertically as well, consequently avoiding sharp temperature fluctuations.

In one female burrow, Kirchshofer found a mother and her seven naked young living in a round nest of shredded grass. The gestation period lasts at least three weeks. During this time the female hastily drags nest materials into the burrow, where she gnaws at them and builds a roundish nest with the fibres. A female field gerbil gives birth to two or three litters a year, with three to seven young in each litter. Other members of the genus produce much larger litters, with twelve to fifteen, or even eighteen, young. "The female is, like all mammal mothers, the source of warmth and food for her young," reports Kirchshofer. "She cleans them and licks their anal areas, which stimulates them to defecate. She turns the young over on their backs when doing this. In addition, she brings them extra food and protects them from other gerbils. If a young gerbil leaves the nest too early, she brings it back. She carries it either by a special "carry grip" on the skin of its neck, or she pulls it back into the nest by its tail. The mother leaves the young when they are four weeks old. They are then independent, although they remain together for a while in a sibling group, probably until the beginning of sexual maturation."

During our expedition through the arid steppes and semi-deserts of Mongolia, we observed the CLAWED JIRD (*Meriones unguiculatus*), which lives together in colonies, and the SOUTHERN MONGOLIAN JIRD (*Meriones meridianus psammophilus*). Both species are active during the day as well as at night. These rodents will often stand on higher ground with their bodies extended, and look out over the surrounding area. Whenever we made even the slightest movement, they beat their hind legs on the sand for a brief period or they made a loud peeping sound and disappeared rapidly into their burrows. Judging by the numerous tracks we found in this area, the sand hills which formed around bushes were totally undermined by these animals. The dense settlements develop very rapidly because the females of both species are able to reproduce from April until September, and, according to mammal specialist M. Stubbe, the young of the first litter are able to reproduce within the same year. The jirds are kept sharply in check by their numerous enemies, which include foxes, steppe foxes, polecats and owls among others. These predators hunt the jird to such an extent that the majority of the jird population does not survive its first winter of life, and only a few animals live to be one year old. In captivity, however, jirds may live over six years. This natural control accounts for the fact that jirds do not cause much damage in cultivated fields and gardens.

Irenäus Eibl-Eibesfeldt was able to make several interesting observations on the body grooming of the PERSIAN JIRD (*Meriones persicus*). The jird began its cleaning procedure by cleaning its head. "Then the animal licked and nibbled at its stomach and flanks. It would also scratch itself and apparently licked its toes at intervals, often holding its feet with its hands. When scratching its stomach or breast, the animal extends its front leg out away from its body. The tail is cleaned with particular care several times a day. Like many mammals, the Persian jird exhibits a definitely socialized cleaning behavior, one that is not limited only to the cleaning of young animals. The males will often clean the females. The females, in turn, invite the males to clean them by running to the males, turning around, and shoving themselves under the males' noses. The males then lick and nibble at the females' chins, snouts, stomachs, and flanks. The females remain motionless. They close their eyes when their head is being licked. When the female has had enough, she springs away, or she wards off further licking by pounding lightly on the ground with her front feet. A short, friendly fight may result."

Fig. 11-48. The beginning of a fight between two Shaw's jirds (*Meriones shawi*): initially the two animals face one another in a threat posture; then, in the subsequent fight, they beat one another with their paws.

Another species, TRISTAM'S JIRD (*Meriones tristami*) is illustrated on page 300.

The GREAT GERBIL (*Rhombomys opimus*; see Color plate, p. 300) is distinguished by its loud, constant whistling which is very similar to that of the little ringed plover. Great gerbils prefer to dig their burrows in drifting sand dunes which may be up to 1 m high. According to Vlasov, one such burrow had a surface area of 358 m^2; the total length of the tunnels, which were built on three different levels, was about 600 m. Vlasov found several kilograms of dried plant remains in the central part of this burrow. We saw this species in the southern Gobi, even near human settlements.

Ludwig Heck wrote the following about the FAT-TAILED MOUSE: "It is quite a curious creature with its grayish-yellow, silky smooth fur and its round head with the large eyes, which make it look like the well-known jerboa. These animals live in the barren areas of Egypt and the other north African countries. The characteristic feature, however, is the thick tail, which gives these animals their name. Sparsely haired with a shimmering flesh-pink color, the fat-tailed mouse looks swollen, as if someone had stepped on it."

LARGE GERBILS, which are active principally at night, feed on bulbs, seeds, green plants, and insects. Only the INDIAN or LARGE INDIAN GERBIL seems to prefer animal matter. According to Walker, these animals will consume not only eggs and young birds, but also their own young. Some species are able to reproduce throughout the year; others have a definite reproductive season. The female has six or eight nipples; she gives birth to between four and eight young, which remain in the nest for three or four weeks. Those species of large gerbils native to South Africa are rigorously controlled by the government because they are carriers or transmitters of bubonic plague; hopefully, the control programs will prevent the spread of this dreaded disease.

The families of BAMBOO RATS (Rhizomyidae) and PALEARCTIC MOLE RATS (Spalacidae) do not have undisputed zoological classifications, although they are usually viewed as distant relatives of the cricetid rodents and the murid rodents. These animals have also sometimes been placed in close conjunction with the porcupinelike rodents, particularly the mole rats, because of their dentition. We do not know for certain whether bamboo rats and palearctic mole rats are more closely related to one another than to other groups. Both groups are so well adapted to a totally subterranean lifestyle that it becomes extremely difficult to determine whether the similar characteristics are due to parallel development or close relationship.

Family: Bamboo rats, by F. Dieterlen

The BAMBOO RATS (Rhizomyidae; BL 16–45 cm, TL 5–15 cm) are sturdy, powerfully built rodents with short legs. The eyes and ears are very small. The skull structure and musculature have been transformed by the diet and life style. These animals have giant incisors and molars. There are sixteen teeth, arranged: $\frac{1 \cdot 0 \cdot 0 \cdot 3}{1 \cdot 0 \cdot 0 \cdot 3}$ or $\frac{1 \cdot 0 \cdot 1 \cdot 2}{1 \cdot 0 \cdot 1 \cdot 2}$. The molars have flat crowns, roots, and a furrowed structure. These animals live in grassy areas with loose and hard soil, in bamboo regions, woods, and similar areas in southeastern Asia and eastern Africa.

There are three genera. The BAMBOO RATS (*Rhizomys*; BL 23–45 cm, TL 5–15 cm; see Color plate, p. 300), are similar to the American pocket gophers. The upper lip is split and the giant, orangish-yellow incisors

Fig. 11-49. 1. African mole rats (genus *Tachyoryctes*); 2. Bamboo rats (genus *Rhizomys*).

are exposed. The tail is not covered with scales. Animals in the northern part of the distribution area have soft, thick hair; those in the southern areas have a thin, coarse coat. There are six species, of which the SUMATRAN BAMBOO RAT (*Rhizomys sumatrensis*) is the largest form. The LESSER BAMBOO RAT (*Cannomys badius*) is from Nepal, Assam (India), Burma, and Indochina.

Bamboo rats, true to their name, prefer to live in bamboo thickets at altitudes of 1000–4000 m. They live between the roots of bamboo bushes and trees. One animal apparently will have several burrows, although it actually lives in only one. The bamboo rat uses its incisors and its feet to dig out the burrow. These rodents eat not only the bamboo roots, but also, like human gourmets, the tender bamboo shoots or sprouts. Usually they eat wherever they happen to be working, but sometimes they will drag the food back to their burrow. Bamboo rats also feed on other plants from the ground surface, and on grasses, seeds, and fruits. The female gives birth to between three and five young per litter. Bamboo rats are hunted and eaten by many local peoples.

The systematic position of the AFRICAN MOLE RATS (*Tachyoryctes*; BL 17–23 cm, TL 5–8 cm, weight 150–300 g) is still uncertain; some zoologists classify them near the murid rodents (family Muridae). The hair is thick, short, and soft. The coloring can be quite varied. Young animals are often black or slate-gray. Adults may be black, brownish, reddish-brown, yellowish-brown, dark gray or white (albino); apparently they may also be piebald. Mole rats within the same population may look quite different. The underside of the body is usually lighter than other parts. There are more than twenty forms, grouped together in a much smaller number of species, including the TANZANIAN MOLE RAT (*Tachyoryctes daemon*; see Color plate, p. 300), and the RWANDA MOLE RAT (*T. ruandae*), which has irregular white flecks on its stomach.

Mole rats live in open areas, in savannas, and on the edges of steppe lands or cattle ranges. They avoid arid regions and apparently do not appear in altitudes of under 1000 m. They have been found at altitudes of up to 3700 m. They live underground, in soft as well as in extremely hard soil; they use their teeth when burrowing through hard soil. The burrow contains a nestroom, with a grass nest, and a specific room for defecation. When the dry season hits Kivu Province (Zaïre), these animals burrow down to a depth of 2 m. In other seasons the burrows may be just under the surface, however, so that cattle and horses often fall through. The presence of the African mole rat can easily be detected by the characteristic piles of earth they throw out of their burrows, piles that are very similar to those made by moles. When digging, the mole rat loosens the earth with its fore feet and teeth, and uses its hind feet to throw the excess dirt to the rear.

Mole rats feed on roots, particularly those of various grasses. However, they also move into croplands, where they eat roots of manioc, sweet potatoes, tea and coffee bushes, and corn. Mole rats are nocturnal. Occasionally they even leave their burrows at night and wander about. During this time they can easily be caught on open flatlands. Man hunts large numbers of mole rats in every area of their range, either by digging up the burrow or by putting snares in the tunnels. In many areas, mole rat meat is a favorite food. We know very little about the reproduction of these animals. The Rwanda mole rat apparently produces between one and three young per litter. Young mole rats, which can sit up or perch, are hairless; they emit squealing sounds. Mole rats can be kept in captivity for months on a diet of tubers, roots, and green plants.

Family: Palearctic mole rats, by F. Dieterlen

PALEARCTIC MOLE RATS (family Spalacidae; L 15–30 cm, weight 130–300 g) are even better adapted to a subterranean lifestyle. There is no outer tail and the inner remnants thereof can be felt only as a firm protuberence. The body is heavy and cylindrical. The neck and head are almost as wide as the body. The feet are short and, at least for a digging animal, rather delicate. There are five digits on each foot, with tiny claws. The eyes are degenerate and functionless. They are very small and are situated under the skin, the only such case among rodents. The only remaining part of the ear in these animals is the cartilaginous opening of the external auditory canal. The palearctic mole rat has a firm row of bristles and tactile hairs along the side of its head. The snout is broad and horny. The skull is also greatly specialized. The skull, for example, is broad, flat and powerful. There are sixteen teeth, arranged: $\frac{1 \cdot 0 \cdot 0 \cdot 3}{1 \cdot 0 \cdot 0 \cdot 3}$. The incisors, which protrude, are wide and strong. The molars have roots and are not in a state of continuous growth. The hair is soft and velvety. It ranges from dark gray to yellowish-gray, and is often shiny. There is one genus (*Spalax*), with three species found in the Balkans, southern Russia, Asia Minor, and along the Mediterranean as far as Libya. These include:

1. The LESSER MOLE RAT (*Spalax leucodon*; L 18–27 cm; see Color plate, p. 300). 2. The GREATER MOLE RAT (*S. microphthalmus*; L 24–51 cm). 3. EHRENHERG'S MOLE RAT (*S. ehrenbergi*), found in Israel and Egypt near the Mediterranean, and in the coastal area (Cyrenaica) in northeastern Libya.

Palearctic mole rats generally live in steppe areas and croplands. They may also occasionally live in wooded areas, and it makes no difference whether the land is particularly low, hilly, or even mountainous. These animals have been found up to an altitude of 2600 m in Turkey. They need a soil suitable for digging and an environment with an annual precipitation of at least 100 mm. Consequently, they avoid desert areas as well as flood plains. These animals are among the most fully developed burrowers in the order of rodents. Their basic digging tool is the wide and rounded head with its powerful incisors, which loosen the soil as well as shoving masses of it to the side like a bulldozer. The fore and hind feet are used to dig out the earth or to push it out of the way; thus,

Fig. 11-50. Palearctic mole rats (genus *Spalax*).

the hind feet push or throw the earth which collects under the body to the rear and out. The earthen mounds, so characteristic of palearctic mole rats, are built from the excess earth thrown out of the burrow.

With the exception of the mating season, palearctic mole rats live in a tunnel system with many branches, all of which are built rather deep down in the ground. This system is equipped with nest rooms, supply rooms, and special defecation chambers. In winter, these systems are built much deeper into the ground; winter burrows of the lesser mole rat have been measured at a depth of 2 m; those of greater mole rats have been measured at 4 m. The horizontal extensions may be 30 m or more. A greater mole rat burrow 169 m in length, with 114 dirt piles, has been uncovered. Generally, the living rooms and supply rooms are found at a greater depth than are the food tunnels, which are usually just under the ground surface and thus run in between the roots of the food plants.

During the mating season, palearctic mole rats construct a particular series of mounds. Nevo has described those structures built by Ehrenberg's mole rats in Israel, where this species reproduces in the moist winter months. These "breeding mounds" are built by the females in the fall, after the first rain. They are over 1 m in diameter and they have a height above the ground surface of about 40 cm. These mounds are usually quite firm; in the center they have a nest chamber measuring about 20 cm in diameter; the young are raised in this room. Tunnels surround the nest chamber. Parts of these tunnels have been expanded into storage rooms or small defecation chambers. The female can use these tunnels to reach the food tunnels or the deeper tunnels which also have rooms. There are several (usually fifteen to twenty) smaller mounds arranged radially around such a female mound; these smaller mounds are built and inhabited by the males, who remain in the area during the mating season.

The summer quarters of Ehrenberg's mole rat is also situated in a mound and includes rooms for sleeping and other purposes. Because mole rats are definitely solitary animals, aside from the mating season, each mound houses only one animal. Mole rats are active in the twilight hours and at night. They may even be found outside their living quarters on the ground surface, but very rarely; this is particularly true of half-grown Ehrenberg's mole rats from March to May. Palearctic mole rats may also be active during the day, particularly during the afternoon. The greater mole rat has even been observed warming itself in the sunlight. The senses of touch and hearing are the most highly developed. The sense of smell is not particularly important.

Palearctic mole rats feed mainly on roots, tubers, and bulbs. The greater mole rat shows a genuine preference for dandelion roots, although these animals will also eat grasses, seedlings, and insects, as well as insect larvae. Their chief enemy is the owl, and mole rat remains have often been found in fur balls left by owls. Whenever a mole rat is attacked by its enemy, it becomes very defensive and can give its attacker several powerful bites while making a hissing threat vocalization. The mating season of the greater and lesser mole rats occurs in March and April. The gestation period lasts about one month, after which time the female gives birth to between one and four, usually two or three, young. A new litter is born every year. Ehrenberg's mole rats mate between November and January. The young, usually two to four, are born from January to March. They weigh 5 g at birth. They are 5 cm long, naked, and helpless. When they are about two weeks old, they acquire their first gray fur coat. The eyes open after two to three weeks. When they are four to six weeks old, they leave the nest.

The agricultural significance of these rodents varies from region to region. In the area around the Mediterranean Sea, these animals apparently are responsible for a great deal of damage, because they eat roots. The greater and lesser mole rats, on the contrary, are generally regarded as not particularly destructive, although they do eat potatoes and bulbs. It is not surprising that many superstitious beliefs have arisen about these unique animals. Some Libyans believe that a person will become blind if he disturbs a mole rat; consequently, they do not attempt to catch these animals. Palearctic mole rats have also served as "archeological helpers" when they carry pieces of objects from earlier civilizations out of their tunnels into the daylight. With such activities, the mole rat unknowingly gives archeologists important indications as to the location of additional digging sites.

Family: Murid rodents, by F. Dieterlen

The MURID RODENTS (family Muridae) are characterized by their lack of specialization and, at the same time, by the incredible number of forms in the family. The few exceptions or special forms do not generally belong to the subfamily Murinae. No mammal family has more species than does the murid rodents. They are the most fertile and the most adaptable mammal group on this earth.

Distinguishing characteristics

They range from mouse-size to rabbit-size, the body length ranging from 5 cm (African native mice) to almost 50 cm (slender-tailed cloud rats). The hair, which generally is short, may be silky-soft or bristly, or anywhere in between. The coloring is usually undistinguished and not readily noticeable. The tail ranges from an average length to very long, often greatly surpassing the body in length. The tail generally is covered with short bristles, although it can also have a thick coat of hair. The scales on the tail are usually arranged like shingles, lying on one another in circles. Some species use the tail as a tool for climbing. Monckton's water rats use their tails as a rudder. The fore feet usually have four digits and a thumb knob which, in some climbing species, is apposable. The hind feet have five digits, of which the outermost may be degenerate as an adaptation to a primarily running or hopping movement. The hind feet are often elongated, although climbing forms may have short, broad hind feet. Some forms (like the water rats) have webbed membranes. The ears are usually normal. There are sixteen teeth, arranged $\frac{1 \cdot 0 \cdot 0 \cdot 3}{1 \cdot 0 \cdot 0 \cdot 3}$ (shrew rats have twelve teeth, some water rats have twelve or even only eight teeth). The molars have roots and are laminate or cuspidate. Females have from four to twenty-four nipples. There are seven subfamilies: Murinae, Cricetomyinae, African tree mice (Dendromurinae), Vlei and karroo rats (Otomyinae), slender-tailed cloud rats (Phloemyinae), shrew rats (Rhynchomyinae), and water rats (Hydromyinae), all of which were originally confined to the Old World and Australia. There are about 100 genera in all, with about 370 species and more than 1500 described subspecies.

Subfamily: Typical mice

The murid rodents are probably one of the most recent animal forms in the evolutionary history of the animal kingdom. We believe that these animals first appeared in the Upper Miocene (10–15 million years ago), descending from hamsterlike forms. We are familiar with seventy living genera within the subfamily Murinae, while only five extinct genera have yet been discovered. The living forms alone have such a wide and complex history of events, all of which led to the development of genera, species, and subspecies, that biologists and zoological systematists are increasingly busy trying to determine the real relationships. The subfamily Murinae offers zoologists the opportunity to study a rodent group at the height of its development. With this subfamily it becomes possible to investigate the adaptive forces and tendencies which enable these animals to maintain a strong population on the earth, in spite of unfavorable circumstances.

What exactly is the difference between "rats" and "mice"? This has long been a popular question. In a very strict sense, we understand "mouse" to mean the house mouse (*Mus musculus*), and "rat" to mean the black rat (*Rattus rattus*). The "mouse" and "rat" belong to different genera, but they are both members of the same subfamily. When we speak of mice and rats we are speaking of a difference in size. We have such a differentiation not only with mice in the strictest sense, but also with the microtine rodents as well as with other rodent groups. Animals that have a body length of less than 13–15 cm are called mice in general speech; animals with a body length of over this amount are referred to as rats. It is obvious that this is merely an artificial division, with no zoological significance.

The only rodents living in Australia and New Guinea belong to the family of murid rodents. According to studies by Rümmler and Tate, the ancestors of these rodents came to Australia and New Guinea from southeastern Asia. The migration took several routes, and there were many successive waves of these animals. There are seventeen genera with from sixty to seventy species, most of which we know very little about. We must confine our discussion here to one genus.

Stick-nest rats

The STICK-NEST RATS (*Leporillus*; BL 19–22 cm, TL 15–17 cm) look like large black rats. The eyes are large. The ears may be very large; they look much like hares' ears (hence the genus name *Leporillus*). These animals live together in large social groups. They make large nests in bushes and other available locations, and they live together in these nests. The shape of the nest reminds one of a small house. There are three species.

Fig. 11-51. 1. Australian native mice (genus *Leggadina*); 2. Giant naked-tailed rats (genus *Uromys*); 3. Stick-nest rats (genus *Leporillus*).

The largest species is the COMMON STICK-NEST RAT (*Leporillus conditor*; BL 21 cm, TL 16 cm; see Color plate, p. 371). Its hair is quite beautiful and downy. The upper side is a soft yellowish-brown, and the underside is gray with reddish-yellow. The large, harelike ears are erect. The upper side of the tail is dark brown to reddish-brown. These animals live in New South Wales and in southern Australia west and north of the Darling River.

Wood-Jones studied the lifestyle of these animals and discovered many interesting facts. The nests which these rodents build from sticks, twigs, and plant stalks, can be almost 1 m tall. For the nest's frame, the stick-nest rats generally use objects already in the area, such as a thin bush. The rats very carefully stack and interweave stick after stick around the frame, so that finally a solid nest is completed and ready for occupation by several rats. The young are born and raised in this nest. Usually there are two or three young per litter. The mother apparently carries her young along with her during all her activities, even when she flees from danger; they hang on to her nipples. The nipples are located well to the rear of her stomach, and thus she can draw her young along after her. This "nipple transport" is quite common among many genera of murid rodents. We must understand that this behavior is both important and valuable; it is used to save all of the young when the female is forced to flee. In other genera, where nipple transport is not a characteristic behavior, the mother must make a separate trip for each of her young.

JONES' STICK-NEST RAT (*Leporillus jonesi*) is almost as large as the stick-nest rat; it looks more like a rabbit. These animals are found only on Franklin Island, off the southern coast of Australia. These stick-nest rats also build large nests, which may have a subterranean connection with the burrows of penguins and shearwaters. Southern short-nosed bandicoots and black tiger snakes also live in these subterranean tunnels so that it is possible to see snakes, penguins, shearwaters, bandicoots, and stick-nest rats all disappear into the same hole when threatened. Wood-Jones was able to observe and describe this extremely rare occurrence. According to Wood-Jones, captive Jones' stick-nest rats can become quite delightful, as they are tame and trusting and do not bite.

Giant naked-tailed rats

The GIANT NAKED-TAILED RATS (*Uromys*; BL 22–33 cm; see Color plate, p. 371) range in size from large to very large. The tail is usually longer than the body. The scales on the tail are arranged in a mosaic pattern, in row after row, as in the mosaic-tailed rats (*Melomys*). Most of the other genera have flat, overlapping scales. There are seven species.

The GIANT NAKED-TAILED RAT (*Uromys anak*; BL 33 cm, TL 38 cm) is particularly large. These animals live in mountain forests on New Guinea and New Britain. Stein gives us the following note: "The natives, who call *Uromys anak* by the rather droll name of *Puradidl*, told me repeatedly that this species lives only in trees where it feeds upon leaves."

Australian native mice

These "giant forms" should be compared to the AUSTRALIAN NATIVE MICE (*Leggadina*; BL 6–9 cm, TL 5–8 cm). The tail is only rarely longer than the body. The hair on the back is usually yellowish-brown or brown-gray. The stomach is usually lighter or even white. There are about six species, including the AUSTRALIAN NATIVE MOUSE (*Leggadina hermannsburgensis*; BL 7–8 cm).

According to the reports from Le Sovef and Burrell, the Australian native mouse is one of the most common forms in central Australia. These animals build subterranean burrows with many branching tunnels, even in hard, stony ground. The entrance is often located under a bush. The entrance tunnel ends in a nest room which contains a nest of dried grass. On the ground above these burrows one can often find a collection of pebbles which are spread about in a rather particular manner. Apparently the mice not only dig up these pebbles in the course of their burrowing activities; they also collect them and bring them together from the surrounding area. An astonishing and enlightening explanation for their behavior suggests that these pebbles serve to catch dew for the mice. The arid, desertlike regions where the Australian native mouse makes its home are without rain and water for much of the year. The nights are cold and the pebbles consequently cool off. In the morning, after the sunrise, the air around the pebble collection rapidly warms up and the small amount of moisture in the air condenses as dew on the cold stones. The mice are apparently able to satisfy their water requirement with this dew.

Australian hopping mice

The back feet of the AUSTRALIAN HOPPING MICE (*Notomys*; BL 10–14.5 cm, TL 9–21 cm) are adapted for hopping movements. The tail is almost always longer than the body. The eyes and ears are large. The hind feet are very large, while the fore feet are relatively small. These animals have gland zones in the neck and throat area. In females these glands may be present but less extensive and less strong, or altogether absent. There are at least ten species, distributed over almost all of Australia. The NECK-POUCHED HOPPING MOUSE (*Notomys cervinus*; BL 10–11 cm, TL 13–16 cm) is lightly and delicately built. The hind feet are very large (3.3–3.5 cm). The ears are large and semi-translucent. The eyes are large and dark. The upper side of the hair is a beautiful light-fawn color. The underside is pure white. These mice have very long vibrissae (over 5 cm long). The tail has a very fine coat of fur, and there is a hair plume at the tip. Males and females both have a small glandular sack on the neck; the opening of this sack has a border of white hairs.

Fig. 11-52. 1. Bandicoot rats (genus *Bandicota*); 2. Australian hopping mice (genus *Notomys*).

This hopping mouse lives in large desert and semi-desert areas in central Australia. Wood-Jones reports that in captivity this animal is quite wonderful, spending the day curled up asleep in its nest, becoming active only toward evening, when it hops about the room. The animal would raise itself upright on its hind legs and then suddenly, without the slightest effort, it would hop onto a bench that was 1 m tall, where it would start to clean itself. These mice can also hop sideways, an ability they share with the North American kangaroo rats.

There are five other genera which are related to the Australian-New Guinean forms. These genera are found exclusively on the Philippine Islands, and we know very little about them. Some twenty genera of typical mice have settled in the remaining areas of southern and south-eastern Asia. There are more than 200 species in all.

Bandicoot rats

The strong and robust BANDICOOT RATS (*Bandicota*; BL 20–36 cm, weight up to 1500 g) comprise one of these genera of typical mice. They range in size from large to very large. The tail is usually somewhat shorter than the body, although in some species it is longer. The upper side of the body is often a dark gray-brown, sometimes with light flecks. The underside is lighter. Females have from twelve to eighteen nipples. There are at least five species, including the LARGE BANDICOOT RAT (*Bandicota indica*; see Color plate, p. 371), and the LESSER OR BENGALI BANDICOOT RAT (*B. bengalensis*).

Lesser bandicoot rat

The LESSER OR BENGALI BANDICOOT RAT has long coarse hairs on its back; these hairs become erect and form something like a mane when the animal is excited. These rats dig large tunnel systems, piling the loose, excess earth in mounds in front of their burrows, similar to the mole. One can often find these burrows in the bushes at the edges of fields, as well as in the banks of rice fields. The animals close off most of the holes they used to throw out excess earth from the inside with dirt. The entrance tunnel of the burrow leads into a central room which may be built at a depth of 60 cm. Several tunnels lead away from this chamber, and one can find a sleeping room and several storage rooms branching off these tunnels. The lesser bandicoot rats often build up regular "corn granaries" from their supplies. In India people often dig up these stores and use them to feed themselves. In many regions, the bandicoot rats are hunted for food.

The lesser bandicoot rat is a carrier of the plague, to which it is particularly susceptible, as well as other diseases. Consequently, like the black rat and the brown rat, these animals are very dangerous to man, especially in large cities.

Rodents living near human settlements are almost always omnivorous. Thus lesser bandicoot rats can live on garbage, feed on green matter, and even attack poultry. These animals are able to swim and dive quite well, abilities that are very useful for life in the wet rice paddies. Usually only one animal lives in each burrow; a high population density can occur however, when many burrows are packed closely together. Females generally give birth to eight to twelve young, all of which are blind and naked at birth. The young are independent after about one month, and after about three months they are sexually mature.

Dao Van Thien and Hoang Trong Cu have studied the closely related LARGER BANDICOOT RAT. This species prefers the dams of irrigation canals in the area around Hanoi (Vietnam), rice fields, and ponds. These animals build burrows similar to those of the lesser bandicoot rat. The larger bandicoot rats are also disease carriers, particularly in China, where they can transmit the seven-day fever, as it is called in rice areas. The rats infect people by depositing their feces in the water of the rice fields. The extent of the role these rodents play in epidemics of the plague is not yet known.

Long-tailed climbing mice

Mice of the genus *Vandeleuria* are excellent climbers. Their feet are especially adapted for climbing. There is probably only one species, the LONG-TAILED CLIMBING MOUSE (*Vandeleuria oleracea*; BL 7–9 cm, TL 9–11 cm). It is chestnut-brown with white undersides. The hair is soft.

Long-tailed climbing mice are nocturnal animals, which live in bushes, trees, and bamboo. Apparently they feed on fruits, buds, and similar matter. The nest is built in tree cavities or between branches, although these animals sometimes use nests vacated by other animals (for example, swallow nests on cliffs). One nest with three young was even discovered in the middle of a thick spider web which was also inhabited by several spiders. The female gives birth to three or four, or six at the most, young per litter.

Pest rats

With their strong bodies, short tails, relatively short heads, and the very powerful incisor teeth which have developed into "digging teeth," the PEST RATS (*Nesokia*) are particularly well equipped for a subterranean way of life. They are average-sized to large. The hair ranges from long and soft to short and coarse; it may even have a few quills. The upper sides are grayish-brown to light brown, sometimes with reddish patches, while the undersides are grayish to white. There is only one species, the PEST RAT (*Nesokia indica*; BL 15–27 cm, TL 9–15 cm), with thirteen subspecies which are distributed from Asia as far as Egypt (possibly by ships from India and Pakistan).

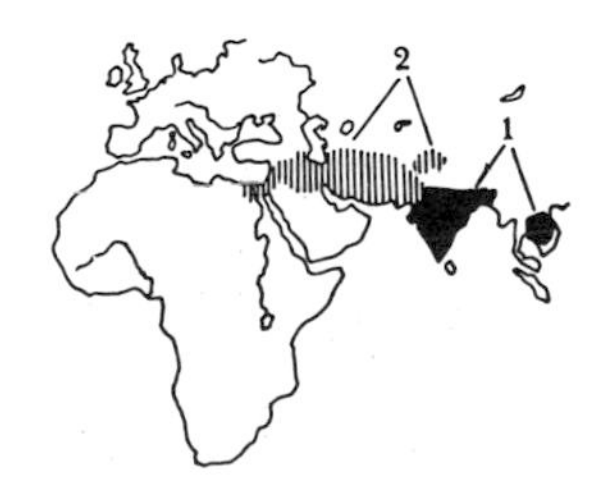

Fig. 11-53. 1. Long-tailed climbing mice (genus *Vandeleuria*); 2. Pest rats (genus *Nesokia*).

Although the pest rat is largely distributed over dry environments, within these environments it lives in areas with a relatively high humidity and warmth. Consequently, these animals are found in moist valleys, river beds, and oases, or near irrigation canals. They build burrows that may be as deep as 60 cm under the surface. They build earthen mounds in front of their exit tunnels. In areas where the plant growth is particularly dense, the animals may come to the surface. They are herbivorous and eat grass, roots, and seeds; as a result, they cause great damage in grain fields. Occasionally people will raid pest rat burrows in order to get these animals' food supplies, just as is the case with the lesser bandicoot rat. The rats themselves are sometimes forced from their burrows by floods or smoke, whereupon they die. The average number of young is four to six. In India and Pakistan the pest rat is recognized as a carrier of the plague.

Rats

The RATS (*Rattus*; BL 10–30 cm) comprise a particularly varied group both in form and lifestyle. The tail is usually longer than the body, although in some cases (brown rat) it is shorter. It is usually sparsely haired. The type and color of hair is quite varied; it ranges from a thick, silky-soft coat to one with quill-like hairs. The coloring ranges from black to grayish, dark brown, reddish-brown, to yellow on the upper side. The underside may be anywhere from gray to completely white. Females have between four and twelve nipples (black rats and brown rats have twelve nipples). In capitivity these rats have a life expectancy of seven years. The number of species is unknown. About 570 different forms have been described. The original area of distribution was eastern and southeastern Asia (see Fig. 11-55).

Fig. 11-54. Rats (genus *Rattus*): The area enclosed by the broken line indicates the approximate pre-Ice Age distribution of these animals. The dark area indicates their distribution before the black rats and brown rats achieved a worldwide distribution (which occurred in the 19th Century).

We would need a long book if we were to adequately describe and discuss the rats, a few species of which have, together with man, managed to conquer the world within a few centuries. Much of the unusual history of these animals would not have occurred without man; on the other hand, human history would have been quite different if the rat had not been present. Only a few rats from the great number in this genus have been involved with man and his living arrangements. Nevertheless, this ability to assimilate is one of the basic characteristics of these animals; no other mammalian genus has this adaptability to the extent that rats do. Many rat species living wild in woods and fields tend to look for man or for human settlements within their original environment even when their house-dwelling relatives are not already present. Generally, however, areas near and around human settlements are occupied by particular species and subspecies. Although many rat species live in trees and are able climbers, the hind feet of these animals are very poorly adapted to an arboreal lifestyle. This is an indication of an important characteristic of the rats: they are not particularly specialized, and they have developed very few special adaptations. Therefore, these animals are able to survive in a wide variety of habitats; habitats like houses, wet rice fields, dry bushes, moist tropical rain forests, subterranean burrows, and treetops, to name just a few.

Fig. 11-55. Brown rat (*Rattus norvegicus*) in Europe.

The original distribution of these animals (the distribution not created by man) is difficult to distinguish today, because most of the world has long been populated by the brown rat and various subspecies of the black rat. Another species, the LITTLE RAT (*Rattus exulans*), was an unwelcome stowaway on boats and ships traveling the Pacific; as a result, it settled on the islands in both the eastern and western halves of the ocean. We know only that large areas of the countries bordering the Mediterranean Sea, the Middle East, India, China, Japan, and all of southeastern Asia to the Philippines, New Guinea, and Australia were once the original habitat of the genus *Rattus*. We are not able to describe the great number of forms in our following discussion, so we will limit ourselves to the brown rat and the black rat, both of which are found all over the world.

The brown rat

The BROWN RAT or NORWAY RAT (*Rattus norvegicus*; BL 22–26 cm, TL 18–22 cm, weight 200–400 g, in extreme cases up to 500 g; see Color plate, p. 371) has a great economic significance. These animals have a stocky, powerful build. There are between 180 and 200 rings on the tail. The tail, which is always shorter than the body, is stubbed when compared with that of the black rat. The ears are short. The fur is usually brownish-gray on the upper side and dirty-white below. There occasionally are black specimens, sometimes with a white patch on the breast and lightercolored fore feet. The white rats (albino rats) used in laboratories are descended from this species.

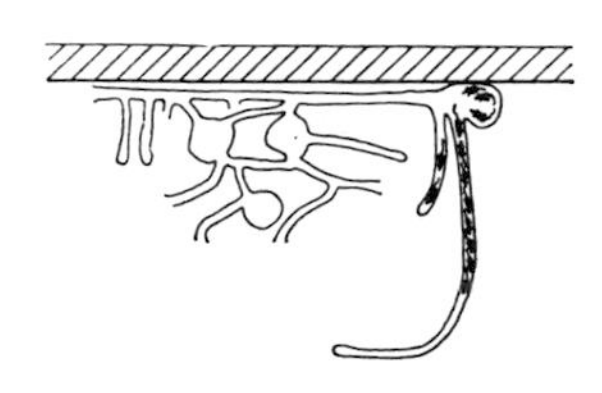

Fig. 11-56. Burrow and branching tunnel system located along a wall, home of a pack of thirteen brown rats—as seen from above.

The original habitat of these animals was the Asian steppes, probably in northern China and Mongolia, where these rats still live in subterranean burrows. Just when the brown rat became involved with man is uncertain; it might have been a few thousand years ago, or perhaps only several hundred. The rat described in 1553 by Konrad Gesner in his animal book was probably a brown rat, so this species was already in central Europe at that time. One often reads of the mass migration of brown rats in 1727, when these animals crossed the Volga River and moved in to "conquer" Europe. However, these narratives are questionable, as the brown rat had probably already appeared in Europe, during the Middle Ages. The only historically justified initial observations of these animals are those from the 18th Century: 1730, in England; 1735, in France; 1750, in eastern Germany; and 1800, in Spain. By 1755 the brown rat was being carried to America in the ships that traveled from Europe across the Atlantic Ocean. Indeed, ships and sea trade were largely responsible for increasing the distribution of these animals, and today brown rats can be found in almost every seaport in the world.

Fig. 11-57. When two brown rats contest a certain spot, they push one another away with the sides of their bodies. As a result, neither rat is able to make sudden movements, and there is no direct attack. Thus these "broadside shovings" do not result in biting fights.

As a ground-dweller, the brown rat, because of its great adaptability, has been able to acclimate itself to any environment that is even reasonably suitable. In buildings, these animals live in cellars, sewage canals, subterranean cavities, stalls, and similar places. In addition to buildings, we can find these animals in rubbish heaps, mulch on compost piles, in wood piles, near water, as well as in pits and dikes, especially wherever there is enough food for these omnivores. Rat burrows dug into the earth have one or more exits, some long, branching tunnels, sleeping rooms, storage rooms, and dead-end tunnels which serve as a last retreat from danger.

The shape and size of the areas where the brown rat is active are largely determined by available food. We must not confuse these activity areas with territories. The latter is the immediate living area of a single animal or a group of animals, and, unlike the activity area, the territory is vigorously defended against all unknown conspecifics. Steiniger describes these often-large activity areas with an example that, at the same time, illustrates the amount of damage brown rats can cause: "At night the animals would make excursions some 3 km long out onto the delta of a north German river, where they would raid the eel nets that were set out. Those eels which were not harmed were able to escape through the gnawed nets at the next high tide. After their meal, the rats traveled the 3 km back to the village where they had their burrows."

In an emergency the brown rat can survive on the simplest food, although these animals are not able to live long without water if the available food does not contain enough fluids. The brown rat is able to adapt itself to water, an ability equalled by only a few of its long-tailed relatives. The brown rat's swimming and diving ability proves useful, for example, when it boards or leaves a ship by the anchor chain. These animals' compatability with water has led to the fact that in some areas they are known as water rats. The brown rat is excellent at catching fish, and it has developed a particular behavior pattern for searching the water in order to find something edible. This is the "searching grasp," as Eibl-Eibesfeldt has named it, whereby the rat makes a sievelike motion with its fore feet in the water current.

Rats generally carry their food back to their burrows, where they consume it. As a result, the burrow contains much edible material which is never eaten. The rats usually show a preference for animal matter, rather than plant material. They will eat not only fish, but also mice, chickens, ducks, and geese. They will even gnaw on young lambs and piglets, killing them indirectly through infection. Rats will even bite large pieces from the fat layer of adult pigs. Brown rats in Hagenbeck's Tierpark (zoo) in Germany even caused the deaths of several elephants when they gnawed into the feet of the pachyderms, creating wounds that would not heal. The fact that these rats occasionally attack helpless people, particularly newborn babies, has been verified by many witnesses.

Brown rats exhibit a behavior much like that of carnivores, particularly when they catch wild birds. In 1945–6 the avian population of the bird sanctuary on the tiny (some 15 hectares in area) island of Nooderoog was almost exterminated by the massive rat population. Some 15,000 rats ate the birds' eggs and caught live seagulls, red shanks, ducks, passerines, and other species. It is interesting to note that rats do not always exhibit behavior characteristic of an animal of prey. We must assume that the brown rat is able to develop certain behaviors above and beyond those in its innate "behavior inventory." In any case, such a behavior is always bound to a particular location and population, although it can be transferred to other members of the pack, particularly the younger animals. This is how what Steiniger calls "tradition" develops, a gen-

Fig. 11-58. The defensive threat posture in the brown rat: a strange rat intrudes on another's territory; it raises up, grinds its incisor teeth, and squeals. The attacker finally yields, and no fight results.

uinely rare phenomenon in the "lower mammals." Such traditions also become important in the social life of the rat, as Steiniger, a highly respected scientist studying the behavior and sociology of the brown rat, will testify.

Within the pack, the sexes do not usually live together in pairs. During her heat, which lasts about six hours, a female is followed by several males, all of which repeatedly mate with her. Steiniger estimates the total number of copulations for one period of heat to be 200–500. The gestation period is twenty-two to twenty-four days, after which six to twelve young, with an average of eight, are born. In rare cases, up to twenty young have been observed. The young are typically altricial; blind and helpless. Their skin is naked, pinkish-red, and has several folds or wrinkles. When they are fifteen days old, the young rats open their eyes; they are fully haired and they begin to show an interest in their immediate surroundings. After twenty-two days, they leave the nest. Their most characteristic behavior is their interest in play, and they romp together or flee from one another. This capability for play is determined by heredity, and it appears in only a few species of the family of mice.

Fig. 11-59. Brown rats in a fight: the opponents beat and kick one another with their paws, as each rat attempts to throw the other. The victor threatens the loser by whetting its teeth and squealing. These fights usually do not result in biting matches.

Males are sexually mature when they are three months old; females achieve this status somewhat later. Brown rats in captivity gave birth to up to seven litters a year. Because the reproductive period of rats in the wild is confined to the warm seasons, or to the moist seasons in the tropics, the reproductive rate of these animals reaches that of rats in captivity only when conditions are especially favorable, if ever. A captive pair of breeding rats can, theoretically, produce more than 800 descendants within one year. In the wild, this number is considerably lower, even when the food supply is adequate.

Steiniger tells us that several females in large packs will often give birth in the same nest room, ". . . the raising of the young is done collectively. One can find young animals from three different age groups all within the same nest. The collective raising of the young insures the security of the next generation. Thus, if one nursing mother is killed, the other mothers raise her young along with their own." The rat's habit of forming packs, and its practice of meeting and facing its varied environments as a member of a pack, appears to be one of the reasons for this animal's success. All of the animals within a pack are usually closely related to one another, and they are generally all descended from an initial breeding pair. However, unrelated rats occasionally join the group, especially if the colony is rather large.

The main base of many rat packs consists of a large family that occupies a certain territory which it defends against all strange packs. According to Telle, forty-five percent of the packs of wild brown rats he found in Germany had more than sixty animals. Giant packs contain over 200 animals. The members of these packs probably recognize each other by a particular smell. In any case, the sense of smell is very highly developed. The nose is the rat's most important sensory organ. Naturally only those animals with a particularly social lifestyle can develop tradition, as this requires a special ability for communication among the animals.

Steiniger has the following description on this phenomenon: "Within the pack there is a large amount of transference of mood, and each example triggers an imitation. If an unfamiliar bait is offered to the group, the first animal to find it usually determines the group acceptance or rejection of the item. When a few members of the pack are the first to approach the bait, and when they do not accept it, then the rest of the pack will show no further interest in it. This was particularly true when the first rat rejected the poisoned bait; the other rats marked it with their urine or feces. Even when local circumstances made it very unpleasant for rats to mark the bait, one would still find droppings on the poisoned material. On the other hand, if the bait is accepted, then all the pack members rapidly share the booty. Consequently, when one puts out poisoned food in particular places, one either has no success at all, or one is 100 percent successful and all the rats die."

The fight against rats has been quite successful in certain areas of Europe and North America since the Second World War. Entire towns, seaports, and even large areas of countryside have been made completely rat-free. However, many of the larger cities of the world, or at least certain parts of these cities, are rat paradises; including, for example, Calcutta and New York City. Brown rats will probably not be exterminated within the next few decades, because they are so adaptable and they are so tough. They play a major role today, as in the past, as carriers of disease, even though many diseases, including the plague, have lost much of their dangerous impact. We must not forget the dangers of this disease, however, as two plagues of the past remind us. In the short time between 1347 and 1352, the black death (bubonic plague) claimed the lives of over twenty-five percent of the population of Europe. Between 1892 and 1918, the plague killed 11 million people in India. Other diseases transmitted by the rat include rabies, typhus, tularemia, and trichinosis, to name a few.

As research animals, however, rats and particularly albino rats, millions of which are used in laboratories today, are also helpful to man (see the discussion on rodents as laboratory animals in Chapter 9). And finally, rats have always had an important role in man's fantasies. We will mention only one such fantasy, which has often appeared in many ancient and modern books on animals. This is the idea of the rat tangle (Ratten König). A rat tangle consists of several rats all connected to one another by their tails. The tails are never fused. They are all entangled and partially knotted, glued together by dirt, encrusted wounds, and the like. Becker and Kemper, in their work on rat tangles, concluded that

Fig. 11-60. Brown rats will clean themselves extensively from time to time.

such "living Gordian knots" occur mainly with half-grown animals. The entanglement probably begins in the nest, where the animals sit close to one another while sleeping or grooming themselves, and the tails become entwined in play, or even without the rats' awareness. When these animals try to separate themselves for the first time and take off in different directions, the tails are pulled tight, the knots strengthened, and the tangle is knitted together. The animals involved in such a tangle soon perish.

The BLACK RAT OR HOUSE RAT (*Rattus rattus*; BL 16–22 cm, TL 17–24 cm, weight 70–300 g; see Color plate, p. 371) may well have lived with man ever since man came into existence. The tail is always longer than the body. The ears are large. The body is more slender than that of the brown rat. Black animals and animals with white breast markings are common, particularly in the subspecies *Rattus rattus rattus*. The center of the distribution area is in southeastern Asia. There are several subspecies, some of which are distributed worldwide, including the HOUSE RAT (*Rattus rattus rattus*), found largely in human dwellings, the ROOF RAT (*R. r. alexandrinus*), and the CORN RAT (*R. r. frugivorus*), whose underside is white. The roof rat and corn rat prefer warm areas. They live in tropical and subtropical areas, in the wild as well as in buildings. All three subspecies may be crossbred successfully, although they very seldom interbreed in the wild.

The black rat

Black rats were present in Europe as early as the Ice Age. Historical documentation shows that these animals have lived in the areas around the Mediterranean for thousands of years. As arboreal climbers, these animals are easily able to live in wooden walls, beams, and in the upper, drier sections of buildings, as well as on ships, particularly the wooden ships from earlier times. More than ninety percent of the rats living on ships are black rats. Some of these animals live on board permanently, others leave the ship as soon as it comes to port, settling in the houses of the town. The earliest explorers must have brought black rats with them to South America, because these animals were there as early as 1540.

Fig. 11-61. Black rat (*Rattus rattus*) in Europe. These rats have an irregular distribution in central Europe; in many areas they are almost or even completely nonexistent.

The black rat appeared in central Europe long before the brown rat; for climatic reasons it occurs only in parts of North America. The number of black rats in central Europe today is much smaller than the number of brown rats. This is probably because wooden houses and attics are much rarer there now than they were earlier. The black rat loses the competition with its larger relative, the brown rat, only in a few areas. The two species encounter one another only in human houses, where the brown rat lives in the lower and basement rooms, while the black rat lives in the building's upper, drier parts, which give it the needed climbing facilities. One can often find black rats in the wild in the areas around the Mediterranean. Sometimes these animals live far away from human settlements, particularly in fruit plantations, in palm trees, or in bushes. Their nests are large and conical. Occasionally these animals will also live in burrows in the ground, in oases near water. The black rat can withstand humidity better than cold.

Fig. 11-62. Like many other rodents, the black rat sometimes holds its food with its forefeet.

The black rat's activity is more related to particular times than is that of the brown rat. These animals are largely nocturnal. Their preferred food is plant material. In the wild they will feed largely on fruit, seeds, and similar items; they eat animals or meat only as additional food. The gestation period lasts twenty-four days, after which the female gives birth to between six and twelve young; the average litter size is eight, although on occasions as many as twenty young have been observed in one litter. The young open their eyes after fifteen days. Their development is very similar to that of the brown rat.

In tropical and subtropical areas the black rat has a greater economic and medical significance than does the brown rat. The black rat, along with its various subspecies, is very closely interconnected with the people of these areas. These animals eat and destroy human supplies to an extent equalled by nearly no other animal. Through its fleas, urine, and dirt, the black rat also brings sickness and disease into the bedroom and dining room. The "plague rats" of the Middle Ages were largely black rats.

House mice

More than 130 forms of HOUSE MICE (*Mus*) have been described, but the number of "viable" species is only about twenty. Included in this number are the LITTLE INDIAN FIELD MOUSE (*Mus booduga*), from India and southeastern Asia, the INDIAN BROWN SPINY MOUSE (*M. platythrix*), from India, and especially the HOUSE MOUSE (*M. musculus*; BL 6–12 cm, TL 6–11 cm, weight 10–30 g; see Color plate, p. 371). The first molar tooth is very large (larger than the second and third together), while the third molar is quite small. The back side of the upper incisors has a sharp-edged elevation (a good identifying characteristic). The hair ranges from silky-smooth to rough and spiny. Females have ten nipples. Breeding forms occur in all color varieties, of which the most widely known is the white mouse (albino). In captivity the life expectancy of these animals is up to six years.

Those forms which are dependent on food gathered by man have moved all over the world from their original home. Depending on climatic conditions, colonies have developed that either live in houses all year long or retreat into houses only during the cold seasons. In certain areas there are also colonies of these animals which remain outside throughout the entire year. Outdoor and indoor colonies may be found in the same area.

The house mouse

All of the common house mice belong to one species. The most important European subspecies are the WESTERN HOUSE MOUSE (*Mus musculus domesticus*), found in western and northwestern Europe, west of the Elbe River, and the NORTHERN HOUSE MOUSE (*M. m. musculus*), found east of the Elbe River in eastern, southeastern, and northeastern Europe. The western house mouse lives largely or even exclusively in houses. The northern house mouse is not so dependent on man; these

Fig. 11-63. House mice (genus *Mus*): the dark area indicates the approximate original distribution of these animals, before they dispersed themselves all over the world (dotted area) in man's wake.

Fig. 11-64. House mouse (*Mus musculus*) in Europe.

Fig. 11-65. Distribution of the northern house mouse (*Mus musculus musculus*) and western house mouse (*M. m. domesticus*), showing contiguous areas.

animals spend at least part of the year outdoors, depending on the weather conditions. These full-time or part-time "boarders," (by which we mean "eating with man") are descended from various wild subspecies. The western house mouse is descended from the BACTRIAN HOUSE MOUSE (*M. m. bactrianus*), a long-tailed, white-bellied subspecies from central Asia, from which various other commensal subspecies are derived. The northern house mouse is descended from the short-tailed, pale-bellied EASTERN HOUSE MOUSE (*M. m. spicilegus*), which is the third most important subspecies in Europe. The eastern house mouse lives outside all year long, and consequently appears only in those areas where the climate allows such a lifestyle, including the Ukraine, Romania, Hungary, and parts of Yugoslavia, lower Austria, and Czechoslovakia. These animals spend the winter in subterranean burrows beneath dirt piles; these burrows are equipped with supply rooms. This behavior sets the eastern house mouse apart from its descendant, the northern house mouse. However, when necessary, the eastern house mouse can also adopt a semi-boarding lifestyle.

As everyone knows, house mice are omnivorous. However, they do show a particular preference for grain and grain products, such as bread. They do not always choose to eat bacon, although they may prefer it as a rare delicacy. In some cases bacon may work particularly well as bait. Wild eastern house mice are the only house mice which store food supplies. House mice living in houses do not hoard food, preferring instead to live near or in man's food supply. Rest and activity periods for these animals may interchange up to twenty times within a twenty-four-hour period, although night is clearly the preferred period of activity. The activity area is sometimes not more than a few square meters, and there may be some mice that never leave this area during their lifetime. House mice are quite capable of executing the most important modes of movement, including running, jumping, climbing, and swimming, although they do not go into the water willingly.

The senses of smell and hearing are the most important in these animals. The nose is used in the search for food and when the animal is following a "scent path" that has been marked with urine. The sense of hearing is especially attuned to picking out high tones. The upper limit of the house mouse's hearing is 40,000 cps. House mice show much less concern over tones from the lower register. This may explain the "musicality" of these animals, including their occasional (unsolicited) appearance when people play music in the evening. The music probably sounds very muffled to the mice, and consequently it does not trigger a flight response. Town musicians have reported mice that were directly in line with a trumpet blast and which remained sitting in a state of "bewilderment." Mice certainly also appear on evenings when no music is being played, only they are usually not noticed so much.

Young house mice are sexually mature after two to three months, or even earlier. Males usually attain this status somewhat sooner than the females. The gestation period lasts twenty to twenty-one days; nursing mothers may have a slightly longer gestation period. The average litter size is six, although the highest extreme in an "average" litter may be up to thirteen young. Larger litters occur very rarely. House mice living in houses may continue to reproduce throughout the winter without any particular reduction in size or frequency. This, of course, leads to a massive propagation. In the wild, on the other hand, there are longer periods between litters. In cold-storage rooms, when the food supply is adequate, mice will raise their young even when constant temperatures of 10°C prevail. Thus, people have found nests with young in cavities within frozen meat that was stored in a cold-storage locker. The northern house mouse is able to effect similar accomplishments in the wild, such as when these animals raise their young in the permafrost areas near the Arctic Ocean. The nests are constructed from all available materials and are built in dark, protected areas. Wild house mice usually have subterranean burrows.

Young house mice are altricial. They are about 2.5 cm long at birth and they weigh just over 1 g. They open their eyes when they are thirteen days old, by which time short, smooth fur has developed all over the body. They begin to eat independently at around the seventeenth day, although they nurse from their mother until they are four weeks old.

House mice usually live together in small groups, which Eibl-Eibesfeldt designates as large families. These groups are organized very much like the brown rat packs. The group occupies a common living area (territory), whose borders are delineated by scent marks. Each animal is able to live alone and build its nest within this area, but it does not occupy a particular territory of its own. Eating and nesting places, escape holes, urine areas, and defecation areas are all used collectively. There is a rank order, particularly among the males; it is not inviolate, however, as the position of the highest-ranking male can be disputed, as can the other positions. Eibl-Eibesfeldt has described the fights, the threat and submissive postures, and many other aspects of the social life of these animals. House mice also practice mutual grooming, especially on areas including the back and neck, where an individual animal is not able to groom itself. Mutual grooming is one of the everyday chores in mouse families.

When a house mouse population begins to become too dense, a particular type of "birth control" comes into effect, as Crowcroft and Rowe found in their investigations. The reproduction is arrested or checked, because many of the females, particularly the adolescents, become infertile. The vagina remains closed, the uterus becomes extremely thin, and the ovaries discontinue their activity. It is possible that the group restlessness which accompanies overpopulation is due to hormonal changes in the female. In any case, this natural birth control appears to be typical of the house mouse.

▻ Fig. 11-66. Long-tailed field mouse cleaning itself. Body grooming is an important part of the daily routine.

The history of the house mouse has been linked with the history of man for thousands of years. These small rodents have very possibly been destroying crops, food stores, and material since man's first appearance on the earth. House mice have also played a role as disease carriers; typhus, spotted fever, tularemia, Salmonella food poisoning, and bubonic plague are only a few of the diseases they have been known to transmit. These animals also assumed a blessed role in ancient times, especially for those who believed in myths and superstition. The ancient Egyptians believed that the heat of the sun during the warmer months begat mice from the mud of the Nile. The Greek philosopher Aristotle believed that mice were produced from the dirt in houses and ships. In Asia Minor, Greece, and Sicily, Apollo was honored for many hundreds of years as the mouse-god, Smintheus; the mouse was highly esteemed, and people kept mice in their temples to help them seek the advice of the gods. Whenever the white mice began to reproduce well, it was regarded as a good sign. The pet name of "little mouse" appeared in Greek drama, and the animal itself was regarded as a symbol of tenderness and sensuality. In medical superstition as well, the mouse was regarded as a helper for afflictions like goiter, epilepsy, cataracts, snake-bite, baldness, constipation, and other conditions.

However, the house mouse has become really valuable to man only in modern times. Around the middle of the 1900s, the house mouse was imported to Europe from Japan as a tame laboratory animal. Since then these animals have served man well in his studies of medicine and of heredity. Two examples of this service are the test for pregnancy and the tests on mice used in cancer research. Biological studies of the house mouse have proved very worthwhile, and the knowledge that we have gained from these tiny rodents more than repays mankind for their parasitic activities.

The Eurasian WOOD AND FIELD MICE (*Apodemus*) have long tails, and range in size from medium to large. The body length is from 8 cm (GEISHA WOOD MOUSE [*Apodemus geisha*], from Japan) to 14 cm (BROAD-TOOTHED FIELD MOUSE [*A. mystacinus*]). The tail is as long as the body or somewhat shorter. The hair is soft and may be a variety of colors. These animals are distributed from Iceland to Morocco, through all of Europe, to eastern Asia, including Japan. There are probably eleven species, of which five are found in Europe.

1. The STRIPED FIELD MOUSE (*Apodemus agrarius*; BL 9.5–12 cm, TL 6.5–8.5 cm, weight 15–25 g; see Color plate. p. 371) has a distinct black stripe on the back; this is their most important characteristic. They live in subterranean burrows equipped with nest and supply rooms. They are active during the day. Females give birth to between four and eight young per litter; the average litter size is six. The gestation period lasts twenty-one to twenty-three days.

2. The BROAD-TOOTHED FIELD MOUSE (*A. mystacinus*; BL 12–15 cm

TL 11.5–14.5 cm, wieight 30–50 g) has a dull grayish-brown upper side, while the underside is grayish-white. These animals have vibrissae which are some 5 cm long. They live in rocky parts of woods and in open areas. They are crepuscular and nocturnal animals. Their principal food consists of seeds. They store food. There is an average of three or four young per litter.

Fig. 11-67. Wood and field mice (genus *Apodemus*).

3. *A. microps* was first discovered and described in 1952, in Czechoslovakia. These animals are similar to the broad-toothed field mouse, but much smaller. They are almost as small as the eastern house mouse. They are found in parts of Czechoslovakia, Hungary, Bulgaria, and Poland. They prefer open landscapes. There is an average of six or seven young per litter.

4. The YELLOW-NECKED FIELD MOUSE (*A. flavicollis*; BL 9–13 cm, TL 9–13 cm, weight 20–40 g; see Color plates, pp. 371 and 321/332) has a sharp distinction between the coloring of the upper and lower sides of the body. These animals have a yellowish hair zone in the throat area; the size and shape of this zone varies considerably, from a wide throat band to a tiny spot.

Fig. 11-68. Striped field mouse (*Apodemus agrarius*).

5. The LONG-TAILED FIELD MOUSE (*A. sylvaticus*; BL 8–11 cm, TL 7–11.5 cm, weight 15–30 g; see Color plates, pp. 371 and 299) is smaller than the preceding species. The line between the coloring of the upper and lower sides of the body is usually indistinct. These animals do not have a throat band, but they do have a long brownish-yellow spot on the breast, which may vary considerably. This spot may extend as far as the underside of the stomach, it may be only 1 mm long, or it may not be present at all.

Fig. 11-69. Broad-toothed field mouse (*Apodemus mystacinus*).

In Europe, the STRIPED FIELD MOUSE lives in a variety of habitats, often humid, like fields or bushes. Occasionally these animals can also be found in drier areas as well as in lightly wooded areas, in parks and gardens, and even in houses. The biological and ecological aspects of these animals have had amazingly little research.

The YELLOW-NECKED FIELD MOUSE and the LONG-TAILED FIELD MOUSE are considered as "twin species" in many areas, because they are found close together. The yellow-necked field mouse does not always have a yellow throat band, so the general name of this animal is not wholly accurate. In many areas, the classification of captive animals into one or the other of these two species is quite difficult. In 1959 Larina was able to crossbreed Caucasian long-tailed field mice and yellow-necked field mice. This success raises several doubts about the independence of these two species, and it is very possible that there may be mixed populations in some areas.

Fig. 11-70. *Apodemus microps*; distribution is largely unknown at present.

Heinrich has given us a clear description of the varying habitats of both species: "The yellow-necked field mouse is clearly linked with wooded habitats or at least habitats where trees and bushes abound. Their optimal environment consists of established deciduous woods,

Fig. 11-71. Yellow-necked field mouse (*Apodemus flavicollis*); the distribution of these animals in eastern and southwestern Europe is not completely known.

Fig. 11-72. Long-tailed field mouse (*Apodemus sylvaticus*) in Europe.

Fig. 11-73. Harvest mouse (*Micromys minutus*).

particularly those comprised of beech and oak trees, although parklands with individual old trees will suffice, as will bushy thickets . . they can also be found in old pine forests. The long-tailed field mouse, on the other hand, clearly prefers open areas. These animals often settle in fields, fallow lands, railroad levees; in short, in all kinds of cultivated prairie lands. Their optimal habitat seems to be open flatlands which have a small amount of brush or which are surrounded by bushes. These animals, then, are better able to find the necessities of life within a wide area than are the yellow-necked field mice. They are also able to adapt to life in a wood or forest. In such cases, the long-tailed field mouse naturally prefers lightly wooded areas with sandy soil, such as those with pine trees or acacia. However, these animals are also able to survive in more dense deciduous forests with humus soil. The long-tailed field mouse cannot exist in forests of old pine and red beech trees where the ground has little or no green growth. These forests are, instead, the preferred habitat of the yellow-necked field mouse. Thus, the habitat of the long-tailed field mouse must contain grasses and low plants. Just as the yellowed-necked field mouse is bound to 'tree and shrub,' so is the long-tailed field mouse linked to 'grass and herb.' The environment of the former species, figuratively speaking, is incomplete without the shade of a green leaf roof. The environment of the latter species is incomplete without the protective covering of some ground vegetation."

Both species, but particularly the long-tailed field mouse, are also able to move into human habitations, especially in the fall and winter. The long-tailed field mouse usually remains in the lower parts of the house, where it feeds on grain, corn, and stored fruit. The yellow-necked field mouse, because of its good climbing ability, prefers the upper stories of houses near wooded areas. These animals show a particular preference for fatty foods, and thus they like to nibble at bacon. In the wild, the yellow-necked field mouse prefers beechnuts, acorns, hazelnuts, chestnuts, pine seeds, and the like. The long-tailed field mouse, on the contrary, feeds largely on the seeds of grasses and weeds, as well as grain, berries, and fruit. These animals also seem to require a small amount of fatty seeds.

Both species store food in the storage chambers of their burrows. The long-tailed field mouse is able to build deep burrows with two entrances. Its nest and supply rooms may occasionally lie between the roots of trees and shrubs. This is much more likely to be the case, however, with the yellow-necked field mouse, which has also been known to build its nest under fallen trees, in rocky crevices, or in burrows abandoned by males. As both species are good climbers, they occasionally use bird nesting boxes for their own nests. Both species are excellent runners and jumpers. The long-tailed field mouse has been called "jumping mouse" in many areas, because it hops on its hind legs when moving at top speed. Both species are also good swimmers. These animals are active mainly during the twilight and nocturnal hours.

Reproduction generally occurs during the warmer seasons, from about March until September. Usually four litters are born during this time, each with an average of five young, eight at the most. Both species have a gestation period of twenty-three days. The young mice are altricial. They are about 3 cm long at birth (not including the tail), and they weigh 2.5 g. Their eyes open around the thirteenth day. They are nursed by the mother for about three weeks, and after about two months they become sexually mature.

The harvest mouse

The HARVEST MOUSE (*Micromys minutus*; BL 5.5–7.5 cm, TL 5–7 cm, weight 5–10 g; see Color plates, pp. 257/258 and 371), the only member of its genus, is adapted to a climbing lifestyle. These animals avoid mountains, and they are rarely seen at altitudes above 2000 m.

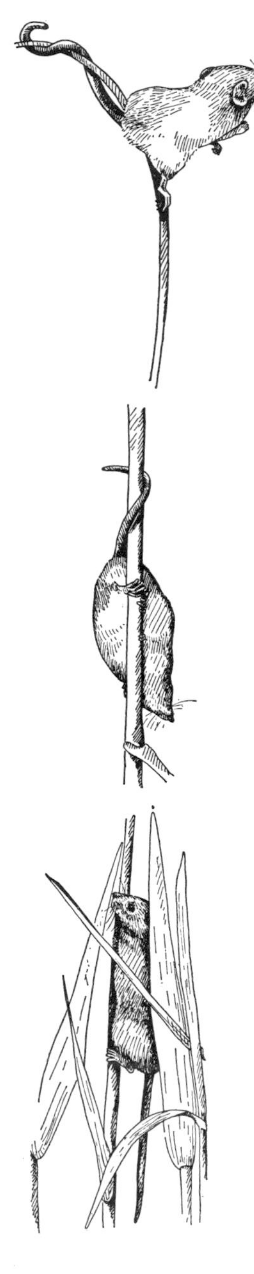

In general, the harvest mouse prefers habitats that are characterized, in the broadest sense, by high grass, thus, overgrown pastures, bulrushes, grain fields, and, in Italy and eastern Asia, rice fields. The original habitat of these animals was moist areas near rivers, ponds, and lakes, which had high, long-lasting grass growth. These animals also like to settle in brushland which has some grassy ground cover. In summer, harvest mice may live in grain or rice fields.

The "grass forest" of fields and prairies gives these tiny rodents everything they need during the warm season. There the harvest mice are able to climb around stalks and tiny branches. There, too, they find the necessary food in the form of grass, seeds, and insects and their larvae. The harvest mice also build their nests in the fall grass. The nests, built slightly above the ground, are round, with one or two openings. The cold seasons of the year, with rain and snow storms, make life in the fall grass impossible, as by that time the grass itself is either not as dense or it has been completely cut down. During these seasons, the harvest mouse settles on the ground or it moves into barns and other buildings, particularly those with food stores, where it spends the winter. These mice, contrary to former belief, do not hibernate. Thus, they are forced to build nests on the ground for this period.

R. Piechocki, F. Frank, and Sleptsov, working in the far East, have been deeply involved in studying the harvest mouse. When climbing in the grasses, these tiny animals spread their toes out and swing their tails around to grip the grass, as support and counterbalance. These animals' slow movements, their motionless "camouflage position," and their habit of suddenly dropping down to the dark ground when threatened, are all measures of defense against enemies. Harvest mice are active principally during the night; to a lesser extent, they are also active during the day. They are able to build their nest in complete darkness, as has been observed with captive animals. The animals intertwine the blades of several neighboring grass stalks. They mesh these blades together and then, on the resultant base, they interlace several other living plants or cut pieces of plants, to build the nestroom. The nest is always lined with particularly fine shredded materials. There, the female gives birth to an average litter of five or six young after a gestation period of twenty-one days. A new nest must be built for the next litter, and so on. When conditions are particularly favorable, the female may give birth to as many as six litters in a year; usually, however, the number is much smaller. The major reproductive season occurs during the warmer months.

Newborn harvest mice weigh just about 1 g. They are only 2 cm long, blind, and naked. When they are three days old, however, they are already able to grip, to some extent. They open their eyes after eight to ten days, and after twelve to thirteen days they leave the nest for the first time and begin to eat. They are weaned at eighteen days. Some observations have shown that these animals go through a certain period when they are extremely playful and unrestrained. Under favorable conditions, the young become sexually mature at five weeks. In any case, they are able to reproduce within the same summer of their birth.

During the gestation period, the female lives in the nest with her young. Normally, though, harvest mice live in very dense groups, particularly in barns during the winter. However, these mice must not be mistakenly classed as social animals. The average life expectancy in the wild may be up to one and one-half years. In captivity, these animals may live to be three or four years old.

African native mice

There are more than thirty genera of typical mice living in Africa. However, only two of these genera also appear in other parts of the world. The AFRICAN NATIVE MICE (*Leggada*; BL 5–9 cm, TL 3.5–7 cm, weight 8–15 g; see distribution, Fig. 11–78) are similar to the house mouse. This genus is very closely related to the genus *Mus*, and the two genera may actually be united. The SMALL AFRICAN NATIVE MOUSE (*Leggada minutoides*; BL 5–6 cm) is one of the smallest mice in the world, and is certainly the smallest mouse in Africa.

Spiny mice

SPINY MICE (*Acomys*; BL 7–12. cm, TL 5–12.5 cm, weight 30–70 g) have backs that are almost completely covered with spiny, oblate hairs. The tail is usually shorter than the body. The color of the upper side of the body ranges from dark grayish-brown to yellow-sand. The underside is gray to completely white. There are two or three young, at most five, per litter. The young are altricial at birth. This genus comprises several species, including the EGYPTIAN SPINY MOUSE (*Acomys cahirinus*), the CRETAN SPINY MOUSE (*A. minous*), and the SINAI SPINY MOUSE (*A. dimidiatus*; see Color plate, p. 371).

◁
Fig. 11-74. The harvest mouse climbs on and around reeds with great agility (top and middle). When in danger, it freezes into a motionless "camouflage position" between the stalks (bottom).

Spiny mice generally live in arid, open habitats, like deserts, prairies, and savannas; some live in dry, wooded areas. In the latter environment these animals were usually caught in rocky landscapes, and occasionally in bushes and undergrowth. The Cretan spiny mouse lives at lower altitudes in rocky brush savannas. Generally, spiny mice avoid altitudes of above 1500 m. In certain areas, including Egypt, these animals also visit human habitations, where they live on foods that man has stored away. Spiny mice are active mainly during the night, although there are also species which are completely diurnal, or which are active during part of the day. These animals live on plant and animal matter, particularly seeds, snails, and insects.

The most significant biological aspect of the spiny mouse, at least of those species studied in Egypt and Crete, is the development of the young at birth. Unlike the young of other mice, spiny mouse young are not naked, blind, and helpless at birth. Newborn spiny mice have a body length of 5–6 cm and they weigh 6–7 g, all of which gives them a rather robust appearance. Their eyes are already open at birth, or they open shortly after. Even though the hair is still sparse, the young are almost able to maintain their body temperatures independently. Thus they need very little body warmth from the mother. When they are three days old they begin to explore their surroundings, although their legs are still somewhat shakey. When threatened, they flee and hide. Their resting place is not a real nest. Some species of spiny mice do not build a nest in any sense of the word. Like many other animals which give birth to well-developed young, the spiny mice have "done away" with the nest-building process. Nevertheless, the mothers are very careful of their young, and they continually carry the young back to the resting place in their mouths, using a special body grip. The mothers painstakingly clean their young and nurse them until they are at least three weeks old. The development of the young proceeds very rapidly. They are usually sexually mature after two to three months. In captivity, these animals breed well when correctly nourished, and can live to be over three years old.

Unfortunately, we know very little about the life of these beautiful, lively creatures in the wild; these animals must certainly be among one of the most interesting rodent groups. We do not know whether the observations of captive animals hold true for wild spiny mice. When compared to the "mouse norm," the gestation period of the female spiny mouse is seen to be fourteen days longer, lasting for thirty-six to thirty-eight days. A female giving birth to a well-developed young is fertile again by the evening of the same day, and she is usually immediately re-impregnated. While the mother raises one litter, she usually carries the next litter in her body, an ability that the spiny mice share with many other typical mice and microtine rodents. Thus, there may be a continuous series of births, up to twelve litters or more before a break of some days or weeks occurs. Some spiny mice give birth to more than twenty litters in their lifetimes and may thus become mothers of fifty or sixty or more offspring.

Most of the young are born between 4 and 8 a.m., after periods of labor which often last several hours. While giving birth, the mother stands on all four feet and raises the back of her body as high as possible; the large young is usually forced out with these motions. Then the mother must turn around in order to lick young and bite off the umbilical cord. Thus, the female spiny mouse behaves, in this instance, much like an

▻
Typical mice: 1. Norway rat (*Rattus norvegicus*); 2. Black rat (*Rattus rattus*); 3. Eastern large bandicoot rat (*Bandicota indica nemorivaga*); 4. House mouse (*Mus musculus*) a. Albino; 5. Long-tailed field mouse (*Apodemus sylvaticus*); 6. Striped field mouse (*A. agrarius*); 7. Yellow-necked field mouse (*A. flavicollis*); 8. Harvest mouse (*Micromys minutus*); 9. Sinai spiny mouse (*Acomys dimidiatus*); 10. Stick-nest rat (*Leporillus conditor*); 11. Giant naked-tailed rat (*Uromys caudimaculatus*); 12. Striped grass mouse (*Lemniscomys striatus*); 13. Multimammate rat (*Mastomys coucha*); 14. Striped field mouse (*Rhabdomys pumilio*). Pouched rat: 15. Gambian pouched rat (*Cricetomys gambianus*).

2
3
1
5
4a
6
7
4
8
9
10
15
13
11
12
14

1
2
3
4
5
6
7

◁
Cloud rats: 1. Slender-tailed cloud rat (*Phloeomys cumingi*). African tree mice: 2. African climbing mouse (*Dendromus insignis*); 3. Fat mouse (*Steatomys pratensis*); 4. Congo forest mouse (*Deomys ferrugineus*). Vlei rats: 5. *Otomys irroratus*. Shrew rats: 6. Shrew rat (*Rhynchomys soricoides*). Water rats: 7. Eastern water rat (*Hydromys chrysogaster*).

ungulate animal and not like other mice, which sit in the nest while giving birth and which are able to bend the backs of their bodies around so that the young come out "frontward," under the mother's abdomen. Such births may take place right in the middle of a pack of these social spiny mice. If one has luck and much patience, one may even observe spiny mice "obstetrics," and see how other females begin to lick the young once it is half out of the mother's body and how they free it from the embryonic pouch so that it can move and breathe. When these "midwives" do not have any young of their own, they attempt to take charge of the newborn young, to nurse it, and to "possess" it. These midwives also attempt to claim the placenta (which many mammals consume), and the mother often is too late in making her claim to this delicacy.

Even those females which have not had young for several months and which are not expecting, help with the birth. Every female spiny mouse which has ever given birth also helps others who have young (midwifery). This behavior has also been observed, in individual cases, in other social animals like the African elephant, the domestic dog, and the South American tamarins. We have long known that birth can be a "social experience" with dolphins, llamas, and elephants. However, the "midwifery" of the spiny mice has not yet been surpassed.

The claim for total possession of the young by the mother or by an "aunt" lasts only for one, or at the most, two or three days. There are fights wherein each side steals the young. Then, finally, all the young also belong to the "aunts" as well as their own mothers; in other words, the young become the common property of every nursing female within the community. Later the question of locality no longer becomes important, so all the young may be nursed by every mother, and they are accepted anywhere. Where several mothers have young that are about the same age, the mothers and young become thoroughly integrated. When, however, the young are of various ages, individual mothers and their young keep more together. In short, any young may be cared for, nursed, and even adopted by any mother of the pack. Thus the young, which are dependent, in any case, on their mother's milk for only six days, will hardly perish even if their mother is lost.

Striped grass mice

The STRIPED GRASS MICE (*Lemniscomys*; BL 9–14 cm, TL 10–15 cm, weight 30–60 g) are characterized by a striped or spotted back. The tail is usually longer than the body. The hair is rather coarse. There are light parts on the upper side of the body, ranging in color from white to yellowish-brown. The underside is white or grayish-white. These mice are found mainly in dry, wooded areas, in prairies, and in savannas. There are six species, which are distinguished from one another by the markings on the back: the SINGLE-STRIPED GRASS MOUSE (*Lemniscomys griselda*) has only a dark center stripe down the back; the NORTH AFRICAN STRIPED GRASS MOUSE (*L. barbarus*) has several dark stripes on the back, between which are lighter stripes, giving the effect of a zebra pattern; in the STRIPED GRASS MOUSE (*L. striatus*; see Color plate, p. 371) the pale, narrow stripes have disintegrated into spots, which make this species look particularly beautiful.

Fig. 11-75. Spiny mice (genus *Acomys*).

All species of striped grass mice, like other grass mice and rats, are active mainly during the day, or at least in the twilight hours. One can see these rodents scurry rapidly across the road in broad daylight. They feed mainly on grass and grain seeds, on fresh plant matter, and, to some extent, on insects. An average of four or five young is born per litter. These animals survive well in captivity. I [Dieterlen] had a pair that lived for two and one-half years, although they did not breed during this time. However, a freshly caught pregnant female gave birth while under my care. The newborn young were only 3–4 cm long and weighed only 2.5 g. One could already see brown pigment, which indicated where the dark stripes would grow in later, on their tiny, hairless backs. Francis Petter and his associates were repeatedly able to breed these animals in Paris. The gestation period lasted twenty-eight days. Females were sexually mature at two months, and they were full-grown at five months.

Striped field mouse

The STRIPED FIELD MOUSE (*Rhabdomys pumilio*; BL 9–13 cm, TL 8–14 cm, weight up to 45 g; see Color plate, p. 371) is the only representative of its genus. Like the members of the genus *Lemniscomys*, striped field mice live in open areas in eastern and southern Africa. On Mt. Kilimanjaro (Tanzania) they appear at altitudes of up to over 4000 m. They are able to live on the ground and to build their nests there, as well as in subterranean burrows. Their basic food consists of green grass; they also eat seeds, insects, and other things. Striped field mice are such strongly diurnal animals that they will even sunbathe on cooler days. Like most African rodents, their mating and reproductive season occurs during the humid months. The female gives birth to an average of four or five young, sometimes up to seven, but rarely more, per litter. The young are fullgrown at three months. Like the GRASS MICE or KUSUS (*Arvicanthis*), striped field mice apparently live together socially. Large families of up to thirty members have been found in one nest. Occasionally these animals move into human habitations, although they are not dependent on these.

Rufous-nosed rat

The RUFOUS-NOSED RAT (*Oenomys hypoxanthus*; BL 13–18 cm, TL 16–20 cm, weight 70–130 g) is well named, as even young rufous-nosed rats have rust-colored hairs on their noses by the time they are but a few days old. The ears and a tiny hair zone at the root of the tail are also reddish. The upper side of the body is brownish-reddish, with an olive-green shimmer, while the undersides are whittish, and often yellowish or orange-colored. This species is probably the only member of its genus. It is found in the virgin forest areas of western Africa, in the Congo-Cameroon region and bordering areas, while certain pocket populations are found east as far as Mt. Kenya.

The ideal habitat of the rufous-nosed rat consists of high, dense plant

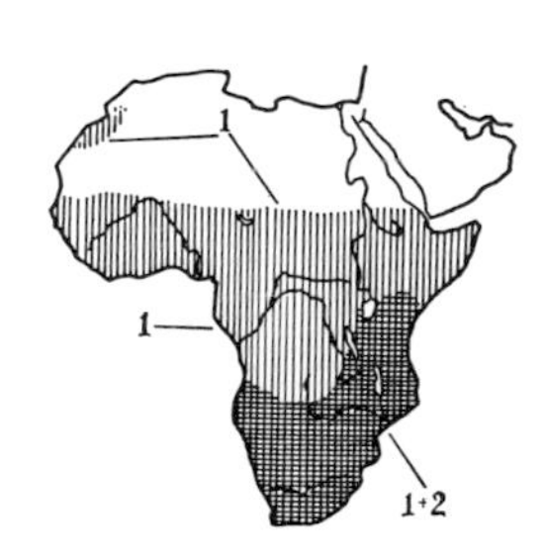

Fig. 11-76. 1. Striped grass mice (genus *Lemniscomys*); 2. Striped field mouse (*Rhabdomys pumilio*).

growth, like that on the edges of woods, in clearings in the middle of wooded areas, secondary woods, or humid savannas. The rats climb around on branches and on the stalks of large grasses. They also descend to the ground rather often, whenever they are unable to find enough food above. Rufous-nosed rats feed mainly on green plant material like delicate grasses, weeds, buds, sprouts, leaves, and bark. The brood nest is a neat ball-shaped structure built between branches or between thick stalks of elephant grass. Externally the nest consists of coarse materials. Internally, however, it is lined with the finest shredded stalks or leaves. Usually these nests have only one entrance, on the side. The female gives birth to an average of two or three young, rarely four, per litter. The reproductive season corresponds with the humid season, at which time the female gives birth to several consecutive litters.

These animals also breed well in captivity; at least they did so in Africa where I [Dieterlen] kept them in a cage for years. Newborn young have a very light, long, brownish-yellow coat of down, under which the hair begins to grow. The newborn young are 5–6 cm long and weigh 5–7 g. They alert their mother with high crying sounds. They open their eyes after seven to eight days. Shortly thereafter they begin to eat green plant material by themselves, and they make their first climbing excursions. Their development lasts a rather long time. A young rufous-nosed rat is not sexually mature until it is four months old. These animals are active both during the day and at night. They climb thin branches with slow, careful movements, an important defensive measure, as they have many flying enemies.

Soft-furred rat

The SOFT-FURRED RATS (*Praomys*; BL 10–13 cm, TL 11–16 cm, weight 30–50 g), like the multimammate rats discussed next, are classified as a subgenus of the rats (*Rattus*) by some zoologists. However, here we will discuss these animals as a separate genus. These mice are relatively large and slender, usually with short hair. They are found especially in the tropical rain forests of Africa, from sea level to an altitude of over 3000 m in the mountains, as well as in densely overgrown savannas. There are between five and ten species, including *Praomys morio*.

These mice are among the most common rodents of the African jungle. One can trap them almost anywhere. They are excellent runners, climbers, and jumpers. They are also good diggers when they make their burrows between the roots of trees. However, one can also find their nests high in the treetrunks, in cavities, in banana groves, or under rocks. Observations of these animals in the wild and in captivity seem to indicate that they live in large family groups. They feed largely on plants, particularly oily and mealy seeds, as well as insects. They eat large quantities of ants.

Fig. 11-77. 1. Rufous-nosed rat (*Oenomys hypoxanthus*); 2. Grass mice (genus *Arvicanthis*).

Outside the rain forest, the reproductive season of members of the genus *Praomys* occurs at about the time of the rainy season. The females may bear several consecutive litters during this period. Usually there are three or four young, six at the most, per litter. Eisentraut was able to breed *P. morio* in captivity, and he has written a description of the raising of the young. Newborn young are 4–5 cm long and weigh 2–3 g. They are naked and helpless. The eyes open after sixteen to nineteen days. The young animals are sexually mature after two to three months. The gestation period lasts twenty-six days.

Multimammate rats

The MULTIMAMMATE RATS (*Mastomys*; BL 9.5–16 cm, TL 9.5–15 cm, weight 30–100 g) are characterized by their unusually high number of nipples, ranging from twelve to twenty-four. The upper side of the body is light grayish-brown to gray-brown or dark brown. The underside is light gray. The many nipples cannot be divided into groups of breast and abdominal nipples, as is the case with other mice; instead they form two longitudinal rows running from front to back. There are at least two species or species groups found in the steppes and savannas of most of subsaharan Africa, with the exception of some regions of the large rain forests. These animals are also found in Morocco. The most important species is *Mastomys coucha* (see Color plate, p. 371).

Fig. 11-78. 1. African native mice (genus *Leggada*) and the multimammate rats (genus *Mastomys*); 2. Soft-furred rats (genus *Praomys*).

The multimammate rats have adapted to a wide variety of habitats within their huge distribution area. They are found mainly in brush-covered areas as well as in areas that have been built up and settled by man. The upper limit of their habitat is found at altitudes of around 2500 m. Multimammate rats thrive in cultivated areas, and they live in fields as well as in human habitations, especially in or near the ground floors of huts. Occasionally one also finds these animals in the upper stories of buildings like warehouses, as they are good climbers. In many areas these rats are unable to move into human habitations or they are forced to give up their attempts to do so when they run into heavy opposition from the black rats already in residence. In other areas, the black rats and the multimammate rats achieve a kind of "coexistence." Unlike the black rats, however, the multimammate rats are able to survive in the wild, and thus they can bide their time before making another attempt to move into the huts.

Multimammate rats spend the daytime in their subterranean burrows and cavities, or in cracks and crevices in rocks and the ground. They do not become active until nightfall. These animals are definitely social, and they prefer to live together in colonies. Burrows that were dug up in Kenya each contained an average of twelve animals; thirty-seven animals were found in one burrow. These rats will also live under garbage piles near human habitations, where they become omnivorous. When they are living in the wild and in fields, multimammate rats feed largely on grain and grass seeds, as well as on tubers, beans, other green materials, and insects. Researchers investigating the stomach contents of these animals found several specimens that were exclusively filled with insects. The multimammate rat is one of the most unpretentious of the African rodents; food procurement and the harsh, arid climate seem to give it no trouble.

The enormous fecundity of these animals is almost unique. Specimens of *Mastomys coucha* found in the most varied regions of Africa had an average of eight to thirteen young or embryos. Researchers found at least two females with nineteen embryos, the record number. The size of the litter depends on the age of the mother; younger females usually have smaller litters than do older members of the same species. When all the living conditions are good, a female multimammate rat may give birth every four weeks. Naturally this reproductive rate sometimes causes massive overpopulations. Researchers in a variety of African areas have found that the reproductive rate is greatest toward the end of the rainy season, when the plant growth offers the largest amount of food and cover. During the dry season, the multimammate rats are one of the last rodents to begin reproducing.

Young multimammate rats are altricial and are only 3–3.5 cm long at birth and they weigh about 2.2 g. The incisor teeth break through the gums around eight days after birth, and the eyes open after sixteen days. The young are weaned after twenty-two days. They are immature for a relatively long period. Generally, multimammate rats are not sexually mature until they are three months old. The population may decline just as rapidly as it developed, if the ground cover is destroyed by fire or drought, or by being harvested. When this happens, the enemies of the multimammate rat, the owls, serval cats, other small cats, and the snakes all have a regular feast. Only the hardiest rats, or the luckiest—those that have a hiding place somewhere—survive to preserve the population until more favorable times.

The multimammate rat is involved with the plague in several parts of Africa. Occasionally the rats themselves carry the plague bacillus, but usually it is carried by the fleas living in their hair. Infected multimammate rats may well transmit the disease as they reproduce so rapidly, move into human settlements from the wild, and move between human settlements. Although these animals are difficult to keep in captivity because of their timidity, their wildness, and their willingness to bite, they are often used as research animals in the study of diseases like the plague and schistosomiasis (or bilharziasis; a disease caused by a trematode). These animals are also used in cancer research, as cancerous growths are common in the proventriculus (glandular stomach) of the South African multimammate rats.

The HARSH-FURRED MICE (*Lophuromys*; BL 10–15 cm, TL 5–12 cm, weight 30–70 g) are quite different from the other typical mice of Africa in many respects, but especially in their hair and coloring. The tail is usually only half as long as the body, although there are also species, for example, the UGANDA HARSH-FURRED MOUSE (*Lophuromys woosnami*), whose tail is almost as long as the body. These animals usually have short, soft, erect hair which looks like fine bristles. The hair color extends down to its base, so the usual basal gray color is missing. The upper side of the body is dark brown, light brown, reddish-yellow, or blackish-red. The underside is multi-colored; it may be orange, for example, or even the beautiful red color of ground paprika. These animals are characterized by a musky odor. There are twenty-one forms, grouped together into about five species. These are found in central Africa from Liberia to Ethiopia and south to Angola and Malawi. The ETHIOPIAN HARSH-FURRED MOUSE (*L. flavopunctatus*) has the largest distribution.

Hanney studied this species in the area around Lake Nyasa, and when he examined stomach contents, he found that the majority of specimens were filled with insect remains and worms. Similar results were obtained in Uganda and in the Congo region. The harsh-furred mouse, like many of the other tropical mice, has its reproductive period during the rainy season, while the dry season constitutes a rest period. The usual number of young is between two and three. The newborn harsh-furred mice are almost hairless; they are 5–6 cm long and weigh about 5 g. There have been numerous indications which suggest that the initial growth of these young mice proceeds very rapidly. Harsh-furred mice in captivity require a diet of insects or meat; otherwise they become very thin and will not breed. They have a lively and combustible temperament; other mice (especially small mice) placed in the same cage are rapidly attacked, killed, and eaten. These mice will also fight among themselves, and not only in captivity, as the large number of injured tails on animals caught in the wild indicates.

Subfamily: Pouched rats

The POUCHED RATS (subfamily Cricetomyinae) have recently been separated from the typical mice. These animals bear some resemblance to the hamster in body characteristics and in lifestyle. Whether the pouched rats should be classified with the murid rodents or whether they belong to the cricetid rodents is still a debated question. There are three genera, with five species: 1. The GIANT POUCHED RATS (*Cricetomys*; BL 29–42 cm, TL 30–48 cm, weight up to 1500 g) are the largest members of this African family. They have large cheek pouches and large ears. There are two species. 2. The LONG-TAILED POUCHED RATS (*Beamys*; BL 12–19 cm, TL 11–16 cm) have two species. 3. The CAPE POUCHED MICE (*Saccostomus*; BL 11–17 cm, TL 4–8 cm, weight 60–120 g) are a grayish color with a variety of hues, usually brown, on the upper side of the body. The underside is white or yellowish-white. There is probably only one species.

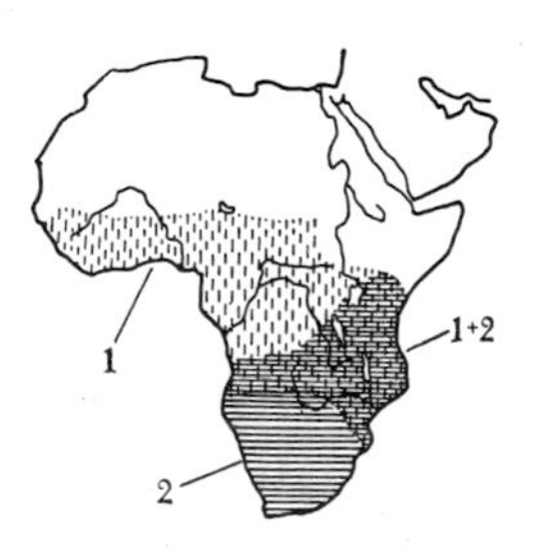

Fig. 11-79. 1. Giant pouched rats (genus *Cricetomys*); 2. Cape pouched mice (genus *Saccostomus*).

The GIANT POUCHED RATS are distributed over almost all of tropical Africa. The GAMBIAN POUCHED RAT (*Cricetomys gambianus*; see Color plate, p. 371) lives in open areas, such as savannas, and dry, wooded areas, according to Genest-Villard, while the CONGO or EMIN'S RAT (*C. emini*) is found in humid tropical rain forests. Both species dig burrows among the roots of large trees, under rocks, or simply on slopes in the woods. The burrows are equipped with between two and six entrances. Tunnels 2–3 m in length lead from the entrances to a centrally located nest room which may be built at a depth of as much as 1 m and which usually also

Fig. 11-80. Harsh-furred mice (genus *Lophuromys*).

serves as a supply room. Only rarely does a burrow have more than one nest chamber. These giant rodents may also make their burrows in termite mounds. The pouched rats use the sleeping chamber as a place to store everything they were able to carry back to the burrow in their cheek pouches from their last excursion, as well as anything they might have been able to carry back in their teeth. This includes seeds, nuts, the pulpy fruit of trees, green plant matter, tubers, and snails and beetles. This "hoarding urge" even carries over to inedible materials. Pens, earrings, pieces of chalk, and other man-made items have been found in these animals' burrows. Investigation of the pouched rats' stores and studies of their stomach contents have led to the conclusion that pouched rats feed largely on plant material, although invertebrate animals seem to make up a fairly consistent part of their diet. Ant and termite remains were also found in their stomachs. The animals that I [Dieterlen] kept in the Congo region preferred to eat nuts and other oily seeds, as well as bananas and similar plant matter. After several weeks, however, when I gave them animal matter for the first time in the form of many living, giant-sized beetles, they rapidly attacked this new treat and consumed it.

Giant pouched rats generally are nocturnal animals; they roam over their territories, looking for food in the dark. It has been determined that in some cases they climb trees; in one case, the rat did so in order to shake unripened fruits down to the ground, where they were later collected. Small objects are stored away in the cheek pouches with amazing rapidity; the edges of the mouth, the tongue, and the cheeks all work together to insure speed. The hands (fore paws) are only used when help with larger objects is needed. Once I acquired a pouched rat that had been killed by a local person; the animal's cheek pouches were so full that they remained open. In this case, the cheek pouches contained 275 Bridelia tree seeds, and each seed was as large as a coffee bean. The cheekpouches are also used as threat and vocal organs. When a pouched rat is highly excited it can blow out its cheek pouches, creating a muffled, repetitive sound.

The nose and the ear seem to be the principle orientation organs. Pouched rats use their own paths; these are narrow runways, 5–10 cm wide, which one can easily find on the floor of the jungle. The animals assume a characteristic posture when running, with the tail held upward in an oblique position. When one is lucky, one can even capture a pouched rat on the road when it has been temporarily blinded. These animals are almost helpless when carried by their tail. Local peoples hunt pouched rats in many ways, with booby traps, by digging up the burrows, and by using dogs to chase them into nets. These animals are eaten almost all over Africa, from Liberia to Mozambique.

Pouched rats that have been caught, particularly young ones, can become very tame. Their affection, the joyful squeaking, their eagerness to lick the human hand, the careful-timid behavior, and the seemingly modest intelligence all make them very loveable creatures. The gestation period, forty-two days, is longer than that of any other mouse species. Between one and four young are born, with an average of two or three. Newborn pouched rats weigh about 20 g, and they are blind and naked. They open their eyes after about three weeks, by which time they weigh almost 100 g. They are weaned after six weeks, and they remain with the mother for some time after that. Burrows with mother, newborn young, and half-grown young—thus with complete colonies—have often been found. The fathers, however, are rarely present. They prefer a solitary lifestyle. Giant pouched rats also live near human settlements, and they will even move into cities when conditions are favorable.

Another noticeable characteristic of the giant pouched rat is a flat, smooth, brownish-yellow insect, numbers of which can be found in the rat's fur. Wild pouched rats almost always have these parasites, although they soon disappear in captive animals. These orthopters are distant relatives of the cockroach, and belong to the genus *Hemimerus*; they are about 1 cm long, wingless, and completely specialized on pouched rat fur. They are not harmful to their host, but they do live on tiny skin flecks or skin fungi, or on food from the pouched rat storage chambers. They may, perhaps, live on both kinds of material.

▻
Garden dormice (*Eliomys quercinus*) live in light deciduous forests. Sometimes one can also find these animals in the fruit trees of gardens belonging to foresters' huts.

The CAPE POUCHED MICE (*Saccostomus campestris*) look very much like the common hamsters with their sturdy bodies, the wide, thick head, the cheek pouches, the short tail, and the small hind feet. The lifestyles of the Cape pouched mice and the common hamsters are also quite similar. Cape pouched mice live in savannas, prairies, and other dry areas. They prefer brush-covered land, grassy plains, and cultivated croplands. In some areas, these animals dig large burrows which have one or two escape holes and one or more nest and supply rooms. The pouched mice, like the hamster, live alone in these burrows, which they leave at night to look for food. Their food consists of seeds, green stuffs, fruit, beans, berries, various insects, termites, ants, and the like. Most of what they collect is stored in the burrow, where it is consumed. In cultivated areas, the Cape pouched mouse shows a preference for grain, and it can cause great damage to grain fields, particularly when, from time to time, large populations of these animals appear, as is the case in South Africa. The average number of young per litter is more than five; nine young have even been found in one litter. Hanney was successful in breeding this species. Newborn Cape pouched mice are sparsely haired and their development is slow. They open their eyes after three weeks, and they are weaned at five weeks of age. These animals do well in captivity, some having lived to be three years old.

Subfamily: Vlei and karroo rats

The subfamily of VLEI AND KARROO RATS (Otomyinae) contains only one genus (*Otomys*), with eleven species, of which seven are distributed throughout South Africa alone. This subfamily is characterized by its dental structure as well as by other features. The incisor teeth have longi-

1
2
3
4
5
6
7

◁
Eurasian dormice:
1. Forest dormouse (*Dryomys nitedula*); 2. Fat dormouse (*Glis glis*);
3. Common dormouse (*Muscardinus avellanarius*);
4. Garden dormouse (*Eliomys quercinus*). African dormice: 5. Common African dormouse (*Graphiurus murinus*); 6. Chinese pygmy dormouse (*Typhlomys cinereus*);
7. Desert dormouse (*Selevinia betpakdalaensis*).

tudinal furrows. The molars consist of transverse lamella, the number of which characterizes each individual tooth. The third molar of the upper jaw is greatly enlarged. These animals can be compared to the "vole type" in body structure and proportions. The body, depending on the species, is from 13 to 20 cm long. The tail is only half or two-thirds as long. The ears are round, relatively large, and half-hidden in the hair. The hair is usually soft, thick, and wooly, and ranges in color from gray to brown or reddish. Most of the species live in open areas, savannas, and prairies. In South Africa and Namibia (South West Africa) some of these animals even live in semi-deserts. Other forms are even found in the cooler mountain forests.

In central Africa, I [Dieterlen] was able to find out several things about the lifestyle of the VLEI RAT (*Otomys irroratus*; see Color plate, p. 372). This animal is a strict vegetarian, and it lives in dense plant growth consisting of grasses and weeds, throughout the year. These animals spend most of their lives above ground, although they also dig burrows. One can easily find their runways and burrows in the vegetation. Vlei and karroo rats must eat large amounts of green plant matter, and, as a result, they are active both during the day and at night. They can cause much damage in tree nurseries by occasionally gnawing off tree bark. These animals are not social, although in some areas they may live in dense settlements. The litter size is not very large. The average number of young per litter is 1.37, which means that most females give birth to only one young. This is compensated for by a continual reproductive period which lasts all year long for every year of the animal's life. Females are reimpregnated shortly after giving birth. In addition, the young are quite well developed at birth, being about 8 cm long and weighing 15–20 g. Their eyes are open, they have a full coat of hair, and they are able to run immediately after birth. Thus, in the subfamily of vlei and karroo rats we find another example of autophagous young within the family of mice (the first example is that of the spiny mouse). The KARROO RAT (*Otomys denti*), from the wooded African mountains, gives birth to the lowest number of young; the average number per litter is 1.07.

Subfamily: African tree mice

The subfamily of AFRICAN TREE MICE (Dendromurinae) includes eight very different genera. These animals live in the rain forests and in the dry areas of all of subsaharan Africa. Most of the forms are small, and their most important characteristic consists of certain distinctions in the shape of the molar teeth. Many scientists classify these animals along with the cricetid rodents (family Cricetidae). These animals were probably a relict group which appeared before the evolutionarily much younger typical mice, and which had many different forms living in a much larger distribution area than it now has. These animals have now adapted themselves to specific habitats, and their body structure and lifestyle have developed somewhat one-sidedly along the lines of this adaptation.

One of the best-known genera of African tree mice is that of the

Fig. 11-81. 1. Congo forest mouse (*Deomys ferrugineus*); 2. Vlei and karroo rats (genus *Otomys*).

AFRICAN CLIMBING MICE (*Dendromus*; BL 6–10 cm, TL 7–12 cm, weight 10–20 g). The hair is usually soft and woolly. The upper side of the body ranges from brownish or reddish to gray, often a long black stripe runs from the head or shoulder area all the way to the rear, although this stripe may occasionally be missing. The underside of the body is white or pale. The fore feet or hands have only three normally developed digits. There are at least three species, from which more than forty-four subspecies have been described. One of these species, the AFRICAN CLIMBING MOUSE (*Dendromus insignis*) is illustrated in the color plate on page 372.

This climbing mouse reminds one both of the common dormouse and, even more, of the northern birch mouse. It lives in open areas which offer good possibilities for climbing. They are particularly prevalent in savannas. African climbing mice have been found at altitudes of over 4000 m in Kenya. They spend most of their lives climbing in the grass, in bushes, in banana groves, and in similar shrubs. They build their nests between stalks or in groups of leaves. The nests are round, and are made of fine stalks or leafy material. Occasionally the climbing mice use abandoned weaverbird nests. Some forms of tree mice dig subterranean burrows, with a room and nest. The climbing mouse can often use its tail as a means of support and for gripping (however imperfectly). Its diet consists mainly of seeds and grains, as well as insects, and occasionally even berries, bird's eggs, and other foods. Up to eight young have been found in the nest of a single climbing mouse, although the average number of young is five. Newborn climbing mice are 2–3 cm long and weigh about 1 g. They are naked, and look much like fetuses. The young go through a long development process after birth. The incisors break through the gums around the fifteenth day of life, and the young mice open their eyes between the twentieth and twenty-fourth day.

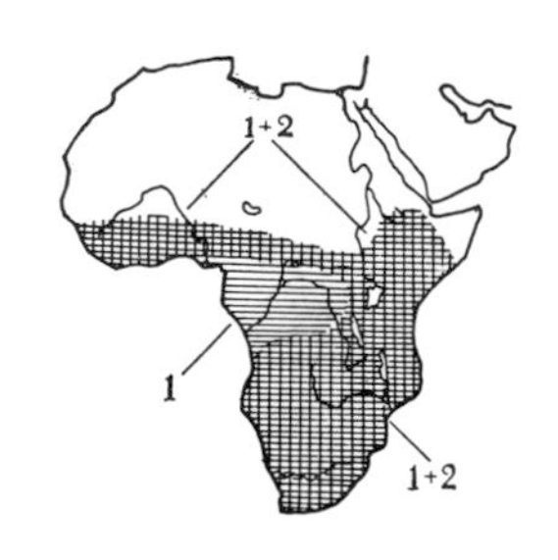

Fig. 11-82. 1. African climbing mice (genus *Dendromus*); 2. Fat mice (genus *Steatomys*).

The FAT MICE (*Steatomys*; BL 7–12 cm, TL 4–5 cm) are distinguished by their rather rotund body shape. They live in savannas and other open or partially rocky, partially open land areas throughout subsaharan Africa, with the exception of the rain forest regions. There are two or three species, including the FAT MOUSE (*Steatomys pratensis*; see color plate, p. 372), with a total of twenty-five subspecies identified.

The fat mice get their name from their ability to store reserves of fat in their bodies. They live partly in deep subterranean burrows which have a nest room that also serves as a supply room. Hanney found 445 plant buds in one *S. pratensis* burrow in Malawi. These animals feed largely on seeds, insects, and green matter. The average litter size is five young. Fat mice are nocturnal; their activity is probably limited only to seasons when food is plentiful, as they estivate for long periods whenever living conditions are unfavorable. This estivation is very similar to the resting period of hibernating animals.

Today we classify another rather extraordinary and varied species with the African tree mice. This is the CONGO FOREST MOUSE (*Deomys *ferrugineus*; BL 12–16 cm, TL 17–21 cm, weight 50–70 g; see Color plate, p. 372). The head is pointed and the molar teeth are cuspidated and tuberculate. The ears are large and the hind feet are rather long. The classification of these animals with the African tree mice is the subject of much debate. They have also been viewed as their own subfamily (Deomyinae).

The only species of this genus is found in the rain forest areas of tropical Africa, where they live on the ground, looking for food mainly during the nocturnal hours. Eighty to ninety percent of their food consists of insects, of which termites—a favorite with other animals, because of their quantity and fat content—make up the major portion. Ants, beetle larvae, worms, snails, frogs, and tree fruit pulp have also been found in these animals' stomachs. Another noticeable characteristic of this species is the limited number of young per litter; only one or two well-developed young are born in each litter. The Congo forest mouse, however, is able to reproduce almost all year long, and as a result it may give birth more often than do other forest-dwelling mice.

Subfamily: Cloud rats

The subfamily of CLOUD RATS (Phloeomyinae) contains some of the largest mouse species in the world. Zoologists have not yet agreed as to the validity of this family, its structure, or even as to the rank of several individual genera within the family. Some of the most important distinguishing characteristics of these animals include their dentition and particular skull or cranial features. The molars tend toward a lamelated structure instead of a tuberculate structure, and the lower molar teeth have a so-called "clover-leaf" pattern. The cloud rats are probably descended from the typical mice. All of these animals are climbers. Their distribution is limited to southeastern Asia. We differentiate six genera, with about twenty species.

Fig. 11-83. 1. Slender-tailed cloud rats (genus *Phloeomys*), the bushy-tailed cloud rat (*Crateromys schadenbergi*), and the shrew rat (*Rhynchomys soricoides*), 2. Water rats (genus *Hydromys*), 3. Monckton's water rat (*Crossomys moncktoni*), 4. Pencil-tailed tree mice (genus *Chiropodomys*).

The SLENDER-TAILED CLOUD RATS (*Phloeomys*) comprise two species, both of which live in the Philippines. One of these species, the SLENDER-TAILED CLOUD RAT (*Phloeomys cumingi*; BL 48–49 cm, TL 20–32 cm; see Color plate, p. 372) is the largest member of the mouse family. Its dentition and skull structure have been greatly transformed. The ears are small and the tail is relatively densely haired. The slender-tailed cloud mouse lives in the northwestern part of the Philippine island of Luzon, apparently in tree cavities.

Although the slender-tailed cloud rat was first discovered and described by Waterhouse, in 1839, and even though several specimens have been caught since then, we still know very little about the biology of this species. Slender-tailed cloud rats have been exhibited and bred in several zoos since 1947, and one born in the National Zoo (Washington, D.C., USA) lived over thirteen years.

The BUSHY-TAILED CLOUD RAT (*Crateromys schadenbergi*; BL 35 cm, TL 35–40 cm) has extremely long, thick hair. The tail also has a thick and bushy coat of hair, a unique characteristic among the mice. The hind feet have strong claws and are adapted to tree life. These animals have relatively small eyes and ears. The hair color varies considerably. Bushy-tailed cloud rats live in trees in the high mountainous areas of northern Luzon (Philippines). They are nocturnal, and feed on buds, bark, and fruits.

The PENCIL-TAILED TREE MICE (*Chiropodomys*; BL 8–13 cm, TL 9.5–15.5 cm) are classified in the same subfamily as the large cloud rats, although the two groups look very different. The tail is not prehensile, and it ends in a fine brush of hair. The feet of these arboreal animals are adapted for climbing. The hind foot is relatively short and wide. The inner tow is apposable. These animals have long vibrissae. There are at least five species, distinguished primarily by body size. One such species is the PENCIL-TAILED TREE MOUSE (*Chiropodomys gliroides*; BL 8–9 cm, TL 10–13 cm, weight 22 g).

These beautiful animals remind one somewhat of a dormouse. Harrison tells us that these animals prefer bamboo thickets on the Malay Peninsula and that they build their nests of leaves in the bamboo stalks. They live in the stalk between two knots; however, they gnaw through the walls of all the bamboo knots so that the whole stalk becomes a giant vertical tunnel, a swaying living area. These animals are apparently quite social, and they may live together in their bamboo towers. Pencil-tailed tree mice do relatively well in captivity, and with some luck and a diet of fruit, rice, and insects, they may also breed. Medway and Harrison have thoroughly studied the reproduction of these animals. Amazingly few young are born per litter. Three litters of only one young each and one litter of two young appeared in Medway's breeding colony. Females studied for possible embryos were found to have an averageof 2.2. The gestation period lasts nineteen to twenty-one days. Newborn pencil-tailed tree mice weigh about 3 g, and are naked, blind, and helpless. Their fur begins to grow in about the fifth day of life. The young are weaned after three weeks. They become full-grown at about three months. Pencil-tailed tree mice do not have definite reproductive seasons; pregnant females have been caught during every season of the year.

Subfamily: Shrew rats

The SHREW RAT (*Rhynchomys soricoides*; BL 19–21.5 cm, TL 14–15 cm; see Color plate, p. 372) is so unique that a special subfamily, the SHREW RATS (Rhynchomyinae), had to be created to encompass it. The dentition is greatly involuted; the upper incisors are very small, and the lower incisors are quite pointed. There are only two small molars in each half of the jaw, and these may possibly no longer serve any function. The total number of teeth is twelve, arranged: $\frac{1 \cdot 0 \cdot 0 \cdot 2}{1 \cdot 0 \cdot 0 \cdot 2}$. The snout area is long, and elongated much like the snout of a shrew. The eyes are small. The hair is short and like thick velvet. Sometimes these animals may have a white spot on their necks.

The shrew rat, which in many respects is quite rare, has occasionally been caught on Monte Data on the island of Luzon (famous for the many rare species of mice found there), in dense bush and humid wooded areas at altitudes between 2000 and 2500 m. The special dentition of these animals suggests that they feed largely on insects. Unfortunately, we know nothing further about their lifestyle.

Subfamily: Water rats

The WATER RATS (Hydromyinae) can be differentiated from the typical mice because of their dentition. The molar teeth in water rats have a much simpler structure and are hollowed out like cups. The third molar may be absent. There are eight genera spread over Australia, New Guinea, and the Philippines; each of these genera exhibits special adaptations to aquatic life.

The Australian–New Guinean WATER RATS (*Hydromys*; BL 15–29 cm, TL 17–29 cm, weight up to 1300 g) belong in this subfamily. These animals have soft, thick hair which is blackish-brown to golden-brown to dark gray on the upper side of the body. The underside is brownish to yellowish-white. The tip of the tail may be of varying lengths; it is white. The head and skull are quite broad, and flattened at the back. The small eyes are situated rather high on the head. The ears are small. The vibrissae are partially directed toward the front. The nostrils can be closed off, an adaptation to life in the water. The body is streamlined, gradually tapering off to the tail, which is particularly firm at its base. The fur is like that of a seal. The hind feet are broad and partially webbed by membranes. These animals have twelve teeth, arranged: $\frac{1 \cdot 0 \cdot 0 \cdot 2}{1 \cdot 0 \cdot 0 \cdot 2}$. There are three species: 1. The EASTERN OR GOLDEN-BELLIED WATER RAT (*Hydromys chrysogaster*; BL 28 cm, rarely up to 39 cm, TL 23 cm; see Color plate, p. 372), from Australia and New Guinea. Both sexes have different coloring. Males often have a golden-yellow abdominal region; this coloring may be linked to the seasons of the year. 2. The NEW BRITAIN WATER RAT (*H. neobritannicus*), found on New Britain Island, northeast of New Guinea. 3. The NEW GUINEA WATER RAT (*H. habbema*), found in the mountains of central New Guinea.

Water rats live, like the beaver, muskrat, or nutria, on the edges of rivers, river deltas, small ocean bays, lakes, and ponds. The water rats build their nests in the ground of river banks, on floating objects, in hollow trees when the habitat is boggy, or even in an old swan's nest depending upon environmental conditions. The burrows built into the bank of a river usually consist of several tunnels which generally contain two rooms, a sleeping room with a nest built out of twigs, grass, and tree bark, and a type of storage chamber. In the evening the water rat leaves its burrow to go out to look for food. Meat is the preferred food on the water rat's varied menu. These animals hunt mussels, crabs, snails, fishes, frogs, and even water birds like young ducks and full-grown coots; they also consume birds' eggs and plant matter.

Whenever water rats find something edible, they carry it to a particular place, a regular "dining area," some stone or tree trunk which sticks up out of the water, and there they eat it. An entire family of water rats was observed as it used such a stone for its base of action, from which the family dived into the water after bullfrogs. Water rats in southeastern Australia have a definite reproductive period which occurs during the southern hemisphere's winter and spring. The females give birth to four or five, or as many as seven, young. Water rats are sexually mature when they weigh about 500 g; it may be almost a year, however, before they are full-grown. Water rat fur is valuable, and has occasionally been substituted for muskrat fur. The water rat population was endangered by excessive hunting, and as a result it became necessary to protect these animals in certain areas. Aside from man, the water rat's chief enemy is the snake.

MONCKTON'S WATER RAT (*Crossomys moncktoni*; BL 20–21 cm, TL 22 cm), from New Guinea is even better adapted to an aquatic lifestyle. The fur is particularly watertight. The external ear is involuted, and only a few minute remains of it are present. The hind feet are webbed with large membranes. The tail is covered with hair and has a bristly seam on the underside that helps it work as a rudder. The bristly seam forks before it reaches the water rat's body. Monckton's water rats are so greatly specialized that the zoologist for whom they are named, Oldfield Thomas Monckton, believed that of all the rodents, these rats were the best adapted to aquatic life.

Superfamily: Dormice, by R. Piechocki

The DORMICE (superfamily Gliroidea) comprise a very old group of closely related rodents. We are familiar with fossilized remains of these animals from as early as the European Tertiary period. There were even some giant dormice which originated on islands in the Mediterranean Sea during the Upper Pliocene.

Distinguishing characteristics

The dormice range from the size of mice to squirrels (BL 6–21 cm, TL 4–16.5 cm). The head is pointed. As they are nocturnal animals, their eyes are large. The sparsely haired ears are usually almost naked. The tail's hair is sparse but evenly distributed on both sides. The fore feet have four distinct and well-developed toes for climbing; the hind feet have five toes also well developed for climbing. Even though these animals resemble the squirrels both in shape and in lifestyle, there is no close relationship between the two groups. There are four molar teeth in each half of the upper and lower jaws, and each of these teeth has permanent roots. There are twenty teeth, arranged: $\frac{1 \cdot 0 \cdot 1 \cdot 3}{1 \cdot 0 \cdot 1 \cdot 3}$ (with the exception of the spiny dormice and the desert dormice). None of the dormice, with the exception of the CHINESE PYGMY DORMOUSE (*Typhlomys cinereus*), which is one of the spiny dormice, has a caecum, a unique occurrence among the rodents. The dormice do not need a caecum because they feed largely on seeds and some animal matter. Those species living in the temperate zones hibernate for several months. There are three families: the TRUE DORMICE (Gliridae), the spiny dormice (Platacanthomyidae; discussed later), and the desert dormice (Seleviniidae; discussed later), with a total of ten genera and some thirty species.

Family: True dormice

All TRUE DORMICE (family Gliridae) have a soft, thick fur which consists mostly of woolly hair. The nasal area is always covered with short hairs; only the nostrils and the split in the upper lip remain hairless. The ears are rounded; they usually protrude above the hair, and are quite easy to see. Dormice are arboreal and primarily nocturnal. Consequently, they have an excellent sense of hearing. The feet are equipped with cushiony growths which help them grip the branches better. These animals vary their diet with the season. After waking from their hibernation, dormice feed on their last seed stores, on tender shoots and fruits which are found on the ground, and on any smaller animals that may be available. During the mating season and the gestation period, these animals feed largely on ripe, juicy fruits, insects, and snails. Just before the birth of the young, and just before hibernation, these animals feed on dry fruits, particularly nuts, which are rich in nutrition.

Subfamily: Eurasian dormice

We distinguish two subfamilies. The EURASIAN DORMICE (Glirinae) live in the temperate zones of the Old World from Great Britain to Japan and from central Sweden to northern Africa and Asia Minor. The AFRICAN DORMICE (Graphiurinae) are distributed over almost all of Africa, from the Sahara to the Cape province. The individual species, which are almost exclusively arboreal, are found in dense deciduous and mixed forests, in wooded prairie lands, and in gardens and park sites. In central Europe these animals may live in favorable mountainous areas at altitudes of up to 2000 m. The availability of food is largely responsible for determining the habitat. Eurasian dormice mate between May and July. The smallest dormouse species, the common dormouse (*Muscardinus avellanarius*), has a gestation period of twenty days; the gestation period of the largest species, the fat dormouse (*Glis glis*), is about one month. The young are born from June to August. The common dormouse gives birth to two litters a year; all the other species probably give birth only once a year. The young, of which there are between two and ten, are nursed for three to four weeks, and they do not become sexually mature until the following year.

The fat dormouse

The largest and most common dormouse is the FAT DORMOUSE (*Glis glis*; BL 13–19 cm, TL 11–15 cm, weight 80–120 g; see Color plate, p. 382), found over almost all of Europe and Asia Minor. In autumn, when the animals have fattened themselves up for hibernation, their weight almost doubles. There are about twelve subspecies, all of which are more or less distinct from one another in size and coloring.

Man has long been familiar with the fat dormouse. In ancient Rome these animals were fattened up and then eaten as a particular delicacy. A. Marcellinus, a Roman historian, angrily reports of how scales were set up along with meals so that fat dormice could be weighed. Notaries present at the meal had to note the weights and they then certified, under seal, the unusually large weights of those fat dormice to be eaten, much to the pride of the host. Varro (116–27 BC) provides us with a description of how these animals were kept in wild breeding areas (*gliarii*): "The glarium must be completely surrounded by a wall made of smooth stone or smooth mortar on the inner wall, so that the animals will not climb out. Inside the pen there should be bushes, those that grow nuts. If these bushes do not bear any fruit, then the animals must be fed acorns and chestnuts. In addition, holes for these animals to bear their young in must be made. They do not need much water, as they drink only a little, and are accustomed to living in arid areas. The fat dormice are fattened up in barrel-like pots like those in country houses; however, the pots used for fattening up these animals are made quite differently from the others. The animals make small pathways and a small hollow in the wall where they store their food. One feeds these animals large amounts of acorns, chestnuts, or other nuts. Consequently they become fat. The barrel-like pot is kept dark."

Fat dormice were eaten not only in ancient times, these animals still appear on the table in many areas. According to Vietinghoff-Riesch, at the beginning of the 19th Century the people of lower Carnolia (Yugoslavia), both rich and poor alike, considered the dormouse, which had a large population in that area, as a particular delicacy. Local people used the fat from these animals for frying (in years when the beechnuts were plentiful), and it was also highly regarded as butter. A century later the fat dormouse was still a prized meal in southeastern Europe, where it was roasted, broiled, and cooked with crackling. The Slovenians in southeastern Austria regard dormouse meat as a delicacy because it tastes of almonds and other nuts. In France, the fat dormouse is still regarded as a gourmet's dish, because its meat is soft, tender, and pleasantly flavored.

Fig. 11-84. Fat dormouse (*Glis glis*) in Europe.

Fat dormice have been held and bred in captivity in recent times, although not as food, but rather so researchers could find out as much as possible about these animals' lifestyle. The following descriptions are based on the work of Lilli Koenig and Vietinghoff-Riesch in 1960, among others.

Like many other nocturnal animals, the fat dormouse has an unusually good sense of hearing (by human standards). When listening, the dormouse alternately flips each of its outer ears back and forth twice per second, while the animal itself remains completely motionless. The sense of smell is equally well developed. Fat dormice are especially able to sense smells indicating food over long distances. The dormouse's most highly developed sense is that of touch. In addition to its vibrissae, which are up to 6 cm long, the fat dormouse has four sparsely haired touch pads on its face, one on its chin, and one on each of the lower arms. When exploring its surroundings, the dormouse stretches its vibrissae ahead of it and feels all objects; at the same time, or at alternating intervals, the animal moves both of its upper lips. In addition, it sniffs and repeatedly sucks in air with a light peeping sound. The very large, dark, forward-directed eyes are especially sensitive to crepuscular vision, so nocturnal orientation is done with the senses of sight and touch.

The fat dormouse is much more exclusively arboreal than the other dormice are; therefore, fat dormice are skilled climbers and can jump as far as 1 m. When climbing, they use their pointed claws to grip the bark. In certain cases the fat dormouse also clings to an object by exerting pressure with the side of the soles of its feet; the sticky secretions of the foot pads help the dormouse to cling simultaneously to all irregularities. These animals, in spite of their rather bulky appearance, are also rather adept at climbing on thin branches. Sometimes the tail is drawn up against the body when such maneuvers are in progress, although usually it serves as a "balancing beam" and helps equalize the animal's weight. When the fat dormouse rests itself, it usually sits directly in the middle of a branch, on all four feet, with its back curved; the tail is either laid across the branch or it hangs free, curved slightly inward.

My [Piechocki] first encounter with a fat dormouse was on the island of Crete, where these "tree cats," as the natives call them, were filling themselves with ripe figs and walnuts during the harvest season. Each morning the nocturnal visitors had vanished without a trace. Only the remains of large meals beneath the plundered trees indicated to the careful observer what had happened. In some years, the population of fat dormice on Crete is so great that no nuts can be harvested. The major distribution area of these animals is in the oak woods, at altitudes of between 800 and 1600 m. Horst Siewert, who was the first to discover the Cretan fat dormouse as a separate subspecies, has the following report: "I spent many evenings in those massive oaks in order to observe the fat dormice and to collect specimens. When the light of the setting sun faded, the branches of the leafy crown of the oak could be seen only as a darker mass against the light of the western sky, and the evening calls of the rock partridge broke the sudden stillness. One could also hear the first peeps of the fat dormice as they began to be active. Long peeping sequences with up to twenty-four sounds, which faded away at the end so the whole vocalization sounded like whimpering, alternated with short peeping sequences. One could also distinguish murmuring or grumbling sounds within the vicinity. Similar sounds could be heard all over the slope, and it became obvious that there were always several animals puttering around in any given tree. At the same time, small shadows would glide like ghosts through holes in the foliage, which, in the meantime, had become quite black. Here a branch would give slightly under the weight of the tiny climber; there a twig moved softly. A peeping sound and a light murmuring always accompanied the lively jumping and hunting that these animals did. Then, after about thirty minutes, one could clearly distinguish the gnawing of sharp teeth, the sounds of empty acorn rinds falling to the ground, the sound of a fruit that was perhaps poorly held, falling to the ground, and the rustling sounds of an animal looking for more food. In the meantime, the rest of the surrounding sounds died down as the whole dormouse population became involved in the search for and consumption of food."

In Germany's deciduous forests the fat dormouse most frequently stays in tall old oak trees. Normally these trees offer it sufficient food and holes in which to live. If there are not enough holes, these animals will occupy birds' nests. Fat dormice do not occur in coniferous forests, especially in the fir sections of the Black Forest. It is remarkable that these animals are never found in mature beech woods, above all in the Swabian Alps. If there are only a few oaks growing in a beech forest and numerous nest holes are available, the fat dormice will occupy only the nest holes in the oak trees.

"It is obvious," reports Löhrl, "that the smooth bark on beech trees makes climbing difficult. The fat dormouse, however, apparently is quite agile in its movements in the crowns of beech trees. One can best observe the motion orientation and climbing patterns and capabilities of these animals, when one forcibly removes a female from her young and releases her some distance away; usually the female rapidly climbs a tree and then moves branch by branch, from tree to tree, until she comes to the nest tree and the nest branch. With this example one can also see that the nest box entrance hole, with its width of 32 mm, is just wide enough to accomodate an adult fat dormouse. However, the animal must enter the nest slowly and with difficulty; it must stretch its back legs to the rear in order to move its thighs through the hole. When the nesting box is made of soft wood, the fat dormice help themselves by widening the entrance hole; when the cavity is built of hardwood, however, the animals make only modest attempts at enlarging the hole." Although fat dormice like to live in artificial holes, and although they will also raise their young in such places, they do not hibernate in these holes, at least not in our latitudes.

Unlike the hamster, the fat dormouse hibernates for a long time. As soon as the fat dormouse has stored away enough fat for a long sleep, and as soon as it feels the time is right, it retreats into a suitable hideout; hollowed-out tree holes, woodpecker holes in large deciduous trees, or even rock crevices may be used for this purpose. When these holes do not have any humus or pulp, the dormice line them with grass and leaves. Fat dormice have often been found sleeping in the mulch of pruned (pollard) willows, as well as in hunting lodges, bee hives, barns, or similar buildings near woods. Most fat dormice dig burrows in the ground, where they hibernate at a depth of between 50 cm and 1 m. Often several animals will sleep closely packed together in one hole. Usually they sleep on their backs with the tail drawn up over the head and abdomen; their eyes are closed, and the pinna is folded over the auditory canal. In this way, the animal manages to reduce its external area so that it needs much less energy. Eisentraut has said about the biological significance of hibernation: "Just as the boiler of a central heating system that is turned on 'low' uses very little fuel and consequently gives out very little warmth to the rooms, so the hibernators, with their decrease in body temperature, need only a minimal amount of nutrition; thus the fat supply stored in these animals' bodies is sufficient to maintain their few active life functions. As a result, each animal is relieved from worrying about where it will get external food."

The earliest the fat dormouse begins its hibernation is by the end of September or the beginning of October; often hibernation begins even later, depending on the availability of food. These animals wake up from their hibernation in the first half of May, by which time they look quite thin. As soon as the dormice have sufficiently regained their weight, the reproductive drive dominates their behavior. Both sexes make a husky "viiiii–viiiii–viiiii" sound which carries well. Squeaking sounds of two to three seconds in length follow immediately upon one another so that there is an average from fifty to one hundred of them, sometimes 300 or more, lasting ten minutes or even longer in the night. According to Lilli Koenig, these sounds are reminiscent of a tracing signal. These long vocalizations, which can be heard out of doors in the summertime within these animals' distribution area, seem to help the partners find one another as each animal gives its location. In addition, fat dormice delineate their territories by depositing smell marks. Whenever a fat dormouse wishes to deposit its "chemical house signs," as Eibl-Eibesfeldt calls the odor marks, it presses the back part of its body firmly against the object to be marked. Then, sniffing, it lowers its nose down to the ground, keeps its tail toward the back in a slight S-curve, presses its heels close to its body, and dribbles forward with a few short steps. Thus, long tracks result; these are sometimes even clearly distinguishable as moist lines of glandular secretions which indicate to other dormice the fact that they are trespassing on occupied territory.

The sexually aroused male courts the female almost unceasingly with soft, vigorous, twittering "zizizi" sounds until the female finally allows the male to mount her. The female begins to isolate herself from the other members of the species only a few days after the copulation; she brings nesting material, preferably green foliage, into her hole. Between three and ten naked and blind young are born after a gestation period of thirty days. The young are nursed until their eyes open, which takes at least thirty days. Because it is rarely possible for anyone to become familiar with the "intimate behavior" that goes on within the fat dormouse's nest in the wild, we will rely on the observations that Lilli Koenig made of her captive animals.

The mother sits over her young with her legs spread wide; the young usually lie on their backs while nursing, held together by the mother's legs as if they were in a box. Sometimes the mother will cover her young in front with her tail. When the nest becomes too warm, the mother lies on her back and allows the nursing young to crawl all around her. When she is disturbed, the mother immediately turns herself over and covers her young protectively. As the young begin to grow larger and livelier, the nursing position becomes more and more uncomfortable for the mother. With her legs outstretched, the mother stands above her rowdy brood like a bridge, while they lift, push, and shove her almost without pause. The mother makes no move to escape, but remains in the middle of the fray, uninvolved, with her eyes half-closed.

The mother often licks her young in the mouth, even in the first few days after birth. The older, larger young eagerly reciprocate this service by lying on their backs, stretching their tongues away from their mouths, and licking the mother's heavily salivated oral cavity for up to one minute or longer; sometimes this action is accompanied by a loud squealing on the part of the young. The mother shuts her eyes and lets the young lick her with pleasure; sometimes she reciprocates the licking, while she firmly holds onto the young with her forefeet. This unusual behavior becomes evident in the young long before other grooming procedures; the young, however, only display this behavior in conjunction with the mother, never among themselves. One can clearly see how eagerly they lick up the mother's saliva; this saliva probably contains important materials which the mother initially gives directly to the young, but which later she allows them to take from her. The mother often licks her young on their mouths, faces, and all possible spots. When the young are licked in the anal region, they give up urine and feces, which the mother then consumes.

During the first few days after birth, the mother leaves the nest only when she is completely undisturbed. Later, once the young begin to grow a coat of hair, the mother makes short excursions; she covers over her young with rubbish before she leaves, however, except when it is particularly hot outside. When she returns to the nest, she approaches it very carefully and then slowly uncovers her young. The mother keeps the father away from the nest until the young have a complete coat of hair; she lets the father into the nest after about the sixteenth day. The male eagerly cleans his young and defends them from all intruders. The young are able to climb at that time; sometimes they follow the father to other nesting boxes, where they remain with him until hunger drives them back to the mother. Often a family will remain together throughout the winter, breaking up as the mating season approaches.

We should note here, with restrictions, that male dormice in the wild may not be as involved with the raising of the young as are the captive dormice which we discussed here. This is because in the wild, the males may often become sexually aroused during the summer when they run after any or every female in heat. Thus, when Lilli Koenig checked her control nesting boxes, she did not find a single adult male in the nest with the young.

As long as fat dormice do not become too numerous, they are, from man's point of view, harmless members of our animal world. The dormice move into fruit gardens and vineyards only in times of poor food supply, when there is a shortage of acorns, beechnuts, and chestnuts. They cause much damage in fruit gardens by biting into unripened fruits as well as eating ripened fruits. According to Heptner, there are an average of thirty fat dormice per hectare of land in the eastern regions of the northern Caucasus; these animals eat about 1 t of pears, about eighteen percent of the total crop. Thus in that area, the fat dormouse is responsible for an estimated annual crop damage of 25,000 t, worth some 3 million rubles. One fat dormouse in the Harz Mountains (Germany) was even able to open jars of preserves, and it ate all the plums contained in them; shortly thereafter, the dormouse was caught and died in a trap, set for what the people thought was a rat.

The common dormouse

The smallest dormouse, the COMMON DORMOUSE (*Muscardinus avellanarius*; BL 6–9 cm, TL 7–7.5 cm, weight 15–40 g; see Color plate, p. 382) is often described as a "young red squirrel" by people who are not familiar with these animals. The animals' weight varies widely depending upon the availability of food. The tail is bushy, unlike tails of the mice, and it has no scales. The black, pearl-like eyes and the short ears give the common dormouse a "doll-like" expression. These animals like warmth.

Fig. 11-85. Common dormouse (*Muscardinus avellanarius*).

The common dormouse prefers to live in hazel bushes, as hazelnuts are an important part of its diet. These animals also live in blackberry bushes and in deciduous forests with a large amount of undergrowth, and, more rarely, in beech or pine woods. They like to settle near paths through the woods or by small clearings in the trees. The easiest way to find a common dormouse is to look for their spherical nests made from grass, leaves, and moss, which are built 1–2 m above ground; one can also look for these animals in bird nesting boxes. The nests of young animals have a diameter of 6–8 cm, as do the nests of single animals. Family nests have a diameter of 12 cm. Usually young born early in the year are born in nests in the ground. Young born later in the year are born in nests in trees. As soon as the young are six or seven weeks old, they become independent. When the temperature drops to below 15–16°C in the fall, the common dormice become inactive and they go into hibernation in holes in the ground, under the leaf cover. Their body temperature sinks to 1°C during this time. When the winter is particularly rough and, as a result, the ground freezes, many hibernating common dormice are killed. The common dormouse begins to be active again in May, after the trees have grown new leaves. These animals usually live through two, or a maximum of four, summers before they die.

The garden dormouse

The GARDEN DORMOUSE (*Eliomys quercinus*; BL 11–17.5 cm, TL 10–13.5 cm; see Color plates, pp. 381 and 382) is somewhat smaller than the fat dormouse. They are easy to identify because of their brightly colored fur. Garden dormice living around the Mediterranean and in northern Africa do not have a uniform color on the undersides of their tails; instead this area is multi-colored. These animals are found in Germany's Black Forest up to rather high altitudes.

Contrary to what one might expect from the name, the garden dormouse does not live in gardens, but rather in light stands of deciduous wood, as well as in pine forests. Occasionally one can also find these animals in alpine huts, in block houses, or in foresters' huts. The nest is built in rock crevices, in tree holes, or, more rarely, in nest boxes. Unlike the other dormice, Löhrl found the garden dormouse rarely, but regularly, hibernating in nest boxes. The nests are not built of leaves like those of the fat dormouse, but rather of cut grasses and moss; they are particularly large and thick.

Fig. 11-86. Garden dormouse (*Eliomys quercinus*).

According to Brehm, garden dormice should be regular monsters, "dumb, raging at night, and untameable." K. Zimmermann, on the contrary, found that even an older garden dormouse will soon begin to relate to people, watch them, react when it is called, and even eat out of one's hand during the daytime. Zimmermann, unexpectedly, found no rank order (of the type observed in rats or in groups of house mice) among the garden dormice, even though large numbers of these animals may live close together near common sleeping quarters or feeding areas. One can bring two strange groups of garden dormice together at any time—with the exception of the mating season—and they will not fight or even become agitated. This would be impossible if one tried it with any species of mouse or microtine rodent.

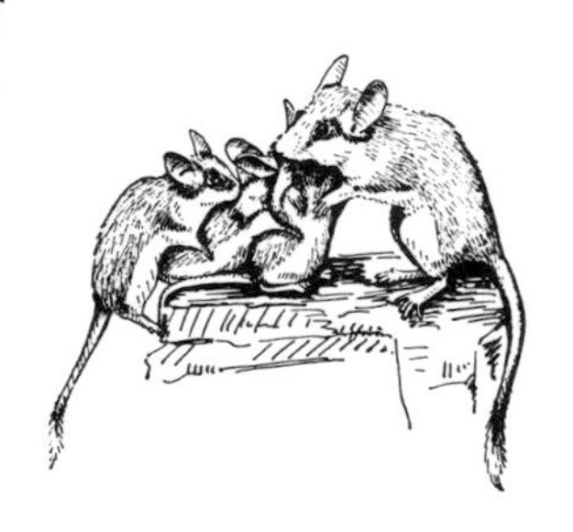

Fig. 11-87. "Caravan" formation in the garden dormouse.

Another distinguishing feature of the garden dormouse, and one that sets it apart from the other species, is the manner in which the mother "leads" her young; this particular behavior is also to be observed in the shrews (*Crocidura leucodon*). According to Kahmann's observations, young garden dormice form "caravans" until they are sixty days old. These caravans consist of groups of two or three young, each of which immediately follows the animal directly in front. The mother leads the caravan, and thus is able to bring her young back to the nest. When a member of the family is attacked, it is able (as are the other dormice) to shed or lose the skin of its tail so that only this tailskin remains in the possession of the enemy. The dormouse then bites off the tail vertebra once the tail has dried and begins to itch; the tail vertebra is not vital to the dormouse.

The forest dormouse

The FOREST DORMOUSE (*Dryomys nitedula*; BL 8–10 cm, TL 8–9 cm; see Color plate p. 382) is an even smaller animal. The ears are small and they protrude above the hair only slightly. The tail is bushy on both its upper and lower side.

Although the forest dormouse is typically an occupant of deciduous forests, it is also found in pine forest of Byelorussia (White Russia) and in mountainous areas. Renate Angermann, who has studied the ecology and biology of this species, reports that forest dormice have been found in rocky meadows at altitudes of 2500–3500 m. This species will also occasionally move into cultivated areas, where it can be found in alpine huts, fruit plantations, gardens, or even regular human habitations. Contrary to what was formerly believed, the forest dormouse prefers animal

Fig. 11-88. 1. Forest dormouse (*Dryomys nitedula*), 2. Betpakdala dormouse (*Selevinia betpakdalaensis*).

matter during the summer; it begins feeding largely on seeds during the fall. We will list the favorite foods of these animals, in order of preference: butterflies, smooth caterpillars, larvae, beatles, sparrow's eggs, ripe sweet berries, acorns and sunflower seeds, and, finally, young sparrows. This preference may be altered depending upon the taste of individual animals.

The forest dormouse prefers to live in free-standing nests which it builds itself or in free holes or bird nesting boxes. In Europe, the female forest dormouse gives birth to one litter a year, usually during the second half of June. There are between two and six young per litter; they become independent four to five weeks after birth. Females living in the evergreen forests of Galilee (Israel), where the forest dormice are active all year long, give birth to two or three litters a year, with an average of three young per litter. When they studied the forest dormice in that area, Nero and Amir were unable to find newborn young only during the months of January and February.

The Japanese dormouse

The JAPANESE DORMOUSE (*Glirulus japonicus*; BL 6.5–8 cm, TL 4–5.5 cm) is the easternmost representative of this family. These animals are very similar to the common dormouse, both in body size and shape. The hair is soft and a faded olive-brown color on the upper side of the body; there are black stripes on the animal's back. Japanese dormice live in the mountain forests on the Japanese islands of Honshu, Shikoku, and Kyushu between altitudes of 400 and 1800 m above sea level. (One animal, thus far considered an exception, was caught at an altitude of 2900 m.)

Japanese dormice feed on fruits, seeds, insects, and birds' eggs. The nest is usually built in the branches of a tree; it is lined with lichen on the outside and bark on the inside. The young are born in June or July; there is an average of three to four young per litter, but in rare cases there may be up to seven. Occasionally, but rarely, a second litter may be born in October. The gestation period lasts for about one month. A dormouse was once found half asleep in the snow of a gorge in the Japanese Alps, in July. The animal was put in a box, where it awoke after some minutes and disappeared. Usually hibernation lasts only for as long as the average monthly temperature remains below 8.8°C.

The mouselike dormouse

The MOUSELIKE DORMOUSE or ASIATIC DORMOUSE (*Myomimus personatus*) was until recently known to us only from central Asia; it has, however, also been found in the rocky areas beyond the Caspian Sea. Surprisingly enough, this species was caught in Bulgaria for the first time in 1959. The mouselike dormouse might well have been discovered twenty-five years earlier, had zoologists been more alert, as R. Angermann recently found a young, unidentified Asiatic dormouse skin among the garden dormice at the Berlin Zoological Museum; this young animal was caught by G. Heinrich, a well-known researcher and traveler, on 14 July, 1935, near Nessebar, north of Burgas, on the Bulgarian part of the Black Sea coast.

The subfamily of AFRICAN DORMICE (Graphiurinae) has one genus,

Subfamily: African dormice

that of the AFRICAN DORMICE (*Graphiurus*; BL 8–16.5 cm, TL 8–13.5 cm), with some twenty species. The members of this subfamily are found in the wooded areas of subsaharan Africa, and from Sudan to the Cape province. The hair is thick and woolly. The tail is relatively short and has a fairly even coat of hair, except on the tip.

The smallest species of this genus is the PYGMY DORMOUSE (*Graphiurus nanus*; BL 8 cm). The COMMON AFRICAN DORMOUSE (*G. murinus*; see Color plate, p. 382) has many subspecies, distributed over wide areas of Africa. African dormice are found in river valleys in southern and eastern Africa, as well as in the rocky sections of arid highlands.

▻

Jumping mice: 1. Northern birch mouse (*Sicista betulina*); 2. Meadow jumping mouse (*Zapus hudsonius*); 3. Woodland jumping mouse (*Napaeozapus insignis*). Jerboas: 4. Pygmy jerboa (*Salpingotus crassicaudatus*); 5. Five-toed dwarf jerboa (*Cardiocranius paradoxus*); 6. Northern three-toed jerboa (*Dipus sagitta*); 7. Desert jerboa (*Jaculus jaculus*); 8. Five-toed jerboa (*Allactaga jaculus*).

Even though the African dormouse is one of the most common nocturnal animals in Africa, we know very little about its lifestyle. If one turns to the available literature on this subject, one will repeatedly find only the observations of G. A. K. Marshall, who noted that pygmy dormice have the unusual habit of using the nests of large communal spiders (*Stegodyphus*) for their dens. Marshall was unable to determine whether this was a standard practice of pygmy dormice, although he was able to trap these animals three times as a result of this habit. Other species build themselves the usual sort of nests, in which they store food supplies. The African dormice feed on seeds, fruits, and insects; they will also eat young reptiles and birds.

Eisentraut and his associates found the subspecies the CAMEROONS DORMOUSE (*Graphiurus murinus haedulus*) in the misty forest belt of Cameroon Mountain, at an altitude between 1700 and 1900 m. The traps were set in old, knotty trees on thick horizontal cavities at a height of about 6–10 m above the ground, a fact that may well support the idea that these animals live almost exclusively in the middle or upper areas of trees. These animals are also active during the day in dark forests. Their vocalizations are apparently loud and shrill. Several litters of between two and five, four on the average, young are born in a year.

Family: Spiny dormice

We know very little about the life of the SPINY DORMICE (family Platacanthomyidae). These animals have a mouselike build. The hair is interspersed with smooth bristles. The tail ends in a brush. These animals have sixteen teeth, arranged: $\frac{1 \cdot 0 \cdot 0 \cdot 3}{1 \cdot 0 \cdot 0 \cdot 3}$. There are two genera, each with one species: 1. the SPINY DORMOUSE (*Platacanthomys lasiurus*; L over 25 cm, TL at least 10 cm of this), which is relatively large. One old female weighed 75 g. The fur on the upper side of the body is reddish-brown. The undersides are whitish. The hair on the back is interspersed with broad, flat quills. The tail is bushy. These animals live in large trees in the Anamalai Mountains on the Malabar coast of southern India, at altitudes of between 600 and 900 m. 2. The CHINESE PYGMY DORMOUSE (*Typhlomys cinereus*; BL 8.5 cm, TL 9.5–13.5 cm; see Color plate, p. 382) is smaller. These animals are found in southeastern China, where they live in dense mountain forests in Fukien Province at an altitude of between 1200 and 2100 m. A somewhat larger subspecies is found in Tonkin (North Vietnam).

1
2
3
4
5
6
7
8

1
2
3
R.B.

The SPINY DORMOUSE is often called the "pepper rat" because it has a preference for eating pepper plants. It also feeds on *Arctocarpas*, and, like the other animals in its environment, it also drinks fermented palm juice. According to Sanderson, this small animal with its large, pointed ears, is unusual insofar as it gnaws at a knothole high in a tree, widening it until it is large enough for the dormouse to live in. The nest is lined with soft plant material from the luxuriant green growth (Epiphyten) found in the tree tops. Spiny dormice are so numerous in many areas that they are hunted down because of the damage they do to pepper plantations. We know even less about the small CHINESE PYGMY DORMOUSE. According to the reports of people living in the areas where they are found, cats will not eat these dormice.

Family: Desert dormice

The BETPAKDALA DORMOUSE (*Selevinia betpakdalaensis*; BL 8 cm, TL 6 cm; see Color plate p. 382 and distribution, Fig. 11-90) is the only representative of the family of DESERT DORMICE (Seleviniidae). These mouse-sized animals look rather plump because the hair is 10 mm long. The ears are large and round. The tail is sparsely haired. The incisors are very large and have deep longitudinal furrows. There are sixteen teeth, arranged: $\frac{1 \cdot 0 \cdot 0 \cdot 3}{1 \cdot 0 \cdot 0 \cdot 3}$. The molars are unusually small and very short. The teeth are so low that they barely protrude from the gums.

The desert dormouse was first discovered as the sole representative of a new rodent family in 1938; it was named after W. A. Selevin, who contributed greatly to the exploration of central Asia, and after the place where it was found. A. N. Formosov has concluded that the Kazakhians, who lived as nomads in the desert, had long been familiar with the desert dormouse, which they called *Kalkan-Kulak* or *Shalman-Kulak*. Both names are very appropriate, pointing out the major characteristics of these animals, their large, round ears. *Kalkan-Kulak* means "shield-ear."

◅ Old World Porcupines: 1. Sumatran thick-spined porcupine (*Thecurus sumatrae*); 2. Crested porcupine (*Hystrix cristata*); 3. Malayan porcupine (*Acanthion brachyura*).

The desert dormouse is able to climb around branches with great agility, although it cannot jump higher than 20 cm. It moves across flat surfaces in an unhurried amble, and when frightened it will make short jumps. In captivity it feeds exclusively on insects, while in the wild it prefers to eat grasshoppers. The desert dormouse is very similar to the southern birch mouse with respect to its harmlessness. It will not bite when one catches it with one's bare hands in the wild. Another important characteristic of these animals is the fact that they do not build nests or burrows during the summer. The individual animals pursue a nomadic existence similar to that of the common hare. Desert dormice begin their hibernation in the fall, when the temperature reaches 5° C.

Superfamily: Dipodoidea, by R. Piechocki

The members of the superfamily Dipodoidea are apparently, from an evolutionary point of view, a very old rodent group. These animals usually have a graceful build, and they range in size from very small to squirrel-sized, at the most. The hair is silky-soft. The elongated hind legs can be quite long. The tail is round and is as long as the body. The upper lip is not cleft, so there is no "harelip." These animals are largely nocturnal; they have relatively large eyes. All of the species within this superfamily hibernate in ground burrows. There are two families: The JUMPING MICE (Zapodidae) and the JERBOAS (Dipodidae), with a total of sixteen genera and some forty-five species.

Family: Jumping mice

The JUMPING MICE (family Zapodidae; BL 5–10 cm, TL 6.5–16 cm, weight 8–25 g) look very much like mice. The hind legs are long, or at least elongated (in birch mice they may be 1.4–1.8 cm long, and in true jumping mice they may be 2.5–3.4 cm long). These animals are good runners and agile climbers. The tail is round, covered with hair, and longer than the body; it is usually pressed against objects the animals has moved or disturbed. The tail can also be used to some extent as a prehensile organ when it is wrapped around plant stalks; it is not strong enough, however, to support the animal in a hanging position. The mating season lasts from May to June. The gestation period lasts four to five weeks. There is only one litter a year, consisting of between two and seven young. Jumping mice build spherical nests on the ground or in small bushes. There are two subfamilies, the birch mice, found in Europe and Asia, and the true jumping mice, found in Asia and North America.

Subfamily: True jumping mice

The subfamily of TRUE JUMPING MICE (Zapodinae) includes: the MEADOW JUMPING MICE (genus *Zapus*), pale brownish-gray animals, found on the prairie lands of North America; the bright-reddish WOODLAND JUMPING MICE (*Napaeozapus insignis*; Color plate, p. 399), also found in North America, where they prefer cool, wooded areas with dense undergrowth; and the CHINESE JUMPING MOUSE (*Eozapus setchuanus*), which lives near riverbeds in cool, high altitude woodland areas. Like all other jumping mice, these forms feed largely on grasses, insects, and berries. American jumping mice will even attack tiny birds. Like the birch mice, the true jumping mice are able to store fat supplies within their bodies and thus are able to survive periods of bad weather, as well as hibernation, without food. Hibernating meadow jumping mice (*Zapus hudsonius*; see Color plate, p. 399) were found sleeping in spherical nests made of grass at a depth of 30–65 cm below the ground surface.

Subfamily: Birch mice

The BIRCH MICE (Sicistinae) are fairly well known in Europe because the northern birch mouse belongs to this group. These animals may be found in eight different flatland habitats, as well as in lower mountainous areas. Within these habitats, birch mice live in pine and deciduous forests, particularly birch or mixed birch forests. They may also be found in wooded bogs, in humid wood and peat meadows with brush cover, in arid regions, and occasionally around grain fields on sandy soil. The NORTHERN BIRCH MOUSE (*Sicista betulina*; see Color plate, p. 399) is found in central Europe; the SOUTHERN BIRCH MOUSE (*Sicista subtilis*) is also found there occasionally. Both of these species can be distinguished from other species because of the distinct black stripe running down the back.

The northern birch mouse

The northern birch mouse is rather small (BL 5–7 cm, TL up to 10 cm). Before hibernation these animals weigh about 12 g; after hibernation the

Fig. 11-89. 1. Meadow jumping mouse (*Zapus hudsonius*), 2. Pacific jumping mouse (*Z. trinotatus*).

Fig. 11-90. 1. Western jumping mouse (*Zapus princeps*), 2. Woodland jumping mouse (*Napaeozapus insignis*).

weight is 6 g. Northern birch mice have a superficial resemblance to the striped field mouse, although the birch mouse's black back stripe begins at the back of the head. The ears are 10–15 mm long; they may be folded up or fully extended. The northern birch mouse can be distinguished from the striped field mouse by two other important points, its mobility and its stubbed nose.

In 1942 Walter von Sanden wrote the following description of the excitement provoked by the capture of northern birch mice. This description appears in his book, *Everything for a Mouse*:

"My living northern birch mouse soon became quite well known, and letters came to me from various places wanting the mouse. I was sent the most beautiful empty transport boxes, free special-delivery bills of lading, and printed labels with the words 'Live animal' in red and green. While the fur and skeleton, and life and death of the birch mouse were being divided among people all over the world, the animal itself lived peacefully and quietly, with the best of appetites, in its silent realm. Its physical condition was excellent. When I lifted it up in my hand, it was noticeably heavier. Like the badger, the birch mouse ate well in the fall and stored fat in its body which would nourish it during the long hibernation."

Today we know that the northern birch mouse has the longest hibernation period of any mammal; its hibernation may last up to eight months. Moreover, the birch mouse falls into a lethargy during the day, and during the activity period as well; such lethargic conditions were previously encountered only in bats. It is no wonder that zoologists are extremely interested in these small mice. In 1964 a northern birch mouse was caught for the first time in the Alps, at an altitude of about 1700 m. Southern birch mice have also been found in Austria, but rarely. Wettsein has suggested that both species are remnants of the Ice Age, and that both are threatened with extinction in the near future.

The northern birch mouse can be very pleasant in captivity. Rokitansky appropriately characterizes the behavior of this unusual animal as follows: "This charming creature—famed from the very beginning—was the delight of all my visitors; even people who hated mice could not resist its charm—I am unable to think of a better expression. It never tried to bite me. I should also like to mention its complete trust in its keeper, even when the latter had something disturbing." When one held it in one's hand, it would eat without fear.

Family: Jerboas

Distinguishing characteristics

The JERBOAS (family Dipodidae; BL 4–15 cm, TL 7–25 cm), with their kangaroo-like shape and their long hind legs—made for jumping—are clearly distinct from the jumping mice. The elongated metacarpal bones of the hind feet are fused to become single metatarsus bones. The hind feet have three to five digits each, depending on the species. The fore feet are small. Jerboas have long, stiff hair bristles on the soles of their hind feet; these prevent the animals from sinking into the fine desert sand, and they also make it easier to throw away excess sand during the digging of the burrow. The tail is long, and it is often pressed into the ground like a handle, for support; the jerboa is then able to sit upright, which enables it to have a wide view of the open land. The tuft of hair at the tip of the tail is almost like a flag; the tail is used as a type of steering apparatus during the animal's jumping, almost its only style of movement. The vibrissae are often as long as the body, and they touch the ground during the jeroboa's leaping runs. The cervical segment of the vertebral column tends to be shortened and—with the exception of the first neck vertebra (atlas)—there may be a complete fusion or a tendency toward such a state. Jerboas have unusually good senses of sight, smell and hearing. The external ears are long and, as a result, these animals can hear the slightest sound during their nocturnal life in deserts and prairies. Studies of the jerboa brain have revealed that the optic center is particularly well developed.

Fig. 11-91. Birch mice (subfamily Sicistinae).

Jerboas are typical inhabitants of arid regions, the prairies and deserts of northern Africa, Arabia, the Middle East, and central Asia. They feed largely on plants and seeds in these areas, although they will also supplement their diet with insects, particularly beetles. They are able to smell their food from some distance away. Desert and semi-desert grasses soon dry out; therefore, jerboas will also eat blossoms, leaves, and thin twigs of permanent plants. In order to reach the often-densely overgrown bushes and trees, some jerboa species will jump several meters and then hold fast to a branch with their incisor teeth and fore feet. The animals use their tails and hind feet as climbing supports, and thus are easily able to reach the fresh food source. Jerboas also occasionally look for their food on the ground, in doing which they may cause much damage when they eat onion plants, a favorite food.

Fig. 11-92. Northern birch mouse (*Sicista betulina*).

We recognize three subfamilies: 1. The JERBOAS (Dipodinae), from Asia, Africa, and Europe, with nine genera, including the NORTHERN THREE-TOED JERBOA (*Dipus*), the COMB-TOED JERBOA (*Paradipus*), the DESERT JERBOA (*Jaculus*), the THICK-TAILED THREE-TOED JERBOA (*Stylodipus*), the FIVE-TOED JERBOA (*Allactaga*), the LITTLE EARTH JERBOA (*Alactagulus*), and the FAT-TAILED JERBOA (*Pygeretmus*); 2. The DWARF JERBOAS (Cardiocraniiae), from central and eastern Asia, with two genera, the FIVE-TOED DWARF JERBOAS (*Cardiocranius*; see Color plate, p. 399), and the PYGMY JERBOAS (*Salpingotus*; see Color plate, p. 399); 3. The LONG-EARED JERBOAS (Euchoreutinae), with only one (particularly noticeable) species from Sinkiang (China) and central Mongolia, (*Euchoreutes naso*). The ears of this animal are three times as long as its head.

Fig. 11-93. Southern birch mouse (*Sicista subtilis*).

The DESERT JERBOA (*Jaculus jaculus*; BL 10–15 cm, TL 15–25 cm, weight 50–70 g; see Color plate p. 399), from northern Africa, Arabia, and the Middle East, is characteristic of the whole family. The hind feet are 5.5–7.5 cm long. Some of the other well-known species include the NORTHERN FIVE-TOED JERBOA (*Allactaga jaculus*; see Color plate, p. 399), found in southern Russia and central Asia, and the LITTLE EARTH JERBOA

Fig. 11-94. Jerboas (subfamily Dipodinae).

(*Alactagulus pygmaeus*), found in the same area. We will focus our discussion on the jerboas from central Asia and Mongolia.

The social behavior and reproductive biology of many central Asian jerboas has, at present, been studied only slightly. With regard to the NORTHERN THREE-TOED JERBOA (*Dipus sagitta*; see Color plate, p. 399), we know only that the gestation period lasts twenty-five to thirty days, and that each of the two or perhaps even three litters a year contains an average of three young. These animals use their own breast and belly hair to line their nests. The DESERT JERBOA, on the other hand, uses camel hair to line its nest. Nursing females always remain by themselves. The young remain in the nest for a long time, even after they are half-grown and are able to look for their food independently.

According to Eibl-Eibesfeldt's observations of the desert jerboa, the male stands fully erect before the female when courting her; then he ducks down and beats his partner on the snout. Finally, the male hops around the female in a half-circle, and challenges her to assume the mating position. The mating ritual is very probably quite different in other species, although at present we have no further information about it.

Fig. 11-95. The northern three-toed jerboa is actually able to walk on two legs when it is moving slowly.

Within the last thirty years, six new jerboa species have been described in central Asia. Two of these species were even found to belong to new genera; these are the COMB-TOED JERBOA (*Paradipus ctenodactylus*) and LICHTENSTEIN'S JERBOA (*Eremodipus lichtensteini*). B. S. Vinogradov has been particularly active in his studies of the biology and classification of the Asiatic jerboas, and we are indebted to him for the following information.

Jerboas use the utmost care in building their burrows, because they are equally sensitive to high and low temperatures. For this reason the entrances to these burrows are built at the greatest possible depth. If the temperature on the ground surface is 45–50°C, it will be only half as high in a burrow which is under a bush and which is 40 cm away from its entrance. At a depth of 50 cm, the animals are almost completely protected from the daily fluctuations in temperature. The comb-toed jerboa, which gets its name from the long hair bristles on its hind feet, is a typical inhabitant of sand dunes, and, as such, it builds its burrows toward the wind, so that the entrance to the burrow is usually rapidly closed off with sand. Every evening, the comb-toed jerboa must dig out another exit hole; this action is also a good safeguard against enemies. Most jerboa species dig out their burrows with their fore legs and their lower incisors, which are particularly long and firmly implanted in the jaw. One northern three-toed jerboa that Vinogradov observed used this method: "It began its initial digging with the fore feet and teeth, and it used its hind feet to throw the excess sand to the rear. When the tunnel became deep enough so that the animal completely disappeared, the jerboa turned around and pushed the excess sand out with its nose and chest. Then it turned itself around and used its hind feet to throw the sand some distance

Fig. 11-96. Courtship of the desert jerboa.

away; the animal used such force in this action that some sand grains were flung as far as 1 m and more."

Jerboas prepare for hibernation by building their living quarters at a depth of as much as 2 m, then lining these rooms with insulating materials. Jerboas begin their hibernation around the end of September or beginning of October, depending on climatic conditions. Five-toed jerboas in the area around the lower Volga hibernate for about five months.

Owls, particularly the eagle owl, are the jerboa's chief enemies. In Mongolia we found five SIBERIAN JERBOAS (*Allactaga sibirica*) in the nest of one eagle owl, all victims of one nocturnal hunting spree. We also found the remains of six members of the same species, as well as a northern three-toed jerboa and a variety of other rodents in the droppings of another eagle owl.

Fig. 11-97. 1. Northern three-toed jerboa (*Dipus sagitta*), 2. Comb-toed jerboa (*Paradipus ctenodactylus*), 3. Lichtenstein's jerboa (*Eremodipus lichtensteini*).

Jerboas are rarely caught in the usual kinds of traps, and, as a result, we could only catch them by jumping out of moving cars at night. It was only by such actions that we were able to study the Mongolian species populations and their relationships at closer range. We found hundreds of jerboas on the shore of (Lake) Char-us-nur in the middle of July. Uhlenhaut, our traveling companion, was able to catch twenty-three northern three-toed jerboas in the beam of his headlights within a half hour. The animals exhibited normal behavior in the spotlight; they jumped around on the ground looking for food, without a flight response; they remained seated when we approached, and allowed themselves to be caught by hand or with a net. We were easily able to collect some little earth jerboas near the Bodoncijn River by using similar methods. However, according to Stubbe, it is much harder to catch the GOBI JERBOA (*Allactaga bullata*), which flees away in large leaps, or the Siberian jerboa with this method.

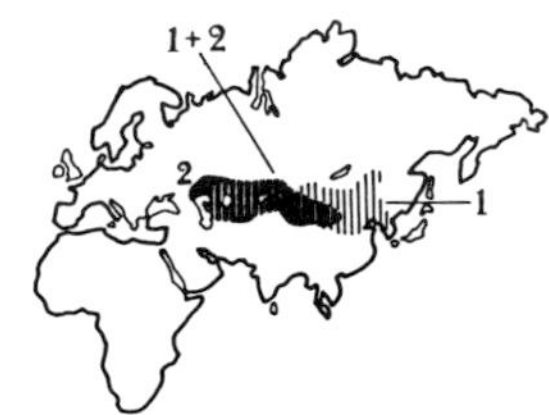

Fig. 11-98. 1. Siberian jerboa (*Allactaga sibirica*), 2. Little earth jerboa (*Alactagulus pygmaeus*).

12 Old World Porcupines, Mole Rats, Rock Rats, and African Cane Rats

Suborder: Porcupines

The Old World porcupines have a special position among the rodents. For this reason we classify them as an independent suborder, that of the PORCUPINES (Hystricomorpha) with the superfamily of porcupines. We will include two other rodent superfamilies within this discussion. Neither of these superfamilies has a distinct systematical classification, and therefore one might as well consider each of them as unique suborders: the mole ratlike rodents (Bathyergoidea) and the cane and rock rats (Petromuroidea).

Family: Old World porcupines, by H.-A. Freye

The superfamily of PORCUPINES (Hystricoidea) consists of only one family, the OLD WORLD PORCUPINES (Hystricidae; BL 35–85 cm, TL 4–23 cm, weight up to 15 kg); these animals should not be confused with the New World porcupines (discussed in Chapter 13), as the two families are not related. The family of Old World porcupines consists of five genera, with twenty-one species. They have a sturdy build, and their bodies are covered with thick, elastic, erectile spines or quills, which may be either long or short. These quills are particularly apparent on the upper part of the back, and because of them the porcupine looks much heavier than it actually is. The limbs are low and the eyes are small. The ears are short (3–5 cm) and are almost hidden between the hairs. Some species have long tails and others have short tails; regardless of its length, the tail always has specially modified bristles or quills. The neck mane is long, loose, bristly, and is intermixed with soft, woolly hair. The soles of the feet are hairless. The thumbs on the fore feet are involuted; the larger toes are complete, but are not of a uniform length. The nostrils are usually S-shaped. The upper lip is cleft; the tip of the nose is stubbed and covered with velvety hairs. The skull of the primordial genus (*Trichys*) is almost flat on top, while the skull of the most highly developed genus (*Hystrix*) is noticeably curved, with many air chambers (pneumatophores). There are many different transitional skull forms between these two extremes. These animals have a strong zygomatic arch. The collar bone is often incomplete. Old World porcupines are herbivorous. They have twenty teeth arranged: $\frac{1 \cdot 0 \cdot 1 \cdot 3}{1 \cdot 0 \cdot 0 \cdot 3}$. The incisors do not have longitudinal grooves; the molars are rooted and have irregular enamel folds which are rapidly worn away. The tongue has several rows of prickly, diagonal horn teeth. The large intestine is simple; it is not fused with the caecum. The lungs have several lobes. Porcupines are able to make growling and snarling sounds. They have a sharp sense of hearing and of smell, but their sense of sight is poorly developed. The gestation period lasts about two months; they can have two or three litters a year; each litter may consist of between one and four, usually two, young. From their birth, the young are precocial; able to eat other food besides their mothers' milk. These animals are limited to the Old World. They are ground-dwellers preferring open woods or steppes. They spend the day either in holes in the ground, which they have either dug themselves or have found, or in rock crevices.

The Old World porcupines are characterized, more than any other mammal group is, by a single body feature; in their case, the quills. Actually, however, the porcupines of the Old and New Worlds are not the only animals with quills; echidnas, tenrecs, and spiny hedgehogs also have them. Perhaps quills were an ancient characteristic of the mammals. However, the long, conspicuous quills of the porcupine are unique.

Even the ancient Greeks and Romans were familiar with the Old World porcupines because of these particular characteristics. Pliny and Aristotle both mentioned these animals in their writings, and Claudius even dedicated a poem to them. Porcupines were often depicted in old Egyptian pictures. Africans used the quills as darts and as arrow tips; the incisors and quills were also made into magic bundles and fetishes.

When we talk about quills, we are actually talking about a variety of different forms: long needles or spears, whose tips are completely rigid; genuine quills with movable tips and rigid shafts, which may be up to 40 cm in length; and quills of over 20 cm long, which are bristly (setaceous), at the tip. The points of these bristly quills are so sharp and rigid that one can easily hurt oneself on contact. Not all of these spines or quills are round; many have up to three longitudinal furrows which make the quills look like hourglasses, kidneys, or even stars, in cross section. "The Asian species of the genera *Acanthion* and *Thecurus*, which have either a very weak mane or no mane at all, as well as all the brush-tailed porcupines of the genus *Atherurus*, have long stilettolike quills all over their bodies in those areas which do not have spines, needles, or bristly quills; these stilettolike quills are between 3 and 7 cm long; they are two or three times wider than they are thick, and they have grooves on one or both sides," writes Erna Mohr. Usually the quills or spines have either light or dark rings, although sometimes they may be completely of one dark color or totally white.

All porcupines have a particular shape at the end of their tails. The short-tailed genera *Hystrix*, *Acanthion*, and *Thecurus* have hollow petal-like cupules called "rattle cupules," which are a variety of lengths (see Fig. 12-6). These are shaken when the animal is aroused, and hit against one another, producing a rattling, clapping sound. When they are new,

▷
Old World porcupines: 1. West African brush-tailed porcupine (*Atherurus africanus*); 2. Bornean long-tailed porcupine (*Trichys lipura*). New World porcupines: 3. North American porcupine (*Erethizon dorsatum*); 4. Prehensile-tailed porcupine (*Coëndou prehensilis*); 5. Upper Amazonian porcupine (*Echinoprocta rufescens*); 6. Thin-spined porcupine (*Chaetomys subspinosus*).

1
2
4
3
5
6
P.B.

these cupules are covered over with continuous cusps. The pointed cover falls off; later, after much use, the pointed cupule also falls off. For this reason one can find animals with both short cupules and long, covered cupules. We should note that as the neck mane of a particular species becomes more developed, these tail rattles become stronger and more powerful. The brush-tailed porcupines (*Atherurus*) have a light-silver tuft of plateletlike bristles on the ends of their long tails; these bristles have a long, slender shaft which is distended at regular intervals along its length, and which, when distended, is as thick as a rice kernel. The long-tailed porcupines (*Trichys*) have smooth, slightly beveled parchmentlike bristles at the end of the tail; these bristles may be up to 20 cm long.

The quills, needles, bristly quills, and platelet quills of newborn porcupines are very short and relatively weak or soft. One can readily recognize the scaly portion of the skin in young animals (compare with echidnas, Vol. X). All of the hair elements grow very rapidly in young animals. A quill or needle that is cracked or broken while growing will grow together once again (cicatrization).

There has been much discussion and confusion about the question of whether or not the porcupine can throw its quills. Aristotle writes that these animals can shoot their "deadly needles like darts" over great distances at hunters and dogs. Thus the European Turks called the porcupines "dart-carrying hedgehogs." Isidore of Seville mentions the hum of quills shot off by the animal, and even Shakespeare, in *Hamlet*, refers to the "terrible porcupine." Certainly these animals, when aroused, do erect their manes and their armored quills, at the same time making a chattering noise with their tail rattles. While these animals shake their bodies, it is possible that a loose quill could be thrown at an opponent. This happens involuntarily, but the quill is thrown with such power that it becomes deeply imbedded in a plank of wood or in the sturdy trousers of a zoo attendant. In his photographic book *Belauschte Tierwelt* ("Eavesdropping on the Animal World"), Berger mentions the story of a farmer who one morning found a porcupine in a springtrap attached to a long chain that had been set for a leopard. The farmer saw, to his amazement, "that several quills were sticking into the high branches of the tree above the trap. The farmer thought that perhaps the porcupine had climbed the tree, trap and all. However, it turned out instead that the animal had thrown the quills with tremendous force, like darts."

Porcupines are very strong. They slide around so that their side or back, with the bristled quills, faces the enemy. Some of the quills are equipped with tiny barbed hooks which can cause painful injuries. The large wild cats, however, are able to kill porcupines without mishap, in spite of the bristled quills, as Petzsch described in 1966. Experienced dogs are also able to overpower these animals.

◁
White-tailed porcupine
(*Hystrix leucura*).

Next to the wild cats and hyenas, the chief enemies of the porcupine are birds of prey, particularly the serpent eagle (*Spilornis*), and pythons. Man, too, feeds on the porcupine in many areas; its meat has been compared with pork. Porcupine meat is occasionally sold in Italy, as well as in Tunisia, Lebanon, and in Asia Minor. South African farmers eat these animals regularly, and some Asian forms are apparently raised like rabbits in their native countries, and eaten. Perhaps the Romans used these animals in their meat pots, and for this reason brought them from northern Africa to Italy, where the crested porcupine (*Hystrix cristata*) still lives.

Fig. 12-1. Asiatic brush-tailed porcupines (*Atherurus*).

In the Pleistocene, during the Ice Age, the porcupine probably had a large population which was distributed over Europe, Asia, and Africa. In Europe, during the interval between the Riss-Würm Ice Ages, these animals were contemporaries of the mammoth, the cave bear, the cave lion, the cave hyena, and the wild horse. The first fossil remains of the porcupine in Europe come from the Middle Miocene (some 15–18 million years ago), while in Asia the first finds are from the Upper Miocene.

Long-tailed porcupines

The oldest porcupines, the LONG-TAILED PORCUPINES (*Trichys*; BL 38–46 cm, TL 17.5–23 cm) have a superficial resemblance to the brown rat. The quills on the body are short and grooved. The end of the tail is a brush of flat, narrow, parchmentlike strips (which frequently disappear sometime during the animal's life, especially in females). These animals are agile climbers, and they get their food from the tops of trees and bushes. There are three species from the southern Malay Peninsula, Borneo, and Sumatra. Included among these are the BORNEAN LONG-TAILED PORCUPINE (*Trichys lipura*; see Color plate, p. 409) and the MALAYAN LONG-TAILED PORCUPINE (*T. fasciculata*), which is often included with the brush-tailed porcupines. They have four front toes, which are connected by a continuous membrane. The hind feet have five toes, which are also connected by a short membrane.

Brush-tailed porcupines

The BRUSH-TAILED PORCUPINES (*Atherurus*; BL 40–54 cm, TL 15–25 cm), like all other porcupines, are outwardly rather difficult to distinguish from one another. The body is covered with quills, of which those in the middle of the back are the longest. There are platelet bristles at the root of the tail, and the tail ends in a long tuft of quills. There are two geographically distinct species groups: 1. African species, with thick, rigid quills on the middle and lower back; these include the WEST AFRICAN BRUSH-TAILED PORCUPINE (*Atherurus africanus*; see Color plate, p. 409), the only longtailed porcupine that has a few air cavities in its slightly arched skull. 2. The Asian species, the ASIATIC BRUSH-TAILED PORCUPINE (*A. macrourus*).

These long-tailed porcupines live in all the tropical forests of their homeland, as well as in low-land forests and wooded islands, up to an altitude of 3000 m. They do not like to dig out burrows themselves, but prefer holes in trees, rock crevices, and other natural burrows, often near banana, manioc, and sweet potato plantations.

Thick-spined porcupines

The THICK-SPINED PORCUPINES (*Thecurus*; BL 42–54 cm, TL 2.5–16 cm) come between these long-tailed forms and the Malayan porcupines. They have a short nasal bone, like the brush-tailed porcupines, but, in common with the Malayan porcupines, they also have the short tail and the rattle cupules. These animals have a superficial resemblance to the maneless porcupines of the genus *Acanthion*; they do, however, have a curved head ("cat's head"). There are three species, including the SUMATRAN THICK-SPINED PORCUPINE (*Thecurus sumatrae*; see Color plate, p. 400).

Malayan porcupines

The MALAYAN PORCUPINES (*Acanthion*) have some characteristics similar to those of the large porcupines (*Hystrix*). However, the neck mane of the Malayan porcupines is not as well developed, and their needles are shorter. There are no long, thin quills or quill bristles on the sides of the body. Consequently, these animals appear to be thinner than the large porcupines. There are five species, including the MALAYAN PORCUPINE (*Acanthion brachyura*; see Color plate, p. 400) and the JAVAN PORCUPINE (*A. javanicum*), which are the oldest animals of this group.

Large porcupines

The LARGE PORCUPINES (*Hystrix*; BL 60–80 cm, TL 12.5–15 cm, weight 17–27 kg) appear to be more powerful than they actually are, because of their thick quill covering. The head, neck, shoulders, limbs and underside of the body are covered with coarse bristles, which become longer toward the rear of the body (30–40 cm). The sides and back half of the body have round needles which are densely packed together. The rattle cupules are larger than those of other porcupines. The skull is greatly distended. The legs are short, with wide feet whose soles are calloused. The thumbs are largely involuted. There are four species, which are difficult to distinguish from one another; they are often crossbred in zoos.

1. The CRESTED PORCUPINE (*Hystrix cristata*; BL 25 cm; see Color plate, p. 400) is the only porcupine species found in Europe. The hind feet are between 10 and 11 cm long; the fore feet are 5 cm long.

2. The WHITE-TAILED PORCUPINE (*H. leucura*; see Color plate, p. 410) has a narrow head. Its snout is covered with hair, and the mane is dark.

3. The EAST AFRICAN PORCUPINE (*H. galeata*) and 4. the SOUTH AFRICAN PORCUPINE (*H. africaeaustralis*). The South African porcupine is black; the tips of the bristles and the needles in the mane, as well as the tail quills, are white. These animals are found in eastern and southern Africa. Their meat is highly prized by local peoples in these areas. South African farmers hunt and trap these animals mercilessly, although they cause only occasional damage to the crops.

Fig. 12-2. African brush-tailed porcupines: 1. West African brush-tailed porcupine (*Atherurus africanus*); 2. Central African brush-tailed porcupine (*A. centrallis*); 3. East African brush-tailed porcupine (*A. turneri*).

All porcupines are chiefly herbivores, although they will also eat meat, like many rodents. In zoos, these animals are fed potatoes, carrots, fruits, greens, black bread, and dog biscuits; one can also give them branches to gnaw on. Wild porcupines eat bulbs, roots, tree bark, thistles, a variety of plants, leaves, and fruit; they will also eat corn (maize), sweet potatoes, pineapples, sugar cane, young cocoa and oil palms, bamboo, melons, onions, and other cultivated crops. These animals are not welcomed in mulberry or rubber plantations, because they gnaw and peel the bark from trees; consequently, they are severely persecuted in Indonesia and South Africa. They seize the food with their incisors and hold it firmly in their "hands" (fore feet). When drinking, they suck up large quantities of water without making a sound. The feces are shaped like long beans, and are usually deposited in particular "dropping areas."

Like all crepuscular animals, porcupines retreat from light (heliophobia), and during the day they keep out of the light. The pupils of their eyes are round. Their sense of hearing is unusually sharp. Erna Mohr reports that a group of sleeping or dozing porcupines can even hear an acorn fall from a tree several meters away. As soon as one animal finds the nut and begins to crack it open, all the other porcupines become active. "They know and recognize the characteristic sound of edible materials falling to the ground." An old Arabian expression, "sharper hearing than a porcupine," is based on this excellent hearing ability. The porcupine also has tactile bristles which are particularly important; these are found on the animal's snout, around its eyes, and in certain areas all over the body. These bristles allow the animals, which live in caves and rock caverns, to find their way by touch. Food is discovered not only with the help of the animal's well-developed sense of smell, but also with the help of its long vibrissae; when the animal is sniffing, these vibrissae are always in motion.

Fig. 12-3. 1. Sumatran thick-spined porcupine (*Thecurus sumatrae*); 2. Bornean thick-spined porcupine (*T. crassispini*); 3. Philippine thick-spined porcupine (*T. pumilis*); 4. Fossil remains of the genus *Thecurus* discovered.

Porcupines adapt to captivity relatively easily and quickly. In 1930, Fischel observed a definite automatic training and learning capability. In 1910, Vosseler described his pet porcupine, which was allowed free run of the house, as a loveable and amusing room-mate. These animals in captivity have a definite emotional need; they will often sleep close to one another, stretched out on their stomachs. Adults take the young animals between them and warm and lick them.

Fig. 12-4. 1. Chinese porcupine (*Acanthion subcristatum*); 2. Himalayan crestless porcupine (*A. hodgsoni*); 3. Kloss' porcupine (*A. klossi*); 4. Malayan porcupine (*A. brachyura*); 5. Javan porcupine (*A. javanicum*).

In spite of their bulky appearance, porcupines have an easy and graceful movement. They use an alternating gait when walking slowly and probably a trot when running. Porcupines can even swim fairly well. Occasionally they may even swim several hundred meters, quietly and cautiously. They show a particular preference for running games and games of motion, whereby they gallop, hop, and jump with ease. Erna Mohr observed four large porcupines in the elephant house at the Munich-Hellabrunn Zoo, Germany that were jumping up and hopping around one another, with their quills slightly raised, just like a group of young dancing girls. The late director of the Leipzig Zoo (Germany), Karl Max Schneider, had a similar experience; he saw an adult porcupine jump through the fence like a ball, accompanied by much hissing. "The circle of spines is raised, as is the long woolly hair on the head. At the conclusion of this performance the animal stamps its feet on the ground. However, in this case the stamping is certainly not a warning signal, but rather an indication of general excitement." Another time Schneider surprised two porcupines hopping around one another in a regular "round dance."

Occasionally these games lead to fights. When the animal assumes a threat posture, not only does it spread its quills and direct these toward its opponent, it also stamps its feet; in addition, it gnashes its teeth and growls and hisses. Porcupines have only a few moderately developed cleaning motions. After eating they often wash off their snouts with one hand. Scratching movements with the hind feet or grooming of the skin with the fore feet have not yet been observed in these animals. Instead, they lick parts of their breast, upper arms, sides, and, more rarely, the side quills. They also have an instinctive shaking movement. When two porcupines lick one another it is a sign of affection, and no part of the body is ignored. Adults do not limit these tender gestures merely to other adults; males and females both lick all the young, which may find it rather difficult to escape from all this family affection.

Fig. 12-5. 1. South African porcupine (*Hystrix africaeaustralis*); 2. East African porcupine (*H. galeata*); 3. Crested porcupine (*H. cristata*); 4. White-tailed porcupine (*H. leucura*).

Males and females also lick one another during mating while they are lying or standing next to each other. When she is ready for copulation, the female flattens her tail against her back, and the male then mounts her with his fore feet. The exact length of gestation is unknown, but is believed to be between six and eight weeks. The shortest time lapse between two normal litters in the Nuremberg Zoo (Germany) was ninety-one days; the shortest time lapse between two litters of brush-tailed porcupines was 110 days. Newborn large porcupines are born with their eyes open; their incisors have broken through the gums and the stomach is covered with short black hair. The back of the body has a longitudinal band of soft black hair, and the longest quills near the shoulders are white. The young animal's forelock is white, and there are five longitudinal white stripes on each side of the body. "After ten days," reports Erna Mohr, "the quills (which initially were soft) are already so sharp and rigid that one can no longer handle the animals without being injured."

Fig. 12-6. Tail of the Bornean thick-spined porcupine with "rattle cupules."

These liverylike stripes, which are readily apparent in all newborn and young animals, disappear after the fourth week in the large porcupines and after the fifth week in the African brush-tailed porcupines. The Asiatic brush-tailed porcupines and the maneless porcupines may retain these stripes for years. The young animals are well cared for by the adults. H. Hediger, director of the Zurich (Switzerland) Zoo, described how a male crested porcupine energetically and impressively defended its young: "It placed its offspring between its fore feet or under its stomach, raised its quills, and made threatening sounds."

Porcupines can become rather old when they are under human care in captivity. In zoos porcupines may live to be almost thirty years old. Several African porcupines of the genus *Hystrix* have lived in captivity for over twenty years. Thick-spined and brush-tailed porcupines have lived for over ten years. One Malayan porcupine lived in the National Zoo (Washington, D.C.) for over twenty-seven years.

Crested porcupine

The CRESTED PORCUPINE (*Hystrix cristata*) is the best-known Old World porcupine, and the most common in European zoos. Several different reports have indicated that this porcupine, present in Italy, might also be found in Greece; however, Jochen Niethammer found, in spite of his exhausting search, that there was no basis for such reports. "Hunters and furriers in Corfu were not familiar with this porcupine." It is very probable that this porcupine was not always a European species, but rather was imported into Italy in ancient times. According to Niethammer, the porcupines of northern Africa and southern Europe are so closely related as to be indistinguishable from one another, even in subspecies classification; thus, according to present knowledge, we can conclude that the bond between these two groups must have been broken only recently. In earlier times, porcupines were far more significant economically than they are today. Their meat was eaten, and man found many varied uses for their quills. Fishermen still use the porcupine's bristles as floats because of their weight and their durability.

The porcupine is the second largest rodent in Europe, after the beaver. These animals prefer dry plains and hilly slopes with substantial brush cover. They show a definite preference for habitats near farms and croplands; they dig out holes or burrows themselves. The animals remain together in these holes all day long during the winter months, although they do not have a genuine hibernation period. They are adept at digging up roots and bulbs; occasionally they will even peel the bark from a tree. They are particularly fond of nuts and fruit, especially grapes, figs, and dates. Occasionally they also eat insects, frogs, reptiles, or other smaller animals. The porcupine adapts readily to human care in spite of its natural shyness, and although these animals prefer twilight when they are wild, in captivity they can become accustomed to the sun and its warmth.

"The porcupines' box, which was lined with layers of hay and straw, was used mainly as a resting place where the animal stayed at night or during the day when it did not want to be disturbed," reports Vosseler of his many experiences with his porcupine friend. "Moreover, as its confidence in its environment grew, it would trot through those rooms of the house which were accessible to it and explore outside with much diligence and conviction. Between these trips, it would stretch out comfortably in front of its master's feet, on a sunny spot, or, in winter, in front of the warm stove. Occasionally it would even trot along behind its master like a dog, licking his boots or scratching to show that it was hungry. It even learned to open doors which had been left ajar, and to get people to open closed doors for it by grunting and scratching. It soon began to take food from the hand, and it chose a particular spot for defecation. Thus, it was "housebroken," as one might put it (this was certainly an innate behavior), and to a certain extent, it learned to adapt itself to its master's wishes."

Family: Mole rats, by F. Dieterlen

We will include another unique rodent group in our discussion of the porcupines; this is the group of MOLE RATLIKE RODENTS (superfamily Bathyergoidea). We do not know anything about these animals' origin or evolution. Their subterranean habits have produced some distinct adaptations, just as with other burrowing rodents. There have been only isolated discoveries of mole rat fossils, and these do not tell us anything about

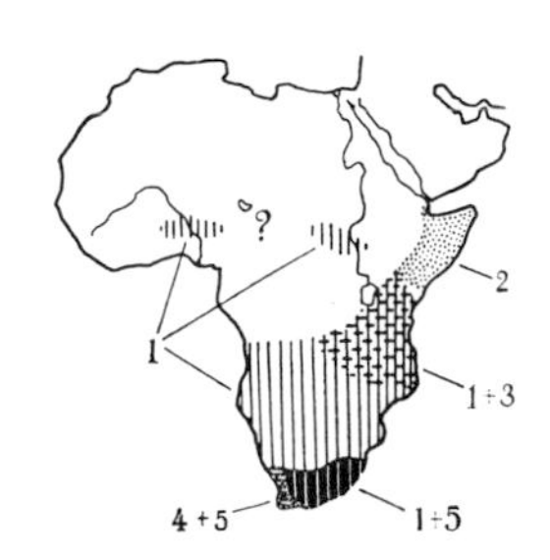

Fig. 12-7. 1. Blesmol (genus *Cryptomys*); 2. Naked sand rat (*Heterocephalus glaber*); 3. Sand rat (genus *Heliophobius*); 4. Cape sand mole (*Bathyergus suillus*); 5. Cape mole rat (*Georhychus capensis*).

possible relationships with other rodent groups. Many zoologists classify these animals along with the cricetid rodents (see Chapter 11), but other zoologists classify them as a unique suborder. Thus we can see that their position in our zoological system is at best uncertain, at least on the basis of our present knowledge.

There is only one family, the MOLE RATS (Bathyergidae; BL 8–33 cm, TL 1–7 cm). These animals are strong and sturdy. The body and skull structure are adapted to subterranean living. The chewing (masseter) muscles are specialized. The eyes, ears, and tail are involuted. Each of the four feet has five digits. There are between twelve and twenty-eight teeth, arranged: $\frac{1\cdot0\cdot2-3\cdot0-3}{1\cdot0\cdot2-3\cdot0-3}$, with the exception of the sand-rats. The incisors are gigantic; the roots of the upper pair may extend back to over the rear molar teeth. The masticating surface of the molars usually has an annular pattern. Today these animals are found in subsaharan Africa. Fossilized mole rats from the Oligocene have been found in Mongolia. There are five genera with, at most, eleven species:

1. The BLESMOL (*Cryptomys*; BL 12–25 cm, TL 1–3 cm) is often identified by a white spot on the head. There are twenty teeth, arranged: $\frac{1\cdot0\cdot2\cdot2}{1\cdot0\cdot2\cdot2}$. The fur color varies from whitish to yellowish to grayish-brown to black. These animals avoid tropical rain forests. Almost fifty different forms have been described, but probably there are between two and five species. These include the TOGO MOLE RAT (*Cryptomys zechi*; see Color plate, p. 426) and the AFRICAN MOLE RAT (*C. hottentotus*).

2. NAKED SAND RATS (*Heterocephalus*). Except for single scattered hairs and the tactile hair zones, they are completely hairless, a unique occurrence among wild rodents. They look like newborn animals. The eyes are tiny and there is no external ear (pinna). They have sixteen teeth. There is only one species, the NAKED SAND RAT (*Heterocephalus glaber*; BL 8–9 cm, TL 3–4 cm, weight 40 g; see Color plate, p. 426). They live in dry steppes and savannas at lower and middle altitudes.

3. SAND RATS (*Heliophobius*; BL 10–20 cm, TL 1.5–4 cm) have soft, thick hair. There may be up to six molar teeth in each half of the jaw, and a total of twenty-eight teeth, unique among rodents. There are three species, including the SILVERY MOLE RAT (*Heliophobius argenteocinereus*).

4. MOLE RATS (*Bathyergus*) have thick, woolly hair. There is only one species, the CAPE SAND MOLE (*Bathyergus suillus*; BL 17.5–33 cm, TL 4–7 cm), which lives in sand dunes.

5. CAPE MOLE RATS (*Georhychus*) have soft, woolly, fluffy hair, which often covers the tiny tail. There is only one species, the CAPE MOLE RAT (*Georhychus capensis*; BL 15–20 cm, TL 1.5–4 cm; see Color plate, p. 426), which is very similar to the blesmol.

The BLESMOL prefers open areas, especially those with sandy ground. Its subterranean burrows are usually close to the surface, often as close as 10–20 cm; in loose soil they may be as deep as 50 cm. Sometimes these animals dig out their burrows in termite mounds. Their burrows consist of tunnels running in all directions, as well as nest and supply rooms. The blesmol digs with its fore feet and its incisors; it pushes the loose earth out of the tunnel mostly by using its hind feet. If the soil is damp, the animal uses it to build a large plug which is erected in front of the tunnel from which the excess earth has been thrown out; this erect plug reminds one of a tiny termite mound. Blesmols prefer to eat the roots of certain grasses as well as the bulbs of a particular species of iris (which happens to be poisonous to cattle). These rodents will even harvest aloe plant blossoms from their tunnel openings, and they will also eat earthworms.

These subterranean mammals may be active at all hours of the day and night, although they seem to prefer the warmer daylight hours. We know very little about the reproduction and the development of the young in these animals. The female African mole rat apparently gives birth to between one and three young. We do not know if these animals have special mating seasons. Their chief enemy, aside from snakes and small carnivores, is the owl; blesmol remains have been found in owls' fur balls, which indicates that the blesmols also leave their nests occasionally at night. Although these rodents frequently build their burrows into small colonies, we can not call them particularly social animals.

The NAKED SAND RAT has the most unusual appearance of all the mole rats; it is one of the oddest mammals known. Sanderson compares this almost completely hairless animal with a "newborn puppy" or even "a shriveled sausage." Evidently, naked sand rats never come out into the daylight. They eat the roots of trees and vines, as well as insects. The small colonies in which they settle can be identified by their peculiar tiny craters, which Dietrich Starck compares with the cones of an ant lion (*Myrmeleon*). He writes: "During the warm afternoon hours, one can see sand fountains some 30 cm high, which are thrown up out of the cone openings at regular intervals of a few seconds. . . . Down in the center of the cone one can see an animal's tail, bending and twisting like a hook. The sand is thrown out by the animal's hind feet, which are visible when the animal is working near the cone opening." While none of the species of this family has been commonly exhibited in captivity, some have survived over two years, and have even bred on occasion.

Superfamily: Cane and rock rats, by D. Heinemann

We turn our discussion now to the African cane rats and rock rats, two rodent families found in Africa which have been tied in with the octodont rodents (Octodontidae) because of definite similarities in body build. However, an evolutionary link between these two African groups and the octodont rodents, which are limited to South America, is highly unlikely for geographical reasons. Like the mole rats, the African cane rats and rock rats are most easily linked to the porcupines, and we classify these two groups together as the superfamily of CANE AND ROCK RATS (Petromuroidea).

The family of AFRICAN CANE RATS (Thryonomyidae) includes only one genus (*Thryonomys*; BL 35–61 cm, TL 7–25 cm, weight 4–7 or up to 9 kg). The head is thickset; the ears are short and round, and protrude only slightly above the hair. The first finger is involuted, and the second, third, and fourth fingers are well developed. The fifth finger is small and almost useless. The toes are longer than the fingers; the first toe is missing. The claws are thick and powerful. The fur is firm and bristly; the hair itself is flat and has no underwool. The tail has scales and a few short bristles. The soles of all four feet are hairless. There are twenty teeth, arranged: $\frac{1 \cdot 0 \cdot 1 \cdot 3}{1 \cdot 0 \cdot 1 \cdot 3}$. The incisors are very strong and wide; they are an orange color, and each upper incisor has three longitudinal grooves. There are several (perhaps six) species, including: 1. The GREAT CANE RAT (*Thryonomys swinderianus*; see Color plate, p. 426), from South Africa, and 2. The LESSER CANE RAT (*Thryonomys gregorianus*), from central Africa.

Cane rats generally live alone in moist, swampy areas, especially in the cane and rush growths along rivers. Evidently these animals do not dig burrows into the earth; instead, they build their den in the densest plant growth available. They also give birth to and raise their young in these nests. There are between two and four young animals in every litter; these are fully developed at birth, with their eyes open, and a thick fur. The young remain in the nest, which has been lined with grasses and broken reeds, until they become accomplished runners.

Cane rats eat the soft parts of grasses and shrubbery, as well as nuts and bark. Sometimes they can cause extensive damage to sugar cane plantations, and in order to keep the number of cane rats in these areas under control, man protects their most important natural enemy, the python. Other enemies of these animals include leopards and various birds of prey. Man, however, is the chief enemy of these animals, and it is man who hunts them not only because they are undesirable competitors, but also because of their delicious meat.

The family of ROCK RATS (Petromuridae) consists of only one species, the ROCK RAT (*Petromus typicus*; BL 14–20 cm, TL 13–18 cm; see Color plate, p. 426). The head is rather flat, and the ears are short. All four feet are narrow; there are four fingers and five toes. The claws are narrow. The hair looks bristly, but it actually is soft and silky; there is no underwool. There are two or three pairs of nipples high on the side of the body. These animals have twenty teeth arranged: $\frac{1 \cdot 0 \cdot 1 \cdot 3}{1 \cdot 0 \cdot 1 \cdot 3}$. The incisors are narrow. These animals are found in rocky cavities and crevices in the stony deserts of the Upper Karroo plateau and in similar South African landscapes. Their bodies are so supple and their ribs so flexible that these rock rats can slip through the narrowest cracks. They usually emerge from their hiding places in the mornings and late afternoons in order to bask in the sun and to look for food. Sometimes they even look for food after sunset. These animals are herbivorous, and they prefer blossoms of certain steppe and desert plants, as well as greens, seeds, and fruits.

The female gives birth to one or two young in late December or early January; the young are rather large at birth, fully developed, and covered with hair.

13 The Cavies

Suborder: Cavies

Of the four suborders which comprise the order of rodents, the suborder of the CAVIES has the greatest evolutionary unity. These animals are found exclusively in the New World; with the exception of a single North American genus, their distribution is limited to Central and South America, the Antilles, and the Bahama Islands. All of these animals share a common dental pattern: $\frac{1.0.1.3}{1\cdot0\cdot1\cdot3}$ However, we will not try to discuss other characteristics of this suborder, because the individual groups and species of these animals have adapted themselves to so many different habitats and habits, and, in so doing, have developed in so many different forms, that any attempt at delineating common characteristics would be meaningless. There are five superfamilies, with twelve families in all:

A. Octodont rodents (superfamily Octodontoidea); 1. Octodont rodents (Octodontidae); 2. Tuco-tucos (Ctenomyidae); 3. Rat chinchillas (Abrocomidae); 4. Spiny rats (Echimyidae); 5. Hutias (Capromyidae); 6. Coypus (Myocastoridae).

B. Viscachas and chinchillas (superfamily Chinchilloidea); 7. Chinchillas (Chinchillidae).

C. Cavies (superfamily Cavioidea); 8. Cavies (Caviidae); 9. Capybaras (Hydrochoeridae); 10. Pacas and agoutis (Dasyproctidae).

D. Pacaranalike rodents (superfamily Dinomyoidea); 11. Pacaranas (Dinomyidae).

E. New World porcupines (superfamily Erethiozontoidea); 12. New World porcupines (Erethizontidae).

Phylogeny, by E. Thenius

Geologically, the oldest known South American rodents come from the Oligocene period (some 35–40 million years ago). These animals are descended from the North American paramyids, and apparently groups of them began to settle on the island chains only after the South American separation. These Oligocene rodents (+*Platypittamys*; see Color plate, pp. 214/215) might very possibly, at least according to their bodily characteristics, be the ancestors of the present-day members of the cavy suborder (with the exception of the New World porcupines).

Superfamily: Octodont rodents, by D. Heinemann

Of all the South American rodents extant today, this evolutionary group is most similar to a range of small to medium-sized rodents which look more or less like rats. However, appearances are deceptive in this case; the internal structures of these animals indicate that they are not related to the ratlike rodents, but instead to the cavies, agoutis, and chinchillas. Consequently, the members of this rodent family are called octodont rodents (Octodontidae), and they are grouped together with five additional families of cavies (all of which have a superficial ratlike appearance) in the superfamily of OCTODONT RODENTS (Octodontoidea).

Family: Octodont rodents

The OCTODONT RODENTS (family Octodontidae; BL 12.5–19.5 cm, TL 4–18 cm) have a relatively large head; the ears are medium sized, round, and sparsely haired. These animals also have long vibrissae. The thumb is involuted. The body hair is usually long, thick, and soft. The hair on the tail increases in length as one moves to the tip of the tail. There are four pairs of nipples. The chewing-surface enamel of the molar teeth has a pattern of grooves shaped like a figure eight, hence the name Octodontidae. There are six genera (grouped together on the distribution map, Fig. 13-1), with a total of eight species, including:

Fig. 13-1. 1. Degu (*Octodon degus*); the other species in the genus *Octodon* are found in generally the same area; 2. Bori (*Octodontomys gliroides*); 3. South American rock rat (*Aconaemys fuscus*).

1. The DEGU (*Octodon degus*; BL 15 cm, TL 12 cm; see Color plate, p. 435), whose body is not particularly adapted to digging activities. The tail is long and slightly bushy. 2. The CURURO (*Spalacopus cyanus*; BL 14–16 cm, TL 4–5 cm; see Color plate, p. 435), which is specially adapted to digging activities. The upper incisors are large and protruding. 3. The SOUTH AMERICAN ROCK RAT (*Aconaemys fuscus*; BL 15–18.5 cm, TL 5.5–7.5 cm; see Color plate, p. 435); its body has special adaptations for burrowing activities. The upper incisors do not protrude. 4. The VISCACHA RAT (*Octomys mimax*; BL 16–17 cm, TL 17–18 cm) has no special adaptations for digging.

The degu is the only octodont rodent whose habits are reasonably well known to us. According to Pöppig, these animals are particularly common in central Chile, where hundreds of them live together in hedges and bushes. "They rarely leave the ground to climb into the lower branches of nearby bushes. They await the approach of their enemy (man) with relative coolness, and then large numbers of them dive into the openings of their burrows which have many branches (with tails erect), only to reappear after a few seconds at another opening." According to Pöppig, the degu's behavior is more similar to the behavior pattern of a squirrel than to that of a rat. "It collects provisions, even when the weather is warm, and it does not hibernate."

Fig. 13-2. 1. Viscacha rat (*Octomys mimax*); 2. *Tympanoctomys barrerae*; 3. Cururo (*Spalacopus cyanus*).

When a degu is caught by its tail and struggles to free itself, the skin of the tail breaks at that point where the tail has been held fast, and the animal then escapes. The resultant injury causes very little bleeding; the degu later bites off that part of the tail which is without skin with its incisors. Unlike the lizard, the tail of the degu does not grow back again.

Degus and other octodont rodents have been kept in recent years in many European and American zoos, where they have bred easily. The young are fully haired and almost completely developed at birth; they open their eyes shortly thereafter.

Family: Tuco-tucos

The family of TUCO-TUCOS (Ctenomyidae) consists of only one genus, the TUCO-TUCOS (*Ctenomys*; BL 17–25 cm, TL 6–11 cm, weight 200–700 g). The head is large and sturdy; the incisors protrude out of the mouth. The eyes and ears are small. The body is compact, and the tail is short. The neck and limbs are short and powerful. All of the digits are equipped with strong digging claws. There are more than twenty species, including KNIGHT'S TUCO-TUCO (*Ctenomys knighti*; see Color plate, p. 435).

The tuco-tucos are similar to the North and Central American pocket gophers (Geomyidae) both in habits and in their bodily adaptations to digging; however, they have no closer relationship to these squirrel-like rodents. Tuco-tucos do not have cheek pouches (the characteristic responsible for the pocket gopher's name). Tuco-tucos live in South America from the Altiplano, the plateau that lies between the Andes mountain chains, as far as Patagonia and Tierra del Fuego, in wastelands and areas of poor vegetation. The number of animals living in these sparsely vegetated—sometimes completely barren—regions is astonishing. In many areas these rodents have created such a large system of interconnecting tunnels and have burrowed through the earth so assiduously that anyone walking through the area will most likely find his feet breaking through the surface, and horses may even break a leg. Tuco-tucos get their name from the alarm call by which some species of this rodent signal the approach of a threat or disturbance; this call can even be heard after the animal has taken refuge underground. Tuco-tucos have air combs on each of their digits (just as do the gundis and pocket gophers), and these combs are particularly useful in shoveling away loose earth.

Fig. 13-3. 1. Distribution of the tuco-tucos (genus *Ctenomys*); 2. Knight's tuco-tuco (*Ctenomys knighti*).

Tuco-tucos eat roots, bulbs, and plant stalks. They also collect provisions which they store in their burrows. The female gives birth once a year to between one and five young, which are fully developed at birth.

The family of RAT CHINCHILLAS (family Abrocomidae) also has only one genus, the RAT CHINCHILLAS (*Abrocoma*; BL 15–25 cm, TL 6–18 cm). The body shape is similar to that of a rat; the head is large and pointed. The eyes and ears are large, while the limbs are short. The hair is thick and soft, like that of the chinchilla, but not as woolly. There are two pairs of nipples. The intestine is very long; the small intestine has a length of 1.5 m, the colon, 1 m, and the caecum, 20 cm. There are two species, including the RAT CHINCHILLA (*Abrocoma cinerea*; see Color plate, p. 435).

Chinchilla rats live in burrows in the ground or in rock cavities; the entrance to their den usually is located under a bush or some rocks. Often one can find these animals living together in colonies. We know

▷
The prehensile-tailed porcupine (*Coëndou prehensilis*) is not related to the Old World porcupines. It is a distant relative of the cavy.

▷ ▷
Capybara mother with young (*Hydrochoerus hydrochaeris*). These capybaras are bred regularly in the Frankfurt zoo.

▷ ▷ ▷
The hare-sized Patagonian cavy (*Dolichotis patagonum*) is closely related to the cavy.

1
2
4
3
5
6
7

Fig. 13-4. 1. Rat chinchilla (*Abrocoma cinerea*); 2. Bennett's rat chinchilla (*A. bennetti*).

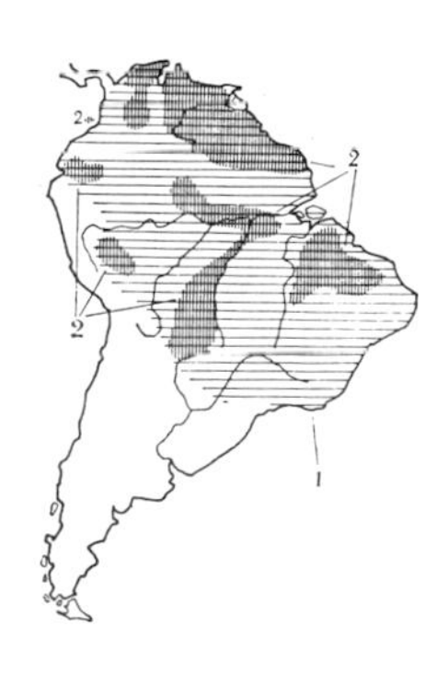

Fig. 13-5. 1. Approximate distribution of all the spiny rats in the genus *Proëchimys*; 2. Approximate distribution of the Cayenne spiny rat (*Proëchimys guyannensis*).

◁ Cane and rock rats: 1. Great cane rat (*Thryonomys swinderianus*); 2. Rock rat (*Petromus typicus*). Mole rats: 3. Zech's or Togo mole rat (*Cryptomys zechi*); 4. Cape mole rat (*Georhychus capensis*); 5. Naked sand rat (*Heterocephalus glaber*). Gundis: 6. Gundi (*Ctenodactylus gundi*); 7. Speke's pectinator (*Pectinator spekei*).

very little about their reproductive pattern. In many areas the local peoples hunt these animals in order to sell the skin at a local market; however, no great price is paid for the fur.

The family of SPINY RATS (Echimyidae; BL 8–48 cm, TL 4.5–43 cm) consists of many species, and is widely distributed throughout South America. The body is shaped like that of a rat. The nose is pointed or somewhat truncated. The eyes and ears are medium-sized. The first digit of the fore foot is stunted. Females have three pairs of nipples. There are two subfamilies:

1. The SPINY RATS (Echimyinae) have bristly hair which often has oblate, pointed, and furrowed quills. 2. The CORO-COROS (Dactylomyinae) have soft hair, without quills.

We will mention only two of the eleven genera in the subfamily of SPINY RATS (Echimyinae). These are: 1. The SPINY RATS (*Proëchimys*), which consists of twelve species, including the CAYENNE SPINY RAT (*Proëchimys guyannensis*; see Color plate, p. 435); and 2. The CRESTED SPINY RATS (*Echimys*), which have ten species, including the ARMORED SPINY RAT (*Echimys armatus*; see Color plate, p. 435). The distributions of additional genera are shown in the maps in this chapter.

We know very little about the habits of the spiny rats. One of the best-known species is the CAYENNE SPINY RAT, one of the most common mammals in many areas of its distribution range. These animals live singly or in pairs in holes in the ground, under tree trunks, or in rock crevices. Because these spiny rats are nocturnal animals, one rarely sees them, in spite of their large numbers.

Cayenne spiny rats and other species of the genus feed on various plant matter. In capitivity they will eat bananas, fresh sweet-corn, pieces of coconut, grain, and a variety of seeds. A female, kept in Brazil by M. Snethlage, a zoologist, gave birth to four young, which opened their eyes immediately and began to move around. Usually there are only two litters a year, and these generally consist of two or three young; however, six young have been observed in a single litter.

Like the degu, many spiny rats easily break and lose their tails when one holds that appendage firmly; the tail usually breaks or splits at a point in the center of the fifth tail vertebra. This ability to lose the tail in danger has certainly saved many spiny rats from death. Almost half of the spiny rats that Snethlage caught had mutilated tails.

The CRESTED SPINY RATS (*Echimys*) and a few other genera of this subfamily are arboreal. The ARMORED SPINY RAT (*Echimys armatus*) incidently, has been introduced on the island of Martinique.

The RATO DE TAQUARA (*Kannabateomys amblyonyx*; BL 25 cm, TL 32 cm; see Color plate, p. 435) is the only species of its genus. Its hair is thick and soft, and has no bristles or quills. Other genera in the subfamily of CORO-COROS (Dactylomyinae) are mentioned in the distribution maps.

In Brazil the rato de taquaro lives in bamboo thickets along river banks. At night these animals climb up the bamboo stalks with great agility, gripping the stalk between the third and fourth digits of their fore feet. Thus, they are able to reach the young, juicy buds, for which they have a particular preference.

A pregnant female was once caught in November. Apparently only one young is born each time.

Three genera of the family of spiny rats once lived on the Antilles, the islands between North and South America. However, these genera died out some 400 to 500 years ago. Remains of bones from these animals were later found in Cuba, Haiti, and Puerto Rico.

Family: Hutias, by Herbert Wendt

The family of HUTIAS (Capromyidae) have a wide variety of forms that once were found on the Antilles. Superficially these animals resemble large, heavy, broad-headed rats. These rodents lived on the Greater Antilles and neighboring islands shortly before and after the Spanish conquest; there were at least six genera, with some fifteen species identified to date. Spaniards of the first West Indian expedition became familiar with some species of these animals; using the native name, they called them *hutias*. Like the ancient solenodons (see Vol. X), these ancient hutias did not have many animal competitors on their Caribbean islands, and thus they survived to the modern era. However, they were hunted by the Indians, and the hutia population on some islands was completely destroyed long before the European discovery of America. When the European arrived, he cleared the forests and brought cats, dogs, and other hunters with him, and consequently most of these Caribbean rodents died out. Only a few animals survived in isolated, barely accessible regions. Probably the Indians and, later, the black slaves transported to the Caribbean, hunted hutias for food, as remains of these presently extinct forms were often found in cooking wastes. In the 1870s, Indian mongoose were introduced into the Antilles to combat the rat population; they also hunted the hutias, which were much less agile and totally unfamiliar with such an enemy, and thus more easily caught. The few remaining hutia forms have been divided into three genera:

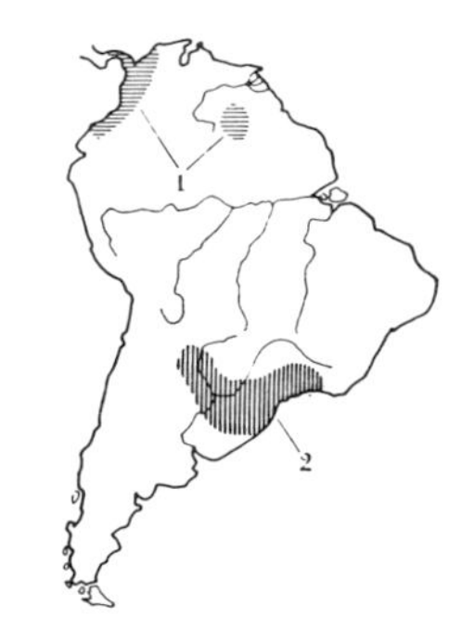

Fig. 13-6. 1. Armored rat (*Hoplomys gymnurus*); 2. Suira (*Euryzygomatomys spinosus*).

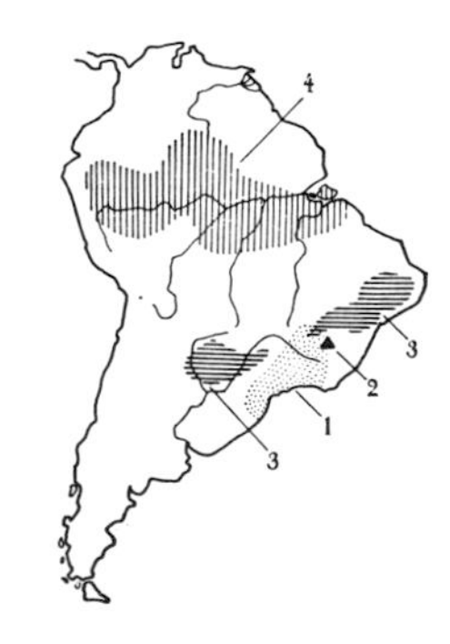

Fig. 13-7. 1. Distribution of the genus *Clyomys*; 2. The spiny rat species *Carterodon sulcidens*; 3. The punare (*Cercomys cunicularis*) was found in this approximate area; 4. *Mesomys hispidus*, another species of spiny rats.

A. The *Capromys* (BL 30–50 cm, TL 15–30 cm, weight 4–7 kg). The hair is coarse, although on the tail it is softer and thinner; there is no underwool. The tail is more or less prehensile. The stomach is divided into three parts by two strictures (an unusual occurrence in rodents). There are two pairs of nipples. These animals are diurnal and are usually arboreal, feeding on fruit, leaves, and other plant matter. There are four species in this genera, including: 1. The PREHENSILE-TAILED HUTIA (*Capromys prehensilis*; see Color plate, p. 438). 2. The BLACK-TAILED HUTIA (∅ *C. melanurus*), which is more delicate. 3. The DWARF HUTIA (∅ *C. nana*), which has a lighter-colored hair. 4. The CUBAN HUTIA (*C. pilorides*), the largest species of the genus. The tail is relatively short, and there can be a variety of fur colorations.

Fig. 13-8. 1. *Isothrix bistriata*, a South American hutia; 2. Arboreal soft-furred spiny rat (*Diplomys caniceps*).

Fig. 13-9. 1. Approximate distribution of the crested spiny rats (*genus Echimys*); 2. Armored spiny rat (*Echimys armatus*); 2a. introduced on Martinique.

Fig. 13-10. 1. Coro-coros (genus *Dactylomys*); 2. Rato de Taquara (*Kannabateomys amblyonyx*); 3. *Thrinacodus albicauda*.

B. The BAHAMAN and JAMAICAN HUTIAS (*Geocapromys*; BL 33–35 cm, TL 3.5–6 cm). The body is heavy and the limbs and tail are short. The hair is short but thick. These animals are largely nocturnal. They live in holes in the ground, in tree holes, or under rocks. They are herbivorous. There are two species extant today, the JAMACIAN HUTIA (∅ *Geocapromys brownii*; see Color plate, p. 438) and the BAHAMAN HUTIA (∅ *G. ingrahami*). Another species, + *G. columbianus*, is extinct.

C. The HISPANIOLAN HUTIAS (*Plagiodontia*; BL 31–40 cm, TL 12–15 cm). The hair is short and thick. The tail, which may be prehensile, is hairless, or almost so, and scaly. These animals are nocturnal. They eat fruits, roots, and other plant matter. There are three species, two of which, + *Plagiodontia ipnaeum* and + *P. spelaeum*, are extinct. The single extant species consists of two subspecies: CUVIER'S HUTIA (∅/+ *P. aedium aedium*), which probably died out within the last decade, and the DOMINICAN HUTIA (∅ *P. a. hylaeum*; see Color plate, p. 438), which is very rare.

Today the CUBAN HUTIA is the most common species in the genus *Capromys*. These animals have found refuge in the large Cienaga de Zapata reserve in south-western Cuba, which also provides a sanctuary for a large and varied bird population. There the hutias live together in pairs. These animals show a particular preference for sunbathing on the leafy branches of high trees; they look almost like bundles of leaves when seen from the ground. When they are disturbed or alarmed, these animals usually make a sharp barking sound and chatter their teeth. When they greet other animals, they make rather melodic chirping sounds. Hutias move about on the ground in an awkward gallop; sometimes they even make short jumps or sit upright.

Cuban hutias have occasionally appeared in zoos. In some cases these animals are not particularly compatible, but in others they are. According to Erna Mohr, they are usually very friendly with their keepers and can be held and carried about without any particular danger; there are, of course, exceptions. Occasionally these animals reproduce in captivity (in zoos and under private care); Cuban hutias in the Berlin and Frankfurt zoos have reproduced, as have animals in the London, Bronx, and National (Washington, D.C.) zoos. According to Bucher and Walker, the gestation period lasts for seventeen to eighteen weeks. The young animals are able to see and to control and direct their own movements at birth; they are also fully haired. In addition they have a prehensile tail which is able to grasp the keeper's finger even before the animal is completely dry. After ten days the young Cuban hutias begin to take solid food, although they are not weaned until after at least one and one-half months. Lee S. Crandall reports that these animals thrive in capitivity on a diet of general plant material. Cuban hutias have lived in captivity over twelve years.

The JAMAICAN HUTIA is not so well off. In Jamaica these animals are found only in a few inaccessible rocky areas of the Blue Mountains. There they are apparently hunted with dogs and even dug up out of their burrows, although legally they have been completely protected since 1922. This species suffered greatly from the importation of the mongoose to Jamaica in 1872. We know almost nothing about the behavior and habits of these animals. How they came to tiny Swan Island, which is over 700 km from Jamaica, is a mystery even today. There are small colonies of this species at the Tacoma (U.S.A.) and National (Washington, D.C.) zoos.

Fig. 13-11. 1 The Jamaican hutia (*Geocapromys brownii*) is found only in two limited mountain areas in Jamaica, and on Swan Island; 2. *G. columbianus* (extinct); 3. The Bahaman hutia (*G. ingrahami*) is found only on a small rocky island in the southern Bahamas.

The HISPANIOLAN HUTIAS, which live on Hispaniola and are even more rare than the Jamaican hutia have an exciting history of discovery. In the beginning of the 19th Century, a French Jesuit found tooth and skeletal remains of these animals in some caves in northern Haiti. Frederic Cuvier, brother of the famous French naturalist Georges Cuvier, was able to describe these animals in 1836, although he believed that the HAITIAN HUTIAS (*Plagiodontia aedium*), as these animals were named, were already extinct. Shortly thereafter, however, fresh skeletons and skins were found on several islands in the Antilles chain. One animal lived in the London Zoo in 1855 and 1856. Basic scientific research within this century has enabled us to know what had happened to these animals. All of the Hispaniolan hutias in the Sierra de Monti Christi, the wooded mountain chain in northern Haiti, had been caught. They were hunted by local peoples and persecuted by the mongoose to such an extent that Cuvier's hutia (*Plagiodontia aedium aedium*) is now believed to be totally extinct.

Fig. 13-12. 1. Black-tailed hutias (*Capromys melanurus*) can be found in parts of these areas; 2. The dwarf hutia (*Capromys nana*) is found only in this area; 3. The prehensile-tailed hutia is found only on the Isle of Pines (Cuba); 4. Distribution of the very rare Haitian hutia (*Plagiodontia aedium*).

In 1923 W. L. Abbott met an old man on the northern coast of the Dominican Republic who offered to catch Hispaniolan hutias if Abbott would pay him five dollars per animal. The old man and his dogs went hunting in the neighborhood of a small lagoon on the sea coast. There they searched all the tree holes, and eventually they came back with eleven Hispaniolan hutias. Most of these animals were pregnant females, each of which gave birth to only one young. Shortly thereafter, Abbott caught two more animals in the same area. These hutias were investigated by Gerrit S. Miller, who in 1927 identified them as a new form. Since then these Dominican hutias (*Plagiodontia aedium hylaeum*) have been caught several times by local people. These animals are also found in Haiti. In 1947, the Bronx Zoo (U.S.A.) received two of these animals from a Haitan fish hatchery; one of them was albino with red eyes. Contrary to previous reports which designated these hutias as arboreal, the two animals in the Bronx Zoo were very poor climbers. Unfortunately, these Hispaniolan hutias were unable to survive more than five weeks of captivity. This species has also been bred in the early 1970s in the Tacoma (U.S.A.) and National (Washington, D.C.) zoos, where they have lived well over five years.

The COYPU or NUTRIA (*Myocastor coypus*; BL 43–63 cm, TL 30–40 cm, weight 7–9 kg, occasionally up to 17 kg; see Color plate, p. 438) is the only species in its genus. (The genus name comes from the Greek words for mouse and beaver.) Many zoologists classify these animals with the family of hutias; however, we will discuss them as a separate family, that of the COYPUS (Myocastoridae).

Family: Coypus, by D. Heinemann

Distinguishing characteristics

Males are larger than the females. The head is sturdy. The ears are small and the vibrissae are long. All four feet are hairless. The digits on the hind feet (with the exception of the fourth and fifth) have webbed membranes. The hair is long and coarse; it covers a thick, soft underwool. The tail is covered with scales and is sparsely covered with hair. It is round when seen in cross section, not wide like the beaver's tail or flattened at the sides like the muskrat's. There are six pairs of nipples, located high on the sides of the body. The sebaceous (fat) glands are situated at the corner of the mouth and near the anus. The incisors are large and orange. The oral cavity behind the incisors is capable of closing. These animals are usually active in the twilight hours. They are found on the banks of rivers and lakes.

The coypu

Spanish-speaking South Americans were the first to name these animals *nutrias*, meaning "fish otter." Their more common English-language popular name, coypu, was originally the Chilean name for these animals. Although these animals have many different names, and when alive are called coypus in most English-language books, their fur is sold around the world under the name of nutria.

Fig. 13-13. Cuban hutia (*Capromys pilorides*).

Coypus live in rivers and lakes in the temperate zones of South America. They dig their burrows in sloping banks; these burrows are usually short, with no branching tunnels, and they generally end in a simple chamber. When the lake shore or river bank is such that the animal is unable to dig its burrow in the ground, it carefully builds a nest of reeds, either on land or in shallow water. Coypus are excellent swimmers, spending most of their time in the water. They can remain underwater for as much as five minutes after a dive. On land these animals have an awkward, clumsy movement, although when the need arises they can run and jump rather quickly for short distances. Coypus are especially shy and fearful; the slightest disturbance will send them scurrying to the shelter of the water, their burrow, or some other hiding place.

Fig. 13-14. Coypu (*Myocastor coypus*).

Usually these animals live together in pairs, although they often form large colonies. The major part of their diet consists of plant matter, particularly water plants, and reeds. According to Walker, coypus also often eat mussels and snails. Their digestive organs are easily able to break down foods that are considered hard to digest. Large amounts of crude fibers are decomposed in the caecum under the influence of bacteria that are able to break down cellulose particles. The end product of these crude fibers is then adapted for use in the animal's body. When eating, coypus take the food in their fore feet and transfer it to their mouths.

Coypus assume a similar posture when cleaning and lubricating their fur with their fore feet. Every area of the body is carefully combed and arranged. At regular intervals the animals wipe the fat glands in the corners of their mouths. Whether or not the sebaceous glands near the anus are also used in lubricating the body has not yet been determined.

Apparently mating takes place in September or October. According to several observers, the female coypu gives birth to two or three litters a year. Animals specially raised on fur farms are able to mate and give birth all year long, although more young are born between February and May than at any other time of the year. The gestation period lasts for between 128 and 132 days, on the average. There are five or six (up to twelve) young in each litter; they are fully developed, able to see, and covered with hair at birth.

The coypu's fur is highly valued because of the particularly thick, fine, and soft underwool, although many people feel that the long, coarse hair of the top coat is unattractive. Consequently, the cover hair and dense hair are separated through a special preparatory process, before the "nutria" fur goes on the market. As has been the case with so many other mammals, the soft fur of the coypu has led to the destruction of these rodents. The coypu population has been so severely reduced by hunting that only a few animals are left in many areas of the distribution range. The number of coypus in central Argentina was so low even in the first decade of the 1800s that, according to Rengger in 1830, a hat-maker from Buenos Aires realized that he would have to start breeding these animals himself. We do not know what happened to this first attempt at raising coypus commercially. However, a century later, people began to breed and raise coypus on a regular basis in fur farms; these began in South America and spread out all over the world.

The herbivorous coypu is much more easily fed than are most other furbearing animals being raised on farms today. Presently coypu fur is "not in style," and commercial farms are therefore not as important now as they once were. Coypus in Germany have been raised on a diet consisting chiefly of potatoes, but including oats, clover, corn (maize or Indian corn), hay, green forage, legumes, turnips, and cabbage. The animal must eat large amounts of foods which are not easily digestible, because its intestinal tract is built for such a purpose.

Like almost all domestic animals, the captive coypu has developed several variations in color; the white and yellowish-white breeding lines are particularly important, as these furs may be dyed any color with relative ease.

The cold European winters often cause the coypus' tails to freeze, as these appendages are hairless; the animals do not seem to notice the discomfort much. A much more dangerous situation arises when the lake, stream, or river where they live freezes over; coypus can not find their way under the ice as easily as do beavers, for example. I [Heinemann] was able to observe coypus in the Münster (Germany) zoo several times

Fig. 13-15. Coypu in threat posture.

as they repeatedly fished blocks of ice several cubic centimeters large out of the water, carried them to their burrows with their teeth, and attempted to build a nest with them. Possibly these animals have an instinctive nest-building reaction to any objects of a certain size which are floating in the water.

Man has tried to settle the coypu in the northern temperate zones as a sport-hunting animal; these attempts, however have been successful only in certain areas of the U.S.A. Most of the coypu colonies in Europe which have been either settled or stocked with animals raised on fur farms have been unsuccessful.

Superfamily: Viscachas and chinchillas, by D. Heinemann

The second superfamily within the suborder of cavies is that of the VISCACHAS AND CHINCHILLAS (Chinchilloidea). This superfamily contains only one family, the CHINCHILLAS (Chinchillidae; BL 22.5–66 cm, TL 7.5–32 cm, weight 0.5–7 kg). The head is large and the snout is wide.

Distinguishing characteristics

The eyes are large; the ears are curved. The body is relatively slender. The fore legs (arms) are short; the fore feet are small and have four digits. The hind legs are long and muscular; the hind feet are long and have either three or four digits. The soles of both the fore and hind feet are hairless. The incisors are narrow. The molars grow continuously. There are three genera, the viscachas (*Lagostomus*), the mountain chinchillas (*Lagidium*), and the chinchillas (*Chinchilla*); there are five species all together.

Fig. 13-16. 1. Viscacha (*Lagostomus maximus*); 2. Former distribution of the short-tailed chinchilla (*Chinchilla chinchilla*); 3. Former distribution of the chinchilla (*Chinchilla laniger*).

The VISCACHA (*Lagostomus maximus*; BL 47–66 cm, TL 15–20 cm, weight 7 kg; see Color plate, p. 438), the largest species in the superfamily, is very possibly the only species of its genus. Another form, with which we are familiar only through a single skeleton, is probably extinct. The viscacha's head is particularly large and bulky. The ears are medium-sized. The hind feet have only three toes each; these are equipped with strong claws. The hair is coarse. The tail has a full coat of hair. These animals live in dry plains, in colonies of fifteen to thirty, sometimes as many as fifty animals. These colonial settlements (called viscacheras) are formed in open areas; each structure consists of a network of subterranean tunnels with many exits. The excess earth is piled into a mound, on top of which the animals deposit any garbage they may push out of their burrows. However, that is not enough; the viscachas bring the strangest-looking objects back to their burrows from their local expeditions: "bones and horns from cattle, stones, thistle stalks, hard clumps of earth, etc., all of which they pile up in a mound, one that might easily fill a wheelbarrow," noted Charles Darwin. He goes on to tell of a man riding over the pampas at night and discovering that he had lost his watch. The next morning he searched out all the viscacha burrows that had been in his line of travel, and sure enough, he found his watch at the entrance to a burrow. The viscacha had found the watch the previous night and carried it back to its burrow. The meaning of this strange behavior has yet to be explained.

Other animals often live along side the viscachas in their viscacheras, as "subtenants." These include the burrowing owl (*Speotyto*; see Vol. VIII), and the miners (*Geositta*; see Vol. IX), as well as lizards and snakes. Some of these "subtenants" dig their own holes in the sides of the viscacha mound, and others move right in with these large rodents.

Viscachas move out onto the pasture lands at night, where they eat grass, roots, and other plant matter. Their meat is edible, and toward the end of 1950s it was even sold commercially on the German market. Viscachas in their native areas are hunted not for their meat but rather because their burrowing activities create a hazard. These animals dig so many tunnels under the earth's surface that horseback riding in these areas can be extremely dangerous. The horse puts its foot in a hole or breaks through the surface, falls, throws its rider, and often remains lying on the ground with a broken leg. For this reason, man has carried out a merciless war of extermination against these animals since 1907, and they have disappeared from large areas of their original distribution range.

The MOUNTAIN CHINCHILLAS (*Lagidium*; BL 32–40 cm, TL 23–32 cm, weight 0.9–1.6 kg; see distribution, Fig. 13-17) are smaller and more delicate than the viscachas. There are three species, one of which, the COMMON MOUNTAIN CHINCHILLA (*Lagidium viscacia*), is pictured in the color plate on page 438. The ears are very large and the vibrissae are rather long. The hind feet have four toes, all of which have weak claws. The hair is thick and soft, except for the upper side of the tail, which is covered with coarse hair. The coloring varies. These animals are diurnal. They live in high mountainous regions (900–5000 m) in dry areas with poor vegetation.

Mountain chinchillas live in rock faults and crevices or in rock caves in the sides of hills. These animals live in social groups ranging from a few animals to groups of up to eighty. There has been no recorded observation of a mountain chinchilla living alone. These rabbit-sized animals climb and jump over, around, and between rocks with an agility that seems almost unbelievable. Their diet consists of the few plants available within the habitat, including grasses, mosses, and lichens. These animals must have water, and they are not to be found in areas without water.

The mating season is in October and November. The gestation period lasts for about three months. Usually only one young is born in each litter. Although mountain chinchillas have been severely persecuted because of their fur and their tasty meat, apparently at present, especially in southern Peru and the La Paz plateau (Bolivia), they are not really threatened with extinction. In 1939 and in 1946 Oliver P. Pearson found that mountain chinchillas were among the most common mammals in Peru.

The closely related CHINCHILLAS (*Chinchilla*) have had a much more difficult time in the wild. These animals are totally extinct in most areas

▷
Octodont rodents: 1. Degu (*Octogon degus*); 2. Cururo (*Spalacopus cyanus*); 3. South American rock rat (*Aconaemys fuscus*). Tuco-tucos: 4. Knight's tuco-tuco (*Ctenomys knighti*). Rat chinchillas: 5. Rat chinchilla (*Abrocoma cinerea*). Spiny rats: 6. Cayenne spiny rat (*Proëchimys guyannensis*); 7. Armored spiny rat (*Echimys armatus*); 8. Rato de Taquara (*Kannabateomys amblyonyx*).

▷▷
Cavies: 1. Wild cavy (*Cavia aperea tschudii*); 2. Guinea pig (*C. a. porcellus*), a and b) short-haired, c) long-haired; 3. Southern mountain cavy (*Microcavia australis*); 4. *Galea musteloides*; 5. Rock cavy (*Kerodon rupestris*).

1
2
3
4
5
6
7
8

1
2a
2b
2c
3
4
5
F.B.

1
2
3
4
5
6
7
8
9
P. B.

1
2
3
4
6
5
7
P.B.

of their original distribution range. Zoologists' opinions differ about the systematic classification of this genus. We will agree with Haltenorth and Cabrera, dividing the genus into two species with a total of three subspecies:

1. The SHORT-TAILED CHINCHILLA (⊕ *Chinchilla chinchilla*) has a large body and relatively short ears and tail. The subspecies + *C. c. chinchilla* (BL 36–38 cm, TL, with end hairs, 14–16 cm) has ears which are 4–5 cm long. These animals probably died out several decades ago. The subspecies ⊕ *C. c. boliviana* (BL 30–32 cm, TL, with end hairs, 14–16 cm), has ears which are 4–5 cm long. The gestation period lasts for 120–128 days. Females give birth to two, occasionally three, litters a year; each litter consists of between one and four young. These animals have a life expectancy of twenty-two years. They are largely extinct in the wild and appear only rarely in captivity.

2. The CHINCHILLA (⊕ *C. laniger*; BL 25–26 cm, TL, with end hairs, 17–18 cm; see Color plate, p. 438) has a small body with long (6 cm) ears and tail. Gestation lasts for 108–111 days. Females give birth to two or three litters a year, and each litter consists of between one and six young. The life expectancy is some eighteen years. These animals are largely extinct in the wild; they are, however, rather common in captivity.

◁
Viscachas and chinchillas: 1. Viscacha (*Lagostomus maximus*; 2. Common mountain chinchilla (*Lagidium viscacia*); 3. Chinchilla (*Chinchilla laniger*). Hutias: 4. Prehensile-tailed hutia (*Capromys prehensilis*); 5. Dominican hutia (*Plagiodontia aedium hylaeum*); 6. Jamaican hutia (*Geocapromys brownii*); 7.Coypu (*Myocastor coypus*), nursing female.

◁◁
Cavies: 1. Patagonian cavy (*Dolichotis patagonum*); 2. Salt-desert cavy (*Pediolagus salinicola*). Capybaras: 3. Capybara (*Hydrochoerus hydrochaeris*). Pacaranas: 4. Pacarana (*Dinomys branickii*). Agoutis: 5. Mountain paca (*Stictomys taczanowskii*); 6. Spotted paca (*Cuniculus paca*); 7. Acouchy (*Myoprocta acouchy exilis*); 8. Sooty agouti (*Dasyprocta fuliginosa*); 9. Orange-rumped agouti (*Dasyprocta aguti*).

In earlier times, chinchillas must have been extremely plentiful in their home range. At the time of the Spanish conquest of South America, these animals lived from the rocky slopes of the Andes Mountains in Peru and Chile, down to the coast below, in colonies of 100 animals or more. They were so unafraid that, as one author from that period tells us, they would run around and between horses' legs, and a traveler could see as many as 1000 animals in one day. These animals live in rock crevices and caverns. Chinchillas are basically crepuscular and nocturnal animals, although they have been observed on bright days, sitting in front of their holes, or climbing and jumping around the rocks with incredible agility.

Their hair is unusually thick, soft, and fine, and it is because of their fur that man has so ruthlessly trapped and hunted these animals. The Indians have always hunted the chinchilla so that they might use its fur for blankets and for clothing. However, their hunting methods did not threaten the chinchilla population to any great extent. The first chinchilla furs appeared on the European market as rarities in the 18th Century. In the 19th Century chinchilla fur was a regular trade article. However, it was not until the turn of the century that the exportation of furs to Europe and North America became such a large concern that many Indian *chinchilleros* supported themselves solely by hunting chinchillas. A good chinchilla fur sold for around thirteen dollars in 1890. In 1905, 216,000 furs were exported from the Chilean harbor of Coquimbo; four years later, however, these animals were already so greatly reduced in numbers that only 27,000 furs could be exported. In the meantime, the price of a first class fur (untouched) rose to forty dollars. Thus it is not surprising to find that the *chinchilleros* hunted the chinchillas in even the most inaccessible rocky lands, in order to find, kill, and sell the remaining survivors. Protective laws, which had been passed by most of the South American states, were of little value. In 1930 a chinchilla fur sold for 200 dollars! However, after that virtually no wild chinchilla skins were available on the open market.

Juan Ignazio Molina, the learned Jesuit priest who in 1782 was responsible for giving the chinchilla its scientific name, was one of the first people to suggest that chinchillas "could be bred at home without difficulty and with limited cost," in order to obtain their fur. Almost 150 years passed before Molina's suggestion was carried out. In the 1920s, some twenty Indian trappers worked for M. F. Chapman, an American mining engineer catching chinchillas in the Andes Mountains so that a breeding stock could be established. These trappers worked for almost three years collecting animals; they finally succeeded in trapping seventeen chinchillas. Chapman brought eleven of these animals to the U.S.A. in 1923. Most of the commercially bred chinchillas today are descended from these eleven animals. In 1934 F. Holst took sixteen short-tailed chinchillas back to his native Norway. However, this species is not yet so commonly bred on fur farms as is its smaller relative. There are probably hundreds of thousands, if not millions, of chinchillas in fur farms throughout the world today.

Fig. 13-17. 1. Peruvian mountain chinchilla (*Lagidium peruanum*); 2. Common mountain chinchilla (*Lagidium viscacia*); 3. Wolffson's mountain chinchilla (*Lagidium wolffsoni*).

These animals are fed good hay, wheat, corn, oats, and green matter, as well as commercial food pellets. They do not need water as long as they can get whatever water is necessary to their survival from green plants or fruits.

Chinchillas are monogamous. They reproduce very slowly, which is why many hopeful breeders of these animals, especially private breeders who can afford only a small capital investment, have been disappointed in their hopes of a rapid financial turnover and have subsequently been forced to give up their breeding ventures. In addition, fur from animals that have been bred commercially is not worth as much as the fur of wild animals, and commercially bred animals, like all domestic animals, come in a variety of shapes and colors. Thus it is extremely difficult, if not impossible, to match 150 furs which all look exactly alike, to make high-quality chinchilla coats.

Recently people have tried to restock the Andes Mountains with chinchillas that have been bred on fur farms. Whether these efforts will be successful depends not only on the animals' abilities to regain their wild characteristics and abilities, but also on the ability of the various South American governments to enforce the laws protecting the chinchilla. These laws have been in effect since 1910, but they have been ignored. If these efforts are not successful, wild chinchillas will disappear from the face of the earth.

Superfamily: Cavies, by D. Heinemann

The families of cavies (Caviidae), capybaras (Hydrochoeridae), and agoutis (Dasyproctidae) are classified together within the superfamily of CAVIES (Cavioidea). These animals are medium-sized to large-sized rodents, their body length ranging from 22.5 to 130 cm. They are adapted for running. The thorax is shaped more or less like a keel. The sternum is narrow. The collarbone is involuted or almost nonexistent. The feet are usually tall and slender (with the exception of the cavies), and have ridges at all the main joints. The side fingers and toes tend to be involuted. The claws are more or less hoof-shaped. The tail is short or stubbed.

Family: Cavies

The CAVIES (family Caviidae) have rootless molar teeth which grow continuously. Rows of these teeth meet toward the front of the mouth. Each fore foot has three digits; each hind foot has four. There are two subfamilies:

Subfamily: Cavies

1. The CAVIES (Caviinae; BL 22.5–33.5 cm). The head is large. The ears and legs are short. The tail is involuted. The nails are short and either dull or sharp. There are four genera (*Cavia*, *Galea*, *Microcavia*, and *Kerodon*). 2. The PATAGONIAN CAVIES (Dolichotinae).

The aperea

The APEREA (*Cavia aperea*) has the largest distribution of all the wild cavies in the genus *Cavia*. The subspecies WILD CAVY (*C. a. tschudii*; see Color plate, p. 436), from southern central Chile, is the evolutionary ancestor of the well-known guinea pig. The wild cavy lives at altitudes of up to 4200 m, while most of the other subspecies of the aperea live in plains and flatlands. These rodents live together in small groups of from five to ten animals. They make their homes in subterranean structures, which they dig out themselves or take over from other animals. They leave their dens at night and, using well-trod pathways, make their way through the tall grass to their feeding places. Wild cavies eat all kinds of grasses, herbs, and other plant matter. Copulation and birth are not confined to any particular season. Gestation, at least in guinea pigs, lasts some sixty or seventy days. Between one and four young are born in every litter; the young are fully developed and are able to run shortly after birth. The mother nurses them for three weeks, and after fifty-five to seventy days they become sexually mature.

The guinea pig

According to the rules suggested by H. Bohlken, the GUINEA PIG (see Color plate, p. 436) should have the scientific name of *Cavia aperea porcellus*. Peruvians kept guinea pigs as domestic animals long before the Spaniards conquered the Incan civilization. When J. J. von Tschudi traveled through Peru during his South American explorations around the middle of the 19th Century, he visited several rather isolated areas where he saw large numbers of guinea pigs being kept loose in the Indians' huts. These animals "ran over and around the faces and bodies of the sleeping Indians all night long." At that time, at least, guinea pig meat was important food for people living in these areas. Some mountain Indians still use these animals for sacrifices.

Guinea pigs were sent to Europe long ago. In 1554 (sixty-two years after Columbus discovered America), Konrad Gesner reported on the "Indian rabbit or pig," so named because Columbus was virtually convinced that his newly discovered continent was part of India.

The "pig" in their popular name comes from the fact that these small animals are round and plump like a pig, and they have a pig-like squeal; the "guinea" is believed to be because many of the first specimens came to England from South America on slave ships from Guinea (Africa), or because of their resemblance to the young of the bush pig (Guinea hog; *Potamochoerus*); there is probably no connection with "Guiana."

Fig. 13-18. 1. Aperea (*Cavia aperea*); 2. Amazonian wild cavy (*C. fulgida*); 3. *C. stolida.*

If one spends some time with guinea pigs, they become quite tame. They will not bite and will let people cradle them or carry them around. Rosl Kirchshofer tells of a guinea pig that she had as a child: "It was familiar with my walk and it would squeal in anticipation, while I was still climbing stairs. Thus my mother always knew that I was home. Otherwise, it only squealed (but in a different tone) when it was hungry and my mother passed by. She was the one who always fed it. It was able to tell the difference between us though . . . . Many people think that guinea pigs are dumb animals. This is possibly because these lovable creatures never defend themselves, but more probably it is because these people never spend any time with guinea pigs and do not know anything about them."

The guinea pig can easily be distinguished from its wild relatives because of the roundness and bulkiness of its head and body. In addition, its fur has a wide range of variations (see Color plate, p. 436). Aside from their natural (wild) color, guinea pigs may be black, gray, brown, yellow, or white with either dark or red eyes. These animals may also have a piebald or even three-toned coloration. The hair of the smooth-haired species corresponds in length and striation to the hair of the wild forms. The curly-haired species have no striations; instead the fur is in a series of curls or rosettes. The long-haired or Angora guinea pigs have an inhibited molt, so the constantly growing hair may reach a considerable length.

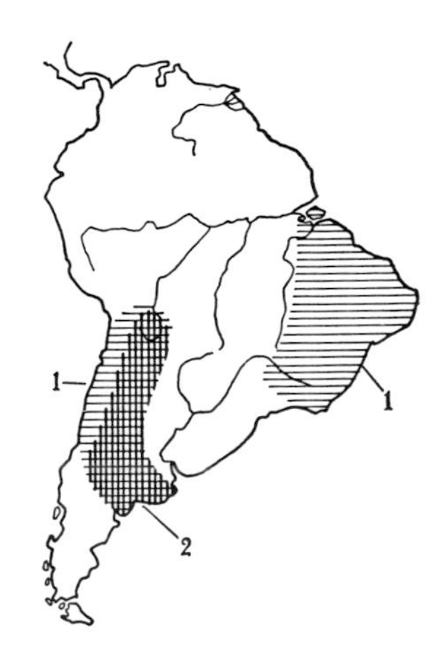

Fig. 13-19. 1. Distribution of the genus *Galea*; 2. *Galea musteloides* (see Color plate, p. 436)

"Domestic animals also have some internal changes," reports Rosl Kirchshofer. "This is because domestic guinea pigs have been fed foods rich in nutritional value, with a high water content, for hundreds of years, while the wild forms have fought a difficult struggle for existence. In many areas of South America, cactus pulp is the only source of water for these wild animals. Consequently, the domestic guinea pigs have a larger stomach and a longer small intestine, although both the caecum and the colon are shorter. Therefore, these animals can take in large amounts of food at one time and they use only a small quantity of cellulose in their digestive processes. Because the food itself is easily digestible, these animals produce very few feces. The wild cavy, on the other hand, is able to take in only a small amount of food at one time. Its food is water-poor but has a high cellulose content, and, as a result, this animal has a small stomach, a short intestine, and a long caecum and large intestine. Because its food is hard to digest, the wild cavy produces many feces."

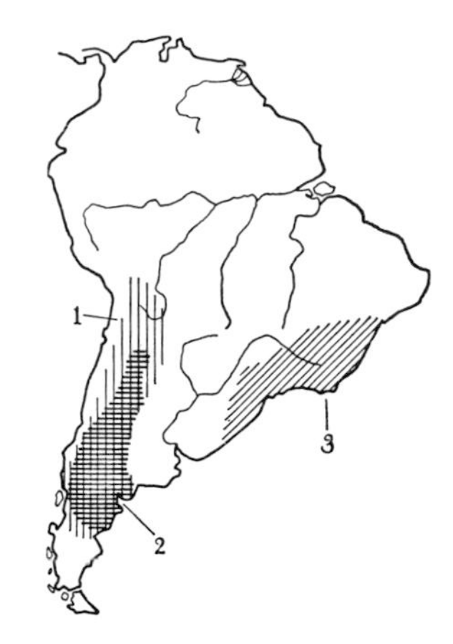

Fig. 13-20. 1. Mountain cavies (genus *Microcavia*); 2. Southern mountain cavy (*Microcavia australis*; see Color plate, p. 436); 3. Rock cavy (*Kerodon rupestris*).

Cavies are unassuming animals. Rosl Kirchshofer has the following advice on the care of these animals as house pets: "A large box (1.0 × 0.5 × 0.3 m) can accommodate one or two animals. A cover for the box is not absolutely necessary, as these animals are not climbers. Peat dust or sawdust will work well as bedding, and sleeping boxes and feeding bowls complete the arrangements. A diet of various vegetables, raw and cooked potatoes, carrots, and grains (oats, barley) is recommended. When the animals are fed juicy foods, they do not need water. However, after the food becomes somewhat 'drier,' the guinea pigs welcome some milk or water.

"Cavies are best kept in pairs. If this is the case, the female usually responds with a litter of two, after a gestation period that lasts nine weeks. The young are completely developed, fully haired, and able to see at birth. The teeth are already fully developed at birth, and although the young nurse from their mother for the first fourteen days, they also often eat some of the plant matter on and after their second day of life.

"I would not recommend keeping two males together. They bite one another and fight over the females. Although they may be tame toward people, the situation is not always so peaceful when the animals are by themselves."

The cavy has an important role as a laboratory animal, along with the rat, mouse, and rabbit. Cavies are especially important to research in the field of medical health, for studies of diet and genetics. They are also important to the development and manufacture of serums and vaccines.

The rock cavies

An important genus of the other genera and species which make up the family of cavies (see Color plate, p. 436) is the ROCK CAVY (*Kerodon*), which has a slender, long-legged body reminiscent of the maras.

There is only one species, the ROCK CAVY (*Kerodon rupestris*; weight 900–1000 g; see Color plate, p. 436). The toe pads have calluses. The nails are short and wide. These animals live in dry, rocky mountainous areas. The female probably has two litters a year, each consisting of one or two young.

Rock cavies live under rocks or in rock crevices; sometimes they will also dig holes in the ground, under rocks. They emerge from their quarters late in the afternoon or in the evening, to look for food. These animals not only move about on the ground like the other cavies, but they also climb cliffs with ease, and they will even climb trees in order to eat the leaves. Ludwig Heck kept rock cavies in the Berlin zoo, and he has discussed these animals' climbing abilities in *Brehms Tierleben*: "These rat-sized rascals are easily able to climb smooth cement or rough glass walls that are 1.20 m high . . . by making several successive diagonal cross-jumps. From a low, pebbly base they can jump over 1 m high onto an overhanging roof of glass." Heck soon gave up trying to prevent his rock cavies from leaving their cage, and let them alone. "They did not go far from their quarters, and people were delighted to find an animal that was not in its cage on the rodent hill."

Subfamily: Patagonian cavies

The PATAGONIAN CAVIES (subfamily Dolichotinae; BL 45–75 cm) are much larger than the cavies. Because these animals look like common hares, they were formerly called "pampas-hares." The ears are long, the legs are long and thin, and the tail is short. The nails on the fore feet (fingernails) are shaped like claws; the toenails are shaped like hooves. There are two genera (*Dolichotis* and *Pediolagus*), each of which has one species. Many recent authors place *Pediolagus* as a subgenus of *Dolichotis*.

Distinguishing characteristics

The PATAGONIAN CAVY (*Dolichotis patagonum*; BL 69–75 cm, TL up to 4.5 cm, weight 9–16 kg; see Color plate, p. 437), which is larger than a hare, looks, at first glance, more like a common hare or even a dwarf-sized antelope than a cavy. It is sometimes called the MARA.

The Patagonian cavy

Patagonian cavies live in dry grasslands or shrubbery. They are rapid runners and accomplished jumpers, often making leaps of 2 m. They either dig wide, deep holes in the ground as burrows or they take over burrows abandoned by other animals. For a long time the Patagonian cavies in the Frankfurt zoo had their burrow next to the fence of their cage; it did not take these animals long to dig under the stone foundation of the cage and make an exit near the visitors' path. These animals did nor remain close to their cage (unlike the rock cavies in the Berlin zoo mentioned above by Heck). Instead they began moving out onto the road, which had a great deal of traffic; consequently, new foundations had to be built around the cages, sunk to a much greater depth in order to prevent the maras from escaping.

The female gives birth to the young in a nest inside the burrow. Usually there are two young per litter, although there may be either one or three. As is the case with all the cavies, the young Patagonian cavies are able to walk immediately; their eyes are open and the young animals look like miniatures of their parents. "Initially, at least, the young are nervous and easily frightened," Erna Mohr says of Patagonian cavies raised in captivity. "The two pairs of nipples are situated on the side of the body. The first pair is toward the front of the body, right behind the elbow and shoulder; the second pair is in the inguinal region, just in front of the knee. The nipples of a nursing female Patagonian cavy are up 3.5 cm long; they are very thin and look rather like clinging leeches. Initially, when the young are small, the mother will lie on her stomach so that they can nurse from her. Later, she sits erect on her haunches, and the young squat at right angles to her in order to nurse."

Fig. 13-21. 1. Salt-desert cavy (*Pediolagus salincola*); 2. Patagonian cavy (*Dolichotis patagonum*), largely extinct in the eastern areas of the indicated distribution.

Erna Mohr observed the Patagonian cavies in Hagenbeck's zoo (Germany), where they are not kept in cages but instead were allowed

Fig. 13-22. Capybara (*Hydrochoerus hydrochaeris*).

a free run of the area, and she noticed that the young of these animals remained with the mother for at least nine months. "Apparently, however, either they were chased away before a new litter arrived or they left of their own accord; in any case, one never saw a nursing female with both newborn and almost fully grown young beside her. Mothers with young do not have as close a bond to the rest of the tribe as do non-nursing females. The latter gather together in large groups to graze and rest; however, in spite of their compatibility, these animals always keep a certain distance between themselves."

The Patagonian cavy cleans itself by licking its sides and apparently by "combing" its fur with its teeth. They wipe their faces like cats do, by moving the inside of the fore leg (arm) from back to front. As a result, some of the hair on the inside of the fore feet eventually deteriorates, and a few bare spots may result.

The Patagonian cavy's usual movement is an ambling walk. "When these animals are in a hurry, they move into a gallop, after a series of small irregular hops. The gallop is rather peculiar, as the movement of the four feet changes quite erratically," according to Erna Mohr.

Patagonian cavies are herbivorous. Their diet in zoos consists of hay, green matter, raw vegetables, and crushed oats. They should also have a supply of drinking water, although they will use it only rarely as long as they are supplied with enough green plant materials. These animals like to have salt blocks. Patagonian cavies are able to survive even the European winters as long as they have a protected burrow. These animals have been successfully raised and bred in many zoos. One Patagonian cavy in captivity in South America lived for almost fourteen years. However, most specimens do not live beyond ten years.

The salt-desert cavy

While the Patagonian cavy lives in the brush pampas of Patagonia (Argentina), the SALT DESERT CAVY (*Pediolagus salinicola*; BL 45 cm; see Color Plate, p. 437) is found in the dry, wintery, wooded brush areas of the western Chaco (Argentina). Hans Krieg found these animals in this area "inevitably, wherever there was no grassy topsoil. Under and between the bushes, which were at most only a few meters high, one could see the hard, barren, and unusually light soil, the top layer of which was, in my opinion, dried mud." Krieg and his associates did not find any of these animals in the salt deserts of central Gran Chaco. This is particularly remarkable, as the species name *salinicola* means living in salty regions. According to Krieg, the salt desert cavy avoids grassy soil, visiting such areas only occasionally to look for food. "Usually these animals are satisfied with only the shoots of bushes and a few coarse weeds and grassy growths (all of which have a low water content) that grow almost all year round over wide stretches of the barren ground." Salt desert cavies have also been successfully raised and bred in several zoos.

Family: Capybaras

The largest living rodent is the CAPYBARA OR CARPINCHO (*Hydrochoerus hydrochaeris*; BL 100–130 cm, shoulder height 50 cm, weight up to 50 kg; see Color plate, p. 437), a member of the family of CAPYBARAS (Hydrochoeridae). The head is large, and the snout is blunt. The upper lip is split. The eyes protrude. The ears are small and rounded. The neck is short. The body is thickset and heavy. The tail is very short. The fore feet each have four fingers; the hind feet have three toes each. All of the digits are connected by short web membranes, and all have short, hooflike nails. The incisors are quite wide (some 2 cm) and white. The molars are rootless and continuously growing. The rows of teeth converge toward the front of the mouth. The hair is long and so sparsely distributed over the body that the skin shows through. Adult males have a large scent-gland area on the bridge of the nose. There are several subspecies, including the PANAMA CAPYBARA or CARPINCHO (*Hydrochoerus hydrochaeris isthmius*), which is smaller and weighs only about 27 kg.

Distinguishing characteristics

The capybara

The name capybara (in Portugese, *capyvara*) comes from the Guaran language, and means "master of the grasses." Spanish-speaking South Americans call this animal the *carpincho*.

Capybaras live near ponds, lakes, rivers, or swamps in woods with thick undergrowth. They live together in families or small herds which rarely have more than twenty animals. However, Goeldi observed colonies of 100 or more on the island of Marajo in the mouth of the Amazon (Brazil). These large rodents are usually crepuscular, when they have not been disturbed. In those areas where they have been hunted by man, they are nocturnal. These animals do not dig burrows; instead they rest in shallow depressions on the ground.

When threatened, the capybara flees in long leaps, although its usual gait is a slow walk. They will often take refuge in the water, as they are excellent swimmers and divers and can remain underwater for several minutes, while moving a considerable distance away from their enemy. They let only their nostrils, eyes, and ears protrude above the surface of the water, just as hippopotamuses do.

Capybaras often stand in water up to their stomachs, eating water plants. They also eat grass and the bark of young trees. Occasionally they may even venture out onto pasture and graze alongside domestic cattle. Sometimes these large rodents even raid watermelon crops, or corn (maize), rice, and sugar cane plantations.

Capybaras are quiet, peaceful animals, and are not nearly as playful as many other social rodents. When they are content, they make soft, whimpering noises. However, they can also produce a shrill, piercing whistle or a short grunting sound.

Females give birth to only one litter a year, after a gestation period of fifteen to eighteen weeks. One litter consists of from two to eight young, which are fully developed at birth. They weigh 900–1350 g five days after birth. Wild capybaras apparently have a life expectancy of eight to ten years; in captivity some have lived over twelve years.

Man is the chief enemy of the capybara, although these animals are also threatened by the jaguar on land and by the alligator in the water. Man hunts these animals because of the damage they occasionally cause to plantations and for their meat, which is not of a very high quality. Man also hunts the capybara for the fat layer under their skin, which apparently has a high iodine content and thus can be used in the preparation of pharmaceutical materials.

The PANAMA CAPYBARA or CARPINCHO is considered a separate species by many zoologists, although these animals differ from the other capybaras mainly in their small size. Herbert Wendt discusses this little-known animal:

"As I was sailing across Lake Gatun in the Panama Canal in 1962, I saw the partially submerged head of an animal that I could not immediately identify, near the island of Barro Colorado. For a short time only the tip of the nose, the eyes, and ears appeared above the surface. I was familiar with capybaras from Brazil; I had observed them several times on river banks and swimming in the water. However, I did not think of finding a capybara in the Panama Canal, and especially a small and much rarer form. As the animal moved out of the water and onto the bank, its angular head and drooping lips showed that it was a Panama carpincho or *piropiro*. The Panama carpinchos seem to have found a final refuge on these islands in Lake Gatun, which in reality are the peaks of mountains that were covered over with water when this man-made lake was created."

Family: Agoutis

The AGOUTIS (family Dasyproctidae; BL 32–79.5 cm, TL 1–7 cm, weight 1–10 kg) are even further removed from the cavies than are the capybaras. These rodents are medium-sized. They have an arched dorsal line, an elevated rump, and a short or stunted tail. The claws are strong and hooflike. There are scent glands near the anal region. There are two subfamilies, with two genera each.

Subfamily: Pacas

The PACAS (subfamily Cuniculinae; BL 60–79.5 cm, TL 2–3 cm, weight 6.3–10 kg) are larger and bulkier than the agoutis. The head is large and distended on the sides. The eyes protrude. The ears are medium-sized. These animals have a powerful body. The limbs are sturdy. The fore feet each have four digits; the hind feet each have five digits. There are two pairs of nipples. The zygomatic arch is compressed and has a cavity in the center which is connected with the oral cavity (a unique occurrence among mammals). There are two genera, which are distinguished by cranial and dental differences; each genus has one species: 1. The SPOTTED PACA (*Cuniculus paca*; see Color plate, p. 437), the larger of the two. It has a coarse fur and no underwool. 2. The MOUNTAIN PACA (*Stictomys taczanowskii*; see Color plate, p. 437), which has somewhat thicker and softer hair.

The SPOTTED PACA lives on wooded or densely overgrown banks of rivers and other bodies of water, especially in flatlands but also in lower and middle mountainous altitudes. These animals are nocturnal, spending the daylight hours in the burrows they dig in river banks or other slopes, between the roots of trees, or under rocks. These burrows usually have several emergency exits. Generally only one animal lives in a burrow; as night falls the individual animals leave their dens to look for food (usually alone) and to go to the nearby water source. They have definite pathways from the burrow to the feeding ground and to the water. Although spotted pacas are land animals, they are excellent swimmers and enjoy going into the water. They also flee to the water as a refuge in times of danger.

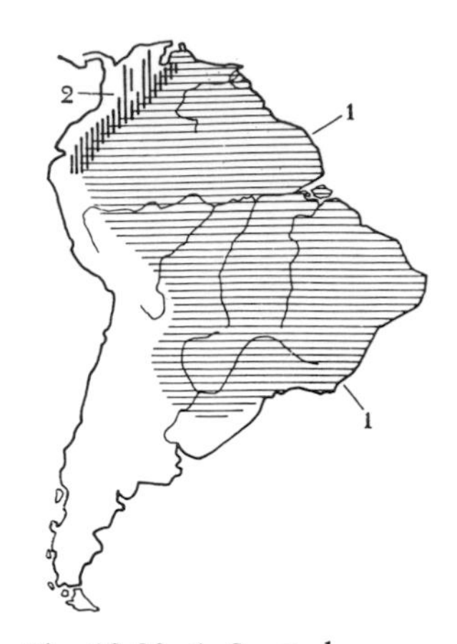

Fig. 13-23. 1. Spotted paca (*Cuniculus paca*); 2. Mountain paca (*Stictomys taczanowskii*).

Females often give birth twice a year, with one young in each litter; twins are rare. Captive pacas have occasionally reproduced, including specimens in the London, National (Washington, D.C.), and Berlin zoos. Several spotted pacas in the National (Washington, D.C.) zoo lived over sixteen years.

Spotted pacas eat the leaves, stalks, and roots of a variety of plants, as well as fallen fruit. They show a particular preference for avocados and mangoes. These animals are not very popular in areas near human settlements because of the large amounts of food they require. They can often cause extensive damage to yam, manioc, vegetable, and sugar cane plantations. Their meat is fatty and very tasty. One often pays more than seventy-five cents for 1 kg of spotted paca meat on the market. It is therefore not surprising that these animals are so ruthlessly hunted by men with guns, traps, and dogs which are specially bred for hunting spotted pacas.

It is very easy to misunderstand an animal's habits and disposition. This is particularly true of nocturnal animals, because their activity period does not coincide with our own, thus making observations of these animals rather difficult. The spotted pacas are an excellent illustration of this point. The fourth edition of *Brehms Tierleben*, which appeared in 1920, still contained Goeldi's description of the spotted paca, which as an exclusively nocturnal animal was "a grumbling and surly philistine capable in all its stupidity of only anger and anxiety."

In 1953 and 1954, I [Heinemann] kept a pair of spotted pacas and a mountain paca, first in my home, and then in a laboratory at the Frankfurt Zoo. I found that these "grumbling philistines" were actually lively and playful creatures, as well as being completely obstinate. At night they jumped over and around one another in their large cage, and playfully stalked each other (these games almost never degenerated into real fights), while making a frightful amount of noise. The male spotted paca often took the aluminum feeding bowl in its teeth and ran around the cage with it, making a lot of noise. Each of the animals had its own particular personality: The male spotted paca liked to be stroked or petted; the mountain paca was unfriendly to us and would snap his teeth at human hands which came too close; the spotted paca female was afraid and would attempt to disentangle herself from one's hands as one picked her up. Once the spotted paca male jumped from its sand bed to the top of a table that was 80 cm high; another time it jumped over a bulkhead that was more than 1 m high. One would not expect such a bulky-looking animal to make jumps of this sort at all.

The mountain paca, "Baby," was trained to recognize certain visual signals. He learned to choose and follow a pathway marked with particular signals, in the process ignoring the three other pathways, also equipped with optical signals, but not the right ones, that were simultaneously presented to him. At the end of the pathway with the correct visual signals, "Baby" always found a dish of food. The animal paid particular attention to his task, and he often worked diligently even after his hunger had been satiated and he was no longer interested in the food. As time went on, the animal learned to differentiate between blue and yellow signals which were of uniform brightness. In the morning I would take "Baby" out of the cage where he lived with the two pacas, and I would carry him to the experimental station in a transport box. Usually he would be waiting at the door of the cage and would immediately step into the transport box, which was sealed with a trap door; "Baby" would even go into the box when the other animals were being fed and when food covered the floor of his cage. When he had been subjected to many successive experimental runs, "Baby" would hide in his transport box, which stood in a corner of the work room, and would not come out until the box was carried back to his cage. A few times I saw "Baby" disappear into his transport box, where he turned around, grabbed the door of the box in his teeth, and closed it behind him. These actions appeared to be particularly intelligent. Actually, however, they were only slight variations of instinctive behavior. Walker has noted that the escape holes of paca burrows are often stopped up with leaves, although he was not able to determine whether the pacas actually did this themselves or whether the leaf piles built up naturally when the burrows were no longer in use. The behavior of my mountain paca seemed to indicate that these animals may have an instinct for closing off the entrances to their burrows, and, in captivity, the den had a wooden door.

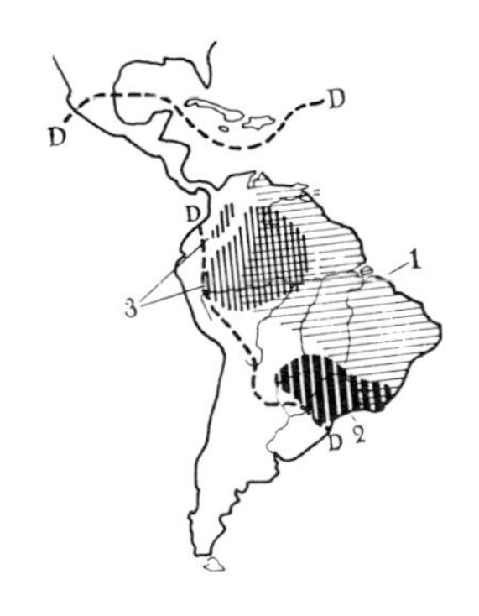

Fig. 13-24. D—D. Approximate distribution of the agoutis (genus *Dasyprocta*). 1. Orange-rumped agouti (*Dasyprocta aguti*); 2. Azara's agouti (*D. azarae*); 3. Sooty agouti (*D. fuliginosa*).

Subfamily: Agoutis

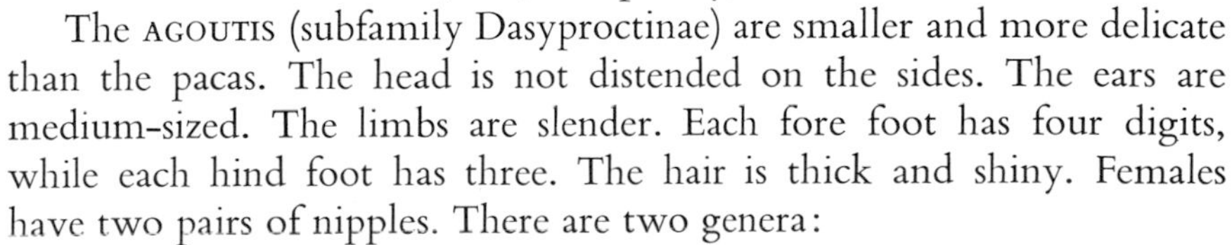

The AGOUTIS (subfamily Dasyproctinae) are smaller and more delicate than the pacas. The head is not distended on the sides. The ears are medium-sized. The limbs are slender. Each fore foot has four digits, while each hind foot has three. The hair is thick and shiny. Females have two pairs of nipples. There are two genera:

A. AGOUTIS (*Dasyprocta*; BL 41.5–62 cm, TL 1–3.5 cm, weight 1.3–4 kg) are the larger animals with shorter tails. There are probably seven species, including the ORANGE-RUMPED AGOUTI (*Dasyprocta aguti*; see Color plate, p. 437), the SOOTY AGOUTI (*D. fuliginosa*; see Color plate, p. 437), and AZARA'S AGOUTI (*D. azarae*), which has an olive-green hair color.

B. ACOUCHYS (*Myoprocta*; BL 32–38 cm, TL 4.5–7 cm) are smaller and have a longer tail. There are two species, including the GUYANA ACOUCHY (*Myoprocta acouchy*).

In contrast to the nocturnal pacas, the agoutis are largely diurnal animals, at least in those areas where they have not been disturbed. They live in moist or marshy woods near lowland rivers, as well as in the dry forests of plateaus, on grassy river banks, in dense thickets, and in open savannas and fields. These animals dig their burrows under rocks, between the roots of trees, or in sloping banks. Apparently, like the pacas, each agouti usually lives alone in its burrow. *Myoprocta acouchy exilis* (see Color plate, p. 437), an acouchy subspecies living in the central Amazon region, reportedly lives in colonies. Well-trodden paths lead from all sides of each agouti burrow to the feeding area. Agoutis are herbivorous; they eat the stalks, leaves, roots, and fruit of a variety of plant species. While eating, they sit on their haunches and hold the food in their fore feet.

Fig. 13-25. M—M. Approximate distribution of the acouchys (genus *Myoprocta*). Shaded area indicates distribution of the acouchy (*Myoprocta acouchi*).

When in danger, the agoutis become motionless, a behavior that can also be observed in the common hare. If these animals are discovered, they race away extremely rapidly, defying pursuit, and take refuge in the hollow centers of fallen tree trunks, often emerging unnoticed from the other end.

Some agouti species seem to have two mating seasons each year. The young are born in a nest of leaves, roots, and hair, after a gestation period of three months. There are usually two young per litter.

According to Walker and Crandall, agoutis may have a life expectancy of ten to twenty years, an astounding age for such a small animal. These animals adapt themselves to captivity well, and with appropriate care they can be bred without difficulty. The fathers are quite friendly with their lively, playful young, as long as they are not sexually mature. Young agoutis raised in captivity are very tame and devoted. These animals, like pacas, can be fed vegetables, fruit, potatoes, and bread.

Man hunts the agouti relentlessly, especially in densely settled areas, because of the great damage it occasionally causes to sugar cane plantations and because agouti meat is particularly tasty.

Superfamily: Pacaranalike rodents, by H. Wendt

The superfamily of PACARANALIKE RODENTS (Dinomyoidea) now contains only one species, although this group had a large and varied membership during the Oligocene, in the Tertiary period (35–40 million years ago).

Extinct giant rodents

Twenty-five years after Columbus discovered America, Gonzalo Fernando de Oviedo y Valdes, one of the first historians to write about the New World, mentioned, in his discussion of West Indian lifestyles, a rather unique animal called the QUEMI, which was found on Haiti and was eaten by the people there. Skull and dental remains of these paca-sized rodents were later found in several caves in Haiti and the Dominican Republic, and in 1929 G. S. Miller gave these extinct animals the scientific name + *Quemisia gravis*. Miller believes that this species died out soon after the arrival of the Spanish, in the first half of the 16th Century.

The quemi belongs to the family + Heptaxodontidae, which is similar to the pacaranas. The other members of this family died out long before the quemi. One species, + *Amblyrhiza inundata*, of which skull and skeletal remains have been discovered on the islands of St. Martin and Anguilla, must have been as large and strong as the North American black bear. We do not know when this giant form died out.

Relatives of this unique Caribbean rodent family were once present on the South American continent. Some giant forms also developed among these animals, such as the genus *Eumegamys*. These animals, however, did not live to see the European's conquest of their environment.

The pacarana

Only one species of this once so varied rodent group, which is related to the octodont rodents, the chinchillas, and the pacas, has survived to the present. This is the PACARANA (*Dinomys branickii*; BL 73–79, TL 20 cm, weight 10–15 kg; see Color plate, p. 437), the only representative of the family Dinomyidae. Its head is wide and somewhat more pointed at the front than is that of the paca. The ears are short and curved. These animals have long vibrissae. The upper lip has a deep cleft. The pacarana is a plantigrade animal. The feet are broad. There are four digits on each foot, and each digit is equipped with a long, powerful claw. The soles of the feet are hairless. The hair is coarse; the individual hairs have a variety of lengths. The tail is densely covered with hair. The incisors are broad and strong. The molars have very high crowns and are probably rootless and continuously growing.

Distinguishing characteristics

The third-largest rodent

The pacarana is the third-largest living rodent, after the beaver and the capybara. These animals are found only in a few widely dispersed areas on the slopes and in the valleys of the virgin forested eastern foothills of the Andes Mountains. Probably the pacarana is quite rare and threatened with extinction. The circumstances surrounding the discovery of this large rodent are very interesting; as is so often the case, chance played an important role. In the 1870s, Count Branicki, a well-known promoter of Warsaw's natural history museum, sent some trained animal collectors to South America, chiefly to investigate some of the rodents of that continent. In 1873 one of these collectors, Constantin Jelski, saw a strange, dog-sized animal with long vibrissae, running about fearlessly in the bright daylight in the garden of the hacienda (country house) of Amablo Mari, near Victoc, in the eastern Peruvian Andes. Jelski killed the animal with his saber, and sent the skin and skull to Warsaw with the comment that apparently the animal was a rarity, as no one in the general area was familiar with it. Peters, a director of the Berlin zoo and museum during the 19th Century, was actively involved in the systematic classification of mammals; he classified the animal as a member of a new genus, according to the information he received from Warsaw.

Fig. 13-26. Pacarana (*Dinomys branickii*).

Thirty years passed and no other pacaranas were seen. The species was believed to have died out. Then in 1904, Goeldi, Director of the

Para (today Belem) museum in Brazil, came across a nest which contained a female pacarana and an almost fully grown young. Like so many other rare mammals, these two pacaranas were brought from the Amazon region to Para in a boat transporting rubber. Other pacaranas came to Para in the following years. William Beebe brought one of these animals to the Bronx Zoo in New York, where it lived for almost seven years on a diet of grains, green matter, fruits, and bread. Pacaranas appear to be unusually even-tempered and peaceful animals. They devour large quantities of food, and have a special preference for fruit from various species of palm. While eating, these animals like to sit on their hind feet and hold the food in their fore feet. They are active mainly at night. Occasionally they will bite through even the strongest wire netting with their powerful incisors.

Margarete Snethlage, who was later to become director of the small zoo in Para, once raised a young animal on milk. The pacarana was familiar with its nurse and always wanted to be near her.

The pacarana disappeared from zoological observation for many years. However, just as the possibility of the animal's being extinct was again suggested, collectors were able to capture several specimens, in the widest range of locations, but especially in the upper courses of the Acre, Purus, and Jurua Rivers in the Amazon region. Some of these specimens appeared in North American and European zoos; however, all of them were in poor condition and died soon. Since 1946, more pacaranas have appeared in zoos, for example in Cologne (Germany) and Basel (Switzerland). A pacarana lived for almost ten years in the Philadelphia (Pennsylvania) zoo. The species has bred in the San Antonio (Texas), Zurich (Switzerland), and National (Washington, D.C.) zoos in recent years.

We know very little about the life of the pacarana in the wild. We do not even know whether these animals' long claws are for digging or for climbing. Pacaranas in captivity have not yet attempted to dig.

My first encounter with a pacarana remains one of the most vivid experiences I [Wendt] had with animals in South America. This large animal, sitting upright on its hind legs, with its coarse hair, heavily bearded muzzle, and long tail, is so unlike the "usual" picture of the South American rodent that it literally leaves you breathless in the first moment. I found the tameness of this animal particularly surprising. Although it had come from the wild to the São Paulo (Brazil) zoo only a short time before, it allowed one to touch and stroke it, and it even tried to come into closer body contact with humans. Pacaranas are contact animals, and their behavior does not seem to include maintaining a certain distance between individual animals. It is true, however, that while some pacaranas are gentle, others are quite vicious. Their behavior seems to be based on individual, rather than species, traits.

The pacarana I observed in the Basel zoo had a similar behavior. It even allowed people with whom it was unfamiliar to pick it up and carry it in their arms, without biting or scratching them. This animal shared its cage with a female two-fingered (two-toed) sloth, and it consistently tried to make body contact with this companion of another species. Initially the sloth defended itself against these attempts and tried to bite, although it was almost never successful in doing so because of its slowness. Later both animals sat side by side on a branch and warmed themselves under the heat lamp. The pacarana in São Paulo and the animal in Basel both liked to climb; perhaps then the long claws are used for climbing trees in order to look for food or for escaping pursuit by fleeing up steep rocks and cliffs.

We know almost nothing about the reproduction of these animals. Possibly the usual number of young in a litter is two; of those born in captivity in 1974, half were litters of one, half, of two; no litters had more than two young. Unfortunately, they do not survive long in captivity, according to the experiences reported to date, with the few exceptions we have noted.

Superfamily: New World porcupines, by D. Heinemann

The last superfamily of cavies and the last in our presentation of the order of rodents is that of the NEW WORLD PORCUPINES (Erethizontoidea), which has only one family, that of the NEW WORLD PORCUPINES (Erethizontidae). It is easy to understand how these arboreal American porcupines were formerly grouped together with the terrestrial Old World porcupines. However, in reality these two rodent families do not have much in common except for their coat of quills.

Distinguishing characteristics

These porcupines are medium to large sized rodents. The body is between 30 and 86 cm long, and the tail length ranges from 7.5 to 45 cm. The head is large and bulky. The muzzle is blunt. The upper lip is split only slightly or not at all. The eyes are small, and the ears are hidden by the animal's hair. The limbs are short and powerful. Each foot has four digits, which are equipped with strong, curved climbing claws. The body and tail hair has been, in part, transformed into short, pointed quills, which often have barbed hooks. There are two pairs of nipples. The molars have roots. The intestinal tract is particularly long (up to over 3 m). There are two subfamilies, which are distinguished by particular cranial characteristics. There are four genera, with eight species:

Fig. 13-27. 1. Approximate distribution of the thin-spined porcupine (*Chaetomys subspinosus*); 2. Approximate distribution of the upper Amazonian porcupine (*Echinoprocta refescens*).

A. The THIN-SPINED PORCUPINES (subfamily Chaetomyinae) have only one species, the THIN-SPINED PORCUPINE (*Chaetomys subspinosus*; BL 43–46 cm, TL 25.5–28 cm; see Color plate, p. 409). The quills are short and undulated. They become more like bristles on the animal's back. The root of the tail is thick, while the tail itself is covered with quills, and on the underside, with short hair. It may be prehensile. These animals live in dense undergrowth on the edges of savannas and cultivated areas. They are usually slow-moving and not particularly lively, although when necessary they are able to move much faster. These porcupines are excellent climbers and jumpers.

B. The subfamily Erethizontinae has three genera:

1. The UPPER AMAZONIAN PORCUPINES (*Echinoprocta*), has only one species, the UPPER AMAZONIAN PORCUPINE (*Echinoprocta rufescens*; see Color plate, p. 409). The quills are short and become strong toward the back. The tail is short and covered with hair; it is not prehensile. Very little is known about the habits of these animals.

2. The PREHENSILE-TAILED PORCUPINES (*Coëndou*; BL 30–60 cm, TL 33–45 cm, weight 900–4310 g). The quills are short and thick. The quills of the wooly prehensile-tailed porcupine and its relatives are hidden in long, soft hair. The tail has no quills, and is hairless on the upper side; it is bent outward and is prehensile. There are five species, including the PREHENSILE-TAILED PORCUPINE (*Coëndou prehensilis*; see Color plates, pp. 409 and 423) and the WOOLLY PREHENSILE-TAILED PORCUPINE (*C. insidiosus*).

3. The NORTH AMERICAN PORCUPINES (*Erethizon*) have only one species, the NORTH AMERICAN PORCUPINE (*Erethizon dorsatum*; BL 64.5–86 cm [the larger extreme being exceptional], TL 14.5–30 cm, weight 3.5–7 kg [exceptionally to 18 kg]; see Color plate, p. 409). The head is thin and the body is powerful. The tail is stubbed. The quills are thick and pointed, with barbed hooks. They are hidden between long, stiff hairs. These animals have an underwool. The under-side of the body has no quills. There are several subspecies.

PREHENSILE-TAILED PORCUPINES (*Coëndou*) are nocturnal, arboreal, and herbivorous rodents, whose movements are slow and cautious. However, they are unusually sure and skillful climbers. Their prehensile tails are unique in that they grip with the upper side (in contrast to the tails of the spider monkeys, marsupial rats, and other prehensile mammals). Wilhem Haake, a zoology professor and former director of the Frankfurt zoo, suggested that the outward-grip posture of these prehensile-tailed porcupines may be related to the general tendency of the tails of nocturnal animals to curl outward. Alfred E. Brehm, in his *Tierleben*, has described the behavior of one of these porcupines when it wants to climb from one branch to another, more distant branch. The animal "holds onto the branch with its hind feet and tail while stretching the front of its body forward horizontally, and attempts to grab onto the selected branch with its fore feet. This position requires a great amount of strength, but the porcupine can maintain it for several minutes, as well as moving back and forth to the side with apparent ease. As soon as the animal is able to fasten its fore feet on the new branch, it looses first its hind feet, then its tail, and swings itself (motivated by its own weight) under the new branch, which it grasps with its tail and hind feet; then it is able to climb up and onto the branch with ease."

Fig. 13-28. Approximate distribution of the prehensile-tailed porcupine (genus *Coëndou*).

Prehensile-tailed porcupines spend the day in the thick, leafy crowns of trees, or in hollow trees, tree stumps, or holes in the ground. Their quill coat serves as very effective protection from most of their enemies.

The quills are loose and fall out easily; because of the barbed hooks, they embed themselves more firmly in the skin of an attacker than in the porcupine's skin. A dog that attacks a porcupine which is quietly sitting on the ground will end up with quills in its nose and mouth; the dog cannot pull these quills out itself, and so, without human aid, it will starve, suffocate, or die of thirst. Often those parts of the body that have been damaged by quills will swell up noticeably. In spite of this, however, many carnivorous animals apparently learn how to overpower the porcupine without being hurt. Azara found porcupine quills in jaguar droppings several times; it seemed as though the quills had passed right through the digestive tract without causing any damage.

Although prehensile-tailed porcupines can force themselves to remain quiet and motionless toward their enemies, they are often rather bellicose among themselves. They bite or snap at members of their own species, and they attempt to stick their quills into one another. Usually these animals live alone. Porcupines can, however, be very friendly and tame toward people. In zoos some of these animals have lived to be almost nine years old.

In general, only one young is born at a time; this newborn is surprisingly large. The young of the prehensile-tailed porcupine (*Coëndou prehensilis*) have a long, reddish fur intermixed with short quills which initially are soft and flexible, but which soon become stiff and rigid.

Habits of the North American porcupine

The NORTH AMERICAN PORCUPINE (*Erethizon dorsatum*) is the only North American representative of the suborder of cavies. These rodents live in extensive forests in the temperate zones. They are nocturnal animals, spending the day in rocky crevices, rock piles, high trees, holes in the ground, or in nests that they build in trees. The North American porcupine is more or less a loner, although one can find several animals together in a particularly favorite spot. These animals do not hibernate, but they do remain in their nests during bad weather.

The North American porcupine is also dependent solely upon its quills for defense when it is attacked. It raises its quills, turns it back to the enemy, and strikes out with its quill-covered tail toward the attacker. Alfred Brehm tells about his experience with a North American porcupine he acquired. He stroked its back to find out the softness of the fur and his hand reached the end of the tail. "I had barely touched it when the animal rapidly snapped its wide, flat tail up, and a stinging pain in my fingertips told me that I had been badly hurt by its defense. There were eighteen quills embedded in my fingertips, so deeply that I was unable to get them out myself." Quills that remain embedded in an attacker's skin, or even quill tips that have broken off under the skin, penetrate further into the body with every movement, because the barbed hooks can only move forward, not back. These quills often move further into the body at a speed that may be as much as 1 mm per hour, and they can be fatal if they reach a vital organ. However, many preda-

Fig. 13-29. North American porcupine (*Erethizon dorsatum*).

tory animals, including the great horned owl (*Bubo virginianus*) are able to overpower the North American porcupine without injury. The fisher (*Martes pennanti*), a mustelid, specializes on porcupines. Some areas of New York State were restocked with fishers, to control the porcupine population.

In spring and summer these animals eat buds, leaves, and thin twigs; in winter they eat pine needles and the bark of trees. Frequently they will strip a tree of so much of its bark that it dies. During the winter, when it is difficult to move from tree to tree because of snow, the North American porcupine often remains at one tree until it has stripped off all the bark. These animals are often ruthlessly hunted in many areas because of the damage they do. However, the loss of trees due to these animals is greatly overestimated. Exact studies of the damage have not been made as yet, but according to the few estimates available so far, no more than 0.7 percent of commercial timber trees have been destroyed by these animals. North American porcupines also make themselves unpopular in logging camps by gnawing canoe paddles, ax handles, saddles, and other objects. Apparently the animals are attracted to these objects by the salty taste which results from the residues of sweat that accumulate with the use of these tools.

The male North American porcupine begins its travels in search of a mate with the advent of winter or shortly thereafter. The mating follows an elaborate foreplay. The female gives birth, almost always to only one young, after a gestation period of 210–217 days. The newborn porcupine weighs 1.5 kg. It is able to climb trees when it is two days old, and is sexually mature in its second year of life.

North American porcupines are not difficult to keep in zoos; however, these slow, nocturnal animals do not have much "visitor appeal." The fence, wall, etc. must be able to withstand their strong incisors, and these arboreal animals must have excellent climbing possibilities available to them if they are to survive. They can be fed crushed oats, corn, and other grains, as well as potatoes, fruit, carrots, and other vegetables, and bread; they also need supplementary vitamins and trace elements. They should be fed fresh pine, maple, and willow branches as often as possible. The longest life expectancy in captivity is seven to ten years. There have not been many breeding successes, but this may be due, in small part, to the fact that many zoos only have single animals.

Phylogeny of the New World porcupines, by E. Thenius

The New World porcupines stand alone in the suborder of cavies. For a long time it was believed that the oldest South American rodents, the genus + *Platypittamys* and their relatives (from which all the other cavies are descended) were the immediate ancestors of the New World porcupines, but now we believe, according to our present knowledge, that these animals are descended, through various transitional forms, from the genus + *Protosteiromys*, which was present in South America during the Oligocene (25–40 million years ago).

14 The Whales (Cetaceans)

Order: Whales (Cetaceans), by E. J. Slijper and D. Heinemann

Of all the mammals, the whales are the furthest removed from the primordial mammalian types, both in habits and body structure. Thus, the fact that Aristotle classified these large fishlike, ocean-going mammals as fish in his systematic classification of the animal world in 400 B.C., is not so surprising. During Aristotle's time, and for several hundreds of years thereafter, an animal's environmental adaptations were generally considered as the basis for its classification. Thus, unlike today, the phylogenetic relationship between certain animals was not investigated or believed important in those times. The whale's former classification as a fish was not the result of ignorance; Aristotle was even able to report in detail the fact that whales have hair, that they breathe with lungs rather than gills, and that they give birth to live young which are then nursed with milk from the mother. In 1693 John Ray was the first to place these animals as mammals, and by 1758 Linné established the subdivision of toothed cetaceans (whales) and baleen whales.

Whales (order Cetacea; L 1.25–33 m, weight 23–136,000 kg) have adapted completely to life in the water; they are not able to survive on land. Their bodies are fish-shaped. The head is large, and the neck is small or is not differentiated at all. The eyes are small; they are involuted in some species. The nostrils, which in whales are called blow holes, are located toward the back of the body, generally at the highest point of the head; toothed whales have only one blow hole. There is no external ear. The fore limbs have been transformed into flippers. The upper arm and forearm are reduced. The "hands" are often lengthened, and the number of digital phalanges is increased. The hind limbs, with the exception of the pelvis (and the thigh bones in right whales), have withered away. The tail has developed into a horizontal fluke which is the major source of locomotion. These animals usually have a dorsal fin made of connective tissue. The body hair has retrogressed so that only a few tactile hairs remain. The dentition either consists of numerous uniform teeth (in the fish-eating toothed whales), or is partially degenerate (as in the sepia-eating toothed whales), or is completely lacking (as in baleen whales). The lower jaw is expanded and does not have a perpendicular branch. The nose bone has been shifted toward the rear. The upper jaw and the intermaxillary bone are expanded and have a strong growth toward the back; these bones have been shifted over (and in the baleen whales, partially under) the frontal bone. The parietal bone is crowded to the side. The vertebral column in the neck is shortened, and the vertebrae may be fused.

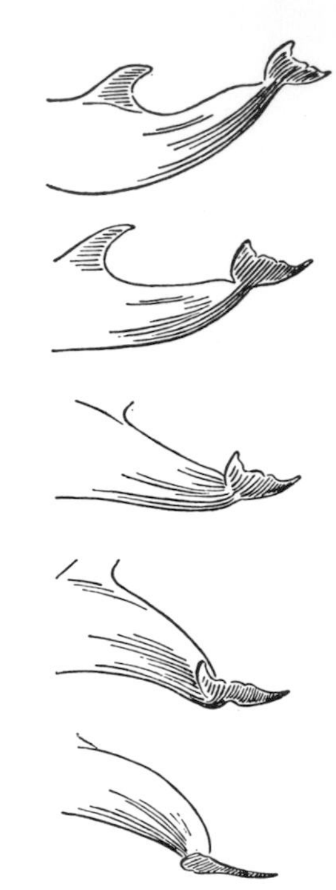

Fig. 14-1. Ten tail movements of a swimming bottle-nosed dolphin (taken from a film made in Marineland Aquarium, Florida). Above: downward motion; Right: upward motion.

There are two suborders, with a total of eleven families, thirty-eight genera, and ninety-two species. These suborders are the BALEEN WHALES (suborder Mystacoceti), with the families of right whales (Balaenidae), gray whales (Eschrichtiidae) and finback whales (Balaenopteridae); and the TOOTHED WHALES (suborder Odontoceti), with the families of sperm whales (Physeteridae), beaked whales (Ziphiidae), gangetic dolphins (Platanistidae), Amazon dolphins (Iniidae), white whales and narwhals (Monodontidae), porpoises (Phocaenidae), long-snouted dolphins (Stenidae), and dolphins (Delphinidae). Another suborder, that of the + Archaeoceti, died out in the pre-Oligocene period (some 30 million years ago).

The whale's fishlike shape

Aside from their torpedolike fish shape, whales have many other internal and external structural adaptations which allow them to live, move, and feed in the water. Their skin is smooth and hairless, with the exception of a few single hairs or hair remains near the jaw, which may serve as tactile hairs. The fore and hind limbs appear in the embryo like those of any other mammal. However, the fore legs always develop into flat, more or less extended flukes. Whereas in a whale embryo that is 20 mm long one can see external indications of the hind feet, these appendages, indications and all, have completely disappeared by the time the embryo is 30 mm long. The blow hole (nostril) is located on the end of the snout in very young (4–5 mm long) whale embryos. (All other mammals have a similar location of the nostril.) However, by the time the embryo has reached a length of 22 cm, the blow hole has been shifted back to the upper part of the head, the same location as in adult whales. The location of the blow hole is probably related to the whale's weight distribution, which differs from that of land mammals, (see Fig. 14-3). The sperm whales are the only members of this order which have their blow hole further up front.

The breast fins or flukes play a very minor role, or no role at all, in the whale's usual swimming motions; the tail provides all the driving power necessary for forward movement. Scientists have been able to study the tail movements of some whale species in the large American seaquariums, and, with the help of photographs, they have established that the tail is moved up and down on an exact vertical line. The real motor drive is produced almost exclusively by the broad, sweeping tail fin, the sides of which are not supported by bones. The longer forward

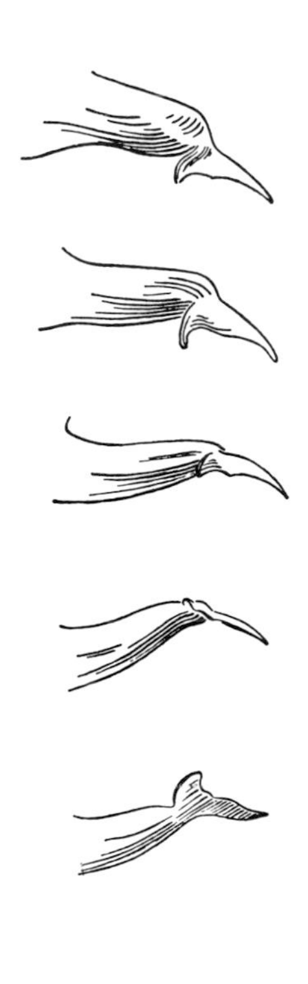

part of the tail provides the muscle power necessary for this driving motion; this part of the tail is a deep oval which is compressed at the sides, and it cuts through the water with very little resistance as the whale moves it up and down.

The speeds that whales can attain are astounding, at least insofar as baleen whales and dolphins are concerned. The slower right whales and gray whales can reach a peak speed of some 11 km/h (7 mph); the humpback whale can reach a speed of up to 18 km/h. The usual speed of these three whales is generally from 3.5 to 5.5 km/h. Sperm whales can maintain a speed of 18 km/h over long distances, and they may travel as fast as 37 km/h over short periods of time. The large finback whales can maintain a general speed of between 22 and 26 km/h, while their peak speed is 50 km/h. The much smaller bottle-nosed dolphins and other dolphins are also able to reach similar speeds; scientists have measured a general velocity of 22 to 26 km/h for these animals. Thus, they, like the large finback whales, are able to keep pace with our modern passenger ships. This becomes particularly remarkable when we consider that in animals of similar build, the larger animal (like the larger machine) usually will be faster than the smaller. However, we can understand this ability of the dolphins and bottle-nosed dolphins if we accept the fact that these animals do not produce any waves or whirls on the surface of the water as they hurry along below; instead, the water layers, which are cut apart at varying speeds by the animal's body, glide along smoothly on each other (laminated streams). Recent studies by M. O. Kramer have shown that the dolphin's elastic skin surface is largely responsible for its ability to swim fast without producing whirls or eddies in the water.

Breathing and diving

One of the greatest difficulties that a vertebrate land animal must overcome in its transition to life in and under water is the adaptation of its breathing to the very different requirements of aquatic living. Unlike fish, which breathe through gills, whales are forced to come to the surface of the water in order to breathe. When it reaches the surface, the whale expels the moist air from its lungs, under great pressure. When the pressure on the exhaled air is released, the air rapidly disperses and, as a result, is cooled to such an extent that the moisture it carries condenses and becomes visible as a white cloud. This cloud of steam, the "blast," can reach a height of 3–4 m with right whales, a height of 2 m with humpback whales, 4–6 m with rorquals, 6 m with blue whales, and 5–8 m with sperm whales. Each whale species has a distinctive steam cloud, and a person acquainted with the types and shapes of clouds can easily distinguish the type of whale. Right whales have a double cloud. Finback whales release a cloud that is shaped somewhat like a pear, and the sperm whale's cloud is directed forward on a diagonal line.

The whole procedure of exhaling and inhaling lasts only for one to two seconds; large whales can exhale some 2000 l of air through their small blow hole, and inhale a similar amount during this short time. Finback whales swimming quietly on the surface breathe every one or two minutes; when they make deep dives, they come up for air only every four to forty minutes. When feeding, finback whales usually do not dive deeper than 10–15 m. Scientists were able to attach a manometer to a harpoon and thereby determined that finback whales can dive as deep as 350 m without difficulty. Sperm whales and bottle-nosed whales normally reach a diving depth of at least 500 m, and they are able to remain under water for between fifty and ninety minutes. Sometimes sperm whales entangle themselves in the telegraph or telephone cables on the ocean floor, and consequently drown miserably. To date, thirteen such cases have been reported. Five times these accidents occurred at depths of some 900 m, and one such accident occurred at a depth of 988 m! At those depths the water pressure is about 100 atmospheres. Dolphins and porpoises probably do not dive any deeper than 25 m. According to Cadenat, bottle-nosed dolphins apparently can dive to 200 m. Man can dive at most 120 m without protection from the water pressure.

One immediately asks why the bodies of the deep-diving whales are not crushed by such enormous pressures. The answer is that the whale's body mass is made up mainly of materials that are hardly able to be crushed; the water content alone (as in man) accounts for two-thirds of the body weight. The lungs, which are filled with air, one might think could constitute a problem. However, the volume of the whale's lungs can be reduced to one-tenth, so that a difference in pressure between the contents of the lungs and the outer surface of the whale does not occur until the animal has gone below a depth of 100 m. The danger of such a difference in pressure is reduced as the volume of the lungs becomes smaller. Thus, it is important that the deep-diving whales take only a small volume of air down with them. For this reason, the lungs of sperm whales, bottle-nosed whales, and finback whales have a weight and volume capacity which is only half as large as that of land mammals. In contrast, lungs of the porpoises and dolphins, neither of which dive very deep, are one and one half to two times larger than those of land mammals.

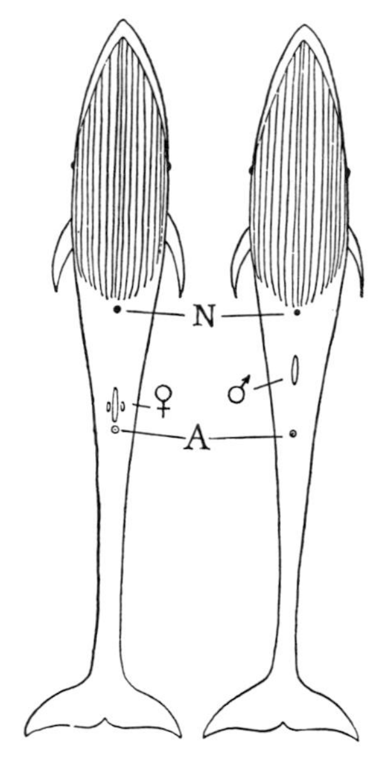

Fig. 14-2. Left, female rorqual; right, male rorqual; both views are of the abdominal side. The longitudinal grooves or furrows extend from the throat area almost as far as the navel (N). The penis slit (♂) is much further forward than the female genital orifice (♀), which is visible near the nipple slits. The part of the body behind the anus (A) is the tail.

This discovery is surprising. One would think that animals which remain underwater for forty, and even ninety, minutes need particularly large lungs in order to carry enough oxygen down with them. However, in reality, the lungs are not the only source of oxygen. In man, the lungs contain only 34% of the oxygen reserve; 41% of the oxygen is found in our blood, where it is combined with the red pigment, hemoglobin; 13% is distributed in the muscles; and 12% can be found in the remaining body tissues. The oxygen in the muscles is loosely combined with myoglobin. In contrast to man, the whale carries only 9% of its oxygen in its lungs. However, it carries 41% in its muscles, 41% in its

Fig. 14-3. Comparative positions of a horse and a porpoise in water. The horse assumes a more sloping or diagonal position (in relation to the water surface) because (due to the position of its lungs) the buoyant force which pushes its body up is located more toward the head than is the gravitational force which pulls its body down.

blood, and 9% in the remaining body tissues. Because of its high myoglobin content, the whale's flesh is much darker than the flesh of most land mammals.

The whale's diving capabilities are not explained solely by the large amount of oxygen it carries in its muscles. We must realize that the metabolic processes of the muscles (which would die without free oxygen) play a more important role when the whale is diving than when is is resting quietly on the surface. Land animals also have an oxygen-releasing process as part of their muscle metabolism; however, this process becomes insignificant when compared to their oxygen utilization process.

The reason deep-diving whales are not affected by the bends can be explained by the limited capacity of their lungs. A man who makes a deep dive without protection against the water pressure dissolves a large amount of the air in his lungs into the blood, because new air is constantly being supplied from the air tank to the blood. If the diver surfaces too rapidly, the resulting lower pressure causes the nitrogen to form bubbles in his blood resulting in the bends which may lead to death. During slow ascent the nitrogen is exhaled through the lungs without harm. Deep-diving whales, however, have such a small quantity of air in their lungs to begin with that only a small amount finds its way into the blood.

With each breath he takes, man exchanges only ten to fifteen percent of the air in his lungs; this is also true of the other land mammals. Whales, however, exchange eighty-five to ninety percent of the air in their lungs with each breath. Thus man must breathe sixteen times a minute, while a dolphin of the same size averages only one to three breaths a minute, in spite of its prolonged diving time. The large sperm and finback whales breathe, on the average, every two minutes (their long diving times have been included in these calculations). The whale's breathing organs have a particularly firm and incompressible structure due to the pressure changes at different water levels and to the fact that these animals empty their lungs almost completely with every breath. The rings of cartilage, which in man keep the trachea and large bronchi firm and open, in whales reach all the way down to the smallest bronchial branches. One can also find elastic vessels in the long tissues, especially in the pleura. Dolphins have rows of interconnecting valves in each of their tiny bronchi; these regulate the animal's breathing during its diving and surfacing, and they can be closed off by the muscular fasciculus. Deep diving baleen whales, bottle-nosed whales, and sperm whales are able to close off every single lung vesicle with a small muscle (sphincter).

Digestion and digestive organs

The stomach of a large baleen whale can hold about 1000 l. This is not very much, when we consider the enormous size of these animals. However, a large stomach is not to be expected in a carnivore. The general structure of the whale's stomach is surprisingly similar to that of an herbivorous mammal. It consists of a large crop or antestomach which has no glandular discharge, a main stomach with its stomach glands which release pepsin and hydrochloric acid, and a third stomach cavity which contains the secretion of those glands in the pyloric area. The crop probably has a role similar to that of the gizzard in birds (see Vol. VII), where the food is crushed between sand and pebbles as the bird's muscles contract. Like birds, whales swallow their food whole; the whales probably use the hard shells of tiny crabs or the bones of the fish they eat for crushing their food, just as the birds use sand and pebbles. Stones and gravel have been found in the stomachs of toothed whales.

As is the case with most carnivores, the whale's small intestine is not located next to the large intestine. Baleen whales and the gangetic dolphins have small caeca. The other toothed whales have no such structure.

Insulation

Water has a thermal conduction capacity which is about twenty-seven times that of air; thus aquatic animals that have their own body heat need special protection that will limit heat loss. A heavy fur coat is not particularly expedient for an animal that spends all of its life in the water (compare this with the seals, Vol. XII). Consequently, whales have completely lost their fur, with the exception of a few tactile hairs, and instead they have developed an insulating layer of fat. The whale's outer skin is thin, only 5–7 mm in the largest species. In addition, the corium or dermis generally reaches a thickness of only a few millimeters, and thus cannot (excepting the skin of the white whale and certain areas of the sperm whale's skin) be made into leather. However, the subcutaneous fatty tissue embedded in the subcutaneous connective tissue under the dermis is well developed. The fat (blubber) layer in right whales is 50 cm thick on the average, but it may reach a thickness of 70 cm. In sperm whales and humpback whales, this fat layer is an average of 12–18 cm thick, and in blue whales and rorquals it is generally 8–14 cm thick, while in sei whales it is only 5–8 cm thick. These figures are only average measurements; the fat layer has a varying degree of thickness in specific areas of the whale's body. Large whales have a thicker blubber layer than do the smaller members of their species. Pregnant females have an especially thick fat layer, while nursing mothers have only a thin one. The thickness of this blubber layer also changes with the seasonal variations in food. When members of a migrating whale species first appear in the polar waters (which have a large supply of food), their fat layers are very thin. However, by the time these whales leave the polar areas, their blubber layers are quite thick. The fat layer of the antarctic baleen whales do not become any thicker after the end of January. If it did these animals would be unable to dissipate the excess heat that is generated by their rapid swimming, and they would perish in the polar circle from an excess of body heat.

▷
Right whales: 1. Atlantic right whale (*Eubalaena glacialis*); 2. Bowhead whale (*Balaena mysticetus*); 3. Pygmy right whale (*Neobalaena marginata*).

▷▷
Gray whales: 1. Gray whale (*Eschrichtius gibbosus*). Finback whales: 2. Sei whale (*Balaenoptera borealis*); 3. Lesser rorqual (*B. acutorostrata*); 4. Common rorqual (*B. physalus*).

▷▷▷
Finback whales: 1. Blue whales (*Balaenoptera musculus*); 2. Bryde's whale (*B. edeni*); 3. Humpback whale (*Megaptera novaeangliae*).

Excretion of salt

Vertebrate animals are not really built for an aquatic life; the salt content of their blood and body tissues is less than that of sea water.

1
2
3

2
3
4
1

3
1
2

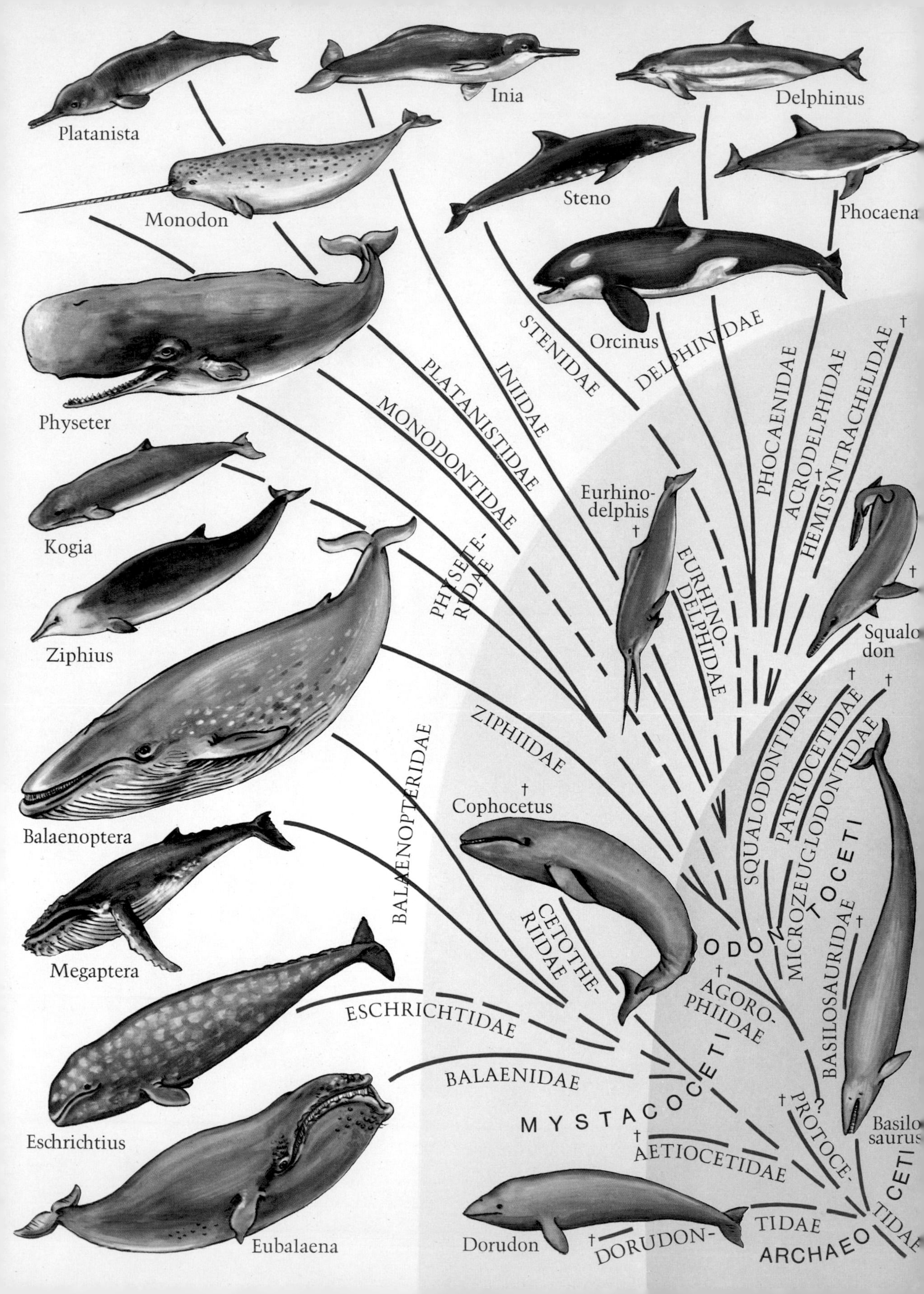
Platanista
Inia
Delphinus
Monodon
Steno
Phocaena
Physeter
Orcinus
STENIDAE
DELPHINIDAE
PLATANISTIDAE
INIIDAE
MONODONTIDAE
PHOCAENIDAE
ACRODELPHIDAE
HEMISYNTRACHELIDAE
Kogia
Eurhinodelphis
EURHINODELPHIDAE
Ziphius
PHYSETERIDAE
Squalodon
ZIPHIIDAE
SQUALODONTIDAE
PATRIOCETIDAE
MICROZEUGLODONTIDAE
Balaenoptera
Cophocetus
BALAENOPTERIDAE
ODONTOCETI
Megaptera
CETOTHERIIDAE
AGOROPHIIDAE
BASILOSAURIDAE
ESCHRICHTIDAE
BALAENIDAE
MYSTACOCETI
PROTOCETIDAE
Eschrichtius
AETIOCETIDAE
Basilosaurus
Eubalaena
Dorudon
DORUDONTIDAE
ARCHAEOCETI

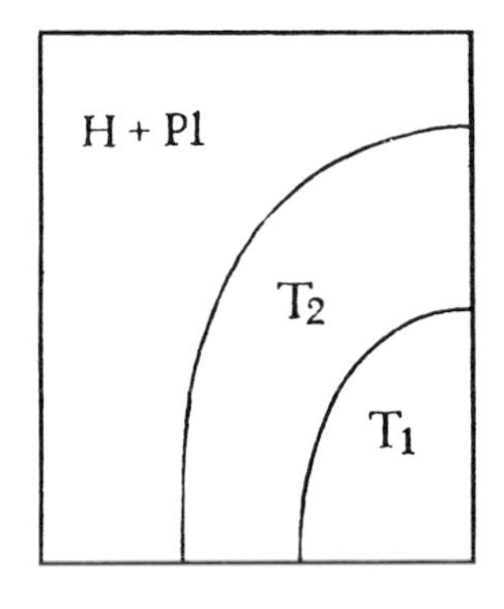

Thus the excess salt which is accumulated from food and water must be eliminated from the body. Whales do not have salt excretory cells like those the bone fishes have on their gills (see Vol. IV). They are not able to excrete salt through their nasal glands like many sea birds (tube-nosed swimmers; see Vol. VII). Whales also do not have perspiration glands, which help many other mammals give off excess salt. The kidneys are the whale's only salt-excreting organs. Whales have very large kidneys relative to those of land mammals; the kidneys of small whales are twice as large as those in land mammals of the same size. In addition, the whale's kidneys are divided into several lobes, the renculi, so that the important exterior surface area is increased. Whales are probably able to excrete large quantities of urine with their kidneys. We do know that the urine of the bottle-nosed dolphin occasionally contains particularly large amounts of salt.

Regulation of blood pressure

Whales have a much larger fluctuation in their blood pressure when they dive and surface or inhale and exhale than do land mammals. The retia mirabilia, which work to maintain a constant blood pressure (and are present in many other animals), are particularly well developed in the whale. These have a spongy looking shape, and consist of a widely branching and inextricable network system of arteries. In whales, these structures are located mainly to the rear of the thorax near the spinal column, but they are also found in the neck, between the ribs, in the vertebral canal, and at the base of the skull. Retia mirabilia made up chiefly of veins are found mostly at the back of the abdominal cavity. Moreover, they are connected to the two large longitudinal veins which run down through the entire vertebral canal, from the ventricle of the brain to the tip of the tail. Apparently the blood in these tiny veins can flow in either direction. These arrangements make it possible for the whale to move its blood "out of the way" and into the retia mirabilia whenever the blood pressure in any particular area rises sharply. As soon as this pressure is reduced, the blood is available once again.

◁ The phylogeny of the whale: A. The extinct suborder +Archaeoceti. B. Baleen whales (Mystacoceti): 1. +Cetotheriidae; 2. Right whales (Balaenidae); 3. Gray whales (Eschrichtiidae); 4. Finback whales (Balaenopteridae). C. Toothed whales (Odontoceti): 1. Fossil families; 2. Beaked whales (Ziphiidae); 3. Sperm whales (Physeteridae); 4. White whales and Narwhals (Monodontidae); 5. River dolphins (Platanistidae and Iniidae); 6. Long-snouted dolphins (Stenidae); 7. Dolphins (Delphinidae); 8. Porpoises (Phocaenidae). Geological periods: T_1 Lower Tertiary; T_2 Upper Tertiary; Pl Ice Age (Pleistocene); H Modern day (Holocene).

Migration

Most whales are ocean-going animals; the few freshwater forms belong to the suborder of toothed whales. All baleen whales, as well as some species of toothed whales, have regular migrations. In 1931, scientists began to study those animals regularly, because of the economic importance of the whale industry; whales were tagged or marked so that scientists could plot their migrational routes and find out more about their way of life. A rust-free metal strip about 27 cm long was used to mark the whale; this strip had a lead top and could be shot at the animal from a rifle (with the help of a hunting cartridge). The marker would usually remain stuck in the whale's back, often for extremely long periods of time. Whales with markers that were shot into them over thirty years ago are still being caught (1970). The men of the ship *William Soresby* shot 5063 markers into whales over a period of six years; about 400 of these markers have been recovered.

Reproduction

Thus far, only a very few people have been able to observe the whale's reproductive behavior. All of the observers agree that the copulation is preceded by an apparently extensive courtship. The partners swim toward one another with their breast flukes. The copulation itself lasts only five to twenty seconds. Humpback whales, rorquals, and sperm whales have been observed in a particular copulation position; the partners face each other, stomach to stomach, and rise out of the water at right angles to the surface to mate. In other cases, the animals swim side by side, each with its underside turned toward the other. Dolphins sometimes swim side by side while the male bends his tail under the female's abdomen and so mates with her in this position.

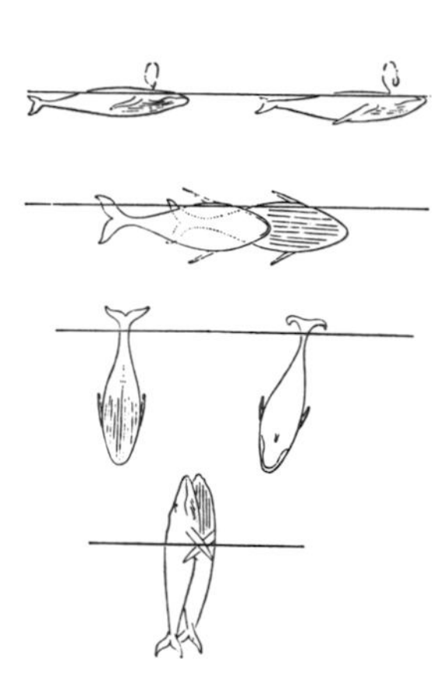

Fig. 14-4. Diagram of the mating process in the humpback whale. The spouting male follows the female; then both whales swim side by side with their stomachs turned toward one another. They "stand on their heads," so that the tail flukes extend out of the water; the two whales rise out of the water, perpendicular to the surface, for the actual mating.

The gestation period in dolphins and porpoises lasts for ten to twelve months. The larger baleen whales have a gestation period of only eleven months. This is particularly noteworthy, since within a related group the larger forms usually have a much longer gestation period than do the smaller animals. The larger toothed whales (in any case, the sperm whale) have a gestation period of sixteen months. The young are fully developed at birth, and look almost exactly like their parents. They are rather large at birth; newborn blue whales are about 7 m long and weigh 2 t. Newborn dolphins, relatively speaking, are even larger; they may be half as long as the mother and may weigh up to one-sixth of her weight. There is almost always only one young; twins appear (almost exactly as in man) in only one out of a hundred births. Triplets are much rarer. In rare occurrences, up to six embryos have been found inside a dead female; however, it is doubtful whether these sextuplets would have developed enough to have been born.

Unlike most of the other larger mammals, whales are born tail first. The birth itself is usually a fairly smooth process, in spite of the size of the young, due to their smooth fish shape. Scientists have been able to observe the birth of some of the smaller whale species such as porpoises, dolphins, spotted dolphins, and bottle-nosed dolphins, in the larger seaquariums. The female bottle-nosed dolphin starts to swim more slowly when her labor pains begin. The other females of the group remain close to the pregnant female, presumably to protect her and to see that she does not get separated from the herd. The umbilical cord is very long, and it cannot be easily broken until the head of the whale calf is completely free; only then does the cord break off, near the calf's stomach. This prevents premature breathing motions by the calf. The mother pushes her young to the surface of the water immediately after birth; there it fills its lungs with air for the first time. The mother is often helped by several other adult members of the herd ("aunts") as she cares for the calf. The placenta appears from one and a half to ten hours after birth; the mother does not eat it.

Blue whales and rorquals nurse their young for between five and seven months; most of the other whales nurse their young for almost a year. The nipples are situated in skin folds on both sides of the genital orifice; the young nurse underwater (a practice we are familiar with in hippopotamuses and sea cows). Whale milk is particularly rich in fats and protein; it is 40–50% fat (compared to between 2 and 17% in land mammals), and only 40–50% water (80–90% in land mammals). The protein content is twice that of land mammals, and the sugar content is 1–2% lower than that of land mammals, whose sugar content is 3–8%. This exceptionally nourishing milk helps the whale calf grow very rapidly; a young blue whale, for example, grows 9 m in seven months, a daily growth of 4.5 cm. During this time, the young whale's weight increases from 2 to 23 t, an increase of about 100 kg daily. Bottle nosed dolphins are sexually mature at five years, although common porpoises reach sexual maturity at fifteen months. The larger whales are sexually mature after four to six years.

The females of some whale species can be re-impregnated immediately after the birth of their young. However, usually female blue whales, rorquals, and sperm whales do not become pregnant until after their young are weaned. For this reason, these species generally give birth only once every two years. If we assume that the large whales have a life expectancy of thirty to forty years (we know very little about the age limits and other factors), then a female can give birth to between ten and twelve calves at the most during her lifetime. This is a rather low birth rate, adapted to the low natural death rate of these animals.

Economic significance

Man has long hunted whales and used various parts of the whale for his needs. The early inhabitants of Alaska hunted whales in 1500 B.C. The first written report of whaling in North America is from A.D. 890. The most important part of the whale for man is the oil which comes from the blubber layer in the subcutaneous connective tissue. Earlier, whale oil was used as lamp oil; today it is used principally in the manufacture of margarine. Whale oil also plays an important role as a raw material in the soap industry and in the production of linoleum and synthetic resins. The oil yield from whaling is only two percent of the world's fat production and only five percent of the production of animal fats. Thus, it cannot be said that whaling has any great significance in the nourishment of an ever increasing human population. The meat of the larger baleen whales is eaten especially by the Japanese, but also in parts of western Europe. The meat of these whales is also used as dog food and, when it has been dried and crushed into meat powder, as cattle feed.

Man uses whale bones to make glue and gelatin or manure (once the bones have been crushed). The baleen whale was formerly important as the supplier of the "whale-bone" (baleen) which was a vital feature of the corset industry. Today we use steel springs and synthetic materials for the same purpose. Additional whaling products include vitamins and hormones from different internal organs, connective tissue fibers (used, for example, in stringing tennis rackets), ivory from the teeth of sperm whales, spermaceti, and ambergris.

History of whaling

In 1583, when the Englishman Jonas Poole attempted to find a passage through the Arctic Ocean to the east, he found a large number of whales; Heemskerk, Barendz, and de Rijp, three Dutchmen who made the same trip in 1596, also found many whales. Although these explorers were unsuccessful in their quest, their reports of a whale kingdom brought whalers from Norway, Denmark, Germany, the Netherlands, and England to the far north. At that time whalers hunted mainly the bowhead whale and the Atlantic right whale. These two right whales are very slow swimmers; their fat layer is so thick that their corpses do not sink. Thus, man was able to hunt these animals with the relatively simple tools available to him at that time. The dead whale was not processed until the whalers were back on land, and so islands like Jan Mayen (in the Atlantic Ocean), Svalbard (Spitsbergen, Norway), and others became whale-processing sites. Later, whalers were able to bind the whale carcass to their ships and thus remove the fat while they were still at sea. Americans began to take an active part in whaling from the 18th Century on. They built whaling stations on the coast and, together with the British, the French, and the Portuguese, began to hunt the humpback whale, the southern right whale, and especially the sperm whale.

The Norwegians and the British have always tried to hunt the finback whale as well. This is a difficult task, as finback whales are much faster swimmers than right whales and their fat layer is so thin that the carcass sinks. In addition, their baleen is smaller and poorer than that of the right whales; for these reasons and because these animals have such a low fat yield, they are much less profitable to hunt than the right whales. Hunting the finback whale first became practical after Svend Foyn developed the harpoon gun with its grenadelike harpoon and after man began to hunt whales with steam-powered ships. The harpoon has an explosive cap just in front of the barb, which explodes inside the whale. If the gun is aimed correctly, the whale may be dead in a matter of seconds. Then the whalers pump air into the whale carcass with a tube, so that it will not sink.

Depletion of the whale population

The whale population in the northern seas was so drastically reduced by man and his whale hunting that whaling in that area is hardly worth the effort. The growth of the natural-gas and steel industries in the 19th Century reduced the demand for whalebone and (temporarily) for whale oil, so that whales had a relatively undisturbed time during the second half of that century. However, around the turn of the century, whaling began to flourish again, especially after 1905, when a new process was developed for hardening fat (the conversion of unsaturated fatty acids to saturated fatty acids), which led to the use of whale oil in the production of margarine. At this time man began to search for new hunting grounds.

James Cook, Sir James C. Ross, James Weddell, and other explorers of the 18th and 19th Centuries had reported huge populations of baleen whales around Antarctica. Thus, at the beginning of this century, whalers began to hunt their prey in these southern waters on a large scale. Initially, the whalers used whaling stations on land for the "industrial" processing of whales, but later floating "kitchens" allowed them to do the work at sea. Whaling in the Antarctic reached its peak in the 1930s. At that time forty-one whaling ships were in use, along with their longboats. The whale population declined steadily as a result of this overkill. Whaling did not reach the same high point after the Second World War, but at the same time other nations, including Russia, Japan, Peru, and Chile joined the first whaling countries in the hunt for these large mammals. As a result, 33,000 whales were killed in the antarctic waters alone in the first few years after the war; 25,000 of these were rorquals. In addition, some 11,000 whales are killed annually by hunters from whaling stations on land.

Protective measure

When we read these figures, we can understand why not only international conservation societies are concerned about the decrease in the whale population, but why judicious governments are attempting to find workable ways of limiting the number of whales being killed. However, most whales are killed on the high seas, outside the territorial waters of individual countries, and consequently any protective measure regarding these animals must be adopted by voluntary international agreement. This results in restrictions on the whaling fleets of those countries that support an agreement to protect these animals; unfortunately, however, not all countries are prepared to support any such agreement at this time. The first covenant of this sort was reached in 1936. Ten years later, on 2 December 1946, representatives of the countries involved in the whaling industry met in Washington, D.C., and established the International Whaling Commission, which now represents eighteen different countries.

International Whaling Commission

This commission has placed many restrictions on whaling. There are areas where one is not allowed to hunt whales at all, while whaling is allowed elsewhere only during definite hunting seasons. Right whales and gray whales are completely protected, as are females with calves. Unfortunately, it is not possible to protect female whales, or even pregnant females, completely; one does not really know the sex of a whale until it is dead and on the rope. Only in sperm whales is the sex of the animal easy to determine, as the females are much smaller than the males. Minimum permissible sizes for the individual species have been established in an attempt to limit the killing of whale calves and female sperm whales; animals that have not reached the minimal size are protected.

Lack of knowledge impedes protection

One of the most difficult jobs of the whaling commission is determining the number of whales that can be killed during each coming hunting season. The commission allowed a total of 15,000 "blue whale units" (whereby two rorquals, six sei whales, or two and one half humpback whales could be killed instead of one blue whale) to be killed per season in the years following the Second World War.

Unfortunately, it became apparent that the hunting quotas were too high and that in some cases the protective measures adopted by the commission were not being followed. There are government inspectors on board every whaling ship, whose job is to make sure that the commission's rules and restrictions are followed. Nevertheless, in recent years these controls do not seem to be working properly. For example, sperm whales that are no more than 11.5 m (38 feet) long may not be killed. Thus the female sperm whales are almost completely protected, as they rarely reach such a length; this protection is particularly important for these polygamous animals. Nonetheless, in 1963, homeward-bound whalers that had been working around Antarctica caught a total of 2004 sperm whales in the Indian Ocean; about half of these sperm whales were females. Each and every one of the hunting statistics indicated a total length for these animals of 11.5 or 11.8 m (38 or 39 feet). The blue whale and the rorqual are presently threatened with extinction.

Whether the protective restrictions are adequate does not depend entirely upon whether they are followed. It is equally important that we be thoroughly familiar with the habits and reproductive behavior of the individual whale species. Only in this way will we be able to predict how this or that precautionary measure will affect the whale population. Such predictions are easy to make when one knows the strength of the existing stock, the number of animals caught annually, the natural death rate, and the number of calves born each year. However, of all these prerequisite figures, we presently have an exact knowledge of only one, the number of animals captured (and killed) annually. One could perhaps ascertain the size of the whale population through an extensive marking program, which would probably take but a few years to do. The natural birth and death rates can be approximately calculated. Professor Ruud from Oslo (Norway), Slijper and his associates, and other research groups are working to develop a reliable way of determining a whale's age. A Commission on Population Dynamics (scientists who investigate the population development of all living beings), working with the latest data, has been able to closely approximate the number of whales killed annually; this commission works under the auspices of the Food and Agriculture Organization (F.A.O.) of the United Nations.

▻ A whale being harpooned (above). Efficient whaling ships and the harpoon gun make it possible for today's whalers to hunt the faster finback whale. Man, with all his new techniques, is now able to kill off these giant sea mammals within a few years, and only the voluntary restriction of whaling will postpone, or even hopefully prevent the total extinction of these animals.

A whaling station on the South African coast (below). Land whaling stations play a minor role compared to the whaling ships, complete with their own cooking stations, which roam the high seas.

Phylogeny of the whale, by E. Thenius

The members of the genera + *Eocetus* and + *Pappocetus* were among the geologically oldest whales of the Eocene (some 50 million years ago); the members of these two genera were already completely adapted to an aquatic life, both in their body shape and in the development of their limbs. The fore limbs had been transformed into flukes; the hind limbs had regressed. The tail carried a fluke, and the ear organs were similar

1♀
1♂
2
Zieger

◅
Sperm whales: 1. Male and female sperm whales (*Physeter catodon*); 2. Pygmy sperm whale (*Kogia breviceps*).

to those found in whales today. Consequently, we must assume that these animals' ancestors moved from land to water much earlier, probably in the Lower Tertiary, or in the Upper Cretaceous period (65–75 million years ago). The ancient insect eaters (+ Protoinsectivora) or the ancient carnivores (+ Creodonta) are the only animals that might possibly be considered as ancestors of the whale. The two major suborders, the baleen whales and the toothed whales, must have separated from each other during the Lower Tertiary. It is even possible that each of these two groups has a different ancestral group (and evolution). Thus Slijper, as the result of his comparative anatomical studies, believes that toothed whales are descended from a long-tailed form of ancient insect-eaters or carnivores that were aquatic. He believes that the baleen whales are descended from a short-tailed form of the same group. Should this hypothesis prove to be correct, we would have to classify baleen whales and toothed whales as two separate orders, instead of suborders.

The oldest known whales, from the Eocene, and their later relatives have been classified as a separate suborder, the + Archaeoceti, because of their numerous primitive characteristics. These animals definitely lived in shallow water. Their fossil remains have been found in the sediments of the former coastlines of northern and western Africa, Europe, North America, and New Zealand. The skull of the Archaeoceti whales seems to be related to the skull of the ancient creodonts. It does not have the asymmetrical structure of the toothed whale skull, nor does it have a shift in the position of the individual bones. The + *Protocetus atavus*, from the Middle Eocene in Egypt, is particularly interesting. Its skull is elongated, the blow hole (nostril) is toward the front of the head (as in land mammals), and the dentition consists of eleven teeth in each half of the jaw (3.1.4.3, as in the ancient carnivores), although the incisors are shaped like canine teeth. The vertebrae, as far as is known, are similar in structure to those of land carnivores. These animals even had sacral vertebra attached to the pelvis; this connection is not present in any other whale. Probably the hind limbs of the genus *Protocetus* were much less involuted than are those of other whales, and we cannot completely dismiss the possibility that these animals may have been able to move about on land from time to time. Although *Protocetus atavus* may not be considered as a direct ancestor of either the baleen whales or the toothed whales, it still gives us a good idea of what the general structure of the ancestors of today's whales may have been. The more highly developed Archaeoceti whales had a largely involuted pelvis, and their hind limbs had no real significance. In addition, the later Archaeoceti whales had particular adaptations which made the possibility of the baleen and toothed whales being descended from these animals highly unlikely, if not impossible. For example, one evolutionary branch of the Archaeoceti whales, the family + Basilosauridae, acquired a snakelike shape through the elongation of its back and lumbar vertebrae. The best known genus of this family is + *Basilosaurus*, which formerly was also called + *Zeuglodon* (see Color plate, p. 466). The Archaeoceti whales died out toward the end of the early Tertiary (some 25 million years ago).

The oldest known toothed whale (+ *Agorophius*) lived about 45 million years ago in the Upper Eocene. The SQUALODONTIDS (family + Squalodontidae) from the Oligomiocene, considered to be the group from which present-day toothed whales are descended, developed from the AGOROPHIDS (family + Agorophiidae). The agorophids are very possibly descended from the ancient Archaeoceti whales, although the present fossil finds are insufficient to support such a hypothesis. Baleen whales, on the contrary, are only known to us from the Oligocene (some 35 million years ago). These cetotheriids (family Cetotheriidae), which we might consider as the ancestors of present-day baleen whales, were small to medium-sized toothless whales; their lower jaw was not as enlarged as those of the right and finback whales. Slijper has noted a series of similarities in body structure between baleen whales and the Archaeoceti whales, and he has suggested that perhaps the Archaeoceti whale is more closely related to the baleen whale than to the toothed whale. This hypothesis has been confirmed by fossil discoveries from the Oligocene. Remains of whales from the genus + *Aetiocetus* (family + Aetiocetidae) were found in sediment layers dating back 25–40 million years; these animals provide a link between the body structure of baleen whales and Archaeoceti whales, and they resembled the gray whale in their appearance.

15 The Baleen Whales

Suborder: Baleen whales, by E. J. Slijper and D. Heinemann

The first major group of whales living today is that of the BALEEN WHALES (Mystacoceti), which we will discuss here as a suborder. A series of discoveries has indicated that the baleen whales are not so closely related to the second large whale group, that of the toothed whales, as was formerly believed. Thus, it is also possible to consider these two groups as separate orders.

Distinguishing characteristics

These animals range in size from large to very large, their length ranging from 6 to 33 m. The head is large and wide. The blow holes (nostrils) are not united into a single opening. The jaws of a newborn calf are toothless, although the embryos do show the rudiments of teeth. The lower jaw surrounds the narrower upper jaw. On each side of the jaw there are between 130 and 400 baleen, thin horny plates suspended from the palate, which strain the whale's food. The skull is symmetrical. The concha and olfactory lobe are poorly developed. The connection between the nasal duct and the larynx is not as solid as that in toothed whales. There are three familes: right whales (Balaenidae), gray whales (Eschrichtiidae), and finback whales (Balaenopteridae), with a total of six genera and twelve species.

One large blue whale weighs as much as twenty-five elephants, 150 head of cattle, or 1600 men, about 130 t. Even the largest prehistoric dinosaur, the brontosaurus, weighed only one fourth as much as a blue whale. The bones and muscles are not equipped to support and carry such a large body mass around on land. As an animal grows larger, its weight increases cubicly (to the third power). However, the carrying ability of the animal's bones and muscles, which must lift its body, is only squared (increased to the second power), because this ability is determined by two factors (width and thickness), rather than three (width, thickness, and length). In other words, an animal whose length is doubled (2) has a weight increase of 800 percent (2^3); the animal's muscles and bones, however, can only support a weight increase of 400 percent (2^2), as long as the body structure remains constant. Thus,

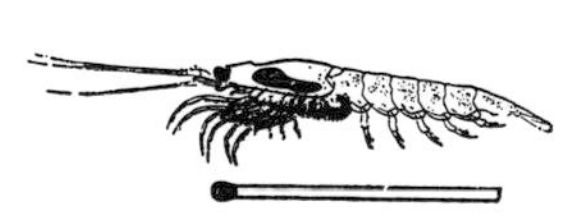

Fig. 15-1. The shrimp (*Euphausia superba*) is only 6 cm long (size comparison with a match stick). Large quantities of these animals form the basic food of the baleen whales.

land animals have definite limits to their growth potential. These limits do not apply to aquatic animals, because the buoyancy of the water balances out the animal's weight. Consequently, the whale does not need to support its body with its muscles or bones.

Diet of the baleen whale

It is hard to believe that the largest animals on earth feed themselves on very minute forms of life. The basic diet of almost all baleen whales consists of plankton. Bryde's whale, however, is an exception, feeding largely on fish. The rorqual also eats fish in certain areas, for example, when it is in the North Atlantic. The shrimp (*Euphausia superba*; see Vol. I), which is about 6 cm long, is the main food of baleen whales around Antarctica. Shrimp can cover the water's surface and its immediate surroundings in massive numbers, making the water look like a glistening reddish-colored soup. The shrimp's orange-red color is due to its supply of carotene, a preliminary stage of vitamin A. Shrimp feed on green diatoms which, like all other plants, are able (with the help of sunlight), to convert carbon dioxide (CO_2) into organic compounds, which animals can use as food. Plant life, however, and animal life with it, does not depend solely on the presence of carbon dioxide; plants need sufficient quantities of oxygen and inorganic salts in order to survive. These life requirements are fulfilled even in the waters of the Antarctic. Cold water contains much larger quantities of dissolved oxygen and carbon dioxide than does warm water, and the inorganic salts are carried to the surface by a warm current, originating in the tropics, which comes to the surface around 53° south latitude. Thus the population of plankton in this area is particularly large. Shrimp, apparently, flourish in very cold water, and for this reason one finds large quantities of shrimp in only two areas: in the zone between 63° south latitude and the Antarctic coast, and in the Weddell Current, which flows out of the Weddell Sea (Antarctica) in a north-easterly direction.

The major baleen whale herds are also found in these two areas during the summer. In the autumn, however, the ice packs move slowly but constantly north, covering the feeding grounds of the baleen whales, which, because they have lungs and must breathe, are unable to live under the ice. A number of these whales probably spend the winter in the open waters along the northern border of the ice floes. The majority of them, however, move north into subtropical and tropical waters in March and April. British and Dutch researchers, working together with officers of commercial ships, have recently come to the conclusion that there are also certain areas in tropical waters where large numbers of baleen whales can be found, due to the rich food supply in these regions. The same areas (which include, among others the Arabian Sea; the Gulf of Aden, Arabia; the Caribbean Sea; and the area around Dakar, Senegal) also have a particularly large population of fish. This supply notwithstanding, the available food in the warmer oceans does not appear to be sufficient for the baleen whale; most of these animals return

▷
River dolphins: 1. Gangetic dolphin (*Platanista gangetica*); 2. Chinese river dolphin (*Lipotes vexillifer*); 3. La Plata dolphin (*Stenodelphis blainvillei*); 4. Amazon dolphin (*Inia geoffrensis*). Beaked whales: 5. Cuvier's beaked whale (*Ziphius cavirostris*).

▷▷
Beaked whales: 1. Baird's beaked whale (*Berardius bairdi*); 2. Layard's whale (*Mesoplodon layardii*); 3. True's beaked whale (*M. mirus*); 4. Sowerby's whale (*M. bidens*); 5. Bottlenose whale (*Hyperoodon ampullatus*).
(Continued on page 481.)

1
2
3
4
5

1
2
3
4
5

8
11
12
9
10
6
7 ♂
7 ♀

23
24
17
18
16
26
25

13
14
19
20
21
15
22

to the colder waters in the spring looking very much thinner. We do not yet know why some baleen whales remain behind in the tropics; the number of whales that do so probably changes every year. Apparently the food supply determines the number of whales remaining behind.

Baleen whales have a large apparatus which allows them to filter large quantities of shrimp and other tiny animals from the water. This apparatus, called baleen, consists of winglike horn plates hanging from the palate, one behind the other, which are equipped with fimbriae (fringes) on the inner border. When the baleen whale feeds, it takes a large quantity of water and floating or swimming plankton into its mouth. Then it closes its mouth and twists the floor of the mouth and the tongue upward, thereby decreasing the area of the oral cavity and forcing the water to the side and through the baleen. The plankton remain in the mouth, hanging on the baleen fibers; then, somehow, the food is moved into the throat and esophagus. The size and form of baleen, and the shape of the fimbriae, differs with each species of baleen whale, according to the main source of food. For example, the hair fibers on the baleen of the sei whale are particularly fine and soft, because this species feeds mainly on small crabs and other plankton; the fimbriae on the baleen of a Bryde's whale are thick and rigid, as this animal's diet consists mainly of fish.

The baleen whales have a much more important role in whaling than do the toothed whales. For this reason, these animals are in danger as the result of overkill on the part of the whaling industry. Thus, in its *Red Data Book*, the International Union of Conservationists and Naturalists, (IUCN), listed seven of the twelve species of baleen whale as being threatened with extinction. These are: the bowhead whale (Ø *Balaena mysticetus*), the Atlantic right whale (Ø *Eubalaena glacialis*), the right whale (Ø *Eubalaena japonica*), the southern right whale (Ø *Eubalaena australis*), the humpback whale (Ø *Megaptera novaeangliae*), the common rorqual (Ø *Balaenoptera physalus*), and both subspecies of the blue whale (Ø *Balaenoptera musculus*). With the exception of the common rorqual, these species are also fully protected under the regulations of the International Whaling Commission. As a result, the populations of the bowhead whale, the Atlantic right whale, and especially the gray whale, have begun to increase.

The RIGHT WHALES (family Balaenidae; L 6–16 m) have smooth, unfurrowed throats and necks. The head is very large (up to two-fifths of the whale's total length) and has a high arch. The baleen are very long and flexible. The lower lip forms a protrusive curve on each side of the mouth. The upper jaw is narrow and has a large curve; when the mouth is closed, it lies between the lower lips. The breast flukes are short and somewhat rounded. These animals have a very thick fat (blubber) layer. They are slow swimmers. All seven neck vertebrae are fused. There are three genera with a total of five species.

(Continued from page 478.)

White whales and narwhals: 6. White whale (*Delphinapterus leucas*); 7. Narwhal (*Monodon monoceros*). Long snouted dolphins: 8. Rough-toothed dolphin (*Steno bredanensis*); 9. West African white dolphin (*Sousa teuszi*). True dolphins: 10. Red Sea bottle-nosed dolphin (*Tursiops aduncus*); 11. Common dolphin (*Delphinus delphis*); 12. Blue-white dolphin (*Stenella caeruleoalba*); 13. White-sided dolphin (*Lagenorhynchus acutus*); 14. Bottle-nosed dolphin (*Tursiops truncatus*); 15. Risso's dolphin (*Grampus griseus*). Right whale dolphins: 16. Northern right whale dolphin (*Lissodelphis borealis*). Pilot and killer whales: 17. Northern pilot whale (*Globicephala melaena*); 18. Pygmy killer whale (*Feresa attenuata*); 19. Irrawaddy dolphin (*Orcaella brevirostris*); 2. Killer whale or orca (*Orcinus orca*); 21. False killer whale (*Pseudorca crassidens*). Commerson's dolphins: 22. Commerson's dolphin (*Cephalorhynchus commersonii*). Porpoises: 23. Common porpoise (*Phocaena phocaena*); 24. Spectacled porpoise (*Phocaena dioptrica*); 25. Dall's harbor porpoise (*Phocaenoides dalli*); 26. Black finless porpoise (*Neophocaena phocaenoides*).

The BOWHEAD WHALE (♀ *Balaena mysticetus*; L up to 18.3 m; see Color plate, p. 463) is the only species of its genus. The length of the head (in older adult males) may be up to two-fifths of the total length. The lower jaw is 4.8–6.1 m long and 3–3.6 m wide. The lower lip is 4.5–6.1 m long on each side and 1.5–1.8 m high. There are more than 300 baleen on each side of the jaw. Each baleen has an average length of 3 m and an average width of 25–30 cm. The longest baleen are 4–4.6 m. These whales do not have a dorsal fin. The tail fin is about 1.5–1.8 m long and 5.5–7.9 m wide. The upper jaw and the tip of the lower jaw have a few short white hairs. These animals are found only in the northernmost waters. Their population was once almost extinct, but it has recovered and continues to grow.

Bowhead whale

Most of our knowledge about the bowhead whale comes from William Scoresby, who published *A Report on the Arctic Area* in 1820. At that time, this whale species was so common in the northern waters that anyone who spoke of "the whale" always meant the bowhead whale. When whaling was carried out with longboats and hand harpoons, the whalers preferred to hunt the bowhead whale, as it was "less active, slower in its movements, and not so agressive as the other similar, larger species," according to Scoresby's descriptions. "In short, they were easier to capture." In addition, the bowhead whale has the longest and most valuable baleen of all whales. At that time, as Harmer has reported, every ton of whalebone (baleen) was worth up to 2250 British pounds; thus the 1.5 t of whalebone that a large bowhead whale carried in its mouth was worth up to 3375 pounds. It is not surprising that larger and larger whaling fleets set out for the Arctic Ocean to hunt these animals and that consequently the bowhead whale population began to diminish even though each whaling ship rarely caught more than seven to ten whales per trip. As early as 1630–40, the bowhead whale began to disappear from the bays of the northern coasts and islands. The whalers were forced to hunt these animals on the high seas and to look for new hunting grounds. The bowhead whale population had decreased to such an extent by the second half of the 19th Century that hunting these animals was no longer profitable. Today this whale is the rarest of all the whale species. A small population of these animals has survived in the Davis Straits (Canada), and in the Bering Sea. Bowhead whales were believed to be fully extinct in the European section of the Arctic Ocean, but recently these whales have occasionally been observed in that area.

The fact that bowhead whales and their relatives have such long baleen is probably due to their particular method of feeding. Apparently, right whales swim through masses of plankton with their mouth open almost all the time; the finback whale, on the other hand, seems to swallow more while it hunts.

The bowhead whale does not seem to travel over long distances. According to Scoresby, the mating takes place at the end of the summer.

▷ An albino bottle-nosed dolphin (*Tursiops truncatus*).

Females with nursing calves can be found most commonly in spring. Scoresby believes that the young are born in February or March, after a gestation period of nine to ten months. The average life expectancy is probably forty years.

The Atlantic right whale and its relatives

The ATLANTIC RIGHT WHALE (♁ *Eubalaena glacialis*; see Color plate, p. 463) and its closest relatives, the PACIFIC RIGHT WHALE (♁ *Eubalaena japonica*) and the SOUTHERN RIGHT WHALE or ICE BALEEN WHALE (♁ *Eubalaena australis*), are all classified together in one genus, and consequently are separated from the bowhead whale. Their length is, at the most, between 18 m (Atlantic right whale) and 21 m (Pacific right whale). The head comprises about one-fourth of the total length. The upper jaw is not as curved as that of the bowhead whale. The baleen are, on the average, 30–60 cm shorter than those of the bowhead whale. These whales do not have white spots on their chins. However, they do have sharply contrasting white areas on their skins, caused by parasites. There is a borny protuberance called a "bonnet" on the tip of the upper jaw. This bonnet is about 20 × 30 cm large, and is usually infested with barnacles and worms. These animals do not have a dorsal fin. Each one of these species has been threatened with extinction, but as the result of protective restrictions, the individual populations are increasing.

The Atlantic right whale and its relatives were formerly rather common within their distribution areas. The Atlantic right whale was found in the North Atlantic from the Bay of Biscay (France) to Norway and Newfoundland. The northern limit of its distribution range bordered the southern limit of the bowhead whale's distribution. The Pacific right whale was commonly found on the western coast of North America from Vancouver Island (Canada) to the Aleutians (Alaska), and on the Asian coast near Kamchatka (U.S.S.R) and in the Sea of Okhotsk (U.S.S.R.). The southern right whale lived in large herds off the coasts of South Georgia (a South Atlantic island under British rule) and South Africa, in the southern Indian Ocean, and off the coasts of Australia and New Zealand. However, these whales have also been hunted to excess. The Vikings began to hunt the Atlantic right whale long ago. The techniques then associated with whaling were probably spread by the Vikings as they made their raids further south. It is not improbable to believe that the Basques learned the whys and hows of whaling from the inhabitants of Normandy. In any case, the Basques hunted the Atlantic right whale in the Bay of Biscay as early as the 11th Century. Later these whalers hunted their prey over the whole North Atlantic, and in 1578, thirty Basque whaling ships appeared off the coast of Newfoundland. Some of the expressions still associated with the whaling industry can be traced back to a Basque origin; the word harpoon, for example, is believed to have come down to us from the Basques. Other peoples also hunted the Atlantic right whale north of Norway, near Iceland, and off the coast of New England. These animals were hunted to such

◁ Underwater photograph of a white whale (*Delphinapterus leucas*). The animal is exhaling; one can see the air escaping in bubbles from the blow hole on the top of the head.

an extent that the herds came together and finally disappeared completely. For a long time after its disappearance, people believed the Atlantic right whale to be extinct. However, in the second half of the 19th Century a few Atlantic right whales were seen in the areas where these animals had once been so common. The hunting of these whales began again, but the limited yield showed that the hunting of these animals would have to be forbidden once more. Today the Atlantic right whale, like all right whales, is fully protected; this means that the member nations of the International Whaling Commission may not hunt or kill these whales. Within the last few years people have been successful in filming some schools or small groups of these whales off the coast of New England. Their vocalizations have also been tape recorded.

The pygmy right whale

The Pacific right whale and the southern right whale are also very rare today, as the result of decades of persecution. The PYGMY RIGHT WHALE (*Neobalaena marginata*; L up to 6 m; see Color plate, p. 463), on the contrary, was probably never very common. Luckily these animals have no economic significance and are therefore, in spite of their rarity, not threatened with extinction. The pygmy right whale is the smallest of the baleen whales. It has a small dorsal fin which is bent to the rear. There are only about 230 baleen on each side of the jaw. The breast flukes are small and narrow; they have four digits. The number of ribs is larger than usual; only the last two back vertebrae have no ribs. The rear ribs are much wider than the front ribs.

The pygmy right whale lives only in southern waters. These whales have been seen off the coast of South Africa, South Australia, New Zealand, and South America. We know nothing about the habits of these animals; in some respects we are not even sure what a pygmy right whale looks like.

Family: Gray whales

The GRAY WHALE (*Eschrichtius gibbosus*; L up to 15 m; see Color plate, p. 464) provides a transition between the right whales and the finback whales, although it is not closely related to either group. Consequently, this species is classified in a special family, that of the GRAY WHALES (*Eschrichtiidae*).

Distinguishing characteristics

These whales are medium-sized with a relatively small head. There are 138–174 baleen on each side of the jaw; these baleen are 45 cm long at the most, and very thick. There are from two to four 1.5 m long furrows on the throat. These whales have more hair on the tip of the upper jaw and along the length of the lower jaw than do the other baleen whales. The breast flukes are about average length and very broad; they have four digits. Gray whales do not have a dorsal fin. The tail fin is heavier than those of right and finback whales. The gestation period lasts eleven to twelve months.

The gray whale is found in the northern Pacific Ocean. Between November and May these animals can be found off the coast of southern California (U.S.A.) and Baja California (Mexico), where they show a

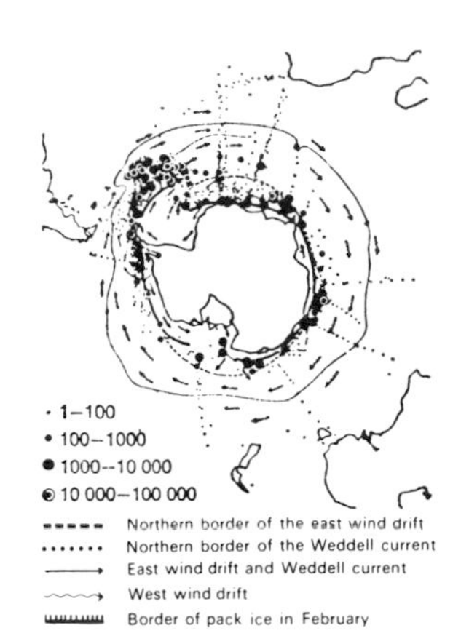

Fig. 15-2.
This is the major distribution area of plankton, large quantities of tiny crustaceans that form the basis of the baleen whale's diet in the Antarctic. Great numbers of baleen whales migrate to this area each summer. The size of the dots indicates the number of plankton that can be caught with one cast of a 100 cm diameter plankton net on the surface of the water.

particular preference for shallow water. In his thorough report on these animals, written in 1874, Scammon reported: "One gray whale lay playing in the surf for half an hour, just like seals do in a rough sea. It rolled itself from one side to the other, with its breast flukes half stretched, and moved through the heavy breakers as they came toward shore. Occasionally it bent its tail fin and made a playful jump; it flung its body out of the water, and landed on the surface again with a loud slap, then it spouted two or three times and dived underwater." Scammon found that the water where this game took place was only 4 m deep. Gray whales are frequently stranded in shallow water. When this happens, the animal usually lies quietly in water that is 60–90 cm deep, which they can do without damaging themselves, and waits for the high tide to free it.

While in their winter quarters off California and Baja California, pregnant females search out shallow lagoons, where they then give birth to their young. Usually only one calf is born, although twins have been observed. The newborn calf is about 4.5 m long.

In the spring, the gray whales move north along the coast. Some of them spend the summer in the coastal regions of British Columbia (Canada). However, most of these animals travel further north to the Bering Sea. There, the herd divides: one half travels westward to the Koryak coast (U.S.S.R.); the other half moves northward through the Bering Straits into the Arctic Ocean, where it then follows the coastline of Siberia or Alaska. The return trip begins toward the end of September or in October, with the whales retracing their previous route. The gray whale population off the Asian coast has a similar behavior pattern; these animals usually spend the winter in the waters off South Korea.

When it is in the arctic waters, the gray whale feeds mainly on small swimming crabs. However, bristle worms (Chaetopoda) and sea snails have also been found in their stomachs. Gray whales apparently take in very little food during their stay in the south. In any case, researchers have found plant remains and even sardines, as well as crabs, in the stomachs of gray whales examined in their winter quarters.

The hunting of the gray whale on the western coast of North America did not reach a large scale until 1851. At that time, according to Scammon's calculations, some 1000 gray whales passed the California coast each day during the migrating season from 15 December to 1 February. Twenty-three years later this number had been reduced to about forty animals a day, as the result of man's merciless persecution of these whales. The gray whale was believed to be extinct in the first decade of the 20th Century. Then, however, it became apparent that the Asian herd of these animals had not been as drastically reduced as had the herd on the western coast of North America. These whales appeared again off the coast of California, and in the winter of 1925–6, forty-two gray whales were killed in that area. The gray whale has been completely protected since 1937. At that time there were only an estimated 250 gray whales in the world, and the population would need a long time to recover its former strength. By 1968 the estimated population of gray whales had increased to approximately 18,300 animals.

Family: Finback whales

The last family of baleen whales is that of the FINBACK WHALES (*Balaenopteridae*; L 9–33 m, weight up to 130,000 kg) which includes the largest whales of all. These whales are more slender than the right whales. The length of the head may be at most, up to one-fourth of the whale's total length. The head itself is flat. The jaws are attenuated. The baleen are shorter, wider, coarser, and less flexible than the baleen of right whales. Finback whales have 70–100 grooves or furrows on their throat and chest. The breast flukes are thin and pointed. These whales have a dorsal fin. The neck vertebrae are not fused. The fat (blubber) layer is thinner in finback whales than in right whales. These whales are extremely fast swimmers. They are sparsely haired on their upper and lower jaws. There are two genera, with a total of six species.

The grooves on the chest and throat of the finback whale are much more numerous and more noticeable than those of the gray whale; the finback whale's grooves are also longer. The length of the baleen and of the throat furrows are both related to the particular manner in which finback whales catch their food. While right whales apparently catch their food by swimming through schools of shrimp with their mouths open, finback whales capture their food in gulps. These gulps are quite productive; the throat furrows between the two sides of the lower jaw and on the breast make it possible for these whales to expand their mouths, and consequently, to take in tremendous amounts of water and plankton. Then, when the whale's mouth is almost closed, the floor of the mouth comes together again and pushes the tongue up against the gums, so that the water is forced out of the mouth at the sides, through the baleen. The plankton are left, hanging on the baleen, just as in right whales.

The blue whale and its relatives

The genus of RORQUALS (*Balaenoptera*) contains five species: 1. The BLUE WHALE (⍭ *Balaenoptera musculus*; average L 24 m, up to 33 m, weight up to 130,000 kg; see Color plate, p. 465) which some zoologists classify as a separate genus (*Sibbaldus*). These whales are the largest animals that have ever lived on the earth. The breast flukes are long (up to one-seventh of the whale's total length) and pointed. The dorsal fin is very small and insignificant. Blue whales have between 70 and 118 long grooves or furrows on their throats. The baleen are black. These animals are distributed worldwide, but they are rare. There is one subspecies, ⍭ *Balaenoptera musculus brevicauda*. 2. The COMMON RORQUAL ⍭ *Balaenoptera physalus*; L 18–24 m; see Color plate, p. 464) has small breast flukes (about one-ninth of the total length). These whales have a high, triangular dorsal fin. Their coloring is not symmetrical. The dark coloration on the left side of the back is much stronger and extends

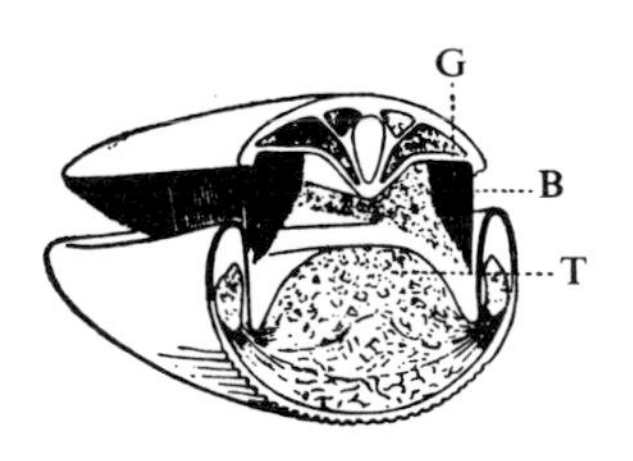

Fig. 15-3. Cross section of the head of a finback whale showing the baleen (B) which hang from the gums (G), and the large tongue (T).

further down the side toward the stomach than does the corresponding coloration on the right side of the back. The right lower jaw is light colored on the outside and dark on the inside; the coloring of the left lower jaw is the opposite of this. The outer edge of the baleen is blue-gray. The front of the baleen is white, while the back is blue-gray. Common rorquals are distributed worldwide. 3. The SEI WHALE (*Balaenoptera borealis*; L 15–18 m; see Color plate, p. 464) has small, narrow breast flukes (only one-eleventh of the total length). These whales have 60–100 relatively short throat furrows. The baleen are particularly fine; most are black, although a few individual baleen may be white. Sei whales have a blue-black coloring on their back, shoulders, and head. The sides of the body are a somewhat lighter color. The underside is covered with white longitudinal stripes of varying width. These whales are distributed worldwide. 4. BRYDE'S WHALE (*Balaenoptera edeni*; L 13 m; see Color plate, p. 465) is very similar to the sei whale, although the Bryde's whale's baleen is much coarser and is white in front, black in back. They have between forty-two and fifty-four quite long throat furrows extending to the navel. Bryde's whale is found in the tropics and subtropics. 5. The LESSER RORQUAL (*Balaenoptera acutorostrata*; L 9 m; see Color plate, p. 464) has breast flukes which are only one-eighth as long as the whale's total length. The dorsal fin is easily noticeable. These whales have a blue-gray coloration on their backs, shoulders, and head. The underside of the body is white; the breast flukes have a white band. There are about fifty throat furrows. Lesser rorquals have up to 235 yellowish-white baleen on each side of their jaws. These whales are distributed worldwide, although they are not common in the tropics.

The individual species of finback whales have slightly different diets, the diet is also affected by changes in season and location. The BLUE WHALE feeds almost exclusively on the small crustacean shrimp (about the length of a match). These shrimp are also the major part of the diet of the COMMON RORQUAL, at least in the Antarctic. The blue whale generally feeds on the adult shrimp, while the common rorqual takes in masses of shrimp larvae when it is in the Antarctic. In the North Atlantic, however, the common rorqual prefers different types of plankton, and it also eats herring and smaller fishes. The blue whale generally does not feed on fish, but it has been known to do so. The SEI WHALE and the LESSER RORQUAL have particularly fine fringes on their baleen, which allow them to catch plankton that are even smaller than the shrimp crustacean. Thus, aside from shrimp, one can find copepod (see Vol. I) in the stomachs of these animals, along with several other smaller plankton. The lesser rorqual also feeds on fish. Allen has reported that lesser rorquals caught off of the New England coast even have a diet consisting mainly of fish: codfish, capelin, whiting, sea salmon, mackerel-like fish, and spiny dogfish were found in the stomachs of these whales. BRYDE'S WHALE is even better adapted to a fish diet, with its particularly thick and rigid baleen. These animals feed mainly on herring and a species of mackerel that is about 30 cm long. Researchers once found fourteen large penguins in the stomach of a Bryde's whale! Olsen believes that these birds probably swam into the whale's open mouth in order to catch the fish there, and that the whale merely swallowed the birds along with the fish.

Common rorquals, blue whales, sei whales, and lesser rorquals all migrate in the spring to the polar regions, which are rich in plankton; in the fall these whales move to warmer regions. In the North Atlantic these migrations (especially of the common rorqual) are much less extensive than the migrations of those animals in the southern hemisphere and in the northern Pacific Ocean. The blue whale makes the deepest penetration into the polar waters; these animals can be found regularly—or could be found—between the ice packs. Common rorquals and sei whales usually remain in ice-free water. Bryde's whale apparently does not travel to the colder polar regions at all, but remains instead between 30° north latitude and 30° south latitude throughout the year.

In Chapter 14 we discussed the hunting of finback whales. The blue whale has suffered the greatest damage as a result of whaling, because it is the largest whale, and therefore it yields the most profit per animal. During the whaling season of 1934–5, 16,500 blue whales and 12,500 common rorquals were killed in the Antarctic, although only 1800 whales of other species were killed at this time. In 1960–1, whalers were able to catch and kill only 1744 blue whales, along with almost 29,000 common rorquals and 10,800 whales of other species. However, even this relative reduction in the number of blue whales killed is frighteningly inadequate when we consider the population count of these animals in the *Red Data Book*. According to the report of the Committee of Three Scientists, there were an estimated 930–2790 blue whales alive in 1961–2 and only 650–1950 in 1963; we can include the population of the subspecies *B. m. brevicauda*, which was estimated to be 2000–3000 at the same time, along with these figures. The blue whale is now completely protected. We can easily see as a result of these figures, however, that the common rorqual will also be threatened with extinction unless a more effective program of protection is put into effect soon. According to the International Whaling Commission the world's population of common rorquals decreased in the eight years from 1955–6 to 1963–4 from some 110,400 to 32,400 animals.

The sei whale was formerly hunted only in Japanese and Korean waters, and in the North Atlantic to any great extent. The whaling in these areas was carried out on a much smaller scale than that in the southern hemisphere. However, since the blue whale and common rorqual have become so much rarer, the whaling industry has become more and more interested in the southern sei whale populations, and the hunting of the sei whale has rapidly increased since 1955. It is to be hoped that this species will not be the next victim of whaling overkill. Bryde's whale and the lesser rorqual do not yet have an important role in the whaling industry.

The mating of the southern blue whales takes place in June or July, during the animals' stay in warmer waters. The gestation period lasts about a year. The calf is nursed for six or seven months. Usually only one calf is born; twins are rare, but they do occur. The blue whale cow does not usually mate again until after the calf is weaned, so she generally gives birth only every second year. Common rorquals have much the same behavior, although their mating season may last seven to eight months. Lesser rorquals also mate in their winter quarters; which are usually in the temperate zones. The gestation period for these whales lasts only ten months. We know very little about the reproduction of the sei whale and Bryde's whale. Newborn finback whales are very large: the blue whale calf weighs 2 t at birth and is over 7 m long; the young lesser rorqual measures at least 2.7 m. Whale calves grow extremely rapidly. The blue whale is sexually mature at four and a half years, while the lesser rorqual attains this status after two and a half years. The young whales, however, continue to grow in size for several years after this time. The average life expectancy for all species is twenty to thirty years.

The humpback whale

The HUMPBACK WHALE (♁ *Megaptera novaeangliae*; L 11.5–15 m; see Color plate, p. 465) is the only representative of its genus. These animals are heavy and bulky. The breast flukes are narrow and very long (up to one-third of the total length). The dorsal fin is similar to that of the common rorqual. There are rows of irregular skin protuberances on the top of the head and the lower jaw; each protuberance has two bristles. Humpback whales have from fourteen to twenty throat furrows; these extend down to the navel and are much further apart than those of the other finback whales. There are about 400 baleen on each side of the jaw; they are not much more than 60 cm long.

While the other finback whales prefer the high seas, the humpback whale is basically a coastal inhabitant. These animals will often swim into harbors and sometimes even down the mouths of large rivers for some distance. Humpback whales in the northern hemisphere mate in April; those in the southern hemisphere mate in September. Scammon has described the process: "The humpback whales are known for their courtship and for their amusing and unusual caresses during the mating season . . . When they lie side to side in the water, the humpback whale couple exchange blows with their breast flukes, blows that are apparently meant to be tender. On quiet days these blows can be heard for miles. The animals also stroke one another with these large, flexible arms, roll themselves occasionally from one side to the other, and make jumps, which are more easily imagined than described, into the air." Humpback whale cows are pregnant for about a year. The newborn calf is 4.5–5 m long; it is nursed by its mother for about five months. The young animals

Fig. 15-4. The humpback whale will often make regular somersaults in the air.

are sexually mature at twenty to twenty-two months, and they reach their full size at ten years. The average life expectancy is estimated to be about thirty years.

In spite of its bulky body shape, the humpback whale is one of the most agile and lively whales. One can often see these animals, whose average weight is 29 t, spring completely out of the water and fall back again with a loud slapping sound.

Their diet consists mainly of plankton, although these animals also eat copelin and other tiny fish. The rich supply of plankton in the waters around Antarctica attracts herds of humpback whales to that area every year. Consequently, the humpback whale is one of the most hunted whale species. In 1930 the number of these animals in the Antarctic region alone was estimated to be about 22,000. In 1965 there were probably fewer than 3000 humpback whales in that area. In addition, there are barely 5000 of these animals in the North Pacific; the number of humpback whales in the North Atlantic has not yet been estimated. However, these animals still seem to be fairly common off the eastern coast of North America.

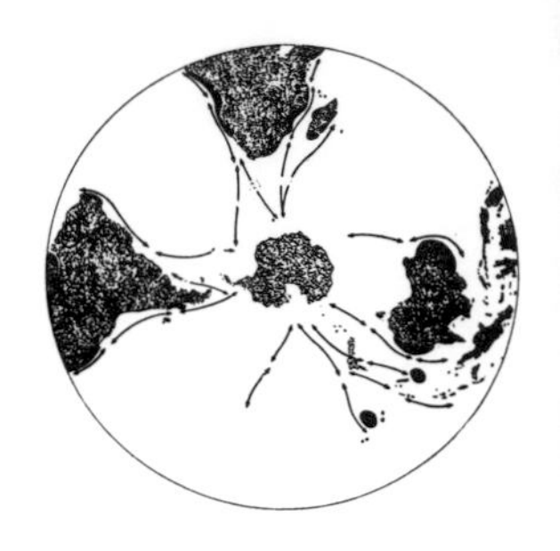

Fig. 15.5. Distribution (black) and migration (lines) of the humpback whale (*Megaptera novaeangliae*) in the Southern Hemisphere.

16 The Toothed Whales

Suborder: Toothed whales, by E. J. Slijper and D. Heinemann

According to our present knowledge, the toothed whales, unlike the baleen whales, are probably not descended from members of the suborder Archaeoceti. For this reason it is also possible to classify the toothed whales as a separate order. However, we will discuss these animals as a suborder of the order of whales (Cetacea), which is how, until recently, they were always classified.

Distinguishing characteristics

The TOOTHED WHALES (suborder Odontoceti) range in length between 1 and 18 m. The males are larger than the females. The nostrils are united in a single blow hole, which is located even further back on the head than in baleen whales (exception: sperm whale). The olfactory mucous membrane and the olfactory lobe in the brain are not present. The brain itself is unusually well developed; the brain of some whale forms can be compared only to the brains of man and the elephant in regard to size and segmentation. Fish-eating members of this suborder usually have more teeth (up to 272 in the Amazon dolphin) than do the sepia-eating toothed whales (bottle-nosed whales, among others, have only two teeth). There are always some teeth. The skull is asymmetric (see Fig. 16-1); we do not know why. The breast flukes always have five digits. There are nine families, which we will group together in four superfamilies: A. Sperm whales (superfamily Physeteroidea) with the families of 1. Sperm whales (Physeteridae) and 2. Beaked whales (Ziphiidae). B. River dolphins (superfamily Platanistoidea), with the families of 3. Gangetic dolphins (Platanistidae), 4. Amazonian dolphins (Iniidae), and 5. La Plata dolphins (Stenodelphidae). C. Narwhals (superfamily Monodontoidea), with the family of 6. White whales and narwhals (Monodontidae). D. Dolphins (superfamily Delphinoidea), with the families of 7. Porpoises (Phocaenidae), 8. Long-snouted dolphins (Stenidae), and 9. Dolphins (Delphinidae). There are about thirty genera.

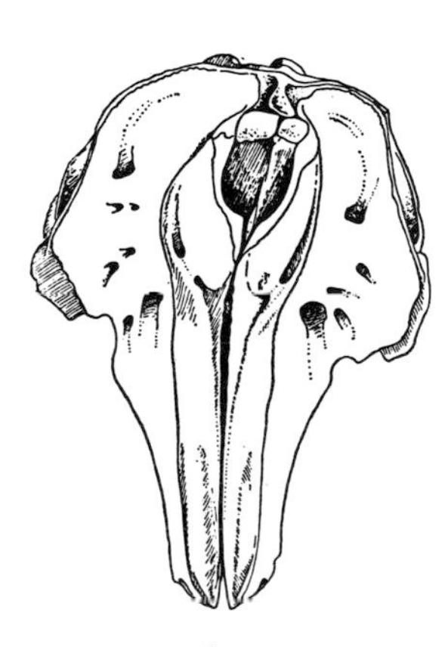

Fig. 16-1. The asymmetric skull of a female narwhal, viewed from below.

Unlike the baleen whales, the suborder of toothed whales includes animals of many different shapes. The toothed whales specialize in limited food sources; however, the individually related groups all hunt different varieties of fish and consequently these groups have developed several very different adaptations to their diets. None of these animals has a sense of smell. We will discuss the development of the brain in toothed whales, and the resultant abilities these animals have, in our description of the dolphin, later in this chapter.

In 1654 Bartholinus described the unusual shape of the larynx in the porpoise. This was the first indication of a characteristic which all toothed whales have in common. Two of the cartilages in the larynx, the epiglottis and the arytenoids are elongated, resulting in a saclike protuberance of the larynx. This elongation is shaped like the beak of a goose, and it extends into the lower part of the nasal cavity, where it is surrounded by a sphincter (circular muscle) which can close it off. As a result, the nasal cavity and the larynx are directly connected to each other, an arrangement that makes it possible for toothed whales to absorb pressure differences while breathing. This connection may also be related to the transmission of sounds in the body.

Generally the smaller toothed whales are classified as "dolphins," without any real regard for their actual relationship and evolution; the larger forms, on the other hand, are generally called "whales." Thus the distinction between dolphins and whales in the popular name generally indicates the size of the animal rather than the classification of its related group.

Superfamily: Sperm whales

The SPERM WHALES (superfamily Physeteroidea) hunt sepia (Cephalopoda, see Vol. III). Because cuttlefish are relatively slow-moving animals with rather soft bodies, the dentition of sepia-eating whales is underdeveloped. The sperm whales, with the exception of Shepherd's beaked whale (*Tasmacetus shepherdi*), have visible teeth only on their lower jaws. The upper jaw often has rudimentary teeth which never break through the gums.

The family of SPERM WHALES (Physeteridae) consists of only two species: 1. The SPERM WHALE (*Physeter catodon*; L, ♂♂ 18–20 m, ♀♀ 11–12.5 m; see Color plates, pp. 466 and 474). The head is very large (over one-third of the whale's total length) and heavy. The lower jaw is long and very thin; it has between eighteen and thirty uniform pointed teeth on each side. The sperm whale's left blow hole is located toward the front of the body. The breast flukes are small. There is no dorsal fin, but there is a row of humps down the last third of the back. The tail fluke is 3.6–4.6 m wide. The gestation period probably lasts for sixteen months. The calf is 4 m long at birth; it is nursed by its mother for about six months. 2. The PYGMY SPERM WHALE (*Kogia breviceps*; L 2.7–4 m; see Color plate, p. 474) has a relatively short head (one-sixth of the total length). The lower jaw is very short; it has nine to fourteen pointed teeth on each side. The breast flukes are small. These animals have a triangular dorsal fin. They have a large distribution range and are fairly common, although we know almost nothing about their way of life.

The unusual shape of the sperm whale's head is not determined by the shape of the skull, but by a massive cushionlike growth on the front of the head. The superior maxillary and the intermaxillary bone together form a flat, elongated basin within which there is a large tissue growth that contains the clear, oily liquid we call spermaceti. When this liquid is exposed to air, it hardens into a soft, white wax. Earlier, spermaceti was believed to be the sperm whale's semen; today, as a result, we call this liquid spermaceti or whale sperm, and our name for the whale species itself denotes a similar connotation. We do not know just what significance this supply of spermaceti has for the whale. The sperm whale's nasal passage does not run vertically from the middle of the head up to the bony blow hole, as is the case with the other toothed whales. Instead it slopes to the front, so the blow hole is found on the left side at the front end of the head.

The sperm whale feeds almost exclusively on cuttlefish. Usually it hunts fishes that are only 1–2 m long, although it can also overpower larger specimens. R. Clarke studied a cuttlefish of the genus *Architeuthis* which was found in the stomach of a sperm whale caught near the Azores. The sepia had not been digested; it was over 10 m long (including the arms) and weighed about 186 kg. The whale that swallowed this giant mollusk was not particularly large; it measured only 14.5 m. One can often find indigestible remains of horny cuttlefish beaks in the sperm whale's intestinal canal; the size of some of these remains shows us that sperm whales must occasionally overpower and swallow sepia of enormous dimensions. The sperm whale's skin is usually covered with large, round scars, left by the suction cups on the arms of the sepia as these animals sought to defend themselves. The sperm whale also eats fish from time to time. Thus, researchers found a 3-m-long herring in a whale's stomach.

The larger species of cuttlefish, those preferred by the sperm whale, are found mainly at the lower ocean depths. As a result, the sperm whale is the second most efficient diver of all the whales, next to the bottle-nosed whale, which also eats cuttlefish. The sperm whale regularly makes dives of between 500 and 1000 m in depth in its search for food.

The intestine of a large sperm whale is about 160 m long. This is surprising, as carnivores generally have a relatively short intestine. Sometimes, but rarely, one can find an abnormal secretion, called amber, in the intestines of these whales. This secretion is a firm gray or black mass which smells rather unpleasant when fresh, but which acquires a pleasantly sweet scent as it grows older. This amber was formerly used as a magic love potion, very much in keeping with the belief, prevalent at that time, that sperm whales have a supply of semen in their heads. Later, amber was used as a raw material in the manufacture of perfume; it still has some significance in that industry. We do not know exactly how the amber is produced; it is probably a pathological, or at least

Fig. 16-2. Bottle-nosed whale (*Hyperoodon ampullatus*; dotted area) and southern bottlenosed whale (*H. planifrons*; black area).

abnormal, growth. The amber pieces can be very large; pieces weighing 400 to 500 kg have been found. The present price is 50–175 dollars for 1 kg of amber, even though the price has been greatly decreased since the development of synthetic perfume bases. One can also find amber floating in the ocean or lying on the shore. The main supply of this material, however, comes from the intestines of dead sperm whales.

Hunting the sperm whale has always been a very lucrative business. The whale's oil, including the spermaceti, is chemically not an oil at all, but rather a waxlike mass. The sperm whale's oil cannot be used in the manufacture of soap and margarine, as are the oils from the baleen whales. Instead, it is used in the production of medicine and cosmetics, as a base for salves and skin creams, makeup, and lipstick. The sperm whale's oil is even used for shoe polish and in the production of lubricating oil. Because man has found so many uses for the oil from these animals, the number of sperm whales killed annually has risen sharply within recent years. Over 20,000 sperm whales were killed in the hunting season of 1959–60.

The sperm whale lives mainly in the oceans of the tropic and temperate zones. Bachelor males may even move into the polar regions. Female sperm whales and young calves are found only in the waters between 40° north latitude and 40° south latitude. There the animals live together in a harem community: groups of females and their calves, together with half-grown young, are all led by a single large sperm whale bull. Hans Hass and his associates were able to approach a herd of sperm whales near the Galapagos Islands, where they were successful in filming the animals underwater. They found that the whales communicated by making squeaking sounds. The leader's position is always strongly challenged during the mating season. Old bulls that have been turned away or driven away from the herd can be very dangerous. Herman Melville has given us a very explicit and poetic portrait of one such animal in his classic tale of the sea, *Moby Dick*.

▻
A group of white-sided dolphins (*Lagenorhynchus acutus*) making simultaneous jumps in the air.

▻▻
The pilot whales (*Globicephala*) show a learning ability and a sense of understanding which is greater than that of all the other dolphins presently kept in captivity. In spite of this fact, however, large numbers of these whales are still being killed and used for dog food and oil.

▻▻▻
The friendly intelligent dolphins are particular favorites of the public whenever they are kept in captivity. Here a bottle-nosed dolphin (*Tursiops truncatus*) jumps out of the water at his trainer's command.

Family: Beaked whales

The BEAKED WHALES (family Ziphiidae; L 4.5–over 9 m) are also sepia-eaters. The snout is long and narrow; in some species it is clearly distinguishable from the head. There are very few teeth in the lower jaw. The teeth of the female do not break through the jaw. These whales have two deep longitudinal grooves on their throats. The breast flukes are small. The dorsal fin is triangular. There are five genera, with about fifteen species, some of which are almost unknown.

The BOTTLE-NOSED WHALE (*Hyperoodon ampullatus*; L, ♂♂ 9.1 m ♀♀ 7.3 m; see Color plate, p. 480) is the best-known beaked whale. The jaw area is beaklike. Older males have a high, steep forehead, which is distended by two bony ridges. These animals have only one pair of large, pointed teeth on the tip of the lower jaw. In older animals these teeth occasionally break through; otherwise they remain hidden in the gums. Sometimes there is a second, smaller pair of teeth behind the first

MARINE
STUDIOS

pair. There are occasionally rows of vestigial teeth in both the upper and lower jaws. The bottle-nosed whale is found in the North Atlantic. It is clearly related to the SOUTHERN BOTTLE-NOSED WHALE (*H. planifrons*), found in the southern hemisphere.

The bottle-nosed whale is found in the North Atlantic especially in the area between the Faeroe Islands (Denmark), Jan Mayen Island (Norway), and Iceland. These animals have been known to move into the Arctic Ocean. They spend the winter in the more southern oceans, as far south as the Cape Verde Islands, and occasionally in the North and Baltic Seas.

Bottle-nosed whales form small packs as well as herds of up to 1000 animals. They are probably polygamous, like the sperm whale. The young calves are born between March and May; they are 3 m long at birth. The bottle-nosed whale is also a sepia-eater, and it makes dives of over 500 m to hunt these animals.

The two species of the genus of GIANT BOTTLE-NOSED WHALES (*Berardius*), BAIRD'S BEAKED WHALE (*Berardius bairdi*; L, ♂♂ up to 12.8 m; see Color plate, p. 480) and ARNOUX'S WHALE (*B. arnouxii*) are found in the northern and southern Pacific Ocean. These whales look very much like the bottle-nosed whale; however, they always have two teeth in each half of the lower jaw. Baird's whale is the largest of the beaked whales; reportedly it has a voice like that of a bellowing bull.

CUVIER'S BEAKED WHALE (*Ziphius cavirostris*; L up to 8 m; see Color plate, p. 479) is sometimes called the GOOSEBEAK WHALE. It is found in almost every ocean of the world. The individual animals of the species may be colored quite differently from one another. As yet we know practically nothing about the habits of these animals.

The genus of BEAKED WHALES or COWFISH (*Mesoplodon*; see Color plate, p. 480) has a variety of forms. There are probably nine species, which are differentiated mainly by the shape and position of the pair of teeth on the lower jaw. The best-known species is SOWERBY'S WHALE (*Mesoplodon bidens*; L up to 4.9 m; see Color plate, p. 480), whose head is slender and pointed, with a sloping forehead. The triangular teeth are found in the front third of the jaw. These whales live in the North Atlantic.

We know almost nothing about the other species of this genus. We are familiar with *Mesoplodon europaeus* through only three individual specimens. LAYARD'S OF STRAP-TOOTHED WHALE (*M. layardii*; L 5 m; see Color plate, p. 480) is distinguished chiefly by the unusual shape of its teeth. The tooth pair is located in the front third of the jaw. The large teeth are flattened at the sides; they grow to the side, out of the mouth, and in a curve around the upper jaw so that the whale can open its mouth only to a slight degree. No one knows the significance of this rather singular growth pattern. These rare whales have thus far been found only in the southern hemisphere.

◁ American scientists are presently studying dolphin vocalizations. Here a microphone is held to the blow hole of a bottle-nosed dolphin (*Tursiops truncatus*); certain whale vocalizations come not from the animal's mouth, but rather from the blow hole on the top of the head.

SHEPHERD'S BEAKED WHALE (*Tasmacetus shepherdi*) represents its own genus. This whale was first described, in 1937, by W. R. B. Oliver, who based his observations on an animal that had been caught near New Zealand. This toothed whale had a pair of large teeth in the front part of the lower jaw, just like the closely related beaked whales. In back of this pair, however, the Shepherd's whale had 26 small but fully developed teeth on each side of the lower jaw. It also had 19 teeth in each half of the upper jaw. We must rely on a newspaper article for a description of the Shepherd's whale's appearance: The head is round and the eyes are relatively large. The back of the animal is black, the sides are spotted with grayish-yellow, and the stomach is white.

Superfamily: River dolphins, by H. Hediger

The RIVER DOLPHINS have a special place among the toothed whales, and consequently we must regard them as a single superfamily (Platanistoidea). These are ancient whales; the shape of their skeleton is reminiscent of certain dolphin fossils.

Their length ranges between 1.5 and 2.5 m. The cranial bones are not quite as crowded together as in other whales. The neck vertebrae are large and are not fused. The sternum is well developed. The head is clearly distinguishable from the body. Both the upper and lower jaws are long, slender and beaklike, with many pointed teeth. River dolphins are fish-eaters. They live in fresh water. There are three families, with four genera, each of which has a single species.

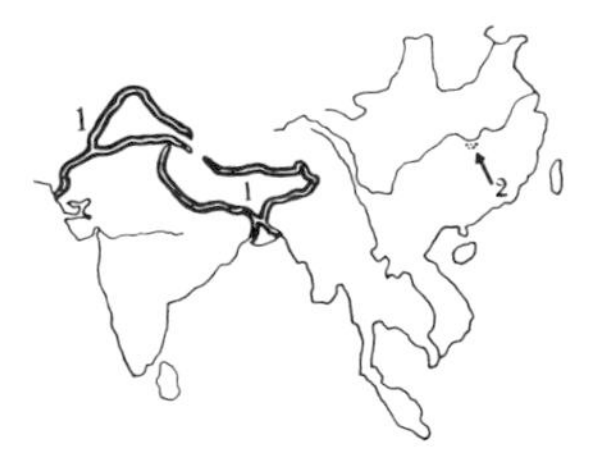

Fig. 16-3. 1. Ganges River or Gangetic dolphin (*Platanista gangetica*), 2. Chinese river dolphin (*Lipotes vexillifer*).

A. The GANGETIC DOLPHINS (family Platanistidae) have only one species, the GANGES RIVER OF GANGETIC DOLPHIN (*Platanista gangetica*; L up to 2.4 m; see Color plate, p. 479). The eyes are tiny, and the dolphin has very poor eyesight. The breast fluke is short and has a broad, fanlike shape. The dorsal fin is low. These dolphins are found only in the Ganges and Indus Rivers, where they are called *susa* by Indians. They dig into the mud for fishes and crabs. Generally a single young is born, after a gestation period of nine months.

B. The AMAZONIAN DOLPHINS (Iniidae) have two species: 1. The AMAZON OF AMAZONIAN DOLPHIN (*Inia geoffrensis*; L up to 2.1 m; see Color plate, p. 479) is usually bluish-gray, but the range of colors of individuals is from pink to black. The eyes are very small, and the ear opening is clearly visible. The breast flukes are large and wide. The tail fin is relatively large. These animals are found in the upper Amazon, where local people call it the *bouto* or *inia*. 2. The CHINESE RIVER DOLPHIN (*Lipotes vexillifer*; L up to 2.2 m; see Color plate, p. 479) has very poor vision. The breast flukes are wide. The dorsal fin is tall and triangular; its flaglike appearance gives the dolphin another popular name, the WHITE FLAG DOLPHIN. These animals are found only in Lake Tung-Ting (Hunan Province, central China).

C. The LA PLATA DOLPHINS (Stenodelphidae) have only one species, the LA PLATA DOLPHIN (*Stenodelphis blainvillei*; L up to 1.5 m; see Color plate, p. 479). The male is usually smaller than the female. They have

Fig. 16-4. Amazon dolphin (*Inia geoffrensis*).

Fig. 16-5. La Plata dolphin (*Stenodelphis blainvellei*).

from fifty to sixty tiny pointed teeth in each half of the jaw. The head is less distinct from the body than in other dolphins. A female usually gives birth to a single young at a time; it may be 45 cm long, and weigh 7 kg.

We do not know very much about the life of the river dolphins. These animals usually live in muddy water, and consequently they have been observed only on rare occasions. Bates has reported that the AMAZON DOLPHIN lives in pairs and that these animals often swim about in shallow water. Amazon dolphins living in American aquariums have shown a similar behavior pattern.

Toothed whales are particularly intelligent, and they have a great learning capacity. Man first noticed and evaluated these accomplishments in salt-water dolphins, and much later in their fresh-water relatives. Observations of fresh-water dolphin behavior is about twenty-five years behind that of the salt-water dolphins. However, scientists are beginning to spend more time and effort in studying these interesting animals. Improvements in aquarium equipment are making it possible to keep delphinids in captivity. In particular, major developments in the creation of inexpensive filtration processes, which permit the use of artificial (man-made) salt water, have been responsible for the more extensive exhibition of marine animals, especially in inland zoos and aquariums.

The Amazon dolphin has been the most widely studied of the river dolphins. These animals are presently being kept in a few American zoos and aquariums. Superficially, these dolphins do not look very much like the salt-water dolphins; they do not have the striking torpedo-like body shape of their salt-water relatives, and their eyes are really very tiny, compared to the large, expressive eyes of the salt-water dolphins.

I was recently able to watch and admire a large number of Amazon dolphins in the Shedd Aquarium in Chicago. These animals were as friendly, as tame, and as eager for affection as a bottle-nosed dolphin (*Tursiops truncatus*) or as any other salt-water dolphin might be. They were unusually receptive to being stroked; they pressed themselves against my hand, and I found that their skin felt like soft leather gloves. The feel of Amazon dolphins' skin is much different from the stiff, plasticlike skin of the salt-water dolphins. The Amazon dolphins also have vibrissae about 1 cm long on their faces, and an elongated, beaklike snout, two other points of difference between these animals and the salt-water dolphins. Amazon dolphins do not make long or high leaps, as do their salt-water relatives; however, these animals undoubtedly have other special skills which we hope to discover in the future.

Superfamily: Narwhals, by E. J. Slijper and D. Heinemann

The NARWHALS (Monodontoidea) are a separate superfamily which contains only one family, that of the WHITE WHALES AND NARWHALS (Monodontidae). There is no dorsal fin. The neck vertebrae are not fused. There are two genera, each with one species.

The white whale

The WHITE WHALE (*Delphinapterus leucas*; L 3.7–4.3 m, up to 5.5 m; see Color plate, p. 480) is also called by its Russian name, BELUKHA or BELUGA. The head is rounded, with only the indications of a beak. The breast flukes are wide and round. There are from eight to ten teeth in each half of the jaw. The young are dark gray; later they become mottled or yellow. After four or five years they become white. Mating and parturition occur from April until June. The nursing period lasts eight months. Males are sexually mature at three years, females, at two years. The average life expectancy is twenty years. These whales are found near the coasts of the Arctic Ocean and in the nearby ocean areas.

White whales live together in families or in small groups of between five and ten animals. They are usually found in coastal regions, in bays, fjords, or in the mouths of rivers. Individual animals sometimes travel south into the North and Baltic Seas. Occasionally these whales will even swim into rivers. One white whale lived in the Rhine River, Germany, for one month during the spring of 1966, where it became quite popular.

A white whale in the Rhine

The whale was first observed on 15 May, near Nijmegen; it appeared again on 18 May, near Duisburg. Initially, the beluga let boats come quite close to it, and people assumed that the animal's extended stay in the polluted waters of the Rhine had somehow caused the whale to become sick or weakened. For fourteen days, Wolfgang Gewalt, director of the Duisburg zoo, tried to catch the whale and remove it from the unhealthy waters and bring it to the Duisburg zoo's aquarium. However, the whale became quite shy and would not let Gewalt's boat approach it. Gewalt reported:

"We were only occasionally able to see the back of the animal and part of its upper body above the surface of the water; in addition we saw the tail fluke two or three times during a particularly powerful diving maneuver. When the whale breathed, its back rose 15–20 cm at most out of the water; on rare occasions it rose 30–40 cm. The head was never raised to such an extent that the eyes or mouth opening could be seen. The beluga did not play around our various boats like a dolphin. In spite of the animal's very great suppleness, it did not have enough mobility to do so. However, the whale did maintain a certain loose contact with the smaller boats, probably out of curiosity, as long as it was not crowded by the crew members of these boats. When darkness fell, the beluga searched out a quiet resting place in which to spend the night, near the river bank. We were usually able to find the whale again on the next morning, not far from this resting place."

All of the attempts to capture this white whale failed. The possibility of drugging the animal with an anesthetic shot, a technique presently used with many other mammals when man wants to transport them to safety, was inapplicable due to the nature of the white whale. Unlike all other mammals, the beluga does not breathe involuntarily; a drug or anesthetic would have made voluntary breathing impossible and the animal, as a result, would have suffocated. For this reason, Gewalt attempted to subdue the white whale with a shot of tranquilizer. This was successful, but the medicine itself was too weak. Gewalt and his associates decided not to attempt a larger dose of the tranquilizer, for fear of accidentally rendering the animal unconscious.

The white whale finally swam further up the Rhine, and on 12 and 13 June, it rested near Bonn. On 14 June it appeared at Rolandseck, near Bad Honnef, 400 km from the sea. Then the animal reversed its direction and swam back downstream very rapidly, reaching the open sea on 16 June at 6:45 p.m. near the Hook of Holland. It had remained in the Rhine for a month.

Fig. 16-6. White whale (*Delphinapterus leucas*). Individual animals will frequently move further south than is indicated here.

We do not know why this whale appeared so far south in the polluted waters of the Rhine River. Man has often attributed such "meandering" behavior to a deficiency or pathological weakness in the animal. Erna Mohr, however, advises us to be careful of such judgements; this behavior might also be due to a certain "active exploratory curiosity" which is present in these animals. External circumstances may affect such southern movements. In any case, at least, eleven white whales moved south from the Arctic into the temperate European waters between 1964 and 1966. Apparently another similar white whale migration occurred in 1902–3.

Man has long tried to keep the white whale in captivity. In 1877, one of these animals was brought from Labrador to the Westminster Aquarium in London; however, the whale died there after four days. In the early 1970s, white whales were exhibited in several aquariums, including New York, Vancouver (British Columbia, Canada), San Diego's Sea-world, and Duisburg (Germany). The white whales in the Duisburg collection had, in 1974, been in captivity for over five years.

The beluga's natural diet consists of crabs, cuttlefish, and fish. These whales prefer bottom-dwelling fishes like flounder and halibut, and consequently they often hunt their food at great depths. W. G. Heptner has reported that these animals also hunt jellyfish and Ctenophora near the water's surface.

The narwhal

The NARWHAL (*Monodon monoceros*; L, without the tusk, 4–4.9 m; see Color plate, p. 480) is closely related to the white whale. The young of these two species are extremely difficult to differentiate. The body is shaped like that of the white whale, but there is no indication of a beak. Adults have only one pair of teeth, which grow horizontally in the upper jaw. In males, the left tooth develops into a straight, screw-shaped tusk between 1.8 and 2.5 m long (sometimes up to 2.7 m long). The male's right tooth and both of the female's teeth remain hidden in the jawbone; in rare cases males have been known to develop two tusks. The lower jaw is toothless. These animals are able to mate and to give birth at any time of the year. The young are blue-gray, darker than the adults. They have two additional pairs of teeth in the upper jaw. A narwhal was exhibited for a short time at the Vancouver (British Columbia, Canada) Public Aquarium.

Fig. 16-7. Narwhal (*Monodon monoceros*).

We do not yet know the significance of the single tusk. It has been suggested that the narwhal uses its tusk to break through the ice cover in order to breathe; possibly the animals use their tusks as rakes when hunting food on the ocean floor. All of these possibilities are very tentative, especially when we remember that the female generally does not have a tusk. It is much more probable that the tusk of the narwhal plays a role in the social or sexual life of these animals, rather like the deer's antlers, the lion's mane, or the nuptual plumage of the male duck.

"Narwhals are fast, agile, harmless animals," says Scoresby. "They are social animals, and they appear in numerous small groups of six or more animals. Often the animals in these groups are all of the same sex." The narwhal has the northernmost distribution of all the whales; it rarely appears further south than 65° north latitude.

Whalers formerly hunted the narwhal relentlessly for the ivory of its teeth and for its oil. The screwlike tusk was believed, by people who did not know its origin, to be the "horn" of the mythical unicorn, and it was believed to have magical and medicinal powers. During the Middle Ages, these tusks were worth their weight in gold. Thus an elector in Saxony once paid 100,000 taler for one narwhal tusk, and Kaiser Karl V was able to pay off a large (and economically dangerous) debt to the Margrave of Bayreuth with two narwhal tusks. Today the narwhal has no great economic significance, and consequently these animals are relatively undisturbed.

Superfamily: Dolphins

The last superfamily within the suborder of toothed whales is that of the DOLPHINS (Delphinoidea) with the three families of porpoises (Phocaenidae), long-snouted dolphins (Stenidae), and dolphins (Delphinidae).

Family: Porpoises, by E. J. Slijper and D. Heinemann

The PORPOISES (family Phocaenidae; L 1.3–1.8 m) are a group of small dolphinlike whales with a worldwide distribution. There is no indication of a beak. The teeth are spatula-shaped. These animals are fish-eaters. The first and second neck vertebrae are fused. There are three genera: *Phocaena* and *Phocaenoides*, which have dorsal fins, and *Neophocaena*, without dorsal fins. There are seven species, including the COMMON PORPOISE (*Phocaena phocaena*; L up to 1.8 m, weight 45–54 kg; see Color plate, p. 480). Its breast flukes are small, and the dorsal fin is medium-sized and triangular. There are between twenty-three and twenty-seven teeth in each half of the jaw. Mating takes place in summer. The gestation period lasts about a year. The newborn young are 70–80 cm long; they are sexually mature by the age of fifteen months. These porpoises are found on the North Atlantic coast, and they are one of the most familiar whales of the European coast. Like the white whale, the common porpoise occasionally travels up rivers. These animals have occasionally been seen in the Elbe, the Meuse, the Schelde, the Seine, and the Thames. They feed primarily on herring (*Clupea*), and on whiting (*Merlangius merlangus*) and sole (*Solea vulgaris*), as well as occasionally on other fishes and crabs. Researchers have even found plant remains in the stomachs of these animals. Porpoise meat used to be regarded very highly; the flesh of the young animals is apparently particularly tender and tasty. Now, however, the porpoise is no longer hunted to any large extent. The SPECTACLED PORPOISE (*Phocaena dioptrica*; see Color plate, p. 480) is found in the South Atlantic. The BLACK FINLESS PORPOISE (*Neophocaena phocaenoides*; L 1.3–1.4 m; see Color plate, p. 480) has no dorsal fin. These animals are found from the coast of the Cape of Good Hope to Japan. These porpoises have been known to swim 1000 km up the Yangtze Kiang River. Their diet consists of fish, crabs, and cuttlefish. Other species are shown in the color plate on page 480.

Family: Long-snouted dolphins, by E. J. Slijper and D. Heinemann

The porpoise is one of the most highly developed of the dolphin superfamily. The LONG-SNOUTED DOLPHINS (family Stenidae), however, have remained quite primitive. This is indicated by the fact that their first two neck vertebrae are only slightly fused. The long-snouted dolphins are small, with a slender head and long, beaklike jaws. The dorsal fin is average-sized and triangular; the point of the triangle is directed toward the back. These animals are generally lightly colored, although one species is rose-colored. There are three genera:

A. The ROUGH-TOOTHED DOLPHINS (*Steno*) have only one species, the ROUGH-TOOTHED DOLPHIN (*Steno bredanensis*; L 2.4 m; see Color plate, p. 480). They have between twenty and twenty-seven teeth in each half of the jaw. We do not know anything about the habits of these animals. They live in the warmer regions of the Atlantic and Indian Oceans.

B. The WHITE DOLPHINS (*Sotalia*). Members of this genus have from twenty-eight to thirty-one teeth in each half of the jaw. Their range includes the Indian Ocean and the coasts of western Africa and eastern South America. They live near rocky coasts and in brackish water; a few live in rivers. There are about eight species, including the BUFFEO NEGRO (*Sotalia fluviatilis*; L 1–1.6 m), which is yellowish-white, dark gray, or black. They are fresh-water dolphins which live in the Amazon. The GUIANAN RIVER OR GUIANA DOLPHIN (*S. guianensis*) is black or brown, with a white underside. These animals inhabit bays and harbors of the eastern South American coast. The division and distinction between additional species of this genus is open to debate.

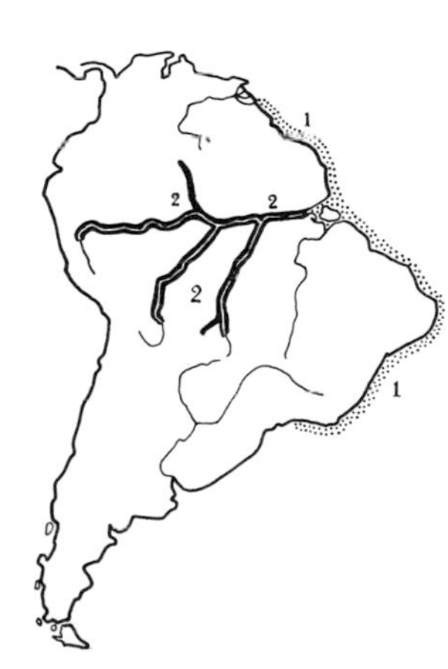

Fig. 16-8. 1. Guianan river dolphin (*Sotalia guianensis*). 2. Buffeo negro (*S. fluviatilis*).

C. The genus *Sousa* (L 1.2–2.5 m). Some scientists place these dolphins in *Sotalia*. Members of this genus have between twenty-three and thirty-seven teeth in each half of the jaw. These animals live near rocky coasts and in brackish water. There are six species, including the WEST AFRICAN WHITE DOLPHIN (*Sousa teuszi*; see Color plate, p. 480), found near the western coast of Africa and in the river deltas of that area. 2. The CHINESE WHITE DOLPHIN (*S. sinensis*). These animals are milk-white with reddish fins. They have black eyes. They are found near the Chinese coast.

For a long time man believed the West African white dolphin's diet consisted solely of plants. The first animal caught in the port of Cameroon had nothing but leaves, grass, and mango fruit in its stomach. Other animals of this species were not caught until 1958 and 1959, when researchers found West African white dolphins in a river delta in Senegal. These specimens had nothing but fish in their stomachs. Probably the plant remains found in the stomach of the first had come from the stomachs of fishes the animal swallowed.

Family: Dolphins

The DOLPHINS (Delphinidae) have the greatest number of species of all the families within the whale order. Their length ranges between 1 and 9 m. The first and second neck vertebrae are fused. The snout is more or less beaklike. The teeth are cone-shaped. Most of these animals have a distinct dorsal fin. The dolphins are basically fish-eaters. There are four subfamilies: 1. Right whale dolphins (Lissodelphinae), 2. True dolphins (Delphininae), 3. Pilot and killer whales (Orcininae), and 4. Commerson's dolphins (Cephalorhynchinae), with about thirteen genera and more than twenty-five species in all.

Ancient dolphin stories, by B. Grzimek

Man has long built fantasies and myths around the dolphin. The Greeks and Romans told many wonderful tales about man's experiences with the dolphin, tales that for over 2000 years have been laughed at by the experts. Now, however, we find that some of these exciting adventures may well have been based on fact.

When Telemachus, son of the famous sailor Odysseus, fell into the sea and nearly drowned, a dolphin rescued him and brought him to shore. For this reason, Odysseus had a dolphin engraved on his ring, and his shield carried a picture of a dolphin. This, at least, is how it was according to the Roman Plutarch. Admittedly, Plutarch lived long after Odysseus; in any case, Plutarch's history is the first report of the dolphin's friendliness and helpfulness to man.

The famous Grecian minstrel and poet Arion, of the late 7th Century, B.C., traveled from town to town in his native Italy, earning much money. The sailors of the ship he took back to Greece robbed him and threw him into the ocean. Dolphins rescued him and brought him to shore. As a result, Arion and the king were able to confront the murderers and were able to punish them.

Near the Grecian city of Milet, on the coast of the country we now call Turkey, there lived a young boy named Dionysios, who liked to swim in the sea with his friends after school. One day a dolphin appeared, hoisted Dionysios on its back, and began to swim out to sea. Dionysios was terrified, but then the dolphin turned around and carried its rider in to shore. Later the dolphin appeared again and again and played with the boy. Unfortunately, the story has a tragic ending. One day the dolphin became stranded on the beach and could not get back to the water. The children who wanted to help the animal could not, because it was too heavy and its skin was too slippery. Consequently, the dolphin died.

The famous Roman scientist Pliny, killed in A.D. 79 by the eruption of Mt. Vesuvius, wrote the story of a dolphin-human friendship in his *Natural History*. The events occurred around the time of Augustus Caesar, in the years when Jesus was still a child. At that time, a flat sea lagoon covered the area we know as Naples, Italy. In this lagoon there lived a dolphin that had been brought over the small piece of connecting land and into the sheltered water by local fishermen. This dolphin became friends with a young boy who had to travel around the lagoon to get to school in Puteoli. The two friends played together, and after a while the boy was allowed to ride on the dolphin's back. Finally, the dolphin carried the boy all the way across the lagoon to Puteoli and school every morning. When the boy was ready to return, the dolphin brought him back home.

Two thousand years ago there was a small town called Hippo, on the African coast where the city of Bizerta, Tunisia, now stands. The children and young men of Hippo were also fond of swimming, and there too a dolphin befriended one of the young boys. The dolphin played with its friend and let him stroke it. It even carried the young boy out to sea and brought him safely back. The other children were frightened of the large animal at first, but they lost their fear within a few days. The friends played together daily. The dolphin was always accompanied by a second dolphin; this second dolphin, however, only remained in the general area, and would not let itself be touched. The news of the dolphin spread rapidly, and all the officials of the province went to Hippo to see the animal. Finally so many people kept coming to Hippo to see the dolphin that there was not enough food to feed them all, and the town fathers decided to kill the dolphin secretly so that the town of Hippo would be left in peace.

We can easily understand why modern scientists originally objected to these stories. Alfred Brehm wrote in the first edition of his *Illustrirten Tierlebens* (Illustrated Animal Life) in 1865: "Unfortunately, we must give up all these beautiful stories by poets and writers of fairy tales; there is no proof whatsoever for these events."

Recent studies of the dolphin, by H. Hediger

However, in this case Brehm was wrong. Perhaps there really were individual people around the Mediterranean Sea who, in ancient times, had a close relationship with these highly developed aquatic mammals which, according to the beliefs of that age, were said to be gifted with human intelligence. This is not as improbable as it sounds; modern researchers have established similar relationships and have had similar experiences with the dolphin. Many of the old beliefs about the dolphin's behavior, which in ancient times arose as the result of unprejudiced observations, have been substantiated recently through exacting experimentation and research. Much of our new knowledge about the dolphin has come about as a result of the aquariums and marinelands in the U.S.A. and other countries all over the world. These aquariums offer us all a unique opportunity to study and observe these small toothed whales.

Man has known for hundreds of years that whale, and especially the smaller toothed whale species, have a particularly large and richly convoluted brain. In an article published in 1942, E. Grünthal noted that, with the dolphins, mammals have reached the primate stage for the second time. Thus, as Grünthal has concluded, the dolphin brain is equipped with a particular primate characteristic in the form of substantia nigra (zona nigra); among other animals, this substantia nigra is present only in the human brain. It does not occur in the brains of the anthropoid apes. In 1962 the brain anatomist G. Pilleri found that the brain forms of some individual whales have "a degree of centralization that far surpasses that of the human brain." According to Pilleri's view, "man's position as the highest point in the rank order of mammals begins to be doubtful today."

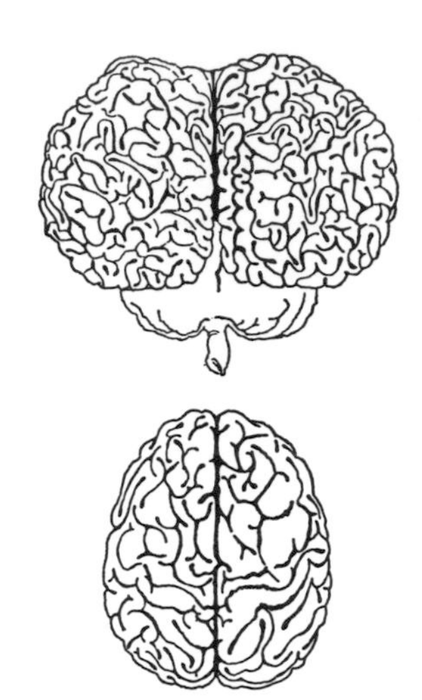

Fig. 16-9. The brain of a dolphin (above), and of a man (below). The larger number of furrows in the dolphin brain is related to its larger volume. However, the number of furrows in the dolphin's brain also indicates the extraordinary faculties of these animals.

In view of these facts, we might well begin to wonder just how the dolphin uses its extremely well-developed brain. This question went unanswered for many hundreds of years; man and the dolphin have long been separated by their opposing environments, and until recently it has not been possible to keep these animals in captivity. During the second half of the 19th Century, people occasionally tried to keep individual whales in large tanks. However, it was not until 1938, with the opening of the Marine Studios in Marineland, Florida, that a biological basis was established for keeping these animals alive in captivity. It became apparent even during the daily visitation periods which are so necessary to the care of dolphins kept captive in such tanks, that they are very easily tamed, that they need companions—even humans are acceptable substitutes—and that they learn quickly. Thus, someone came up with the idea of training dolphins so that they could put on a real show before the paying public. This was more a case of a circuslike production, rather than scientific research. Such training, however, can be adapted to scientific use.

When I [Hediger] visited Marineland in 1951, the male bottle-nosed dolphin (*Tursiops truncatus*) called "Flippy" was the most gifted of trainer Adolf Frohn's pupils. Flippy had been born in the Oceanarium, and at the time of my visit he was 2.10 m long and weighed 105 kg. When I approached his basin, Flippy came to meet me, as far as the salt-water enclosure allowed. He stretched his head and neck to the side, toward me, as far as possible, and, even though I had never stroked a living whale before, I understood instinctively that the animal wanted to be gently scratched. The valvelike breathing opening on the dolphin's head seemed unusual and strange to the touch.

Fig. 16-10. Flippy, the bottle-nosed dolphin, rapidly learned to pull a bell rope.

When Flippy raised his large, antelopelike, and expressive eyes out of the water, one could always see the gelatinous "tears" which are secreted from the inner corners of the eye. These "tears" were 5–10 cm long and so viscous that they could not be wiped away with the hand. Flippy would close his dangerous-looking teeth on a human hand with such care that one was never afraid of being hurt. I came away with the impression of an incredibly tame and affectionate contact animal.

Within a few weeks, Flippy learned to jump through a tire covered with paper, to catch balls and sticks in the air and to return them, to ring a bell by pulling a rope, to twist himself around his center axis in the water, and many other tricks. He even let himself be harnessed to a surfboard, and would pull people around behind him on a ride into the neighboring lagoon. Flippy's commands for these tricks were given to him through specific movements and through calls; I was able to see that in many cases the command word alone sufficed. Flippy needed only a very short time to master a trick. This is apparently true of all trained dolphins. Even apes usually need a much longer period of time to learn whatever their trainer wants them to do.

The training performances of the dolphins and the other animals in the Marine Studios aquarium were greatly increased during the year after my visit. The Pacific white-sided dolphins (*Lagenorhynchus obliquidens*) in the Marineland of the Pacific (California) are particularly impressive when a group of them leaps high out of the water in formation: in groups of three, four, five, or six, these animals take their underwater position at their trainer's command and wait for the starting signal. As soon as the starting signal is given, all of these dolphins (each of which is highly conscious of the others) swim rapidly through the water for a certain distance. Then they all jump, rising above the water's surface together in the same instant. They glide through the air for several meters in a parallel formation, and then dive back into the water, head first. The exact timing and coordination that this effort requires, leads us to believe that the dolphins were able to understand a large part of the task and that they carried it out with particular care and concentration.

Recently a pilot whale (*Globicephala sieboldii*) or a false killer whale (*Pseudorca crassidens*) was observed turning a somersault at the same time the Pacific white-sided dolphins made their jump. Thus it is possible that other whale species may soon be included in the training programs heretofore used principally on dolphins.

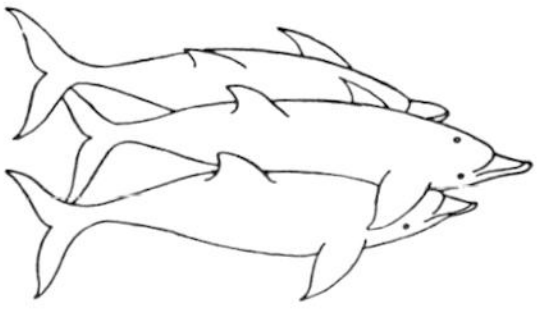

The pilot whale has shown that it has a particularly high learning capacity. Its intelligence is apparent in the animal's high comprehension level, which surpasses even that of dolphins. An animal trainer is able to make the pilot whale understand new techniques or tricks with a minimum of indication. These round-headed animals differentiate and answer vocal sounds, visual signals, and gestures made by their trainer in the quickest and most subtle possible way. There is a close emotional bond between man and whale in this case, a bond that one can feel but cannot describe or analyze. I can see no reason not to mention this important fact.

Fig. 16-11. Two bottle-nosed dolphins hold a wounded conspecific so that its blow hole remains above the water's surface. Dolphins have also rescued drowning people in a similar manner.

One example of the many tricks learned by the pilot whale which will perhaps clarify this astounding learning ability is the following. When the trainer so commanded, the animal beat its horizontal tail fluke against the water surface until the trainer gave it the signal to stop. Superficially, this action might remind one of a well-trained circus elephant beating a drum with its tail. However, the training procedure in each of these cases is quite different. The trainer teaches the elephant this trick by holding the animal's tail in his hand and continuously beating the drum with it, until the elephant understands what is required. This method is not applicable with the pilot whale. No one could hold the tail fin and beat it forcefully against the water surface, since the pilot whale itself weighs nearly 1000 kg. Instead, the trainer must wait until the whale more or less happens to beat its tail against the surface by chance. Then he quickly rewards the whale with food and with applause, making the animal realize through voluntary and involuntary indications that it should beat its tail in response to a given signal.

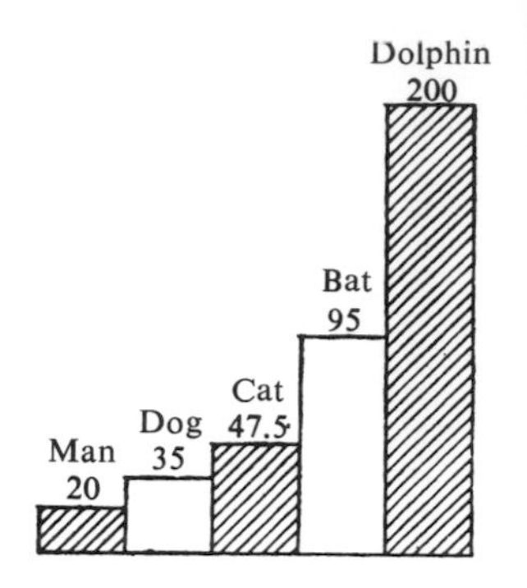

Fig. 16-12. The upper hearing limits of various mammals: Man, 20 kHg; Dog, 35 kHg; Cat, 47.5 kHg; Bat, 95 kHg; Dolphin, 200 kHg.

Unless the animals understand what it is that their trainer wants, these training exercises will not be successful in either the elephant or the pilot whale. However, pilot whales are able to reach this sort of understanding more easily, more quickly, and even more elegantly than elephants or chimpanzees engaged in similar learning procedures.

The close and trusting animal-man relationship has a practical importance in the behavior of a dolphin which has not completely adapted to its new environment, when the water is removed from the dolphin's pool so that the pool can be cleaned. Most whales become highly uncomfortable in shallow water, particularly when the water surface keeps sinking; consequently the animals in the draining pool panic easily, and this panic can result in a senseless racing around the pool which, in turn, can lead to injury. This dangerous situation can be relieved when the keeper climbs into the pool as the water drains, and reassures the animals with words and touch. Later, these clever creatures adapt naturally to the regular pool cleanings.

Why do dolphins help drowning people?, by B. Grzimek

The wide range of observations made possible through marinelands and aquariums helps us to understand and accept the dolphin's astonishing intelligence. Thus we must view the ancient legends about these animals in a much different light. Many of these stories which were believed to be either myth or fantasy, apparently either were true or, at least, were based on a true occurrence.

An attorney's wife went swimming off a deserted beach on the coast of Florida in 1943. She swam some distance from shore and suddenly noticed that she was being pulled under by the current. The woman was terrified. She began to swallow water and to lose consciousness. All of the sudden she was pushed from below and from the side; something pushed her further and further in toward the shore. She landed on the beach, but was too exhausted to turn around and thank her rescuers.

When she finally did turn around, there was no one in sight, but dolphins were swimming a couple of meters offshore. A man who had witnessed the last part of the incident ran to the woman and told her that first he had seen her body floating like a corpse, but then he saw a dolphin pushing her in to shore.

That the woman was indeed rescued by a dolphin is generally understood. However, this does not necessarily mean that the animal wanted to rescue the woman. There have also been reports of dolphins that have worked hard pushing a sodden mattress to the nearest shore, where they have shoved it up onto the beach. Why do they do this? It has been noticed that dolphins, as well as certain other whale species, make sure that sick or wounded conspecifics (members of their own species) keep their blow holes above the water. Occasionally these animals also help animals of another species, especially in aquariums. Consequently, we can understand how this apparently innate behavior can also occasionally be invoked when a human being swims in the ocean. When the conditions are particularly favorable, the dolphin may well rescue a drowning person.

Do dolphins have a language?, by H. Hediger

Among the many tricks that dolphins have learned, we should mention the sounds that they produce on command. Whales have an incredibly wide range of vocalizations, some of which are within the range of human hearing, some of which are beyond that range. Like bats, dolphins produce ultrasonic sounds which they use for echo sounding or echolocation, by which they find direction and locate obstacles and prey. However, these whales also use sound to indicate their moods and to communicate with one another, as is the case with most animals capable of producing vocalizations.

John C. Lilly, a neuroresearcher, has long been involved with dolphin vocalizations, and he has high expectations for his studies. Lilly has established that dolphins are able to imitate human speech. However, the dolphin's rendition is not as clear and distinct as that of a well-trained parrot. Moreover, dolphins often speak so rapidly that we humans cannot understand a word. In view of the extremely high development of the dolphin brain and these whales' unusual intelligence, however, we can expect that as time goes by, their imitation of human speech will surpass that which we are familiar with in parrots and other mocking birds.

Dr. Lilly is convinced that one day it will be possible for him to converse with his dolphins in English. To date, however, Lilly has not proved that the dolphin, unlike all other animals, except man, is able to produce genuine two-word sentences. This two-word sentence is the first prerequisite for a real language.

In addition, and related to their highly developed ability to vocalize, dolphins have an extremely well-refined sense of hearing. They can hear sounds with a frequency of 200,000 vibrations per second. Man is only able to hear sounds with a maximum frequency of 20,000 vibrations per second.

W. H. Dawbin, a whale researcher, once told me [Hediger] about an unusual method of catching dolphins, used by the people of the Solomon Islands. They sail out on the ocean in a fleet of canoes. In each canoe there are two rowers; a third man sits in the middle of the canoe with a number of peculiarly shaped stones. When a herd (school) of dolphins is discovered, the canoes enclose the herd in a wide ring, one that can be as much as 10–15 km from the coast. Then the men in the middle of the canoes bend over and strike the stones together underwater. The resultant sound is apparently unbearable to the dolphins. The natives use their widespread "sound net" to drive the animals toward shore, where they are forced into a bay; there they either die in the shallow water or they become so confused that they do not resist as they are carried on to land, where they rapidly die. The sound of the stones underwater seems to be absolutely intolerable for these animals. There is presently no explanation for the cause of death in these animals other than the acceptance of the idea that they undergo a total nervous breakdown; this is also Dr. Dawbin's explanation for this occurrence.

If this explanation had been published in a book some twenty or thirty years ago, it would never have been accepted today. One would have believed such an idea to be legend or a sailor's yarn rather than a subject for scientific research. Today we are somewhat better equipped to receive the idea. The observations regarding such behavior were first published in 1966, and the behavior itself is presently being investigated by zoologists at the University of Sydney (Australia). We are anxiously awaiting the results of their study.

Overview of the dolphin species, by E. J. Slijper and D. Heinemann

We will classify the genus of RIGHT WHALE DOLPHINS (*Lissodelphis*) as a separate subfamily (Lissodelphinae), distinct from the other dolphins. The snout is protrusive and curves downward. There is no dorsal fin. This genus has two species, the NORTHERN OR PACIFIC RIGHT WHALE DOLPHIN (*Lissodelphis borealis*; L 2.5 m, see Color plate, p. 480), found in the North Pacific, and PERON'S DOLPHIN (*L. peronii*; L 1.8 m), found in southern oceans. The upper head, back, and tail fin are black; the rest of the body is white.

Subfamily: True dolphins

The majority of dolphins belong in the subfamily of TRUE DOLPHINS (Delphininae; L 1.2–4 m). The first and second neck vertebrae are fused. Each jaw has from twenty-two to fifty-two teeth (with the exception of the genus *Grampus*). There are six genera, with a total of about thirty species.

The SPOTTED DOLPHINS (*Stenella*; L 1.2–3 m, weight, in larger species, up to 165 kg) include a wide variety of generally smaller dolphins. They have between thirty-seven and fifty-two tiny, pointed teeth in each half of the jaw. The beak is elongated and narrow. Ten species have been described, but the validity of their classification is open to debate.

The BLUE-WHITE DOLPHIN (*Stenella caeruleoalba*; see Color plate, p. 480) is small but powerfully built. The animal from which the species was described was only 1 m long. These dolphins are found off the southeastern coast of South America, near the mouth of the La Plata River (Argentina).

The DOLPHINS (genus *Delphinus*) also encompass a wide variety of forms. The status of some of these animals as independent species is also subject to debate. Consequently we will mention only one species, the COMMON DOLPHIN (*Delphinus delphis*; L 1.5–2.5 m, weight up to 75 kg; see color plate p. 480) found all over the world in warm and temperate waters. The beaklike jaw is relatively long, thin, and pointed. It is clearly distinguishable from the forehead. The dorsal fin is medium sized, as are the breast flukes, which are pointed. There are forty to fifty very tiny teeth in each half of the jaw.

The TRUE DOLPHIN has a very large distribution range, and often appears in large herds. This is the most common species in the Mediterranean Sea, and therefore these animals are probably responsible for the many stories, legends, and fables told about the dolphin in ancient times.

The COMMON DOLPHIN is quite well known, and well loved, because it often follows ships in larger or smaller groups, and entertains sailors and travelers alike with its lively and playful antics. "The glistening bodies sail rapidly through the air in curves of several meters distance, plunging headfirst into the water, only to shoot back out again, always the same game, over and over," writes Pechuel-Loesche. He describes how some members of the herd make vertical jumps, and how, standing erect or in a curved position, they dance around the water surface on their tail fins. When a ship appears in their field of vision, the dolphins race toward it: "They encircle the ship with a wide curve, and jump in front of it and along the sides; then they turn around and show off their best tricks."

The largest species of the true dolphins is RISSO'S DOLPHIN (*Grampus griseus*; L 3.7–4 m; see Color plate, p. 480). They do not have a beak, and they have only three to seven teeth in each half of the lower jaw. The upper jaw has no teeth. These dolphins are sepia eaters. The breast flukes are fairly long and narrow. The dorsal fin is high and pointed. These animals are found all over the world.

Risso's dolphin lives alone or in small herds (schools) of fewer than twelve animals. We know very little about the habits of these dolphins, as is the case with many other toothed whales. One Risso's dolphin that became world famous was called "Pelorus Jack." Pelorus Jack was protected by the New Zealand Parliament because for decades he swam and played around ships in the Pelorus Sound, between Nelson and Wellington (New Zealand). He even swam in front of the ships and showed them the way.

The BOTTLE-NOSED DOLPHIN (*Tursiops truncatus*; L 1.75–3.6 m, weight

The bottle-nosed dolphin

150–200 kg; see Color plate, p. 480) is the dolphin species most commonly kept in aquariums and marinelands. The observations of the dolphin's learning ability and vocalization, etc., discussed earlier in this chapter, were all initially made with this species. Their beak is short and easily distinguishable. They have twenty to twenty-six teeth in each half of the jaw. These dolphins, found all over the world, are fish-eaters. The demarcation of other individual species within the genus has not yet been finally decided. Bottle-nosed dolphins, some of which have increased the popularity of their species and other delphinids by becoming movie and television "stars," have been bred in captivity. One born in the Marine Studios (Florida) in 1947, lived there for over twenty years. In recent years this species has bred in several aquariums, including the Brookfield (Chicago) zoo's which uses artificial (man-made) salt water. A close relative, the RED SEA BOTTLE-NOSED DOLPHIN (*T. aduncus*) is pictured on page 480.

Next to the common dolphin, the bottle-nosed dolphin is one of the best known and most beloved whales. These dolphins are particularly common on the eastern coast of the U.S.A. The animals in that region were hunted relentessly in earlier times. Between 15 November 1884 and 15 May 1885, 1268 bottle-nosed dolphins were killed off Cape Hatteras (North Carolina) for their oil. Today the bottle-nosed dolphin and other dolphin species no longer have a large economic significance, at least not after they are dead. The U.S.S.R. has even forbidden the hunting or killing of these animals, because their brain is so much like that of man.

The bottle-nosed dolphin mates in spring or summer. The gestation period apparently lasts for ten to twelve months, and the young are nursed for about sixteen months. These animals' readiness to help wounded or sick conspecifics was also evident in captive bottle-nosed dolphins in three different situations; each time, two dolphins carried another dolphin, that had been floating helplessly about, to the surface, where it was able to breathe. The two helpers stuck their heads under the breast flukes of the wounded animal; however, as they were unable to breathe in such a position, the two rescuers occasionally had to let go of the other animal.

Fig. 16-13. Genus *Lagenorhynchus.*

Other true dolphins

We group five to seven dolphin species, again, with only tentative parameters around some groups, together in the WHITE-SIDED DOLPHINS genus (*Lagenorhynchus*; L 1.5–3 m). They are found in the Pacific, Indian, and South Atlantic Oceans. These animals all have a short snout which is not clearly distinguishable from the head. The breast flukes are relatively large and pointed. The dorsal fin is relatively high and pointed; it curves to the back.

The dolphins of this genus live together in large or small herds. Herds of 1000 to 1500 WHITE-SIDED DOLPHINS (*Lagenorhynchus acutus*; see Color plate, p. 480) and WHITE-BEAKED DOLPHINS (*L. albirostris*) have been observed on several occasions. The PACIFIC WHITE-SIDED DOLPHIN (*L. obliquidens*), which is not uncommon off the western coast of the U.S.A. has much the same role in the aquariums of California as the bottle-nosed dolphin has in the aquariums in the eastern U.S.A.

Subfamily: Commerson's dolphin

The COMMERSON'S OR BLACK-AND-WHITE DOLPHINS (genus *Cephalorhynchus*; L 1.2–1.8 m) are classified as a separate subfamily (Cephalorhynchinae), and are thus distinct from the true dolphins. The beak formation is not particularly apparent. The shape of the flukes differs with each individual species. These dolphins have black and white body markings which are easily distinguishable, and which give them the popular name SKUNK DOLPHINS. They live in the colder waters of the southern hemisphere. Their diet consists of sepia and crab. We know virtually nothing about habits of these animals. There are four species, COMMERSON'S DOLPHIN (*Cephalorhynchus commersonii*; see Color plate, p. 480), the WHITE-LIPPED DOLPHIN (*C. heavisidii*), the WHITE-HEADED DOLPHIN (*C. albifrons*), and the HECTOR DOLPHIN (*C. hectori*).

Subfamily: Pilot and killer whales

The subfamily Orcininae includes both smaller and the very largest dolphin species. This subfamily consists of the five genera of the PILOT AND KILLER WHALES. These animals do not have a beaklike snout as do many other dolphins.

The PILOT WHALES OR BLACKFISH (genus *Globicephala*; L 3.6–8.5 m) have a globular forehead which is curved outward; as a result, these animals have one of the most unusual shapes of all the dolphins. The breast flukes are long and narrow. The dorsal fin has a long base, is low, and points to the rear. There are seven to eleven teeth in each half of the jaw. These whales are sepia eaters. They are found in all of the world's warm and temperate oceans. These whales are generally black; some species have white or gray markings. The division into species is uncertain; we will list them in three species: 1. NORTHERN PILOT WHALE OR ATLANTIC BLACKFISH (*Globicephala melaena*; Color plates, pp. 480 and 498), found predominantly in the North Atlantic. 2. INDIAN PILOT WHALE OR SHORT-FINNED BLACKFISH (*G. macrorhyncha*), found from the coast of the eastern U.S.A. to South Africa. 3. PACIFIC PILOT WHALE (*G. sieboldii*).

The pilot whale

Pilot whales live in herds, and one can often find schools of several hundred animals. Apparently these whales have some innate urge to blindly follow runaway members of the same species. The inhabitants of the Faeroe Islands (Denmark) take advantage of this urge when they hunt the pilot whale: they wound only one or two whales with harpoons. These injured animals rapidly swim away, driven by pain and fear, and the rest of the herd immediately follows the runaway. When the whalers are able to force the wounded "pacemakers" toward the beach, the whale herd becomes stranded and the whalers are easily able to kill their prey. Such strandings are probably due largely to the failure of echo-sounding or echolocation waves to return to the whale in shallow or muddy water.

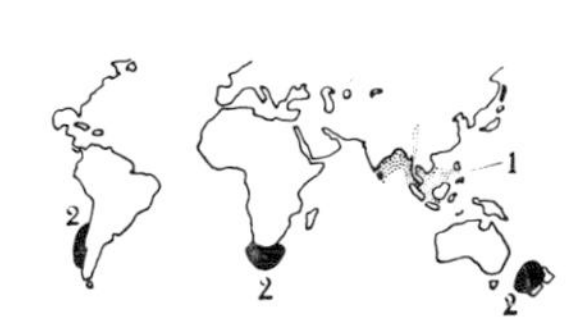

Fig. 16-14. 1. Irrawaddy dolphin (*Orcaella brevirostris*), 2. Genus *Cephalorhynchus*.

Behavior such as that we have just described seems to us to be particularly stupid. In reality, however, the pilot whale is not stupid; of all the dolphins presently in captivity, the pilot whale is the fastest learner, as well as the animal with the greatest apparent understanding.

Unfortunately, the pilot whale is still being slaughtered in great numbers, in spite of its particular friendliness to man. From 1955 to 1960 alone, an average of 4780 pilot whales were killed every year off the coast of Newfoundland. The meat of these animals was frozen and sold in Canada and the U.S.A as dog food. The fat was manufactured into oil right at the hunting site.

The PYGMY KILLER WHALE (*Feresa attenuata*; L 2.35 m; see Color plate, p. 480) is barely known to man at the present time. There are ten to thirteen teeth in each half of the jaw. Only four animals have been caught to date; these specimens were found in the northern Pacific, in the southern Pacific, and in the southern Atlantic Oceans.

The Irrawaddy dolphin

Another species of this subfamily has adapted itself, at least partially, to living in fresh water. This is the IRRAWADDY DOLPHIN (*Orcaella brevirostris*; L 2–2.15 m; see Color plate, p. 480). The head is round and similar to that of the pilot whale. The breast flukes are wide and triangular. The dorsal fin is small. There are twelve to nineteen teeth in each half of the jaw. These animals are fish-eaters.

The Irrawaddy dolphin lives in the coastal waters of Southeast Asia, from the Gulf of Bengal to Borneo. These animals have been found upstream in the Irrawaddy River as far as 1400 kilometers. There they live together in small schools, and often accompany the riverboats. Their fat is used in various areas of India as a medicine for rheumatism.

The false killer whale

The FALSE KILLER WHALE (*Pseudorca crassidens*; L, ♂♂ up to 6 m, ♀♀ up to 4.5 m, weight up to 1350 kg; see Color plate, p. 480) gets its name from its similarity to the killer whale, to which it is closely related. The body is more slender and the head more graceful than those of the killer whale. The breast flukes are tapered more to a point, while the dorsal fin is smaller and curves more to the rear than the killer whale's. There are 8–12 teeth in each half of the jaw. These animals are found all over the world, except in the polar seas.

Richard Owen was the first to describe the false killer whale in 1846; he based his descriptions on a skull that had been found in a swamp in eastern England. Owen believed that these animals were extinct. Sixteen years after this initial description, however, more than 100 false killer whales unexpectedly appeared in the Kiel Bay (Germany). During the next decades, large herds of these whales were stranded on beaches all over the world. Apparently the false killer whale lives on the high seas and only rarely comes close to the coast. Consequently, the behavior of these animals is not adapted to the dangers of shallow water, such as the misdirection of the echosounding waves.

The false killer whale prefers a diet of sepia or cuttlefish, but it will also hunt fish, particularly cod. These animals have recently been introduced into aquariums, although they are still rare exhibits.

The killer whale, or orca, by David R. Martinez and E. Klinghammer

The KILLER WHALE OR ORCA (*Orcinus orca*; L, ♂♂ up to over 9 m, ♀♀ up to 6 m, weight up to 8 t; see Color plate, p. 480) is the largest, most powerful, fleetest member of the delphinid family. Killer whales have a striking black and white coloration and prominent dorsal fins, which in old bulls can be 2 m high. The dorsal fin can be used to determine the sex of adult orcas, for it is triangular in bulls and sickle-shaped in cows. Each jaw half contains ten to fourteen conical interlocking teeth. The species is distributed in every ocean throughout the world, and wherever it occurs it is at the top of the ecological food chain, with no natural enemies.

Man has long been intrigued by killer whales, and descriptions of this giant dolphin made over the past four centuries provide absorbing reading for those who take the time to scan the literature. The earliest reports are from medieval Japan, one of the first definite references dating from 1708 (others, as old as the 9th Century, are somewhat ambiguous). These old descriptions deal with the prominent dorsal fin and with predatory behavior, which is perhaps the most dramatic aspect of killer whale life and one which, unfortunately, has been glamorized and exaggerated in both popular and scientific writing until recently.

Writers in the 19th and 20th Centuries produced dramatic accounts of the attacking behavior of the killer whale, which was once known under such bynames as *Tyrannus balaenarum* and *Formidabilis balaenarum hostis*. Some of the early scientific names given to killer whales included such provocative ones as *Delphinus gladiator* and *Orca destructor*. In 1904 Daniel G. Elliot wrote that "from sheer love of slaughter more creatures are killed in their forays than can be devoured. They delight in blood and rapine...." Another writer suggested that the killer whale uses the "long and pointed dorsal fin ... for aggressive purposes, to rip up the belly of a whale!" The killer whale was depicted as a ravenous monster taking joy in tearing apart huge marine mammals such as walruses, seals, and even the giant baleen whales, which are much larger than themselves. Vivid accounts filled the literature, even that of scientific journals. A 19th Century naturalist wrote of one killer whale 6.4 m long whose stomach contained thirteen porpoises and fourteen seals, with a fifteenth seal in the whale's mouth! The stomach contained only fragments of so many seals and porpoises, but someone later quoted this naturalist as finding entire *bodies* of seals and porpoises, an exaggerated falsehood that crept into even highly regarded scientists' works.

One of the most sensationally reported incidents was that experienced by the Scott expedition to Antarctica in 1905. Two dogs leashed to the expedition ship were on an ice floe at least 75 cm thick, and several killer whales swam about the floe. The whales disappeared underwater, and a photographer moved to the edge of the floe to get some pictures of the giant delphinids. Suddenly the ice floe was knocked about in the sea and fragmented; the whales had come up from beneath and were breaking up the thick ice floe by butting it with their heads! The dogs and photographer survived, and in numerous later reports the inference was drawn that the voracious whales were after the man but, by luck or Providence, could not get him. Actually, it is far more probable that these whales were trying to get the dogs, but to report that would make the entire episode far less dramatic. Of still greater scientific significance is the fact that the behavioral pattern of submerging, approaching from below, and coming up quickly is a species-typical pattern found in killer whales throughout the world as they attack various kinds of small prey.

Reliable, skilled observers in the 1950s and 1960s delivered blow-by-blow accounts of killer whale predation on marine mammals, and these showed that *Orcinus orca* is capable of sophisticated group maneuvers varying with different prey. For example, one group of fifteen to twenty of them attacked some 100 dolphins off Baja California (Mexico) by swimming in gradually narrowing circles about the dolphins. Small dolphins typically respond to danger by bunching together, so what would normally be adaptive behavior only aided the attempts of the killer whales to seize the dolphins. When the small dolphins were tightly crowded, one of the killer whales rushed into the group and fed on some of them while the other whales remained circling. The whale returned to the circling group as a second one swam into the crowded dolphins. In this manner nearly the entire group of dolphins was eaten.

A different approach is used with larger whales. These are attacked *en masse* and differentially: from three to forty killer whales attack at once, several grabbing the lower jaw and tearing out chunks from this part of the body; another group bites into the prey's pectoral fins, immobilizing the animal; one killer whale may even cover the prey's blow hole. At some point the struggling victim stick its tongue out (or has it forced out by the whales at its mouth), and when it does so the orcas tear out the tongue. The killer whales usually feed on the softer body parts (i.e., tongue, lips, parts of the jaw) and leave the rest.

The accuracy of the above and other descriptions by modern observers on predatory behavior in killer whales can scarcely be doubted, but whether these episodes are *representative* of most feeding encounters in the species is an entirely different question. Beneath all the emphasis on mass attacks on large marine mammals lurks an implicit assumption that *Orcinus orca* specializes in gorging itself on such animals. As late as 1959, a member of the prestigious Paris natural history museum wrote the following words about the killer whale in a totally serious work on cetaceans: "It is a voracious carnivore and will tackle all manner of quarry: seals, penguins, dolphins and porpoises (it is nearly a cannibal!), and large fishes. . . . All who have seen the tactics employed by killers acknowledge that these bloodthirsty beasts maneuvre in a most astute way, or, if you like, display a really astonishing instinct for hunting in packs."

Is this what the "killer" whale is all about? What was missing from all literature prior to the late 1950s and the 1960s was analysis of the stomach contents of a large, representative sample of killer whales. The fact that a group of killer whales overwhelms a large baleen whale does not tell us how many other baleen whales *got away* by fleeing or fighting off the killer whales. Similarly, the fact that some observers have seen these whales feeding on large marine mammals does not mean they will eat practically nothing else. Within the last decade, studies have indeed been published on prey found within the stomachs of a great many killer whales, and these studies have cast a new light on killer whale feeding patterns. What results is less exciting and sensational but is far more realistic than what we believed in the past. Not only do killer whales not confine themselves to marine mammals, but mammals are not even the most important component of their diet in general! Two scientists, surveying Japanese catch records on over 500 killer whales caught over a ten-year period, found that fishes and cephalopods comprised about two-thirds of the stomach contents, no matter how large the killer whale. Cetaceans comprised only twenty to thirty-three percent of the stomach contents, and almost all of these consisted of small dolphins less than 2.5 m long. A Norwegian investigator with data on over 1400 killer whales came to the conclusion that the killer whale does not attack large cetaceans that are healthy. Throughout a thirty-year period there were no reports among Norwegian whalers of killer whales attacking large whales under normal conditions. Furthermore, other scientists are publishing papers on baleen whales with scar signs from killer whales: in other words, they were baleen whales that got away!

The picture we now have of killer whales is that of a highly intelligent, curious animal that usually feeds on fishes but sometimes will take a weakened marine mammal (or less often a healthy one) and may even gorge itself on occasion. They are group-living animals with a highly developed social organization and an obvious need for companionship. Their reaction toward humans is a docile one, even in the wild; almost every case of killer whales reacting violently toward human beings arises from some sort of provocation on the part of the people involved. As part of my studies on killer whale behavior, I [Martinez] was in a small motorboat in Puget Sound off the southwestern coast of Canada. Our small craft was in radio contact with a larger one 16 km away (also part of the research team). Both boats were surrounded by groups of perhaps twenty killer whales, including infants, juveniles, cows, and bulls. As other observers had noted, I saw that a killer whale group is structured in such a way that the cows and young are in the middle and the bulls are on the periphery or even a considerable distance away from the main group. The bulls, with their huge dorsal fins and powerful bodies, seem to determine the movements of the group as a whole, signaling intentions to move by slapping the water with their flukes or pectoral fins or even by leaping out of the water and landing against the surface broadside (a maneuver called breaching), which creates a splash so loud it can be heard for 8 km! Vocalizations, not heard on the surface, probably include additional signals. Our small boat was surrounded by these mighty delphinids for several hours, and although it would have required no great effort on their part to capsize it, the whales did not disturb us in any way and merely seemed curious about our presence. We were as close as 5 m from some of the group members, and some of them may well have swum right beneath our little boat. The entire group remained in the same area throughout the time we observed them, until at one point one of the peripheral males created a loud splash against the water, whereupon the whole group disappeared from view. At that same instant our companion boat radioed that their whales also submerged! Neither group was sighted again, both because killer whales are very durable underwater swimmers and because an obscuring fog entered the areas where we were. Although these two groups—perhaps we should say subgroups—were separated by a great distance, both stopped, and finally disappeared in unison. One could hardly attribute this to sheer coincidence; it seems virtually certain that they utilized their vocalizations to communicate and thus to coordinate their activities.

Few of us can observe wild killer whales at such close quarters and under such natural conditions.

A seemingly well substantiated case of a killer whale attack on a small boat in the Pacific, west of Hawaii, occurred on 15 June 1972. Dougal Robertson and his family were rammed in their boat from below and they saw killer whales before their boat went down. He and his family had seen killer whales in captivity at the Wometco Seaquarium in Miami, Florida earlier and said that their identification was positive. The killer whales did not return and attack the people in their small lifeboat. Whether this attack was directed at this boat because of its similarity to one from which they had been hunted at another time, or whether there is some other reason for the attack could not be ascertained by Robertson.

On the other hand, Ted Griffin, who captured wild killer whales in Puget Sound for various seaquaria would enter the water with his diver colleagues when the animals were surrounded by the net in which they were captured. They would actually touch the whales and cut them free when they became entangled in the net—yet they were never attacked. Griffin also mentioned how he had rowed in a small boat in Puget Sound among killer whales and how he had never been afraid of them, nor had he ever been attacked. Eskimos and Indians along the west coast of Canada also seem to have no fear of orcas as far as we know.

In 1972, on the other hand, a man in a black wet suit lying on a surfboard and paddling offshore along the coast of central California was apparently mistaken for a sea lion and was bitten in the leg. The young man turned to see a large black head, he hit the animal, it let go and disappeared. Doctors, who were also experienced divers, showed him pictures of various sea animals and he identified the animal as a killer whale. Unlike sharks, the whale did not press the attack, and left. The man recovered. This seems to be the first authenticated report of a killer whale attack on a human being. The circumstances seem to indicate that the animal mistook the person for its natural prey, and let go as soon as it discovered its error. Such cases of mistaken identity suggest caution in the presence of killer whales, yet the rarity of attacks on humans do not warrant any fear of them. However, since the 1960s, killer whales have been captured and displayed at numerous aquariums and marinelands in North America, Europe, and Japan, and they provide us with at least a glimpse of their total being. Presently (1974) there are nearly twenty different establishments with captive (usually performing) killer whales. A number of these, such as the Vancouver Public Aquarium, in British Columbia, offer a realistic, educational, and informative commentary on the killer whale performance; these institutions are helping to place the killer whale in proper perspective and are giving the spectators a meaningful presentation. Others persist in glamorizing and inappropriately sensationalizing the species (comparing them with wolves and tigers, something which does little to dispel myths about all three animals), and only the complaints of the public (or their refusal to patronize this kind of performances) would put an end to this practice.

A great deal about the life of killer whales is still unknown or poorly understood. Very little is known about their reproductive behavior, for example. Although killer whales have been kept in captivity for over ten years, none has been captive bred. Apparently the gestation period lasts about one year, and newborn calves are 2.1 m long. They are brilliant orange on those parts of the body that are white in older animals; their distinctive coloration may serve in facilitating recognition of the infants by the other members of the group. Unlike many other whales, female killer whales can probably conceive again shortly after giving birth, which would enable them to bear young every year. It appears that one area off the southwestern coast of Canada is a sort of "killer whale nursery," for the same adults appear there year after year, each time with new infants.

There is also very little known about any migration behavior in killer whales. So far it has not been possible to mark individuals and follow them for long periods of time. New marking techniques may enable future researchers to carry out an investigation of this nature.

Like the wolf, the shark, and the grizzly bear (to name only a few), the killer whale has been one of those maligned animals that have been greatly misunderstood. For centuries we have been describing it as a savage, "malicious" carnivore that kills for "sheer joy," and only now are we gaining a deeper insight into the place of the killer whale in the oceanic animal kingdom. Those who see captive ones at the better establishments and those fortunate enough to see them in their natural element are always impressed by them.

While we can see why *Orcinus orca* has been called killer whale, it seems quite inappropriate to single out one of many predators who must all kill their prey by including the word "killer" in its name. One common name in German even refers to it as "murder whale" (Mordwal), as if hunting one's natural food has anything to do with murder. Another name in German, Schwertwal (sword whale), refers to the high dorsal fin of the males, which is a much better term. Perhaps we could all help to put the killer whale's place in nature more into its proper perspective by simply calling it orca.

Currently (1974) plans are underway to set aside Puget Sound in the State of Washington, U.S.A. as a killer whale or orca sanctuary. However, opposition to this proposal has so far prevented this plan from becoming a reality. The species is presently not endangered in the world, although it is feared that continued hunting them for live capture may cause them to abandon Puget Sound as part of their range.

Systematic Classification

With the exception of man (family Hominidae), fossil forms are not listed. Page numbers refer to the main article; numbers in parentheses refer to illustrations or distribution maps of species not mentioned in the text. Species and subspecies without page numbers are not mentioned in the text, nor illustrated or shown on maps. Species and subspecies marked with ⊕ are endangered; those marked with ∅ are extinct.

Order Primates—Continued from Volume X

Suborder Monkeys (Simiae)

Infraorder Old World Monkeys (Catarrhina)

Superfamily Apes and Men (Hominoidea)

Family Anthropid Apes (Pongidae)—continuation 51

Genus *Pan* 19ff

Chimpanzee, *P. troglodytes* (Blumenbach, 1799) 19ff

Bonobo, ⊕ *P. paniscus* (Schwarz, 1929) 47

Family Man (Homididae)

Subfamily Ape-Man (∅ Australopithecinae)

Genus Ape-man (∅ *Australopithecus*)

∅ *A. africanus* (Dart, 1925) 53

∅ *A. robustus* (Broom, 1938) 54

Subfamily Man (Homininae)

Genus Man (*Homo*)

Java Man, ∅ *H. erectus* (Dubois, 1894)

Java Man, ∅ *H. e. erectus* (Dubois, 1894)

Peking Man, ∅ *H. e. pekinensis* (Black, 1927)

Mauretania Man, ∅ *H. e. mauritanicus* (Arambourg, 1954)

Heidelberg Man, ∅ *H. e.* (?) *heidelbergensis* Schoetensack, 1908

Human or Man, *H. sapiens* Linné, 1758 56

Neanderthal Man, ∅ *H. s. neanderthalensis* King, 1864

Cro-Magnon Man, ∅ *H. s.* "*fossilis*" Lartet, 1869

Modern Man, *H. s. sapiens* (Linné, 1758) 55

Order Flying Lemurs (Dermoptera)

Family Flying Lemurs (Cynocephalidae) 64ff

Genus *Cynocephalus* 64ff

Temminck Flying Lemur, *C. temminckii* (Waterhouse, 1838) 64f

Philippine Flying Lemur, *C. volans* (Linné, 1758) 64ff

Order Bats (Chiroptera)

Suborder Fruit Bats (Megachiroptera)

Family Fruit Bats (Pteropidae) 97f

Subfamily Flying Foxes (Pteropinae) 97

Genus *Dobsonia*

Genus *Eidolon*

Straw-colored bat, *E. helvum* (Kerr, 1792) 97f

E. h. helvum (Kerr, 1792)

E. h. dupreanum (Schlegel and Pollen, 1867)

E. h. sabaeum (K. Andersen, 1907)

Genus Rousette Bats (*Rousettus*) 98

Egyptian Fruit Bat, *R. aegyptiacus* (Geoffroy, 1810) 98f

R. a. aegyptiacus (E. Geoffroy, 1810) 98

R. a. leachi (A. Smith, 1829) 98

R. a. occidentalis (Eisentraut, 1959) 98

Angolan Fruit Bat, *R. angolensis* (Bocage, 1889) 98f

R. a. angolensis (Bocage, 1898) 98

R. a. smithi Thomas, 1908 98

R. a. ruwenzorii Eisentraut, 1965 98

Genus *Myonycteris* 99

Collared Fruit Bat, *M. torquata* (Dobson, 1878) 99

Genus Flying Foxes (*Pteropus*) 100

Tonga Flying Fox, *P. tonganus* Quoy and Gaimard, 1830

P. niger (Kerr, 1792)

P. alecto gouldi Peters, 1867

Indian Flying Fox, *P. giganteus* (Brünnich, 1782) 100f

P. g. giganteus (Brünnich, 1782) 100

P. g. leucocephalus Hodgson, 1835 100

Red-necked Fruit Bat, *P. vampyrus* (Linné, 1758) 100

P. v. vampyrus (Linné, 1758) 100
Gray-headed Flying Fox, *P. poliocephalus* Temminck, 1825 100
P. subniger (Kerr, 1792) 100
Rufous Flying Fox, *P. rufus* E. Geoffroy, 1803 100
P. capistratus Peters, 1876

Subfamily Epauletted Fruit Bats (Epomophorinae) 102f
Genus *Epomops*
Franquet's Epauletted Fruit Bat, *E. franqueti* (Tomes, 1860) 102
Büttikofer's Epauletted Fruit Bat, *E. büettikoferi* (Matschie, 1899) 102
Genus *Epomophorus*
Wahlberg's Epauletted Fruit Bat, *E. wahlbergi* (Sundevall, 1846) 103
E. w. wahlbergi (Sundevall, 1846)
E. w. haldemani (Halowell, 1846) 103
Genus *Micropteropus*
Dwarf Epauletted Fruit Bat, *M. pusillus* (Peters, 1867) 103
Genus *Hypsignathus*
Hammer-headed Fruit Bat, *H. monstrosus* H. Allen, 1861 107
Genus *Scotonycteris*
Zenker's Fruit Bat, *S. zenkeri* Matschie, 1894 107
Snake-toothed Bat, *S. ophiodon* Pohle, 1943 107

Subfamily Short-nosed Fruit Bats (Cynopterinae) 108
Genus *Cynopterus*
Indian Short-nosed Fruit Bat, *C. sphinx* (Vahl, 1797) 108

Subfamily Tube-nosed Fruit Bats (Nyctimeninae) 108
Genus *Nyctimene*
Large Tube-nosed Fruit Bat, *N. major* (Dobson, 1877)
Large-headed Tube-nosed Fruit Bat, *N. cephalotes* (Pallas, 1767)
Queensland Tube-nosed Fruit Bat, *N. robinsoni* Thomas, 1904 108
Genus Lesser Tube-nosed Fruit Bats (*Paranyctimene*) 108

Family Long-tongued Bats (Macroglossidae) 108
Genus *Eonycteris*
Dobson's Long-tongued Dawn Bat, *E. spelaea* (Dobson, 1871) 109
Genus *Macroglossus*
M. lagochilus Matschie, 1899 109
Dwarf Long-tongued Fruit Bat, *M. minimus* (E. Geoffroy, 1810) 109
Genus *Megaloglossus*
African Long-tongued Fruit Bat, *M. woermanni* Pagenstecher, 1885 109
Genus *Notopteris*
Genus *Nesonycteris*

Family Harpy Fruit Bats (Harpyionycteridae) 110
Genus *Harpyionycteris*
Whitehead's Harpy Fruit Bat, *H. whiteheadi* Thomas, 1886 110

Suborder Insectivorous Bats (Microchiroptera)

Superfamily Emballonuroidea

Family Rat-tailed Bats (Rhinopomatidae) 121
Genus *Rhinopoma*
Larger Rat-tailed Bat, *R. microphyllum* (Brünnich, 1782)
Lesser Rat-tailed Bat, *R. hardwickei* Gray, 1831 122

Family Sac-winged Bats (Emballonuridae) 122

Subfamily Emballonurinae
Genus Old World Sheath-tailed Bats (*Emballonura*)
Genus *Coleura*
Genus Sac-winged Bats (*Balantiopteryx*)
Genus *Rhynchonycteris*
Proboscis Bat, *R. naso* (Weid, 1820) 122
Genus Sheath-tailed Bats (*Saccopteryx*) 122
Two-lined Sheath-tailed Bat, *S. bilineata* (Temminck, 1839) 122
Genus Tomb Bats (*Taphozous*) 129
Subgenus *Taphozous*
Tomb Bat, *T. (Taphozous) perforatus* E. Geoffroy, 1818 129
Mauritian Tomb Bat, *T. (Taphozous) mauritianus* E. Geoffroy, 1818 129
Subgenus *Liponycteris*
Naked-bellied Tomb Bat, *T. (Liponycteris) nudiventris* Cretzschmar, 1826 129
Subgenus Pouch-bearing Bats (*Saccolaimus*)

Subfamily Diclidurinae
Genus *Diclidurus*
Ghost Bat, *D. albus* Weid, 1820

Family Bulldog Bats (Noctilionidae) 129
Genus *Noctilio*

Mexican Bulldog Bat, *N. leporinus* (Linné, 1758) 129f
Southern Bulldog Bat, *N. labialis* Kerr, 1792 130

Superfamily Megadermatoidea

Family Slit-faced Bats (Nycteridae) 130
Genus *Nycteris*
Great Slit-faced Bat, *N. grandis* Peters, 1871 130
Geoffroy's Slit-faced Bat, *N. thebaica* E. Geoffroy, 1818 130
Hispid Slit-faced Bat, *N. hispida* (Schreber, 1775) 130
Javanese Slit-faced Bat, *N. javanica* Geoffroy, 1813 130

Family Large-winged Bats (Megadermatidae) 130
Genus *Lavia*
African Yellow-winged Bat, *L. frons.* (E. Geoffroy, 1810) 131
Genus False Vampires (*Megaderma*)
Subgenus *Cardioderma*
Heart-nosed False Vampire, *M.* (*Cardioderma*) *cor* Peters, 1872 131
Subgenus *Lyroderma*
Indian False Vampire, *M.* (*Lyroderma*) *lyra* E. Geoffroy, 1810 131
Subgenus *Megaderma*
Malayan False Vampire, *M.* (*Megaderma*) *spasma* (Linné, 1758) 131
Genus *Macroderma*
Australian Giant False Vampire Bat, *M. gigas* (Dobson, 1880) 131

Superfamily Horseshoe Bat Relatives (Rhinolophoidea)

Family Horseshoe Bats (Rhinolophidae) 131
Genus *Rhinolophus*
Maclaud's Horseshoe Bat, *R. maclaudi* Pousargues, 1897 132
Lesser Horseshoe Bat, *R. hipposideros* (Bechstein, 1800) 132
Dent's Horseshoe Bat, *R. denti* Thomas, 1904 132
Lander's Horseshoe Bat, *R. landeri* Martin, 1838 132
R. alcyone Temminck, 1852 132
Greater Horseshoe Bat, *R. ferrumequinum* Schreber, 1775 132
Hildebrandt's Horseshoe Bat, *R. hildebrandtii* Peters, 1878
Mediterranean Horseshoe Bat, *R. euryale* Blasius, 1853
Blasius Horseshoe Bat, *R. blasii* Peters, 1866
Mehely's Horseshoe Bat, *R. mehelyi* Matschie, 1901

Family Old World Leaf-nosed Bats (Hipposideridae) 131f
Genus *Hipposideros*
Commerson's Leaf-nosed Bat, *H. commersoni* (E. Geoffroy, 1813) 132
Angolan Leaf-nosed Bat, *H. c. gigas* (Wagner, 1845) 132
South African Lesser Leaf-nosed Bat, *H. caffer* (Sundevall, 1846) 132f
Greater Himalayan Leaf-nosed Bat, *H. armiger* (Hodgson, 1835) 139
H. jonesi Hayman, 1947
Genus *Asellia*
Trident Leaf-nosed Bat, *A. tridens* (E. Geoffroy, 1813) 139
Genus *Anthops*
Flower-faced Bat, *A. ornatus* (Thomas, 1888) 139
Genus *Triaenops*
Persian Leaf-nosed Bat, *T. persicus* Dobson, 1871 139

Superfamily Phyllostomoidea

Family Leaf-nosed Bats (Phyllostomidae) 139

Subfamily Mustache Bats (Chilonycterinae) 139
Genus *Chilonycteris* 139
Mustache Bat, *C. personata* Wagner, 1843
Genus *Pteronotus*
Suapure Naked-backed Bat, *P. suapurensis* (J. A. Allen, 1904) 139
Naked-backed Bat, *P. davyi* Gray, 1838 139

Subfamily Big-eared Leaf-nosed Bats (Phyllostominae) 140
Genus *Lonchorrhina*
Tome's Long-eared Bat, *L. aurita* Tomes, 1863 140
Genus *Macrophyllum*
Long-legged Bat, *M. macrophyllum* (Wied, 1821) 140
Genus Spear-nosed Bats (*Phyllostomus*)
Spear-nosed Bat, *P. hastatus* Pallas, 1767 140
Pale Spear-nosed Bat, *P. discolor* Wagner, 1843 140
Genus *Trachops*
Fringe-lipped Bat, *T. cirrhosus* (Spix, 1823)
Genus *Vampyrum*
Linné's False Vampire Bat, *V. spectrum* (Linné, 1758) 140

Subfamily Long-tongued Bats (Glossophaginae) 141
Genus *Glossophaga*
Long-tongued Bat, *G. soricina* (Pallas, 1766) 141

Genus *Platalina*
P. genovensium Thomas, 1928
Genus *Choeronycteris*
Mexican Long-nosed Bat, *C. mexicana* Tschudi, 1844 141
Genus *Musonycteris*
Banana Bat, *M. harrisoni* (Schaldach and McLaughlin, 1960) 141

Subfamily Phyllonycterinae
Genus *Phyllonycteris*

Subfamily Short-tailed Leaf-nosed Bats (Carolliinae)
Genus *Carollia*
Seba's Short-tailed Bat, *C. perspicillata* (Linné, 1758) 141

Subfamily Yellow-shouldered Bats (Sturnirinae) 141
Genus *Sturnira*
Yellow-shouldered Bat, *S. lilium* (E. Geoffroy, 1810) 142

Subfamily Red Fruit-eating Bats (Stenoderminae) 141
Genus *Ectophylla*
White Bat, *E. alba* H. Allen, 1892
Genus *Artibeus*
Dwarf Fruit-eating Bat, *A. nanus* Andersen, 1906 142
Big Fruit-eating Bat, *A. lituratus* (Olfers, 1818) 142
Jamaican Fruit-eating Bat, *A. jamaicensis* Leach, 1821 142
Genus *Uroderma*
Tent-building Bat, *U. bilobatum* Peters, 1866 142
Genus *Centurio*
Wrinkle-faced Bat, *C. senex* Gray, 1842 142

Family Vampire Bats (Desmodontidae) 142f
Genus *Desmodus*
Vampire Bat, *D. rotundus* (E. Geoffroy, 1810) 143
Genus *Diaemus*
White-winged Vampire Bat, *D. youngi* (Jentink, 1893) 143
Genus *Diphylla*
Hairy-legged Vampire Bat, *D. ecaudata* Spix, 1823 143

Superfamily Vespertilionid Bats (Vespertilionoidea) 143

Family Funnel-eared Bats (Natalidae) 143
Genus *Natalus*
Funnel-eared Bat, *N. stramineus* Gray, 1838 143

Family Smokey Bats (Furipteridae) 143
Genus *Furipterus*
Smokey Bat, *F. horreus* (F. Cuvier, 1828) 143

Family Disk-winged Bats (Thyropteridae)
Genus *Thyroptera*
Spix's Disk-winged Bat, *T. tricolor* Spix, 1823 144
Honduran Disk-winged Bat, *T. discifera* (Lichtenstein and Peters, 1854) 144

Family Sucker-footed Bats (Myzopodidae) 144
Genus *Myzopoda*
Golden Bat, *M. aurita* (Milne-Edwards and Grandidier, 1878) 144

Family Vespertilionid Bats (Vespertilionidae) 144

Subfamily Vespertilioninae
Genus *Tylonycteris*
Flat-headed Bat, *T. pachypus* (Temminck, 1835) 144
Genus Mouse-eared Bats, (*Myotis*)
Large Mouse-eared Bat or European Little Brown Bat, *M. myotis* (Borkhausen, 1797)
Mediterranean Bat, *M. oxygnathus* (Monticelli, 1885)
Bechstein's Bat, *M. bechsteini* (Kuhl, 1818) 144
Pond Bat, *M. dasycneme* (Boie, 1825) 144
Water Bat, *M. daubentoni* (Kuhl, 1819) 144
Long-fingered Bat, *M. capaccinii* (Bonaparte, 1837)
Geoffroy's Bat, *M. emarginatus* (E. Geoffroy, 1806)
Natterer's Bat, *M. nattereri* (Kuhl, 1819) 144
Whiskered Bat, *M. mystacinus* (Kuhl, 1819) 144
Welwitsch's Bat, *M. welwitschii* (Gray, 1866)
Genus *Pizonyx*
Fish-eating Bat, *P. vivesi* (Menegaux, 1901)
Genus Pipistrelles (*Pipistrellus*) 144
Common Pipistrelle, *P. pipistrellus* (Schreber, 1774) 144
Nathusius' Pipistrelle, *P. nathusii* (Keyserling and Blasius, 1839)
Kuhl's Pipistrelle, *P. kuhli* (Kuhl, 1819)
Savi's Pipistrelle, *P. savii* (Bonaparte, 1837)
African Banana Bat, *P. nanus* (Peters, 1852) 145
West African Pipistrelle, *P. nanulus* Thomas, 1904
Genus Noctule Bats (*Nyctalus*) 145
Common Noctule Bat, *N. noctula* (Schreber, 1774) 145
Lesser Noctule, *N. leisleri* (Kuhl, 1818) 145
Giant Noctule, *N. lasiopterus* (Schreber, 1780) 145
Genus Big Brown Bats (*Eptesicus*)
Serotine Bat, *E. serotinus* (Schreber, 1774) 145

Big Brown Bat, *E. fuscus* (Pal. de Beauvois, 1796) 145
White-winged Serotine Bat, *E. tenuipinnis* (Peters, 1872) 145
Northern Bat, *E. nilssoni* (Keyserling and Blasius, 1839) 145
Genus *Vespertilio*
Particolored Bat, *V. murinus* Linné, 1758
Genus Butterfly Bats (*Glauconycteris*)
Genus *Lasiurus*
Red Bat, *L. borealis* (Müller, 1776) 145
Hoary Bat, *L. cinereus* (Pal. de Beauvois, 1796) 145
Genus *Barbastella*
Barbastelle, *B. barbastellus* (Schreber, 1774) 145
Genus *Plecotus*
Long-eared Bat, *P. auritus* (Linné, 1758) 145
Southern Long-eared Bat, *P. austriacus* (Fischer, 1829) 146

Subfamily Miniopterinae
Genus *Miniopterus*
Long-winged Bat, *M. schreibersi* (Kuhl, 1819) 146

Subfamily Murininae
Genus Tube-nosed Bats (*Murina*)
Genus Hairy-winged Bats (*Harpiocephalus*)

Subfamily Nyctophilinae
Genus *Antrozous*
Pallid Bat, *A. pallidus* (Le Conte, 1856) 146

Family New Zealand Short-tailed Bats (Mystacinidae) 146
Genus *Mystacina*
New Zealand Short-tailed Bat, *M. tuberculata* (Gray, 1843) 146

Family Free-tailed Bats (Molossidae) 146
Genus *Molossops*
Mexican Dog-faced Bat, *M. malagai* (Villa, 1955)
Genus *Tadarida*
Brazilian Free-tailed Bat, *T. brasiliensis* (I. Geoffroy, 1824)
Mexican Free-tailed Bat, *T. brasiliensis mexicana* (Saussure, 1860) 147
European Free-tailed Bat, *T. teniotis* (Rafinesque, 1814) 147
Braided Free-tailed Bat, *T. limbata* (Peters, 1852) 147
Angolan Free-tailed Bat, *T. condylura* (A. Smith, 1833)
Genus *Mops*
M. leonis (Thomas, 1908)
Genus *Cheiromeles*
Naked Bat, *C. torquatus* Horf, 1824 147
Necklace Hairless Bat, *C. parvidens* Miller and Hollister, 1921 147
Genus *Eumops*
Californian Greater Mastiff Bat, *E. perotis californicus* (Merriam, 1890)
Genus *Molossus*
Red Velvety Free-tailed Bat, *M. rufus* Geoffroy, 1805
Giant Velvety Free-tailed Bat, *M. major* (Kerr, 1792)

Order Edentates (Edentata)

Suborder Xenarthra

Infraorder Cingulata

Family Armadillos (Dasypodidae) 154f
Tribe Dasypodini
Genus Armadillos (*Dasypus*)
Kappler's Armadillo, *D. kappleri* Krauss, 1862 157
Nine-banded Armadillo, *D. novemcinctus* Linné, 1758 157ff
Seven-banded Armadillo, *D. septemcinctus* Linné, 1758 157
D. pilosus (Fitzinger, 1856) 157f
Tribe Three-banded Armadillos (Tolypeutini)
Genus Three-banded Armadillos (*Tolypeutes*)
La Plata Three-banded Armadillo, *T. matacus* (Desmarest, 1804) 160f
Three-banded Armadillo, *T. tricinctus* (Linné, 1758) 160
Tribe Priodontini
Genus *Priodontes*
Giant Armadillo, ⌀ *P. giganteus* (Geoffroy, 1803) 161f
Genus Eleven-banded Armadillos (*Cabassous*)
Eleven-banded Armadillo, *C. unicinctus* (Linné, 1758) 161
Spiny Armadillo, *C. hispidus* (Burmeister, 1854) 161
C. lugubris (Gray, 1873) 161
C. loricatus (Wagner, 1855) 161
Tribe Euphractini

Subtribe Euphractina
Genus *Euphractus*
Six-banded Armadillo, *E. sexcinctus* (Linné, 1758) 163f
Hairy Armadillo, *E. villosus* (Desmarest, 1804) 163ff
Furry Armadillo, *E. nationi* (Thomas, 1894) 163
Long-haired Armadillo, *E. vellerosus* (Gray, 1865) 163
Pygmy Armadillo, *E. pichiy* (Desmarest, 1804) 163f
Subtribe Chlamyphorina 165
Genus *Chlamyphorus*
Lesser Pichiciago, ∅ *C. truncatus* Harlan, 1825 165f
Genus *Burmeisteria*
Greater Pichiciago, ∅ *B. retusa* (Burmeister, 1863) 165f

Infraorder Armorless Edentates (Pilosa)

Family Sloths (Bradypodidae) 166ff
Genus Two-fingered Sloths (*Choloepus*) 169
Hoffmann's Two-fingered Sloth, *C. hoffmanni* Peters, 1858 169f
Common or South American Two-fingered Sloth, *C. didactylus* (Linné, 1758) 169ff
Genus Three-fingered Sloths (*Bradypus*) 171f
South American Three-fingered Sloth, *B. tridactylus* Linné, 1758 172
B. cuculliger Wagler, 1831 172
Necklace Sloth, *B. torquatus* Illiger, 1811 172

Family Anteaters (Myrmecophagidae) 172ff
Genus *Myrmecophaga*
Giant Anteater, *M. tridactyla* Linné, 1758 177ff
Genus *Tamandua*
Tamandua or Collared Anteater, *T. tetradactyla* (Linné, 1758) 179f
Genus *Cyclopes*
Two-toed or Silky Anteater, *C. didactylus* (Linné, 1758) 180f

Order Pangolins (Pholidota)

Family Pangolins (Manidae)
Genus Pangolins (*Manis*) 183ff
Subgenus *Phataginus*
White-bellied Tree Pangolin, *M.* (*Phataginus*) *tricuspis* Rafinesque, 1820 182
Subgenus *Uromanis*
Long-tailed Tree Pangolin, *M.* (*Uromanis*) *tetradactyla* Linné, 1766 183ff
Subgenus *Smutsia*
Giant Pangolin, *M.* (*Smutsia*) *gigantea* Illiger, 1815 183
Cape Pangolin, *M.* (*Smutsia*) *temmincki* Smuts, 1832 183
Subgenus *Phatages*
Indian Pangolin, *M.* (*Phatages*) *crassicaudata* Gray, 1827 183
Subgenus *Manis*
Chinese Pangolin, *M.* (*Manis*) *pentadactyla* Linné, 1758 183
Subgenus *Paramanis*
Malayan Pangolin, *M.* (*Paramanis*) *javanica* Desmarest, 1822 183

Order Rodents (Rodentia)

Suborder Sciurid or Squirrel-like Rodents (Sciuromorpha)

Superfamily Mountain Beavers (Aplodontoidea)

Family Mountain Beavers or Sewellels (Aplodontidae) 201
Genus *Aplodontia*
Mountain Beaver or Sewellel, *A. rufa* (Rafinesque, 1817) 201ff

Superfamily Sciurid Rodents or Typical Squirrels (Sciuroidea) 203

Family Tree Squirrels or Typical Squirrels (Sciuridae) 203ff

Subfamily Ground and Tree Squirrels (Sciurinae) 204f
Tribe Northern Ground Squirrels (Marmotini) 205
Genus Marmots (*Marmota*) 205
Alpine Marmot, *M. marmota* (Linné, 1758) 205ff
True Alpine Marmot or European Marmot, *M. m. marmota* (Linné, 1758)
M. m. latirostris Kratochvil, 1960
Bobac Marmot, *M. bobak* (Müller, 1776) 205
Mongolian Marmot or Tarbagan, *M. b. sibirica* (Radde, 1862)
Altai Marmot, *M. b. baibacina* (Brandt, 1843)

Long-tailed Marmot, *M. caudata* (Jacquemont, 1844) 205
Afghan Marmot, *M. c. dichrous* (Anderson, 1875) 206
Hoary Marmot, *M. caligata* (Eschscholtz, 1829) 206
M. c. vancouverensis Swarth, 1911
Kamchatken Marmot, *M. c. camtschatica* (Pallas, 1811) 206
Himalayan Marmot, *M. himalayana* (Hodgson, 1841)
Tien Shan Marmot, *M. menzbieri* (Kashkarov, 1925)
Yellow-bellied Marmot, *M. flaviventris* (Audubon and Bachman, 1841) 206
Woodchuck, *M. monax* (Linné, 1758) 206
Genus Prairie Dogs (*Cynomys*) 225ff
Black-tailed Prairie Dog, *C. ludovicianus* (Ord, 1815) 225ff
White-tailed Prairie Dog, *C. gunnisoni* (Baird, 1855) 225ff
Utah Prairie Dog, ∅ *C. gunnisoni parvidens* J. A. Allen, 1905
Genus Ground Squirrels and Sousliks (*Citellus*) 231f
European Souslik, *C. citellus* (Linné, 1776) 231
Spotted Souslik, *C. suslicus* (Güldenstaedt, 1770) 232
Large-toothed or Aral Yellow Souslik, *C. fulvus* (Lichtenstein, 1823) 232
Mohave Ground Squirrel, *C. mohavensis* (Merriam, 1889)
Ring-tailed Ground Squirrel, *C. annulatus* (Audubon and Bachman, 1842)
Little Souslik, *C. pygmaeus* (Pallas, 1779) 232
Long-tailed Souslik, *C. eversmanni* (Brandt, 1841) 232
Amur Souslik, *C. e. menzbieri* Ognev, 1937 232
Red Souslik, *C. rufescens* (Keyserling and Blasius, 1840)
Nelson's Antelope Squirrel, *C. nelsoni* (Merriam, 1893)
Arctic Ground Squirrel, *C. undulatus* (Pallas, 1779) 232
Thirteen-lined Ground Squirrel, *C. tridecemlineatus* (Mitchill, 1821) 232
Franklin's Ground Squirrel, *C. franklinii* (Sabine, 1822) 232
California Ground Squirrel, *C. beecheyi* (Richardson, 1829) 232
Round-tailed Ground Squirrel, *C. tereticaudus* (Baird, 1858) 233
Golden-mantled Ground Squirrel, *C. lateralis* (Say, 1823) 233
Rock Squirrel, *C. variegatus* (Erxleben, 1777)
Townsend's Ground Squirrel, *C. townsendii* (Bachman, 1839)
Washington Ground Squirrel, *C. washingtoni* A. H. Howell, 1938
Belding's Ground Squirrel, *C. beldingi* (Merriam, 1888)
Uinta Ground Squirrel, *C. armatus* (Kennicott, 1863)
Mexican Ground Squirrel, *C. mexicanus* (Erxleben, 1777)
Richardson's Ground Squirrel, *C. richardsonii* (Sabine, 1822)
Spotted Ground Squirrel, *C. spilosoma* (Bennett, 1833)
Harris' Antelope Ground Squirrel, or Antelope Squirrel, *C. harrisi* (Audubon and Bachman, 1854)
White-tailed Antelope Squirrel, *C. leucurus* (Merriam, 1889) 233
Genus Old World and Western Chipmunks (*Eutamias*)
Burunduk, *E. sibiricus* (Laxmann, 1769) 238f
Genus Chipmunks, or Eastern Chipmunks (*Tamias*) 241
Subgenus *Tamias*
Eastern Chipmunk, *T.* (*Tamias*) *striatus* (Linné, 1758) 242
Subgenus *Neotamias*
Alpine Chipmunk, *T.* (*Neotamias*) *alpinus* Merriam, 1893 242
Least Chipmunk, *T.* (*Neotamias*) *minimus* Bachman, 1839 242f
Yellow Pine Chipmunk, *T.* (*Neotamias*) *amoenus* J. A. Allen, 1890 242
Colorado Chipmunk, *T.* (*Neotamias*) *quadrivittatus* (Say, 1823) 242
Townsend's Chipmunk, *T.* (*Neotamias*) *townsendii* Bachman, 1839 242
Tribe Bristly Ground Squirrels (Xerini) 243
Genus *Xerus*
X. rutilis (Cretzschmar, 1826)
South African Ground Squirrel, *X. inauris* (Zimmermann, 1780)
Genus *Atlantoxerus*
North African Ground Squirrel, *A. getulus* (Linné, 1758) 243
Genus *Spermophilopsis*
Long-clawed Ground Squirrel, *S. leptodactylus* (Lichtenstein, 1823) 243f
Tribe Red Squirrels or Chickarees (Tamiasciurini) 244
Genus Rock Squirrels (*Sciurotamias*) 244
Père David's Rock Squirrel, *S. davidianus* (Milne-Edwards, 1867) 244

Genus Chickarees (*Tamiasciurus*)
North American Red Squirrel, *T. hudsonicus* (Erxleben, 1777) 244
Chickaree, *T. douglasii* (Bachman, 1839) 244
Tribe Tree Squirrels (Sciurini) 244f
Genus Red Squirrels (*Sciurus*)
Red Squirrel, *S. vulgaris* Linné, 1758 245f
Siberian Red Squirrel, *S. v. exalbidus* Pallas, 1779
Central European Red Squirrel, *S. v. fuscoater* Altum, 1876
British Red Squirrel, *S. v. leucourus* Kerr, 1792
Gray Squirrel, *S. carolinensis* Gmelin, 1788 255
Eastern Fox Squirrel, *S. niger* Linné, 1758
∅ *S. n. cinereus* Linné, 1758
∅ *S. n. vulpinus* Gmelin, 1788
Kaibab Squirrel, ∅ *S. kaibabensis* Merriam, 1904
Western Gray Squirrel, *S. griseus* Ord, 1818
Arizona Gray Squirrel, *S. arizonensis* Coues, 1867
Abert's Squirrel, *S. aberti* Woodhouse, 1853
Apache Squirrel, *S. apache* J. A. Allen, 1893
Caucasian Squirrel, *S. anomalus* Schreber, 1785 255
Subgenus *Guerlinguetus*
Brazilian Squirrel, *S.* (*Guerlinguetus*) *aestuans* Linné, 1766
Genus Dwarf Tree Squirrels (*Microsciurus*) 260
Dwarf Tree Squirrel, *M. alfari* (Allen, 1895) 260
Genus *Rheithrosciurus*
Groove-toothed Squirrel, *R. macrotis* (Gray, 1856) 260
Tribe Asian Squirrels (Callosciurini) 260
Genus *Callosciurus*
Swinhoe's or Asiatic Striped Squirrel, *C. swinhoei* (Milne-Edwards, 1874) 260
Spotted Squirrel, *C. notatus* (Boddaert, 1785) 260
Genus Oriental Pygmy Squirrels (*Nannosciurus*) 260
Whitehead's Dwarf Squirrel, *N. whiteheadi* (Thomas, 1887) 260
Brown Dwarf Squirrel, *N. melanotis* (S. Müller, 1844) 260
Genus *Rhinosciurus*
Long-nosed Squirrel, *R. laticaudatus* (S. Müller, 1844) 261
Genus *Hyosciurus*
Long-snouted Squirrel, *H. heinrichi* Tate and Archbold, 1935 261
Genus *Dremomys*
Perny's Long-nosed Squirrel, *D. pernyi* (Milne-Edwards, 1867) 261
Genus *Lariscus*
Malayan Black-striped Squirrel, *L. insignis* (F. Cuvier, 1821) 261
Bornean Black-striped Squirrel, ∅ *L. hosei* (Thomas, 1892) 261
Tribe African Tree Squirrels (Funambulini)
Genus Palm Squirrels (*Funambulus*) 261
Indian Palm Squirrel, *F. palmarum* (Linné, 1766) 261
Genus African Striped Squirrels (*Funisciurus*) 261
Western African Striped Squirrel, *F. lemniscatus* (Le Conte, 1857) 261
Congo Striped Squirrel, *F. congicus* (Kuhl, 1820)
Genus African Bush Squirrels (*Paraxerus*) 261
Mantled African Bush Squirrel, *P. palliatus* (Peters, 1852) 261
Boehm's African Bush Squirrel, *P. boehmi* (Reichenow, 1886) 262
P. flavivittis (Peters, 1852) 262
Genus *Myosciurus*
African Pygmy Squirrel, *M. pumilio* (Le Conte, 1857) 262
Tribe Indo-Malayan Giant Squirrels (Ratufini)
Genus *Ratufa*
Malayan Giant Squirrel, *R. bicolor* (Sparrmann, 1778) 262
Indian Giant Squirrel, *R. indica* (Erxleben, 1777) 262
Tribe Oil Palm Squirrels (Protoxerini) 262
Genus *Protoxerus*
Oil Palm Squirrel, *P. stangeri* (Waterhouse, 1843) 262
Genus *Epixerus*
Wilson's Palm Squirrel, *E. wilsoni* (Du Chaillu, 1860) 262
Ebien, or African Palm Squirrel, ∅ *E. ebii* (Temminck, 1853) 262
Genus Sun Squirrels (*Heliosciurus*) 262
West African Sun Squirrel, *H. gambianus* (Ogilby, 1835) 263
H. undulatus (True, 1892) 263

Subfamily Flying Squirrels (Pteromyinae) 263
Genus *Hylopetes*
Indo-Malayan Flying Squirrel, *H. lepidus* (Horsfield, 1824) 264
Genus *Petaurista*
Common Giant Flying Squirrel, *P. petaurista* (Pallas, 1766) 264
Formosan Giant Flying Squirrel, *P. grandis* (Swinhoe, 1862) 264
Japanese Giant Flying Squirrel, *P. leucogenys* (Temminck, 1827)
Genus *Pteromys*
European Flying Squirrel, *P. volans* (Linné, 1758) 264

Genus *Eupetaurus*
Woolly Flying Squirrel, *E. cinereus* Thomas, 1888 265
Genus *Glaucomys*
Southern Flying Squirrel, *G. volans* (Linné, 1758) 265f
Northern Flying Squirrel, *G. sabrinus* (Shaw, 1801) 265f

Superfamily Pocket Gophers (Geomyoidea) 268
Family Pocket Gophers (Geomyidae) 268f
Genus Eastern Pocket Gophers (*Geomys*)
Plains or Eastern Pocket Gopher, *G. bursarius* (Shaw, 1800) 270f
Texas Pocket Gopher, *G. personatus* True, 1889
Southeastern Pocket Gopher, *G. pinetis* Rafinesque, 1817
Genus *Thomomys*
Western Pocket Gopher, *T. bottae* (Eydoux and Gervais, 1836) 271
Southern Pocket Gopher, *T. umbrinus* (Richardson, 1829)
Northern Pocket Gopher, *T. talpoides* (Richardson, 1828)
Oregon Pocket Gopher, *T. mazama* Merriam, 1897
Sierra Pocket Gopher, *T. monticola* J. A. Allen, 1893
Genus Large Pocket Gophers, or Tuzas (*Orthogeomys*)
Tuza, *O. grandis* (Thomas, 1893) 271
Genus Yellow Pocket Gophers (*Cratogeomys*)
Yellow Pocket Gopher, *C. castanops* (Baird, 1852) 272

Family Heteromyid Rodents or Kangaroo Mice (Heteromyidae) 272f
Genus Pocket Mice (*Perognathus*) 272ff
Bailey's Pocket Mouse, *P. baileyi* Merriam, 1894 273
Silky Pocket Mouse, *P. flavus* Baird, 1855 273
Plains Pocket Mouse, *P. flavescens* Merriam, 1889
Wyoming Pocket Mouse, *P. fasciatus* Wied-Neuwied, 1839
Hispid Pocket Mouse, *P. hispidus* Baird, 1858
Great Basin Pocket Mouse, *P. parvus* (Peale, 1848)
California Pocket Mouse, *P. californicus* Merriam, 1889 273
Genus *Microdipodops*
Pale Kangaroo Mouse, *M. pallidus* Merriam, 1901 273
Dark Kangaroo Mouse, *M. megacephalus* Merriam, 1891
Genus Kangaroo Rats (*Dipodomys*)
Ord's Kangaroo Rat, *D. ordii* Woodhouse, 1853 273f
Big-eared Kangaroo Rat, ∅ *D. elephantinus* (Grinnell, 1919) 273
Texas Kangaroo Rat, ∅ *D. elator* Merriam, 1894 273
Desert Kangaroo Rat, *D. deserti* Stephens, 1887 273
Pacific Kangaroo Rat, *D. agilis* Gambel, 1848 273
Merriam Kangaroo Rat, *D. merriami* Mearns, 1890
Narrow-faced Kangaroo Rat, *D. venustus* Merriam, 1904
Genus *Heteromys*
Goldman's Spiny Pocket Mouse, *H. goldmani* Merriam, 1902
South American Forest Mouse, *H. anomalus* (Thompson, 1815) 275
Genus *Liomys*
Painted Spiny Pocket Mouse, *L. pictus* (Thomas, 1893) 275
Mexican Spiny Pocket Mouse, *L. irroratus* (Gray, 1868) 275

Superfamily Beaver-like Rodents (Castoroidea) 275ff

Family Beavers (Castoridae) 275f
Genus *Castor*
Beaver, *C. fiber* Linné, 1758 276ff
Scandinavian Beaver, *C. f. fiber* Linné, 1758 276
Elbe Beaver, ∅ *C. f. albicus* Matschie, 1907 276
Rhone Beaver, ∅ *C. f. galliae* Geoffroy, 1803 276
Polish or Byelorussian Beaver, *C. f. vistulanus* Matschie, 1907 276
Ural Beaver, *C. f. pohlei* Serebrennikov, 1929 276
Mongolian Beaver, *C. f. birulai* Serebrennikov, 1929 276
Canadian Beaver, *C. f. canadensis* Kuhl, 1820 276
Michigan Beaver, *C. f. michiganensis* Bailey, 1913 276
Newfoundland Beaver, *C. f. caecator* Bangs, 1913 276
Rio Grande Beaver, *C. f. frondator* Mearns, 1897 276
Golden-bellied Beaver, *C. f. subauratus* Taylor, 1912 276

Superfamily Gundis (Ctenodactyloidea) 286

Family Gundis (Ctenodactylidae) 286f
Genus *Massoutiera*

Genus *Felovia*
Genus Gundis (*Ctenodactylus*)
Gundi, *C. gundi* (Rothmann, 1776) 287
Genus *Pectinator*
Speke's Pectinator, *P. spekei* Blyth, 1856 289

Superfamily Scaly-tailed Squirrel-like Rodents (Anomaluroidea) 289

Family Scaly-tailed Squirrels (Anomaluridae) 289

Subfamily Scaly-tailed Squirrels (Anomalurinae) 289
Genus *Anomalurus*
Pel's Scaly-tailed Squirrel, *A. peli* (Schlegel and S. Müller, 1843) 290
Pygmy Scaly-tailed Squirrel, *A. pusillus* Thomas, 1887 290
Jackson's Scaly-tailed Squirrel, *A. jacksoni* De Winton, 1898 291ff
Neave's Scaly-tailed Squirrel, *A. neavei* Dollmann, 1909 291
Gabon Scaly-tailed Squirrel, *A. chrysophoenus* Dubois, 1888 291
Fraser's Scaly-tailed Squirrel, *A. fraseri* Waterhouse, 1843 291
Belden's Scaly-tailed Squirrel, *A. beldeni* Du Chaillu, 1860 291
Beecroft's Scaly-tailed Squirrel, *A. beecrofti* Fraser, 1853
A. erythronotus Milne-Edwards, 1879 291f

Subfamily Flightless Scaly-tailed Squirrels (Zenkerellinae)
Genus African Small Flying Squirrels (*Idiurus*) 291
Zenker's Small Flying Squirrel, *I. zenkeri* Matschie, 1894 291
Large-eared Small Flying Squirrel, *I. macrotis* Miller, 1898 291
Lang's Small Flying Squirrel, *I. langi* J. A. Allen, 1922 291
Genus *Zenkerella*
Flightless Scaly-tailed Squirrel, *Z. insignis* Matschie, 1898 290f

Superfamily Springhaas (Pedetoidea) 294

Family Springhaas, Spring Hares or Jumping Hares (Pedetidae) 294
Genus Springhaas (*Pedetes*)
East African Springhaas, *P. surdaster* Thomas, 1902 295
Cape Springhaas, *P. cafer* (Pallas, 1779) 295

Suborder Myomorphs or Mouse-like Rodents (Myomorpha)

Superfamily Murid Rodents (Muroidea)

Family Cricetid Rodents (Cricetidae) 296

Subfamily Cricetinae
Tribe New World Mice (Hesperomyini) 301
Genus Aquatic Rats (*Ichthyomys*)
Genus White-footed Mice (*Peromyscus*) 301
Piñon Mouse, *P. truei* (Shufeldt, 1885) 301
Golden Mouse, *P. nuttalli* (Harlan, 1832) 301f
Deer Mouse, *P. maniculatus* (Wagner, 1845) 301
White-footed Mouse, *P. leucopus* (Rafinesque, 1818) 301f
P. l. tornillo Mearns, 1896
Genus *Baiomys*
Pygmy Mouse, *B. taylori* (Thomas, 1887) 302
Genus Grasshopper Mice (*Onychomys*) 302
Northern Grasshopper Mouse, *O. leucogaster* (Wied-Neuwied, 1841) 302
Southern Grasshopper Mouse, *O. torridus* (Coues, 1874) 302
Genus American Harvest Mice (*Reithrodontomys*) 302
Western Harvest Mouse, *R. megalotis* (Baird, 1858) 302
Salt Marsh Harvest Mouse, ∅ *R. raviventris* Dixon, 1908
Genus Rice Rats (*Oryzomys*) 302f
Rice Rat, *O. palustris* (Harlan, 1837) 303
Genus Vesper Mice (*Calomys*)
Vesper Mouse, *C. musculinus* (Thomas, 1913) 303
Genus Wood Rats (*Neotoma*) 304
White-throated Wood Rat, *N. albigula* Hartley, 1894 304
Bushy-tailed Wood Rat, *N. cinerea* (Ord, 1815) 304
Eastern Wood Rat, *N. floridana* (Ord, 1818)
Dusky-footed Wood Rat, *N. fuscipes* Baird, 1858
Genus Cotton Rats (*Sigmodon*) 304f
Hispid Cotton Rat, *S. hispidus* Say and Ord, 1825 304
Least Wood Rat, *S. minimus* Mearns, 1894
Yellow-nosed Wood Rat, *S. ochrognathus* Bailey, 1902
Tribe Hamsters (Cricetini) 305
Genus *Calomyscus*
Mouse-like Hamster, *C. bailwardi* Thomas, 1905 305f
Genus Dwarf Hamsters (*Phodopus*) 306
Striped Hairy-footed Hamster, *P. sungorus* (Pallas, 1773) 306
Roborovsky's Dwarf Hamster, *P. roborovskii* (Satunin, 1903) 306f

Genus *Allocricetulus*
Eversmann's Dwarf Hamster, *A. eversmanni* (Brandt, 1859) 306
Short Dwarf Hamster, *A. curtatus* (G. Allen, 1925) 306f
Genus Ratlike Hamsters (*Cricetulus*) 306
Striped Hamster, *C. barabensis* (Pallas, 1773) 306
Chinese Hamster, *C. griseus* Milne-Edwards, 1867 306
Long-tailed Hamster, *C. longicaudatus* (Milne-Edwards, 1867) 306f
Migratory Hamster, *C. migratorius* (Pallas, 1773) 306
Tibetan Hamster, *C. lama* Bonhote, 1905
Genus *Tscherskia*
Rat-like Hamster, *T. triton* (De Winton, 1899) 307
Genus *Cricetus*
Common Hamster, *C. cricetus* (Linné, 1758) 309
Rumanian Hamster, *C. c. nehringi* Matschie, 1901
Genus Golden Hamsters (*Mesocricetus*)
Golden Hamster, *M. auratus* (Waterhouse, 1839) 315
Tribe White-tailed Rats (Mystromyini) 316
Genus *Mystromys* 315f
White-tailed Rat, *M. albicaudatus* (A. Smith, 1834) 316
Long-tailed Rat, *M. longicaudatus* Noack, 1887 316
Tribe Mole Mice (Myospalacini) 316
Genus Zokors (*Myospalax*) 316
Manchurian Zokor, *M. psilurus* (Milne-Edwards, 1874) 316
Altai Zokor, *M. myospalax* (Laxmann, 1773) 316f
Transbaikal Zokor, *M. aspalax* (Pallas, 1776) 316

Subfamily Malagasy Rats (Nesomyinae) 318
Genus *Hypogeomys*
Votsotsa, *H. antimena* Grandidier, 1869 318
Genus *Macrotarsomys*
M. bastardi Milne-Edwards and Grandidier, 1898 318
Genus *Nesomys*
Lamberton's Malagasy Rat, *N. lambertoni* Grandidier, 1928 318
Genus *Brachytarsomys*
B. albicauda Günther, 1875 318
Genus *Eliurus*
E. minor Forsyth Major, 1896 318
Genus *Brachyuromys* 318

Subfamily Maned Rats (Lophiomyinae) 318
Genus Maned Rats (*Lophiomys*)
Maned Rat, *L. imhausi* Milne-Edwards, 1867 318f

Subfamily Microtine Rodents (Microtinae) 319

Tribe Lemmings (Lemmini)
Genus Collared Lemmings (*Dicrostonyx*) 323
Arctic Lemming, *D. torquatus* (Pallas, 1779) 323
Greenland Collared Lemming, *D. groenlandicus* (Traill, 1823) 323
Collared Lemming, *D. hudsonicus* (Pallas, 1778)
Genus Bog Lemmings (*Synaptomys*) 323
Southern Bog Lemming, *S. cooperi* Baird, 1858 323f
Northern Bog Lemming, *S. borealis* (Richardson, 1828) 323
Genus *Myopus*
Wood Lemming, *M. schisticolor* (Lilljeborg, 1844) 324
Genus Brown Lemmings (*Lemmus*) 324f
Siberian Lemming, *L. sibiricus* (Kerr, 1792)
Amur Lemming, *L. amurensis* Vinogradov, 1924
Norway Lemming, *L. lemmus* (Linné, 1758) 324ff
American Brown Lemming, *L. trimucronatus* (Richardson, 1825)
Black-footed Lemming, *L. nigripes* (True, 1894)
Tribe Voles (Microtini) 328
Genus Red-backed Voles (*Clethrionomys*)
Bank Vole, *C. glareolus* (Schreber, 1780) 328f
Gapper's Red-backed Vole, *C. gapperi* (Vigors, 1830)
Western Red-backed Vole, *C. occidentalis* (Merriam, 1890)
Northern Red-backed Vole, *C. rutilus* (Pallas, 1779) 330
Large-toothed Red-backed Vole, *C. rufocanus* (Sundevall, 1846) 330
Genus *Eothenomys*
Père David's Vole, *E. melanogaster* (Milne-Edwards, 1872) 330
Genus Western Voles (*Alticola*) 330
High-mountain Vole, *A. strelzowi* (Kastschenko, 1900) 330
Royle's High-mountain Vole, *A. roylei* Gray, 1842 330
Genus Water Voles (*Arvicola*)
European Water Vole, *A. terrestris* (Linné, 1758) 331f
Iberian Water Vole, *A. sapidus* Miller, 1908 331
Genus Muskrats (*Ondatra*)
Newfoundland Muskrat, *O. obscura* (Bangs, 1894) 332
Muskrat, *O. zibethica* (Linné, 1766) 332
Genus *Neofiber*
Round-tailed Muskrat, *N. alleni* True, 1884 333
Genus Pine Voles (*Pitymys*) 335
Southern European Pine Vole, *P. savii* (de Selys-Longchamps, 1838)

North Italian Pine Vole, *P. multiplex* (Fatio, 1905)
Mediterranean Pine Vole, *P. duodecimcostatus* (de Selys-Longchamps, 1839)
European Pine Vole, *P. subterraneus* (de Selys-Longchamps, 1836) 335f
American Pine Vole, *P. pinetorum* (Le Conte, 1830) 335
Genus Voles (*Microtus*) 336
Common Vole, *M. arvalis* (Pallas, 1779) 336ff
Gunther's Vole, *M. guentheri* (Danford and Alston, 1880)
Eastern Meadow Mouse, *M. pennsylvanicus* (Ord, 1815) 336
Block Island Vole, ∅ *M. p. provectus* Bangs, 1908
Beach Vole, ∅ *M. breweri* (Baird, 1858)
Yellow-cheeked Vole, ∅ *M. xanthognathus* (Leach, 1815)
California Vole, *M. californicus* (Peale, 1848)
Mexican Vole, *M. mexicanus* (Saussure, 1861)
Prairie Vole, *M. ochrogaster* (Wagner, 1842)
Rock Vole, *M. chrotorrhinus* (Miller, 1894)
Townsend's Vole, *M. townsendi* (Bachman, 1839)
Creeping Vole, *M. oregoni* (Bachman, 1839)
Long-tailed Vole, *M. longicaudus* (Merriam, 1888)
Singing Vole, *M. miurus* Osgood, 1901
Richardson's or American Vole, *M. richardsoni* (de Kay, 1842)
Field Vole, *M. agrestis* (Linné, 1761) 336
Tundra Vole, *M. oeconomus* (Pallas, 1776) 336
Brandt's Vole, *M. brandti* (Radde, 1861) 336
Snow Vole, *M. nivalis* (Martins, 1842) 336
Genus *Lagurus*
Steppe Lemming, *L. lagurus* Pallas, 1773 341
Genus *Prometheomys*
Long-clawed Mole Vole, *P. schaposchnikowi* Satunin, 1901 341f
Tribe Mole Lemmings (Ellobiini) 342
Genus Mole Lemmings (*Ellobius*)
Mole Lemming, *E. talpinus* (Pallas, 1770) 342f
Afghan Mole Lemming, *E. fuscocapillus* (Blyth, 1843) 342

Subfamily Gerbils (Gerbillinae) 343
Genus Gerbils (*Gerbillus*) 343
Field Gerbil, *G. nanus garamantis* Lataste, 1881 343f
Large North African Gerbil, *G. campestris* Levaillant, 1857 343
Genus Jirds (*Meriones*) 343
Clawed Jird, *M. unguiculatus* (Milne-Edwards, 1867)
Southern Mongolian Jird, *M. meridianus psammophilus* (Milne-Edwards, 1871)
Persian Jird, *M. persicus* (Blanford, 1875)
Shaw's Jird, *M. shawi* (Duvernoy, 1842)
Tristram's Jird, *M. tristrami* Thomas, 1892
Genus Great Gerbils (*Rhombomys*)
Great Gerbil, *R. opimus* (Lichtenstein, 1823) 343
Genus Fat-tailed Mice (*Pachyuromys*)
Fat-tailed Mouse, *P. duprasi* Lataste, 1880 343
Genus Large Gerbils (*Tatera*) 343
East African Gerbil, *T. vicina* (Peters, 1878) 343
Indian or Large Gerbil, *T. indica* (Hardwicke, 1807) 343
Genus Small Naked-soled Gerbils (*Taterillus*)
Small Naked-soled Gerbil, *T. emini* (Thomas, 1892) 343

Family Bamboo Rats (Rhizomyidae) 364f
Genus *Rhizomys*
Sumatran Bamboo Rat, *R. sumatrensis* (Raffles, 1821) 347
Genus *Cannomys*
Lesser Bamboo Rat, *C. badius* (Hodgson, 1841) 347
Genus African Mole Rats (*Tachyoryctes*) 347f
Tanzanian Mole Rat, *T. daemon* Thomas, 1909 347
Rwanda Mole Rat, *T. ruandae* Lönnberg and Gyldenstolpe, 1925 347f

Family Palearctic Mole Rats (Spalacidae) 348
Genus *Spalax*
Lesser Mole Rat, *S. leucodon* Nordmann, 1840 348ff
Greater Mole Rat, *S. microphthalmus* Güldenstaedt, 1770 348ff
Ehrenberg's Mole Rat, *S. ehrenbergi* Nehring, 1898 348ff

Family Murid Rodents (Muridae) 350

Subfamily Murinae
Genus Stick-nest Rats (*Leporillus*) 351f
Common Stick-nest Rat, *L. conditor* (Sturt, 1848) 352
Jones' Stick-nest Rat, *L. jonesi* Thomas, 1921 352
Genus Giant Naked-tailed Rats (*Uromys*) 352
Giant Naked-tailed Rat, *U. anak* Thomas, 1907 353
U. caudimaculatus (Krefft, 1867)
Genus Mosaic-tailed Rats (*Melomys*) 352
Genus Australian Native Mice (*Leggadina*)
Australian Native Mouse, *L. hermannsburgensis* (Waite, 1896) 353
Genus Australian Hopping Mice (*Notomys*) 353
Neck-pouched Hopping Mouse, *N. cervinus* (Gould, 1853) 353

Genus Bandicoot Rats (*Bandicota*) 354
Lesser or Bengali Bandicoot Rat, *B. bengalensis* (Gray, 1835) 354
Large Bandicoot Rat, *B. indica* (Bechstein, 1800) 354f
Eastern Large Bandicoot Rat, *B. i. nemorivaga* (Hodgson, 1836)
Genus *Vandeleuria*
Long-tailed Climbing Mouse, *V. oleracea* (Bennett, 1832) 355
Genus Pest Rats (*Nesokia*)
Pest Rat, *N. indica* (Gray, 1830) 355
Genus *Tokudaia*
Ryukyu Rat, ∅ *T. osimensis muenninki* (Johnson, 1946)
Genus *Rattus* 356
Little Rat, *R. exulans* (Peale, 1848) 356
Brown Rat or Norway Rat, *R. norvegicus* (Berkenhout, 1769) 357ff
Black Rat or House Rat, *R. rattus* (Linné, 1758) 361f
House Rat, *R. r. rattus* (Linné, 1758) 361
Roof Rat, *R. r. alexandrinus* (E. Geoffroy and Audouin, 1829) 361
Corn Rat, *R. r. frugivorus* (Rafinesque, 1814) 361
Genus *Mus* 362
Little Indian Field Mouse, *M. booduga* (Gray, 1837) 362
Indian Brown Spiny Mouse, *M. platythrix* Bennett, 1832 362
House Mouse, *M. musculus* Linné, 1758 362f
Western House Mouse, *M. m. domesticus* Rutty, 1772 362
Northern House Mouse, *M. m. musculus* Linné, 1758 362
Bactrian House Mouse, *M. m. bactrianus* Blyth, 1846 363
Eastern House Mouse, *M. m. spicilegus* Petenyi, 1882 362
Genus Wood and Field Mice (*Apodemus*) 365
Geisha Wood Mouse, *A. geisha* (Thomas, 1905)
Striped Field Mouse, *A. agrarius* (Pallas, 1771)
Broad-toothed Field Mouse, *A. mystacinus* (Danford and Alston, 1877) 365
A. microps Kratochvil and Rosicky, 1952 366
Yellow-necked Field Mouse, *A. flavicollis* (Melchior, 1834) 366f
Long-tailed Field Mouse, *A. sylvaticus* (Linné, 1758) 366f
Genus *Micromys*
Harvest Mouse, *M. minutus* (Pallas, 1771) 368f
Genus African Native Mice (*Leggada*) 369
Small African Native Mouse, *L. minutoides* (A. Smith, 1834) 369
Genus Spiny Mice (*Acomys*) 369f
Egyptian Spiny Mouse, *A. cahirinus* (Desmarest, 1819) 369
Cretan Spiny Mouse, *A. minous* Bate, 1906 369
Sinai Spiny Mouse, *A. dimidiatus* (Cretzschmar, 1826) 369
Genus Striped Grass Mice (*Lemniscomys*) 373
Single-striped Grass Mouse, *L. griselda* (Thomas, 1904) 373
North African Striped Grass Mouse, *L. barbarus* (Linné, 1767) 373
Striped Grass Mouse, *L. striatus* (Linné, 1758) 374
Genus *Rhabdomys*
Striped Field Mouse, *R. pumilio* (Sparrman, 1784) 374
Genus Grass Mice or Kusus (*Arvicanthis*)
Genus *Oenomys*
Rufous-nosed Rat, *O. hypoxanthus* (Pucheran, 1855) 374f
Genus Soft-furred Rats (*Praomys*) 375
Cameroon Soft-furred Rat, *P. morio* (Trouessart, 1881) 375f
Genus Multimammate Rats (*Mastomys*) 376
M. coucha (A. Smith, 1834) 376f
Genus Harsh-furred Mice (*Lophuromys*) 377
Uganda Harsh-furred Mouse, *L. woosnami* (Thomas, 1906) 377
Ethiopian Harsh-furred Mouse, *L. flavopunctatus* Thomas, 1888 378

Subfamily Pouched Rats (Cricetomyinae) 378
Genus Giant Pouched Rats (*Cricetomys*) 378ff
Emin's or Congo Pouched Rat, *C. emini* Wroughton, 1910 378
Gambian Pouched Rat, *C. gambianus* Waterhouse, 1840 378
Genus Long-tailed Pouched Rats (*Beamys*)
Genus Cape Pouched Mice (*Saccostomus*)
Cape Pouched Mouse, *S. campestris* 380

Subfamily Vlei and Karroo Rats (Otomyinae) 380
Genus *Otomys*
Vlei Rat, *O. irroratus* (Brants, 1827) 383
Karroo Rat, *O. denti* Thomas, 1906 383

Subfamily African Tree Mice (Dendromurinae) 383
Genus African Climbing Mice (*Dendromus*)
African Climbing Mouse, *D. insignis* (Thomas, 1903) 384
Genus Fat Mice (*Steatomys*)
Fat Mouse, *S. pratensis* Peters, 1846 384
Genus *Deomys*
Congo Forest Mouse, *D. ferrugineus* Thomas, 1888 384f

Subfamily Cloud Rats (Phloeomyinae) 385
Genus *Phloeomys*
Slender-tailed Cloud Rat, *P. cumingi* (Waterhouse, 1839) 385
Genus *Crateromys*
Bushy-tailed Cloud Rat, *C. schadenbergi* (Meyer, 1895) 385f
Genus Pencil-tailed Tree Mice (*Chiropodomys*)
Pencil-tailed Tree Mouse, *C. gliroides* (Blyth, 1856) 386

Subfamily Shrew Rats (Rhynchomyinae) 386
Genus *Rhynchomys*
Shrew Rat, *R. soricoides* Thomas, 1895 386

Subfamily Water Rats (Hydromyinae) 387
Genus Water Rats (*Hydromys*) 387
Eastern or Golden-bellied Water Rat, *H. chrysogaster* E. Geoffroy, 1804 387
New Britain Water Rat, *H. neobritannicus* Tate and Archbold, 1935 387
New Guinea Water Rat, *H. habbema* Tate and Archbold, 1941 387
Genus *Crossomys*
Monckton's Water Rat, *C. moncktoni* Thomas, 1907 388

Superfamily Dormice (Gliroidea) 388ff

Family True Dormice (Gliridae) 388f

Subfamily Eurasian Dormice (Glirinae) 389
Genus *Glis*
Fat Dormouse, *G. glis* (Linné, 1766) 389ff
Genus *Muscardinus*
Common Dormouse, *M. avellanarius* (Linné, 1758)
Genus *Eliomys*
Garden Dormouse, *E. quercinus* (Linné, 1766) 395
Genus *Dryomys*
Forest Dormouse, *D. nitedula* (Pallas, 1779) 396f
Genus *Glirulus*
Japanese Dormouse, *G. japonicus* (Schinz, 1845) 397
Genus *Myomimus*
Mouse-like Dormouse or Asiatic Dormouse, *M. personatus* Ognev, 1924 397

Subfamily African Dormice (Graphiurinae) 397f
Genus African Dormice (*Graphiurus*) 398
Pygmy Dormouse, *G. nanus* (de Winton, 1897) 398
Common African Dormouse, *G. murinus* (Desmarest, 1822) 398
Cameroons Dormouse, *G. m. haedulus* Dollman, 1912 398

Family Spiny Dormice (Platacanthomyidae) 398
Genus *Platacanthomys*
Spiny Dormouse, *P. lasiurus* Blyth, 1859 398
Genus *Typhlomys*
Chinese Pygmy Dormouse, *T. cinereus* Milne-Edwards, 1877 398

Family Desert Dormice (Seleviniidae) 401
Genus *Selevinia*
Betpakdala Dormouse, *S. betpakdalaensis* Belosludov and Bashanav, 1938 401

Superfamily Dipodoidea 410f

Family Jumping Mice (Zapodidae) 402

Subfamily True Jumping Mice (Zapodinae) 402
Genus Meadow Jumping Mice (*Zapus*)
Meadow Jumping Mouse, *Z. hudsonius* (Zimmermann, 1780) 402
Pacific Jumping Mouse, *Z. trinotatus* Rhoads, 1895
Western Jumping Mouse, *Z. princeps* J. A. Allen, 1893
Genus *Napaeozapus*
Woodland Jumping Mouse, *N. insignis* (Miller, 1891) 402
Genus *Eozapus*
Chinese Jumping Mouse, *E. setchuanus* (Pousargues, 1896) 402

Subfamily Birch Mice (Sicistinae) 402
Genus Birch Mice (*Sicista*)
Northern Birch Mouse, *S. betulina* (Pallas, 1779) 402f
Southern Birch Mouse, *S. subtilis* (Pallas, 1773) 402

Family Jerboas (Dipodidae) 403

Subfamily Jerboas (Dipodinae)
Genus Northern Three-toed Jerboas (*Dipus*)
Northern Three-toed Jerboa, *D. sagitta* (Pallas, 1773) 405f
Genus Desert Jerboas (*Jaculus*)
Desert Jerboa, *J. jaculus* (Linné, 1758) 404f
Genus Comb-toed Jerboas (*Paradipus*)
Comb-toed Jerboa, *P. ctenodactylus* (Vinogradov, 1929) 404f
Genus *Eremodipus*
Lichtenstein's Jerboa, *E. lichtensteini* (Vinogradov, 1927) 405
Genus Thick-tailed Three-toed Jerboas (*Stylodipus*)
Genus Fat-tailed Jerboas (*Pygeretmus*)
Genus Five-toed Jerboas (*Allactaga*)
Northern Five-toed Jerboa, *A. jaculus* (Pallas, 1779) 404

Siberian Jerboa, *A. sibirica* (Forster, 1778) 406
Gobi Jerboa, *A. bullata* G. Allen, 1925 406
Genus Jerboas (*Alactagulus*)
Little Earth Jerboa, *A. pygmaeus* (Pallas, 1779) 404

Subfamily Dwarf Jerboas (Cardiocraniinae)
Genus Five-toed Dwarf Jerboas (*Cardiocranius*)
Satunin's Pygmy Jerboa, *C. paradoxus* Satunin, 1903
Genus *Salpingotus*
Pygmy Jerboa *S. crassicauda* Vinogradov, 1924 404

Subfamily Long-eared Jerboas (Euchoreutinae)
Genus *Euchoreutes*
Long-eared Jerboa, *E. naso* Sclater, 1891

Suborder Porcupines (Hystricomorpha)

Superfamily Porcupine-like Rodents (Hystricoidea) 407

Family Old World Porcupines (Hystricidae) 407
Genus Long-tailed Porcupines (*Trichys*) 412
Malayan Long-tailed Porcupine, *T. fasciculata* (Shaw, 1801) 412
Bornean Long-tailed Porcupine, *T. lipura* Günther, 1876 412
Genus Brush-tailed Porcupines (*Atherurus*) 412
Central African Brush-tailed Porcupine, *A. centralis* Thomas, 1895 413
East African Brush-tailed Porcupine, *A. turneri* St. Leger, 1932
Asiatic Brush-tailed Porcupine, *A. macrourus* (Linné, 1758) 412
West African Brush-tailed Porcupine, *A. africanus* Gray, 1842 412
Genus Thick-spined Porcupines (*Thecurus*) 412
Bornean Thick-spined Porcupine, *T. crassispinis* (Günther, 1876)
Philippine Thick-spined Porcupine, *T. pumilis* (Günther, 1879)
Sumatran Thick-spined Porcupine, *T. sumatrae* Lyon, 1907 413
Genus Malayan Porcupines (*Acanthion*)
Malayan Porcupine, *A. brachyura* (Linné, 1758) 413
Javan Porcupine, *A. javanicum* F. Cuvier, 1822 413
Chinese Porcupine, *A. subcristatum* (Swinhoe, 1870)
Himalayan Crestless Porcupine, *A. hodgsoni* Gray, 1847
Kloss' Porcupine, *A. klossi* Thomas, 1916
Genus Large Porcupines (*Hystrix*) 413
Crested Porcupine, *H. cristata* Linné, 1758 413ff
White-tailed Porcupine, *H. leucura* Sykes, 1831 413
South African Porcupine, *H. africaeaustralis* Peters, 1852 413
East African Porcupine, *H. galeata* Thomas, 1893 413

Superfamily Mole Ratlike Rodents (Bathyergoidea) 416

Family Mole Rats (Bathyergidae) 417
Genus Blesmols (*Cryptomys*) 417f
Togo Mole Rat, *C. zechi* (Matschie, 1900) 417
African Mole Rat, *C. hottentotus* (Lesson, 1826) 417f
Genus Naked Sand Rats (*Heterocephalus*)
Naked Sand Rat, H. *glaber* Rüppell, 1842 417f
Genus Sand Rats (*Heliophobius*) 417
Silvery Mole Rat, *H. argenteocinereus* Peters, 1846 417
Genus Mole Rats (*Bathyergus*) 417
Cape Sand Mole, *B. suillus* (Schreber, 1782) 417
Genus Cape Mole Rats (*Georhychus*)
Cape Mole Rat, *G. capensis* (Pallas, 1778) 417

Superfamily Cane and Rock Rats (Petromuroidea) 419

Family African Cane Rats (Thryonomyidae) 419
Genus *Thryonomys*
Great Cane Rat, *T. swinderianus* (Temminck, 1827) 419
Lesser Cane Rat, *T. gregorianus* (Thomas, 1894) 419

Family Rock Rats (Petromuridae) 419
Genus *Petromus*
Rock Rat, *P. typicus* A. Smith, 1831 419

Suborder Cavies (Caviomorpha)

Superfamily Octodont Rodents (Octodontoidea) 420f

Family Octodont Rodents (Octodontidae) 420f
Genus South American Bush Rats (*Octodon*)
Degu, *O. degus* (Molina, 1782) 421
Genus *Octodontomys*
Bori, *O. gliroides* (Gervais and d'Orbigny, 1844)
Genus *Spalacopus*
Cururo, *S. cyanus* (Molina, 1782) 421

Genus *Aconaemys*
South American Rock Rat, *A. fuscus* (Waterhouse, 1841) 421
Genus *Octomys*
Viscacha Rat, *O. mimax* Thomas, 1920 421
Genus *Tympanoctomys*
T. barrerae (Lawrence, 1941)

Family Tuco-tucos (Ctenomyidae) 422
Genus Tuco-tucos (*Ctenomys*)
Knight's Tuco-tuco, *C. knighti* Thomas, 1919 422

Family Rat Chinchillas (Abrocomidae)
Genus Rat Chinchillas (*Abrocoma*)
Rat Chinchilla, *A. cinerea* Thomas, 1919 422
Bennett's Rat Chinchilla, *A. bennetti* Waterhouse, 1837

Family Spiny Rats (Echimyidae) 427

Subfamily Spiny Rats (Echimyinae)
Genus Spiny Rats (*Proëchimys*)
Cayenne Spiny Rat, *P. guyannensis* (E. Geoffroy, 1803) 427
Genus Crested Spiny Rats (*Echimys*) 427
Armored Spiny Rat, *E. armatus* (I. Geoffroy, 1838) 427
Genus *Hoplomys*
Armored Rat, *H. gymnurus* (Thomas, 1897)
Genus *Euryzygomatomys*
Suira, *E. spinosus* (G. Fischer, 1814)
Genus *Clyomys*
C. laticeps (Thomas, 1909)
Genus *Carterodon*
C. sulcidens (Lund, 1841)
Genus *Cercomys*
Punare, *C. cunicularis* Cuvier, 1829
Genus *Mesomys*
M. hispidus (Desmarest, 1817)
Genus *Lonchothrix*
L. emiliae Thomas, 1920
Genus *Isothrix*
Toro, *I. bistriata* Wagner, 1845
Genus *Diplomys*
Arboreal Soft-furred Spiny Rat, *D. caniceps* (Günther, 1876)

Subfamily Coro-coros (Dactylomyinae)
Genus Coro-coros (*Dactylomys*)
D. dactylinus (Desmarest, 1817)
Genus *Kannabateomys*
Rato de Taquara, *K. amblyonyx* Wagner, 1845 427f
Genus *Thrinacodus*
T. albicauda Günther, 1879

Family Hutias (Capromyidae) 428
Genus *Capromys* 428
Cuban Hutia, *C. pilorides* (Say, 1822) 428f
Black-tailed Hutia, ∅ *C. melanurus* Peters, 1864 428
Dwarf Hutia, ∅ *C. nana* G. M. Allen, 1917 428
Prehensile-tailed Hutia, *C. prehensilis* Poeppig, 1824 428
Genus Bahaman and Jamaican Hutias (*Geocapromys*)
Jamaican Hutia, ∅ *G. brownii* (Fischer, 1830) 429f
Bahaman Hutia, ∅ *G. ingrahami* (J. A. Allen, 1891) 429
† *G. columbianus* (Chapman, 1892) 429
Genus Hispaniolan Hutias (*Plagiodontia*) 429f
† *P. ipnaeum* Johnson, 1948 429
† *P. spelaeum* Miller, 1929 429
Haitian Hutia, ∅ *P. aedium* F. Cuvier, 1836
Cuvier's Hutia, ∅ / † *P. aedium aedium* F. Cuvier, 1836 429f
Dominican Hutia, ∅ *P. aedium hylaeum* Miller, 1927 429f

Family Coypus (Myocastoridae) 431
Genus *Myocastor*
Coypu or Nutria, *M. coypus* (Molina, 1782) 430ff

Superfamily Viscachas and Chinchillas (Chinchilloidea) 433

Family Chinchillas (Chinchillidae) 433
Genus Viscachas (*Lagostomus*)
Viscacha, *L. maximus* (Desmarest, 1817) 433f
Genus Mountain Chinchillas (*Lagidium*) 433f
Common Mountain Chinchilla, *L. viscacia* (Molina, 1782) 434
Peruvian Mountain Chinchilla, *L. peruanum* Meyen, 1933
Wolffson's Mountain Chinchilla, *L. wolffsoni* (Thomas, 1907)
Genus Chinchillas (*Chinchilla*)
Chinchilla, *C. laniger* (Molina, 1782) 439ff
Short-tailed Chinchilla, ∅ *C. chinchilla* (Lichtenstein, 1830) 439ff
† *C. c. chinchilla* (Lichtenstein, 1830) 439
∅ *C. c. boliviana* Brass, 1911 439

Superfamily Cavy-like Rodents (Cavioidea) 441

Family Cavies (Caviidae) 441

Subfamily Cavies (Caviinae)
Genus Guinea Pigs (*Cavia*)
Aperea, *C. aperea* Erxleben, 1777 441
Wild Cavy, *C. a. tschudii* (Fitzinger, 1857) 441
Guinea Pig, *C. a. porcellus* (Linné, 1758) 441f
Amazon Cavy, *C. fulgida* Wagler, 1831
C. stolida Thomas, 1926

Genus *Galea*
G. musteloides Meyen, 1833
Genus Mountain Cavies (*Microcavia*)
Southern Mountain Cavy, *M. australis* (Geoffroy and d'Orbigny, 1833)
Genus Rock Cavies (*Kerodon*)
Rock Cavy, *K. rupestris* (Wied, 1820) 443

Subfamily Patagonian Cavies (Dolichotinae) 444
Genus *Dolichotis*
Patagonian Cavy or Mara, *D. patagonum* (Zimmermann, 1780) 444f
Genus *Pediolagus*
Salt-desert Cavy, *P. salinicola* (Burmeister, 1875) 445

Family Capybaras (Hydrochoeridae) 446
Genus *Hydrochoerus*
Capybara or Carpincho, *H. hydrochaeris* (Linné, 1766) 446f
Panama Capybara, *H. h. isthmius* Goldman, 1912 446f

Family Agoutis (Dasyproctidae) 447

Subfamily Pacas (Cuniculinae)
Genus *Cuniculus*
Spotted Paca, *C. paca* (Linné, 1758) 447f
Genus *Stictomys*
Mountain Paca, *S. taczanowskii* (Stolzmann, 1865) 447

Subfamily Agoutis (Dasyproctinae) 449
Genus Agoutis (*Dasyprocta*)
Orange-rumped Agouti, *D. aguti* (Linné, 1766) 449
Sooty Agouti, *D. fuliginosa* Wagler, 1832 449
Azara's Agouti, *D. azarae* Lichtenstein, 1823 449
Genus Acouchys (*Myoprocta*) 450
Guyana Acouchy, *M. acouchi* (Erxleben, 1777) 450
Amazonian Acouchy, *M. a. exilis* (Wagler, 1851) 450

Superfamily Pacaranalike Rodents (Dinomyoidea) 450

Family Pacaranas (Dinomyidae) 451
Genus Pacaranas (*Dinomys*)
Pacarana, *D. branickii*, Peters, 1873 451ff

Superfamily New World Porcupines (Erethizontoidea) 453

Family New World Porcupines (Erethizontidae)

Subfamily Thin-spined Porcupines (Chaetomyinae)
Genus *Chaetomys*
Thin-spined Porcupine, *C. subspinosus* (Olfers, 1818) 453

Subfamily Erethizontinae
Genus Upper Amazonian Porcupines (*Echinoprocta*)
Upper Amazonian Porcupine, *E. rufescens* (Gray, 1865) 454
Genus Prehensile-tailed Porcupines (*Coëndou*)
Prehensile-tailed Porcupine, *C. prehensilis* (Linné, 1758) 454f
Woolly Prehensile-tailed Porcupine, *C. insidiosus* (Kuhl, 1820) 454
Genus *Erethizon*
North American Porcupine, *E. dorsatum* (Linné, 1758) 454ff

Order Whales (Cetacea)

Suborder Baleen Whales (Mystacoceti)

Family Right Whales (Balaenidae) 481
Genus *Balaena*
Bowhead Whale, ∅ *B. mysticetus* Linné, 1758 482f
Genus *Eubalaena*
Atlantic Right Whale, ∅ *E. glacialis* (Borowski, 1781) 485f
Pacific Right Whale, ∅ *E. japonica* (Lacépède, 1818) 485f
Southern Right Whale or Ice Baleen Whale, ∅ *E. australis* (Desmoulins, 1822) 485
Genus *Neobalaena*
Pygmy Right Whale, *N. marginata* (Gray, 1846) 486

Family Gray Whales (Eschrichtiidae) 486
Genus *Eschrichtius*
Gray Whale, *E. gibbosus* (Erxleben, 1777) 486f

Family Finback Whales (Balaenopteridae) 488
Genus Rorquals (*Balaenoptera*) 488
Blue Whale, ∅ *B. musculus* (Linné, 1758) 488ff
∅ *B. m. brevicauda* Zemsky and Boronin, 1964 488
Common Rorqual, ∅ *B. physalus* (Linné, 1758) 488ff
Sei Whale, *B. borealis* Lesson, 1828 489ff
Bryde's Whale, *B. edeni* Anderson, 1878 489ff
Lesser Rorqual, *B. acutorostrata* Lacépède, 1804 489f
Genus *Megaptera*
Humpback Whale, ∅ *M. novaeangliae* (Borowski, 1781) 491f

Suborder Toothed Whales (Odontoceti)

Superfamily Sperm Whales (Physeteroidea)

Family Sperm Whales (Physeteridae) 494f
- Genus *Physeter*
  - Sperm Whale, *P. catodon* Linné, 1758 494ff
- Genus *Kogia*
  - Pygmy Sperm Whale, *K. breviceps* (Blainville, 1838) 494

Family Beaked Whales (Ziphiidae) 496
- Genus Bottle-nosed Whales (*Hyperoodon*)
  - Bottle-nosed Whale, *H. ampullatus* (Forster, 1770) 496f
  - Southern Bottle-nosed Whale, *H. planifrons* Flower, 1882 501
- Genus Giant Bottle-nosed Whales (*Berardius*) 501
  - Baird's Beaked Whale, *G. bairdi* Stejneger, 1883 501
  - Arnoux's Whale, *B. arnouxii* Duvernoy, 1851 501
- Genus *Ziphius*
  - Cuvier's Beaked Whale or Goosebeak Whale, *Z. cavirostris* G. Cuvier, 1823 501
- Genus Beaked Whales or Cowfish (*Mesoplodon*)
  - Sowerby's Whale, *M. bidens* (Sowerby, 1804) 501
  - Gervais' Whale, *M. europaeus* (Gervais, 1852) 501
  - Layard's or Strap-toothed Whale, *M. layardii* (Gray, 1865) 501
  - True's Beaked Whale, *M. mirus* True, 1913
- Genus *Tasmacetus*
  - Shepherd's Beaked Whale, *T. shepherdi* Oliver, 1937 502

Superfamily River Dolphins (Platanistoidea) 502

Family Gangetic Dolphins (Platanistidae) 502
- Genus *Platanista*
  - Gangetic or Ganges River Dolphin, *P. gangetica* (Lebeck, 1801) 502

Family Amazonian Dolphins (Iniidae) 502
- Genus *Inia*
  - Amazon or Amazonian Dolphin, *I. geoffrensis* (Blainville, 1817) 502f
- Genus *Lipotes*
  - Chinese River or White Flag Dolphin, *L. vexillifer* G. S. Miller, 1918 502

Family La Plata Dolphins (Stenodelphinidae) 502
- Genus *Stenodelphis*
  - La Plata Dolphin, *S. blainvillei* (Gervais, 1844) 502

Superfamily Narwhals (Monodontoidea) 503

Family White Whales and Narwhals (Monodontidae) 503
- Genus *Delphinapterus*
  - White Whale or Beluga or Belukha, *D. leucas* (Pallas, 1776) 504f
- Genus *Monodon*
  - Narwhal, *M. monoceros* Linné, 1758 505f

Superfamily Dolphins (Delphinoidea) 506

Family Porpoises (Phocaenidae) 506
- Genus *Phocaena*
  - Common Porpoise, *P. phocaena* (Linné, 1758) 506
  - Pacific Harbor Porpoise, *P. sinus* Norris and McFarland, 1958
  - Spectacled Porpoise, *P. dioptrica* Lahille, 1912 507
  - Burmeister's Porpoise, *P. spinipinnis* Burmeister, 1865
- Genus *Phocaenoides*
  - Dall's Harbor Porpoise, *P. dalli* (True, 1885)
- Genus *Neophocaena*
  - Black Finless Porpoise, *N. phocaenoides* (G. Cuvier, 1829) 507

Family Long-snouted Dolphins (Stenidae) 507
- Genus Rough-toothed Dolphins (*Steno*)
  - Rough-toothed Dolphin, *S. bredanensis* (Lesson, 1828) 507
- Genus White Dolphins (*Sotalia*) 507
  - Buffeo Negro, *S. fluviatilis* (Gervais, 1855) 507
  - Guianan River or Guiana Dolphin, *S. guianensis* (P. J. v. Beneden, 1863) 507
- Genus *Sousa*
  - West African White Dolphin, *S. teuszi* (Kükenthal, 1892) 507f
  - Chinese White Dolphin, *S. sinensis* (F. Cuvier, 1835) 507

Family Dolphins (Delphinidae) 508

Subfamily Right Whale Dolphins (Lissodelphinae) 514
- Genus Right Whale Dolphins (*Lissodelphis*) 514
  - Northern Right Whale Dolphin, *L. borealis* (Peale, 1848) 514
  - Peron's Dolphin, *L. peronii* (Lacépède, 1804) 514

Subfamily True Dolphins (Delphininae) 514f
- Genus Spotted Dolphins (*Stenella*) 514
  - Blue-White Dolphin, *S. caeruleoalba* (Mayen, 1833) 515

Narrow-snouted Dolphin, *S. attenuata* (Gray, 1846)
Bridled Dolphin, *S. frontalis* (G. Cuvier, 1829)
S. pernettensis (Blainville, 1817)
S. clymene (Gray, 1850)
Genus Dolphins (*Delphinus*)
Common Dolphin, *D. delphis* Linné, 1758 515
Genus *Grampus*
Risso's Dolphin, *G. griseus* (G. Cuvier, 1812) 515
Genus *Tursiops*
Bottle-nosed Dolphin, *T. truncatus* (Montagu, 1821) 515f
Red Sea Bottle-nosed Dolphin, *T. aduncus* (Hemprich and Ehrenberg, 1832)
Gill's Dolphin, *T. gillii* Dall, 1873
Genus *Lagenodelphis*
L. hosei Fraser, 1956
Genus *Lagenorhynchus*
White-sided Dolphin, *L. acutus* (Gray, 1828) 516
White-beaked Dolphin, *L. albirostris* Gray, 1846 516
Pacific White-sided Dolphin, *L. obliquidens* Gill, 1865 517

Subfamily Commerson's Dolphins (Cephalorhynchinae) 517

Genus Commerson's or Black-and-white Dolphins or Skunk Dolphins (*Cephalorhynchus*)
Commerson's Dolphin, *C. commersonii* (Lacépède, 1804) 517

Subfamily Pilot and Killer Whales (Orcininae) 517

Genus Pilot Whales or Blackfish (*Globicephala*) 517
Northern Pilot Whale or Atlantic Blackfish, *G. melaena* (Traill, 1809) 517
Indian Pilot Whale or Short-finned Blackfish, *G. macrorhyncha* (Gray, 1846) 517
Pacific Pilot Whale, *G. sieboldii* (Gray, 1848) 517
Genus *Feresa*
Pygmy Killer Whale, *F. attenuata* Gray, 1875 518
Genus *Orcaella*
Irrawaddy Dolphin, *O. brevirostris* (Owen, 1866) 518
Genus *Pseudorca*
False Killer Whale, *P. crassidens* (Owen, 1846) 518
Genus *Orcinus*
Common Killer Whale or Orca, *O. orca* (Linné, 1758) 519ff

On the Zoological Classification and Names

For many years, zoologists and botanists have tried to classify animals and plants into a system which would be a survey of the abundance of forms in fauna and flora. Such a system, of course, may be established under very different aspects. Since Charles Darwin, his predecessors, and his successors have found that all creatures have evolved out of common ancestors, species of animals and plants have been classified according to their natural relationships. Our knowledge about the phylogeny, and thus the relationship of each living being to the other, is augmented every year by new discoveries and insights. Old ideas are replaced with more recent and more appropriate ones. Therefore, the natural classification of the animal kingdom (and the plant kingdom) is subject to changes. Furthermore, the opinions of zoologists, who are working on the classification of animals into the various groups, are anything but uniform. These differences and changes are usually insignificant. The classification of vertebrates into the classes of fish, amphibians, reptiles, birds, and mammals has been fixed for many decades. Only the Cyclostomata were recently separated from the fish and all other classes of vertebrates as the "jawless" Agnatha (comp. Vol. 4).

The animal kingdom has been split into several subkingdoms and these were again divided into further sections, subsections, and so on. The scale of the most important systematic categories follows in a descending rank order:

Kingdom
Subkingdom
Phylum
Subphylum
Class
Subclass
Superorder
Order
Suborder
Infraorder
Family
Subfamily
Tribe
Genus
Subgenus
Species
Subspecies

The scientific names of the animals and their spelling follow the international rules for the zoological nomenclature as agreed upon by the XV International Congress for Zoology and are obligatory for all zoological publications. The name of the genus, which is a Latin or Latinized noun, is singular and capitalized. After the name of the genus follows the name of the species and of the subspecies. The names of the species and subspecies may be nouns or adjectives, and they are spelled in the lower case. The name of a subgenus, which is formed in the same manner as a genus, may be added in brackets following the name of the genus. The names of the tribes, subfamilies, families, and superfamilies are plural capitalized nouns. They are formed from the name of a given genus by adding to the principal word the endings -ini for the tribe, -inae for the subfamily, -idae for the family, and -oidea for the superfamily. The names of the authors who were the first to describe and to name a species, subspecies, or group of animals should be cited with the year of this naming at least once in each scientific publication. The name of the author and year are not enclosed in brackets when the species or subspecies is classified as belonging to the same genus with which the author had originally classified it. They are in brackets when another genus name is used in the present publication. The scientific names of the genus, subgenus, species, and subspecies are supposed to be printed with different letters, usually italics.

ANIMAL DICTIONARY

For scientific names of species see the German-English-French-Russian section of this dictionary or the index.

1. English—German—French—Russian

In most cases names of subspecies are formed by putting an adjective or geographical specification before the name of species. These English names of subspecies will, as a rule, not appear in this part of the zoological dictionary.

ENGLISH NAME	GERMAN NAME	FRENCH NAME	RUSSIAN NAME
Acouchi	Acouchi	Acouchi	Акучи
Acouchis	Geschwänzte Agutis	Acouchis	Хвостатые агути
Afghan Mole Lemming	Südlicher Mull-Lemming	Rat taupe d'Afghan	Афганская слепушонка
African Cane Rats	Rohrratten		Триономииды, Тростниковые крысы
– Climbing Mice	Aalstrich-Klettermäuse	Souris des bananiers	
– Dormice	Afrikanische Bilche Pinselschwanzbilche	Graphiures	Африканские сони, Кистехвостые сони
– Ground Squirrels	– Borstenhörnchen	Écureuils foisseurs	Африканские земляные белки
– Long-tongued Fruit Bat	Afrikanischer Langzungen-Flughund	Mégaloglosse de Woermann	Африканский длинноязычный крылан
– Mole Rat	Hottentotten-Graumull	Rat taupe africain	Готтентотский землекоп
– Mole Rats	Afrikanische Maulwurfsratten	Rats taupes africains	Африканские кротовидные крысы
– Native Mice	– Kleinmäuse	Souris d'Afrique	Африканские мыши-малютки
– Palm Squirrel	Wilsons Riesenhörnchen	Écureuil de Wilson	Гигантская белка Вильсона
– Porcupine	Afrikanisches Stachelschwein	Porc-épic d'Afrique	Африканский дикобраз
– Pygmy Squirrel	– Zwerghörnchen	Écureuil nain du Gabon	Африканская карликовая белка
– Small Flying Squirrels	Gleitbilche	Anomalures nains	Сони-летяги
– Striped Squirrels	Rotschenkelhörnchen	Écureuils rayés d'Afrique	
– Tree Mice	Baummäuse	Dendromurinés	Африканские древесные мыши
– Tree Squirrels	Palmenhörnchen		Пальмовые белки
– Yellow-winged Bat	Gelbflüglige Großblattnase	Mégaderme à ailes orangées	
Agoutis	Agutis, Eigentliche Agutis, Stummelschwanzagutis	Dasyproctidés, Dasyproctinés, Agoutis	Агутиевые, Агути, Куцые агути
Ai	Dreifingerfaultier	Bradype	Ай
Alpine Chipmunk	Gebirgs-Chipmunk	Néotamia de montagne	Горный чипмунк
– Marmot	Alpenmurmeltier	Marmotte des Alpes	Альпийский сурок
Amazon Dolphins	Inias	Iniidés	Амазонские дельфины
Amazonian Dolphin	Amazonas-Delphin	Inie de Geoffroy	Амазонская иния
American Harvest Mice	Amerikanische Erntemäuse	Souris des moissons d'Amérique	
– Leaf-nosed Bats	Blattnasen		Листоносы
Andean Swamp Rat	Baumwollratte	Sigmodon velu	
Angola Fruit Bat	Angola-Flughund	Roussette d'Angola	Ангольский крылан
Ant-eaters	Ameisenbären	Fourmiliers	Муравьеды
Anthropoid Apes	Menschenaffen	Singes anthropomorphes	Человекоподобные
Apara	Dreibinden-Kugelgürteltier	Apar de Buffon	Трехпоясный броненосец
Ape-man	Vormensch	Australopithèque	Австралопитски
Ape-men	Vormenschen	Australopithèques	Австралопитеки
Aperea	Aperea	Cobaye sauvage	Аперея
Aral Yellow Souslik	Gelbziesel	Spermophile jaune	Суслик-песчаник
Arboreal Soft-furred Spiny Rat	Graukopf-Baumstachelratte	Rat épineux à tête grise	
Arctic Ground Squirrel	Parry-Ziesel	Souslik de Parry	Американский длиннохвостый суслик
– Lemming	Halsbandlemming	Lemming arctique	Копытный лемминг
– Right Whale	Grönlandwal	Baleine boréale	Гренландский кит
Arizona Grasshopper Mouse	Südliche Grashüpfermaus	Onchomys du sud	
Armadillos	Gürteltiere, Weichgürteltiere	Tatous	Броненосцы
Armoured Spiny Rat	Sania	Échimys armé	
– Rat	Lanzenratte	Rat armé	Гимнуровая крыса
Arnoux's Whale	Südlicher Schwarzwal	Baleine d'Arnoux	Южный плавун
Asiatic Brush-tailed Porcupine	Langschwanz-Quastenstachler	Athérure à longue queue	Малайский кистехвост
– Dormouse	Dünnschwanz-Baumschläfer	Myomime à queue fine	Мышевидная соня
– Long-tongued Fruit Bat	Zwerg-Langzungen-Flughund	Macroglosse minime	Малый длинноязычный крылан
Atlantic Right Whale	Nordkaper	Baleine noire	Бискайский кит
Australian Ghost Bat	Australische Gespenstfledermaus	Macroderme d'Australie	
– Hopping Mice	Australische Hüpfmäuse		Австралийские прыгающие мыши
– Native Mice	– Kleinmäuse	Souris d'Australie	Австралийские малые мыши

ENGLISH NAME	GERMAN NAME	FRENCH NAME	RUSSIAN NAME
– Native Mouse	Hermannsburg-Zwergmaus	– d'Hermannsburg	
– Water Rats	Schwimmratten	Rats d'eau d'Australie	
Azara Agouti	Azara-Aguti	Agouti d'Azara	Азарская агути
Bactrian House Mouse	Baktrische Maus		Центральноазиатская домовая мышь
Bahaman Hutia	Bahama-Ferkelratte	Rat des Bahamas	Багамская крыса
Bailey's Pocket Mouse	Bailey-Taschenmaus	Souris à poche de Bailey	Мешетчатый тушканчик Бейлейя
Baird's Beaked Whale	Baird-Wal	Baleine de Baird	Северный плавун
Baleen Whales	Bartenwale	Baleines	Усатые киты
Bamboo Rats	Wurzelratten, Bambusratten	Rats des bambous	Ризомииды, Бамбуковые крысы
Banana Bat	Bananenfledermaus, Bananen-Zwergfledermaus	Chauve-souris des bananes, Pipistrelle naine à ailes brunes	Банановый листонос, Банановый нетопырь
Bandicoot Rat	Bandikutratte	Bandicoot rat d'Inde	Индийская бандикутовая крыса
– Rats	Bandikutratten	– rats	Бандикутовые крысы
Bank Vole	Rötelmaus	Campagnol roussâtre	Рыжая полевка
Barbastelle	Mopsfledermaus	Barbastelle d'Europe	Европейская широкоушка
Bats	Fledertiere	Chiroptères	Рукокрылые
Beaked Whales	Schnabelwale	Ziphiidés	Клюворылые
Beaver	Biber	Castor	Бобр
Beavers	Biberartige, Biber	Castors	Бобровые, Бобры
Bechstein's Bat	Bechstein-Fledermaus	Vespertilion de Bechstein	Ночница Бехштейна
Beecroft's Scaly-tailed Squirrel	Beecroft-Dornschwanzhörnchen	Anomalure de Beecroft	Шипохвостая белка Бикрофта
Belden's Scaly-tailed Squirrel	Belden-Dornschwanzhörnchen	– de Belden	Шипохвостая белка Бельдена
Bengali Bandicoot Rat	Indische Maulwurfsratte	Bandicoot rat du Bengale	Бенгальская бандикутовая крыса
Bent-winged Bats		Minioptères	Длиннокрылы
Betpakdala Dormice	Salzkrautbilche	Souris de Selevin	Селевиниевые
Big Brown Bat	Große Braune Fledermaus	Sérotine de maison	
– – Bats	Breitflügel-Fledermäuse	Sérotines	Кожаны
Big-eared Kangaroo Rat	Großohr-Känguruhratte	Rat kangourou à grandes oreilles	Большеухий мешетчатый прыгун
– Leaf-nosed Bats	Eigentliche Blattnasen	Phyllostomes	Настоящие листоносы
Birch Mice	Streifen-Hüpfmäuse	Sicistinés, Sicistes	Мышовки
Black Flying Fox	Schwarzer Flughund	Roussette noire	
– Rat	Hausratte	Rat commun	Черная крыса
– Right Whale	Nordkaper	Baleine noire	Бискайский кит
Black-tailed Hutia	Schwarzschwanz-Baumratte	Rat à queue noire	Чернохвостая кубинская крыса
– Prairie Dog	Schwarzschwanz-Präriehund	Cynomys social	Чернохвостая луговая собачка
Blesmols	Bleßmulle	Rat taupes du Cap	
Blood-sucking Bat	Gemeiner Vampir	Vampire d'Azara	Большой кровосос
Blue Whale	Blauwal	Balénoptère bleu	Синий кит
Blue-white Dolphin	Blau-Weißer Delphin	Dauphin bleu-blanc	Однополосый продельфин
Bobac Marmot	Bobak	Marmotte Bobac	Степной сурок
Boehm's African Bush Squirrel	Boehm-Hörnchen	Écureuil de Boehm	
Bori	Bori	Bori	Бори
Boris	Pinselschwanzratten	Boris	
Bottle-nosed Dolphin	Großer Tümmler	Tursiops tronqué	Североатлантическая афалина
– Dolphins	Tümmler	Tursiops	Афалины
– Whale	Nördlicher Entenwal	Hypérodon du nord	Высоколобый бутылконос
– Whales	Entenwale	Baleines à bec	Бутылконосы
Bowhead	Grönlandwal	Baleine boréale	Гренландский кит
Braided Free-tailed Bat	Borten-Fledermaus	Tataride bordée	
Brandt's Vole	Brandt's Steppenwühlmaus	Campagnol de Brandt	Полевка Брандта
Brazilian Squirrel	Brasilhörnchen	Écureuil de la Guyane	Бразильская белка
Bristly Ground Squirrels	Borstenhörnchen		Земляные белки
Broad-toothed Field Mouse	Felsenmaus	Mulot rupestre	Горная мышь
Brown Dwarf Squirrel	Braunes Zwerghörnchen	Écureuil pygmée brun	
– Lemming	Berglemming	Lemming des toundras	Норвежский лемминг
– Lemmings	Echte Lemminge		Настоящие лемминги
– Rat	Wanderratte	Rat surmulot	Пасюк
Brush-tailed Porcupines	Quastenstachler	Athérures	Кистехвосты
Bryde's Whale	Brydewal	Balénoptère de Bryde	Полосатик Брайда
Buffeo Negro	Amazonas-Sotalia	Sotalie	Амазонский речной дельфин
Bulldog Bats	Hasenmäuler	Noctilions	Зайцеротые летучие мыши, Зайцероты
Burmeister's Porpoise	Burmeister-Schweinswal	Marsouin spinipenne	Морская свинья Бурмейстера

ENGLISH NAME	GERMAN NAME	FRENCH NAME	RUSSIAN NAME
Bushtail Wood Rat	Buschschwanzratte		Серый кустовый хомячок
Bushy-tailed Cloud Rat	Schadenbergs Borkenratte	Rat d'ecorce de Schadenberg	Древесная крыса Шаденберга
Buttikofer's Epauletted Fruit Bat	Büttikofer-Epauletten-Flughund	Épomophore de Buttikofer	Крылан Бютикофера
California Ground Squirrel	Kalifornischer Ziesel	Spermophile de Californie	Калифорнийский суслик
– Pocket Mouse	Kalifornische Taschenmaus	Souris à poche de Californie	Калифорнийский мешетчатый тушканчик
Canadian Beaver	Kanadischer Biber	Castor de Canada	Канадский бобр
Cane and Rock Rats	Felsenrattenartige		Петромиидовые
Cape Blesmole	Kap-Bleßmull	Rat taupe du Cap	Капский пескорой
– Mole Rat	Kap-Strandgräber	Fouisseur	
– Pangolin	Steppen-Schuppentier	Pangolin de Temminck	Степной ящер
– Pouched Mice	Kurzschwanz-Hamsterratten		Короткохвостые хомяковидные крысы
– Spring Haas	Südafrikanischer Springhase	Lièvre sauteur d'Afrique du sud	Кафрский долгоног
Capybara	Capybara	Capybara	Капибара
Caucasian Squirrel	Kaukasisches Eichhörnchen	Écureuil du Caucase	Кавказская белка
Cavies	Meerschweinchen, Eigentliche Meerschweinchen	Caviidés, Caviinés	Морские свинки, Настоящие морские свинки
Cavy-like Rodents	Meerschweinchenverwandte		Свинкообразные
Cayenne Spiny Rat	Cayenneratte	Rat de Guyane	Гвианская крыса
Chickaree	Rothörnchen	Écureuil d'Hudson	Канадская белка
Chickarees	Nordamerikanische Rothörnchen		Североамериканские красные белки
Chimpanzee	Schimpanse	Chimpansé	Обыкновенный шимпанзе
Chimpanzees	Schimpansen	Chimpansés	Шимпанзе
China Bamboo Rat	Chinesische Bambusratte	Rat des bambous Chinois	Китайская ризомиида
Chinchilla	Langschwanz-Chinchilla	Chinchilla à longue queue	Длиннохвостая шиншилла
– Rat	Chinchillaratte	Rat-chinchilla cendré	Шиншилловая крыса
Chinchillas	Chinchillas i. e. S.	Chinchillas	Шиншиллы
Chinese Pangolin	Chinesisches Ohren-Schuppentier	Pangolin à queue courte	Индостанский панголин
– Pygmy Dormouse	Chinesische Zwergschlafmaus	Souris naine de Chine	Китайская карликовая соня
– Ratlike Hamster	Chinesischer Zwerghamster	Hamster nain de Chine	Китайский хомячок
– River Dolphin	– Flußdelphin	Dauphin d'eau douce de Chine	Китайский речной дельфин
– White Dolphin	– Weißer Delphin	Sotalie de Chine	Китайский бледный речной дельфин
Club-footed Bat	Bambus-Fledermaus	Vespertilion du bambou	Бамбуковая летучая мышь
Collared Fruit Bats	Kragenflughunde	Myonyctères à collier	
– Lemming	Halsbandlemming	Lemming arctique	Копытный лемминг
– Lemmings	Halsbandlemminge		Копытные лемминги
Colorado Chipmunk	Colorado-Chipmunk	Néotamia du Colorado	Колорадский чипмунк
Coloured Bat	Bunte Lanzennase	Phyllostome coloré	
Comb-toed Jerboa	Kammzehen-Springmaus		Гребнепалый тушканчик
Commerson's Dolphin	Commerson-Delphin	Dauphin de Commerson	Трезубцевый дельфин Коммерсона
– Dolphins	Schwarz-Weiß-Delphine	Dauphins de Commerson	Трезубцевые дельфины
– Leaf-nosed Bat	Riesen-Rundblattnase	Phyllorine de Commerson	Большой ложный подковонос
Common Bats	Mausohr-Fledermäuse	Murins	Ночницы
– Dolphin	Delphin	Dauphin commun	Североатлантический дельфин-белобочка
– Dormouse	Haselmaus	Muscardin	Орешниковая соня
– Hamster	Feldhamster	Hamster d'Europe	Обыкновенный хомяк
– Killer Whale	Schwertwal	Orque épaulard	Косатка
– Mouse	Hausmaus	Souris domestique	Домовая мышь
– Noctule	Großer Abendsegler	Noctule	Рыжая вечерница
– Pipistrelle	Zwergfledermaus	Vespertilion pipistrelle	Нетопырь-карлик
– Porpoise	Schweinswal	Marsouin commun	Морская свинья
– Rorqual	Finnwal	Rorqual commun	Сельдяной полосатик
– Vole	Feldmaus	Campagnol des champs	Обыкновенная полевка
Congo Forest Mouse	Insektenessende Waldmaus	Rat à manteau roux	Насекомоядная лесная мышь
Corn Rat	Fruchtratte	– des fruits	
Coro-coro	Eigentliche Fingerratte	Dactylomys	
Cotton Rats	Baumwollratten	Sigmodons	Хлопковые хомячки
Coypu	Sumpfbiber	Ragondin	Нутрия
Crested Hamster	Mähnenratte	Hamster d'Imhause	Гривистая крыса
– Spiny Rats	Kammstachelratten	Échimys	
Cretan Spiny Mouse	Kreta-Stachelmaus	Souris épineuse de Crête	Критская ежовая мышь
Cricetid Rats and Mice	Wühler	Cricétidés	Хомячьи
Cro-Magnon Man	Cro-Magnon-Mensch	Homme de Cro-Magnon	Кроманьонец
Cuban Hutia	Hutiaconga	Rat poilé	Хутия-конга
Cururo	Cururo	– bleu	Куруро

ENGLISH NAME	GERMAN NAME	FRENCH NAME	RUSSIAN NAME
Dall's Harbour Porpoise	Dall-Hafenschweinswal	Marsouin de Dall	Северная белокрылая свинья
Dark Flying Fox	Rauchgrauer Flughund	Roussette foncée	
Daubenton's Bat	Wasserfledermaus	Vespertilion de Daubenton	Водяная ночница
Deer Mouse	Hirschmaus	Souris du soir	
Degu	Degu	Dègue du Chili	Дегу
Dent's Horseshoe Bat	Dent-Hufeisennase	Rhinolophe de Dent	
Desert Dormouse	Salzkrautbilch	Souris du désert	Боялычная соня
– Jerboa	Wüstenspringmaus	Gerboise du steppe	Песчаный тушканчик
– Kangaroo Rat	Wüsten-Känguruhratte	Rat kangourou du désert	Пустынный мешетчатый прыгун
Disk-winged Bats	Amerikanische Haftscheiben-Fledermäuse	Thyropteridés	Американские присосковые летучие мыши
Dobson's Long-tongued Dawn Bat	Höhlen-Langzungen-Flughund	Eonyctère des cavernes	Пещерный крылан
Dolphins	Delphine, Delphine i. e. S.	Delphinidés, Dauphins	Дельфины, Дельфины-белобочки
Dormice	Bilche	Loirs	Сони
Dwarf Epauletted Fruit Bat	Zwerg-Epauletten-Flughund	Petit Microptère	Карликовый эполетовый крылан
– Hamster	Dshungarischer Zwerghamster	Hamster nain de Djoungarie	Джунгарский хомячок
– Hamsters	Kurschwänziger Zwerghamster	Hamsters nains à queue courte	Мохноногие хомячки
– Jerboas	Herzschädel-Springmäuse	Cardiocraninés	Карликовые тушканчики
– Tree Squirrel	Zwerghörnchen	Microsciure nain	
– Tree Squirrels	Neuweltliche Zwerghörnchen	Microsciures	Американские карликовые белки
Earth Vole	Kleine Wühlmaus	Campagnol souterrain	Подземная полевка
Eastern Chipmunk	Streifenbackenhörnchen	Tamia strié	Американский бурундук
– Chipmunks	Chipmunks	Tamias	Чипмунки
– House Mouse	Ährenmaus		Средиземноморская домовая мышь
– Meadow Mouse	Wiesenwühlmaus	Campagnol de Pennsylvania	Пенсильванская полевка
– Pocket Gopher	Flachland-Taschenratte	Gaufre à poche	Равнинный гофер
– Pocket Gophers	Flachland-Taschenratten	Gaufres à poche	
Edentates	Zahnlose	Édentés	Неполнозубые
Egyptian Fruit Bat	Ägyptischer Flughund	Roussette d'Égypte	Египетский крылан
– Spiny Mouse	Ägyptische Stachelmaus	Souris épineuse	Египетская ежовая мышь
Ehrenberg's Mole Rat	Ehrenberg-Blindmaus	Spalax d'Ehrenberg	Слепыш Эренбера
Elbe Beaver	Elbebiber	Castor de l'Elbe	Эльбский бобр
Elephant-eared Kangaroo Rat	Großohr-Känguruhratte	Rat kangourou à grandes oreilles	Большеухий мешетчатый прыгун
Eleven-banded Armadillo	Großes Nacktschwanzgürteltier	Tatou à onze bandes	Большой голохвостый броненосец
– Armadillos	Nacktschwanzgürteltiere	Tatous à onze bandes	Голохвостые броненосцы
El Salvador Sheath-tailed Bat	Zweistreifen-Taschenfledermaus	Saccopteryx à deux raies	Двухполосая мешкокрылая мышь
Emin's Rat	Emin-Riesenhamsterratte	Rat géant d'Emin	
Epauletted Bat	Franquet-Epauletten-Flughund	Épomophore de Franquet	Крылан Франкета
– Fruit Bats	Epauletten-Flughunde	Épomophores	Эполетовые крыланы
Eurasian Dormice	Eigentliche Bilche	Glirinés	Настоящие сони
European and American Tree Squirrels	Eichhörnchen	Écureuils	Белки
– Flying Squirrel	Gewöhnliches Gleithörnchen	Palatouche	Обыкновенная летяга
– Free-tailed Bat	Bulldogg-Fledermaus	Tataride bouledogue	Широкоухий складчатогуб
– Little Brown Bat	Mausohr	Vespertilion murin	Большая ночница
– Souslik	Ziesel	Souslik d'Europe	Серый суслик
Eversman's Dwarf Hamster	Eversmann-Zwerghamster	Hamster nain d'Eversman	Хомячок Эверсманна
– Souslik	Langschwänziger Ziesel	Spermophile d'Eversman	Длиннохвостый суслик
False Killer Whale	Kleiner Schwertwal	Pseudorque à dents épaisses	Черная косатка
– Vampire Bats	Falsche Vampire	Faux Vampires	
– Vampires	Eigentliche Großblattnasen	Mégadermes	Лироносы
Fat Dormouse	Siebenschläfer	Loir gris	Соня-полчок
– Mouse	Fettmaus	Rat adipeux	Жирная древесная мышь
Fat-tailed Mouse	Dickschwanzmaus	Souris à grosse queue	Толстохвостая песчанка
Field Gerbil	Nordafrikanische Rennmaus	Gerbille champêtre	Североафриканская песчанка
– Vole	Erdmaus	Campagnol agreste	Темная полевка
Finback Whales	Furchenwale, Finnwale	Rorquals	Полосатики, Настоящие полосатики
Finless Black Porpoise	Indischer Schweinswal	Marsouin de l'Inde	Бесперая морская свинья
Fish-eating Bat	Großes Hasenmaul	Noctilio pêcheuse	Обыкновенный зайцерот
Five-toed Dwarf Jerboas	Fünfzehen-Zwergspringmäuse		Пятипалые карликовые тушканчики
– Jerboa	Pferdespringer	Gerboise à cinq doigts	Большой тушканчик
Flat-haired Mouse		Souris aux cheveux plats	Бурая иглистая мышь

ENGLISH NAME	GERMAN NAME	FRENCH NAME	RUSSIAN NAME
Flat-headed Bat	Bambus-Fledermaus	Vespertilion du bambou	Бамбуковая летучая мышь
Flightless Scaly-tailed Squirrel	Dornschwanzbilch	Zenkerelle	Шипохвостая соня
– – Squirrels	Dornschwanzbilche	Zenkerelles	Шипохвостые сони
Flower-faced Bat	Blumennasen-Fledermaus	Anthops orné	Цветконосый ложный подковонос
Flying Foxes	Flederhunde, Langnasen-Flughunde, Eigentliche Flughunde	Megachiroptères, Roussettes	Летучие собаки, Длинноносые крыланы
– Lemurs	Riesengleiter, Riesengleitflieger	Cynocephalidés	Шерстокрыловые, Шерстокрылы
– Squirrels	Gleithörnchen	Écureuils volants	Летяги
Forest Dormouse	Baumschläfer	Lérotin	Лесная соня
Franklin's Ground Squirrel	Franklin-Ziesel	Spermophile de Franklin	Суслик Франклина
Fraser's Scaly-tailed Squirrel	Fraser-Dornschwanzhörnchen	Anomalure de Fraser	Шипохвостая белка Фразера
Free-tailed Bats	Bulldogg-Fledermäuse, Faltlippen-Fledermäuse	Molossidés, Tatarides	Бульдоговые летучие мыши, Складчатогубы
Frosted Bats	Zweifarbige Fledermäuse	Sérotines bicolores	Двухцветные кожаны
Fruit Bats	Flederhunde, Flughunde i. e. S.	Megachiroptères, Pteropidés	Летучие собаки, Крыланы
Funnel-eared Bat	Trichterohr	Vespertilion à couleur de paille	
– Bats	Trichterohren	Natalidés	
Gambia Pouched Rat	Gambia-Riesenhamsterratte	Rat géant de Gambia	Гамбийская хомяковидная крыса
Gangetic Dolphin	Ganges-Delphin	Dauphin du Gange	Сусук
– Dolphins	Ganges-Delphine	Platanistidés	Гангские дельфины
Garden Dormouse	Gartenschläfer	Lérot	Садовая соня
Geisha Wood Mouse	Geishamaus	Mulot de Geisha	Японская мышь
Geoffroy's Slit-faced Bat	Geoffroy-Schlitznase	Nyctère de Geoffroy	
Gerbils	Rennmäuse, Eigentliche Rennmäuse	Gerbillinés, Gerbilles	Песчанки, Африканские песчанки
Gervais' Whale	Gervais-Zweizahnwal	Mésoplodon de Gervais	Европейский ремнезуб
Ghost Bat	Gespenst-Fledermaus	Diclidure de fantôme	
Giant Armadillo	Riesengürteltier	Tatou géant	Гигантский броненосец
– Bottle-nosed Whales	Schwarzwale		Плавуны
– False Vampire	Australische Gespenstfledermaus	Macroderme d'Australie	
– Flying Squirrel	Taguan	Écureuil volant géant	Тагуан
– Fruit Bat	Indischer Flughund	Roussette géante	Летучая лисица
– Malabar Squirrel	Riesenhörnchen	Écureuil géant	Малайская гигантская белка
– Naked-tailed Rat	Gebirgs-Mosaikschwanz-Riesenratte	Rat à queue en mosaique des montagnes	
– – Rats	Mosaikschwanz-Riesenratten	Rats à queue en mosaique	
– Noctule	Riesen-Abendsegler	Noctule géante	Гигантская вечерница
– Pangolin	Riesen-Schuppentier	Grand Pangolin	Гигантский ящер
– Pouched Rats	Riesenhamsterratten	Rats géants	Большие хомяковидные крысы
– Tube-nosed Fruit Bat	Großer Röhrennasen-Flughund	Nyctimène géant	Большой ночной крылан
– Velvety Free-tailed Bat	Große Samt-Fledermaus	Molosse géante	
Gill's Dolphin	Gill-Tümmler	Dauphin de Gill	Афалина Гилля
Gobi Jerboa	Gobi-Springmaus	Gerboise du Gobi	Монгольский земляной заяц
Golden Bat	Madagassische Haftscheiben-Fledermaus	Vespertilion doré	Золотистая летучая мышь
– Bats	– Haftscheiben-Fledermäuse	Myzopodidés	Мадагаскарские присосковые летучие мыши
– Hamster	Syrischer Goldhamster	Hamster doré	Переднеазиатский хомяк
Golden-mantled Ground Squirrel	Goldmantelziesel	Spermophile à manteau doré	Золотистоспинннный суслик
Golden Mouse	Goldmaus	Souris dorée	Золотистый мышевидный хомячок
Goldman's Spiny Pocket Mouse	Goldmann-Stacheltaschenmaus	– à poche de Goldman	Иглистый тушканчик Гольдмана
Goosebeak Whale	Cuvier-Schnabelwal	Baleine du bec des oies	Настоящий клюворыл
Grasshopper Mice	Grashüpfermäuse	Onychomys	
Great Ant-eater	Großer Ameisenbär	Grand Fourmilier	Трехпалый муравьед
– Bat	– Abendsegler	Noctule	Рыжая вечерница
– Cane Rat	Große Rohrratte	Aulacode	Большая тростниковая крыса
– False Vampire	– Spießblattnase	Faux Vampire commun	Большой вампир
– Gerbil	– Rennmaus	Gerbille géante	Большая песчанка
– Himalayan Leaf-nosed Bat	Himalaja-Rundblattnase	Phyllorine d'Himalaye	
Slit faced Bat	Große Schlitznase	Grand Nyctère	
Greater Horseshoe Bat	– Hufeisennase	Grand Rhinolophe obscur	Большой подковонос
– Mole Rat	Ostblindmaus	Spalax oriental	Обыкновенный слепыш
– Pichiciago	Burmeister-Gürtelmull	Grand Chlamyphore	Плащеносец Бурмейстера
Greenland Collared Lemming	Grönländischer Halsbandlemming		Гренландский лемминг
– Right Whale	Grönlandwal	Baleine boréale	Гренландский кит

ENGLISH NAME	GERMAN NAME	FRENCH NAME	RUSSIAN NAME
Grey-headed Flying Fox	Graukopf-Flughund	Roussette à tête cendrée	Сероголовый восточно-австралийский крылан
Grey Squirrel	Grauhörnchen	Écureuil gris	Американская серая белка
– Whale	Grauwal	Baleine grise	Серый кит
– Whales	Grauwale	Eschrichtiidés	Серые киты
Groove-toothed Squirrel	Borneo-Hörnchen	Reithrosciure de Borneo	Борнеоская белка
Ground and Tree Squirrels	Erd- und Baumhörnchen		Земляные и древесные беличьи
– Squirrels	Ziesel	Sousliks	Суслики
Guano Bat	Guano-Fledermaus	Tataride de Mexique	
Guinea Pig	Hausmeerschweinchen	Cobaye	Домашняя морская свинка
– Pigs	Meerschweinchen i. e. S.	Cobayes	
Gundi	Gundi	Goundi	Гунди
Gundis	Kammfingerartige, Kammfinger, Gundis	Goundis	Гребнепаловые, Гребнепалые крысы
Hairy Armadillo	Pelzgürteltier, Braunzottiges Borstengürteltier	Tatou poilu, Tatou velu	Мохнатый броненосец, Волосатый броненосец
Hairy-armed Bat	Rauharm-Fledermaus	Noctule de Leisler	Малая вечерница
Hairy-legged Vampire Bat	Kleiner Blutsauger	Diphylle sans queue	Малый кровосос
Haitian Hutia	Cuviers Zaguti	Plagiodonte d'Haiti	Загути Кювье
Hammer-headed Fruit Bat	Hammerkopf	Hypsignathe monstrueux	Молотоголовый крылан
Hamsters	Hamster	Hamsters	Хомяки
Harpy Fruit Bats	Spitzzahn-Flughunde	Harpionyctères	Крыланы-гарпии
Harsh-furred Mice	Afrikanische Bürstenhaarmäuse	Rats à pelage en brosse	
Harvest Mouse	Eurasiatische Zwergmaus	Souris des moissons	Мышь-малютка
Heart-nosed False Vampire	Herznasenfledermaus	Mégaderme du cœur	
Heidelberg Man	Heidelberger Frühmensch	Pithécanthrope d'Heidelberg	Гейдельбергский прачеловек
High-mountain Vole	Gebirgsmaus	Campagnol des montagnes	Плоскочерепная полевка
– Voles	Gebirgswühlmäuse		Каменные полевки
Hildebrandt's Horseshoe Bat	Hildebrandt-Hufeisennase	Rhinolophe d'Hildebrandt	Подковонос Гильдебрандта
Hispid Slit-faced Bat	Rauhhaar-Schlitznase	Nyctère hérissé	
Hoary Bat	Weißgraue Fledermaus	Chauve-souris cendrée	Бело-серый кожан
– Marmot	Eisgraues Murmeltier	Marmotte grise	Североамериканский серый сурок
Hoffmann's Two-toed Sloth	Hoffmann-Zweifingerfaultier	Unau d'Hoffmann	Двупалый ленивец Гоффманна
Honduran Disk-winged Bat	Honduras-Haftscheiben-Fledermaus	Vespertilion d'Honduras	
Horseshoe Bats	Hufeisennasen, Eigentliche Hufeisennasen	Rhinolophidés, Rhinolophes	Подковоносы
Horseshoe-nosed Bats	Hufeisennasen-Verwandte	Rhinolophes	Подковоносовые
House Bat	Große Braune Fledermaus	Sérotine de maison	
House-building Jerboa Rat	Langohr-Häschenratte		Длинноухая зайцевидная мышь
– Jerboa Rats	Australische Häschenratten		Австралийские зайцевидные мыши
House Mice	Mäuse i. e. S.	Souris	Мыши
– Mouse	Hausmaus	– domestique	Домовая мышь
House Rat	Hausratte, Hausratte i. e. S.	Rat commun, Rat noir	Черная крыса, Северо-западная черная крыса
Humpback Whale	Buckelwal	Mégaptère	Горбач
Hutias	Baum- und Ferkelratten		Ежовые крысы
Ice Baleen Whale	Südlicher Glattwal	Baleine australe	Австралийский кит
Indian False Vampire	Lyra-Fledermaus	Mégaderme lyre	Индийский ложный вампир
– Flying Fox	Indischer Flughund	Roussette géante	Летучная лисица
– Giant Squirrel	Königsriesenhörnchen	Écureuil d'Inde	Индийская гигантская белка
– Pangolin	Vorderindisches Schuppentier	Pangolin indien	Переднеиндийский ящер
– Pilot Whale	Indischer Grindwal	Globicéphale d'Inde	Индийская гринда
Indo-Malayan Flying Squirrel	Javanisches Gleithörnchen	Écureuil volant de Java	Яванская летяга
– Giant Squirrels	Indomalaiische Riesenhörnchen		Индомалайские гигантские белки
Indonesian Porcupine	Zwergstachelschwein	Porc-épic d'Indonésie	Карликовый дикобраз
– Porcupines	Insel-Stachelschweine	Porcs-épic d'Indonésie	Индонезийские дикобразы
Insect-eating Bats	Fledermäuse	Chauves-souris insectivores	Летучие мыши
Insectivorous Bats	Fledermäuse	– insectivores	Летучие мыши
Irawady Dolphin	Irawadi-Delphin	Orcaelle d'Irawadi	Иравадийский дельфин
Jackson's Scaly-tailed Squirrel	Jackson-Dornschwanzhörnchen	Anomalure de Jackson	Шипохвостая белка Джексона
Jamaican Hutia	Jamaika-Ferkelratte	Rat jamaique	Ямайская крыса
Japanese Dormouse	Japanischer Schläfer	Glirule de Japon	Японская соня
Java Man	Frühmensch, Javanischer Frühmensch	Pithécanthrope de Java	Прачеловек, Питекантроп
Javanes Porcupine	Java-Stachelschwein	Acanthion de Java	Яванский дикобраз

ENGLISH NAME	GERMAN NAME	FRENCH NAME	RUSSIAN NAME
– Slit-faced Bat	Java-Hohlnase	Nyctère de Java	Яванская никтерида
Javelin Bat	Lanzennase	Phyllostome fer de lance	Копьенос
Jerboa	Erdhase	Gerboise lièvre	Тарбаганчик
Jerboas	Springmäuse, Eigentliche Springmäuse	Gerboises, Dipodinés	Тушканчики, Трехпалые тушканчики
Jird	Mongolische Rennmaus	Mérione de Monglie	Когтистая песчанка
Jirds	Sandmäuse	Mériones	Малые песчанки
Jone's House-building Jerboa Rat	Kurzohr-Häschenratte		Короткоухая зайцевидная мышь
Jumping Mice	Hüpfmäuse	Zapodidés	Мышовки и прыгунчики
Kangaroo Mice	Taschenmäuse, Känguruhmäuse	Souris kangourou	Мешетчатые крысы, Мешетчатые прыгунчики
– Rats	Taschenspringer	Rats kangourou	Мешетчатые прыгуны
Kappler's Armadillo	Kappler-Weichgürteltier	Tatou de Kappler	Броненосец Капплера
Knight's Tuco-tuco	Knight-Tukotuko	Ctenomys de Knight	Тукотуко
Lamberton's Malagasy Rat	Lamberton-Inselratte	Rat de Lamberton	Мадагаскарская крыса Ламбертона
Lander's Horseshoe Bat	Lander-Hufeisennase	Rhinolophe de Lander	
Lang's Small Flying Squirrel	Lang-Gleitbilch	Anomalure nain de Lang	Соня-летяга Ланга
La Plata Dolphin	La-Plata-Delphin	Dauphin de la Plata	Лаплатский дельфин
Large Fruit Bats	Flughunde i. e. S.	Pteropidés	Крыланы
Large Gerbil	Indische Nacktsohlen-Rennmaus	Gerbille d'Inde	Индийская песчанка
– Gerbils	Nacktsohlen-Rennmäuse		Голоступые песчанки
– Porcupines	Eigentliche Stachelschweine	Porcs-épic	Настоящие дикобразы
Larger Rat-tailed Bat	Ägyptische Klappnase	Rhinopôme microphylle	Египетский ланцетонос
Large-eared Small Flying Squirrel	Großohr-Gleitbilch	Anomalure nain à longues oreilles	Ушастая соня-летяга
Large-headed Tube-nosed Fruit Bat	Großkopf-Röhrennasen-Flughund	Nyctimène à grosse tête	
Large-toothed Red-backed Vole	Graurötelmaus	Campagnol gris-roux	Красно-серая полевка
Large-winged Bats	Großblattnasen	Mégadermes	Мегадермы
Layard's Whale	Layard-Wal	Mésoplodon de Layard	Ремнезуб Лэйярда
Least Chipmunk	Kleiner Chipmunk	Néotamia nain	Малый чипмунк
– Sac-winged Bats	Glattnasige Freischwanz-Fledermäuse		Плосконосые свободнохвосты
Leisler's Bat	Rauharm-Fledermaus	Noctule de Leisler	Малая вечерница
Lemming Mice	Lemmingmäuse		Лемминговые мыши
Lemmings	Lemminge	Lemmings	Лемминги
Lesser Bamboo Rat	Kleine Bambusratte	Petit Rat des bambous	
– Cane Rat	– Rohrratte	Rat de Grégorian	Малая тростниковая крыса
– Horseshoe Bat	– Hufeisennase	Petit Rhinolophe	Малый подковонос
– Mole Rat	Westblindmaus	Spalax occidental	Малый слепыш
– Noctule	Rauharm-Fledermaus	Noctule de Leisler	Малая вечерница
– Pichiciago	Gürtelmull	Chlamyphore tronqué	Аргентинский плащеносец
– Rat-tailed Bat	Hardwicke-Klappnase	Rhinopôme d'Hardwick	Ланцетонос Хардвика
– Rorqual	Zwergwal	Petit Rorqual	Малый остромордый полосатик
Libyan Jird	Rotschwänzige Rennmaus	Mérione de Lybie	Краснохвостая песчанка
Lichtenstein's Jerboa	Lichtensteins Springmaus	Gerboise de Lichtenstein	Тушканчик Лихтенштейна
Little Ant eater	Zwergameisenbär	Fourmilier didactyle	Двупалый муравьед
– Collared Fruit Bat	Schmalkragen-Flughund	Myonyctère à collier	
– Souslik	Zwergziesel	Souslik nain	Малый суслик
Long-clawed Ground Squirrel	Zieselmaus	Spermophile leptodactyle	Тонкопалый суслик
– Mole Vole	Prometheus-Maus		Прометеева полевка
Long-eared Bat	Braunes Langohr	Oreillard commun	Обыкновенный ушан
– Bats	Großohren	Oreillards	Ушаны
– Jerboas	Riesenohr-Springmäuse		Длинноухие тушканчики
Long-legged Bat	Langbein-Fledermaus	Macrophylle	
Long-nosed Squirrel	Langnasenhörnchen	Écureuil à nez long	Длинноносая белка
Long-snouted Squirrel	Ferkelhörnchen	– d'Heinrich	
Long-tailed Cinchilla	Langschwanz-Chinchilla	Chinchilla à longue queue	Длиннохвостая шиншилла
– Climbing Mice	Langschwänzige Indische Baummäuse	Souris à longue queue	Индийские пальмовые мыши
– Climbing Mouse		– à longue queue	Пальмовая мышь
– Field Mouse	Feld-Waldmaus	Mulot sylvestre	Лесная мышь
– Fruit Bats		Notoptères à queue longue	
– Hamster	Langschwanz-Zwerghamster	Hamster nain à longue queue	Длиннохвостый хомячок
– Marmot	Langschwänziges Murmeltier	Marmotte à longue queue	Длиннохвостый сурок
– Pouched Rats	Kleine Hamsterratten		Малые хомяковидные крысы
– Porcupine	Malaiischer Pinselstachler	Porc-épic à longue queue	Борнеоский дикобраз
– Rat	Langschwänziger Hamster	Hamster à longue queue	Длиннохвостый хомяк
Long-tongued Bat	Spitzmaus-Langzüngler	Glossophage de Pallas	Глоссофага

ENGLISH NAME	GERMAN NAME	FRENCH NAME	RUSSIAN NAME
– Bats	Langzungen-Fledermäuse, Langzungen	Glossophages	Длинноязычные вампиры, Глоссофаги
– Fruit Bats	Langzungen-Flughunde	Macroglosses	Длинноязычные крыланы
Long-winged Bat	Langflügel-Fledermaus	Minioptère à longues ailes	Длиннокрыл Шрейбера
– Bats	Langflügel-Fledermäuse	Minioptères	Длиннокрылы
Maclaud's Bat	Maclaud-Hufeisennase	Rhinolophe de Maclaud	Подковонос Маклауда
Malagasy Rats	Madagaskarratten	Rats de Madagascar	Мадагаскарские крысы
Malayan False Vampire	Malaiischer Falscher Vampir	Mégaderme spasmę	Малайский ложный вампир
– Flying Lemur	Temminck-Gleitflieger		Шерстокрыл Темминка
– Pangolin	Javanisches Schuppentier	Pangolin javanais	Яванский ящер
– Porcupine	Kurzschwanz-Stachelschwein	Acanthion de Malaysie	
Man	Menschen, Menschen i. e. S., Mensch	Hominidés, Homininés, Homo sapiens	Гоминиды, Люди, Человек
Manchurian Zokor	Chinesischer Blindmull	Zokor de Mandchourie	Северокитайский цокор
Mantled African Bush Squirrel	Rotschwanzhörnchen	Écureuil d'Afrique à manteau	
Mara	Mara	Mara	Патагонская свинка
Maras	Maras	Dolichotinés	Мары
Marmots	Murmeltiere	Marmottes	Сурки
Mauretania Man	Nordafrikanischer Frühmensch	Pithécanthrope de Mauretanie	Североафриканский прачеловек
Mauritian Tomb Bat	Mauritanischer Grabflatterer	Taphien de Maurice	
Meadow Jumping Mice	Feldhüpfmäuse	Zapodes des prées	
– – Mouse	Wiesenhüpfmaus	Zapode du Canada	Североамериканский луговой прыгунчик
Mexican Bulldog Bat	Großes Hasenmaul	Noctilio pêcheuse	Обыкновенный зайцерот
– Free-tailed Bat	Guano-Fledermaus	Tataride de Mexique	
Mexican Long-nosed Bat	Langnasen-Fledermaus	Chauve-souris de Mexique	Длинноносый листонос
– Spiny Pocket Mouse	Mexikanische Stacheltaschen-maus	Souris épineuse à poche de Mexique	Мексиканский иглистый тушканчик
Michigan Beaver	Waldbiber	Castor de Michigan	Мичиганский бобр
Migratory Hamster	Grauer Zwerghamster	Hamster migrateur	Серый хомячок
Moco	Moko	Cobaye des roches	Горная морская свинка
Mole Lemming	Nördlicher Mull-Lemming	Rat taupe	Обыкновенная слепушонка
– Lemmings	Mull-Lemminge	Rats taupes	Слепушонки
– Mice	Blindmulle		Цокоры
– Rats	Sandgräber, Blindmäuse	– – africains, Rats taupes	Землекопы, Слепыши
Monckton's Water Rat	Moncktons Schwimmratte		Австралийская крыса Монктона
Mongolian Beaver	Mongolischer Biber	Castor de Mongolie	Монгольский бобр
Monkeys	Affen	Singes	Обезьяны
Mountain Beaver	Stummelschwanzhörnchen	Castor de montagne	Горный бобр
– Beavers	Stummelschwanzhörnchen	Castors de montagne	Горные бобры
– Cavies	Zwergmeerschweinchen		Карликовые свинки
– Cinchilla	Cuvier-Hasenmaus	Lagostome des montagnes	Пушак
– Chinchillas	Hasenmäuse	Lagostomes des montagnes	Пушаки
– Paca	Bergpaka	Paca des montagnes	Горная пака
– Viscachas	Hasenmäuse	Lagostomes des montagnes	Пушаки
Mouse-eared Bats	Mausohr-Fledermäuse	Murins	Ночницы
Mouselike Hamster	Mausartiger Zwerghamster		Мышевидный хомячок
Multimammate Rats	Vielzitzenmäuse	Rats à mamelles multiples	
Muskrat	Bisamratte	Rat musqué	Ондатра
Muskrats	Bisamratten	Ondatras	Мускусные крысы
Mustache Bats	Kinnblatt-Fledermäuse, Schnurrbart-Fledermäuse	Chilonycterinés	Листобороды
Naked-backed Bat	Kleine Nacktrücken-Fledermaus	Vespertilion à dos nu	
Naked Bat	Nacktfledermaus	Cheiromèle nue	Голая летучая мышь
Naked-bellied Tomb Bat	Nacktbäuchiger Grabflatterer	Taphien à ventre nu	Голобрюхая летучая мышь
Naked Sand Rat	Nacktmull	Rat nu de sable	Голая крыса
Naked-soled Gerbils	Kleine Nacktsohlen-Rennmäuse		Малые голоступые песчанки
Narrow-snouted Dolphin	Schlankdelphin	Dauphin douteux	Стройный продельфин
Narwhal	Narwal	Narval	Нарвал
Natterer's Bat	Fransenfledermaus	Vespertilion de Natterer	Ночница Наттерера
Neanderthal Man	Neandertaler	Homme de Néanderthale	Неандерталец
Neave's Scaly-tailed Squirrel	Neave-Dornschwanzhörnchen	Anomalure de Neave	
Necklace Hairless Bat	Halsband-Fledermaus	Cheiromèle à collier	
– Sloth	Kragenfaultier	Bradype à collier	Ошейниковый ленивец
Nelson's Souslik	Nelson-Antilopenziesel	Souslik de Nelson	Калифорнийский антилоповый суслик
Neotropical False Vampire Bats	Falsche Vampire	Faux Vampires	
– Fruit-eating Bats		Artibées	Тупомордые листоносы
Newfoundland Beaver	Neufundland-Biber	Castor de Terre-Neuve	Ньюфаундлендский бобр
New World Mice	Neuweltmäuse	Souris du Nouveau Monde	Американские мышевидные хомячки
– – Porcupines	Baumstachlerartige, Baumstachler	Erethizontidés	Дикобразы Нового Света, Иглошерсты

ENGLISH NAME	GERMAN NAME	FRENCH NAME	RUSSIAN NAME
– Zealand Short-tailed Bat	Neuseeland-Fledermaus	Mystacine tubercule	Новозеландская летучая мышь
– – Short-tailed Bats	– Fledermäuse	Mystacinidés	Новозеландские летучие мыши
Nine-banded Armadillo	Neunbindengürteltier	Tatou à neuf bandes	Девятипоясный броненосец
Noctule Bats	Abendsegler	Noctules	Вечерницы
North Altai Zokor	Blindmull	Rat taupe	Алтайский цокор
– American Porcupine	Urson	Porc-épic nord américain	Североамериканский иглошерст
– – Porcupines	Nordamerikanische Baumstachler	Porcs-épic nord américains	Североамериканские иглошерсты
– – Red Squirrel	Rothörnchen	Écureuil d'Hudson	Канадская белка
Northern Bat	Nordische Fledermaus	Sérotine boréale	Северный кожанок
– Birch Mouse	Birkenmaus	Siciste de bouleaux	Лесная мышовка
– Bog Lemming	Nördliche Lemmingmaus	Lemming du nord	Северная лемминговая мышь
– Flying Squirrel	Nördliches Gleithörnchen	Écureuil volant du nord	
– Grasshopper Mouse	Nördliche Grashüpfermaus	Onychomys du nord	
– Ground Squirrels	Erdhörnchen		Сурковые
– House Mouse	Nördliche Hausmaus		Североевропейская домовая мышь
– Pilot Whale	Gewöhnlicher Grindwal	Grinde	Северная гринда
– Red-backed Vole	Polarrötelmaus	Campagnol boréal	Красная полевка
– Right Whale Dolphin	Nördlicher Glattdelphin	Dauphin du nord	Северный китовидный дельфин
– Three-toed Jerboa	Rauhfuß-Springmaus	Gerboise à pattes rugueuses	Мохноногий тушканчик
Norway Lemming	Berglemming	Lemming des toundras	Норвежский лемминг
– Rat	Wanderratte	Rat surmulot	Пасюк
Nutria	Sumpfbiber	Ragondin	Нутрия
Nutrias	Biberratten	Ragondins	Бобровые крысы
Octodont Rodents	Trugrattenartige, Trugratten	Octodontidés	Осьмизубовые, Осьмизубые
Oil-palm Squirrel	Ölpalmenhörnchen	Grand Écureuil de Stanger	Масличная белка
Old World Fruit Bats	Flughunde i. e. S.	Pteropidés	Крыланы
– – Leaf-nosed Bats	Rundblattnasen		Ложные подковоносы
– – Monkeys	Schmalnasen	Catarrhiniens	Узконосые обезьяны или Обезьяны Старого Света
– – Porcupines	Altwelt-Stachelschweine	Hystricidés	Дикобразы Старого Света
– – Rats and Mice	Mäuse	Souris et Rats de l'Ancien Monde	Мышиные
Orange-rumped Agouti	Goldaguti	Agouti doré	Золотистая агути
Ord's Kangaroo Rat	Ord-Känguruhratte	Rat kangourou d'Ord	Мешетчатый прыгун Орда
Oriental Giant Squirrel	Riesenhörnchen	Écureuil géant	Малайская гигантская белка
– Pygmy Squirrels	Asiatische Zwerghörnchen	Écureuils pygmées	Азиатские карликовые белки
Paca	Paka	Paca	Пака
Pacarana	Pakarana	Pacarana	Длиннохвостая пака
Pacaranas	Pakaranaartige, Pakaranas	Pacaranas	Диномииды
Pacas	Pakas	Cuniculinés	Паки
Pacific Harbour Porpoise	Pazifischer Hafenschweinswal	Marsouin du Pacifique	Северотихоокеанская морская свинья
– Kangaroo Rat	Pazifik-Känguruhratte	Rat kangourou du Pacifique	Проворный мешетчатый прыгун
– Pilot Whale	Pazifischer Grindwal	Globicéphale de Siebold	Черная гринда
– Right Whale	Nordpazifik-Glattwal	Baleine de Siebold	Японский кит
– White-sided Dolphin	Weißstreifendelphin	Lagenorhynque de Gill	Тихоокеанский белобокий дельфин
Painted Spiny Pocket Mouse	Gemalte Stacheltaschenmaus	Souris épineuse à poche	
Palaearctic Mole Rats	Blindmäuse	Rats taupes	Слепыши
Pale Kangaroo Mouse	Blasse Känguruhmaus	Souris kangourou pâle	Бледный мешетчатый прыгунчик
Pallid Bat	– Fledermaus	Oreillard pâle	Бледная летучая мышь
Palm Squirrel	Palmenhörnchen	Écureuil des palmes	Пальмовая белка
– Squirrels	Gestreifte Palmenhörnchen	Écureuils des palmes	
Pangolins	Schuppentiere	Pholidotes, Manidés	Панголины, Ящеры
Particoloured Bat	Zweifarbige Fledermaus	Petite Chauve-souris murine	Двухцветный кожан
– Bats	– Fledermäuse	Sérotines bicolores	Двухцветные кожаны
Peking Man	Peking-Frühmensch	Pithécanthrope de Péking	Синантроп
Pel's Scaly-tailed Squirrel	Pel-Dornschwanzhörnchen	Anomalure de Pel	Шипохвостая белка Пеля
Pencil-tailed Tree Mouse	Malaiische Pinselschwanz-Baummaus	Souris d'arbre de Malaysie	Малайская древесная крыса
Pere David's Vole	Schwarzbauch-Wühlmaus	Campagnol de Père David	Давидова полевка
– – Voles	Père-Davids-Wühlmäuse		Давидовы полевки
Peron's Dolphin	Südlicher Glattdelphin	Dauphin de Peron	Южный китовидный дельфин

ENGLISH NAME	GERMAN NAME	FRENCH NAME	RUSSIAN NAME
Persian Jird	Persische Wüstenmaus	Mérione de Perse	Персидская песчанка
– Leaf-nosed Bat	Dreiblatt-Fledermaus	Triaenops de Perse	Персидский ложный подковонос
Pest Rat	Kurzschwanz-Maulwurfsratten	Rat à queue courte	Индийская земляная крыса
Phatagin	Langschwanz-Schuppentier	Pangolin tétradactyle	Длиннохвостый ящер
Philippines Flying Lemur	Philippinen-Gleitflieger		Кагуан
Pichicies	Gürtelmulle	Chlamyphores	Плащеносные броненосцы
Pilot Whales	Grindwale	Globicéphales	Гринды
Pine Vole	Kiefernwühlmaus	Campagnol du pin	Американская боровая полевка
– Voles	Kleine Wühlmäuse		Малые полевки
Piñon Mouse	Pinjonmaus	Souris de True	
Pipistrelles	Zwergfledermäuse	Pipistrelles	Нетопыри
Pocket Gophers	Taschennager, Taschenratten	Gaufres à poche	Мешетчатые грызуны, Гоферы
– Mice	Eigentliche Taschenmäuse	Souris à poche	Мешетчатые тушканчики
Pond Bat	Teichfledermaus	Verspertilion des marais	Прудовая ночница
Porcupine	Gewöhnliches Stachelschwein	Porc-épic	Обыкновенный дикобраз
Porcupine-like Rodents	Stachelschweinverwandte		Дикобразовые
Porpoises	Schweinswale	Phocénidés	Морские свиньи
Prairie Dogs	Präriehunde	Chiens des prairies	Луговые собачки
Prehensile-tailed Porcupine	Greifstachler i. e. S.	Porc-épic préhensile	Бразильский коэнду
– Hutia	Hutiacarabali	Rat à queue préhensile	Хутия-карабали
Primates	Herrentiere	Primates	Приматы
Proboscis Bat	Nasenfledermaus	Rhynchonyctère	Носатая летучая мышь
Puffing Pig	Schweinswal	Marsouin commun	Морская свинья
Pygmy Armadillo	Zwerggürteltier	Tatou nain	Карликовый броненосец
– Chimpanzee	Bonobo	Chimpansé nain	Бонобо
– Dormouse	Zwergschläfer	Graphiure nain	Карликовая африканская соня
Pygmy Jerboas	Koslows Zwergspringmäuse		Трехпалые карликовые тушканчики
– Killer	Zwerggrindwal		Карликовая косатка
– Mouse	Amerikanische Zwergmaus		Карликовый мышевидный хомячок
– Right Whale	Zwergglattwal	Baleine naine	Карликовый гладкий кит
– Scaly-tailed Squirrel	Zwerg-Dornschwanzhörnchen	Anomalure nain	Карликовая шипохвостая белка
– Sperm Whale	Zwergpottwal	Cachalot pygmée	Карликовый кашалот
Queensland Tube-nosed Fruit Bat	Robinson-Röhrennasen-Flughund	Nyctimène de Robinson	
Rabbit Rat	Westliche Erntemaus	Souris occidentale des moissons	
Rat Chinchillas	Chinchillaratten	Abrocomidés	Шиншилловые крысы
Ratlike Hamster	Daurischer Zwerghamster, Rattenartiger Zwerghamster	Hamster nain de Daourie	Даурский хомячок, Крысовидный хомяк
– Hamsters	Graue Zwerghamster	Hamsters nains gris	Хомячки
Rats	Eigentliche Ratten	Rats proprement dits	Крысы
Rat-tailed Bats	Mausschwanz-Fledermäuse, Klappnasen	Rhinopômes	Ланцетоносы
Red-backed Scaily-tailed Squirrel	Rotrücken-Dornschwanz-hörnchen	Anomalure à dos rouge	
– Voles	Rötelmäuse	Campagnols	Лесные полевки
Red Bat	Rote Fledermaus	Chauve-souris boréale	Рыжий кожан
Red-fruit Bats	Fruchtvampire		Плодовые вампиры
Red-grey Bat	Fransenfledermaus	Vespertilion de Natterer	Ночница Наттерера
Red-necked Fruit Bat	Kalong	Roussette à cou rouge	Калонг
Red Sea Bottle-nosed Dolphin	Rotmeer-Tümmler	Tursiops du Mer Rouge	Красноморская афалина
– Souslik	Rotziesel	Spermophile rouge	Большой суслик
– Squirrel	Eichhörnchen	Écureuil commun	Обыкновенная белка
– Squirrels	Rothörnchen		Красные белки
– Velvety Free-tailed Bat	Rote Samt-Fledermaus	Molosse rouge	Рыжая бульдоговая мышь
Rhone Beaver	Rhonebiber	Castor du Rhône	Ронский бобр
Rice Rat	Sumpf-Reisratte	Rat du riz	Рисовая мышь
– Rats	Reisratten	Rats du riz	
Right Whale Dolphins	Glattdelphine	Lissodelphininés	Бесперые дельфины
– Whales	Glattwale	Baleines	Гладкие киты
Rio-Grande Beaver	Rio-Grande-Biber	Castor de Rio-Grande	Бобр Рио-Гранде
Risso's Dolphin	Rundkopfdelphin	Dauphin de Risso	Серый дельфин
River Dolphins	Flußdelphinartige		Речные дельфины
Roborowsky's Dwarf Hamster	Roborowski-Zwerghamster	Hamster nain de Roborowsky	Хомячок Роборовского
Rock Cavies	Bergmeerschweinchen	Cobayes des roches	Горные свинки
– Cavy	Moko	Cobaye des roches	Горная морская свинка
– Rat	Felsenratte	Rat typique	Горная мышь
– Rats	Afrikanische Felsenratten		Африканские горные мыши

ENGLISH NAME	GERMAN NAME	FRENCH NAME	RUSSIAN NAME
– Squirrel	Père-Davids-Felsenhörnchen	Écureuil des rochers	Давидова белка
– Squirrels	Chinesische Rothörnchen		Китайские красные белки
Rodents	Nagetiere	Rongeurs	Грызуны
Roof Rat	Hausratte, Alexandriner Hausratte	Rat commun, Rat d'Aléxandrie	Черная крыса, Александрийская черная крыса
Rorquals	Furchenwale, Finnwale	Rorquals	Полосатики, Настоящие полосатики
Round-tailed Ground Squirrel	Rundschwanzziesel	Spermophile à queue ronde	Круглохвостый суслик
– Muskrat	Florida-Wasserratte	Rat d'eau de Floride	Флоридская водяная крыса
Roussette Bats	Höhlenflughunde	Roussettes	Ночные крыланы
Royle's High-mountain Vole	Mongolische Gebirgsmaus	Campagnol de Royle	Серебристая полевка
Rufous Flying Fox	Roter Flughund	Roussette rougeâtre	Мадагаскарская летучая собака
Rufous-nosed Rat	Afrikanische Rotnasenratte	Rat à museau roux	Африканская красноносая крыса
Sac-winged Bats	Glattnasen-Freischwänze, Freischwänze		Свободнохвостые летучие мыши, Эмбаллонуры
Sagebrush Vole	Steppenlemming		Степная пеструшка
– Voles	Steppenlemminge		Степные пеструшки
Salt-desert Cavy	Kleine Mara	Cobaye halophile	Малая мара
Scaly Ant-eaters	Schuppentiere	Manidés	Ящеры
Scaly-tailed Squirrels	Dornschwanzhörnchen, Eigentliche Dornschwanzhörnchen	Anomaluridés, Anomalures	Шипохвостые белки-летяги, Шипохвостые белки
Seba's Short-tailed Bat	Brillen-Blattnase	Vespertilion à nez plat	Очковый листонос
Sei Whale	Seiwal	Balénoptère boréal	Сайдяной кит
Serotine Bat	Spätfliegende Fledermaus	Sérotine commune	Поздний кожан
Seven-banded Armadillo	Siebenbindengürteltier	Tatou à sept bandes	Семипоясный броненосец
Sheath-tailed Bats	Glattnasen-Freischwänze, Taschenfledermäuse	Saccopteryx	Свободнохвостые летучие мыши, Мешкокрылые летучие мыши
Shepherd's Beaked Whale	Shepherd-Wal	Tasmacète de Shepherd	Новозеландский клюворыл Шеперда
Short Dwarf Hamster	Mongolischer Zwerghamster	Hamster nain de Mongolie	Монгольский хомячок
Short-faced Bats	Fruchtvampire		Плодовые вампиры
Short-nosed Fruit Bat	Indischer Kurznasen-Flughund	Cynoptère à nez court	Обыкновенный коротконосый крылан
– Fruit Bats	Kurznasen-Flughunde	Cynoptères	Коротконосые крыланы
Short-tailed Leaf-nosed Bats	Kurzschwanz-Blattnasen	Vespertilions à nez plat	Короткохвостые вампиры
– Chinchilla	Kurzschwanz-Chinchilla	Chinchilla à queue courte	Короткохвостая шиншилла
Shrew Rat	Nasenratte	Rat au nez	
– Rats	Nasenratten	Rhynchomyinés	
Sibbold's Rorqual	Blauwal	Balénoptere bleu	Синий кит
Sibirian Jerboa	Sibirische Springmaus	Gerboise de Sibérie	Тушканчик-прыгун
Silky Pocket Mouse	Seiden-Taschenmaus	Souris soyeux à poche	Шелковистый мешетчатый тушканчик
Sinai Spiny Mouse	Sinai-Stachelmaus	– épineuse de Sinai	Бледная ежовая мышь
Six-banded Armadillo	Weißborsten-Gürteltier	Tatou à six bandes	Шестипоясный броненосец
– Armadillos	Borstengürteltiere	Tatous à six bandes	
Slender-tailed Cloud Rat	Gescheckte Riesenborkenratte	Rat d'ecorce tacheté	Пятнистая древесная крыса
Slit-faced Bats	Schlitznasen	Nyctères	Никтериды
Small African Native Mouse	Afrikanische Zwergmaus	Souris naine d'Afrique	Африканская мышь-малютка
– Naked-soled Gerbil	Kleine Nacktsohlen-Rennmaus	Gerbille d'Emin	
Smoky Agouti	Mohrenaguti		Аспидная агути
– Bats	Stummeldaumen	Furipteridés	
Snake-toothed Fruit Bat	Schlangenzahn-Flughund	Scotonyctère à dent du serpent	
Snow Vole	Schneemaus	Campagnol des neiges	Снежная полевка
Snuffer	Schweinswal	Marsouin commun	Морская свинья
Selden Hamsters	Mittelhamster		Средние хомяки
South African Lesser Leaf-nosed Bat	Gewöhnliche Rundblattnase	Phyllorine de Cafrérie	
– – Porcupine	Südafrikanisches Stachelschwein	Porc-épic d'Afrique du sud	Южноафриканский дикобраз
– American Bush Rats	Strauchratten	Octodons	Кустарниковые крысы
– – Rock Rat	Südamerikanische Felsenratte	Rat foncu	
Southern Birch Mouse	Streifenmaus	Siciste des steppes	Степная мышовка
– Bog Lemming	Südliche Lemmingmaus	Lemming du sud	Южная лемминговая мышь
– Bottle-nosed Whale	Südlicher Entenwal	Hypérodon du sud	Плосколобый бутылконос
– Bulldog Bat	Kleines Hasenmaul	Noctilio du sud	Малый зайцерот
– Flying Squirrel	Assapan	Écureuil volant du sud	Североамериканская летяга
– Grasshopper Mouse	Südliche Grashüpfermaus	Onychomys du sud	

ENGLISH NAME	GERMAN NAME	FRENCH NAME	RUSSIAN NAME
– Jird	Mittagsrennmaus	Mérione du sud	Полуденная песчанка
– Long-eared Bat	Graues Langohr	Oreillard du sud	
– Mountain Cavy	Südliches Zwergmeerschweinchen		Южная карликовая свинка
– Right Whale	Südlicher Glattwal	Baleine australe	Австралийский кит
Sowerby's Whale	Sowerby-Zweizahnwal	Mésoplodon de Sowerby	Атлантический ремнезуб
Spear-nosed Bats	Lanzennasen	Phyllostomes	
Spectacled Porpoise	Brillenschweinswal	Marsouin des lunettes	Очковая морская свинья
Speke's Pectinator	Speke-Kammfinger	Pectinator de Speke	
Sperm Whale	Pottwal	Cachalot macrocéphale	Кашалот
– Whales	Pottwale	Physétéridés	Кашалоты
Spiny Armadillo	Kleines Nacktschwanzgürteltier	Tatou épineux	Малый голохвостый броненосец
– Dormice	Stachelbilche		Шипохвостые сони
– Dormouse	Südindischer Stachelbilch		Южноиндийская соня
– Mice	Stachelmäuse	Souris épineuses	Ежовые мыши
– Rats	Stachelratten, Stachelratten i. e. S.	Échimyidés, Échimyinés	Цепкохвостые щетинистые крысы, Щетинистые крысы
Spix's Disk-winged Bat	Dreifarbige Haftscheiben-Fledermaus	Vespertilion tricolore	Трехцветная летучая мышь
Spotted Dolphins	Fleckendelphine	Dauphins tachetés	Продельфины
– Souslik	Perlziesel	Souslik tacheté	Крапчатый суслик
Spring Haas	Springhasenartige, Springhasen, Ostafrikanischer Springhase	Lièvres sauteurs, Lièvre sauteur d'Afrique de l'est	Долгоноговые, Долгоноги, Восточноафриканский долгоног
Squirrel-like Rodents	Hörnchenverwandte		Белкоподобные
Straw-coloured Bat	Palmenflughund	Roussette paillée	Пальмовый крылан
Striped Field Mouse	Brandmaus, Afrikanische Striemen-Grasmaus	Mulot rayé, Rat rayé du champ	Полевая мышь
– Grass Mice	Afrikanische Streifen-Grasmäuse	Rats rayés	Африканские полосатые мыши
Suapure Naked-backed Bat	Große Nacktrücken-Fledermaus	Vespertilion de Suapuré	
Sucker-footed Bats	Madagassische Haftscheiben-Fledermäuse	Myzopodidés	Мадагаскарские присосковые летучие мыши
Suira	Suira	Rat épineux	
Sumatran Bamboo Rat	Sumatra-Bambusratte	– des bambous de Sumatra	Суматранская ризомиида
Sun Squirrel	Graufußhörnchen	Écureuil de Gambie	
Tamandua	Tamandua	Tamandua à quatre doigts	Четырехпалый муравьед
Tent-building Bat	Gelbohr-Fledermaus	Vespertilion bilobé	
Texas Kangaroo Rat	Texas-Känguruhratte	Rat kangourou de Texas	Техасский мешетчатый прыгун
Thin-spined Porcupine	Borsten-Baumstachler	Porc-épic épineux	Крысохвостый дикобраз
Thirteen-striped Ground Squirrel	Streifenziesel	Spermophile à treize bandes	Полосатый американский суслик
Three-banded Armadillo	Kugelgürteltier	Tatou à trois bandes	Шаровидный броненосец
– Armadillos	Kugelgürteltiere	Tatous à trois bandes	
Three-pointed Pangolin	Weißbauch-Schuppentier	Tricuspide	Белобрюхий ящер
Three-toed Sloth	Dreifingerfaultier	Bradype	Ай
– Sloths	Dreifingerfaultiere	Bradypes	Трехпалые ленивцы
Tibetan Ratlike Hamster	Tibetanischer Zwerghamster	Hamster nain du Tibet	Тибетский хомячок
Tiny Hutia	Zwergbaumratte	Rat nain	Карликовая кубинская крыса
Tomb Bat	Grabflatterer	Taphien perforé	
– Bats	Grabflatterer	Thapiens	Могильные летучие мыши
Tome's Long-eared Bat	Schwertnase	Vespertilion de Tome	Малый вампир
Tonga Flying Fox	Tonga-Flughund	Roussette de Tonga	Тонгский крылан
Toothed Cetaceans	Zahnwale	Odontocétés	Зубатые киты
Townsend's Chipmunk	Townsend-Chipmunk	Néotamia de Townsend	Чипмунк Тоунсенда
Transbaikal Zokor	Daurischer Blindmull	Zokor de Transbaikalie	Даурский цокор
Tree Porcupines	Greifstachler	Porcs-épic préhensiles	Коэнду
Tree-Rat	Bambus-Fingerratte	Rat du bambou	
Tree Sloths	Faultiere	Bradypodidés	Ленивцы
– Squirrels	Baumhörnchen		Настоящие белки
Tricoloured Squirrel	Dreifarbenhörnchen	Écureuil à trois couleurs	Индонезийская великолепная белка
Trident Leaf-nosed Bat	Dreizack-Blattnase	Asellia à trois endentures	Трехзубчатоносый ложный подковонос
Tristram's Jird	Tristram-Wüstenmaus	Mériones de Tristram	
True Dolphins	Eigentliche Delphine	Delphininés	Настоящие дельфины
– Jumping Mice	– Hüpfmäuse	Zapodinés	Прыгунчики
True's Beaked Whale	True-Wal	Mésoplodon de True	Североатлантический ремнезуб
True Vampire	Gemeiner Vampir	Vampire d'Azara	Большой кровосос
Tube-nosed Fruit Bats	Röhrennasen-Flughunde	Myctimènes	Трубконосые крыланы
Tuco-tucos	Kammratten, Tukotukos	Tuco-tucos	Ктеномииды
Tundra Vole	Nordische Wühlmaus	Campagnol nordique	Полевка-экономка

ENGLISH NAME	GERMAN NAME	FRENCH NAME	RUSSIAN NAME
Tuza	Hamsterratte	Gaufre	Хомячий гофер
Tuzas	Riesentaschenratten	Gaufres	
Two-toed Ant-eater	Zwergameisenbär	Fourmilier didactyle	Двупалый муравьед
– Sloths	Zweifingerfaultiere	Unaus	Двупалые ленивцы
Typical Squirrels	Hörnchen	Écureuils	Беличьи
Unau	Unau	Unau commun	Унау
Upper Amazonian Porcupine	Bergstachler	Porc-épic rougeâtre	Горный иглошерст
Ural Beaver	Uralbiber	Castor d'Ural	Уральский бобр
Vampire	Gemeiner Vampir	Vampire d'Azara	Большой кровосос
– Bats	Gemeine Vampire	Vampires	
Vampires	Echte Vampire	–	Кровососы
Velvety Free-tailed Bats	Samt-Fledermäuse	Molosses	
Vespertilionid Bats	Glattnasen-Verwandte, Glattnasen-Fledermäuse	Vespertilionidés	Гладконосые летучие мыши, Обыкновенные летучие мыши
Viscacha	Viscacha	Lagostome des Pampas	Вискаша
– Rat	Viscacharatte	Rat minime	Вискашевидная крыса
Viscachas and Chinchillas	Chinchillaartige, Chinchillas	Chinchillidés	Шиншилловые, Шиншиллы
Vlei and Karroo Rats	Afrikanische Lamellenzahnratten	Otomyinés	Африканские ушастые крысы
Vole Rat	Schermaus	Campagnol terrestre	Водяная крыса
Voles	Eigentliche Wühlmäuse, Feldmäuse		Полевки, Серые полевки
Votsotsa	Votsotsa		Вотсотса
Wahlberg's Epauletted Fruit Bat	Wahlberg-Epauletten-Flughund	Épomophore de Wahlberg	Крылан Вальберга
Water Bat	Wasserfledermaus	Vespertilion de Daubenton	Водяная ночница
– Hogs	Riesennager	Hydrochéridés	Водосвинки
– Rats	Schwimmratten	Hydromyinés	Австралийские водяные крысы
– Vole	West-Schermaus	Campagnol amphibie	Водяная крыса
– Voles	Schermäuse		Водяные крысы
Welwitsch's Bat	Welwitsch-Fledermaus	Vespertilion de Welwitsch	
West African Brush-tailed Porcupine	Westafrikanischer Quastenstachler	Athérure africain	Африканский кистехвост
– – White Dolphin	Kamerunfluß-Delphin	Sotalie de Teusz	Камерунский речной дельфин
Western African Striped Squirrel	Westliches Rotschenkelhörnchen	Écureuil rayé d'Afrique occidental	
Western Chipmunks	Streifenhörnchen	Eutamias	Бурундуки
– Harvest Rat	Westliche Erntemaus	Souris occidentale des moissons	
– House Mouse	– Hausmaus		Западноевропейская домовая мышь
– Pocket Gopher	Gebirgs-Taschenratte	Gaufre à poche des montagnes	Горный гофер
Whalebone Whales	Bartenwale	Baleines	Усатые киты
Whales	Waltiere	Cétacés	Китообразные
Whiskered Bat	Bartfledermaus	Vespertilion à moustaches	Усатая ночница
White and Narwhales	Gründelwale	Monodontidés	Белухи
White-beaked Dolphin	Weißschnauzendelphin	Lagénorhynque à rostre blanc	Беломордый дельфин
White-footed Mice	Weißfußmäuse	Souris à pattes blanches	
– Mouse	Weißfußmaus	Souris à pattes blanches	Оленья мышь
Whitehead's Dwarf Squirrel	Whitehead-Zwerghörnchen	Écureuil pygmée de Whitehead	Белка Уайтхеда
– Harpy Fruit Bat	– Spitzzahn-Flughund	Harpionyctère de Whitehead	
White-sided Dolphin	Weißseitendelphin	Lagénorhynque à bec pointu	Белобокий дельфин
White-tailed Antelope Squirrel	Weißschwanz-Antilopenziesel	Spermophile d'antilope à queue blanche	Белохвостый антилоповый суслик
– Porcupine	– Stachelschwein	Porc-épic à queue blanche	Индийский дикобраз
– Prairie Dog	– Präriehund	Cynomys de Gunnison	Белохвостая луговая собачка
– Rat	Weißschwänziger Hamster	Hamster à queue blanche	Белохвостый хомяк
Whitethroat Wood Rat	Wüstenratte	Rat des steppes	Пустынный кустовый хомячок
White Whale	Weißwal	Dauphin blanc	Белуха
White-winged Vampire Bat	Weißschwingen-Vampir	Vampire d'ailes blanches	
Wild Cavy	Aperea, Tschudi-Meerschweinchen	Cobaye sauvage	Аперея, Перуанская морская свинка
Wood and Field Mice	Wald- und Feldmäuse	Mulots	Лесные и полевые мыши
Woodchuck	Waldmurmeltier	Monax	Лесной сурок
Woodland Jumping Mice	Waldhüpfmäuse	Zapodes des bois	
– – Mouse	Waldhüpfmaus	Zapode des bois	Лесной прыгунчик
Wood Lemming	Waldlemming	Lemming des forêts	Лесной лемминг
– Rats	Buschratten	Rats des bois	Кустовые хомячки
Woolly Flying Squirrel	Fels-Gleithörnchen	Écureuil volant cendré	Каменная летяга
– Prehensile-tailed Porcupine	Wollgreifstachler	Porc-épic laineux	Мохнатый цепкохвост
Wrinkled-faced Bat	Greisengesicht	Vespertilion ridé	

ENGLISH NAME	GERMAN NAME	FRENCH NAME	RUSSIAN NAME
Wrinkled-lipped Bats	Bulldogg-Fledermäuse	Molossidés	Бульдоговые летучие мыши
Xenarthra	Nebengelenktiere	Xénarthrés	
Yellow-bellied Marmot	Gelbbäuchiges Murmeltier	Marmotte à ventre fauve	Желтобрюхий сурок
Yellow-eared Bat	Gelbohr-Fledermaus	Vespertilion bilobée	
Yellow-haired Fruit Bat	Palmenflughund	Rousette paillée	Пальмовый крылан
Yellow-necked Field Mouse	Gelbhalsmaus	Mulot fauve	Желтогорлая мышь
Yellow Pine Chipmunk	Gelber Fichten-Chipmunk	Néotamia jaune	Желтый чипмунк
Yellow Pocket Gopher	Mexikanische Taschenratte	Rat à poche mexicain	Мексиканский гофер
– – Gophers	Tuzas	Rats à poche	
Yellow-shouldered Bat	Gelbschulter-Blattnase	Sturnire fleur-de-lys	
– Bats	– Blattnasen	Sturnires	
Yellow-winged Bats	Großblattnasen	Mégadermes	Мегадермы
Zagoutis	Zagutis	Plagiodontes	Загути
Zenker's Small Flying Squirrel	Zenker-Gleitbilch	Anomalure nain de Zenker	Соня-летяга Ценкера
– Fruit Bat	– Flughund	Scotonyctère de Zenker	Крылан Ценкера
Zokors	Blindmulle		Цокоры

2. German—English—French—Russian

Unterartnamen werden meist aus den Artnamen durch Voranstellen von Eigenschaftswörtern oder geographischen Bezeichnungen gebildet. In diesem Teil des Tierwörterbuchs sind so gebildete deutsche Unterartnamen sowie die wissenschaftlichen Unterartnamen in der Regel nicht aufgeführt.

GERMAN NAME	ENGLISH NAME	FRENCH NAME	RUSSIAN NAME
Aalstrich-Klettermäuse	African Climbing Mice	Souris des bananiers	
Abendsegler	Noctule Bats	Noctules	Вечерницы
Abrocoma cinerea	Chinchilla Rat	Rat-chinchilla cendré	Шиншилловая крыса
Abrocomidae	Rat Chinchillas	Abrocomidés	Шиншилловые крысы
Acanthion brachyura	Malayan Porcupine	Acanthion de Malaysie	
– *javanicum*	Javanese Porcupine	– de Java	Яванский дикобраз
Acomys	Spiny Mice	Souris épineuses	Ежовые мыши
– *cahirinus*	Egyptian Spiny Mouse	– épineuse	Египетская ежовая мышь
– *dimidiatus*	Sinai Spiny Mouse	– – de Sinai	Бледная ежовая мышь
– *minous*	Cretan Spiny Mouse	– – de Crête	Критская ежовая мышь
Aconaemys fuscus	South American Rock Rat	Rat foncu	
Acouchi	Acouchi	Acouchi	Акучи
Ägyptische Klappnase	Larger Rat-tailed Bat	Rhinopôme microphylle	Египетский ланцетонос
– Stachelmaus	Egpytian Spiny Mouse	Souris épineuse	Египетская ежовая мышь
Ägyptischer Flughund	– Fruit Bat	Roussette d'Égypte	Египетский крылан
Ährenmaus	Eastern House Mouse		Средиземноморская домовая мышь
Affen	Monkeys	Singes	Обезьяны
Afrikanische Bilche	African Dormice		Африканские сони
– Borstenhörnchen	– Ground Squirrels	Écureuils foisseurs	Африканские земляные белки
– Bürstenhaarmäuse	Harsh-furred Mice	Rats à pelage en brosse	
– Felsenratten	Rock Rats		Афирканские горные мыши
– Kleinmäuse	African Native Mice	Souris d'Afrique	Африканские мыши-малютки
– Lamellenzahnratten	Vlei and Karroo Rats	Otomyinés	Африканские ушастые крысы
– Maulwurfsratten	African Mole Rats	Rats taupes africains	Африканские кротовидные крысы
– Rotnasenratte	Rufous-nosed Rat	Rat à museau roux	Африканская красноносая крыса
– Streifen-Grasmäuse	Striped Grass Mice	Rats rayés	Африканские полосатые мыши
– Striemen-Grasmaus	– Field Mouse	Rat rayé du champ	
– Tüpfel-Grasmäuse	– Grass Mice	Rats rayés	Африканские полосатые мыши
– Zwergmaus	Small African Native Mouse	Souris naine d'Afrique	Африканская мышь-малютка
Afrikanischer Langzungen-Flughund	African Long-tongued Fruit Bat	Mégaloglosse de Woermann	Африканский длинноязычный крылан
Afrikanisches Stachelschwein	– Porcupine	Porc-épic d'Afrique	Африканский дикобраз
– Zwerghörnchen	– Pygmy Squirrel	Écureuil nain du Gabon	Африканская карликовая белка
Agutis	Agoutis	Dasyproctidés	Агутиевые
Alactagulus pygmaeus	Jerboa	Gerboise lièvre	Тарбаганчик
Alexandriner Hausratte	Roof Rat	Rat d'Aléxandrie	Александрийская черная крыса

GERMAN NAME	ENGLISH NAME	FRENCH NAME	RUSSIAN NAME
Allactaga bullata	Gobi Jerboa	Gerboise du Gobi	Монгольский земляной заяц
– *jaculus*	Five-toed Jerboa	– à cinq doigts	Большой тушканчик
– *sibirica*	Sibirian Jerboa	– de Sibérie	Тушканчик-прыгун
Allocricetulus curtatus	Short Dwarf Hamster	Hamster nain de Mongolie	Монгольский хомячок
– *eversmanni*	Eversman's Dwarf Hamster	– – d'Eversman	Хомячок Эверсманна
Alpenmurmeltier	Alpine Marmot	Marmotte des Alpes	Альпийский сурок
Alticola	High-mountain Voles		Каменные полевки
– *roylei*	Royle's High-mountain Vole	Campagnol de Royle	Серебристая полевка
– *strelzowi*	High-mountain Vole	– des montagnes	Плоскочерепная полевка
Altwelt-Stachelschweine	Old World Porcupines	Hystricidés	Дикобразы Старого Света
Amazonas-Delphin	Amazonian Dolphin	Inie de Geoffroy	Амазонская иния
Amazonas-Sotalia	Buffeo Negro	Sotalie	Амазонский речной дельфин
Ameisenbären	Ant-eaters	Fourmiliers	Муравьеды
Amerikanische Erntemäuse	American Harvest Mice	Souris des moissons d'Amérique	
Amerikanische Haftscheiben-Fledermäuse	Disk-winged Bats	Thyropteridés	Американские присосковые летучие мыши
– Zwergmaus	Pygmy Mouse		Карликовый мышевидный хомячок
Angola-Flughund	Angola Fruit Bat	Roussette d'Angola	Ангольский крылан
Anomaluridae	Scaly-tailed Squirrels	Anomaluridés	Шипохвостые белки-летяги
Anomalurus beecofti	Beecroft's Scaly-tailed Squirrel	Anomalure de Beecroft	Шипохвостая белка Бикрофта
– *beldeni*	Belden's Scaly-tailed Squirrel	– de Belden	Шипохвостая белка Бельдена
– *erythronotus*	Red-backed Scaly-tailed Squirrel	– à dos rouge	
– *fraseri*	Fraser's Scaly-tailed Squirrel	– de Fraser	Шипохвостая белка Фразера
– *jacksoni*	Jackson's Scaly-tailed Squirrel	– de Jackson	Шипохвостая белка Джексона
– *neavei*	Neave's Scaly-tailed Squirrel	– de Neave	
– *peli*	Pel's Scaly-tailed Squirrel	– de Pel	Шипохвостая белка Пеля
– *pusillus*	Pygmy Scaly-tailed Squirrel	– nain	Карликовая шипохвостая белка
Anthops ornatus	Flower-faced Bat	Anthops orné	Цветконосый ложный подковонос
Antrozous pallidus	Pallid Bat	Oreillard pâle	Бледная летучая мышь
Aperea	Aperea	Cobaye sauvage	Аперея
Aplodontia rufa	Mountain Beaver	Castor de montagne	Горный бобр
Aplodontidae	– Beavers	Castors de montagne	Горные бобры
Apodemus	Wood and Field Mice	Mulots	Лесные и полевые мыши
– *agrarius*	Striped Field Mouse	Mulot rayé	Полевая мышь
– *flavicollis*	Yellow-necked Field Mouse	– fauve	Желтогорлая мышь
– *geisha*	Geisha Wood Mouse	– de Geisha	Японская мышь
– *mystacinus*	Broadtoothed Field Mouse	– rupestre	Горная мышь
– *sylvaticus*	Long-tailed Field Mouse	– sylvestre	Лесная мышь
Artibeus	Neotropical Fruit-eating Bats	Artibées	Тупомордые листоносы
– *jamaicensis*	Mexican Fruit-eating Bat	Artibée de la Jamaique	Ямайский тупомордый листонос
– *lituratus*	Big Fruit-eating Bat	Grande Artibée	Большой тупомордый листонос
– *nanus*	Dwarf Fruit-eating Bat	Artibée naine	Карликовый тупомордый листонос
Arvicola sapidus	– Vole	Campagnol amphibie	Западноевропейская водяная крыса
– *terrestris*	Vole Rat	– terrestre	Водяная крыса
Asellia tridens	Trident Leaf-nosed Bat	Asellia à trois endentures	Трехзубчатоносый ложный подковонос
Asiatische Zwerghörnchen	Oriental Pygmy Squirrels	Écureuils pygmées	Азиатские карликовые белки
Assapan	Southern Flying Squirrel	Écureuil volant du sud	Североамериканская летяга
Atherurus africanus	West African Brush-tailed Porcupine	Athérure africain	Африканский кистехвост
– *macrourus*	Asiatic Brush-tailed Porcupine	– à longue queue	Малайский кистехвост
Australische Gespenstfledermaus	Australian Ghost Bat	Macroderme d'Australie	
– Häschenratten	House-building Jerboa Rats		Австралийские зайцевидные мыши
– Hüpfmäuse	Australian Hopping Mice		Австралийские прыгающие мыши
– Kleinmäuse	– Native Mice	Souris d'Australie	Австралийские малые мыши
Australopithecus	Ape-man	Australopithèque	Австралопитеки

GERMAN NAME	ENGLISH NAME	FRENCH NAME	RUSSIAN NAME
Azara-Aguti	Azara Agouti	Agouti d'Azara	Азарская агути
Bahama-Ferkelratte	Bahaman Hutia	Rat des Bahamas	Багамская крыса
Bailey-Taschenmaus	Bailey's Pocket Mouse	Souris à poche de Bailey	Мешетчатый тушканчик Бейлейя
Baiomys taylori	Pygmy Mouse		Карликовый мышевидный хомячок
Baird-Wal	Baird's Beaked Whale	Baleine de Baird	Северный плавун
Baktrische Maus	Bactrian House Mouse		Центральноазиатская домовая мышь
Balaena mysticetus	Bowhead	Baleine boréale	Гренландский кит
Balaenidae	Right Whales	Baleines	Гладкие киты
Balaenoptera	Rorquals	Rorquals	Настоящие полосатики
– *acutorostrata*	Lesser Rorqual	Petit Rorqual	Малый остромордый полосатик
– *borealis*	Sei Whale	Balénoptère boréal	Сайдяной кит
– *edeni*	Bryde's Whale	– de Bryde	Полосатик Брайда
– *musculus*	Blue Whale	– bleu	Синий кит
– *physalus*	Common Rorqual	Rorqual commun	Сельдяной полосатик
Balaenopteridae	Finback Whales	Rorquals	Полосатики
Balantiopteryx	Least Sac-winged Bats		Плосконосые свободнохвосты
Bambus-Fingerratte	Tree-Rat	Rat du bambou	
Bambus-Fledermaus	Flat-headed Bat	Vespertilion du bambou	Бамбуковая летучая мышь
Bambusratten	Bamboo Rats	Rats des bambous	Бамбуковые крысы
Bananenfledermaus	Banana Bat	Chauve-souris des bananes	Банановый листонос
Bananenfledermäuse	– Bats	Chauves-souris des bananes	Банановые листоносы
Bananen-Zwergfledermaus	– Bat	Pipistrelle naine à ailes brunes	Банановый нетопырь
Bandicota bengalensis	Bengali Bandicoot Rat	– rat du Bengale	Бенгальская бандикутовая крыса
– *indica*	Bandicoot Rat	– – d'Inde	Индийская бандикутовая крыса
Bandikutratte	– Rat	– – d'Inde	Индийская бандикутовая крыса
Bandikutratten	– Rats	– rats	Бандикутовые крысы
Barbastella	Barbastelles	Barbastelles	Широкоушки
– *barbastellus*	Barbastelle	Barbastelle d'Europe	Европейская широкоушка
Bartenwale	Baleen Whales	Baleines	Усатые киты
Bartfledermaus	Whiskered Bat	Vespertilion à moustaches	Усатая ночница
Bathyergidae	Mole Rats	Rats taupes africains	Землекопы
Bathyergus suillus	Cape Mole Rat	Fouisseur	
Baumhörnchen	Tree Squirrels		Настоящие белки
Baummäuse	African Tree Mice	Dendromurinés	Африканские древесные мыши
Baum- und Ferkelratten	Hutias		Ежовые крысы
Baumschläfer	Forest Dormouse	Lérotin	Лесная соня
Baumstachler	New World Porcupines	Erethizontidés	Иглошерсты
Baumwollratte	Andean Swamp Rat	Sigmodon velu	
Baumwollratten	Cotton Rats	Sigmodons	Хлопковые хомячки
Beamys	Long-tailed Pouched Rats		Малые хомяковидные крысы
Bechstein-Fledermaus	Bechstein's Bat	Vespertilion de Bechstein	Ночница Бехштейна
Beecroft-Dornschwanzhörnchen	Beecroft's Scaly-tailed Squirrel	Anomalure de Beecroft	Шипохвостая белка Бикрофта
Belden-Dornschwanzhörnchen	Belden's Scaly-tailed Squirrel	– de Belden	Шипохвостая белка Бельдена
Beluga	White Whale	Dauphin blanc	Белуха
Berardius	Giant Bottle-nosed Whales		Плавуны
– *arnouxii*	Arnoux's Whale	Baleine d'Arnoux	Южный плавун
– *bairdi*	Baird's Beaked Whale	– de Baird	Северный плавун
Berglemming	Norway Lemming	Lemming des toundras	Норвежский лемминг
Bergmeerschweinchen	Rock Cavies	Cobayes des roches	Горные свинки
Bergpaka	Mountain Paca	Paca des montagnes	Горная пака
Bergstachler	Upper Amazonian Porcupine	Porc-épic rougeâtre	Горный иглошерст
Bergviscachas	Mountain Chinchillas	Lagostomes des montagnes	Пушаки
Biber	Beavers, Beaver	Castors, Castor	Бобры, Бобр
Biberhörnchen	Mountain Beaver	Castor de montagne	Горный бобр
Biberratten	Nutrias	Ragondins	Бобровые крысы
Bilch	Fat Dormouse	Loir gris	Соня-полчок
Bilche	Dormice	Loirs	Сони
Birkenmaus	Northern Birch Mouse	Siciste de bouleaux	Лесная мышовка
Bisamratte	Muskrat	Rat musqué	Ондатра
Bisamratten	Muskrats	Ondatras	Мускусные крысы
Blasse Fledermaus	Pallid Bat	Oreillard pâle	Бледная летучая мышь
– Känguruhmaus	Pale Kangaroo Mouse	Souris kangourou pâle	Бледный мешетчатый прыгунчик

GERMAN NAME	ENGLISH NAME	FRENCH NAME	RUSSIAN NAME
Blattnasen	American Leaf-nosed Bats		Листоносы
Blauwal	Blue Whale	Balénoptère bleu	Синий кит
Blau-Weißer Delphin	Blue-white Dolphin	Dauphin bleu-blanc	Однополосый продельфин
Bleßmulle	Blesmols	Rats taupes du Cap	
Blindmäuse	Palaearctic Mole Rats	– taupes	Слепыши
Blindmull	North Altai Zokor	Rat taupe	Алтайский цокор
Blindmulle	Mole Mice		Цокоры
Blumennasen-Fledermaus	Flower-faced Bat	Anthops orné	Цветконосый ложный подковонос
Bobak	Bobac Marmot	Marmotte Bobac	Степной сурок
Boehm-Hörnchen	Boehm's African Bush Squirrel	Écureuil de Boehm	
Bonobo	Pygmy Chimpanzee	Chimpansé nain	Бонобо
Bori	Bori	Bori	Бори
Borkenratten		Phloeomyinés	Азиатские древесные крысы
Borneo-Hörnchen	Groove-toothed Squirrel	Reithrosciure de Borneo	Борнеоская белка
Borsten-Baumstachler	Thin-spined Porcupine	Porc-épic épineux	Крысохвостый дикобраз
Borstengürteltiere	Six-banded Armadillos	Tatous à six bandes	
Borstenhörnchen	Bristly Ground Squirrels		Земляные белки
Borten-Fledermaus	Braided Free-tailed Bat	Tataride bordée	
Brachytarsomys albicauda		Souris à pattes courtes	Коротколапая мадагаскарская крыса
Bradypodidae	Tree Sloths	Bradypodidés	Ленивцы
Bradypus	Three-toed Sloths	Bradypes	Трехпалые ленивцы
– *cuculliger*		Bradype à capuchon	Капюшонный ленивец
– *torquatus*	Necklace Sloth	– à collier	Ошейниковый ленивец
– *tridactylus*	Three-toed Sloth	Bradype	Ай
Brandmaus	Striped Field Mouse	Mulot rayé	Полевая мышь
Brandts Steppenwühlmaus	Brandt's Vole	Campagnol de Brandt	Полевка Брандта
Brasilhörnchen	Brazilian Squirrel	Écureuil de la Guyane	Бразильская белка
Braunes Langohr	Long-eared Bat	Oreillard commun	Обыкновенный ушан
– Zwerghörnchen	Brown Dwarf Squirrel	Écureuil pygmée brun	
Braunzottiges Borstengürteltier	Hairy Armadillo	Tatou velu	Волосатый броненосец
Breitflügel-Fledermäuse	Big Brown Bats	Sérotines	Кожаны
Brillen-Blattnase	Seba's Short-tailed Bat	Vespertilion à nez plat	Очковый листонос
Brillenschweinswal	Spectacled Porpoise	Marsouin des lunettes	Очковая морская свинья
Brydewal	Bryde's Whale	Balénoptère de Bryde	Полосатик Брайда
Buckelwal	Humpback Whale	Mégaptère	Горбач
Bulldogg-Fledermaus	European Free-tailed Bat	Tataride bouledogue	Широкоухий складчатогуб
Bulldogg-Fledermäuse	Free-tailed Bats	Molossidés	Бульдоговые летучие мыши
Bunte Lanzennase	Coloured Bat	Phyllostome coloré	
Burmeister-Gürtelmull	Greater Pichiciago	Grand Chlamyphore	Плащеносец Бурмейстера
Burmeister-Schweinswal	Burmeister's Porpoise	Marsouin spinipenne	Морская свинья Бурмейстера
Burmeisteria retusa	Greater Pichiciago	Grand Chlamyphore	Плащеносец Бурмейстера
Burunduk	Burunduk		Бурундук
Buschratten	Wood Rats	Rats des bois	Кустовые хомячки
Buschschwanzratte	Bushtail Wood Rat		Серый кустовый хомячок
Büttikofer-Epauletten-Flughund	Buttikofer's Epauletted Fruit Bat	Épomophore de Buttikofer	Крылан Бютикофера
Cabassous	Eleven-banded Armadillos	Tatous à onze bandes	Голохвостые броненосцы
– *hispidus*	Spiny Armadillo	Tatou épineux	Малый голохвостый броненосец
– *unicinctus*	Eleven-banded Armadillo	– à onze bandes	Большой голохвостый броненосец
Callosciurus notatus	Tricoloured Squirrel	Écureuil à trois couleurs	Индонезийская великолепная белка
Calomyscus bailwardi	Mouselike Hamster		Мышевидный хомячок
Cannomys badius	Lesser Bamboo Rat	Petit Rat des bambous	
Capromyidae	Hutias		Ежовые крысы
Capromys melanurus	Black-tailed Hutia	Rat à queue noire	Чернохвостая кубинская крыса
– *nana*	Tiny Hutia	– nain	Карликовая кубинская крыса
– *pilorides*	Cuban Hutia	– poilé	Хутия-конга
– *prehensilis*	Prehensile-tailed Hutia	– à queue préhensile	Хутия-карабали
Capybara	Capybara	Capybara	Капибара
Cardiocraniinae	Dwarf Jerboas	Cardiocraninés	Карликовые тушканчики
Cardiocranius	Five-toed Dwarf Jerboas		Пятипалые карликовые тушканчики
Carollia perspicillata	Seba's Short-tailed Bat	Vespertilion à nez plat	Очковый листонос
Carolliinae	Short-tailed Leaf-nosed Bats	Vespertilions à nez plat	Короткохвостые вампиры
Castor fiber	Beaver	Castor	Бобр
– – *albicus*	Elbe Beaver	– de l'Elbe	Эльбский бобр
– – *birulai*	Mongolian Beaver	– de Mongolie	Монгольский бобр

GERMAN NAME	ENGLISH NAME	FRENCH NAME	RUSSIAN NAME
– – *caecator*	Newfoundland Beaver	– de Terre-Neuve	Ньюфаундлендский бобр
– – *canadensis*	Canadian Beaver	– de Canada	Канадский бобр
– – *frondator*	Rio-Grande Beaver	– de Rio-Grande	Бобр Рио-Гранде
– – *galliae*	Rhone Beaver	– du Rhône	Ронский бобр
– – *michiganensis*	Michigan Beaver	– de Michigan	Мичиганский бобр
– – *pohlei*	Ural Beaver	– d'Ural	Уральский бобр
– – *subauratus*	Golden-bellied Beaver		Золотистый бобр
Castoridae	Beavers	Castors	Бобры
Castoroidea	Beavers	Castors	Бобровые
Catarrhina	Old World Monkeys	Catarrhiniens	Узконосые обезьяны или Обезьяны Старого Света
Cavia	Guinea Pigs	Cobayes	
– *aperea*	Aperea	Cobaye sauvage	Аперея
– – *porcellus*	Guinea Pig	Cobaye	Домашняя морская свинка
– – *tschudii*	Wild Cavy	– sauvage	Перуанская морская свинка
Caviidae	Cavies	Caviidés	Морские свинки
Caviinae	Cavies	Caviinés	Настоящие морские свинки
Caviomorpha	Cavy-like Rodents		Свинкообразные
Cayenneratte	Cayenne Spiny Rat	Rat de Guyane	Гвианская крыса
Centurio senex	Wrinkled-faced Bat	Vespertilion ridé	
Cephalorhynchus	Commerson's Dolphins	Dauphins de Commerson	Трезубцевые дельфины
– *commersonii*	– Dolphin	Dauphin de Commerson	Трезубцевый дельфин Коммерсона
Cetacea	Whales	Cétacés	Китообразные
Chaetomys subspinosus	Thin-spined Porcupine	Porc-épic épineux	Крысохвостый дикобраз
Cheiromeles parvidens	Necklace Hairless Bat	Cheiromèle à collier	
– *torquatus*	Naked Bat	– nue	Голая летучая мышь
Chilonycterinae	Mustache Bats	Chilonycterinés	Листобороды
Chinchilla	Chinchillas	Chinchillas	Шиншиллы
– *chinchilla*	Short-tailed Chinchilla	Chinchilla à queue courte	Короткохвостая шиншилла
– *laniger*	Chinchilla	– à longue queue	Длиннохвостая шиншилла
Chinchillaartige	Viscachas and Chinchillas		Шиншилловые
Chinchillaratte	Chinchilla Rat	Rat-chinchilla cendré	Шиншилловая крыса
Chinchillaratten	Rat Chinchillas	Abrocomidés	Шиншилловые крысы
Chinchillas	Viscachas and Chinchillas	Chinchillidés	Шиншиллы
– i. e. S.	Chinchillas	Chinchillas	Шиншиллы
Chinchillidae	Viscachas and Chinchillas	Chinchillidés	Шиншиллы
Chinchilloidea	Viscachas and Chinchillas		Шиншилловые
Chinesische Bambusratte	China Bamboo Rat	Rat des bambous Chinois	Китайская ризомиида
– Rothörnchen	Rock Squirrels		Китайские красные белки
– Zwergschlafmaus	Chinese Pygmy Dormouse	Souris naine de Chine	Китайская карликовая соня
Chinesischer Blindmull	Manchurian Zokor	Zokor de Mandchourie	Северокитайский цокор
– Flußdelphin	Chinese River Dolphin	Dauphin d'eau douce de Chine	Китайский речной дельфин
– Weißer Delphin	– White Dolphin	Sotalie de Chine	Китайский бледный речной дельфин
– Zwerghamster	– Ratlike Hamster	Hamster nain de Chine	Китайский хомячок
Chinesisches Ohren-Schuppentier	– Pangolin	Pangolin à queue courte	Индостанский панголин
Chipmunks	Eastern Chipmunks	Tamias	Чипмунки
Chiropodomys gliroides	Pencil-tailed Tree Mouse	Souris d'arbre de Malaysie	Малайская древесная крыса
Chiroptera	Bats	Chiroptères	Рукокрылые
Chlamyphorina	Pichicies	Chlamyphores	Плащеносные броненосцы
Chlamyphorus truncatus	Lesser Pichiciago	Chlamyphore tronqué	Аргентинский плащеносец
Choeronycteris	Mexican Long-nosed Bats	Chauves-souris du Mexique	Длинноносые листоносы
– *mexicana*	– – Bat	Chauve-souris du Mexique	Длинноносый листонос
Choloepus	Two-toed Sloths	Unaus	Двупалые ленивцы
– *didactylus*	Unau	Unau commun	Унау
– *hoffmanni*	Hoffmann's Two-toed Sloth	– d'Hoffmann	Двупалый ленивец Гоффманна
Citellus	Ground Squirrels	Sousliks	Суслики
– *beecheyi*	California Ground Squirrel	Spermophile de Californie	Калифорнийский суслик
– *citellus*	European Souslik	Souslik d'Europe	Серый суслик
– *eversmanni*	Eversman's Souslik	Spermophile d'Eversman	Длиннохвостый суслик
– *franklinii*	Franklin's Ground Squirrel	– de Franklin	Суслик Франклина
– *fulvus*	Aral Yellow Souslik	– jaune	Суслик-песчаник
– *lateralis*	Golden-mantled Ground Squirrel	– à manteau doré	Золотистоспинный суслик
– *leucurus*	White-tailed Antelope Squirrel	– d'aptilope à queue blanche	Белохвостый антилоповый суслик
– *nelsoni*	Nelson's Souslik	Souslik de Nelson	Калифорнийский антилоповый суслик

GERMAN NAME	ENGLISH NAME	FRENCH NAME	RUSSIAN NAME
– *pygmaeus*	Little Souslik	– nain	Малый суслик
– *rufescens*	Red Souslik	Spermophile rouge	Большой суслик
– *suslicus*	Spotted Souslik	Souslik tacheté	Крапчатый суслик
– *tereticaudus*	Round-tailed Ground Squirrel	Spermophile à queue ronde	Круглохвостый суслик
– *tridecemlineatus*	Thirteen-striped Ground Squirrel	– à treize bandes	Полосатый американский суслик
– *undulatus*	Arctic Ground Squirrel	Souslik de Parry	Американский длинно-хвостый суслик
Clethrionomys	Red-backed Voles	Campagnols	Лесные полевки
– *glareolus*	Bank Vole	Campagnol roussâtre	Рыжая полевка
– *rufocanus*	Large-toothed Red-backed Vole	– gris-roux	Красно-серая полевка
– *rutilus*	Northern Red-backed Vole	– boréal	Красная полевка
Clyomys laticeps	Spiny Rat	Rat à grosse tête	
Coëndou	Tree Porcupine	Porcs-épic préhensiles	Коэнду
– *insidiosus*	Woolly Prehensile-tailed Porcupine	Porc-épic laineux	Мохнатый цепкохвост
– *prehensilis*	Prehensile-tailed Porcupine	– préhensile	Бразильский коэнду
Colorado-Chipmunk	Colorado Chipmunk	Néotamia du Colorado	Колорадский чипмунк
Commerson-Delphin	Commerson's Dolphin	Dauphin de Commerson	Трезубцевый дельфин Коммерсона
Crateromys schadenbergi	Bushy-tailed Cloud Rat	Rat d'ecorce de Schadenberg	Древесная крыса Шаденберга
Cratogeomys	Yellow Pocket Gophers	Rats à poche	
– *castanops*	– – Gophers	Rat à poche mexicain	Мексиканский гофер
Cricetidae	Cricetid Rats and Mice	Cricétidés	Хомячьи
Cricetinae		Cricétinés	Хомяки
Cricetini	Hamsters	Hamsters	Хомяки
Cricetomyinae		Cricétomyinés	Хомяковидные крысы
Cricetomys	Giant Pouched Rats	Rats géants	Большие хомяковидные крысы
– *emini*	Emin's Rat	Rat géant d'Emin	
– *gambianus*	Gambia Pouched Rat	– – de Gambia	Гамбийская хомяковидная крыса
Cricetulus	Ratlike Hamsters	Hamsters nains gris	Хомячки
– *barabensis*	– Hamster	Hamster nain de Daourie	Даурский хомячок
– *griseus*	Chinese Ratlike Hamster	– – de Chine	Китайский хомячок
– *lama*	Tibetan Ratlike Hamster	– – du Tibet	Тибетский хомячок
– *longicaudatus*	Long-tailed Hamster	– – à longue queue	Длиннохвостый хомячок
– *migratorius*	Migratory Hamster	– migrateur	Серый хомячок
Cricetus cricetus	Common Hamster	– d'Europe	Обыкновенный хомяк
Cro-Magnon-Mensch	Cro-Magnon Man	Homme de Cro-Magnon	Кроманьонец
Crossomys moncktoni	Monckton's Water Rat		Австралийская крыса Монктона
Cryptomys hottentotus	African Mole Rat	Rat taupe africain	Готтентотский землекоп
Ctenodactylidae	Gundis	Goundis	Гребнепалые крысы
Ctenodactyloidea	Gundis	Goundis	Гребнепаловые
Ctenodactylus	Gundis	Goundis	
– *gundi*	Gundi	Goundi	Гунди
Ctenomyidae	Tuco-tucos	Tuco-tucos	Ктеномииды
Ctenomys	Tuco-tucos	Tuco-tucos	
– *knighti*	Knight's Tuco-tuco	Ctenomys de Knight	Тукотуко
Cuniculinae	Pacas	Cuniculinés	Паки
Cuniculus paca	Paca	Paca	Пака
Cururo	Cururo	Rat bleu	Куруро
Cuvier-Hasenmaus	Mountain Chinchilla	Lagostome des montagnes	Пушак
Cuvier-Schnabelwal	Goosebeak Whale	Baleine du bec des oies	Настоящий клюворыл
Cuviers Zaguti	Haitian Hutia	Plagiodonte d'Haiti	Загути Кювье
Cyclopes didactylus	Two-toed Ant-eater	Fourmilier didactyle	Двупалый муравьед
Cynocephalidae	Flying Lemurs	Cynocephalidés	Шерстокрылы
Cynocephalus temminckii	Malayan Flying Lemur		Шерстокрыл Темминка
– *volans*	Philippines Flying Lemur		Кагуан
Cynomys	Prairie Dogs	Chiens des prairies	Луговые собачки
– *gunnisoni*	White-tailed Prairie Dog	Cynomys de Gunnison	Белохвостая луговая собачка
– *ludovicianus*	Black-tailed Prairie Dog	– social	Чернохвостая луговая собачка
Cynopterinae	Short-nosed Fruit Bats	Cynoptères	Коротконосые крыланы
Cynopterus sphinx	– – Bat	Cynoptère à nez court	Обыкновенный коротконосый крылан
Dachratte	House Rat	Rat commun	Черная крыса
Dactylomys dactylinus	Coro-coro	Dactylomys	
Dall-Hafenschweinswal	Dall's Harbour Porpoise	Marsouin de Dall	Северная белокрылая свинья
Dasypodidae	Armadillos	Tatous	Броненосцы
Dasyprocta	Agoutis	Agoutis	Куцые агути

GERMAN NAME	ENGLISH NAME	FRENCH NAME	RUSSIAN NAME
– *aguti*	Orange-rumped Agouti	Agouti doré	Золотистая агути
– *azarae*	Azara Agouti	– d'Azara	Азарская агути
– *fuliginosa*	Smoky Agouti		Аспидная агути
Dasyproctidae	Agoutis	Dasyproctidés	Агутиевые
Dasyproctinae	Agoutis	Dasyproctinés	Агути
Dasypus	Armadillos	Tatous	
– *kappleri*	Kappler's Armadillo	Tatou de Kappler	Броненосец Капплера
– *novemcinctus*	Nine-banded Armadillo	– à neuf bandes	Девятипоясный броненосец
– *pilosus*	Hairy Armadillo	– poilu	Мохнатый броненосец
– *septemcinctus*	Seven-banded Armadillo	– à sept bandes	Семипоясный броненосец
Daurischer Blindmull	Transbaikal Zokor	Zokor de Transbaikalie	Даурский цокор
– Zwerghamster	Ratlike Hamster	Hamster nain de Daourie	Даурский хомячок
Degu	Degu	Dègue du Chili	Дегу
Delphin	Common Dolphin	Dauphin commun	Североатлантический дельфин-белобочка
Delphinapterus leucas	White Whale	– blanc	Белуха
Delphine	Dolphins	Delphinidés	Дельфины
– i. e. S.	Dolphins	Dauphins	Дельфины-белобочки
Delphinidae	Dolphins	Delphinidés	Дельфины
Delphininae	True Dolphins	Delphininés	Настоящие дельфины
Delphinus	Dolphins	Dauphins	Дельфины-белобочки
– *delphis*	Common Dolphin	Dauphin commun	Североатлантический дельфин-белобочка
Dendromurinae	African Tree Mice	Dendromurinés	Африканские древесные мыши
Dendromus	– Climbing Mice	Souris des bananiers	
Dent-Hufeisennase	Dent's Horseshoe Bat	Rhinolophe de Dent	
Deomys ferrugineus	Congo Forest Mouse	Rat à manteau roux	Насекомоядная лесная мышь
Dermoptera	Flying Lemurs		Шерстокрыловые
Desmodontidae	Vampires	Vampires	Кровососы
Desmodus	Vampire Bats	Vampires	
– *rotundus*	Vampire	Vampire d'Azara	Большой кровосос
Diaemus youngi	White-winged Vampire Bat	– d'ailes blanches	
Dickschwanzmaus	Fat-tailed Mouse	Souris à grosse queue	Толстохвостая песчанка
Diclidurus albus	Ghost Bat	Diclidure de fantôme	
Dicrostonyx	Collared Lemmings		Копытные лемминги
– *groenlandicus*	Greenland Collared Lemming		Гренландский лемминг
– *torquatus*	Arctic Lemming	Lemming arctique	Копытный лемминг
Dinomyidae	Pacaranas	Pacaranas	Диномииды
Dinomyoidea	Pacaranas	Pacaranas	
Dinomys	Pacaranas	Pacaranas	
– *branickii*	Pacarana	Pacarana	Длиннохвостая пака
Diphylla ecaudata	Hairy-legged Vampire Bat	Diphylle sans queue	Малый кровосос
Diplomys caniceps	Arboreal Soft-furred Spiny Rat	Rat épineux à tête grise	
Dipodidae	Jerboas	Gerboises	Тушканчики
Dipodinae	Jerboas	Dipodinés	Трехпалые тушканчики
Dipodomys	Kangaroo Rats	Rats kangourou	Мешетчатые прыгуны
– *agilis*	Pacific Kangaroo Rat	Rat kangourou du Pacifique	Проворный мешетчатый прыгун
– *deserti*	Desert Kangaroo Rat	– – du désert	Пустынный мешетчатый прыгун
– *elator*	Texas Kangaroo Rat	– – de Texas	Техасский мешетчатый прыгун
– *elephantinus*	Big-eared Kangaroo Rat	– – à grandes oreilles	Большеухий мешетчатый прыгун
– *ordii*	Ord's Kangaroo Rat	– – d'Ord	Мешетчатый прыгун Орда
Dipus sagitta	Northern Three-toed Jerboa	Gerboise à pattes rugueuses	Мохноногий тушканчик
Dobsonia	Bare-backed Fruit Bats	Chauve-souris à dos nu	
Dögling	Bottle-nosed Whale	Hypérodon du nord	Высоколобый бутылконос
Dolichotinae	Maras	Dolichotinés	Мары
Dolichotis patagonum	Mara	Mara	Патагонская свинка
Dornschwanzbilch	Flightless Scaly-tailed Squirrel	Zenkerelle	Шипохвостая соня
Dornschwanzbilche	– – Squirrels	Zenkerelles	Шипохвостые сони
Dornschwanzhörnchen	Scaly-tailed Squirrels	Anomaluridés	Шипохвостые белки-летяги
Dreibinden-Kugelgürteltier	Apara	Apar de Buffon	Трехпоясный броненосец
Dreiblatt-Fledermaus	Persian Leaf-nosed Bat	Triaenops de Perse	Персидский ложный подковонос
Dreifarbenhörnchen	Tricoloured Squirrel	Écureuil à trois couleurs	Индонезийская великолепная белка
Dreifarbige Haftscheiben-Fledermaus	Spix's Disk-winged Bat	Vespertilion tricolore	Трехцветная летучая мышь
Dreifingerfaultier	Three-toed Sloth	Bradype	Ай

GERMAN NAME	ENGLISH NAME	FRENCH NAME	RUSSIAN NAME
Dreifingerfaultiere	– Sloths	Bradypes	Трехпалые ленивцы
Dreizack-Blattnase	Trident Leaf-nosed Bat	Asellia à trois endentures	Трехзубчатоносый ложный подковонос
Dryomys nitedula	Forest Dormouse	Lérotin	Лесная соня
Dshungarischer Zwerghamster	Dwarf Hamster	Hamster nain de Djoungarie	Джунгарский хомячок
Dünnschwanz-Baumschläfer	Asiatic Dormouse	Myomime à queue fine	Мышевидная соня
Echimyidae	Spiny Rats	Échimyidés	Цепкохвостые щетинистые крысы
Echimyinae	Spiny Rats	Échimyinés	Щетинистые крысы
Echimys	Crested Spiny Rats	Échimys	
– *armatus*	Armoured Spiny Rat	– armé	
Echinoprocta rufescens	Upper Amazonian Porcupine	Porc-épic rougeâtre	Горный иглошерст
Echte Lemminge	Brown Lemmings		Настоящие лемминги
– Vampire	Vampires	Vampires	Кровососы
Edentata	Edentates	Édentés	Неполнозубые
Ehrenberg-Blindmaus	Ehrenberg's Mole Rat	Spalax d'Ehrenberg	Слепыш Эренберга
Eichhörnchen	European and American Tree Squirrels, Red Squirrel	Écureuils, Écureuil commun	Белки, Обыкновенная белка
Eichkätzchen	Red Squirrel	Écureuil commun	Обыкновенная белка
Eidolon helvum	Straw-coloured Bat	Roussette paillée	Пальмовый крылан
Eigentliche Agutis	Agoutis	Dasyproctinés	Агути
– Bilche	Eurasian Dormice	Glirinés	Настоящие сони
– Blattnasen	Big-eared Leaf-nosed Bats	Phyllostomes	Настоящие листоносы
– Delphine	True Dolphins	Delphininés	Настоящие дельфины
– Dornschwanzhörnchen	Scaly-tailed Squirrels	Anomalures	Шипохвостые белки
– Fingerratte	Coro-coro	Dactylomys	
– Flughunde	Flying Foxes	Roussettes	Летучие собаки
– Großblattnasen	False Vampires	Mégadermes	Лироносы
– Hufeisennasen	Horseshoe Bats	Rhinolophes	Подковоносы
– Hüpfmäuse	True Jumping Mice	Zapodinés	Прыгунчики
– Meerschweinchen	Cavies	Caviinés	Настоящие морские свинки
– Ratten	Rats	Rats proprement dits	Крысы
– Rennmäuse	Gerbils	Gerbilles	Африканские песчанки
Eigentliche Springmäuse	Jerboas	Dipodinés	Трехпалые тушканчики
– Stachelschweine	Large Porcupines	Porcs-épic	Настоящие дикобразы
– Taschenmäuse	Pocket Mice	Souris à poche	Мешетчатые тушканчики
– Wühler		Cricétinés	Хомяки
– Wühlmäuse	Voles		Полевки
Eisgraues Murmeltier	Hoary Marmot	Marmotte grise	Североамериканский серый сурок
Elbebiber	Elbe Beaver	Castor de l'Elbe	Эльбский бобр
Eliomys quercinus	Garden Dormouse	Lérot	Садовая соня
Ellobiini	Mole Lemmings	Rats taupes	Слепушонки
Ellobius	– Lemmings	– taupes	Слепушонки
– *fuscocapillus*	Afghan Mole Lemming	Rat taupe d'Afghan	Афганская слепушонка
– *talpinus*	Mole Lemming	Rat taupe	Обыкновенная слепушонка
Emballonura	Sac-winged Bats		Эмбаллонуры
Emballonuridae	– Bats		Свободнохвостые летучие мыши
Emin-Riesenhamsterratte	Emin's Rat	Rat géant d'Emin	
Entenwale	Bottle-nosed Whales	Baleines à bec	Бутылконосы
Eonycteris	Dawn Bats	Eonyctères	Пещерные крыланы
– *spelaea*	Dobson's Long-tongued Dawn Bat	Eonyctère des cavernes	Пещерный крылан
Eothenomys	Pere David's Voles		Давидовы полевки
– *melanogaster*	– – Vole	Campagnol de Père David	Давидова полевка
Epauletten-Flughunde	Epauletted Fruit Bats	Épomophores	Эполетовые крыланы
Epixerus wilsoni	African Palm Squirrel	Écureuil de Wilson	Гигантская белка Вильсона
Epomophorinae	Epauletted Fruit Bats	Épomophores	Эполетовые крыланы
Epomophorus wahlbergi	Wahlberg's Epauletted Fruit Bat	Épomophore de Wahlberg	Крылан Вальберга
Epomops	Epauletted Fruit Bats	Épomophores	Эполетовые крыланы
– *buettikoferi*	Buttikofer's Epauletted Fruit Bat	Épomophore de Buttikofer	Крылан Бютикофера
– *franqueti*	Epauletted Bat	– de Franquet	Крылан Франкета
Eptesicus	Big Brown Bats	Sérotines	Кожаны
– *fuscus*	– – Bat	Sérotine de maison	
– *nilssoni*	Northern Bat	– boréale	Северный кожанок
– *serotinus*	Serotine Bat	– commune	Поздний кожан
Erdhase	Jerboa	Gerboise lièvre	Тарбаганчик
Erdhörnchen	Northern Ground Squirrels		Сурковые
Erd- und Baumhörnchen	Ground and Tree Squirrels		Земляные и древесные беличьи

GERMAN NAME	ENGLISH NAME	FRENCH NAME	RUSSIAN NAME
Erdmaus	Field Vole	Campagnol agreste	Темная полевка
Eremodipus lichtensteini	Lichtenstein's Jerboa	Gerboise de Lichtenstein	Тушканчик Лихтенштейна
Erethizon dorsatum	North American Porcupine	Porc-épic nord américain	Североамериканский иглошерст
Erethizontidae	New World Porcupines	Erethizontidés	Иглошерсты
Erethizontoidea	– – Porcupines		Дикобразы Нового Света
Eschrichtiidae	Grey Whales	Eschrichtiidés	Серые киты
Eschrichtius gibbosus	– Whale	Baleine grise	Серый кит
Eubalaena australis	Ice Baleen Whale	– australe	Австралийский кит
– *glacialis*	Atlantic Right Whale	– noire	Бискайский кит
– *japonica*	Pacific Right Whale	– de Siebold	Японский кит
Euchoreutinae	Long-eared Jerboas		Длинноухие тушканчики
Eupetaurus cinereus	Woolly Flying Squirrel	Écureuil volant cendré	Каменная летяга
Euphractus	Six-banded Armadillos	Tatous à six bandes	
– *pichiy*	Pygmy Armadillo	Tatou nain	Карликовый броненосец
– *sexcinctus*	Six-banded Armadillo	– à six bandes	Шестипоясный броненосец
– *villosus*	Hairy Armadillo	– velu	Волосатый броненосец
Eurasiatische Zwergmaus	Harvest Mouse	Souris des moissons	Мышь-малютка
Europäisch-Nordafrikanisches Stachelschwein	Porcupine	Porc-épic	Обыкновенный дикобраз
Euryzygomatomys spinosus	Suira	Rat épineux	
Eutamias	Western Chipmunks	Eutamias	Бурундуки
– *sibiricus*	Burunduk		Бурундук
Eversmann-Ziesel	Eversman's Souslik	Spermophile d'Eversman	Длиннохвостый суслик
Eversmann-Zwerghamster	– Dwarf Hamster	Hamster nain d'Eversman	Хомячок Эверсманна
Falsche Vampire	False Vampire Bats	Faux Vampires	
– Vampir-Fledermaus	Great False Vampir	– Vampire commun	Большой вампир
Faltlippen-Fledermäuse	Free-tailed Bats	Tatarides	Складчатогубы
Faultiere	Tree Sloths		Ленивцы
Feldhamster	Common Hamster	Hamster d'Europe	Обыкновенный хомяк
Feld-Hausmaus	Northern House Mouse		Североевропейская домовая мышь
Feldhüpfmäuse	Meadow Jumping Mice	Zapodes des prées	
Feldmaus	Common Vole	Campagnol des champs	Обыкновенная полевка
Feldmäuse	Voles		Серые полевки
Feld-Waldmaus	Long-tailed Field Mouse	Mulot sylvestre	Лесная мышь
Felsenmaus	Broadtoothed Field Mouse	– rupestre	Горная мышь
Felsenmeerschweinchen	Rock Cavies	Cobayes des roches	Горные свинки
Felsen-Moko	– Cavy	Cobaye des roches	Горная морская свинка
Felsenratte	– Rat	Rat typique	Горная мышь
Felsenrattenartige	Cane and Rock Rats		Петромиидовые
Fels-Gleithörnchen	Woolly Flying Squirrel	Écureuil volant cendré	Каменная летяга
Feresa attenuata	Pygmy Killer		Карликовая косатка
Ferkelhörnchen	Long-snouted Squirrel	– d'Heinrich	
Fettmaus	Fat Mouse	Rat adipeux	Жирная древесная мышь
Finnwal	Common Rorqual	Rorqual commun	Сельдяной полосатик
Finnwale	Rorquals	Rorquals	Настоящие полосатики
Fischfledermaus	Mexican Bulldog Bat	Noctilio pêcheuse	Обыкновенный зайцерот
Flachland-Taschenratte	Eastern Pocket Gopher	Gaufre à poche	Равнинный гофер
Fleckendelphine	Spotted Dolphins	Dauphins tachetés	Продельфины
Flederhunde	Fruit Bats	Megachiroptères	Летучие собаки
Fledermäuse	Insectivorous Bats	Chauves-souris insectivores	Летучие мыши
Fledertiere	Bats	Chiroptères	Рукокрылые
Florida-Wasserratte	Round-tailed Muskrat	Rat d'eau de Floride	Флоридская водяная крыса
Flugfuchs	Indian Flying Fox	Roussette géante	Летучая лисица
Flugfüchse	Flying Foxes	Roussettes	Летучие собаки
Flughunde	Fruit Bats	Megachiroptères	Летучие собаки
– i. e. S.	– Bats	Pteropidés	Крыланы
Flußdelphinartige	River Dolphins		Речные дельфины
Franklin-Ziesel	Franklin's Ground Squirrel	Spermophile de Franklin	Суслик Франклина
Franquet-Epauletten-Flughund	Epauletted Bat	Épomophore de Franquet	Крылан Франкета
Fransenfledermaus	Natterer's Bat	Vespertilion de Natterer	Ночница Наттерера
Fraser-Dornschwanzhörnchen	Fraser's Scaly-tailed Squirrel	Anomalure de Fraser	Шипохвостая белка Фразера
Freischwänzige	Sac-winged Bats		Эмбаллонуры
Fruchtratte	Corn Rat	Rat des fruits	
Fruchtvampire	Red-fruit Bats		Плодовые вампиры
Frühmensch	Java Man	Pithécanthrope de Java	Прачеловек
Funambulini	African Tree Squirrels		Пальмовые белки
Funambulus	Palm Squirrels	Écureuils des palmes	
– *palmarum*	– Squirrel	Écureuil des palmes	Пальмовая белка
Fünfzehen-Zwergspringmäuse	Five-toed Dwarf Jerboas		Пятипалые карликовые тушканчики
Funisciurus	African Striped Squirrels	Écureuils rayés d'Afrique	

GERMAN NAME	ENGLISH NAME	FRENCH NAME	RUSSIAN NAME
– *lemniscatus*	Western African Striped Squirrel	Ecureuil rayé d'Afrique occidental	
Furchenwale	Finback Whales	Rorquals	Полосатики
Furipteridae	Smoky Bats	Furipteridés	
Gabelkrallenlemming	Arctic Lemming	Lemming arctique	Копытный лемминг
Gambia-Riesenhamsterratte	Gambia Pouched Rat	Rat géant de Gambia	Гамбийская хомяковидная крыса
Ganges-Delphin	Gangetic Dolphin	Dauphin du Gange	Сусук
Ganges-Delphine	– Dolphins	Platanistidés	Гангские дельфины
Gänseschnabelwal	Goosebeak Whale	Baleine du bec des oies	Настоящий клюворыл
Gartenschläfer	Garden Dormouse	Lérot	Садовая соня
Gebirgs-Chipmunk	Alpine Chipmunk	Néotamia de montagne	Горный чипмунк
Gebirgsmaus	High-mountain Vole	Campagnol des montagnes	Плоскочерепная полевка
Gebirgs-Mosaikschwanz-Riesenratte	Giant Naked-tailed Rat	Rat à queue en mosaique des montagnes	
Gebirgs-Taschenratte	Western Pocket Gopher	Gaufre à poche des montagnes	Горный гофер
Gebirgswühlmäuse	– Voles		Каменные полевки
Geishamaus	Geisha Wood Mouse	Mulot de Geisha	Японская мышь
Gelbbäuchiges Murmeltier	Yellow-bellied Marmot	Marmotte à ventre fauve	Желтобрюхий сурок
Gelber Fichten-Chipmunk	Yellow Pine Chipmunk	Néotamia jaune	Желтый чипмунк
Gelbflüglige Großblattnase	African Yellow-winged Bat	Mégaderme à ailes orangées	
Gelbhalsmaus	Yellow-necked Field Mouse	Mulot fauve	Желтогорлая мышь
Gelbohr-Fledermaus	Tent-building Bat	Vespertilion bilobé	
Gelbschulter-Blattnase	Yellow-shouldered Bat	Sturnire fleur-de-lys	
Gelbschulter-Blattnasen	– Bats	Sturnires	
Gelbziesel	Aral Yellow Souslik	Spermophile jaune	Суслик-песчаник
Gemalte Stacheltaschenmaus	Painted Spiny Pocket Mouse	Souris épineuse à poche	
Gemeine Vampire	Vampire Bats	Vampires	
Gemeiner Vampir	Vampire	Vampire d'Azara	Большой кровосос
Geocapromys brownii	Jamaican Hutia	Rat jamaique	Ямайская крыса
– *ingrahami*	Bahama Hutia	– des Bahamas	Багамская крыса
Geoffroy-Schlitznase	Geoffroy's Slit-faced Bat	Nyctère de Geoffroy	
Geomyidae	Pocket Gophers	Gaufres à poche	Гоферы
Geomyoidea	– Gophers	– à poche	Мешетчатые грызуны
Geomys	Eastern Pocket Gophers	– à poche	
– *bursarius*	– – Gopher	Gaufre à poche	Равнинный гофер
Georhychus capensis	Cape Blesmol	Rat taupe du Cap	Капский пескорой
Gerbillinae	Gerbils	Gerbillinés	Песчанки
Gerbillus	Gerbils	Gerbilles	Африканские песчанки
Gerbillus nanus	Field Gerbil	Gerbille champêtre	Североафриканская песчанка
Gervais-Zweizahnwal	Gervais' Whale	Mésoplodon de Gervais	Европейский ремнезуб
Gescheckte Riesenborkenratte	Slender-tailed Cloud Rat	Rat d'ecorce tacheté	Пятнистая древесная крыса
Geschwänzte Agutis	Acouchis	Acouchis	Хвостатые агути
Gespenst-Fledermaus	Ghost Bat	Diclidure de fantôme	
Gestreifte Palmenhörnchen	Palm Squirrels	Écureuils des palmes	
Gewöhnliche Rundblattnase	South African Lesser Leaf-nosed Bat	Phyllorine de Cafrérie	
Gewöhnlicher Grindwal	Northern Pilot Whale	Grinde	Северная гринда
Gewöhnliches Gleithörnchen	European Flying Squirrel	Palatouche	Обыкновенная летяга
– Stachelschwein	Porcupine	Porc-épic	Обыкновенный дикобраз
Gill-Tümmler	Gill's Dolphin	Dauphin de Gill	Афалина Гилля
Glattdelphine	Right Whale Dolphins	Lissodelphininés	Бесперые дельфины
Glattnasen-Fledermäuse	Vespertilionid Bats	Vespertilionidés	Обыкновенные летучие мыши
Glattnasen-Freischwänze	Sac-winged Bats		Свободнохвостые летучие мыши
Glattnasen-Verwandte	Vespertilionid Bats		Гладконосые летучие мыши
Glattnasige Freischwanz-Fledermäuse	Least Sac-winged Bats		Плосконосые свободнохвосты
Glattwale	Right Whales	Baleines	Гладкие киты
Glaucomys sabrinus	Northern Flying Squirrel	Écureuil volant du nord	
– *volans*	Southern Flying Squirrel	– – du sud	Североамериканская летяга
Gleitbilche	African Small Flying Squirrels	Anomalures nains	Сони-летяги
Gleithörnchen	Flying Squirrels	Écureuils volants	Летяги
Gliridae	Dormice	Loirs	Сони
Glirinae	Eurasian Dormice	Glirinés	Настоящие сони
Glirulus japonicus	Japanese Dormouse	Glirule de Japon	Японская соня
Glis glis	Fat Dormouse	Loir gris	Соня-полчок
Globicephala	Pilot Whales	Globicéphales	Гринды
– *macrorhyncha*	Indian Pilot Whale	Globicéphale d'Inde	Индийская гринда

GERMAN NAME	ENGLISH NAME	FRENCH NAME	RUSSIAN NAME
– *melaena*	Northern Pilot Whale	Grinde	Северная гринда
– *sieboldii*	Pacific Pilot Whale	Globicéphale de Siebold	Черная гринда
Glossophaga	Long-tongued Bats	Glossophages	Глоссофаги
– *soricina*	– Bat	Glossophage de Pallas	Глоссофага
Glossophaginae	– Bats	Glossophages	Длинноязычные вампиры
Gobi-Springmaus	Gobi Jerboa	Gerboise du Gobi	Монгольский земляной заяц
Goldaguti	Orange-rumped Agouti	Agouti doré	Золотистая агути
Goldbiber	Golden-bellied Beaver		Золотистый бобр
Goldman-Stacheltaschenmaus	Goldman's Spiny Pocket Mouse	Souris à poche de Goldman	Иглистый тушканчик Гольдмана
Goldmantelziesel	Golden-mantled Ground Squirrel	Spermophile à manteau doré	Золотистоспинный суслик
Goldmaus	Golden Mouse	Souris dorée	Золотистый мышевидный хомячок
Grabflatterer	Tomb Bats, Tomb Bat	Taphiens, Taphien perforé	Могильные летучие мыши
Grabfledermäuse	– Bats	Taphiens	Могильные летучие мыши
Gramper	Risso's Dolphin	Dauphin de Risso	Серый дельфин
Grampus griseus	– Dolphin	– de Risso	Серый дельфин
Graphiurinae	African Dormice		Африканские сони
Graphiurus	– Dormice	Graphiures	Кистехвостые сони
– *nanus*	Pygmy Dormouse	Graphiure nain	Карликовая африканская соня
Grashüpfermäuse	Grasshopper Mice	Onychomys	
Graue Zwerghamster	Ratlike Hamsters	Hamsters nains gris	Хомячки
Grauer Zwerghamster	Migratory Hamster	Hamster migrateur	Серый хомячок
Graues Langohr	Southern Long-eared Bat	Oreillard du sud	
Graufußhörnchen	Sun Squirrel	Écureuil de Gambie	
Grauhörnchen	Grey Squirrel	– gris	Американская серая белка
Graukopf-Baumstachelratte	Arboreal Soft-furred Spiny Rat	Rat épineux à tête grise	
Graukopf-Flughund	Grey-headed Flying Fox	Roussette à tête cendrée	Сероголовый восточно-австралийский крылан
Graurötelmaus	Large-toothed Red-backed Vole	Campagnol gris-roux	Красно-серая полевка
Grauwal	Grey Whale	Baleine grise	Серый кит
Grauwale	– Whales	Eschrichtiidés	Серые киты
Grauziesel	Little Souslik	Souslik nain	Малый суслик
Greifstachler	Tree Porcupines	Porcs-épic préhensiles	Коэнду
– i. e. S.	Prehensile-tailed Porcupine	Porc-épic préhensile	Бразильский коэнду
Greisengesicht	Wrinkled-faced Bat	Vespertilion ridé	
Grindwale	Pilot Whales	Globicéphales	Гринды
Grönländischer Halsbandlemming	Greenland Collared Lemming		Гренландский лемминг
Grönlandwal	Bowhead	Baleine boréale	Гренландский кит
Großblattnasen	Large-winged Bats	Mégadermes	Мегадермы
Große Braune Fledermaus	Big Brown Bat	Sérotine de maison	
– Hufeisennase	Greater Horseshoe Bat	Grand Rhinolophe obscur	Большой подковонос
– Mara	Mara	Mara	Патагонская свинка
– Nacktrücken-Fledermaus	Suapure Naked-backed Bat	Vespertilion de Suapuré	
– Rennmaus	Great Gerbil	Gerbille géante	Большая песчанка
– Rohrratte	– Cane Rat	Aulacode	Большая тростниковая крыса
– Samt-Fledermaus	Giant Velvety Free-tailed Bat	Molosse géante	
– Schlitznase	Great Slit-faced Bat	Grand Nyctère	
– Spießblattnase	– False Vampire	Faux Vampire commun	Большой вампир
– Waldmaus	Yellow-necked Field Mouse	Mulot fauve	Желтогорлая мышь
Großer Abendsegler	Common Noctule	Noctule	Рыжая вечерница
– Ameisenbär	Great Ant-eater	Grand Fourmilier	Трехпалый муравьед
– Röhrennasen-Flughund	Giant Tube-nosed Fruit Bat	Nyctimène géant	Большой ночной крылан
– Tümmler	Bottle-nosed Dolphin	Tursiops tronqué	Североатлантическая афалина
Großes Hasenmaul	Mexican Bulldog Bat	Noctilio pêcheuse	Обыкновенный зайцерот
– Nacktschwanzgürteltier	Eleven-banded Armadillo	Tatou à onze bandes	Большой голохвостый броненосец
Großkopf-Röhrennasen-Flughund	Large-headed Tube-nosed Fruit Bat	Nyctimène à grosse tête	
Großohren	Long-eared Bats	Oreillards	Ушаны
Großohr-Gleitbilch	Large-eared Small Flying Squirrel	Anomalure nain à longues oreilles	Ушастая соня-летяга
Großohr-Känguruhratte	Big-eared Kangaroo Rat	Rat kangourou à grandes oreilles	Большеухий мешетчатый прыгун
Gründelwale	White and Narwhales	Monodontidés	Белухи
Guano-Fledermaus	Mexican Free-tailed Bat	Tataride de Mexique	
Guayana-Delphin	Guyanian River Dolphin	Sotalie de la Guyane	Гвианский речной дельфин
Gundi	Gundi	Goundi	Гунди
Gürtelmaus	Lesser Pichiciago	Chlamyphore tronqué	Аргентинский плащеносец

GERMAN NAME	ENGLISH NAME	FRENCH NAME	RUSSIAN NAME
Gürtelmull	– Pichiciago	– tronqué	Аргентинский плащеносец
Gürtelmulle	Pichicies	Chlamyphores	Плащеносные броненосцы
Gürteltiere	Armadillos	Tatous	Броненосцы
Hackee	Eastern Chipmunk	Tamia strié	Американский бурундук
Halsband-Fledermaus	Necklace Hairless Bat	Cheiromèle à collier	
Halsbandlemming	Arctic Lemming	Lemming arctique	Копытный лемминг
Halsbandlemminge	Collared Lemmings		Копытные лемминги
Hammerkopf	Hammer-headed Fruit Bat	Hypsignathe monstrueux	Молотоголовый крылан
Hamster	Hamsters, Common Hamster	Hamsters, Hamster d'Europe	Хомяки, Обыкновенный хомяк
Hamsterratte	Tuza	Gaufre	Хомячий гофер
Hamsterratten		Cricétomyinés	Хомяковидные крысы
Hardwicke-Klappnase	Lesser Rat-tailed Bat	Rhinopôme d'Hardwick	Ланцетонос Хардвика
Harpyionycteridae	Harpy Fruit Bats	Harpionyctères	Крыланы-гарпии
Harpyionycteris	– – Bats	Harpionyctères	
– *whiteheadi*	Whitehead's Harpy Fruit Bat	Harpionyctère de Whitehead	
Haselmaus	Common Dormouse	Muscardin	Орешниковая соня
Hasenmäuler	Bulldog Bats	Noctilions	Зайцеротые летучие мыши, Зайцероты
Hasenmäuse	Mountain Chinchillas	Lagostomes des montagnes	Пушаки
Haus-Hausmaus	Western House Mouse		Западноевропейская домовая мышь
Hausmaus	House Mouse	Souris domestique	Домовая мышь
Hausmeerschweinchen	Guinea Pig	Cobaye	Домашняя морская свинка
Hausratte	House Rat	Rat commun	Черная крыса
– i. e. S.	– Rat	– noir	Северозападная черная крыса
Heidelberger Frühmensch	Heidelberg Man	Pithécanthrope d'Heidelberg	Гейдельбергский прачеловек
Heliosciurus gambianus	Sun Squirrel	Écureuil de Gambie	
Hermannsburg-Zwergmaus	Australian Native Mouse	Souris d'Hermannsburg	
Herrentiere	Primates	Primates	Приматы
Herznasenfledermaus	Heart-nosed False Vampire	Mégaderme du cœur	
Herzschädel-Springmäuse	Dwarf Jerboas	Cardiocraninés	Карликовые тушканчики
Hesperomyini	New World Mice	Souris du Nouveau Monde	Американские мышевидные хомячки
Heterocephalus glaber	Naked Sand Rat	Rat nu de sable	Голая крыса
Heteromyidae	Kangaroo Mice	Souris kangourou	Мешетчатые крысы
Heteromys goldmani	Goldman's Spiny Pocket Mouse	– à poche de Goldman	Иглистый тушканчик Гольдмана
Hildebrandt-Hufeisennase	Hildebrandt's Horseshoe Bat	Rhinolophe d'Hildebrandt	Подковонос Гильдебрандта
Himalaja-Rundblattnase	Great Himalayan Leaf-nosed Bat	Phyllorine d'Himalaye	
Hipposideridae	Old World Leaf-nosed Bats		Ложные подковоносы
Hipposideros armiger	Great Himalayan Leaf-nosed Bat	Phyllorine d'Himalaye	
– *caffer*	South African Lesser Leaf-nosed Bat	– de Cafrérie	
Hipposideros commersoni	Commerson's Leaf-nosed Bat	Phyllorhine de Commerson	Большой ложный подковонос
Hirschmaus	Deer Mouse	Souris du soir	
Hirschmäuse	White-footed Mice	– à pattes blanches	
Hoffmann-Zweifingerfaultier	Hoffmann's Two-toed Sloth	Unau d'Hoffmann	Двупалый ленивец Гофманна
Höhlen-Flughunde	Roussette Bats	Roussettes	Ночные крыланы
Höhlen-Langzungen-Flughund	Dobson's Long-tongued Dawn Bat	Eonyctère des cavernes	Пещерный крылан
Hohlnasen	Slit-faced Bats	Nyctères	Никтериды
Hominidae	Man	Hominidés	Гоминиды
Homininae	Man	Homininés	Люди
Homo	Man	Homo	Люди
– *erectus*	Java Man	Pithécanthrope de Java	Прачеловек
– – *erectus*	– Man	– de Java	Питекантроп
– – *heidelbergensis*	Heidelberg Man	– d'Heidelberg	Гейдельбергский прачеловек
– – *mauritanicus*	Mauretania Man	– de Mauretanie	Североафриканский прачеловек
– – *pekinensis*	Peking Man	– de Péking	Синантроп
– *sapiens*	Man	Homo sapiens	Человек
– – *»fossilis«*	Cro-Magnon Man	Homme de Cro-Magnon	Кроманьонец
– – *neanderthalensis*	Neanderthal Man	– de Néanderthale	Неандерталец
– – *sapiens*	Man	Homo sapiens	Современный человек
Honduras Haftscheiben-Fledermaus	Honduran Disk-winged Bat	Vespertilion d'Honduras	
Hoplomys gymnurus	Armoured Rat	Rat armé	Гимнуровая крыса
Hörnchen	Typical Squirrels	Écureuils	Беличьи
Hörnchenartige		Écureuils	Белкообразные

GERMAN NAME	ENGLISH NAME	FRENCH NAME	RUSSIAN NAME
Hörnchenverwandte	Squirrel-like Rodents		Белкоподобные
Hottentotten-Graumull	African Mole Rat	Rat taupe africain	Готтентотский землекоп
Hudsonhörnchen	Chickaree	Écureuil d'Hudson	Канадская белка
Hufeisennasen	Horseshoe Bats	Rhinolophidés	Подковоносы
Hufeisennasen-Verwandte	Horseshoe-nosed Bats	Rhinolophes	Подковоносовые
Hüpfmäuse	Jumping Mice	Zapodidés	Мышовки и прыгунчики
Hutiacarabali	Prehensile-tailed Hutia	Rat à queue préhensile	Хутия-карабали
Hutiaconga	Cuban Hutia	– poilé	Хутия-конга
Hutiasata	Black-tailed Hutia	– à queue noire	Чернохвостая кубинская крыса
Hydrochoeridae	Water Hogs	Hydrochéridés	Водосвинки
Hydrochoerus hydrochaeris	Capybara	Capybara	Капибара
Hydromyinae	Water Rats	Hydromyinés	Австралийские водяные крысы
Hydromys	Australian Water Rats	Rats d'eau d'Australie	
Hylopetes lepidus	Indo-Malayan Flying Squirrel	Écureuil volant de Java	Яванская летяга
Hyosciurus heinrichi	Long-snouted Squirrel	– d'Heinrich	
Hyperoodon	Bottle-nosed Whales	Baleines à bec	Бутылконосы
– *ampullatus*	– Whale	Hypérodon du nord	Высоколобый бутылконос
– *planifrons*	Southern Bottle-nosed Whale	– du sud	Плосколобый бутылконос
Hypogeomys antimena	Votsotsa		Вотсотса
Hypsignathus monstrosus	Hammer-headed Fruit Bat	Hypsignathe monstrueux	Молотоголовый крылан
Hystricidae	Old World Porcupines	Hystricidés	
Hystricomorpha	Porcupine-like Rodents		Дикобразовые
Hystrix	Large Porcupines	Porcs-épic	Настоящие дикобразы
– *africaeaustralis*	South African Porcupine	Porc-épic d'Afrique du sud	Южноафриканский дикобраз
– *cristata*	Porcupine	Porc-épic	Обыкновенный дикобраз
– *galeata*	African Porcupine	– d'Afrique	Африканский дикобраз
– *leucura*	White-tailed Porcupine	– à queue blanche	Индийский дикобраз
Idiurus	African Small Flying Squirrels	Anomalures nains	Сони-летяги
– *langi*	Lang's Small Flying Squirrel	Anomalure nain de Lang	Соня-летяга Ланга
– *macrotis*	Large-eared Small Flying Squirrel	– – à longues oreilles	Ушастая соня-летяга
– *zenkeri*	Zenker's Small Flying Squirrel	– – de Zenker	Соня-летяга Ценкера
Indische Maulwurfsratte	Bengali Bandicoot Rat	Bandicoot rat du Bengale	Бенгальская бандикутовая крыса
– Nacktsohlen-Rennmaus	Large Gerbil	Gerbille d'Inde	Индийская песчанка
– Pestratte	Bengali Bandicoot Rat	Bandicoot rat du Bengale	Бенгальская бандикутовая крыса
Indischer Flughund	Indian Flying Fox	Roussette géante	Летучая лисица
– Grindwal	– Pilot Whale	Globicéphale d'Inde	Индийская гринда
– Kurznasen-Flughund	Short-nosed Fruit Bat	Cynoptère à nez court	Обыкновенный коротконосый крылан
– Schweinswal	Finless Black Porpoise	Marsouin de l'Inde	Бесперая морская свинья
Indomalaiische Riesenhörnchen	Indomalayan Giant Squirrels		Индомалайские гигантские белки
Inia	Amazonian Dolphin	Inie de Geoffroy	Амазонская иния
Inia geoffrensis	– Dolphin	– de Geoffroy	Амазонская иния
Inias	Amazon Dolphins	Iniidés	Амазонские дельфины
Iniidae	Amazon Dolphins	Iniidés	Амазонские дельфины
Insektenessende Waldmaus	Congo Forest Mouse	Rat à manteau roux	Насекомоядная лесная мышь
Insel-Stachelschweine	Indonesian Porcupines	Porcs-épic d'Indonesie	Индонезийские дикобразы
Irawadi-Delphin	Irawady Dolphin	Orcaelle d'Irawadi	Иравадийский дельфин
Jackson-Dornschwanzhörnchen	Jackson's Scaly-tailed Squirrel	Anomalure de Jackson	Шипохвостая белка Джексона
Jaculus jaculus	Desert Jerboa	Gerboise du steppe	Песчаный тушканчик
Jamaika-Ferkelratte	Jamaican Hutia	Rat jamaique	Ямайская крыса
Japanischer Schläfer	Japanese Dormouse	Glirule de Japon	Японская соня
Java-Hohlnase	Javanese Slit-faced Bat	Nyctère de Java	Яванская никтерида
Javanischer Frühmensch	Java Man	Pithécanthrope de Java	Питекантроп
Javanisches Gleithörnchen	Indo-Malayan Flying Squirrel	Écureuil volant de Java	Яванская летяга
– Schuppentier	Malayan Pangolin	Pangolin javanais	Яванский ящер
Java-Stachelschwein	Javanese Porcupine	Acanthion de Java	Яванский дикобраз
Jelarang	Oriental Giant Squirrel	Écureuil géant	Малайская гигантская белка
Kalifornische Taschenmaus	California Pocket Mouse	Souris à poche de Californie	Калифорнийский мешетчатый тушканчик
Kalifornischer Ziesel	– Ground Squirrel	Spermophile de Californie	Калифорнийский суслик
Kalong	Red-necked Fruit Bat	Roussette à cou rouge	Калонг
Kamerunfluß-Delphin	West African White Dolphin	Sotalie de Teusz	Камерунский речной дельфин
Kammfinger	Gundis	Goundis	Гребнепалые крысы
Kammfingerartige	Gundis	Goundis	Гребнепаловые
Kammratten	Tuco-tucos	Tuco-tucos	Ктеномииды

GERMAN NAME	ENGLISH NAME	FRENCH NAME	RUSSIAN NAME
Kammstachelratten	Crested Spiny Rats	Échimys	
Kammzehen-Springmaus	Comb-toed Jerboa		Гребнепалый тушканчик
Kanadischer Biber	Canadian Beaver	Castor de Canada	Канадский бобр
Känguruhmäuse	Kangaroo Mice	Souris kangourou	Мешетчатые прыгунчики
Kannabateomys amblyonyx	Tree-Rat	Rat du bambou	
Kap-Bleßmull	Cape Blesmol	– taupe du Cap	Капский пескорой
Kappler-Weichgürteltier	Kappler's Armadillo	Tatou de Kappler	Броненосец Капплера
Kap-Strandgräber	Cape Mole Rat	Fouisseur	
Kapuzenfaultier		Bradype à capuchon	Капюшонный ленивец
Kaukasisches Eichhörnchen	Caucasian Squirrel	Écureuil du Caucase	Кавказская белка
Kerodon	Rock Cavies	Cobayes des roches	Горные свинки
– *rupestris*	– Cavy	Cobaye des roches	Горная морская свинка
Kiefernwühlmaus	Pine Vole	Campagnol du pin	Американская боровая полевка
Kinnblatt-Fledermäuse	Mustache Bats	Chilonycterinés	Листобороды
Klappnasen	Rat-tailed Bats	Rhinopômes	
Kleine Bambusratte	Lesser Bamboo Rat		
– Chinchilla	Chinchilla	Chinchilla à longue queue	Длиннохвостая шиншилла
– Fünfzehen-Springmaus	Jerboa	Gerboise lièvre	Тарбаганчик
– Hamsterratten	Long-tailed Pouched Rats		Малые хомяковидные крысы
– Hufeisennase	Lesser Horseshoe Bat	Petit Rhinolophe	Малый подковонос
– Mara	Salt-desert Cavy	Cobaye halophile	Малая мара
– Nacktrücken-Fledermaus	Naked-backed Bat	Vespertilion à dos nu	
– Nacktsohlen-Rennmaus	Small Naked-soled Gerbil	Gerbille d'Emin	
– Nacktsohlen-Rennmäuse	Naked-soled Gerbils		Малые голостопые песчанки
– Rohrratte	Lesser Cane Rat	Rat de Grégorian	Малая тростниковая крыса
– Waldmaus	Long-tailed Field Mouse	Mulot sylvestre	Лесная мышь
– Wühlmaus	Earth Vole	Campagnol souterrain	Подземная полевка
– Wühlmäuse	Pine Voles		Малые полевки
Kleiner Blutsauger	Hairy-legged Vampire Bat	Diphylle sans queue	Малый кровосос
– Chipmunk	Least Chipmunk	Néotamia nain	Малый чипмунк
– Schwertwal	False Killer Whale	Pseudorque à dents épaisses	Черная косатка
Kleines Hasenmaul	Southern Bulldog Bat	Noctilio du sud	Малый зайцерот
– Nacktschwanzgürteltier	Spiny Armadillo	Tatou épineux	Малый голохвостый броненосец
Knight-Tukotuko	Knight's Tuco-tuco	Ctenomys de Knight	Тукотуко
Kogia breviceps	Pygmy Sperm Whale	Cachalot pygmée	Карликовый кашалот
Königsriesenhörnchen	Indian Giant Squirrel	Écureuil d'Inde	Индийская гигантская белка
Koslows Zwergspringmäuse	Pygmy Jerboas		Трехпалые карликовые тушканчики
Kragenfaultier	Necklace Sloth	Bradype à collier	Ошейниковый ленивец
Kragenflughunde	Collared Fruit Bats	Myonyctéres à collier	
Kreta-Stachelmaus	Cretan Spiny Mouse	Souris épineuse de Crête	Критская ежовая мышь
Kugelgürteltier	Three-banded Armadillo	Tatou à trois bandes	Шаровидный броненосец
Kurzfuß-Inselratte		Souris à pattes courtes	Коротколапая мадагаскарская крыса
Kurznasen-Flughunde	Short-nosed Fruit Bats	Cynoptères	Коротконосые крыланы
Kurzohr-Häschenratte	Jone's House-building Jerboa Rat		Короткоухая зайцевидная мышь
Kurzschwanz-Blattnasen	Short-tailed Leaf-nosed Bats	Vespertilions à nez plat	Короткохвостые вампиры
Kurzschwanz-Chinchilla	– Chinchilla	Chinchilla à queue courte	Короткохвостая шиншилла
Kurzschwanz-Hamsterratten	Cape Pouched Mice		Короткохвостые хомяковидные крысы
Kurzschwanz-Maulwurfsratte	Pest Rat	Rat à queue courte	Индийская земляная крыса
Kurzschwanz-Stachelschwein	Malayan Porcupine	Acanthion de Malaysie	
Kurzschwänzige Zwerghamster	Dwarf Hamsters	Hamsters nains à queue courte	Мохноногие хомячки
Lagenorhynchus acutus	White-sided Dolphin	Lagénorhynque à bec pointu	Белобокий дельфин
– *albirostris*	White-beaked Dolphin	– à rostre blanc	Беломордый дельфин
– *obliquidens*	Pacific White-sided Dolphin	– de Gill	Тихоокеанский белобокий дельфин
Lagidium	Mountain Chinchillas	Lagostomes des montagnes	Пушаки
– *viscacia*	– Chinchilla	Lagostome des montagnes	Пушак
Lagostomus maximus	Viscacha	– des Pampas	Вискаша
Lagurus	Sagebrush Voles		Степные пеструшки
– *lagurus*	– Vole		Степная пеструшка
Lamberton-Inselratte	Lamberton's Malagasy Rat	Rat de Lamberton	Мадагаскарская крыса Ламбертона
Lander-Hufeisennase	Lander's Horseshoe Bat	Rhinolophe de Lander	
Langbein-Fledermaus	Long-legged Bat	Macrophylle	
Langflügel-Fledermaus	Long-winged Bat	Minioptère à longues ailes	Длиннокрыл Шрейбера
Langflügel-Fledermäuse	– Bats	Minioptères	Длиннокрылы

GERMAN NAME	ENGLISH NAME	FRENCH NAME	RUSSIAN NAME
Lang-Gleitbilch	Lang's Small Flying Squirrel	Anomalure nain de Lang	Соня-летяга Ланга
Langnasen-Fledermaus	Mexican Long-nosed Bat	Chauve-souris de Mexique	Длинноносый листонос
Langnasen-Flughunde	Flying Foxes	Roussettes	Длинноносые крыланы
Langnasenhörnchen	Long-nosed Squirrel	Écureuil à nez long	Длинноносая белка
Langohr-Häschenratte	House-building Jerboa Rat		Длинноухая зайцевидная мышь
Langschnabeldelphine	Long-snouted Dolphins	Sténidés	Длинноклювые дельфины
Langschwanz-Chinchilla	Chinchilla	Chinchilla à longue queue	Длиннохвостая шиншилла
Langschwanz-Quastenstachler	Asiatic Brush-tailed Porcupine	Athérure à longue queue	Малайский кистехвост
Langschwanz-Schuppentier	Phatagin	Pangolin tétradactyle	Длиннохвостый ящер
Langschwanz-Zwerghamster	Long-tailed Hamster	Hamster nain à longue queue	Длиннохвостый хомячок
Langschwänzige Indische Baummäuse	– Climbing Mice	Souris à longue queue	Индийские пальмовые мыши
Langschwänziger Hamster	– Rat	Hamster à longue queue	Длиннохвостый хомяк
– Ziesel	Eversman's Souslik	Spermophile d'Eversman	Длиннохвостый суслик
Langschwänziges Murmeltier	Long-tailed Marmot	Marmotte à longue queue	Длиннохвостый сурок
Langzungen	Long-tongued Bats	Glossophages	Глоссофаги
Langzungen-Fledermäuse	– Bats	Glossophages	Длинноязычные вампиры
Langzungen-Flughunde	– Fruit Bats	Macroglosses	Длинноязычные крыланы
Lanzennase	Javelin Bat	Phyllostome fer de lance	Копьенос
Lanzennasen	Spear-nosed Bats	Phyllostomes	
Lanzenratte	Armoured Rat	Rat armé	Гимнуровая крыса
La-Plata-Delphin	La Plata Dolphin	Dauphin de la Plata	Лаплатский дельфин
Lasiurus borealis	Red Bat	Chauve-souris boréale	Рыжий кожан
– *cinereus*	Hoary Bat	– cendrée	Бело-серый кожан
Lavia frons	African Yellow-winged Bat	Mégaderme à ailes orangées	
Layard-Wal	Layard's Whale	Mésoplodon de Layard	Ремнезуб Лэйярда
Leggada	African Native Mice	Souris d'Afrique	Африканские мыши-малютки
– *minutoides*	Small African Native Mouse	– naine d'Afrique	Африканская мышь-малютка
Leggadina	Australian Native Mice	– d'Australie	Австралийские малые мыши
– *hermannsburgensis*	– – Mouse	– d'Hermannsburg	
Lemminge	Lemmings	Lemmings	Лемминги
Lemmingmäuse	Lemming Mice		Лемминговые мыши
Lemmini	Lemmings	Lemmings	Лемминги
Lemmus	Brown Lemmings		Настоящие Лемминги
– *lemmus*	Norway Lemming	Lemming des toundras	Норвежский лемминг
Lemniscomys	Striped Grass Mice	Rats rayés	Африканские полосатые мыши
Leopardenziesel	Thirteen-striped Ground Squirrel	Spermophile à treize bandes	Полосатый американский суслик
Leporillus	House-building Jerboa Rats		Австралийские зайцевидные мыши
– *conditor*	– – Rat		Длинноухая зайцевидная мышь
– *jonesi*	Jone's House-building Jerboa Rat		Короткоухая зайцевидная мышь
Lichtensteins Springmaus	Lichtenstein's Jerboa	Gerboise de Lichtenstein	Тушканчик Лихтенштейна
Liomys irroratus	Mexican Spiny Pocket Mouse	Souris épineuse à poche de Mexico	Мексиканский иглистый тушканчик
– *pictus*	Painted Spiny Pocket Mouse	Souris épineuse à poche	
Lipotes vexillifer	Chinese River Dolphin	Dauphin d'eau douce de Chine	Китайский речной дельфин
Lissodelphininae	Right Whale Dolphins	Lissodelphininés	Бесперые дельфины
Lissodelphis borealis	Northern Right Whale Dolphin	Dauphin du nord	Северный китовидный дельфин
– *peroni*	Peron's Dolphin	– de Peron	Южный китовидный дельфин
Ljutaga	European Flying Squirrel	Palatouche	Обыкновенная летяга
Lonchorrhina	Sword-nosed Bats		
– *aurita*	Tome's Long-eared Bat	Vespertilion de Tome	Малый вампир
Lophiomys imhausi	Crested Hamster	Hamster d'Imhause	Гривистая крыса
Lophuromys	Harsh-furred Mice	Rats à pelage en brosse	
Lyra-Fledermaus	Indian False Vampire	Mégaderme lyre	Индийский ложный вампир
Maclaud-Hufeisennase	Maclaud's Bat	Rhinolophe de Maclaud	Подковонос Маклауда
Macroderma gigas	Australian Ghost Bat	Macroderme d'Australie	
Macroglossidae	Long-tongued Fruit Bats	Macroglosses	Длинноязычные крыланы
Macroglossus	– – Bats	Macroglosses	
– *minimus*	Asiatic Long-tongued Fruit Bat	Macroglosse minime	Малый длинноязычный крылан
Macrophyllum macrophyllum	Long-legged Bat	Macrophylle	
Madagaskarratten	Malagasy Rats	Rats de Madagascar	Мадагаскарские крысы

GERMAN NAME	ENGLISH NAME	FRENCH NAME	RUSSIAN NAME
Madagassische Haftscheiben-Fledermäuse	Sucker-footed Bats, Golden Bat	Myzopodidés, Vespertilion doré	Мадагаскарские присосковые летучие мыши, Золотистая летучая мышь
Mähnenratte	Crested Hamster	Hamster d'Imhause	Гривистая крыса
Malaiische Pinselschwanz-Baummaus	Pencil-tailed Tree Mouse	Souris d'arbre de Malaysie	Малайская древесная крыса
Malaiischer Falscher Vampir	Malayan False Vampire	Mégaderme spasme	Малайский ложный вампир
– Pinselstachler	Long-tailed Porcupine	Porc-épic à longue queue	Борнеоский дикобраз
Manidae	Pangolins	Manidés	Ящеры
Manis	Scaly Ant-eaters	Manidés	Ящеры
– *crassicaudata*	Indian Pangolin	Pangolin indien	Переднеиндийский ящер
– *gigantea*	Giant Pangolin	Grand Pangolin	Гигантский ящер
– *javanica*	Malayan Pangolin	Pangolin javanais	Яванский ящер
– *pentadactyla*	Chinese Pangolin	– à queue courte	Индостанский панголин
– *temmincki*	Cape Pangolin	– de Temminck	Степной ящер
– *tetradactyla*	Phatagin	– tétradactyle	Длиннохвостый ящер
– *tricuspis*	Three-pointed Pangolin	Tricuspide	Белобрюхий ящер
Mara	Mara	Mara	Патагонская свинка
Maras	Maras	Dolichotinés	Мары
Marmota	Marmots	Marmottes	Сурки
– *bobak*	Bobac Marmot	Marmotte Bobac	Степной сурок
– *caligata*	Hoary Marmot	– grise	Североамериканский серый сурок
– *caudata*	Long-tailed Marmot	– à longue queue	Длиннохвостый сурок
– *flaviventris*	Yellow-bellied Marmot	– à ventre fauve	Желтобрюхий сурок
– *marmota*	Alpine Marmot	– des Alpes	Альпийский сурок
– *monax*	Woodchuck	Monax	Лесной сурок
Marmotini	Northern Ground Squirrels		Сурковые
Mastomys	Multimammate Rats	Rats à mamelles multiples	
Mauritanischer Grabflatterer	Mauritian Tomb Bat	Taphien de Maurice	
Mausartiger Zwerghamster	Mouselike Hamster		Мышевидный хомячок
Mäuse	Old World Rats and Mice	Souris et Rats de l'Ancien Monde	Мышиные
– i. e. S.	House Mice	Souris	Мыши
Mausohr	European Little Brown Bat	Vespertilion murin	Большая ночница
Mausohr-Fledermäuse	Mouse-eared Bats	Murins	Ночницы
Mausschwanz-Fledermäuse	Rat-tailed Bats	Rhinopômes	Ланцетоносы
Meerschweinchen	Cavies	Caviidés	Морские свинки
Meerschweinchen i. e. S.	Guinea Pigs	Cobayes	
Meerschweinchenverwandte	Cavy-like Rodents		Свинкообразные
Megachiroptera	Fruit Bats	Megachiroptères	Летучие собаки
Megaderma	False Vampires	Mégadermes	Лироносы
– *cor*	Heart-nosed False Vampire	Mégaderme du cœur	
– *lyra*	Indian False Vampire	– lyre	Индийский ложный вампир
– *spasma*	Malayan False Vampire	– spasme	Малайский ложный вампир
Megadermatidae	Large-winged Bats	Mégadermes	Мегадермы
Megaloglossus woermanni	African Long-tongued Fruit Bat	Mégaloglosse de Woermann	Африканский длинноязычный крылан
Megaptera novaeangliae	Humpback Whale	Mégaptère	Горбач
Mensch	Man	Homo sapiens	Человек
– der Jetztzeit	Man	Homo sapiens	Современный человек
Menschen	Man	Hominidés, Homo	Гоминиды, Люди
– i. e. S.	Man	Homininés	Люди
Menschenaffen	Anthropoid Apes	Singes anthropomorphes	Человекоподобные
Meriones	Jirds	Mériones	Малые песчанки
– *libycus*	Libyan Jird	Mérione de Lybie	Краснохвостая песчанка
– *meridianus*	Southern Jird	– du sud	Полуденная песчанка
– *persicus*	Persian Jird	– de Perse	Персидская песчанка
Meriones tristrami	Tristram's Jird	Mérione de Tristram	
– *unguiculatus*	Jird	– de Mongolie	Когтистая песчанка
Mesocricetus	Golden Hamsters		Средние хомяки
– *auratus*	– Hamster	Hamster doré	Переднеазиатский хомяк
Mesoplodon bidens	Sowerby's Whale	Mésoplodon de Sowerby	Атлантический ремнезуб
– *europaeus*	Gervais' Whale	– de Gervais	Европейский ремнезуб
– *layardi*	Layard's Whale	– de Layard	Ремнезуб Лэйярда
– *mirus*	True's Beaked Whale	– de True	Североатлантический ремнезуб
Mexikanische Stacheltaschenmaus	Mexican Spiny Pocket Mouse	Souris épineuse à poche de Mexico	Мексиканский иглистый тушканчик
– Taschenratte	Yellow Pocket Gopher	Rat à poche mexicain	Мексиканский гофер
Microcavia	Mountain Cavies		Карликовые свинки
– *australis*	Southern Mountain Cavy		Южная карликовая свинка
Microchiroptera	Insectivorous Bats	Chauves-souris insectivores	Летучие мыши
Microdipodops	Kangaroo Mice	Souris kangourou	Мешетчатые прыгунчики

GERMAN NAME	ENGLISH NAME	FRENCH NAME	RUSSIAN NAME
– *pallidus*	Pale Kangaroo Mouse	– – pâle	Бледный мешетчатый прыгунчик
Micromys minutus	Harvest Mouse	– des moissons	Мышь-малютка
Micropteropus pusillus	Dwarf Epauletted Fruit Bat	Petit Microptère	Карликовый эполетовый крылан
Microsciurus	– Tree Squirrels	Microsciures	Американские карликовые белки
– *alfari*	– – Squirrel	Microsciure nain	
Microtini	Voles		Полевки
Microtus	Voles		Серые полевки
– *agrestis*	Field Vole	Campagnol agreste	Темная полевка
– *arvalis*	Common Vole	– des champs	Обыкновенная полевка
– *brandti*	Brandt's Vole	– de Brandt	Полевка Брандта
– *nivalis*	Snow Vole	– des neiges	Снежная полевка
– *oeconomus*	Tundra Vole	– nordique	Полевка-экономка
– *pennsylvanicus*	Eastern Meadow Mouse	– de Pennsylvania	Пенсильванская полевка
Miniopterinae	Bent-winged Bats	Minioptères	Длиннокрылы
Miniopterus	Long-winged Bats	Minioptères	Длиннокрылы
– *schreibersi*	– Bat	Minioptère à longues ailes	Длиннокрыл Шрейбера
Mittagsrennmaus	Southern Jird	Mérione du sud	Полуденная песчанка
Mittelhamster	Golden Hamsters		Средние хомяки
Mohrenaguti	Smoky Agouti		Аспидная агути
Moko	Rock Cavy	Cobaye des roches	Горная морская свинка
Molossidae	Free-tailed Bats	Molossidés	Бульдоговые летучие мыши
Molossus	Velvety Free-tailed Bats	Molosses	
– *major*	Giant Velvety Free-tailed Bat	Molosse géante	
– *rufus*	Red Velvety Free-tailed Bat	– rouge	Рыжая бульдоговая мышь
Moncktons Schwimmratte	Monckton's Water Rat		Австралийская крыса Монктона
Mongolische Gebirgsmaus	Royle's High-mountain Vole	Campagnol de Royle	Серебристая полевка
– Rennmaus	Jird	Mérione de Mongolie	Когтистая песчанка
Mongolischer Biber	Mongolian Beaver	Castor de Mongolie	Монгольский бобр
– Zwerghamster	Short Dwarf Hamster	Hamster nain de Mongolie	Монгольский хомячок
Monodon monoceros	Narwhal	Narval	Нарвал
Monodontidae	White and Narwhales	Monodontidés	Белухи
Moorlemminge	Lemming Mice		Лемминговые мыши
Mopsfledermaus	Barbastelle	Barbastelle d'Europe	Европейская широкоушка
Mopsfledermäuse	Barbastelles	Barbastelles	Широкоушки
Mosaikschwanz-Riesenratten	Giant Naked-tailed Rats	Rats à queue en mosaique	
Mull-Lemminge	Mole Lemmings	– taupes	Слепушонки
Muridae	Old World Rats and Mice	Souris et Rats de l'Ancien Monde	Мышиные
Murmeltiere	Marmots	Marmottes	Сурки
Mus	House Mice	Souris	Мыши
– *musculus*	– Mouse	– domestique	Домовая мышь
– – *bactrianus*	Bactrian House Mouse		Центральноазиатская домовая мышь
– – *domesticus*	Western House Mouse		Западноевропейская домовая мышь
– – *musculus*	Northern House Mouse		Североевропейская домовая мышь
– – *spicilegus*	Eastern House Mouse		Средиземноморская домовая мышь
– *platythrix*	Flat-haired Mouse	Souris aux cheveux plats	Бурая иглистая мышь
Muscardinus avellanarius	Common Dormouse	Muscardin	Орешниковая соня
Musonycteris	Banana Bats	Chauves-souris des bananes	Банановые листоносы
– *harrisoni*	– Bat	Chauve-souris des bananes	Баноновый листонос
Myocastor coypus	Nutria	Ragondin	Нутрия
Myocastoridae	Nutrias	Ragondins	Бобровые крысы
Myomimus personatus	Asiatic Dormouse	Myomime à queue fine	Мышевидная соня
Myonycteris	Collared Fruit Bats	Myonyctères à collier	
Myonycteris torquata	Little Collared Fruit Bat	Myonyctère à collier	
Myoprocta	Acouchis	Acouchis	Хвостатые агути
– *acouchi*	Acouchi	Acouchi	Акучи
Myopus schisticolor	Wood Lemming	Lemming des forêts	Лесной лемминг
Myosciurus pumilio	African Pygmy Squirrel	Écureuil nain du Gabon	Африканская карликовая белка
Myospalacini	Mole Mice		Цокоры
Myospalax	Zokors		Цокоры
– *aspalax*	Transbaikal Zokor	Zokor de Transbaikalie	Даурский цокор
– *myospalax*	North Altai Zokor	Rat taupe	Алтайский цокор
– *psilurus*	Manchurian Zokor	Zokor de Mandchourie	Северокитайский цокор
Myotis	Mouse-eared Bats	Murins	Ночницы
– *bechsteini*	Bechstein's Bat	Vespertilion de Bechstein	Ночница Бехштейна

GERMAN NAME	ENGLISH NAME	FRENCH NAME	RUSSIAN NAME
– *dasycneme*	Pond Bat	– des marais	Прудовая ночница
– *daubentoni*	Water Bat	– de Daubenton	Водяная ночница
– *myotis*	European Little Brown Bat	– murin	Большая ночница
– *mystacinus*	Whiskered Bat	– à moustaches	Усатая ночница
– *nattereri*	Natterer's Bat	– de Natterer	Ночница Наттерера
– *welwitschii*	Welwitsch's Bat	– de Welwitsch	
Myrmecophaga tridactyla	Great Ant-eater	Grand Fourmilier	Трехпалый муравьед
Myrmecophagidae	Ant-eaters	Fourmiliers	Муравьеды
Mystacina tuberculata	New Zealand Short-tailed Bat	Mystacine tubercule	Новозеландская летучая мышь
Mystacinidae	– – – Bats	Mystacinidés	Новозеландские летучие мыши
Mystacoceti	Baleen Whales	Baleines	Усатые киты
Mystromys albicaudatus	White-tailed Rat	Hamster à queue blanche	Белохвостый хомяк
– *longicaudatus*	Long-tailed Rat	– à longue queue	Длиннохвостый хомяк
Myzopoda aurita	Golden Bat	Vespertilion doré	Золотистая летучая мышь
Myzopodidae	Sucker-footed Bats	Myzopodidés	Мадагаскарские присосковые летучие мыши
Nacktbäuchiger Grabflatterer	Naked-bellied Tomb Bat	Taphien à ventre nu	Голобрюхая летучая мышь
Nacktfledermaus	Naked Bat	Cheiromèle nue	Голая летучая мышь
Nacktmull	– Sand Rat	Rat nu de sable	Голая крыса
Nacktrücken-Fledermäuse	Naked-backed Bats		
Nacktschwanzgürteltiere	Eleven-banded Armadillos	Tatous à onze bandes	Голохвостые броненосцы
Nacktsohlen-Rennmäuse	Large Gerbils		Голоступые песчанки
Nagetiere	Rodents	Rongeurs	Грызуны
Nannosciurus	Oriental Pygmy Squirrels	Écureuils pygmées	Азиатские карликовые белки
– *melanotis*	Brown Dwarf Squirrel	Écureuil pygmée brun	
– *whiteheadi*	Whitehead's Dwarf Squirrel	– – de Whitehead	Белка Уайтхеда
Napaeozapus	Woodland Jumping Mice	Zapodes des bois	
– *insignis*	– – Mouse	Zapode des bois	Лесной прыгунчик
Narwal	Narwhal	Narval	Нарвал
Nasenfledermaus	Proboscis Bat	Rhynchonyctère	Носатая летучая мышь
Nasenratte	Shrew Rat	Rat au nez	
Nasenratten	– Rats	Rhynchomyinés	
Natalidae	Funnel-eared Bats	Natalidés	
Natalus stramineus	– Bat	Vespertilion à couleur de paille	
Neandertaler	Neanderthal Man	Homme de Néanderthale	Неандерталец
Neave-Dornschwanzhörnchen	Neave's Scaly-tailed Squirrel	Anomalure de Neave	
Nebengelenktiere	Xenarthra	Xénarthrés	
Nelson-Antilopenziesel	Nelson's Souslik	Souslik de Nelson	Калифорнийский антилоповый суслик
Neobalaena marginata	Pygmy Right Whale	Baleine naine	Карликовый гладкий кит
Neofiber alleni	Round-tailed Muskrat	Rat d'eau de Floride	Флоридская водяная крыса
Neophocaena phocaenoides	Finless Black Porpoise	Marsouin de l'Inde	Бесперая морская свинья
Neotoma	Wood Rats	Rats des bois	Кустовые хомячки
– *albigula*	Whitethroat Wood Rat	Rat des steppes	Пустынный кустовый хомячок
– *cinerea*	Bushtail Wood Rat		Серый кустовый хомячок
Nesokia indica	Pest Rat	– à queue courte	Индийская земляная крыса
Nesomyinae	Malagasy Rats	Rats de Madagascar	Мадагаскарские крысы
Nesomys lambertoni	Lamberton's Malagasy Rat	Rat de Lamberton	Мадагаскарская крыса Ламбертона
Neufundland-Biber	Newfoundland Beaver	Castor de Terre-Neuve	Ньюфаундлендский бобр
Neunbindengürteltier	Nine-banded Armadillo	Tatou à neuf bandes	Девятипоясный броненосец
Neuseeland-Fledermaus	New Zealand Short-tailed Bat	Mystacine tubercule	Новозеландская летучая мышь
Neuseeland-Fledermäuse	– – – Bats	Mystacinidés	Новозеландские летучие мыши
Neuweltliche Zwerghörnchen	Dwarf Tree Squirrels	Microsciures	Американские карликовые белки
Neuweltmäuse	New World Mice	Souris du Nouveau Monde	Американские мышевидные хомячки
Neuweltratten	New World Mice	Souris du Nouveau Monde	Американские мышевидные хомячки
Nilflughund	Egyptian Fruit Bat	Rousette d'Égypte	Египетский крылан
Noctilio	Bulldog Bats	Noctilions	Зайцероты
– *labialis*	Southern Bulldog Bat	Noctilio du sud	Малый зайцерот
– *leporinus*	Mexican Bulldog Bat	– pêcheuse	Обыкновенный зайцерот
Noctilionidae	Bulldog Bats	Noctilions	Зайцеротые летучие мыши
Nordafrikanische Rennmaus	Field Gerbil	Gerbille champêtre	Североафриканская песчанка

GERMAN NAME	ENGLISH NAME	FRENCH NAME	RUSSIAN NAME
Nordafrikanischer Frühmensch	Mauretania Man	Pithécanthrope de Mauretanie	Североафриканский прачеловек
Nordamerikanische Baumstachler	North American Porcupines	Porcs-épic nord américains	Североамериканские иглошерсты
– Rothörnchen	Chickarees		Североамериканские крысные белки
Nordamerikanisches Zwerggleithörnchen	Southern Flying Squirrel	Écureuil volant du sud	Североамериканская летяга
Nordische Fledermaus	Northern Bat	Sérotine boréale	Северный кожанок
– Wühlmaus	Tundra Vole	Campagnol nordique	Полевка-экономка
Nordkaper	Atlantic Right Whale	Baleine noire	Бискайский кит
Nördliche Grashüpfermaus	Northern Grasshopper Mouse	Onychomys du nord	
– Hausmaus	– House Mouse		Североевропейская домовая мышь
– Lemmingmaus	– Bog Lemming	Lemming du nord	Северная лемминговая мышь
Nördlicher Entenwal	Bottle-nosed Whale	Hypérodon du nord	Высоколобый бутылконос
– Glattdelphin	Northern Right Whale Dolphin	Dauphin du nord	Северный китовидный дельфин
– Mull-Lemming	Mole Lemming	Rat taupe	Обыкновенная слепушонка
Nördliches Gleithörnchen	Northern Flying Squirrel	Écureuil volant du nord	
Nordpazifik-Glattwal	Pacific Right Whale	Baleine de Siebold	Японский кит
Notomys	Australian Hopping Mice		Австралийские прыгающие мыши
Notopteris	Long-tailed Fruit Bats	Notoptères à queue longue	
Nutria	Nutria	Ragondin	Нутрия
Nyctalus	Noctule Bats	Noctules	Вечерницы
– *lasiopterus*	Giant Noctule	Noctule géante	Гигантская вечерница
– *leisleri*	Lesser Noctule	– de Leisler	Малая вечерница
– *noctula*	Common Noctule	Noctule	Рыжая вечерница
Nycteridae	Slit-faced Bats	Nyctères	Никтериды
Nycteris	– Bats	Nyctères	
– *grandis*	Great Slit-faced Bat	Grand Nyctère	
– *hispida*	Hispid Slit-faced Bat	Nyctère hérissé	
– *javanica*	Javanese Slit-faced Bat	– de Java	Яванская никтерида
– *thebaica*	Geoffroy's Slit-faced Bat	– de Geoffroy	
Nyctimene	Tube-nosed Fruit Bats	Nyctimènes	
– *cephalotes*	Large-headed Tube-nosed Fruit Bat	Nyctimène à grosse tête	
– *major*	Giant Tube-nosed Fruit Bat	– géant	Большой ночной крылан
– *robinsoni*	Queensland Tube-nosed Fruit Bat	– de Robinson	
Nyctimeninae	Tube-nosed Fruit Bats	Nyctimènes	Трубконосые крыланы
Octodon	South American Bush Rats	Octodons	Кустарниковые крысы
– *degus*	Degu	Dègue du Chili	Дегу
Octodontidae	Octodont Rodents	Octodontidés	Осьмизубые
Octodontoidea	– Rodents		Осьмизубовые
Octodontomys	Boris	Boris	
– *gliroides*	Bori	Bori	Бори
Octomys mimax	Viscacha Rat	Rat minime	Вискашевидная крыса
Odontoceti	Toothed Cetaceans	Odontocétés	Зубатые киты
Oenomys hypoxanthus	Rufous-nosed Rat	Rat à museau roux	Африканская красноносая крыса
Ohrenratten	Vlei and Karroo Rats	Otomyinés	Африканские ушастые крысы
Ölpalmenhörnchen	Oil-palm Squirrel	Grand Écureuil de Stanger	Масличная белка
Ondatra	Muskrats	Ondatras	Мускусные крысы
– *zibethica*	Muskrat	Rat musqué	Ондатра
Onychomys	Grasshopper Mice	Onychomys	
– *leucogaster*	Northern Grasshopper Mouse	– du nord	
– *torridus*	Southern Grasshopper Mouse	– du sud	
Orcaella brevirostris	Irawady Dolphin	Orcaelle d'Irawadi	Иравадийский дельфин
Orcinus orca	Common Killer Whale	Orque épaulard	Косатка
Ord-Känguruhratte	Ord's Kangaroo Rat	Rat kangourou d'Ord	Мешетчатый прыгун Орда
Orthogeomys	Tuzas	Gaufres	
– *grandis*	Tuza	Gaufre	Хомячий гофер
Oryzomys	Rice Rats	Rats du riz	
– *palustris*	– Rat	Rat du riz	Рисовая мышь
Ostafrikanischer Springhase	Spring Haas	Lièvre sauteur d'Afrique de l'est	Восточноафриканский долгоног
Ostblindmaus	Greater Mole Rat	Spalax oriental	Обыкновенный слепыш
Östlicher Chipmunk	Eastern Chipmunk	Tamia strié	Американский бурундук
Otomyinae	Vlei and Karroo Rats	Otomyinés	Африканские ушастые крысы

GERMAN NAME	ENGLISH NAME	FRENCH NAME	RUSSIAN NAME
Pachyuromys duprasi	Fat-tailed Mouse	Souris à grosse queue	Толстохвостая песчанка
Paka	Paca	Paca	Пака
Pakarana	Pacarana	Pacarana	Длиннохвостая пака
Pakaranas	Pacaranas	Pacaranas	Диномииды
Pakas	Pacas	Cuniculinés	Паки
Palmenflughund	Straw-coloured Bat	Roussette paillée	Пальмовый крылан
Palmenhörnchen	African Tree Squirrels, Palm Squirrel	Écureuil des palmes	Пальмовые белки, Пальмовая белка
Pan	Chimpanzees	Chimpansés	Шимпанзе
– *paniscus*	Pygmy Chimpanzee	Chimpansé nain	Бонобо
– *troglodytes*	Chimpanzee	Chimpansé	Обыкновенный шимпанзе
Paradipus ctenodactylus	Comb-toed Jerboa		Гребнепалый тушканчик
Paraxerus boehmi	Boehm's African Bush Squirrel	Écureuil de Boehm	
– *palliatus*	Mantled African Bush Squirrel	– – à manteau	
Parry-Ziesel	Arctic Ground Squirrel	Souslik de Parry	Американский длинно-хвостый суслик
Pazifik-Känguruhratte	Pacific Kangaroo Rat	Rat kangourou du Pacifique	Проворный мешетчатый прыгун
Pazifischer Grindwal	Pacific Pilot Whale	Globicéphale de Siebold	Черная гринда
– Hafenschweinswal	– Harbour Porpoise	Marsouin du Pacifique	Северотихоокеанская морская свинья
Pectinator spekei	Speke's Pectinator	Pectinator de Speke	
Pedetes	Spring Haas	Lièvres sauteurs	Долгоноги
– *cafer*	Cape Spring Haas	Lièvre sauteur d'Afrique du sud	Кафрский долгоног
– *surdaster*	Spring Haas	– – – de l'est	Восточноафриканский долгоног
Pedetidae	– Haas	Lièvres sauteurs	Долгоноги
Pedetoidea	– Haas	– sauteurs	Долгоноговые
Pediolagus salinicola	Salt-desert Cavy	Cobaye halophile	Малая мара
Peking-Frühmensch	Peking Man	Pithécanthrope de Péking	Синантроп
Pel-Dornschwanzhörnchen	Pel's Scaly-tailed Squirrel	Anomalure de Pel	Шипохвостая белка Пеля
Pelzgürteltier	Hairy Armadillo	Tatou poilu	Мохнатый броненосец
Père-Davids-Felsenhörnchen	Rock Squirrel	Écureuil des rochers	Давидова белка
Père-Davids-Wühlmäuse	Pere David's Voles		Давидовы полевки
Perlziesel	Spotted Souslik	Souslik tacheté	Крапчатый суслик
Perognathus	Pocket Mice	Souris à poche	Мешетчатые тушканчики
– *baileyi*	Bailey's Pocket Mouse	– – – de Bailey	Мешетчатый тушканчик Бейлеия
– *californicus*	California Pocket Mouse	– – – de Californie	Калифорнийский мешетчатый тушканчик
– *flavus*	Silky Pocket Mouse	– soyeux à poche	Шелковистый мешетчатый тушканчик
Peromyscus	White-footed Mice	– à pattes blanches	
– *leucopus*	– Mouse	– à pattes blanches	Оленья мышь
– *maniculatus*	Deer Mouse	– du soir	
– *nuttalli*	Golden Mouse	– dorée	Золотистый мышевидный хомячок
– *truei*	Piñon Mouse	– de True	
Persische Wüstenmaus	Persian Jird	Mérione de Perse	Персидская песчанка
Petaurista petaurista	Giant Flying Squirrel	Écureuil volant géant	Тагуан
Petromuridae	Rock Rats		Африканские горные мыши
Petromuroidea	Cane and Rock Rats		Петромиидовые
Petromus typicus	Rock Rat	Rat typique	Горная мышь
Pferdespringer	Five-toed Jerboa	Gerboise à cinq doigts	Большой тушканчик
Philippinen-Gleitflieger	Philippines Flying Lemur		Кагуан
Phloeomyinae		Phloeomyinés	Азиатские древесные крысы
Phloeomys cumingi	Slender-tailed Cloud Rat	Rat d'ecorce tacheté	Пятнистая древесная крыса
Phocaena dioptrica	Spectacled Porpoise	Marsouin des lunettes	Очковая морская свинья
– *phocaena*	Common Porpoise	– commun	Морская свинья
– *sinus*	Pacific Harbour Porpoise	– du Pacifique	Северотихоокеанская морская свинья
– *spinipinnis*	Burmeister's Porpoise	– spinipenne	Морская свинья Бурмейстера
Phocaenidae	Porpoises	Phocénidés	Морские свиньи
Phocaenoides dalli	Dall's Harbour Porpoise	Marsouin de Dall	Северная белокрылая свинья
Phodopus	Dwarf Hamsters	Hamsters nains à queue courte	Мохноногие хомячки
– *roborovskii*	Roborowsky's Dwarf Hamster	Hamster nain de Roborowsky	Хомячок Роборовского
– *sungorus*	Dwarf Hamster	– – de Djoungarie	Джунгарский хомячок
Pholidota	Pangolins	Pholidotes	Панголины
Phyllostomidae	American Leaf-nosed Bats		Листоносы
Phyllostominae	Big-eared Leaf-nosed Bats	Phyllostomes	Настоящие листоносы

GERMAN NAME	ENGLISH NAME	FRENCH NAME	RUSSIAN NAME
Phyllostomus	Spear-nosed Bats	Phyllostomes	
– discolor	Coloured Bat	Phyllostome coloré	
– hastatus	Javelin Bat	– fer de lance	Копьенос
Physeter catodon	Sperm Whale	Cachalot macrocéphale	Кашалот
Physeteridae	– Whales	Physétéridés	Кашалоты
Pinjonmaus	Piñon Mouse	Souris de True	
Pinselschwanzbilche	African Dormice	Graphiures	Кистехвостые сони
Pinselschwanzratten	Boris	Boris	
Pipistrellus	Pipistrelles	Pipistrelles	Нетопыри
– nanus	Banana Bat	Pipistrelle naine à ailes brunes	Банановый нетопырь
– pipistrellus	Common Pipistrelle	Vespertilion pipistrelle	Нетопырь-карлик
Pitymys	Pine Voles		Малые полевки
– pinetorum	– Vole	Campagnol du pin	Американская боровая полевка
– subterraneus	Earth Vole	– souterrain	Подземная полевка
Plagiodontia	Zagoutis	Plagiodontes	Загути
– aedium	Haitian Hutia	Plagiodonte d'Haiti	Загути Кювье
Platacanthomyidae	Spiny Dormice		Шипохвостые сони
Platacanthomys lasiurus	– Dormouse		Южноиндийская соня
Platanista gangetica	Gangetic Dolphin	Dauphin du Gange	Сусук
Platanistidae	– Dolphins	Platanistidés	Гангские дельфины
Platanistoidea	River Dolphins		Речные дельфины
Plecotus	Long-eared Bats	Oreillards	Ушаны
– auritus	– Bat	Oreillard commun	Обыкновенный ушан
– austriacus	Southern Long-eared Bat	– du sud	
Polarrötelmaus	Northern Red-backed Vole	Campagnol boréal	Красная полевка
Pongidae	Anthropoid Apes	Singes anthropomorphes	Человекоподобные
Pottwal	Sperm Whale	Cachalot macrocéphale	Кашалот
Pottwale	– Whales	Physétéridés	Кашалоты
Präriehunde	Prairie Dogs	Chiens des prairies	Луговые собачки
Primates	Primates	Primates	Приматы
Priodontes giganteus	Giant Armadillo	Tatou géant	Гигантский броненосец
Proëchimys guyannensis	Cayenne Spiny Rat	Rat de Guyane	Гвианская крыса
Prometheomys schaposchnikovi	Long-clawed Mole Vole		Прометеева полевка
Prometheus-Maus	– – Vole		Прометеева полевка
Protoxerus stangeri	Oil-palm Squirrel	Grand Écureil de Stanger	Масличная белка
Pseudorca crassidens	False Killer Whale	Pseudorque à dents épaisses	Черная косатка
Pteromyinae	Flying Squirrels	Écureuils volants	Летяги
Pteromys volans	European Flying Squirrel	Palatouche	Обыкновенная летяга
Pteronotus davyi	Naked-backed Bat	Vespertilion à dos nu	
– suapurensis	Suapure Naked-backed Bat	– de Suapuré	
Pteropidae	Fruit Bats	Pteropidés	Крыланы
Pteropinae	Flying Foxes	Roussettes	Длинноносые крыланы
Pteropus	– Foxes	Roussettes	Летучие собаки
– giganteus	Indian Flying Fox	Roussette géante	Летучая лисица
– niger	Black Flying Fox	– noire	
– poliocephalus	Grey-headed Flying Fox	– à tête cendrée	Сероголовый восточно-австралийский крылан
– rufus	Rufous Flying Fox	– rougeâtre	Мадагаскарская летучая собака
– subniger	Dark Flying Fox	– foncée	
– tonganus	Tonga Flying Fox	– de Tonga	Тонгский крылан
– vampyrus	Red-necked Fruit Bat	– à cou rouge	Калонг
Quastenstachler	Brush-tailed Porcupines	Athérures	Кистехвосты
Rattenartiger Zwerghamster	Ratlike Hamster		Крысовидный хомяк
Rattus	Rats	Rats proprement dits	Крысы
– norvegicus	Brown Rat	Rat surmulot	Пасюк
– rattus	House Rat	– commun	Черная крыса
– – alexandrinus	Roof Rat	– d'Aléxandrie	Александрийская черная крыса
– – frugivorus	Corn Rat	– des fruits	
– – rattus	House Rat	– noir	Северозападная черная крыса
Ratufa bicolor	Oriental Giant Squirrel	Écureuil géant	Малайская гигантская белка
– indica	Indian Giant Squirrel	– d'Inde	Индийская гигантская белка
Ratufini	Indomalayan Giant Squirrels		Индомалайские гигантские белки
Rauchgrauer Flughund	Dark Flying Fox	Roussette foncée	
Rauharm-Fledermaus	Lesser Noctule	Noctule de Leisler	Малая вечерница
Rauhfuß-Springmaus	Northern Three-toed Jerboa	Gerboise à pattes rugueuses	Мохноногий тушканчик
Rauhhaar-Schlitznase	Hispid Slit-faced Bat	Nyctère hérissé	
Reisratten	Rice Rats	Rats du riz	

GERMAN NAME	ENGLISH NAME	FRENCH NAME	RUSSIAN NAME
Reithrodontomys	American Harvest Mice	Souris des moissons d'Amérique	
Reithrodontomys megalotis	Rabbit Rat	Souris occidentale des moissons	
Rennmäuse	Gerbils	Gerbillinés	Песчанки
Rhabdomys pumilio	Striped Field Mouse	Rat rayé du champ	
Rheithrosciurus macrotis	Groove-toothed Squirrel	Reithrosciure de Borneo	Борнеоская белка
Rhinolophidae	Horseshoe Bats	Rhinolophidés	Подковоносы
Rhinolophoidea	Horseshoe-nosed Bats	Rhinolophes	Подковоносовые
Rhinolophus	Horseshoe Bats	Rhinolophes	Подковоносы
– *denti*	Dent's Horseshoe Bat	Rhinolophe de Dent	
– *ferrumequinum*	Greater Horseshoe Bat	Grand Rhinolophe obscur	Большой подковонос
– *hildebrandtii*	Hildebrandt's Horseshoe Bat	Rhinolophe d'Hildebrandt	Подковонос Гильдебрандта
– *hipposideros*	Lesser Horseshoe Bat	Petit Rhinolophe	Малый подковонос
– *landeri*	Lander's Horseshoe Bat	Rhinolophe de Lander	
– *maclaudi*	Maclaud's Bat	– de Maclaud	Подковонос Маклауда
Rhinopoma	Rat-tailed Bats	Rhinopômes	
– *hardwickei*	Lesser Rat-tailed Bat	Rhinopôme d'Hardwick	Ланцетонос Хардвика
– *microphyllum*	Larger Rat-tailed Bat	– microphylle	Египетский ланцетонос
Rhinopomatidae	Rat-tailed Bats	Rhinopômes	Ланцетоносы
Rhinosciurus laticaudatus	Long-nosed Squirrel	Écureuil à nez long	Длинноносая белка
Rhizomyidae	Bamboo Rats	Rats des bambous	Ризомииды
Rhizomys	– Rats	– des bambous	Бамбуковые крысы
– *sinensis*	China Bamboo Rat	Rat des bambous Chinois	Китайская ризомиида
– *sumatrensis*	Sumatran Bamboo Rat	– – – de Sumatra	Суматранская ризомиида
Rhombomys opimus	Great Gerbil	Gerbille géante	Большая песчанка
Rhonebiber	Rhone Beaver	Castor du Rhône	Ронский бобр
Rhynchomyinae	Shrew Rats	Rhynchomyinés	
Rhynchomys soricoides	– Rat	Rat au nez	
Rhynchonycteris naso	Proboscis Bat	Rhynchonyctère	Носатая летучая мышь
Riesen-Abendsegler	Giant Noctule	Noctule géante	Гигантская вечерница
Riesengleiter	Flying Lemurs		Шерстокрыловые
Riesengleitflieger	– Lemurs	Cynocephalidés	Шерстокрылы
Riesengürteltier	Giant Armadillo	Tatou géant	Гигантский броненосец
Riesenhamsterratten	– Pouched Rats	Rats géants	Большие хомяковидные крысы
Riesenhörnchen	Oriental Giant Squirrel	Écureuil géant	Малайская гигантская белка
Riesennager	Water Hogs	Hydrochéridés	Водосвинки
Riesenohr-Springmäuse	Long-eared Jerboas		Длинноухие тушканчики
Riesen-Rundblattnase	Commerson's Leaf-nosed Bat	Phyllorine de Commerson	Большой ложный подковонос
Riesen-Schuppentier	Giant Pangolin	Grand Pangolin	Гигантский ящер
Riesentaschenratten	Tuzas	Gaufres	
Rio-Grande-Biber	Rio-Grande Beaver	Castor de Rio-Grande	Бобр Рио-Гранде
Robinson-Röhrennasen-Flughund	Queensland Tube-nosed Fruit Bat	Nyctimène de Robinson	
Roborowski-Zwerghamster	Roborowsky's Dwarf Hamster	Hamster nain de Roborowsky	Хомячок Роборовского
Rodentia	Rodents	Rongeurs	Грызуны
Röhrennasen-Flughunde	Tube-nosed Fruit Bats	Nyctimènes	Трубконосые крыланы
Rohrratten	African Cane Rats		Трионимииды, Тростниковые крысы
Rote Fledermaus	Red Bat	Chauve-souris boréale	Рыжий кожан
– Samt-Fledermaus	– Velvety Free-tailed Bat	Molosse rouge	Рыжая бульдоговая мышь
Rötelmaus	Bank Vole	Campagnol roussâtre	Рыжая полевка
Rötelmäuse	Red-backed Voles	Campagnols	Лесные полевки
Roter Flughund	Rufous Flying Fox	Roussette rougeâtre	Мадагаскарская летучая собака
Rothörnchen	Red Squirrels, Chickaree	Écureuil d'Hudson	Красные белки, Канадская белка
Rotmeer-Tümmler	– Sea Bottle-nosed Dolphin	Tursiops du Mer Rouge	Красноморская афалина
Rotrücken-Dornschwanzhörnchen	Red-backed Scaly-tailed Squirrel	Anomalure à dos rouge	
Rotschenkelhörnchen	African Striped Squirrels	Écureuils rayés d'Afrique	
Rotschwanzhörnchen	Mantled African Bush Squirrel	Écureuil d'Afrique à manteau	
Rotschwänzige Rennmaus	Libyan Jird	Mérione de Lybie	Краснохвостая песчанка
Rotziesel	Red Souslik	Spermophile rouge	Большой суслик
Rousettus	Roussette Bats	Roussettes	Ночные крыланы
– *aegyptiacus*	Egyptian Fruit Bat	Roussette d'Égypte	Египетский крылан
– *angolensis*	Angola Fruit Bat	– d'Angola	Ангольский крылан
Rundblattnasen	Old World Leaf-nosed Bats		Ложные подковоносы
Rundkopfdelphin	Risso's Dolphin	Dauphin de Risso	Серый дельфин
Rundschwanzziesel	Round-tailed Ground Squirrel	Spermophile à queue ronde	Круглохвостый суслик
Saccopteryx	Sheath-tailed Bats	Saccopteryx	Мешкокрылые летучие мыши
– *bilineata*	El Salvador Sheath-tailed Bat	– à deux raies	Двухполосая мешкокрылая мышь

GERMAN NAME	ENGLISH NAME	FRENCH NAME	RUSSIAN NAME
Saccostomus	Cape Pouched Mice		Короткохвостые хомяковидные крысы
Salpingotus	Pygmy Jerboas		Трехпалые карликовые тушканчики
Salzkrautbilch	Desert Dormouse	Souris du désert	Боялычная соня
Salzkrautbilche	Betpakdala Dormice	– de Selevin	Селевиниевые
Samt-Fledermäuse	Velvety Free-tailed Bats	Molosses	
Sandgräber	Mole Rats	Rats taupes africains	Землекопы
Sandmäuse	Jirds	Mériones	Малые песчанки
Sania	Armoured Spiny Rat	Échimys armé	
Schadenbergs Borkenratte	Bushy-tailed Cloud Rat	Rat d'ecorce de Schadenberg	Древесная крыса Шаденберга
Schermaus	Vole Rat	Campagnol terrestre	Водяная крыса
Schermäuse	Water Voles		Водяные крысы
Schildwurf	Lesser Pichiciago	Chlamyphore tronqué	Аргентинский плащеносец
Schimpanse	Chimpanzee	Chimpansé	Обыкновенный шимпанзе
Schimpansen	Chimpanzees	Chimpansés	Шимпанзе
Schläfer	Dormice	Loirs	Сони
Schlangenzahn-Flughund	Snake-toothed Fruit Bat	Scotonyctère à dent du serpent	
Schlankdelphin	Narrow-snouted Dolphin	Dauphin douteux	Стройный продельфин
Schlichtziesel	European Souslik	Souslik d'Europe	Серый суслик
Schlitznasen	Slit-faced Bats	Nyctères	Никтериды
Schmalkragen-Flughund	Little Collared Fruit Bat	Myonyctère à collier	
Schmalnasen	Old World Monkeys	Catarrhiniens	Узконосые обезьяны или Обезьяны Старого Света
Schnabelwale	Beaked Whales	Ziphiidés	Клюворылые
Schneemaus	Snow Vole	Campagnol des neiges	Снежная полевка
Schnellwühler	African Mole Rats	Rats taupes africains	Африканские кротовидные крысы
Schnurrbartmaus	Broadtoothed Field Mouse	Mulot rupestre	Горная мышь
Schuppentiere	Pangolins, Scaly Ant-eaters	Pholidotes, Manidés	Панголины, Ящеры
Schwarzbauchhamster	Common Hamster	Hamster d'Europe	Обыкновенный хомяк
Schwarzbauch-Wühlmaus	Pere David's Vole	Campagnol de Père David	Давидова полевка
Schwarzer Flughund	Black Flying Fox	Roussette noire	
Schwarzschwanz-Baumratte	Black-tailed Hutia	Rat à queue noire	Чернохвостая кубинская крыса
Schwarzschwanz-Präriehund	– Prairie Dog	Cynomys social	Чернохвостая луговая собачка
Schwarzwale	Giant Bottle-nosed Whales		Плавуны
Schwarz-Weiß-Delphine	Commerson's Dolphins	Dauphins de Commerson	Трезубцевые дельфины
Schweinswal	Common Porpoise	Marsouin commun	Морская свинья
Schweinswale	Porpoises	Phocénidés	Морские свиньи
Schwertnase	Tome's Long-eared Bat	Vespertilion de Tome	Малый вампир
Schwertwal	Common Killer Whale	Orque épaulard	Косатка
Schwimmratten	Water Rats, Australian Water Rats	Hydromyinés, Rats d'eau d'Australie	Австралийские водяные крысы
Sciuridae	Typical Squirrels	Écureuils	Беличьи
Sciurinae	Ground and Tree Squirrels		Земляные и древесные беличьи
Sciurini	Tree Squirrels		Настоящие белки
Sciuroidea		Écureuils	Белкообразные
Sciuromorpha	Squirrel-like Rodents		Белкоподобные
Sciurotamias	Rock Squirrels		Китайские красные белки
– *davidianus*	– Squirrel	Écureuil des rochers	Давидова белка
Sciurus	European and American Tree Squirrels	Écureuils	Белки
– *aestuans*	Brazilian Squirrel	Écureuil de la Guyane	Бразильская белка
– *anomalus*	Caucasian Squirrel	– du Caucase	Кавказская белка
– *carolinensis*	Grey Squirrel	– gris	Американская серая белка
– *vulgaris*	Red Squirrel	– commun	Обыкновенная белка
Scotonycteris ophiodon	Snake-toothed Fruit Bat	Scotonyctère à dent du serpent	
– *zenkeri*	Zenker's Fruit Bat	– de Zenker	Крылан Ценкера
Sechsbinden-Gürteltier	Six-banded Armadillo	Tatou à six bandes	Шестипоясный броненосец
Seiden-Taschenmaus	Silky Pocket Mouse	Souris soyeux à poche	Шелковистый мешетчатый тушканчик
Seiwal	Sei Whale	Balénoptère boréal	Сайдяной кит
Selevinia betpakdalaensis	Desert Dormouse	Souris du désert	Боялычная соня
Seleviniidae	Betpakdala Dormice	– de Selevin	Селевиниевые
Shepherd-Wal	Shepherd's Beaked Whale	Tasmacète de Shepherd	Новозеландский клюворыл Шеперда
Sibirische Springmaus	Sibirian Jerboa	Gerboise de Sibérie	Тушканчик-прыгун
Sicista	Birch Mice	Sicistes	Мышовки
– *betulina*	Northern Birch Mouse	– de bouleaux	Лесная мышовка
– *subtilis*	Southern Birch Mouse	– des steppes	Степная мышовка

GERMAN NAME	ENGLISH NAME	FRENCH NAME	RUSSIAN NAME
Sicistinae	Birch Mice	Sicistinés	Мышовки
Siebenbindengürteltier	Seven-banded Armadillo	Tatou à sept bandes	Семипоясный броненосец
Siebenschläfer	Fat Dormouse	Loir gris	Соня-полчок
Sigmodon	Cotton Rats	Sigmondons	Хлопковые хомячки
– hispidus	Andean Swamp Rat	Sigmodon velu	
Simiae	Monkeys	Singes	Обезьяны
Sinai-Stachelmaus	Sinai Spiny Mouse	Souris épineuse de Sinai	Бледная ежовая мышь
Sotalia fluviatilis	Buffeo Negro	Sotalie	Амазонский речной дельфин
– guianensis	Guyanian River Dolphin	– de la Guyane	Гвианский речной дельфин
Sousa sinensis	Chinese White Dolphin	– de Chine	Китайский бледный речной дельфин
– teuszi	West African White Dolphin	– de Teusz	Камерунский речной дельфин
Sowerby-Zweizahnwal	Sowerby's Whale	Mésoplodon de Sowerby	Атлантический ремнезуб
Spalacidae	Palaearctic Mole Rats	Rats taupes	Слепыши
Spalacopus cyanus	Cururo	Rat bleu	Куруро
Spalax	Mole Rats	Rats taupes	Слепыши
– ehrenbergi	Ehrenberg's Mole Rat	Spalax d'Ehrenberg	Слепыш Эренберга
– leucodon	Lesser Mole Rat	– occidental	Малый слепыш
– microphthalmus	Greater Mole Rat	– oriental	Обыкновенный слепыш
Spätfliegende Fledermaus	Serotine Bat	Sérotine commune	Поздний кожан
Speke-Kammfinger	Speke's Pectinator	Pectinator de Speke	
Spermophilopsis leptodactylus	Long-clawed Ground Squirrel	Spermophile leptodactyle	Тонкопалый суслик
Spitzmaus-Langzüngler	Long-tongued Bat	Glossophage de Pallas	Глоссофага
Spitzschnauzendelphine	Beaked Whales	Ziphiidés	Клюворылые
Spitzschnauzenratte	Shrew Rat	Rat au nez	
Spitzzahn-Flughunde	Harpy Fruit Bats	Harpionyctères	Крыланы-гарпии
Springhasen	Spring Haas	Lièvres sauteurs	Долгоноги
Springhasenartige	– Haas	– sauteurs	Долгоноговые
Springmäuse	Jerboas	Gerboises	Тушканчики
Springnager	Jerboas	Gerboises	Тушканчики
Stachelbilche	Spiny Dormice		Шипохвостые сони
Stachelmäuse	– Mice	Souris épineuses	Ежовые мыши
Stachelratten	– Rats	Échimyidés	Цепкохвостые щетинистые крысы
Stachelratten i. e. S.	– Rats	Échimyinés	Щетинистые крысы
Stachelschweine	Old World Porcupines	Hystricidés	Дикобразы Старого Света
Stachelschweinverwandte	Porcupine-like Rodents		Дикобразовые
Steatomys	Fat Mice	Rats adipeux	
– pratensis	– Mouse	Rat adipeux	Жирная древесная мышь
Stenella	Spottes Dolphins	Dauphins tachetés	Продельфины
– attenuata	Narrow-snouted Dolphin	Dauphin douteux	Стройный продельфин
– caeruleoalba	Blue-white Dolphin	– bleu-blanc	Однополосый продельфин
Stenidae	Long-snouted Dolphins	Sténidés	Длинноклювые дельфины
Stenodelphis blainvillei	La Plata Dolphin	Dauphin de la Plata	Лаплатский дельфин
Stenoderminae	Red-fruit Bats		Плодовые вампиры
Steppenbirkenmaus	Southern Birch Mouse	Siciste des steppes	Степная мышовка
Steppenlemming	Sagebrush Vole		Степная пеструшка
Steppenlemminge	– Voles		Степные пеструшки
Steppenmurmeltier	Bobac Marmot	Marmotte Bobac	Степной сурок
Steppen-Schuppentier	Cape Pangolin	Pangolin de Temminck	Степной ящер
Stictomys taczanowskii	Mountain Paca	Paca des montagnes	Горная пака
Strauchratten	South American Bush Rats	Octodons	Кустарниковые крысы
Streifenbackenhörnchen	Eastern Chipmunk	Tamia strié	Американский бурундук
Streifenhörnchen	Western Chipmunks	Eutamias	Бурундуки
Streifen-Hüpfmäuse	Birch Mice	Sicistinés, Sicistes	Мышовки
Streifenmaus	Southern Birch Mouse	Siciste des steppes	Степная мышовка
Streifenziesel	Thirteen-striped Ground Squirrel	Spermophile à treize bandes	Полосатый американский суслик
Stummeldaumen	Smoky Bats	Furipteridés	
Stummelschwanzagutis	Agoutis	Agoutis	Куцые агути
Stummelschwanzhörnchen	Mountain Beavers	Castors de montagne	Горные бобры, Горный бобр
Sturnira lilium	Yellow-shouldered Bat	Sturnire fleur-de-lys	
Sturnirinae	– Bats	Sturnires	
Südafrikanischer Springhase	Cape Spring Haas	Lièvre sauteur d'Afrique du sud	Кафрский долгоног
Südafrikanisches Stachelschwein	South African Porcupine	Porc-épic d'Afrique du sud	Южноафриканский дикобраз
Südamerikanische Felsenratte	– American Rock Rat	Rat foncu	
Südindischer Stachelbilch	Spiny Dormouse		Южноиндийская соня
Südliche Grashüpfermaus	Southern Grasshopper Mouse	Onychomys du sud	
– Lemmingmaus	– Bog Lemming	Lemming du sud	Южная лемминговая мышь

GERMAN NAME	ENGLISH NAME	FRENCH NAME	RUSSIAN NAME
Südlicher Entenwal	– Bottle-nosed Whale	Hypérodon du sud	Плосколобый бутылконос
– Glattdelphin	Peron's Dolphin	Dauphin de Peron	Южный китовидный дельфин
– Glattwal	Ice Baleen Whale	Baleine australe	Австралийский кит
– Mull-Lemming	Afghan Mole Lemming	Rat taupe d'Afghan	Афганская слепушонка
– Schwarzwal	Arnoux's Whale	Baleine d'Arnoux	Южный плавун
Südliches Zwergmeerschweinchen	Southern Mountain Cavy		Южная карликовая свинка
Suira	Suira	Rat épineux	
Sumatra-Bambusratte	Sumatran Bamboo Rat	– des bambous de Sumatra	Суматранская ризомиида
Sumpfbiber	Nutria	Ragondin	Нутрия
Sumpf-Reisratte	Rice Rat	Rat du riz	Рисовая мышь
Synaptomys	Lemming Mice		Лемминговые мыши
– *borealis*	Northern Bog Lemming	Lemming du nord	Северная лемминговая мышь
– *cooperi*	Southern Bog Lemming	– du sud	Южная лемминговая мышь
Syrischer Goldhamster	Golden Hamster	Hamster doré	Переднеазиатский хомяк
Tachyoryctes	African Mole Rats	Rats taupes africains	Африканские кротовидные крысы
Tadarida	Free-tailed Bats	Tatarides	Складчатогубы
– *brasiliensis mexicana*	Mexican Free-tailed Bat	Tataride de Mexique	
– *limbata*	Braided Free-tailed Bat	– bordée	
– *teniotis*	European Free-tailed Bat	– bouledogue	Широкоухий складчатогуб
Taguan	Giant Flying Squirrel	Écureuil volant géant	Тагуан
Tamandua	Tamandua	Tamandua à quatre doigts	Четырехпалый муравьед
Tamandua tetradactyla	Tamandua	– – – doigts	Четырехпалый муравьед
Tamias	Eastern Chipmunks	Tamias	Чипмунки
– *alpinus*	Alpine Chipmunk	Néotamia de montagne	Горный чипмунк
– *amoenus*	Yellow Pine Chipmunk	– jaune	Желтый чипмунк
– *minimus*	Least Chipmunk	– nain	Малый чипмунк
– *quadrivittatus*	Colorado Chipmunk	– du Colorado	Колорадский чипмунк
– *striatus*	Eastern Chipmunk	Tamia strié	Американский бурундук
– *townsendii*	Townsend's Chipmunk	Néotamia de Townsend	Чипмунк Тоунсенда
Tamiasciurini	Red Squirrels		Красные белки
Tamiasciurus	Chickarees		Североамериканские красные белки
– *hudsonicus*	Chickaree	Écureuil d'Hudson	Канадская белка
Taphozous	Tomb Bats	Taphiens	Могильные летучие мыши
– *mauritianus*	Mauritian Tomb Bat	Taphien de Maurice	
– *nudiventris*	Naked-bellied Tomb Bat	– à ventre nu	Голобрюхая летучая мышь
– *perforatus*	Tomb Bat	– perforé	
Taschenfledermäuse	Sheath-tailed Bats	Saccopteryx	Мешкокрылые летучие мыши
Taschenmäuse	Kangaroo Mice	Souris kangourou	Мешетчатые крысы
Taschennager	Pocket Gophers	Gaufres à poche	Мешетчатые грызуны
Taschenratten	– Gophers	– à poche	Гоферы
Taschenspringer	Kangaroo Rats	Rats kangourou	Мешетчатые прыгуны
Tasmacetus shepherdi	Shepherd's Beaked Whale	Tasmacète de Shepherd	Новозеландский клюворыл Шеперда
Tatera	Large Gerbils		Голоступые песчанки
– *indica*	– Gerbil	Gerbille d'Inde	Индийская песчанка
Taterillus	Naked-soled Gerbils		Малые голоступые песчанки
– *emini*	Small Naked-soled Gerbil	Gerbille d'Emin	
Teichfledermaus	Pond Bat	Vespertilion des marais	Прудовая ночница
Temminck-Gleitflieger	Malayan Flying Lemur		Шерстокрыл Темминка
Texas-Känguruhratte	Texas Kangaroo Rat	Rat kangourou de Texas	Техасский мешетчатый прыгун
Thecurus	Indonesian Porcupines	Porcs-épic d'Indonésie	Индонезийские дикобразы
– *pumilis*	– Porcupine	Porc-épic d'Indonésie	Карликовый дикобраз
Thomomys	Western Pocket Gophers	Gaufres à poche des montagnes	
– *bottae*	– – Gopher	Gaufre à poche des montagnes	Горный гофер
Thryonomyidae	African Cane Rats		Триономииды
Thryonomys	– – Rats		Тростниковые крысы
– *gregorianus*	Lesser Cane Rat	Rat de Grégorian	Малая тростниковая крыса
– *swinderianus*	Great Cane Rat	Aulacode	Большая тростниковая крыса
Thyroptera discifera	Honduran Disk-winged Bat	Vespertilion d'Honduras	
– *tricolor*	Spix's Disk-winged Bat	– tricolore	Трехцветная летучая мышь
Thyropteridae	Disk-winged Bats	Thyropteridés	Американские присосковые летучие мыши
Tibetanischer Zwerghamster	Tibetan Ratlike Hamster	Hamster nain du Tibet	Тибетский хомячок
Tolypeutes	Three-banded Armadillos	Tatous à trois bandes	
– *matacus*	– Armadillo	Tatou à trois bandes	Шаровидный броненосец

GERMAN NAME	ENGLISH NAME	FRENCH NAME	RUSSIAN NAME
– *tricinctus*	Apara	Apar de Buffon	Трехпоясный броненосец
Tolypeutini	Three-banded Armadillos	Tatous à trois bandes	
Tonga-Flughund	Tonga Flying Fox	Roussette de Tonga	Тонгский крылан
Townsend-Chipmunk	Townsend's Chipmunk	Néotamia de Townsend	Чипмунк Тоунсенда
Triaenops persicus	Persian Leaf-nosed Bat	Triaenops de Perse	Персидский ложный подковонос
Trichterohr	Funnel-eared Bat	Vespertilion à couleur de paille	
Trichterohren	– Bats	Natalidés	
Trichys fasciculata	Long-tailed Porcupine	Porc-épic à longue queue	Борнеоский дикобраз
Tristram-Wüstenmaus	Tristram's Jird	Mériones de Tristram	
True-Wal	True's Beaked Whale	Mésoplodon de True	Североатлантический ремнезуб
Trugratten	Octodont Rodents	Octodontidés	Осьмизубые
Trugrattenartige	Octodont Rodents		Осьмизубовые
Tscherskia triton	Ratlike Hamster		Крысовидный хомяк
Tschudi-Meerschweinchen	Wild Cavy	Cobaye sauvage	Перуанская морская свинка
Tukotukos	Tuco-tucos	Tuco-tucos	
Tümmler	Bottle-nosed Dolphins	Tursiops, Tursiops tronqué	Афалины, Североатлантическая афалина
Tursiops	Bottle-nosed Dolphins	Tursiops	Афалины
– *aduncus*	Red Sea Bottle-nosed Dolphin	– du Mer Rouge	Красноморская афалина
– *gillii*	Gill's Dolphin	Dauphin de Gill	Афалина Гилля
– *truncatus*	Bottle-nosed Dolphin	Tursiops tronqué	Североатлантическая афалина
Tuzas	Yellow Pocket Gophers	Rats à poche	
Tylonycteris pachypus	Flat-headed Bat	Vespertilion du bambou	Бамбуковая летучая мышь
Typhlomys cinereus	Chinese Pygmy Dormouse	Souris naine de Chine	Китайская карликовая соня
Umber-Fledermaus	Northern Bat	Sérotine boréale	Северный кожанок
Unau	Unau	Unau commun	Унау
Unechter Schwertwal	False Killer Whale	Pseudorque à dents épaisses	Черная косатка
Uralbiber	Ural Beaver	Castor d'Ural	Уральский бобр
Uroderma bilobatus	Tent-building Bat	Vespertilion bilobé	
Uromys	Giant Naked-tailed Rats	Rats à queue en mosaique	
– *anak*	– – Rat	Rat à queue en mosaique des montagnes	
Urson	North American Porcupine	Porc-épic nord américain	Североамериканский иглошерст
Vampyrum	False Vampire Bats	Faux Vampires	
– *spectrum*	Great False Vampire	– Vampire commun	Большой вампир
Vandeleuria	Long-tailed Climbing Mice	Souris à longue queue	Индийские пальмовые мыши
– *oleracea*	– – Mouse	– – – queue	Пальмовая мышь
Vespertilio	Particoloured Bats	Sérotines bicolores	Двухцветные кожаны
– *murinus*	– Bat	Petite Chauve-souris murine	Двухцветный кожан
Vespertilionidae	Vespertilionid Bats	Vespertilionidés	Обыкновенные летучие мыши
Vespertilionoidea	Vespertilionid Bats		Гладконосые летучие мыши
Vielzitzenmäuse	Multimammate Rats	Rats à mamelles multiples	
Vielzitzenratten	– Rats	– – – multiples	
Viscacha	Viscacha	Lagostome des Pampas	Вискаша
Viscacharatte	– Rat	Rat minime	Вискашевидная крыса
Vorderindisches Schuppentier	Indian Pangolin	Pangolin indien	Переднеиндийский ящер
Vormensch	Ape-man	Australopithèque	Австралопитеки
Votsotsa	Votsotsa		Вотсотса
Wahlberg-Epauletten-Flughund	Wahlberg's Epauletted Fruit Bat	Épomophore de Wahlberg	Крылан Вальберга
Waldbiber	Michigan Beaver	Castor de Michigan	Мичиганский бобр
Waldhüpfmaus	Woodland Jumping Mouse	Zapode des bois	Лесной прыгунчик
Waldlemming	Wood Lemming	Lemming des forêts	Лесной лемминг
Wald- und Feldmäuse	– and Field Mice	Mulots	Лесные и полевые мыши
Waldmurmeltier	Woodchuck	Monax	Лесной сурок
Waldwühlmaus	Bank Vole	Campagnol roussâtre	Рыжая полевка
Wale	Whales	Cétacés	Китообразные
Waltiere	Whales	Cétacés	Китообразные
Wanderratte	Brown Rat	Rat surmulot	Пасюк
Wasserfledermaus	Water Bat	Vespertilion de Daubenton	Водяная ночница
Wasserschwein	Capybara	Capybara	Капибара
Weichgürteltiere	Armadillos	Tatous	
Weißbauch-Schuppentier	Three-pointed Pangolin	Tricuspide	Белобрюхий ящер
Weißborsten-Gürteltier	Six-banded Armadillo	Tatou à six bandes	Шестипоясный броненосец
Weißfußmaus	White-footed Mouse	Souris à pattes blanches	Оленья мышь
Weißgraue Fledermaus	Hoary Bat	Chauve-souris cendrée	Бело-серый кожан

GERMAN NAME	ENGLISH NAME	FRENCH NAME	RUSSIAN NAME
Weißschnauzendelphin	White-beaked Dolphin	Lagénorhynque à rostre blanc	Беломордый дельфин
Weißschwanz-Antilopenziesel	White-tailed Antelope Squirrel	Spermophile d'antilope à queue blanche	Белохвостый антилоповый суслик
Weißschwanz-Präriehund	– Prairie Dog	Cynomys de Gunnison	Белохвостая луговая собачка
Weißschwanz-Stachelschwein	– Porcupine	Porc-épic à queue blanche	Индийский дикобраз
Weißschwänziger Hamster	– Rat	Hamster à queue blanche	Белохвостый хомяк
Weißschwingen-Vampir	White-winged Vampire Bat	Vampire d'ailes blanches	
Weißseitendelphin	White-sided Dolphin	Lagénorhynque à bec pointu	Белобокий дельфин
Weißstreifendelphin	Pacific White-sided Dolphin	– de Gill	Тихоокеанский белобокий дельфин
Weißwal	White Whale	Dauphin blanc	Белуха
Welwitsch-Fledermaus	Welwitsch's Bat	Vespertilion de Welwitsch	
Westafrikanischer Quastenstachler	Western African Brush-tailed Porcupine	Athérure africain	Африканский кистехвост
Westblindmaus	Lesser Mole Rat	Spalax occidental	Малый слепыш
Westliche Erntemaus	Rabbit Rat	Souris occidentale des moissons	
– Hausmaus	Western House Mouse		Западноевропейская домовая мышь
Westliches Rotschenkelhörnchen	– African Striped Squirrel	Écureuil rayé d'Afrique occidental	
West-Schermaus	Water Vole	Campagnol amphibie	Западноевропейская водяная крыса
Whitehead-Spitzzahn-Flughund	Whitehead's Harpy Fruit Bat	Harpionyctère de Whitehead	
Whitehead-Zwerghörnchen	– Dwarf Squirrel	Écureuil pygmée de Whitehead	Белка Уайтхеда
Wiesenhüpfmaus	Meadow Jumping Mouse	Zapode du Canada	Североамериканский луговой прыгунчик
Wiesenwühlmaus	Eastern Meadow Mouse	Campagnol de Pennsylvania	Пенсильванская полевка
Wilsons Riesenhörnchen	African Palm Squirrel	Écureuil de Wilson	Гигантская белка Вильсона
Wollgreifstachler	Woolly Prehensile-tailed Porcupine	Porc-épic laineux	Мохнатый цепкохвост
Wood-Chuck	Woodchuck	Monax	Лесной сурок
Wühler	Cricetid Rats and Mice	Cricétidés	Хомячьи
Wurzelratten	Bamboo Rats	Rats des bambous	Ризомииды
Wüsten-Fledermaus	Pallid Bat	Oreillard pâle	Бледная летучая мышь
Wüsten-Känguruhratte	Desert Kangaroo Rat	Rat kangourou du désert	Пустынный мешетчатый прыгун
Wüstenmäuse	Jirds	Mériones	Малые песчанки
Wüstenratte	Whitethroat Wood Rat	Rat des steppes	Пустынный кустовый хомячок
Wüstenspringmaus	Desert Jerboa	Gerboise du steppe	Песчаный тушканчик
Xenarthra	Xenarthra	Xénarthrés	
Xerini	Bristly Ground Squirrels		Земляные белки
Xerus	African Ground Squirrels	Écureuils foisseurs	Африканские земляные белки
Zagutis	Zagoutis	Plagiodontes	Загути
Zahnarme	Edentates	Édentés	Неполнозубые
Zahnlose	Edentates	Édentés	Неполнозубые
Zahnwale	Toothed Cetaceans	Odontocétés	Зубатые киты
Zapodidae	Jumping Mice	Zapodidés	Мышовки и прыгунчики
Zapodinae	True Jumping Mice	Zapodinés	Прыгунчики
Zapus	Meadow Jumping Mice	Zapodes des prées	
– hudsonius	– – Mouse	Zapode du Canada	Североамериканский луговой прыгунчик
Zenkerella	Flightless Scaly-tailed Squirrels	Zenkerelles	Шипохвостые сони
– insignis	– – Squirrel	Zenkerelle	Шипохвостая соня
Zenker-Flughund	Zenker's Fruit Bat	Scotonyctère de Zenker	Крылан Ценкера
Zenker-Gleitbilch	– Small Flying Squirrel	Anomalure nain de Zenker	Соня-летяга Ценкера
Ziesel	Ground Squirrels, European Souslik	Sousliks, Souslik d'Europe	Суслики, Серый суслик
Zieselmaus	Long-clawed Ground Squirrel	Spermophile leptodactyle	Тонкопалый суслик
Ziphiidae	Beaked Whales	Ziphiidés	Клюворылые
Ziphius cavirostris	Goosebeak Whale	Baleine du bec des oies	Настоящий клюворыл
Zokor	North Altai Zokor	Rat taupe	Алтайский цокор
Zweifarbige Fledermaus	Particoloured Bat	Petite Chauve-souris murine	Двухцветный кожан
– Fledermäuse	– Bats	Sérotines bicolores	Двухцветные кожаны
Zweifingerfaultiere	Two-toed Sloths	Unaus	Двупалые ленивцы
Zweistreifen-Taschenfledermaus	El Salvador Sheath-tailed Bat	Saccopteryx à deux raies	Двухполосая мешкокрылая мышь
Zwergameisenbär	Two-toed Ant-eater	Fourmilier didactyle	Двупалый муравьед
Zwergbaumratte	Tiny Hutia	Rat nain	Карликовая кубинская крыса
Zwerg-Dornschwanzhörnchen	Pygmy Scaly-tailed Squirrel	Anomalure nain	Карликовая шипохвостая белка

GERMAN NAME	ENGLISH NAME	FRENCH NAME	RUSSIAN NAME
Zwerg-Epauletten-Flughund	Dwarf Epauletted Fruit Bat	Petit Microptére	Карликовый эполетовый крылан
Zwergfledermaus	Common Pipistrelle	Vespertilion pipistrelle	Нетопырь-карлик
Zwergfledermäuse	Pipistrelles	Pipistrelles	Нетопыри
Zwergglattwal	Pygmy Right Whale	Baleine naine	Карликовый гладкий кит
Zwerggrindwal	– Killer		Карликовая косатка
Zwerggürteltier	– Armadillo	Tatou nain	Карликовый броненосец
Zwerghörnchen	Dwarf Tree Squirrel	Microsciure nain	
Zwerg-Langzungen-Flughund	Asiatic Long-tongued Fruit Bat	Macroglosse minime	Малый длинноязычный крылан
Zwergmara	Salt-desert Cavy	Cobaye halophile	Малая мара
Zwergmeerschweinchen	Mountain Cavies		Карликовые свинки
Zwergpottwal	Pygmy Sperm Whale	Cachalot pygmée	Карликовый кашалот
Zwergschimpanse	– Chimpanzee	Chimpansé nain	Бонобо
Zwergschläfer	– Dormouse	Graphiure nain	Карликовая африканская соня
Zwergstachelschwein	Indonesian Porcupine	Porc-épic d'Indonésie	Карликовый дикобраз
Zwergwal	Lesser Rorqual	Petit Rorqual	Малый остромордый полосатик
Zwergziesel	Little Souslik	Souslik nain	Малый суслик

3. French—German—English—Russian

Dans la plupart des cas, les noms des sous-espèces sont formés en ajoutant au nom de l'espèce un adjectif ou une désignation géographique. Dans cette partie du dictionnaire zoologique, les noms français des sous-espèces formés de cette manière ne seront en général pas indiqués.

FRENCH NAME	GERMAN NAME	ENGLISH NAME	RUSSIAN NAME
Abrocomidés	Chinchillaratten	Rat Chinchillas	Шиншилловые крысы
Acanthion de Java	Java-Stachelschwein	Javenese Porcupine	Яванский дикобраз
– de Malaysie	Kurzschwanz-Stachelschwein	Malayan Porcupine	
Acouchi	Acouchi	Acouchi	Акучи
Acouchis	Geschwänzte Agutis	Acouchis	Хвостатые агути
Agouti d'Azara	Azara-Aguti	Azara Agouti	Азарская агути
– doré	Goldaguti	Orange-rumped Agouti	Золотистая агути
Agoutis	Stummelschwanzagutis	Agoutis	Куцые агути
Anomalure à dos rouge	Rotrücken-Dornschwanzhörnchen	Red-backed Scaly-tailed Squirrel	
Anomalure de Beecroft	Beecroft-Dornschwanzhörnchen	Beecroft's Scaly-tailed Squirrel	Шипохвостая белка Бикрофта
– de Belden	Belden-Dornschwanzhörnchen	Belden's Scaly-tailed Squirrel	Шипохвостая белка Бельдена
– de Fraser	Fraser-Dornschwanzhörnchen	Fraser's Scaly-tailed Squirrel	Шипохвостая белка Фразера
– de Jackson	Jackson-Dornschwanzhörnchen	Jackson's Scaly-tailed Squirrel	Шипохвостая белка Джексона
– de Neave	Neave-Dornschwanzhörnchen	Neave's Scaly-tailed Squirrel	
– de Pel	Pel-Dornschwanzhörnchen	Pel's Scaly-tailed Squirrel	Шипохвостая белка Пеля
– nain	Zwerg-Dornschwanzhörnchen	Pygmy Scaly-tailed Squirrel	Карликовая шипохвостая белка
– – à longues oreilles	Großohr-Gleitbilch	Large-eared Small Flying Squirrel	Ушастая соня-летяга
– – de Lang	Lang-Gleitbilch	Lang's Small Flying Squirrel	Соня-летяга Ланга
– – de Zenker	Zenker-Gleitbilch	Zenker's Small Flying Squirrel	Соня-летяга Ценкера
Anomalures	Eigentliche Dornschwanzhörnchen	Scaly-tailed Squirrels	Шипохвостые белки
– nains	Gleitbilche	African Small Flying Squirrels	Сони-летяги
Anomaluridés	Dornschwanzhörnchen	Scaly-tailed Squirrels	Шипохвостые белки-летяги
Anthops orné	Blumennasen-Fledermaus	Flower-faced Bat	Цветконосый ложный подковонос
Apar de Buffon	Dreibinden-Kugelgürteltier	Apara	Трехпоясный броненосец
Artibées		Neotropical Fruit-eating Bats	Тупомордые листоносы
Asellia à trois endentures	Dreizack-Blattnase	Trident Leaf-nosed Bat	Трехзубчатоносый ложный подковонос
Athérure africain	Westafrikanischer Quastenstachler	West African Brush-tailed Porcupine	Африканский кистехвост
– à longue queue	Langschwanz-Quastenstachler	Asiatic Brush-tailed Porcupine	Малайский кистехвост
Athérures	Quastenstachler	Brush-tailed Porcupines	Кистехвосты

FRENCH NAME	GERMAN NAME	ENGLISH NAME	RUSSIAN NAME
Aulacode	Große Rohrratte	Great Cane Rat	Большая тростниковая крыса
Australopithèques	Vormenschen	Ape-men	Австралопитеки
Baleine à Bosse	Buckelwal	Humpback Whale	Горбач
– australe	Südlicher Glattwal	Ice Baleen Whale	Австралийский кит
– boréale	Grönlandwal	Bowhead	Гренландский кит
– d'Arnoux	Südlicher Schwarzwal	Arnoux's Whale	Южный плавун
– de Baird	Baird-Wal	Baird's Beaked Whale	Северный плавун
– de Siebold	Nordpazifik-Glattwal	Pacific Right Whale	Японский кит
– du bec des oies	Cuvier-Schnabelwal	Goosebeak Whale	Настоящий клюворыл
– franche	Grönlandwal	Bowhead	Гренландский кит
– grise	Grauwal	Grey Whale	Серый кит
– naine	Zwergglattwal	Pygmy Right Whale	Карликовый гладкий кит
– noire	Nordkaper	Atlantic Right Whale	Бискайский кит
Baleines	Bartenwale, Glattwale	Baleen Whales, Right Whales	Усатые киты, Гладкие киты
– à bec	Entenwale	Bottle-nosed Whales	Бутылконосы
Balénoptère bleu	Blauwal	Blue Whale	Синий кит
– boréale	Seiwal	Sei Whale	Сайдяной кит
– de Bryde	Brydewal	Bryde's Whale	Полосатик Брайда
Bandicoot rat d'Inde	Bandikutratte	Bandicoot Rat	Индийская бандикутовая крыса
– – du Bengale	Indische Maulwurfsratte	Bengali Bandicoot Rat	Бенгальская бандикутовая крыса
– rats	Bandikutratten	Bandicoot Rats	Бандикутовые крысы
Barbastelle d'Europe	Mopsfledermaus	Barbastelle	Европейская широкоушка
Barbastelles	Mopsfledermäuse	Barbastelles	Широкоушки
Bori	Bori	Bori	Бори
Boris	Pinselschwanzratten	Boris	
Bradype	Dreifingerfaultier	Three-toed Sloth	Ай
– à capuchon	Kapuzenfaultier		Капюшонный ленивец
– à collier	Kragenfaultier	Necklace Sloth	Ошейниковый ленивец
Bradypes	Dreifingerfaultiere	Three-toed Sloths	Трехпалые ленивцы
Cachalot macrocéphale	Pottwal	Sperm Whale	Кашалот
– pygmée	Zwergpottwal	Pygmy Sperm Whale	Карликовый кашалот
Campagnol agreste	Erdmaus	Field Vole	Темная полевка
– amphibie	West-Schermaus	Water Vole	Западноевропейская водняная крыса
– boréal	Polarrötelmaus	Northern Red-backed Vole	Красная полевка
– de Brandt	Brandt's Steppenwühlmaus	Brandt's Vole	Полевка Брандта
– de Pennsylvania	Wiesenwühlmaus	Eastern Meadow Mouse	Пенсильванская полевка
– de Père David	Schwarzbauch-Wühlmaus	Pere David's Vole	Давидова полевка
– de Royle	Mongolische Gebirgsmaus	Royle's High-mountain Vole	Серебристая полевка
– des champs	Feldmaus	Common Vole	Обыкновенная полевка
– des montagnes	Gebirgsmaus	High-mountain Vole	Плоскочерепная полевка
– des neiges	Schneemaus	Snow Vole	Снежная полевка
Campagnol du pin	Kiefernwühlmaus	Pine Vole	Американская боровая полевка
– gris-roux	Graurötelmaus	Large-toothed Red-backed Vole	Красно-серая полевка
– nordique	Nordische Wühlmaus	Tundra Vole	Полевка-экономка
– roussâtre	Rötelmaus	Bank Vole	Рыжая полевка
– souterrain	Kleine Wühlmaus	Earth Vole	Подземная полевка
– terrestre	Schermaus	Vole Rat	Водяная крыса
Capybara	Capybara	Capybara	Капибара
Cardiocraninés	Herzschädel-Springmäuse	Dwarf Jerboas	Карликовые тушканчики
Castor	Biber	Beaver	Бобр
– de Canada	Kanadischer Biber	Canadian Beaver	Канадский бобр
– de l'Elbe	Elbebiber	Elbe Beaver	Эльбский бобр
– de Michigan	Waldbiber	Michigan Beaver	Мичиганский бобр
– de Mongolie	Mongolischer Biber	Mongolian Beaver	Монгольский бобр
– de montagne	Stummelschwanzhörnchen	Mountain Beaver	Горный бобр
– de Rio-Grande	Rio-Grande-Biber	Rio-Grande Beaver	Бобр Рио-Гранде
– de Terre-Neuve	Neufundland-Biber	Newfoundland Beaver	Ньюфаундлендский бобр
– d'Ural	Uralbiber	Ural Beaver	Уральский бобр
– du Rhône	Rhonebiber	Rhone Beaver	Ронский бобр
Castors	Biberartige, Biber	Beavers	Бобровые, Бобры
Catarrhiniens	Schmalnasen	Old World Monkeys	Узконосые обезьяны или Обезьяны Старого Света
Caviidés	Meerschweinchen	Cavies	Морские свинки
Caviinés	Eigentliche Meerschweinchen	Cavies	Настоящие морские свинки
Cétacés	Waltiere	Whales	Китообразные
Chauve-souris boréale	Rote Fledermaus	Red Bat	Рыжий кожан
– cendrée	Weißgraue Fledermaus	Hoary Bat	Бело-серый кожан
– de Mexique	Langnasen-Fledermaus	Mexican Long-nosed Bat	Длинноносый листонос

FRENCH NAME	GERMAN NAME	ENGLISH NAME	RUSSIAN NAME
Chauves-souris des bananes	Bananenfledermäuse	Banana Bats	Банановые листоносы
– des tombeaux	Grabflatterer	Tomb Bats	Могильные летучие мыши
– frugivores	Flederhunde	Fruit Bats	Летучие собаки
– insectivores	Fledermäuse	Insectivorous Bats	Летучие мыши
Cheiromèle à collier	Halsband-Fledermaus	Necklace Hairless Bat	
– nue	Nacktfledermaus	Naked Bat	Голая летучая мышь
Chiens des prairies	Präriehunde	Prairie Dogs	Луговые собачки
Chilonycterinés	Kinnblatt-Fledermäuse	Mustache Bats	Листобороды
Chimpansé	Schimpnase	Chimpanzee	Обыкновенный шимпанзе
– nain	Bonobo	Pygmy Chimpanzee	Бонобо
Chimpansés	Schimpansen	Chimpanzees	Шимпанзе
Chinchilla à longue queue	Langschwanz-Chinchilla	Chinchilla	Длиннохвостая шиншилла
– à queue courte	Kurzschwanz-Chinchilla	Short-tailed Chinchilla	Короткохвостая шиншилла
Chinchillas	Chinchillas i. e. S.	Chinchillas	Шиншиллы
Chinchillidés	Chinchillas	Viscachas and Chinchillas	Шиншиллы
Chiroptères	Fledertiere	Bats	Рукокрылые
Chlamyphores	Gürtelmulle	Pichicies	Плащеносные броненосцы
Chlamyphore tronqué	Gürtelmull	Lesser Pichiciago	Аргентинский плащеносец
Cobaye	Hausmeerschweinchen	Guinea Pig	Домашняя морская свинка
– des roches	Moko	Rock Cavy	Горная морская свинка
– halophile	Kleine Mara	Salt-desert Cavy	Малая мара
– sauvage	Aperea, Tschudi-Meerschweinchen	Aperea, Wild Cavy	Аперея, перуанская морская свинка
Cobayes	Meerschweinchen i. e. S.	Guinea Pigs	
– des roches	Bergmeerschweinchen	Rock Cavies	Горные свинки
Cricétidés	Wühler	Cricetid Rats and Mice	Хомячьи
Cricétinés	Eigentliche Wühler		Хомяки
Cricétomyinés	Hamsterratten		Хомяковидные крысы
Ctenomys de Knight	Knight-Tukotuko	Knight's Tuco-tuco	Тукотуко
Cuniculinés	Pakas	Pacas	Паки
Cynocephalidés	Riesengleitflieger	Flying Lemurs	Шерстокрылы
Cynomys de Gunnison	Weißschwanz-Präriehund	White-tailed Prairie Dog	Белохвостая луговая собачка
– social	Schwarzschwanz-Präriehund	Black-tailed Prairie Dog	Чернохвостая луговая собачка
Cynoptère à nez court	Indischer Kurznasen-Flughund	Short-nosed Fruit Bat	Обыкновенный коротконосый крылан
Cynoptères	Kurznasen-Flughunde	– Fruit Bats	Коротконосые крыланы
Dactylomys	Eigentliche Fingerratte	Coro-coro	
Dasyproctidés	Agutis	Agoutis	Агутиевые
Dasyproctinés	Eigentliche Agutis	Agoutis	Агути
Dauphin blanc	Weißwal	White Whale	Белуха
– bleu-blanc	Blau-Weißer Delphin	Blue-white Dolphin	Однополосый продельфин
– commun	Delphin	Common Dolphin	Североатлантический дельфин-белобочка
– d'eau douce de Chine	Chinesischer Flußdelphin	Chinese River Dolphin	Китайский речной дельфин
Dauphin de Commerson	Commerson-Delphin	Commerson's Dolphin	Трезубцевый дельфин Коммерсона
– de Gill	Gill-Tümmler	Gill's Dolphin	Афалина Гилля
– de la Plata	La-Plata-Delphin	La Plata Dolphin	Лаплатский дельфин
– de Peron	Südlicher Glattdelphin	Peron's Dolphin	Южный китовидный дельфин
– de Risso	Rundkopfdelphin	Risso's Dolphin	Серый дельфин
– douteux	Schlankdelphin	Narrow-snouted Dolphin	Стройный продельфин
– du Gange	Ganges-Delphin	Gangetic Dolphin	Сусук
– du nord	Nördlicher Glattdelphin	Northern Right Whale Dolphin	Северный китовидный дельфин
Dauphins	Delphine i. e. S.	Dolphins	Дельфины-белобочки
– de Commerson	Schwarz-Weiß-Delphine	Commerson's Dolphins	Трезубцевые дельфины
– tachetés	Fleckendelphine	Spotted Dolphins	Продельфины
Dègue du Chili	Degu	Degu	Дегу
Delphinidés	Delphine	Dolphins	Дельфины
Delphininés	Eigentliche Delphine	True Dolphins	Настоящие дельфины
Dendromurinés	Baummäuse	African Tree Mice	Африканские древесные мыши
Diclidure de fantôme	Gespenst-Fledermaus	Ghost Bat	
Diphylle sans queue	Kleiner Blutsauger	Hairy-legged Vampire Bat	Малый кровосос
Dipodinés	Eigentliche Springmäuse	Jerboas	Трехпалые тушканчики
Dolichotinés	Maras	Maras	Мары
Échimyidés	Stachelratten	Spiny Rats	Цепкохвостые щетинистые крысы
Échimyinés	– i. e. S.	– Rats	Щетинистые крысы

FRENCH NAME	GERMAN NAME	ENGLISH NAME	RUSSIAN NAME
Échimys	Kammstachelratten	Crested Spiny Rats	
– armé	Sania	Armoured Spiny Rat	
Écureuil à nez long	Langnasenhörnchen	Long-nosed Squirrels	Длинноносая белка
– à trois couleurs	Dreifarbenhörnchen	Tricoloured Squirrel	Индонезийская великолепная белка
– commun	Eichhörnchen	Red Squirrel	Обыкновенная белка
– d'Afrique à manteau	Rotschwanzhörnchen	Mantled African Bush Squirrel	
– d'Heinrich	Ferkelhörnchen	Long-snouted Squirrel	
– d'Hudson	Rothörnchen	Chickaree	Канадская белка
– d'Inde	Königsriesenhörnchen	Indian Giant Squirrel	Индийская гигантская белка
– de Boehm	Boehm-Hörnchen	Boehm's African Bush Squirrel	
– de Gambie	Graufußhörnchen	Sun Squirrel	
– de la Guyane	Brasilhörnchen	Brazilian Squirrel	Бразильская белка
– de Wilson	Wilsons Riesenhörnchen	African Palm Squirrel	Гигантская белка Вильсона
– des palmes	Palmenhörnchen	Palm Squirrel	Пальмовая белка
– des rochers	Père-Davids-Felsenhörnchen	Rock Squirrel	Давидова белка
– du Caucase	Kaukasisches Eichhörnchen	Caucasian Squirrel	Кавказская белка
– géant	Riesenhörnchen	Oriental Giant Squrrel	Малайская гигантская белка
– gris	Grauhörnchen	Grey Squirrel	Американская серая белка
– nain du Gabon	Afrikanisches Zwerghörnchen	African Pygmy Squirrel	Африканская карликовая белка
– pygmée brun	Braunes Zwerghörnchen	Brown Dwarf Squirrel	
– – de Whitehead	Whitehead-Zwerghörnchen	Whitehead's Dwarf Squirrel	Белка Уайтхеда
– rayé d'Afrique occidental	Westliches Rotschenkelhörnchen	Western African Striped Squirrel	
– volant cendré	Fels-Gleithörnchen	Wolly Flying Squirrel	Каменная летяга
– – de Java	Javanisches Gleithörnchen	Indo-Malayan Flying Squirrel	Яванская летяга
– – du nord	Nördliches Gleithörnchen	Northern Flying Squirrel	
– – du sud	Assapan	Southern Flying Squirrel	Североамериканская летяга
– – géant	Taguan	Giant Flying Squirrel	Тагуан
Écureuils	Hörnchenartige, Hörnchen, Eichhörnchen	Typical Squirrels, European and American Tree Squirrels	Белкообразные, Беличьи, Белки
– des palmes	Gestreifte Palmenhörnchen	Palm Squirrels	
– foisseurs	Afrikanische Borstenhörnchen	African Ground Squirrels	Африканские земляные белки
– pygmées	Asiatische Zwerghörnchen	Oriental Pygmy Squirrels	Азиатские карликовые белки
– rayés d'Afrique	Rotschenkelhörnchen	African Striped Squirrels	
– volants	Gleithörnchen	Flying Squirrels	Летяги
Édentés	Zahnlose	Edentates	Неполнозубые
Eonyctère des cavernes	Höhlen-Langzungen-Flughund	Dobson's Long-tongued Dawn Bat	Пещерный крылан
Épomophore de Buttikofer	Büttikofer-Epauletten-Flughund	Buttikofer's Epauletted Fruit Bat	Крылан Бютикофера
– de Franquet	Franquet-Epauletten-Flughund	Epauletted Bat	Крылан Франкета
– de Wahlberg	Wahlberg-Epauletten-Flughund	Wahlberg's Epauletted Fruit Bat	Крылан Вальберга
– monstreux	Hammerkopf	Hammer-headed Fruit Bat	Молотоголовый крылан
Épomophores	Epauletten-Flughunde	Epauletted Fruit Bats	Эполетовые крыланы
Erethizontidés	Baumstachler	New World Porcupines	Иглошерсты
Eschrichtiidés	Grauwale	Grey Whales	Серые киты
Eutamias	Streifenhörnchen	Western Chipmunks	Бурундуки
Faux Vampire commun	Große Spießblattnase	Great False Vampire	Большой вампир
– Vampires	Falsche Vampire	False Vampire Bats	
Fers-à-chevaux	Hufeisennasen	Horseshoe Bats	Подковоносы
Fouisseur	Kap-Strandgräber	Cape Mole Rat	
Fourmilier didactyle	Zwergameisenbär	Two-toed Ant-eater	Двупалый муравьед
Fourmiliers	Ameisenbären	Ant-eaters	Муравьеды
Furipteridés	Stummeldaumen	Smoky Bats	
Gaufre	Hamsterratte	Tuza	Хомячий гофер
– à poche	Flachland-Taschenratte	Eastern Pocket Gopher	Равнинный гофер
– – – des montagnes	Gebirgs-Taschenratte	Western Pocket Gopher	Горный гофер
Gaufres	Riesentaschenratten	Tuzas	
– à poche	Taschennager, Taschenratten, Flachland-Taschenratten	Pocket Gophers, Eastern Pocket Gophers	Мешетчатые грызуны, Гоферы
– – – des montagnes	Gebirgs-Taschenratten	Western Pocket Gophers	
Gerbille champêtre	Nordafrikanische Rennmaus	Field Gerbil	Североафриканская песчанка
Gerbille d'Emin	Kleine Nacktsohlen-Rennmaus	Small Naked-soled Gerbil	
– d'Inde	Indische Nacktsohlen-Rennmaus	Large Gerbil	Индийская песчанка
– géante	Große Rennmaus	Great Gerbil	Большая песчанка

FRENCH NAME	GERMAN NAME	ENGLISH NAME	RUSSIAN NAME
Gerbilles	Eigentliche Rennmäuse	Gerbils	Африканские песчанки
Gerbillinés	Rennmäuse	Gerbils	Песчанки
Gerboise à cinq doigts	Pferdespringer	Five-toed Jerboa	Большой тушканчик
– à pattes rugueuses	Rauhfuß-Springmaus	Northern Three-toed Jerboa	Мохноногий тушканчик
– de Lichtenstein	Lichtensteins Springmaus	Lichtenstein's Jerboa	Тушканчик Лихтенштейна
– de Sibérie	Sibirische Springmaus	Sibirian Jerboa	Тушканчик-прыгун
– du Gobi	Gobi-Springmaus	Gobi Jerboa	Монгольский земляной заяц
– du steppe	Wüstenspringmaus	Desert Jerboa	Песчаный тушканчик
– lièvre	Erdhase	Jerboa	Тарбаганчик
Gerboises	Springmäuse	Jerboas	Тушканчики
Glirinés	Eigentliche Bilche	Eurasian Dormice	Настоящие сони
Glirule de Japon	Japanischer Schläfer	Japanese Dormouse	Японская соня
Globicéphale de Siebold	Pazifischer Grindwal	Pacific Pilot Whale	Черная гринда
– d'Inde	Indischer Grindwal	Indian Pilot Whale	Индийская гринда
Globicéphales	Grindwale	Pilot Whales	Гринды
Glossophage de Pallas	Spitzmaus-Langzüngler	Long-tongued Bat	Глоссофага
Glossophages	Langzungen-Fledermäuse, Langzungen	– Bats	Длинноязычные вампиры, Глоссофаги
Goundi	Gundi	Gundi	Гунди
Goundis	Kammfingerartige, Kammfinger, Gundis	Gundis	Гребнепаловые, Гребнепалые крысы
Grand Chlamyphore	Burmeister-Gürtelmull	Greater Pichiciago	Плащеносец Бурмейстера
– Écureuil de Stanger	Ölpalmenhörnchen	Oil-palm Squirrel	Масличная белка
– Fer-à-cheval	Große Hufeisennase	Greater Horseshoe Bat	Большой подковонос
– Fourmilier	Großer Ameisenbär	Great Ant-eater	Трехпалый муравьед
– Nyctère	Große Schlitznase	– Slit-faced Bat	
– Pangolin	Riesen-Schuppentier	Giant Pangolin	Гигантский ящер
– Rhinolophe obscur	Große Hufeisennase	Greater Horseshoe Bat	Большой подковонос
Graphiure nain	Zwergschläfer	Pygmy Dormouse	Карликовая африканская соня
Graphiures	Pinselschwanzbilche	African Dormice	Кистехвостые сони
Grinde	Gewöhnlicher Grindwal	Northern Pilot Whale	Северная гринда
Hamster à longue queue	Langschwänziger Hamster	Long-tailed Rat	Длиннохвостый хомяк
– à queue blanche	Weißschwänziger Hamster	White-tailed Rat	Белохвостый хомяк
– d'Europe	Feldhamster	Common Hamster	Обыкновенный хомяк
– d'Imhause	Mähnenratte	Crested Hamster	Гривистая крыса
– doré	Syrischer Goldhamster	Golden Hamster	Переднеазиатский хомяк
– migrateur	Grauer Zwerghamster	Migratory Hamster	Серый хомячок
– nain à longue queue	Langschwanz-Zwerghamster	Long-tailed Hamster	Длиннохвостый хомячок
– – de Chine	Chinesischer Zwerghamster	Chinese Ratlike Hamster	Китайский хомячок
– – de Daourie	Daurischer Zwerghamster	Ratlike Hamster	Даурский хомячок
– – de Djoungarie	Dshungarischer Zwerghamster	Dwarf Hamster	Джунгарский хомячок
– – de Mongolie	Mongolischer Zwerghamster	Short Dwarf Hamster	Монгольский хомячок
– – de Roborowsky	Roborowski-Zwerghamster	Roborowsky's Dwarf Hamster	Хомячок Роборовского
– – d'Eversman	Eversmann-Zwerghamster	Eversman's Dwarf Hamster	Хомячок Эверсманна
– – du Tibet	Tibetanischer Zwerghamster	Tibetan Ratlike Hamster	Тибетский хомячок
Hamsters	Hamster	Hamsters	Хомяки
– nains à queue courte	Kurzschwänzige Zwerghamster	Dwarf Hamsters	Мохноногиче хомячки
– – gris	Graue Zwerghamster	Ratlike Hamsters	Хомячки
Harpionyctère de Whitehead	Whitehead-Spitzzahn-Flughund	Whitehead's Harpy Fruit Bat	
Harpionyctères	Spitzzahn-Flughunde	Harpy Fruit Bats	Крыланы-гарпии
Hominidés	Menschen	Man	Гоминиды
Homme de Cro-Magnon	Cro-Magnon-Mensch	Cro-Magnon Man	Кроманьонец
– de Néanderthale	Neandertaler	Neanderthal Man	Неандерталец
Homo sapiens	Mensch	Man	Человек, Современный человек
Hydrochéridés	Riesennager	Water Hogs	Водосвинки
Hydromyinés	Schwimmratten	– Rats	Австралийские водяные крысы
Hypérodon du nord	Nördlicher Entenwal	Bottle-nosed Whale	Высоколобый бутылконос
– du sud	Südlicher Entenwal	Southern Bottle-nosed Whale	Плосколобый бутылконос
Hypsignathe monstrueux	Hammerkopf	Hammer-headed Fruit Bat	Молотоголовый крылан
Hystricidés	Altwelt-Stachelschweine	Old World Porcupines	Дикобразы Старого Света
Inie de Goeffroy	Amazonas-Delphin	Amazonian Dolphin	Амазонская иния
Iniidés	Inias	Amazon Dolphins	Амазонские дельфины
Lagénorhynque à bec pointu	Weißseitendelphin	White-sided Dolphin	Белобокий дельфин
– à rostre blanc	Weißschnauzendelphin	White-beaked Dolphin	Беломордый дельфин
– de Gill	Weißstreifendelphin	Pacific White-sided Dolphin	Тихоокеанский белобокий делфин
Lagostome des montagnes	Cuvier-Hasenmaus	Mountain Cinchilla	Пушак
– des Pampas	Viscacha	Viscacha	Вискаша
Lagostomes des montagnes	Hasenmäuse	Mountain Chinchillas	Пушаки
Lemming arctique	Halsbandlemming	Arctic Lemming	Копытный лемминг
– commun	Berglemming	Norway Lemming	Норвежский лемминг

FRENCH NAME	GERMAN NAME	ENGLISH NAME	RUSSIAN NAME
– des forêts	Waldlemming	Wood Lemming	Лесной лемминг
– des montagnes	Berglemming	Norway Lemming	Норвежский лемминг
– des toundras	Berglemming	– Lemming	Норвежский лемминг
– du nord	Nördliche Lemmingmaus	Northern Bog Lemming	Северная лемминговая мышь
– du sud	Südliche Lemmingmaus	Southern Bog Lemming	Южная лемминговая мышь
Lemmings	Lemminge	Lemmings	Лемминги
Lérot	Gartenschläfer	Garden Dormouse	Садовая соня
Lérotin	Baumschläfer	Forest Dormouse	Лесная соня
Lièvre sauteur d'Afrique de l'est	Ostafrikanischer Springhase	Spring Haas	Восточноафриканский долгоног
– – – du sud	Südafrikanischer Springhase	Cape Spring Haas	Кафрский долгоног
Lièvres sauteurs	Springhasenartige, Springhasen	Spring Haas	Долгоноговые, Долгоноги
Lissodelphininés	Glattdelphine	Right Whale Dolphins	Бесперые дельфины
Loir gris	Siebenschläfer	Fat Dormouse	Соня-полчок
Loirs	Bilche	Dormice	Сони
Macroderme d'Australie	Australische Gespenstfledermaus	Australian Ghost Bat	
Macroglosse minime	Zwerg-Langzungen-Flughund	Asiatic Long-tongued Fruit Bat	Малый длинноязычный крылан
Macroglosses	Langzungen-Flughunde	Long-tongued Fruit Bats	Длинноязычные крыланы
Macrophylle	Langbein-Fledermaus	Long-legged Bat	
Manidés	Schuppentiere	Pangolins, Scaly Ant-eaters	Ящеры
Mara	Mara	Mara	Патагонская свинка
Marmotte à longue queue	Langschwänziges Murmeltier	Long-tailed Marmot	Длиннохвостый сурок
– à ventre fauve	Gelbbäuchiges Murmeltier	Yellow-bellied Marmot	Желтобрюхий сурок
– Bobac	Bobak	Dobac Marmot	Степной сурок
– des Alpes	Alpenmurmeltier	Alpine Marmot	Альпийский сурок
– grise	Eisgraues Murmeltier	Hoary Marmot	Североамериканский серый сурок
Marmottes	Murmeltiere	Marmots	Сурки
Marsouin commun	Schweinswal	Common Porpoise	Морская свинья
– de Dall	Dall-Hafenschweinswal	Dall's Harbour Porpoise	Северная белокрылая свинья
– de l'Inde	Indischer Schweinswal	Finless Black Porpoise	Бесперая морская свинья
– des lunettes	Brillenschweinswal	Spectacled Porpoise	Очковая морская свинья
– du Pacifique	Pazifischer Hafenschweinswal	Pacific Harbour Porpoise	Северотихоокеанская морская свинья
– spinipenne	Burmeister-Schweinswal	Burmeister's Porpoise	Морская свинья Бурмейстера
Megachiroptères	Flederhunde	Fruit Bats	Летучие собаки
Mégaderme à ailes orangées	Gelbflügelige Großblattnase	African Yellow-winged Bat	
– du cœur	Herznasenfledermaus	Heart-nosed False Vampire	
– feuille	Gelbflügelige Großblattnase	African Yellow-winged Bat	
– lyre	Lyra-Fledermaus	Indian False Vampire	Индийский ложный вампир
– spasme	Malaiischer Falscher Vampir	Malayan False Vampire	Малайский ложный вампир
Mégadermes	Großblattnasen, Eigentliche Großblattnasen	Large-winged Bats, False Vampires	Мегадермы, Лироносы
Mégaloglosse de Woermann	Afrikanischer Langzungen-Flughund	African Long-tongued Fruit Bat	Африканский длинноязычный крылан
Mégaptère	Buckelwal	Humpback Whale	Горбач
Mérione de Lybie	Rotschwänzige Rennmaus	Libyan Jird	Краснохвостая песчанка
– de Mongolie	Mongolische Rennmaus	Jird	Когтистая песчанка
– de Perse	Persische Wüstenmaus	Persian Jird	Персидская песчанка
– du sud	Mittagsrennmaus	Southern Jird	Полуденная песчанка
Mériones	Sandmäuse	Jirds	Малые песчанки
– de Tristram	Tristram-Wüstenmaus	Tristram's Jird	
Mésoplodon de Gervais	Gervais-Zweizahnwal	Gervais' Whale	Европейский ремнезуб
– de Layard	Layard-Wal	Layard's Whale	Ремнезуб Лэйярда
– de Sowerbye	Sowerby-Zweizahnwal	Sowerby's Whale	Атлантический ремнезуб
– de True	True-Wal	True's Beaked Whale	Североатлантический ремнезуб
Microchiroptères	Fledermäuse	Insectivorous Bats	Летучие мыши
Microsciure nain	Zwerghörnchen	Dwarf Tree Squirrel	
Microsciures	Neuweltliche Zwerghörnchen	– – Squirrels	Американские карликовые белки
Minioptère à longues ailes	Langflügel-Fledermaus	Long-winged Bat	Длиннокрыл Шрейбера
Minioptères	Langflügel-Fledermäuse	Bent-winged Bats, Long-winged Bats	Длиннокрылы
Molosse géante	Große Samt-Fledermaus	Giant Velvety Free-tailed Bat	
– rouge	Rote Samt-Fledermaus	Red Velvety Free-tailed Bat	Рыжая бульдоговая мышь
Molosses	Samt-Fledermäuse	Velvety Free-tailed Bats	
Molossidés	Bulldogg-Fledermäuse	Free-tailed Bats	Бульдоговые летучие мыши

FRENCH NAME	GERMAN NAME	ENGLISH NAME	RUSSIAN NAME
Monax	Waldmurmeltier	Woodchuck	Лесной сурок
Monodontidés	Gründelwale	White and Narwhales	Белухи
Mulot de Geisha	Geishamaus	Geisha Wood Mouse	Японская мышь
– fauve	Gelbhalsmaus	Yellow-necked Field Mouse	Желтогорлая мышь
– rayé	Brandmaus	Striped Field Mouse	Полевая мышь
– rupestre	Felsenmaus	Broadtoothed Field Mouse	Горная мышь
– sylvestre	Feld-Waldmaus	Long-tailed Field Mouse	Лесная мышь
Mulots	Wald- und Feldmäuse	Wood and Field Mice	Лесные и полевые мыши
Murin	Mausohr	European Little Brown Bat	Большая ночница
Murins	Mausohr-Fledermäuse	Mouse-eared Bats	Ночницы
Muscardin	Haselmaus	Common Dormouse	Орешниковая соня
Myomime à queue fine	Dünnschwanz-Baumschläfer	Asiatic Dormouse	Мышевидная соня
Myonyctère à collier	Schmalkragen-Flughund	Little Collared Fruit Bat	
Myonyctères à collier	Kragenflughunde	Collared Fruit Bats	
Mystacine tubercule	Neuseeland-Fledermaus	New Zealand Short-tailed Bat	Новозеландская летучая мышь
Mystacinidés	Neuseeland-Fledermäuse	– – – Bats	Новозеландские летучие мыши
Myzopodidés	Madagassische Haftscheiben-Fledermäuse	Sucker-footed Bats	Мадагаскарские присосковые летучие мыши
Narval	Narwal	Narwhal	Нарвал
Natalidés	Trichterohren	Funnel-eared Bats	
Néotamia de montagne	Gebirgs-Chipmunk	Alpine Chipmunk	Горный чипмунк
– de Townsend	Townsend-Chipmunk	Townsend's Chipmunk	Чипмунк Тоунсенда
– de Colorado	Colorado-Chipmunk	Colorado Chipmunk	Колорадский чипмунк
– jaune	Gelber Fichten-Chipmunk	Yellow Pine Chipmunk	Желтый чипмунк
– nain	Kleiner Chipmunk	Least Chipmunk	Малый чипмунк
Noctilio du sud	Kleines Hasenmaul	Southern Bulldog Bat	Малый зайцерот
– pêcheuse	Großes Hasenmaul	Mexican Bulldog Bat	Обыкновенный зайцерот
Noctilions	Hasenmäuler	Bulldog Bats	Зайцеротые летучие мыши, Зайцероты
Noctule	Großer Abendsegler	Common Noctule	Рыжая вечерница
– de Leisler	Rauharm-Fledermaus	Lesser Noctule	Малая вечерница
– géante	Riesen-Abendsegler	Giant Noctule	Гигантская вечерница
Noctules	Abendsegler	Noctule Bats	Вечерницы
Notoptères à queue longue		Long-tailed Fruit Bats	
Nyctère de Geoffroy	Geoffroy-Schlitznase	Geoffroy's Slit-faced Bat	
– de Java	Java-Hohlnase	Javanese Slit-faced Bat	Яванская никтерида
– de la Thébaide	Geoffroy-Schlitznase	Geoffroy's Slit-faced Bat	
– hérissé	Rauhhaar-Schlitznase	Hispid Slit-faced Bat	
Nyctères	Schlitznasen	Slit-faced Bats	Никтериды
Nyctimène à grosse tête	Großkopf-Röhrennasen-Flughund	Large-headed Tube-nosed Fruit Bat	
– de Robinson	Robinson-Röhrennasen-Flughund	Queensland Tube-nosed Fruit Bat	
– géant	Großer Röhrennasen-Flughund	Giant Tube-nosed Fruit Bat	Большой ночной крылан
Nyctimènes	Röhrennasen-Flughunde	Tube-nosed Fruit Bats	Трубконосые крыланы
Octodons	Strauchratten	South American Bush Rats	Кустарниковые крысы
Octodontidés	Trugratten	Octodont Rodents	Осьмизубые
Odontocétés	Zahnwale	Toothed Cetaceans	Зубатые киты
Ondatras	Bisamratten	Muskrats	Мускусные крысы
Onychomys	Grashüpfermäuse	Grasshopper Mice	
– du nord	Nördliche Grashüpfermaus	Northern Grasshopper Mouse	
– du sud	Südliche Grashüpfermaus	Southern Grasshopper Mouse	
Orcaelle d'Irawadi	Irawadi-Delphin	Irawady Dolphin	Иравaддийский дельфин
Oreillard Barbastelle	Mopsfledermaus	Barbastelle	Европейская широкоушка
– commun	Braunes Langohr	Long-eared Bat	Обыкновенный ушан
– du sud	Graues Langohr	Southern Long-eared Bat	
– pâle	Blasse Fledermaus	Pallid Bat	Бледная летучая мышь
Oreillards	Großohren	Long-eared Bats	Ушаны
Orque épaulard	Schwertwal	Common Killer Whale	Косатка
Otomyinés	Afrikanische Lamellenzahnratten	Vlei and Karroo Rats	Африканские ушастые крысы
Paca	Paka	Paca	Пака
– des montagnes	Bergpaka	Mountain Paca	Горная пака
Pacarana	Pakarana	Pacarana	Длиннохвостая пака
Pacaranas	Pakranaartige, Pakaranas	Pacaranas	Диномииды
Palatouche	Gewöhnliches Gleithörnchen	European Flying Squirrel	Обыкновенная летяга
Pangolin à queue courte	Chinesisches Ohren-Schuppentier	Chinese Pangolin	Индостанский панголин
– de Temminck	Steppen-Schuppentier	Cape Pangolin	Степной ящер
– indien	Vorderindisches Schuppentier	Indian Pangolin	Переднеиндийский ящер
– javanais	Javanisches Schuppentier	Malayan Pangolin	Яванский ящер
– tétradactyle	Langschwanz-Schuppentier	Phatagin	Длиннохвостый ящер
Pectinator de Speke	Speke-Kammfinger	Speke's Pectinator	

FRENCH NAME	GERMAN NAME	ENGLISH NAME	RUSSIAN NAME
Petite Chauve-souris murine	Zweifarbige Fledermaus	Particoloured Bat	Двухцветный кожан
Petit Fer-à-cheval	Kleine Hufeisennase	Lesser Horseshoe Bat	Малый подковонос
– Microptère	Zwerg-Epauletten-Flughund	Dwarf Epauletted Fruit Bat	Карликовый эполетовый крылан
– Rhinolophe	Kleine Hufeisennase	Lesser Horseshoe Bat	Малый подковонос
– Rorqual	Zwergwal	Lesser Rorqual	Малый остромордый полосатик
Phloeomyinés	Borkenratten		Азиатские древесные крысы
Phocénidés	Schweinswale	Porpoises	Морские свиньи
Pholidotes	Schuppentiere	Pangolins	Панголины
Phyllorine de Cafrérie	Gewöhnliche Rundblattnase	South African Lesser Leaf-nosed Bat	
– de Commerson	Riesen-Rundblattnase	Commerson's Leaf-nosed Bat	Большой ложный подковонос
– d'Himalaye	Himalaja-Rundblattnase	Great Himalayan Leaf-nosed Bat	
Phyllostome coloré	Bunte Lanzennase	Coloured Bat	
– fer de lance	Lanzennase	Javelin Bat	Копьенос
Phyllostomes	Eigentliche Blattnasen, Lanzennasen	Big-eared Leaf-nosed Bats, Spear-nosed Bats	Настоящие листоносы
Physétéridés	Pottwale	Sperm Whales	Кашалоты
Pipistrelle naine à ailes brunes	Bananen-Zwergfledermaus	Banana Bat	Банановый нетопырь
Pipistrelles	Zwergfledermäuse	Pipistrelles	Нетопыри
Pithécanthrope d'Heidelberg	Heidelberger Frühmensch	Heidelberg Man	Гейдельбергский прачеловек
– de Java	Frühmensch, Javanischer Frühmensch	Java Man	Прачеловек, Питекантроп
– de Mauretanie	Nordafrikanischer Frühmensch	Mauretania Man	Североафриканский прачеловек
– de Péking	Peking-Frühmensch	Peking Man	Синантроп
Plagiodonte d'Haiti	Cuviers Zaguti	Haitian Hutia	Загути Кювье
Plagiodontes	Zagutis	Zagoutis	Загути
Platacanthomyidés	Stachelbilche	Spiny Dormice	Шипохвостые сони
Platanistidés	Ganges-Delphine	Gangetic Dolphins	Гангские дельфины
Porc-épic	Gewöhnliches Stachelschwein	Porcupine	Обыкновенный дикобраз
– à longue queue	Malaiischer Pinselstachler	Long-tailed Porcupine	Борнеоский дикобраз
– à queue blanche	Weißschwanz-Stachelschwein	White-tailed Porcupine	Индийский дикобраз
– d'Afrique	Afrikanisches Stachelschwein	African Porcupine	Африканский дикобраз
– – du sud	Südafrikanisches Stachelschwein	South African Porcupine	Южноафриканский дикобраз
– d'Indonésie	Zwergstachelschwein	Indonesian Porcupine	Карликовый дикобраз
– épineux	Borsten-Baumstachler	Thin-spined Porcupine	Крысохвостый дикобраз
– laineux	Wollgreifstachler	Wolly Prehensile-tailed Porcupine	Мохнатый цепкохвост
– nord américain	Ur[illegible]n	North American Porcupine	Североамериканский иглошерст
– préhensile	Greifstachler i. e. S.	Prehensile-tailed Porcupine	Бразильский коэнду
– rougeâtre	Bergstachler	Upper Amazonian Porcupine	Горный иглошерст
Porcs-épic	Eigentliche Stachelschweine	Large Porcupines	Настоящие дикобразы
– nord américains	Nordamerikanische Baumstachler	North American Porcupines	Североамериканские иглошерсты
– préhensiles	Greifstachler	Tree Porcupines	Коэнду
– d'Indonésie	Insel-Stachelschweine	Indonesian Porcupines	Индонезийские дикобразы
Primates	Herrentiere	Primates	Приматы
Pseudorque à dents épaisses	Kleiner Schwertwal	False Killer Whale	Черная косатка
Pteropidés	Flughunde i. e. S.	Fruit Bats	Крыланы
Ragondin	Sumpfbiber	Nutria	Нутрия
Ragondins	Biberratten	Nutrias	Бобровые крысы
Rat adipeux	Fettmaus	Fat Mouse	Жирная древесная мышь
– à manteau roux	Insektenessende Waldmaus	Congo Forest Mouse	Насекомоядная лесная мышь
– à museau roux	Afrikanische Rotnasenratte	Rufous-nosed Rat	Африканская красноносая крыса
– à poche mexicain	Mexikanische Taschenratte	Yellow Pocket Gopher	Мексиканский гофер
– à queu courte	Kurzschwanz-Maulwurfsratte	Pest Rat	Индийская земляная крыса
Rat à queue en mosaique des montagnes	Gebirgs-Mosaikschwanz-Riesenratte	Giant Naked-tailed Rat	
– à queue noire	Schwarzschwanz-Baumratte	Black-tailed Hutia	Чернохвостая кубинская крыса
– à queue préhensile	Hutiacarabali	Prehensile-tailed Hutia	Хутия-карабали
– armé	Lanzenratte	Armoured Rat	Гимнуровая крыса
– au nez	Nasenratte	Shrew Rat	
– bleu	Cururo	Cururo	Куруро
– commun	Hausratte	House Rat	Черная крыса

FRENCH NAME	GERMAN NAME	ENGLISH NAME	RUSSIAN NAME
– d'Aléxandrie	Alexandriner Hausratte	Roof Rat	Александрийская черная крыса
– d'eau	Schermaus	Water Vole	Водяная крыса
– – de Floride	Florida-Wasserratte	Round-tailed Muskrat	Флоридская водяная крыса
– d'ecorce de Schadenberg	Schadenbergs Borkenratte	Bushy-tailed Cloud Rat	Древесная крыса Шаденберга
– – tacheté	Gescheckte Riesenborkenratte	Slender-tailed Cloud Rat	Пятнистая древесная крыса
– de Grégorian	Kleine Rohrratte	Lesser Cane Rat	Малая тростниковая крыса
– de Guyane	Cayenneratte	Cayenne Spiny Rat	Гвианская крыса
– de Lamberton	Lamberton-Inselratte	Lamberton's Malagasy Rat	Мадагаскарская крыса Ламбертона
– des Bahamas	Bahama-Ferkelratte	Bahaman Hutia	Багамская крыса
– des bambous Chinois	Chinesische Bambusratte	China Bamboo Rat	Китайская ризомиида
– – – de Sumatra	Sumatra-Bambusratte	Sumatran Bamboo Rat	Суматранская ризомиида
– des fruits	Fruchtratte	Corn Rat	
– des steppes	Wüstenratte	Whitethroat Wood Rat	Пустынный кустовый хомячок
– du bambou	Bambus-Fingerratte	Tree-Rat	
– du riz	Sumpf-Reisratte	Rice Rat	Рисовая мышь
– épineux	Suira	Suira	
– – à tête grise foncu	Graukopf-Baumstachelratte	Arboreal Soft-furred Spiny Rat	
	Südamerikanische Felsenratte	South American Rock Rat	
– géant de Gambia	Gambia-Riesenhamsterratte	Gambia Pouched Rat	Гамбийская хомяковидная крыса
– – d'Emin	Emin-Riesenhamsterratte	Emin's Rat	
– jamaique	Jamaika-Ferkelratte	Jamaican Hutia	Ямайская крыса
– kangourou à grandes oreilles	Großohr-Känguruhratte	Big-eared Kangaroo Rat	Большеухий мешетчатый прыгун
– – de Texas	Texas-Känguruhratte	Texas Kangaroo Rat	Техасский мешетчатый прыгун
– – d'Ord	Ord-Känguruhratte	Ord's Kangaroo Rat	Мешетчатый прыгун Орда
– – du désert	Wüsten-Känguruhratte	Desert Kangaroo Rat	Пустынный мешетчатый прыгун
– – du Pacifique	Pazifik-Känguruhratte	Pacific Kangaroo Rat	Проворный мешетчатый прыгун
– minime	Viscacharatte	Viscacha Rat	Вискашевидная крыса
– musqué	Bisamratte	Muskrat	Ондатра
– nain	Zwergbaumratte	Tiny Hutia	Карликовая кубинская крыса
– noir	Hausratte, Hausratte i. e. S.	House Rat	Черная крыса, Северо-западная черная крыса
– nu de sable	Nacktmull	Naked Sand Rat	Голая крыса
– poilé	Hutiaconga	Cuban Hutia	Хутия-конга
– rayé du champ	Afrikanische Striemen-Grasmaus	Striped Field Mouse	
– surmulot	Wanderratte	Brown Rat	Пасюк
– taupe	Blindmull, Nördlicher Mull-Lemming	North Altai Zokor, Mole Lemming	Алтайский цокор, Обыкновенная слепушонка
– – africain	Hottentotten-Graumull	African Mole Rat	Готтентотский землекоп
– – d'Afghan	Südlicher Mull-Lemming	Afghan Mole Lemming	Афганская слепушонка
– – du Cap	Kap-Bleßmull	Cape Blesmol	Капский пескорой
– typique	Felsenratte	Rock Rat	Горная мышь
Rat-chinchilla cendré	Chinchillaratte	Chinchilla Rat	Шиншилловая крыса
Rats à mamelles multiples	Vielzitzenmäuse	Multimammate Rats	
– à pelage en brosse	Afrikanische Bürstenhaarmäuse	Harsh-furred Mice	
– à poche	Tuzas	Yellow Pocket Gophers	
– à queue en mosaique	Mosaikschwanz-Riesenratten	Giant Naked-tailed Rats	
– d'eau d'Australie	Schwimmratten	Australian Water Rats	
– de Madagascar	Madagaskarratten	Malagasy Rats	Мадагаскарские крысы
– des bambous	Wurzelratten, Bambusratten	Bamboo Rats	Ризомииды, Бамбуковые крысы
– des bois	Buschratten	Wood Rats	Кустовые хомячки
– du riz	Reisratten	Rice Rats	
– géants	Riesenhamsterratten	Giant Pouched Rats	Большие хомяковидные крысы
– kangourou	Taschenspringer	Kangaroo Rats	Мешетчатые прыгуны
– proprement dits	Eigentliche Ratten	Rats	Крысы
– rayés	Afrikanische Streifen-Grasmäuse	Striped Grass Mice	Африканские полосатые мыши
Rats taupes	**Blindmäuse, Mull-Lemminge**	**Palaearctic Mole Rats, Mole Lemmings**	**Слепыши, Слепушонки**
– – africains	**Sandgräber, Afrikanische Maulwurfsratten**	**Mole Rats, African Mole Rats**	**Землекопы, Африканские кротовидные крысы**
Reithrosciure de Borneo	**Borneo-Hörnchen**	**Groove-toothed Squirrel**	**Борнеоская белка**
Rhinolophe de Dent	**Dent-Hufeisennase**	**Dent's Horseshoe Bat**	

FRENCH NAME	GERMAN NAME	ENGLISH NAME	RUSSIAN NAME
– de Lander	Lander-Hufeisennase	Lander's Horseshoe Bat	
– de Maclaud	Maclaud-Hufeisennase	Maclaud's Bat	Подковонос Маклауда
– d'Hildebrandt	Hildebrandt-Hufeisennase	Hildebrandt's Horseshoe Bat	Подковонос Гильдебрандта
– unifer	Große Hufeisennase	Greater Horseshoe Bat	Большой подковонос
Rhinolophes	Hufeisennasen-Verwandte, Eigentliche Hufeisennasen	Horseshoe-nosed Bats, Horseshoe Bats	Подковоносовые, Подковоносы
Rhinolophidés	Hufeisennasen	Horseshoe Bats	Подковоносы
Rhinopôme d'Hardwick	Hardwicke-Klappnase	Lesser Rat-tailed Bat	Ланцетонос Хардвика
– microphylle	Ägyptische Klappnase	Larger Rat-tailed Bat	Египетский ланцетонос
Rhinopômes	Mausschwanz-Fledermäuse, Klappnasen	Rat-tailed Bats	Ланцетоносы
Rhynchomyinés	Nasenratten	Shrew Rats	
Rhynchonyctère	Nasenfledermaus	Proboscis Bat	Носатая летучая мышь
Rongeurs	Nagetiere	Rodents	Грызуны
Rorqual commun	Finnwal	Common Rorqual	Сельдяной полосатик
Rorquals	Furchenwale, Finnwale	Finback Whales, Rorquals	Полосатики, Настоящие полосатики
Roussette à cou rouge	Kalong	Red-necked Fruit Bat	Калонг
– à tête cendrée	Graukopf-Flughund	Grey-headed Flying Fox	Сероголовый восточно-австралийский крылан
– d'Angola	Angola-Flughund	Angola Fruit Bat	Ангольский крылан
– d'Égypte	Ägyptischer Flughund	Egyptian Fruit Bat	Египетский крылан
– de Geoffroy	– Flughund	– – Bat	Египетский крылан
– de Tonga	Tonga-Flughund	Tonga Flying Fox	Тонгский крылан
– foncée	Rauchgrauer Flughund	Dark Flying Fox	
– géante	Indischer Flughund	Indian Flying Fox	Летучая лисица
– noire	Schwarzer Flughund	Black Flying Fox	
– paillée	Palmenflughund	Straw-coloured Bat	Пальмовый крылан
– rougeâtre	Roter Flughund	Rufous Flying Fox	Мадагаскарская летучая собака
– vulgaire	Kalong	Red-necked Fruit Bat	Калонг
Roussettes	Langnasen-Flughunde, Eigentliche Flughunde, Höhlenflughunde	Flying Foxes, Roussette Bats	Длинноносые крыланы, Летучие собаки, Ночные крыланы
Saccopteryx	Taschenfledermäuse	Sheath-tailed Bats	Мешкокрылые летучие мыши
– à deux raies	Zweistreifen-Taschenfledermaus	El Salvador Sheath-tailed Bat	Двухполосая мешкокрылая мышь
Scotonyctère à dent du serpent	Schlangenzahn-Flughund	Snake-toothed Fruit Bat	
– de Zenker	Zenker-Flughund	Zenker's Fruit Bat	Крылан Ценкера
Sérotine boréale	Nordische Fledermaus	Northern Bat	Северный кожанок
– commune	Spätfliegende Fledermaus	Serotine Bat	Поздний кожан
– de maison	Große Braune Fledermaus	Big Brown Bat	
Sérotines	Breitflügel-Fledermäuse	– – Bats	Кожаны
– bicolores	Zweifarbige Fledermäuse	Particoloured Bats	Двухцветные кожаны
Siciste de bouleaux	Birkenmaus	Northern Birch Mouse	Лесная мышовка
– des steppes	Streifenmaus	Southern Birch Mouse	Степная мышовка
Sicistes	Streifen-Hüpfmäuse	Birch Mice	Мышовки
Sigmodons	Baumwollratten	Cotton Rats	Хлопковые хомячки
Sigmodon velu	Baumwollratte	Andean Swamp Rat	
Singes	Affen	Monkeys	Обезьяны
– anthropomorphes	Menschenaffen	Anthropoid Apes	Человекоподобные
Sotalie	Amazonas-Sotalia	Buffeo Negro	Амазонский речной дельфин
– de Chine	Chinesischer Weißer Delphin	Chinese White Dolphin	Китайский бледный речной делфин
– de la Guyane	Guayana-Delphin	Guyanian River Dolphin	Гвианский речной дельфин
– de Teusz	Kamerunfluß-Delphin	West African White Dolphin	Камерунский речной дельфин
Souris	Mäuse i. e. S.	House Mice	Мыши
– à grosse queue	Dickschwanzmaus	Fat-tailed Mouse	Толстохвостая песчанка
– à longue queue	Langschwänzige Indische Baummäuse	Long-tailed Climbing Mice	Индийские пальмовые мыши
– à pattes blanches	Weißfußmäuse, Weißfußmaus	White-footed Mice, White-footed Mouse	Оленья мышь
– – courtes	Kurzfuß-Inselratte		Коротколапая мадагаскарская крыса
– à poche	Eigentliche Taschenmäuse	Pocket Mice	Мешетчатые тушканчики
– – – de Bailey	Bailey-Taschenmaus	Bailey's Pocket Mouse	Мешетчатый тушканчик Бейлейя
– – – de Californie	Kalifornische Taschenmaus	California Pocket Mouse	Калифорнийский мешетчатый тушканчик
Souris à poche de Goldman	Goldman-Stacheltaschenmaus	Goldman's Spiny Pocket Mouse	Иглистый тушканчик Гольдмана

FRENCH NAME	GERMAN NAME	ENGLISH NAME	RUSSIAN NAME
– aux cheveux plats		Flat-haired Mouse	Бурая иглистая мышь
– d'Afrique	Afrikanische Kleinmäuse	African Native Mice	Африканские мыши-малютки
– d'arbre de Malaysie	Malaiische Pinselschwanz-Baummaus	Pencil-tailed Tree Mouse	Малайская древесная крыса
– d'Australie	Australische Kleinmäuse	Australian Native Mice	Австралийские малые мыши
– d'Hermannsburg	Hermannsburg-Zwergmaus	– – Mouse	
– de Selevin	Salzkrautbilche	Betpakdala Dormice	Селевиниевые
– de True	Pinjonmaus	Piñon Mouse	
– des bananiers	Aalstrich-Klettermäuse	African Climbing Mice	
– des bois	Hirschmaus	Deer Mouse	
– des moissons	Eurasiatische Zwergmaus	Harvest Mouse	Мышь-малютка
– – d'Amérique	Amerikanische Erntemäuse	American Harvest Mice	
– domestique	Hausmaus	House Mouse	Домовая мышь
– dorée	Goldmaus	Golden Mouse	Золотистый мышевидный хомячок
– du désert	Salzkrautbilch	Desert Dormouse	Боялычная соня
– du Nouveau Monde	Neuweltmäuse	New World Mice	Американские мышевидные хомячки
– du soir	Hirschmaus	Deer Mouse	
– épineuse	Ägyptische Stachelmaus	Egyptian Spiny Mouse	Египетская ежовая мышь
– – à poche	Gemalte Stacheltaschenmaus	Painted Spiny Pocket Mouse	
– – – – de Mexique	Mexikanische Stacheltaschenmaus	Mexican Spiny Pocket Mouse	Мексиканский иглистый тушканчик
– – de Crête	Kreta-Stachelmaus	Cretan Spiny Mouse	Критская ежовая мышь
– – de Sinai	Sinai-Stachelmaus	Sinai Spiny Mouse	Бледная ежовая мышь
– épineuses	Stachelmäuse	Spiny Mice	Ежовые мыши
– et Rats de l'Ancien Monde	Mäuse	Old World Rats and Mice	Мышиные
– kangourou	Taschenmäuse, Känguruhmäuse	Kangaroo Mice	Мешетчатые крысы, Мешетчатые прыгунчики
– – pâle	Blasse Känguruhmaus	Pale Kangaroo Mouse	Бледный мешетчатый прыгунчик
– naine d'Afrique	Afrikanische Zwergmaus	Small African Native Mouse	Африканская мышь-малютка
– – de Chine	Chinesische Zwergschlafmaus	Chinese Pygmy Dormouse	Китайская карликовая соня
– occidentale des moissons	Westliche Erntemaus	Rabbit Rat	
– soyeux à poche	Seiden-Taschenmaus	Silky Pocket Mouse	Шелковистый мешетчатый тушканчик
Souslik de Nelson	Nelson-Antilopenziesel	Nelson's Souslik	Калифорнийский антилоповый суслик
– de Parry	Parry-Ziesel	Arctic Ground Squirrel	Американский длиннохвостый суслик
– d'Europe	Ziesel	European Souslik	Серый суслик
– nain	Zwergziesel	Little Souslik	Малый суслик
– tacheté	Perlziesel	Spotted Souslik	Крапчатый суслик
Sousliks	Ziesel	Ground Squirrels	Суслики
Spalax d'Ehrenberg	Ehrenberg-Blindmaus	Ehrenberg's Mole Rat	Слепыш Эренберга
– occidental	Westblindmaus	Lesser Mole Rat	Малый слепыш
– oriental	Ostblindmaus	Greater Mole Rat	Обыкновенный слепыш
Spermophile à manteau doré	Goldmantelziesel	Golden-mantled Ground Squirrel	Золотистоспинный суслик
– à queue ronde	Rundschwanzziesel	Round-tailed Ground Squirrel	Круглохвостый суслик
– à treize bandes	Streifenziesel	Thirteen-striped Ground Squirrel	Полосатый американский суслик
– d'antilope à queue blanche	Weißschwanz-Antilopenziesel	White-tailed Antelope Squirrel	Белохвостый антилоповый суслик
– de Californie	Kalifornischer Ziesel	California Ground Squirrel	Калифорнийский суслик
– de Franklin	Franklin-Ziesel	Franklin's Ground Squirrel	Суслик Франклина
– d'Eversman	Langschwänziger Ziesel	Everman's Souslik	Длиннохвостый суслик
– jaune	Gelbziesel	Aral Yellow Souslik	Суслик-песчаник
– leptodactyle	Zieselmaus	Long-clawed Ground Squirrel	Тонкопалый суслик
– rouge	Rotziesel	Red Souslik	Большой суслик
Sténidés	Langschnabeldelphine	Long-snouted Dolphins	Длинноклювые дельфины
Sturnire fleur-de-lys	Gelbschulter-Blattnase	Yellow-shouldered Bat	
Sturnires	Gelbschulter-Blattnasen	– Bats	
Tamandua à quatre doigts	Tamandua	Tamandua	Четырехпалый муравьед
Tamias	Chipmunks	Eastern Chipmunks	Чипмунки
Tamia strié	Streifenbackenhörnchen	Chipmunk	Американский бурундук
Taphien à ventre nu	Nacktbäuchiger Grabflatterer	Naked-bellied Tomb Bat	Голобрюхая летучая мышь
– de Maurice	Mauritanischer Grabflatterer	Mauritian Tomb Bat	
– perforé	Grabflatterer	Tomb Bat	
Taphiens	Grabflatterer	– Bats	Могильные летучие мыши

FRENCH NAME	GERMAN NAME	ENGLISH NAME	RUSSIAN NAME
Tasmacète de Shepherd	**Shepherd-Wal**	**Shepherd's Beaked Whale**	**Новозеландский клюворыл Шеперда**
Tataride bordée	Borten-Fledermaus	Braided Free-tailed Bat	
– bouledogue	Bulldogg-Fledermaus	European Free-tailed Bat	Широкоухий складчатогуб
– de Mexique	Guano-Fledermaus	Mexican Free-tailed Bat	
Tatarides	Faltlippen-Fledermäuse	Free-tailed Bats	Складчатогубы
Tatou à neuf bandes	Neunbindengürteltier	Nine-banded Armadillo	Девятипоясный броненосец
– à onze bandes	Großes Nacktschwanzgürteltier	Eleven-banded Armadillo	Большой голохвостый броненосец
– à sept bandes	Siebenbindengürteltier	Seven-banded Armadillo	Семипоясный броненосец
– à six bandes	Weißborsten-Gürteltier	Six-banded Armadillo	Шестипоясный броненосец
– à trois bandes	Kugelgürteltier	Three-banded Armadillo	Шаровидный броненосец
– de Kappler	Kappler-Weichgürteltier	Kappler's Armadillo	Броненосец Капплера
– épineux	Kleines Nacktschwanzgürteltier	Spiny Armadillo	Малый голохвостый броненосец
– géant	Riesengürteltier	Giant Armadillo	Гигантский броненосец
– nain	Zwerggürteltier	Pygmy Armadillo	Карликовый броненосец
– poilu	Pelzgürteltier	Hairy Armadillo	Мохнатый броненосец
– velu	Braunzottiges Borstengürteltier	– Armadillo	Волосатый броненосец
Tatous	Gürteltiere, Weichgürteltiere	Armadillos	Броненоцы
– à onze bandes	Nacktschwanzgürteltiere	Eleven-banded Armadillos	Голохвостые броненосцы
– à six bandes	Borstengürteltiere	Six-banded Armadillos	
– à trois bandes	Kugelgürteltiere	Three-banded Armadillos	
Thyropteridés	Amerikanische Haftscheiben-Fledermäuse	Disk-winged Bats	Американские присосковые летучие мыши
Triaenops de Perse	Dreiblatt-Fledermaus	Persian Leaf-nosed Bat	Персидский ложный подковонос
Tricuspide	Weißbauch-Schuppentier	Three-pointed Pangolin	Белобрюхий ящер
Tuco-tucos	Kammratten, Tukotukos	Tuco-tucos	Ктеномииды
Tursiops	Tümmler	Bottle-nosed Dolphins	Афалины
– du Mer Rouge	Rotmeer-Tümmler	Red Sea Bottle-nosed Dolphin	Красноморская афалина
– tronqué	Großer Tümmler	Bottle-nosed Dolphin	Североатлантическая афалина
Unau commun	Unau	Unau	Унау
– d'Hoffmann	Hoffmann-Zweifingerfaultier	Hoffmann's Two-toed Sloth	Двупалый ленивец Гоффманна
Unaus	Zweifingerfaultiere	Two-toed Sloths	Двупалые ленивцы
Vampire d'ailes blanches	Weißschwingen-Vampir	White-winged Vampire Bat	
– d'Azara	Gemeiner Vampir	Vampire	Большой кровосос
Vampires	Echte Vampire, Gemeine Vampire	Vampires, Vampire Bats	Кровососы
Vespertilion à couleur de paille	Trichterohr	Funnel-eared Bat	
– à dos nu	Kleine Nacktrücken-Fledermaus	Naked-backed Bat	
– à moustaches	Bartfledermaus	Whiskered Bat	Усатая ночница
– à nez plat	Brillen-Blattnase	Seba's Short-tailed Bat	Очковый листонос
– bilobé	Gelbohr-Fledermaus	Tent-building Bat	
– Dasycnême	Teichfledermaus	Pond Bat	Прудовая ночница
– de Bechstein	Bechstein-Fledermaus	Bechstein's Bat	Ночница Бехштейна
– de Daubenton	Wasserfledermaus	Water Bat	Водяная ночница
– de Natterer	Fransenfledermaus	Natterer's Bat	Ночница Наттерера
– de Suapuré	Große Nacktrücken-Fledermaus	Suapure Naked-backed Bat	
– de Tome	Schwertnase	Tome's Long-eared Bat	Малый вампир
– de Welwitsch	Welwitsch-Fledermaus	Welwitsch's Bat	
– des marais	Teichfledermaus	Pond Bat	Прудовая ночница
– d'Honduras	Honduras-Haftscheiben-Fledermaus	Honduran Disk-winged Bat	
– doré	Madagassische Haftscheiben-Fledermaus	Golden Bat	Золотистая летучая мышь
– du bambou	Bambus-Fledermaus	Flat-headed Bat	Бамбуковая летучая мышь
– murin	Mausohr	European Little Brown Bat	Большая ночница
– pipistrelle	Zwergfledermaus	Common Pipistrelle	Нетопырь-карлик
– ridé	Greisengesicht	Wrinkled-faced Bat	
– tricolore	Dreifarbige Haftscheiben-Fledermaus	Spix's Disk-winged Bat	Трехцветная летучая мышь
Vespertilionidés	Glattnasen-Fledermäuse	Vespertilionid Bats	Обыкновенные летучие мыши
Vespertilions à nez plat	Kurzschwanz-Blattnasen	Short-tailed Leaf-nosed Bats	Короткохвостые вампиры
Xénarthrés	Nebengelenktiere	Xenarthra	
Zapode des bois	Waldhüpfmaus	Woodland Jumping Mouse	Лесной прыгунчик
– du Canada	Wiesenhüpfmaus	Meadow Jumping Mouse	Североамериканский луговой прыгунчик
Zapodes des bois	Waldhüpfmäuse	Woodland Jumping Mice	
– des prées	Feldhüpfmäuse	Meadow Jumping Mice	
Zapodidés	Hüpfmäuse	Jumping Mice	Мышовки и прыгунчики
Zapodinés	Eigentliche Hüpfmäuse	True Jumping Mice	Прыгунчики
Zenkerelles	Dornschwanzbilche	Flightless Scaly-tailed Squirrels	Шихпохвостые сони

FRENCH NAME	GERMAN NAME	ENGLISH NAME	RUSSIAN NAME
Ziphiidés	Schnabelwale	Beaked Whales	Клюворылые
Zokor	Blindmull	North Altai Zokor	Алтайский цокор
– de Mandchourie	Chinesischer Blindmull	Manchurian Zokor	Северокитайский цокор
– de Transbaikalie	Daurischer Blindmull	Transbaikal Zokor	Даурский цокор

4. Russian—German—English—French

Названия подвидов отличаются от видовых чаще всего лишь дополнительным прилагательным, главным образом географического характера. Такие русские названия подвидов как правило не включены в данную часть зоологического словаря.

RUSSIAN NAME	GERMAN NAME	ENGLISH NAME	FRENCH NAME
Австралийская крыса Монктона	Moncktons Schwimmratte	Monckton's Water Rat	
Австралийские водяные крысы	Schwimmratten	Water Rats	Hydromyinés
Австралийские зайцевидные мыши	Australische Häschenratten	House-building Jerboa Rats	
Австралийские малые мыши	– Kleinmäuse	Australian Native Mice	Souris d'Australie
Австралийские прыгающие мыши	– Hüpfmäuse	– Hopping Mice	
Австралийский кит	Südlicher Glattwal	Ice Baleen Whale	Baleine australe
Австралопитеки	Vormenschen	Ape-men	Australopithèques
Агути	Eigentliche Agutis	Agoutis	Dasyproctinés
Агутиевые	Agutis	Agoutis	Dasyproctidés
Азарская агути	Azara-Aguti	Azara Agouti	Agouti d'Azara
Азиатские древесные крысы	Borkenratten		Phloeomyinés
Азиатские карликовые белки	Asiatische Zwerghörnchen	Oriental Pygmy Squirrels	Écureuils pygmées
Ай	Dreifingerfaultier	Three-toed Sloth	Bradype
Акучи	Acouchi	Acouchi	Acouchi
Александрийская черная крыса	Alexandriner Hausratte	Roof Rat	Rat d'Aléxandrie
Алтайский цокор	Blindmull	North Altai Zokor	Rat taupe
Альпийский сурок	Alpenmurmeltier	Alpine Marmot	Marmotte des Alpes
Амазонская иния	Amazonas-Delphin	Amazonian Dolphin	Inie de Geoffroy
Амазонские дельфины	Inias	Amazon Dolphins	Iniidés
Амазонский речной дельфин	Amazonas-Sotalia	Buffeo Negro	Sotalie
Американская боровая полевка	Kiefernwühlmaus	Pine Vole	Campagnol du pin
Американская серая белка	Grauhörnchen	Grey Squirrel	Écureuil gris
Американские карликовые белки	Neuweltliche Zwerghörnchen	Dwarf Tree Squirrels	Microsciures
Американские мышевидные хомячки	Neuweltmäuse	New World Mice	Souris du Nouveau Monde
Американские присосковые летучие мыши	Amerikanische Haftscheiben-Fledermäuse	Disk-winged Bats	Thyropterides
Американский бурундук	Streifenbackenhörnchen	Eastern Chipmunk	Tamia strié
Американский длиннохвостый суслик	Parry-Ziesel	Arctic Ground Squirrel	Souslik de Parry
Ангольский крылан	Angola-Flughund	Angola Fruit Bat	Roussette d'Angola
Аперея	Aperea	Aperea	Cobaye sauvage
Аргентинский плащеносец	Gürtelmull	Lesser Pichiciago	Chlamyphore tronqué
Аспидная агути	Mohrenaguti	Smoky Agouti	
Атлантический ремнезуб	Sowerby-Zweizahnwal	Sowerby's Whale	Mésoplodon de Sowerby
Афалина Гилля	Gill-Tümmler	Gill's Dolphin	Dauphin de Gill
Афалины	Tümmler	Bottle-nosed Dolphins	Tursiops
Афганская слепушонка	Südlicher Mull-Lemming	Afghan Mole Lemming	Rat taupe d'Afghan
Африканская карликовая белка	Afrikanisches Zwerghörnchen	African Pygmy Squirrel	Écureuil nain du Gabon
Африканская красноносая крыса	Afrikanische Rotnasenratte	Rufous nosed Rat	Rat à museau roux
Африканская мышь-малютка	– Zwergmaus	Small African Native Mouse	Souris naine d'Afrique
Африканские горные мыши	– Felsenratten	Rock Rats	

RUSSIAN NAME	GERMAN NAME	ENGLISH NAME	FRENCH NAME
Африканские древесные мыши	Baummäuse	African Tree Mice	Dendromurinés
Африканские земляные белки	Afrikanische Borstenhörnchen	– Ground Squirrels	Écureuils foisseurs
Африканские кротовидные крысы	– Maulwurfsratten	– Mole Rats	Rats taupes africains
Африканские мыши-малютки	– Kleinmäuse	– Native Mice	Souris d'Afrique
Африканские песчанки	Eigentliche Rennmäuse	Gerbils	Gerbilles
Африканские полосатые мыши	Afrikanische Streifen-Grasmäuse	Striped Grass Mice	Rats rayés
Африканские сони	– Bilche	African Dormice	
Африканские ушастые крысы	Afrikanische Lamellenzahnratten	Vlei and Karroo Rats	Otomyinés
Африканский дикобраз	Afrikanisches Stachelschwein	African Porcupine	Porc-épic d'Afrique
Африканский длинноязычный крылан	Afrikanischer Langzungen-Flughund	African Long-tongued Fruit Bat	Mégaloglosse de Woermann
Африканский кистехвост	Westafrikanischer Quastenstachler	West African Brush-tailed Porcupine	Athérure africain
Багамская крыса	Bahama-Ferkelratte	Bahaman Hutia	Rat des Bahamas
Бамбуковая летучая мышь	Bambus-Fledermaus	Flat-headed Bat	Vespertilion du bambou
Бамбуковые крысы	Bambusratten	Bamboo Rats	Rats des bambous
Банановые листоносы	Bananenfledermäuse	Banana Bats	Chauves-souris des bananes
Банановый листонос	Bananenfledermaus	– Bat	Chauve-souris des bananes
Банановый нетопырь	Bananen-Zwergfledermaus	– Bat	Pipistrelle naine à ailes brunes
Бандикутовые крысы	Bandikutratten	Bandicoot Rats	Bandicoot rats
Беличьи	Hörnchen	Typical Squirrels	Écureuils
Белка Уайтхеда	Whitehead-Zwerghörnchen	Whitehead's Dwarf Squirrel	Écureuil pygmée de Whitehead
Белки	Eichhörnchen	European and American Tree Squirrels	Écureuils
Белкообразные	Hörnchenartige		Écureuils
Белкоподобные	Hörnchenverwandte	Squirrel-like Rodents	
Белобокий дельфин	Weißseitendelphin	White-sided Dolphin	Lagénorhynque à bec pointu
Белобрюхий ящер	Weißbauch-Schuppentier	Three-pointed Pangolin	Tricuspide
Беломордый дельфин	Weißschnauzendelphin	White-beaked Dolphin	Lagénorhynque à rostre blanc
Бело-серый кожан	Weißgraue Fledermaus	Hoary Bat	Chauve-souris cendrée
Белохвостая луговая собачка	Weißschwanz-Präriehund	White-tailed Prairie Dog	Cynomys de Gunnison
Белохвостый антилоповый суслик	Weißschwanz-Antilopenziesel	– Antelope Squirrel	Spermophile d'antilope à queue blanche
Белохвостый хомяк	Weißschwänziger Hamster	– Rat	Hamster à queue blanche
Белуха	Weißwal	White Whale	Dauphin blanc
Белухи	Gründelwale	– and Narwhales	Monodontidés
Бенгальская бандикутовая крыса	Indische Maulwurfsratte	Bengali Bandicoot Rat	Bandicoot rat du Bengale
			Marsouin de l'Inde
Бесперая морская свинья	Indischer Schweinswal	Finless Black Porpoise	
Бесперые дельфины	Glattdelphine	Right Whale Dolphins	Lissodelphininés
Бискайский кит	Nordkaper	Atlantic Right Whale	Baleine noire
Бледная ежовая мышь	Sinai-Stachelmaus	Sinai Spiny Mouse	Souris épineuse de Sinai
Бледная летучая мышь	Blasse Fledermaus	Pallid Bat	Oreillard pâle
Бледный мешетчатый прыгунчик	– Känguruhmaus	Pale Kangaroo Mouse	Souris kangourou pâle
Бобр	Biber	Beaver	Castor
Бобр Рио-Гранде	Rio-Grande-Biber	Rio-Grande Beaver	– de Rio-Grande
Бобровые	Biberartige	Beavers	Castors
Бобровые крысы	Biberratten	Nutrias	Ragondins
Бобры	Biber	Beavers	Castors
Большая ночница	Mausohr	European Little Brown Bat	Vespertilion murin
Большая песчанка	Große Rennmaus	Great Gerbil	Gerbille géante
Большая тростниковая крыса	– Rohrratte	– Cane Rat	Aulacode
Большеухий мешетчатый прыгун	Großohr-Känguruhratte	Big-eared Kangaroo Rat	Rat kangourou à grandes oreilles
Большие хомяковидные крысы	Riesenhamsterratten	Giant Pouched Rats	Rats géants
Большой вампир	Große Spießblattnase	Great False Vampire	False Vampire commun
Большой голохвостый броненосец	Großes Nacktschwanzgürteltier	Eleven-banded Armadillo	Tatou à onze bandes
Большой кровосос	Gemeiner Vampir	Vampire	Vampire d'Azara
Большой ложный подковонос	Riesen-Rundblattnase	Commerson's Leaf-nosed Bat	Phyllorine de Commerson
Большой ночной крылан	Großer Röhrennasen-Flughund	Giant Tube-nosed Fruit Bat	Nyctimène géant
Большой подковонос	Große Hufeisennase	Greater Horseshoe Bat	Grand Rhinolophe obscur

RUSSIAN NAME	GERMAN NAME	ENGLISH NAME	FRENCH NAME
Большой суслик	Rotziesel	Red Souslik	Spermophile rouge
Большой тушканчик	Pferdespringer	Five-toed Jerboa	Gerboise à cinq doigts
Бонобо	Bonobo	Pygmy Chimpanzee	Chimpansé nain
Бори	Bori	Bori	Bori
Борнеоская белка	Borneo-Hörnchen	Groove-toothed Squirrel	Reithrosciure de Borneo
Борнеоский дикобраз	Malaiischer Pinselstachler	Long-tailed Porcupine	Porc-épic à longue queue
Боялычная соня	Salzkrautbilch	Desert Dormouse	Souris du désert
Бразильская белка	Brasilhörnchen	Brazilian Squirrel	Écureuil de la Guyane
Бразильский коэнду	Greifstachler i. e. S.	Prehensile-tailed Porcupine	Porc-épic préhensile
Броненосец Капплера	Kappler-Weichgürteltier	Kappler's Armadillo	Tatou de Kappler
Броненосцы	Gürteltiere	Armadillos	Tatous
Бульдоговые летучие мыши	Bulldogg-Fledermäuse	Free-tailed Bats	Molossidés
Бурая иглистая мышь		Flat-haired Mouse	Souris aux cheveux plats
Бурундук	Burunduk	Burunduk	
Бурундуки	Streifenhörnchen	Western Chipmunks	Eutamias
Бутылконосы	Entenwale	Bottle-nosed Whales	Baleines à bec
Вечерницы	Abendsegler	Noctule Bats	Noctules
Вискаша	Viscacha	Viscacha	Lagostome des Pampas
Вискашевидная крыса	Viscacharatte	– Rat	Rat minime
Водосвинки	Riesennager	Water Hogs	Hydrochéridés
Водяная крыса	Schermaus	Vole Rat	Campagnol terrestre
Водяная ночница	Wasserfledermaus	Water Bat	Vespertilion de Daubenton
Водяные крысы	Schermäuse	– Voles	
Волосатый броненосец	Braunzottiges Borstengürteltier	Hairy Armadillo	Tatou velu
Восточноафриканский долгоног	Ostafrikanischer Springhase	Spring Haas	Lièvre sauteur d'Afrique de l'est
Вотсотса	Votsotsa	Votsotsa	
Высоколобый бутылконос	Nördlicher Entenwal	Bottle-nosed Whale	Hypérodon du nord
Гамбийская хомяковидная крыса	Gambia-Riesenhamsterratte	Gambia Pouched Rat	Rat géant de Gambia
Гангские дельфины	Ganges-Delphine	Gangetic Dolphins	Platanistidés
Гвианская крыса	Cayenneratte	Cayenne Spiny Rat	– de Guyane
Гвианский речной дельфин	Guayana-Delphin	Guyanian River Dolphin	Sotalie de la Guyane
Гейдельбергский прачеловек	Heidelberger Frühmensch	Heidelberg Man	Pithécanthrope d'Heidelberg
Гигантская белка Вильсона	Wilsons Riesenhörnchen	African Palm Squirrel	Écureuil de Wilson
Гигантская вечерница	Riesen-Abendsegler	Giant Noctule	Noctule géante
Гигантский броненосец	Riesengürteltier	– Armadillo	Tatou géant
Гигантский ящер	Riesen-Schuppentier	– Pangolin	Grand Pangolin
Гимнуровая крыса	Lanzenratte	Armoured Rat	Rat armé
Гладкие киты	Glattwale	Right Whales	Baleines
Гладконосые летучие мыши	Glattnasen-Verwandte	Vespertilionid Bats	
Глоссофага	Spitzmaus-Langzüngler	Long-tongued Bat	Glossophage de Pallas
Глоссофаги	Langzungen	– Bats	Glossophages
Голая крыса	Nacktmull	Naked Sand Rat	Rat nu de sable
Голая летучая мышь	Nacktfledermaus	– Bat	Cheiromèle nue
Голобрюхая летучая мышь	Nacktbäuchiger Grabflatterer	Naked-bellied Tomb Bat	Taphien à ventre nu
Голоступые песчанки	Nacktsohlen-Rennmäuse	Large Gerbils	
Голохвостые броненосцы	Nacktschwanzgürteltiere	Eleven-banded Armadillos	Tatous à onze bandes
Гоминиды	Menschen	Man	Hominidés
Горбач	Buckelwal	Humpback Whale	Mégaptère
Горная морская свинка	Moko	Rock Cavy	Cobaye des roches
Горная мышь	Felsenratte, Felsenmaus	Rock Rat, Broadtoothed Field Mouse	Rat typique, Mulot rupestre
Горная пака	Bergpaka	Mountain Paca	Paca des montagnes
Горные бобры	Stummelschwanzhörnchen	– Beavers	Castors de montagne
Горные свинки	Bergmeerschweinchen	Rock Cavies	Cobayes des roches
Горный бобр	Stummelschwanzhörnchen	Mountain Beaver	Castor de montagne
Горный гофер	Gebirgs-Taschenratte	Western Pocket Gopher	Gaufre à poche des montagnes
Горный иглошерст	Bergstachler	Upper Amazonian Porcupine	Porc-épic rougeâtre
Горный чипмунк	Gebirgs-Chipmunk	Alpine Chipmunk	Néotamia de montagne
Готтентотский землекоп	Hottentotten-Graumull	African Mole Rat	Rat taupe africain
Гоферы	Taschenratten	Pocket Gophers	Gaufres à poche
Гребнепаловые	Kammfingerartige	Gundis	Goundis
Гребнепалые крысы	Kammfinger	Gundis	Goundis
Гребнепалый тушканчик	Kammzehen-Springmaus	Comb-toed Jerboa	
Гренландский кит	Grönlandwal	Bowhead	Baleine boréale
Гренландский лемминг	Grönländischer Halsbandlemming	Greenland Collared Lemming	
Гривистая крыса	Mähnenratte	Crested Hamster	Hamster d'Imhause
Гринды	Grindwale	Pilot Whales	Globicéphales
Грызуны	Nagetiere	Rodents	Rongeurs
Гунди	Gundi	Gundi	Goundi

RUSSIAN NAME	GERMAN NAME	ENGLISH NAME	FRENCH NAME
Давидова белка	Père-Davids-Felsenhörnchen	Rock Squirrel	Écureuil des rochers
Давидова полевка	Schwarzbauch-Wühlmaus	Pere David's Vole	Campagnol de Père David
Давидовы полевки	Père-Davids-Wühlmäuse	– – Voles	
Даурский хомячок	Daurischer Zwerghamster	Ratlike Hamster	Hamster nain de Daourie
Даурский цокор	– Blindmull	Transbaikal Zokor	Zokor de Transbaikalie
Двупалые ленивцы	Zweifingerfaultiere	Two-toed Sloths	Unaus
Двупалый ленивец Гоффманна	Hoffmann-Zweifingerfaultier	Hoffmann's Two-toed Sloth	Unau d'Hoffmann
Двупалый муравьед	Zwergameisenbär	Two-toed Ant-eater	Fourmilier didactyle
Двухполосая мешкокрылая мышь	Zweistreifen-Taschenfledermaus	El Salvador Sheath-tailed Bat	Saccopteryx à deux raies
Двухцветные кожаны	Zweifarbige Fledermäuse	Particoloured Bats	Sérotines bicolores
Двухцветный кожан	– Fledermaus	– Bat	Petite Chauve-souris murine
Девятипоясный броненосец	Neunbindengürteltier	Nine-banded Armadillo	Tatou à neuf bandes
Дегу	Degu	Degu	Dègue du Chili
Дельфины	Delphine	Dolphins	Delphinidés
Дельфины-белобочки	– i. e. S.	Dolphins	Dauphins
Джунгарский хомячок	Dshungarischer Zwerghamster	Dwarf Hamster	Hamster nain de Djoungarie
Дикобразовые	Stachelschweinverwandte	Porcupine-like Rodents	
Дикобразы Нового Света	Baumstachlerartige	New World Porcupines	
Диномииды	Pakaranas	Pacaranas	Pacaranas
Длинноклювые дельфины	Langschnabeldelphine	Long-snouted Dolphins	Sténidés
Длиннокрыл Шрейбера	Langflügel-Fledermaus	Long-winged Bat	Miniioptère à longues ailes
Длиннокрылы	Langflügel-Fledermäuse	Bent-winged Bats	Minioptères
Длинноносая белка	Langnasenhörnchen	Long-nosed Squirrel	Écureuil à nez long
Длинноносые крыланы	Langnasen-Flughunde	Flying Foxes	Roussettes
Длинноносый листонос	Langnasen-Fledermaus	Mexican Long-nosed Bat	Chauve-souris de Mexique
Длинноухая зайцевидная мышь	Langohr-Häschenratte	House-building Jerboa Rat	
Длинноухие тушканчики	Riesenohr-Springmäuse	Long-eared Jerboas	
Длиннохвостая пака	Pakarana	Pacarana	Pacarana
Длиннохвостая шиншилла	Langschwanz-Chinchilla	Chinchilla	Chinchilla à longue queue
Длиннохвостый сурок	Langschwänziges Murmeltier	Long-tailed Marmot	Marmotte à longue queue
Длиннохвостый суслик	Langschwänziger Ziesel	Eversman's Souslik	Spermophile d'Eversman
Длиннохвостый хомяк	– Hamster	Long-tailed Rat	Hamster à longue queue
Длиннохвостый хомячок	Langschwanz-Zwerghamster	– Hamster	– nain à longue queue
Длиннохвостый ящер	Langschwanz-Schuppentier	Phatagin	Pangolin tétradactyls
Длинноязычные вампиры	Langzungen-Fledermäuse	Long-tongued Bats	Glossophages
Длинноязычные крыланы	Langzungen-Flughunde	– Fruit Bats	Macroglosses
Долгоноги	Springhasen	Spring Haas	Lièvres sauteurs
Долгоноговые	Springhasenartige	– Haas	– sauteurs
Домашняя морская свинка	Hausmeerschweinchen	Guinea Pig	Cobaye
Домовая мышь	Hausmaus	House Mouse	Souris domestique
Древесная крыса Шаденберга	Schadenbergs Borkenratte	Bushy-tailed Cloud Rat	Rat d'ecorce de Schadenberg
Европейская широкоушка	Mopsfledermaus	Barbastelle	Barbastelle d'Europe
Европейский ремнезуб	Gervais-Zweizahnwal	Gervais' Whale	Mésoplodon de Gervais
Египетская ежовая мышь	Ägyptische Stachelmaus	Egyptian Spiny Mouse	Souris épineuse
Египетский крылан	Ägyptischer Flughund	– Fruit Bat	Roussette d'Égypte
Египетский ланцетонос	Ägyptische Klappnase	Larger Rat-tailed Bat	Rhinopôme microphylle
Ежовые крысы	Baum- und Ferkelratten	Hutias	
Ежовые мыши	Stachelmäuse	Spiny Mice	Souris épineuses
Желтобрюхий сурок	Gelbbäuchiges Murmeltier	Yellow-bellied Marmot	Marmotte à ventre fauve
Желтогорлая мышь	Gelbhalsmaus	Yellow-necked Field Mouse	Mulot fauve
Желтый чипмунк	Gelber Fichten-Chipmunk	Yellow Pine Chipmunk	Néotamie jaune
Жирная древесная мышь	Fettmaus	Fat Mouse	Rat adipeux
Загути	Zagutis	Zagoutis	Plagiodontes
Загути Кювье	Cuviers Zaguti	Haitian Hutia	Plagiodonte d'Haiti
Зайцероты	Hasenmäuler	Bulldog Bats	Noctilions
Зайцеротые летучие мыши	Hasenmäuler	– Bats	Noctilions
Западноевропейская водяная крыса	West-Schermaus	Water Vole	Campagnol amphibie
Западноевропейская домовая мышь	Westliche Hausmaus	Western House Mouse	
Землекопы	Sandgräber	Mole Rats	Rats taupes africains
Земляные белки	Borstenhörnchen	Bristly Ground Squirrels	
Земляные и древесные беличьи	Erd- und Baumhörnchen	Ground and Tree Squirrels	
Золотистая агути	Goldaguti	Orange-rumped Agouti	Agouti doré
Золотистая летучая мышь	Madagassische Haftscheiben-Fledermaus	Golden Bat	Vespertilion doré
Золотистоспинный суслик	Goldmantelziesel	Golden-mantled Ground Squirrel	Spermophile à manteau doré

RUSSIAN NAME	GERMAN NAME	ENGLISH NAME	FRENCH NAME
Золотистый бобр	Goldbiber	Golden-bellied Beaver	
Золотистый мышевидный хомячок	Goldmaus	Golden Mouse	Souris dorée
Зубатые киты	Zahnwale	Toothed Cetaceans	Odontocétés
Иглистый тушканчик Гольдмана	Goldman-Stacheltaschenmaus	Goldman's Spiny Pocket Mouse	Souris à poche de Goldman
Иглошерсты	Baumstachler	New World Porcupines	Erethizontidés
Индийская бандикутовая крыса	Bandikutratte	Bandicoot Rat	Bandicoot rat d'Inde
Индийская гигантская белка	Königsriesenhörnchen	Indian Giant Squirrel	Écureuil d'Inde
Индийская гринда	Indischer Grindwal	– Pilot Whale	Globicéphale d'Inde
Индийская земляная крыса	Kurzschwanz-Maulwurfsratte	Pest Rat	Rat à queue courte
Индийская песчанка	Indische Nacktsohlen-Rennmaus	Large Gerbil	Gerbille d'Inde
Индийские пальмовые мыши	Langschwänzige Indische Baummäuse	Long-tailed Climbing Mice	Souris à longue queue
Индийский дикобраз	Weißschwanz-Stachelschwein	White-tailed Porcupine	Porc-épic à queue blanche
Индийский ложный вампир	Lyra-Fledermaus	Indian False Vampire	Mégaderme lyre
Индомалайские гигантские белки	Indomalaiische Riesenhörnchen	Indomalayan Giant Squirrels	
Индонезийская великолепная белка	Dreifarbenhörnchen	Tricoloured Squirrel	Écureuil à trois couleurs
Индостанский панголин	Chinesisches Ohren-Schuppentier	Chinese Pangolin	Pangolin à queue courte
Иравадийский дельфин	Irawadi-Delphin	Irawady Dolphin	Orcaelle d'Irawadi
Кавказская белка	Kaukasisches Eichhörnchen	Caucasian Squirrel	Écureuil du Caucase
Кагуан	Philippinen-Gleitflieger	Philippines Flying Lemur	
Калифорнийский антилоповый суслик	Nelson-Antilopenziesel	Nelson's Souslik	Souslik de Nelson
Калифорнийский мешетчатый тушканчик	Kalifornische Taschenmaus	California Pocket Mouse	Souris à poche de Californie
Калифорнийский суслик	Kalifornischer Ziesel	– Ground Squirrel	Spermophile de Californie
Калонг	Kalong	Red-necked Fruit Bat	Roussette à cou rouge
Каменная летяга	Fels-Gleithörnchen	Woolly Flying Squirrel	Écureuil volant cendré
Каменные полевки	Gebirgswühlmäuse	High-mountain Voles	
Камерунский речной дельфин	Kamerunfluß-Delphin	West African White Dolphin	Sotalie de Teusz
Канадская белка	Rothörnchen	Chickaree	Écureuil d'Hudson
Канадский бобр	Kanadischer Biber	Canadian Beaver	Castor de Canada
Капибара	Capybara	Capybara	Capybara
Капский пескорой	Kap-Bleßmull	Cape Blesmol	Rat taupe du Cap
Капюшонный ленивец	Kapuzenfaultier		Bradype à capuchon
Карликовая африканская соня	Zwergschläfer	Pygmy Dormouse	Graphiure nain
Карликовая косатка	Zwerggrindwal	– Killer	
Карликовая кубинская крыса	Zwergbaumratte	Tiny Hutia	Rat nain
Карликовая шипохвостая белка	Zwerg-Dornschwanzhörnchen	Pygmy Scaly-tailed Squirrel	Anomalure nain
Карликовые свинки	Zwergmeerschweinchen	Mountain Cavies	
Карликовые тушканчики	Herzschädel-Springmäuse	Dwarf Jerboas	Cardiocraninés
Карликовый броненосец	Zwerggürteltier	Pygmy Armadillo	Tatou nain
Карликовый гладкий кит	Zwergglattwal	– Right Whale	Baleine naine
Карликовый дикобраз	Zwergstachelschwein	Indonesian Porcupine	Porc-épic d'Indonésie
Карликовый кашалот	Zwergpottwal	Pygmy Sperm Whale	Cachalot pygmée
Карликовый мышевидный хомячок	Amerikanische Zwergmaus	– Mouse	
Карликовый эполетовый крылан	Zwerg-Epauletten-Flughund	Dwarf Epauletted Fruit Bat	Petit Microptère
Кафрский долгоног	Südafrikanischer Springhase	Cape Spring Haas	Lièvre sauteur d'Afrique du sud
Кашалот	Pottwal	Sperm Whale	Cachalot macrocéphale
Кашалоты	Pottwale	– Whales	Physétéridés
Китайская карликовая соня	Chinesische Zwergschlafmaus	Chinese Pygmy Dormouse	Souris naine de Chine
Китайская ризомиида	– Bambusratte	China Bamboo Rat	Rat des bambous Chinois
Китайские красные белки	– Rothörnchen	Rock Squirrels	
Китайский бледный речной дельфин	Chinesischer Weißer Delphin	Chinese White Dolphin	Sotalie de Chine
Китайский речной дельфин	– Flußdelphin	– River Dolphin	Dauphin d'eau douce de Chine
Китайский хомячок	– Zwerghamster	– Ratlike Hamster	Hamster nain de Chine
Китообразные	Waltiere	Whales	Cétacés

RUSSIAN NAME	GERMAN NAME	ENGLISH NAME	FRENCH NAME
Кистехвостые сони	Pinselschwanzbilche	African Dormice	Graphiures
Клюворылые	Schnabelwale	Beaked Whales	Ziphiidés
Когтистая песчанка	Mongolische Rennmaus	Jird	Mérione de Mongolie
Кожаны	Breitflügel-Fledermäuse	Big Brown Bats	Sérotines
Колорадский чипмунк	Colorado-Chipmunk	Colorado Chipmunk	Néotamia du Colorado
Копытные лемминги	Halsbandlemminge	Collared Lemmings	
Копытный лемминг	Halsbandlemming	Arctic Lemming	Lemming arctique
Копьенос	Lanzennase	Javelin Bat	Phyllostome fer de lance
Коротконосые крыланы	Kurznasen-Flughunde	Short-nosed Fruit Bats	Cynoptères
Короткопалая мадагаскарская крыса	Kurzfuß-Inselratte		Souris à pattes courtes
Короткоухая зайцевидная мышь	Kurzohr-Häschenratte	Jone's House-building Jerboa Rat	
Короткохвостая шиншилла	Kurzschwanz-Chinchilla	Short-tailed Chinchilla	Chinchilla à queue courte
Короткохвостые вампиры	Kurzschwanz-Blattnasen	– Leaf-nosed Bats	Vespertilions à nez plat
Короткохвостые хомяковидные крысы	Kurzschwanz-Hamsterratten	Cape Pouched Mice	
Косатка	Schwertwal	Common Killer Whale	Orque épaulard
Крапчатый суслик	Perlziesel	Spotted Souslik	Souslik tacheté
Красная полевка	Polarrötelmaus	Northern Red-backed Vole	Campagnol boréal
Красноморская афалина	Rotmeer-Tümmler	Red Sea Bottle-nosed Dolphin	Tursiops du Mer Rouge
Красно-серая полевка	Graurötelmaus	Large-toothed Red-backed Vole	Campagnol gris-roux
Краснохвостая песчанка	Rotschwänzige Rennmaus	Libyan Jird	Mérione de Lybie
Красные белки	Rothörnchen	Red Squirrels	
Критская ежовая мышь	Kreta-Stachelmaus	Cretan Spiny Mouse	Souris épineuse de Crête
Кровососы	Echte Vampire	Vampires	Vampires
Кроманьонец	Cro-Magnon-Mensch	Cro-Magnon Man	Homme de Cro-Magnon
Круглохвостый суслик	Rundschwanzziesel	Round-tailed Ground Squirrel	Spermophile à queue ronde
Крылан Бютикофера	Büttikofer-Epauletten-Flughund	Buttikofer's Epauletted Fruit Bat	Épomophore de Buttikofer
Крылан Вальберга	Wahlberg-Epauletten-Flughund	Wahlberg's Epauletted Fruit Bat	– de Wahlberg
Крылан Франкета	Franquet-Epauletten-Flughund	Epauletted Bat	– de Franquet
Крылан Ценкера	Zenker-Flughund	Zenker's Fruit Bat	Scotonyctère de Zenker
Крыланы	Flughunde i. e. S.	Fruit Bats	Pteropidés
Крыланы-гарпии	Spitzzahn-Flughunde	Harpy Fruit Bats	Harpionyctères
Крысовидный хомяк	Rattenartiger Zwerghamster	Ratlike Hamster	
Крысохвостый дикобраз	Borsten-Baumstachler	Thin-spined Porcupine	Porc-épic épineux
Крысы	Eigentliche Ratten	Rats	Rats proprement dits
Ктеномииды	Kammratten	Tuco-tucos	Tuco-tucos
Куруро	Cururo	Cururo	Rat bleu
Кустарниковые крысы	Strauchratten	South American Bush Rats	Octodons
Кустовые хомячки	Buschratten	Wood Rats	Rats des bois
Куцые агути	Stummelschwanzagutis	Agoutis	Agoutis
Ланцетонос Хардвика	Hardwicke-Klappnase	Lesser Rat-tailed Bat	Rhinopôme d'Hardwick
Ланцетоносы	Mausschwanz-Fledermäuse	Rat-tailed Bats	Rhinopômes
Лаплатский дельфин	La-Plata-Delphin	La Plata Dolphin	Dauphin de la Plata
Лемминги	Lemminge	Lemmings	Lemmings
Лемминговые мыши	Lemmingmäuse	Lemming Mice	
Ленивцы	Faultiere	Tree Sloths	Bradypodidés
Лесная мышовка	Birkenmaus	Northern Birch Mouse	Siciste de bouleaux
Лесная мышь	Feld-Waldmaus	Long-tailed Field Mouse	Mulot sylvestre
Лесная соня	Baumschläfer	Forest Dormouse	Lérotin
Лесной лемминг	Waldlemming	Wood Lemming	Lemming des forêts
Лесной прыгунчик	Waldhüpfmaus	Woodland Jumping Mouse	Zapode des bois
Лесной сурок	Waldmurmeltier	Woodchuck	Monax
Лесные и полевые мыши	Wald- und Feldmäuse	Wood and Field Mice	Mulots
Лесные полевки	Rötelmäuse	Red-backed Voles	Campagnols
Летучая лисица	Indischer Flughund	Indian Flying Fox	Roussette géante
Летучие мыши	Fledermäuse	Insectivorous Bats	Chauves-souris insectivores
Летучие собаки	Flederhunde, Eigentliche Flughunde	Fruit Bats, Flying Foxes	Megachiroptères, Roussettes
Летяги	Gleithörnchen	Flying Squirrels	Écureuils volants
Лироносы	Eigentliche Großblattnasen	False Vampires	Mégadermes
Листобороды	Kinnblatt-Fledermäuse	Mustache Bats	Chilonycterinés
Листоносы	Blattnasen	American Leaf-nosed Bats	
Ложные подковоносы	Rundblattnasen	Old World Leaf-nosed Bats	
Луговые собачки	Präriehunde	Prairie Dogs	Chiens des prairies
Люди	Menschen, Menschen i. e. S.	Man	Homininés, Homo
Мадагаскарская крыса Ламбертона	Lamberton-Inselratte	Lamberton's Malagasy Rat	Rat de Lamberton
Мадагаскарская летучая собака	Roter Flughund	Rufous Flying Fox	Roussette rougeâtre
Мадагаскарские крысы	Madagaskarratten	Malagasy Rats	Rats de Madagascar

RUSSIAN NAME	GERMAN NAME	ENGLISH NAME	FRENCH NAME
Мадагаскарские присосковые летучие мыши	Madagassische Haftscheiben-Fledermäuse	Sucker-footed Bats	Myzopodidés
Малая вечерница	Rauharm-Fledermaus	Lesser Noctule	Noctule de Leisler
Малая мара	Kleine Mara	Salt-desert Cavy	Cobaye halophile
Малая тростниковая крыса	– Rohrratte	Lesser Cane Rat	Rat de Grégorian
Малайская древесная крыса	Malaiische Pinselschwanz-Baummaus	Pencil-tailed Tree Mouse	Souris d'arbre de Malaysie
Малайская гигантская белка	Riesenhörnchen	Oriental Giant Squirrel	Écureuil géant
Малайский кистехвост	Langschwanz-Quastenstachler	Asiatic Brush-tailed Porcupine	Athérure à longue queue
Малайский ложный вампир	Malaiischer Falscher Vampir	Malayan False Vampire	Mégaderme spasme
Малые голоступые песчанки	Kleine Nacktsohlen-Rennmäuse	Naked-soled Gerbils	
Малые песчанки	Sandmäuse	Jirds	Mériones
Малые полевки	Kleine Wühlmäuse	Pine Voles	
Малые хомяковидные крысы	– Hamsterratten	Long-tailed Pouched Rats	
Малый вампир	Schwertnase	Tome's Long-eared Bat	Vespertilion de Tome
Малый голохвостый броненосец	Kleines Nacktschwanzgürteltier	Spiny Armadillo	Tatou épineux
Малый длинноязычный крылан	Zwerg-Langzungen-Flughund	Asiatic Long-tongued Fruit Bat	Macroglosse minime
Малый зайцерот	Kleines Hasenmaul	Southern Bulldog Bat	Noctilio du sud
Малый кровосос	Kleiner Blutsauger	Hairy-legged Vampire Bat	Diphylle sans queue
Малый остромордый полосатик	Zwergwal	Lesser Rorqual	Petit Rorqual
Малый подковонос	Kleine Hufeisennase	– Horseshoe Bat	– Rhinolophe
Малый слепыш	Westblindmaus	– Mole Rat	Spalax occidental
Малый суслик	Zwergziesel	Little Souslik	Souslik nain
Малый чипмунк	Kleiner Chipmunk	Least Chipmunk	Néotamia nain
Мары	Maras	Maras	Dolichotinés
Масличная белка	Ölpalmenhörnchen	Oil-palm Squirrel	Grand Écureuil de Stanger
Мегадермы	Großblattnasen	Large-winged Bats	Mégadermes
Мексиканский гофер	Mexikanische Taschenratte	Yellow Pocket Gophers	Rat à poche mexicain
Мексиканский иглистый тушканчик	– Stacheltaschenmaus	Mexican Spiny Pocket Mouse	Souris épineuse à poche de Mexico
Мешетчатые грызуны	Taschennager	Pocket Gophers	Gaufres à poche
Мешетчатые крысы	Taschenmäuse	Kangaroo Mice	Souris kangourou
Мешетчатые прыгунчики	Känguruhmäuse	– Mice	– kangourou
Мешетчатые прыгуны	Taschenspringer	– Rats	Rats kangourou
Мешетчатые тушканчики	Eigentliche Taschenmäuse	Pocket Mice	– à poche
Мешетчатый прыгун Орда	Ord-Känguruhratte	Ord's Kangaroo Rat	Rat kangourou d'Ord
Мешетчатый тушканчик Бейлейя	Bailey-Taschenmaus	Bailey's Pocket Mouse	Souris à poche de Bailey
Мешкокрылые летучие мыши	Taschenfledermäuse	Sheath-tailed Bats	Saccopteryx
Мичиганский бобр	Waldbiber	Michigan Beaver	Castor de Michigan
Могильные летучие мыши	Grabflatterer	Tomb Bats	Taphiens
Молотоголовый крылан	Hammerkopf	Hammer-headed Fruit Bat	Hypsignathe monstrueux
Монгольский бобр	Mongolischer Biber	Mongolian Beaver	Castor de Mongolie
Монгольский земляной заяц	Gobi-Springmaus	Gobi Jerboa	Gerboise du Gobi
Монгольский хомячок	Mongolischer Zwerghamster	Short Dwarf Hamster	Hamster nain de Mongolie
Морская свинья	Schweinswal	Common Porpoise	Marsouin commun
Морская свинья Бурмейстера	Burmeister-Schweinswal	Burmeister's Porpoise	– spinipenne
Морские свинки	Meerschweinchen	Cavies	Caviidés
Морские свиньи	Schweinswale	Porpoises	Phocénidés
Мохнатый броненосец	Pelzgürteltier	Hairy Armadillo	Tatou poilu
Мохнатый цепкохвост	Wollgreifstachler	Woolly Prehensile-tailed Porcupine	Porc-épic laineux
Мохноногие хомячки	Kurzschwänzige Zwerghamster	Dwarf Hamsters	Hamsters nains à queue courte
Мохноногий тушканчик	Rauhfuß-Springmaus	Northern Three-toed Jerboa	Gerboise à pattes rugueuses
Муравьеды	Ameisenbären	Ant-eaters	Fourmiliers
Мускусные крысы	Bisamratten	Muskrats	Ondatras
Мышевидная соня	Dünnschwanz-Baumschläfer	Asiatic Dormouse	Myomime à queue fine
Мышевидный хомячок	Mausartiger Zwerghamster	Mouselike Hamster	
Мыши	Mäuse i. e. S.	House Mice	Souris
Мышиные	Mäuse	Old World Rats and Mice	– et Rats de l'Ancien Monde
Мышовки	Streifen-Hüpfmäuse	Birch Mice	Sicistinés, Sicistes
Мышовки и прыгунчики	Hüpfmäuse	Jumping Mice	Zapodidés
Мышь-малютка	Eurasiatische Zwergmaus	Harvest Mouse	Souris des moissons
Нарвал	Narwal	Narwhal	Narval
Насекомоядная лесная мышь	Insektenessende Waldmaus	Congo Forest Mouse	Rat à manteau roux

RUSSIAN NAME	GERMAN NAME	ENGLISH NAME	FRENCH NAME
Настоящие белки	Baumhörnchen	Tree Squirrels	
Настоящие дельфины	Eigentliche Delphine	True Dolphins	Delphininés
Настоящие лемминги	Echte Lemminge	Brown Lemmings	
Настоящие листоносы	Eigentliche Blattnasen	Big-eared Leaf-nosed Bats	Phyllostomes
Настоящие морские свинки	– Meerschweinchen	Cavies	Caviinés
Настоящие полосатики	Finnwale	Rorquals	Rorquals
Настоящий клюворыл	Cuvier-Schnabelwal	Goosebeak Whale	Baleine du bec des oies
Неандерталец	Neandertaler	Neanderthal Man	Homme de Néanderthale
Неполнозубые	Zahnlose	Edentates	Édentés
Нетопыри	Zwergfledermäuse	Pipistrelles	Pipistrelles
Нетопырь-карлик	Zwergfledermaus	Common Pipistrelle	Vespertilion pipistrelle
Никтериды	Schlitznasen	Slit-faced Bats	Nyctères
Новозеландская летучая мышь	Neuseeland-Fledermaus	New Zealand Short-tailed Bat	Mystacine tubercule
Новозеландские летучие мыши	Neuseeland-Fledermäuse	– – – Bats	Mystacinidés
Новозеландский клюворыл Шеперда	Shepherd-Wal	Shepherd's Beaked Whale	Tasmacète de Shepherd
Норвежский лемминг	Berglemming	Norway Lemming	Lemming des toundras
Носатая летучая мышь	Nasenfledermaus	Proboscis Bat	Rhynchonyctère
Ночница Бехштейна	Bechstein-Fledermaus	Bechstein's Bat	Vespertilion de Bechstein
Ночница Наттерера	Fransenfledermaus	Natterer's Bat	– de Natterer
Ночницы	Mausohr-Fledermäuse	Mouse-eared Bats	Murins
Ночные крыланы	Höhlenflughunde	Roussette Bats	Roussettes
Нутрия	Sumpfbiber	Nutria	Ragondin
Ньюфаундлендский бобр	Neufundland-Biber	Newfoundland Beaver	Castor de Terre-Neuve
Обезьяны	Affen	Monkeys	Singes
Обыкновенная белка	Eichhörnchen	Red Squirrel	Écureuil commun
Обыкновенная летяга	Gewöhnliches Gleithörnchen	European Flying Squirrel	Palatouche
Обыкновенная полевка	Feldmaus	Common Vole	Campagnol des champs
Обыкновенная слепушонка	Nördlicher Mull-Lemming	Mole Lemming	Rat taupe
Обыкновенные летучие мыши	Glattnasen-Fledermäuse	Vespertilionid Bats	Vespertilionidés
Обыкновенный дикобраз	Gewöhnliches Stachelschwein	Porcupine	Porc-épic
Обыкновенный зайцерот	Indischer Kurznasen-Flughund	Short-nosed Fruit Bat	Cynoptère à nez court
Обыкновенный коротконосый крылан	Großes Hasenmaul	Mexican Bulldog Bat	Noctilio pêcheuse
Обыкновенный слепыш	Ostblindmaus	Greater Mole Rat	Spalax oriental
Обыкновенный ушан	Braunes Langohr	Long-eared Bat	Oreillard commun
Обыкновенный хомяк	Feldhamster	Common Hamster	Hamster d'Europe
Обыкновенный шимпанзе	Schimpanse	Chimpanzee	Chimpansé
Однополосый продельфин	Blau-Weißer Delphin	Blue-white Dolphin	Dauphin bleu-blanc
Оленья мышь	Weißfußmaus	White-footed Mouse	Souris à pattes blanches
Ондатра	Bisamratte	Muskrat	Rat musqué
Орешниковая соня	Haselmaus	Common Dormouse	Muscardin
Осьмизубовые	Trugrattenartige	Octodont Rodents	
Осьмизубые	Trugratten	– Rodents	Octodontidés
Очковая морская свинья	Brillenschweinswal	Spectacled Porpoise	Marsouin des lunettes
Очковый листонос	Brillen-Blattnase	Seba's Short-tailed Bat	Vespertilion à nez plat
Ошейниковый ленивец	Kragenfaultier	Necklace Sloth	Bradype à collier
Пака	Paka	Paca	Paca
Паки	Pakas	Pacas	Cuniculinés
Пальмовая белка	Palmenhörnchen	Palm Squirrel	Écureuil des palmes
Пальмовая мышь		Long-tailed Climbing Mouse	Souris à longue queue
Пальмовые белки	Palmenhörnchen	African Tree Squirrels	
Пальмовый крылан	Palmenflughund	Straw-coloured Bat	Roussette paillée
Панголины	Schuppentiere	Pangolins	Pholidotes
Пасюк	Wanderratte	Brown Rat	Rat surmulot
Патагонская свинка	Mara	Mara	Mara
Пенсильванская полевка	Wiesenwühlmaus	Eastern Meadow Mouse	Campagnol de Pennsylvania
Переднеазиатский хомяк	Syrischer Goldhamster	Golden Hamster	Hamster doré
Переднеиндийский ящер	Vorderindisches Schuppentier	Indian Pangolin	Pangolin indien
Персидская песчанка	Persische Wüstenmaus	Persian Jird	Mérione de Perse
Персидский ложный подковонос	Dreiblatt-Fledermaus	– Leaf-nosed Bat	Triaenops de Perse
Перуанская морская свинка	Tschudi-Meerschweinchen	Wild Cavy	Cobaye sauvage
Песчанки	Rennmäuse	Gerbils	Gerbillinés
Песчаный тушканчик	Wüstenspringmaus	Desert Jerboa	Gerboise du steppe
Петромиидовые	Felsenrattenartige	Cane and Rock Rats	
Пещерный крылан	Höhlen-Langzungen-Flughund	Dobson's Long-tongued Dawn Bat	Eonyctère des cavernes
Питекантроп	Javanischer Frühmensch	Java Man	Pithécanthrope de Java

RUSSIAN NAME	GERMAN NAME	ENGLISH NAME	FRENCH NAME
Плавуны	Schwarzwale	Giant Bottle-nosed Whales	
Плащеносец Бурмейстера	Burmeister-Gürtelmull	Greater Pichiciago	Grand Chlamyphore
Плащеносные броненосцы	Gürtelmulle	Pichicies	Chlamyphores
Плодовые вампиры	Fruchtvampire	Red-fruit Bats	
Плосколобый бутылконос	Südlicher Entenwal	Southern Bottle-nosed Whale	Hypérodon du sud
Плосконосые свободнохвосты	Glattnasige Freischwanz-Fledermäuse	Least Sac-winged Bats	
Плоскочерепная полевка	Gebirgsmaus	High-mountain Vole	Campagnol des montagnes
Подковонос Гильдебрандта	Hildebrandt-Hufeisennase	Hildebrandt's Horseshoe Bat	Rhinolophe d'Hildebrandt
Подковонос Маклауда	Maclaud-Hufeisennase	Maclaud's Bat	– de Maclaud
Подковоносовые	Hufeisennasen-Verwandte	Horseshoe-nosed Bats	Rhinolophes
Подковоносы	Hufeisennasen	Horseshoe Bats	Rhinolophidés, Rhinolophes
Подземная полевка	Kleine Wühlmaus	Earth Vole	Campagnol souterrain
Поздний кожан	Spätfliegende Fledermaus	Serotine Bat	Sérotine commune
Полевая мышь	Brandmaus	Striped Field Mouse	Mulot rayé
Полевка Брандта	Brandt's Steppenwühlmaus	Brandt's Vole	Campagnol de Brandt
Полевка-экономка	Nordische Wühlmaus	Tundra Vole	– nordique
Полевки	Eigentliche Wühlmäuse	Voles	
Полосатик Брайда	Brydewal	Bryde's Whale	Balénoptère de Bryde
Полосатики	Furchenwale	Finback Whales	Rorquals
Полосатый американский суслик	Streifenziesel	Thirteen-striped Ground Squirrel	Spermophile à treize bandes
Полуденная песчанка	Mittagsrennmaus	Southern Jird	Mérione du sud
Прачеловек	Frühmensch	Java Man	Pithécanthrope de Java
Приматы	Herrentiere	Primates	Primates
Проворный мешетчатый прыгун	Pazifik-Känguruhratte	Pacific Kangaroo Rat	Rat kangourou du Pacifique
Продельфины	Fleckendelphine	Spotted Dolphins	Dauphins tachetés
Прометеева полевка	Prometheus-Maus	Long-clawed Mole Vole	
Прудовая ночница	Teichfledermaus	Pond Bat	Vespertilion des marais
Прыгунчики	Eigentliche Hüpfmäuse	True Jumping Mice	Zapodinés
Пустынный кустовый хомячок	Wüstenratte	Whitethroat Wood Rat	Rat des steppes
Пустынный мешетчатый прыгун	Wüsten-Känguruhratte	Desert Kangaroo Rat	Rat kangourou du désert
Пушак	Cuvier-Hasenmaus	Mountain Cinchilla	Lagostome des montagnes
Пятипалые карликовые тушканчики	Fünfzehen-Zwergspringmäuse	Five-toed Dwarf Jerboas	
Пятнистая древесная крыса	Gescheckte Riesenborkenratte	Slender-tailed Cloud Rat	Rat d'ecorce tacheté
Равнинный гофер	Flachland-Taschenratte	Eastern Pocket Gopher	Gaufre à poche
Ремнезуб Лэйярда	Layard-Wal	Layard's Whale	Mésoplodon de Layard
Речные дельфины	Flußdelphinartige	River Dolphins	
Ризомииды	Wurzelratten	Bamboo Rats	Rats des bambous
Рисовая мышь	Sumpf-Reisratte	Rice Rat	Rat du riz
Ронский бобр	Rhonebiber	Rhone Beaver	Castor du Rhône
Рукокрылые	Fledertiere	Bats	Chiroptères
Рыжая бульдоговая мышь	Rote Samt-Fledermaus	Red Velvety Free-tailed Bat	Molosse rouge
Рыжая вечерница	Großer Abendsegler	Common Noctule	Noctule
Рыжая полевка	Rötelmaus	Bank Vole	Campagnol roussâtre
Рыжий кожан	Rote Fledermaus	Red Bat	Chauve-souris boréale
Садовая соня	Gartenschläfer	Garden Dormouse	Lérot
Сайдяной кит	Seiwal	Sei Whale	Balénoptère boréal
Свинкообразные	Meerschweinchenverwandte	Cavy-like Rodents	
Свободнохвостые летучие мыши	Glattnasen-Freischwänze	Sac-winged Bats	
Северная белокрылая свинья	Dall-Hafenschweinswal	Dall's Harbour Porpoise	Marsouin de Dall
Северная гринда	Gewöhnlicher Grindwal	Northern Pilot Whale	Grinde
Северная лемминговая мышь	Nördliche Lemmingmaus	– Bog Lemming	Lemming du nord
Северный китовидный дельфин	Nördlicher Glattdelphin	– Right Whale Dolphin	Dauphin du nord
Северный плавун	Baird-Wal	Baird's Beaked Whale	Baleine de Baird
Северный кожанок	Nordische Fledermaus	Northern Bat	Sérotine boréale
Североамериканская летяга	Assapan	Southern Flying Squirrel	Écureuil volant du sud
Североамериканские красные белки	Nordamerikanische Rothörnchen	Chickarees	
Североамериканский иглошерст	Urson	North American Porcupine	Porc-épic nord américain
Североамериканский луговой прыгунчик	Wiesenhüpfmaus	Meadow Jumping Mouse	Zapode de Canada

RUSSIAN NAME	GERMAN NAME	ENGLISH NAME	FRENCH NAME
Североамериканский серый сурок	Eisgraues Murmeltier	Hoary Marmot	Marmotte grise
Североатлантическая афалина	Großer Tümmler	Bottle-nosed Dolphin	Tursiops tronqué
Североатлантический дельфин-белобочка	Delphin	Common Dolphin	Dauphin commun
Североатлантический ремнезуб	True-Wal	True's Beaked Whale	Mésoplodon de True
Североафриканская песчанка	Nordafrikanische Rennmaus	Field Gerbil	Gerbille champêtre
Североафриканский прачеловек	Nordafrikanischer Frühmensch	Mauretania Man	Pithécanthrope de Mauretanie
Североевропейская домовая мышь	Nördliche Hausmaus	Northern House Mouse	
Северозападная черная крыса	Hausratte i. e. S.	House Rat	Rat noir
Северокитайский цокор	Chinesischer Blindmull	Manchurian Zokor	Zokor de Mandchourie
Северотихоокеанская морская свинья	Pazifischer Hafenschweinswal	Pacific Harbour Porpoise	Marsouin du Pacifique
Селевиниевые	Salzkrautbilche	Betpakdala Dormice	Souris de Selevin
Сельдяной полосатик	Finnwal	Common Rorqual	Rorqual commun
Семипоясный броненосец	Siebenbindengürteltier	Seven-banded Armadillo	Tatou à sept bandes
Серебристая полевка	Mongolische Gebirgsmaus	Royle's High-mountain Vole	Campagnol de Royle
Сероголовый восточно-австралийский крылан	Graukopf-Flughund	Grey-headed Flying Fox	Roussette à tête cendrée
Серые киты	Grauwale	Grey Whales	Eschrichtiidés
Серые полевки	Feldmäuse	Voles	
Серый дельфин	Rundkopfdelphin	Risso's Dolphin	Dauphin de Risso
Серый кит	Grauwal	Grey Whale	Buleine grise
Серый кустовый хомячок	Buschschwanzratte	Bushtail Wood Rat	
Серый суслик	Zwiesel	European Souslik	Souslik d'Europe
Серый хомячок	Grauer Zwerghamster	Migratory Hamster	Hamster migrateur
Синантроп	Peking-Frühmensch	Peking Man	Pithécanthrope de Péking
Синий кит	Blauwal	Blue Whale	Balénoptère bleu
Складчатогубы	Faltlippen-Fledermäuse	Free-tailed Bats	Tatarides
Слепушонки	Mull-Lemminge	Mole Lemmings	Rats taupes
Слепыш Эренберга	Ehrenberg-Blindmaus	Ehrenberg's Mole Rat	Spalax d'Ehrenberg
Слепыши	Blindmäuse	Palaearctic Mole Rats	Rats taupes
Снежная полевка	Schneemaus	Snow Vole	Campagnol des neiges
Современный человек	Mensch der Jetztzeit	Man	Homo sapiens
Сони	Bilche	Dormice	Loirs
Сони-летяги	Gleitbilche	African Small Flying Squirrels	Anomalures nains
Соня-летяга Ланга	Lang-Gleitbilch	Lang's Small Flying Squirrel	Anomalure nain de Lang
Соня-летяга Ценкера	Zenker-Gleitbilch	Zanker's Small Flying Squirrel	– – de Zenker
Соня-полчок	Siebenschläfer	Fat Dormouse	Loir gris
Средиземноморская домовая мышь	Ährenmaus	Eastern House Mouse	
Средние хомяки	Mittelhamster	Golden Hamsters	
Степная мышовка	Streifenmaus	Southern Birch Mouse	Siciste des steppes
Степная пеструшка	Steppenlemming	Sagebrush Vole	
Степной сурок	Bobak	Bobac Marmot	Marmotte Bobac
Степной ящер	Steppen-Schuppentier	Cape Pangolin	Pangolin de Temminck
Степные пеструшки	Steppenlemminge	Sagebrush Voles	
Стройный продельфин	Schlankdelphin	Narrow-snouted Dolphin	Dauphin douteux
Суматранская ризомиида	Sumatra-Bambusratte	Sumatran Bamboo Rat	Rat des bambous de Sumatra
Сурки	Murmeltiere	Marmots	Marmottes
Сурковые	Erdhörnchen	Northern Ground Squirrels	
Суслик-песчаник	Gelbziesel	Aral Yellow Souslik	Spermophile jaune
Суслик Франклина	Franklin-Ziesel	Franklin's Ground Squirrel	– de Franklin
Суслики	Ziesel	Ground Squirrels	Sousliks
Сусук	Ganges-Delphin	Gangetic Dolphin	Dauphin du Gange
Тагуан	Taguan	Giant Flying Squirrel	Écureuil volant géant
Тарбаганчик	Erdhase	Jerboa	Gerboise lièvre
Темная полевка	Erdmaus	Field Vole	Campagnol agreste
Техасский мешетчатый прыгун	Texas-Känguruhratte	Texas Kangaroo Rat	Rat kangourou de Texas
Тибетский хомячок	Tibetanischer Zwerghamster	Tibetan Ratlike Hamster	Hamster nain du Tibet
Тихоокеанский белобокий дельфин	Weißstreifendelphin	Pacific White-sided Dolphin	Lagenorhynque de Gill
Толстохвостая песчанка	Dickschwanzmaus	Fat-tailed Mouse	Souris à grosse queue
Тонгский крылан	Tonga-Flughund	Tonga Flying Fox	Roussette de Tonga
Тонкопалый суслик	Zieselmaus	Long-clawed Ground Squirrel	Spermophile leptodactyle
Трезубцевые дельфины	Schwarz-Weiß-Delphine	Commerson's Dolphins	Dauphins de Commerson

RUSSIAN NAME	GERMAN NAME	ENGLISH NAME	FRENCH NAME
Трезубцевый дельфин Коммерсона	Commerson-Delphin	– Dolphin	Dauphin de Commerson
Трехзубчатоносый ложный подковонос	Dreizack-Blattnase	Trident Leaf-nosed Bat	Asselia à trois endentures
Трехпалые карликовые тушканчики	Koslows Zwergspringmäuse	Pygmy Jerboas	
Трехпалые ленивцы	Dreifingerfaultiere	Three-toed Sloths	Bradypes
Трехпалые тушканчики	Eigentliche Springmäuse	Jerboas	Dipodinés
Трехпалый муравьед	Großer Ameisenbär	Great Ant-eater	Grand Fourmilier
Трехпоясный броненосец	Dreibinden-Kugelgürteltier	Apara	Apar de Buffon
Трехцветная летучая мышь	Dreifarbige Haftscheiben-Fledermaus	Spix's Disk-winged Bat	Vespertilion tricolore
Триономииды	Rohrratten	African Cane Rats	
Тростниковые крысы	Rohrratten	– – Rats	
Трубконосые крыланы	Röhrennasen-Flughunde	Tube-nosed Fruit Bats	Nyctimènes
Тукотуко	Knight-Tukotuko	Knight's Tuco-tuco	Ctenomys de Knight
Тупомордые листоносы		Neotropical Fruit-eating Bats	Artibées
Тушканчик Лихтенштейна	Lichtensteins Springmaus	Lichtenstein's Jerboa	Gerboise de Lichtenstein
Тушканчик-прыгун	Sibirische Springmaus	Sibirian Jerboa	– de Sibérie
Тушканчики	Springmäuse	Jerboas	Gerboises
Узконосые обезьяны или Обезьяны Старого Света	Schmalnasen	Old World Monkeys	Catarrhiniens
Унау	Unau	Unau	Unau commun
Уральский бобр	Uralbiber	Ural Beaver	Castor d'Ural
Усатая ночница	Bartfledermaus	Whiskered Bat	Vespertilion à moustaches
Усатые киты	Bartenwale	Baleen Whales	Baleines
Ушаны	Großohren	Long-eared Bats	Oreillards
Ушастая соня-летяга	Großohr-Gleitbilch	Large-eared Small Flying Squirrel	Anomalure nain à longues oreilles
Флоридская водяная крыса	Florida-Wasserratte	Round-tailed Muskrat	Rat d'eau de Floride
Хвостатые агути	Geschwänzte Agutis	Acouchis	Acouchis
Хлопковые хомячки	Baumwollratten	Cotton Rats	Sigmodons
Хомяковидные крысы	Hamsterratten		Cricétomyinés
Хомяки	Hamster, Eigentliche Wühler	Hamsters	Hamsters, Cricétinés
Хомячий гофер	Hamsterratte.	Tuza	Gaufre
Хомячки	Graue Zwerghamster		Hamsters nains gris
Хомячьи	Wühler	Cricetid Rats and Mice	Cricétidés
Хомячок Роборовского	Roborowski-Zwerghamster	Roborowsky's Dwarf Hamster	Hamster nain de Roborowsky
Хомячок Эверсманна	Eversmann-Zwerghamster	Eversman's Dwarf Hamster	– – d'Eversman
Хутия-карабали	Hutiacarabali	Prehensile-tailed Hutia	Rat à queue préhensile
Хутия-конга	Hutiaconga	Cuban Hutia	– poilé
Цветконосый ложный подковонос	Blumennasen-Fledermaus	Flower-faced Bat	Anthops orné
Центральноазиатская домовая мышь	Baktrische Maus	Bactrian House Mouse	
Цепкохвостые щетинистые крысы	Stachelratten	Spiny Rats	Échimyidés
Цокоры	Blindmulle	Mole Mice	
Человек	Mensch	Man	Homo sapiens
Человекоподобные	Menschenaffen	Anthropoid Apes	Singes anthropomorphes
Черная гринда	Pazifischer Grindwal	Pacific Pilot Whale	Globicéphale de Siebold
Черная косатка	Kleiner Schwertwal	False Killer Whale	Pseudorque à dents épaisses
Черная крыса	Hausratte	House Rat	Rat commun
Чернохвостая кубинская крыса	Schwarzschwanz-Baumratte	Black-tailed Hutia	– à queue noire
Чернохвостая луговая собачка	Schwarzschwanz-Präriehund	– Prairie Dog	Cynomys social
Четырехпалый муравьед	Tamandua	Tamandua	Tamandua à quatre doigts
Чипмунк Таунсенда	Townsend-Chipmunk	Townsend's Chipmunk	Néotamia de Townsend
Чипмунки	Chipmunks	Eastern Chipmunks	Tamias
Шаровидный броненосец	Kugelgürteltier	Three-banded Armadillo	Tatou à trois bandes
Шелковистый мешетчатый тушканчик	Seiden-Taschenmaus	Silky Pocket Mouse	Souris soyeux à poche
Шерстокрыл Темминка	Temminck-Gleitflieger	Malayan Flying Lemur	
Шерстокрыловые	Riesengleiter	Flying Lemurs	
Шерстокрылы	Riesengleitflieger	– Lemurs	Cynocephalidés
Шестипоясный броненосец	Weißborsten-Gürteltier	Six-banded Armadillo	Tatou à six bandes
Шимпанзе	Schimpansen	Chimpanzees	Chimpansés
Шиншилловая крыса	Chinchillaratte	Chinchilla Rat	Rat-chinchilla cendré
Шиншилловые	Chinchillaartige	Viscachas and Chinchillas	
Шиншилловые крысы	Chinchillaratten	Rat Chinchillas	Abrocomidés
Шиншиллы	Chinchillas, Chinchillas i. e. S.	Viscachas and Chinchillas	Chinchillidés, Chinchillas
Шипохвостая белка Бельдена	Belden-Dornschwanzhörnchen	Belden's Scaly-tailed Squirrel	Anomalure de Belden

RUSSIAN NAME	GERMAN NAME	ENGLISH NAME	FRENCH NAME
Шипохвостая белка Бикрофта	Beecroft-Dornschwanzhörnchen	Beecroft's Scaly-tailed Squirrel	– de Beecroft
Шипохвостая белка Джексона	Jackson-Dornschwanzhörnchen	Jackson's Scaly-tailed Squirrel	– de Jackson
Шипохвостая белка Пеля	Pel-Dornschwanzhörnchen	Pel's Scaly-tailed Squirrel	– de Pel
Шипохвостая белка Фразера	Fraser-Dornschwanzhörnchen	Fraser's Scaly-tailed Squirrel	– de Fraser
Шипохвостая соня	Dornschwanzbilch	Flightless Scaly-tailed Squirrel	Zenkerelle
Шипохвостые белки	Eigentliche Dornschwanzhörnchen	Scaly-tailed Squirrels	Anomalures
Шипохвостые белки-летяги	Dornschwanzhörnchen	– Squirrels	Anomaluridés
Шипохвостые сони	Stachelbilche, Dornschwanzbilche	Spiny Dormice, Flightless Scaly-tailed Squirrels	Zenkerelles
Широкоухий складчатогуб	Bulldogg-Fledermaus	European Free-tailed Bat	Tataride bouledogue
Широкоушки	Mopsfledermäuse	Barbastelles	Barbastelles
Щетинистые крысы	Stachelratten i. e. S.	Spiny Rats	Échimyinés
Эльбский бобр	Elbebiber	Elbe Beaver	Castor de l'Elbe
Эмбаллонуры	Freischwänzige	Sac-winged Bats	
Эполетовые крыланы	Epauletten-Flughunde	Epauletted Fruit Bats	Épomophores
Южная карликовая свинка	Südliches Zwergmeerschweinchen	Southern Mountain Cavy	
Южная лемминговая мышь	Südliche Lemmingmaus	– Bog Lemming	Lemming du sud
Южноафриканский дикобраз	Südafrikanisches Stachelschwein	South African Porcupine	Porc-épic d'Afrique du sud
Южноиндийская соня	Südindischer Stachelbilch	Spiny Dormouse	
Южный китовидный дельфин	Südlicher Glattdelphin	Peron's Dolphin	Dauphin de Peron
Южный плавун	– Schwarzwal	Arnoux's Whale	Baleine d'Arnoux
Яванская летяга	Javanisches Gleithörnchen	Indo-Malayan Flying Squirrel	Écureuil volant de Java
Яванская никтерида	Java-Hohlnase	Javanese Slit-faced Bat	Nyctère de Java
Яванский дикобраз	Java-Stachelschwein	– Porcupine	Acanthion de Java
Яванский ящер	Javanisches Schuppentier	Malayan Pangolin	Pangolin javanais
Ямайская крыса	Jamaika-Ferkelratte	Jamaican Hutia	Rat jamaique
Японская мышь	Geishamaus	Geisha Wood Mouse	Mulot de Geisha
Японская соня	Japanischer Schläfer	Japanese Dormouse	Glirule de Japon
Японский кит	Nordpazifik-Glattwal	Pacific Right Whale	Baleine de Siebold
Ящеры	Schuppentiere	Pangolins, Scaly Ant-eaters	Manidés

Conversion Tables of Metric to U.S. and British Systems

U.S. Customary to Metric		*Metric to U.S. Customary*	
To convert	*Multiply by*	*To convert*	*Multiply by*
— Length —			
in. to mm.	25.4	mm. to in.	0.039
in. to cm.	2.54	cm. to in.	0.394
ft. to m.	0.305	m. to ft.	3.281
yd. to m.	0.914	m. to yd.	1.094
mi. to km.	1.609	km. to mi.	0.621
— Area —			
sq. in. to sq. cm.	6.452	sq. cm. to sq. in.	0.155
sq. ft. to sq. mi.	0.093	sq. m. to sq. ft.	10.764
sq. yd. to sq. m.	0.836	sq. m. to sq. yd.	1.196
sq. mi. to ha.	258.999	ha. to sq. mi.	0.004
— Volume —			
cu. in. to cc.	16.387	cc. to cu. in.	0.061
cu. ft. to cu. m.	0.028	cu. m. to cu. ft.	35.315
cu. yd. to cu. m.	0.765	cu. m. to cu. yd.	1.308
— Capacity (liquid) —			
fl. oz. to liter	0.03	liter to fl. oz.	33.815
qt. to liter	0.946	liter to qt.	1.057
gal. to liter	3.785	liter to gal.	0.264
— Mass (weight) —			
oz. avdp. to g.	28.35	g. to oz. avdp.	0.035
lb. avdp. to kg.	0.454	kg. to lb. avdp.	2.205
ton to t.	0.907	t. to ton	1.102
l. t. to t.	1.016	t. to l. t.	0.984

Abbreviations

U.S. Customary

avdp.—avoirdupois
ft.—foot, feet
gal.—gallon(s)
in.—inch(es)
lb.—pound(s)
l. t.—long ton(s)
mi.—mile(s)
oz.—ounce(s)
qt.—quart(s)
sq.—square
yd.—yard(s)

Metric

cc.—cubic centimeter(s)
cm.—centimeter(s)
cu.—cubic
g.—gram(s)
ha.—hectare(s)
kg.—kilogram(s)
m.—meter(s)
mm.—millimeter(s)
t.—metric ton(s)

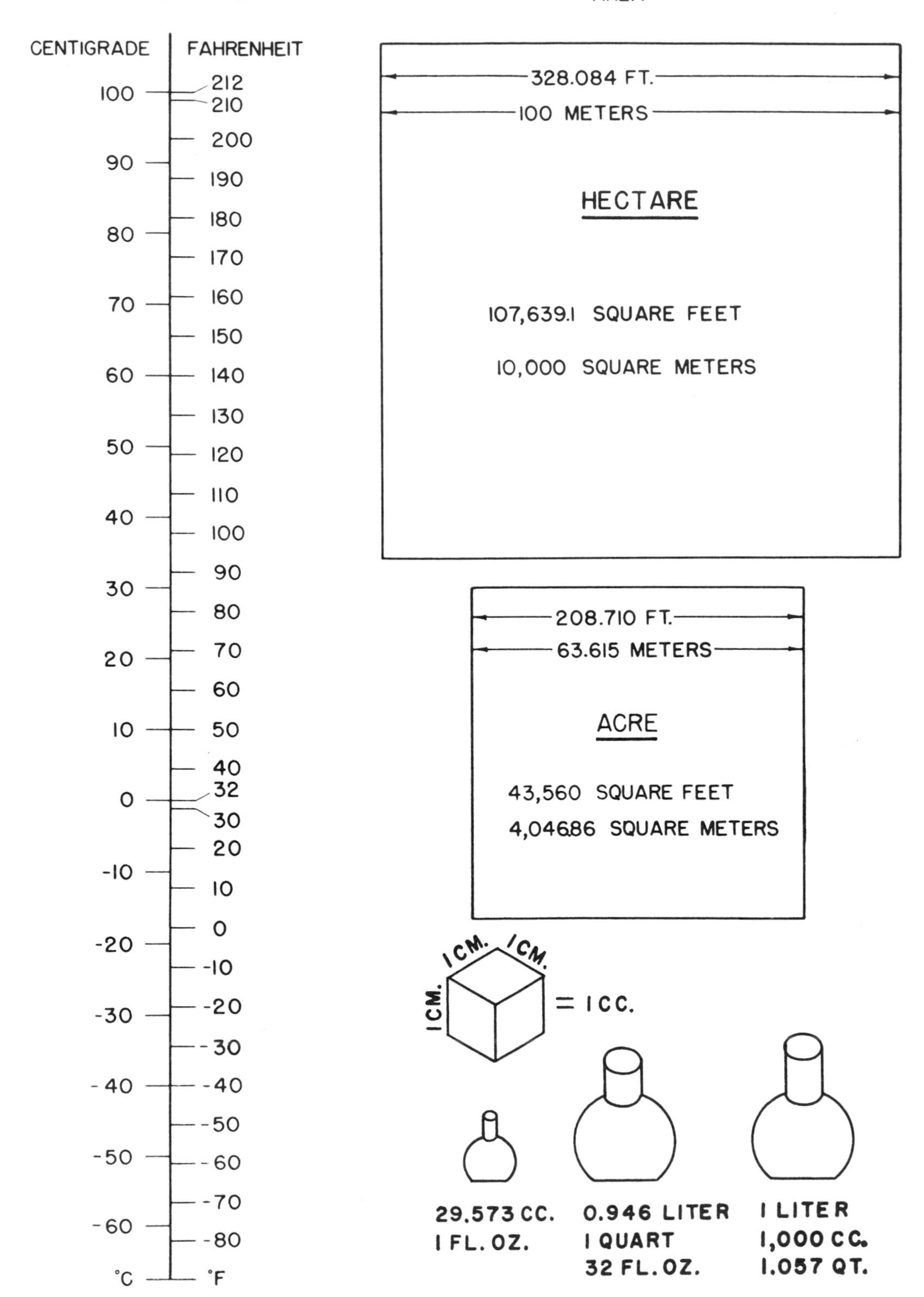

TEMPERATURE
CENTIGRADE
FAHRENHEIT
100
212
210
200
90
190
180
80
170
70
160
150
60
140
130
50
120
110
40
100
90
30
80
20
70
60
10
50
40
32
0
30
20
-10
10
0
-20
-10
-30
-20
-30
-40
-40
-50
-50
-60
-70
-60
-80
°C
°F
AREA
328.084 FT.
100 METERS
HECTARE
107,639.1 SQUARE FEET
10,000 SQUARE METERS
208.710 FT.
63.615 METERS
ACRE
43,560 SQUARE FEET
4,046.86 SQUARE METERS
1CM.
1CM.
1CM.
= 1CC.
29.573 CC.
1 FL. OZ.
0.946 LITER
1 QUART
32 FL. OZ.
1 LITER
1,000 CC.
1.057 QT.

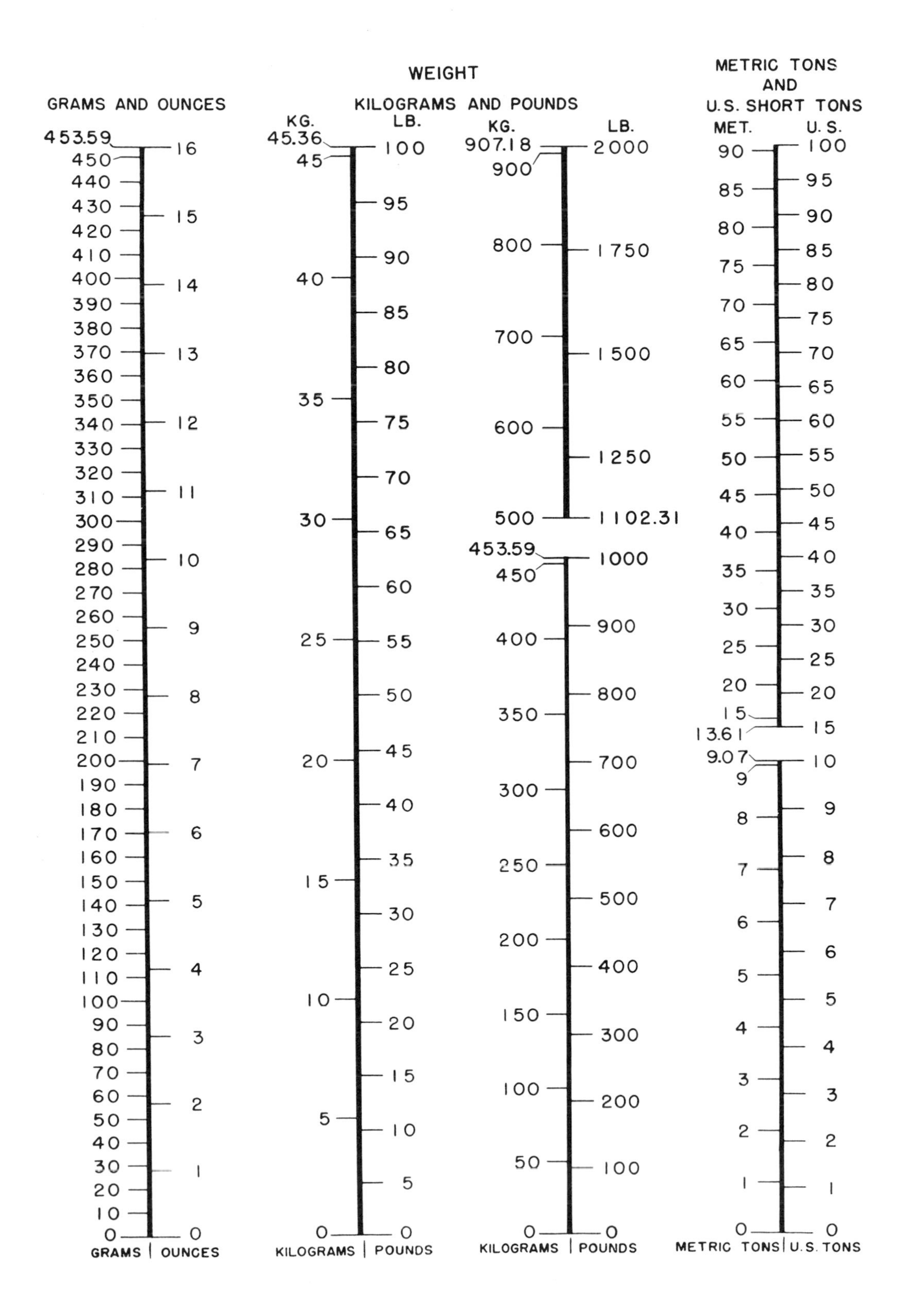

WEIGHT
GRAMS AND OUNCES
453.59
16
450
440
430
15
420
410
400
14
390
380
370
13
360
350
340
12
330
320
310
11
300
290
280
10
270
260
9
250
240
230
8
220
210
200
7
190
180
170
6
160
150
140
5
130
120
110
4
100
90
3
80
70
60
2
50
40
30
1
20
10
0
0
GRAMS | OUNCES
KILOGRAMS AND POUNDS
KG.
LB.
45.36
100
45
95
40
90
85
35
80
75
30
70
65
60
25
55
50
20
45
40
15
35
30
10
25
20
5
15
10
5
0
0
KILOGRAMS | POUNDS
KG.
LB.
907.18
2000
900
800
1750
700
1500
600
1250
500
1102.31
453.59
1000
450
400
900
800
350
700
300
600
250
500
200
400
150
300
100
200
50
100
0
0
KILOGRAMS | POUNDS
METRIC TONS
AND
U.S. SHORT TONS
MET.
U.S.
90
100
85
95
80
90
75
85
70
80
65
75
60
70
55
65
50
60
45
55
40
50
35
45
30
40
25
35
20
30
15
25
13.61
20
9.07
15
9
10
8
9
7
8
6
7
5
6
4
5
3
4
2
3
1
2
0
1
0
METRIC TONS | U.S. TONS

LENGTH: MILLIMETERS AND INCHES

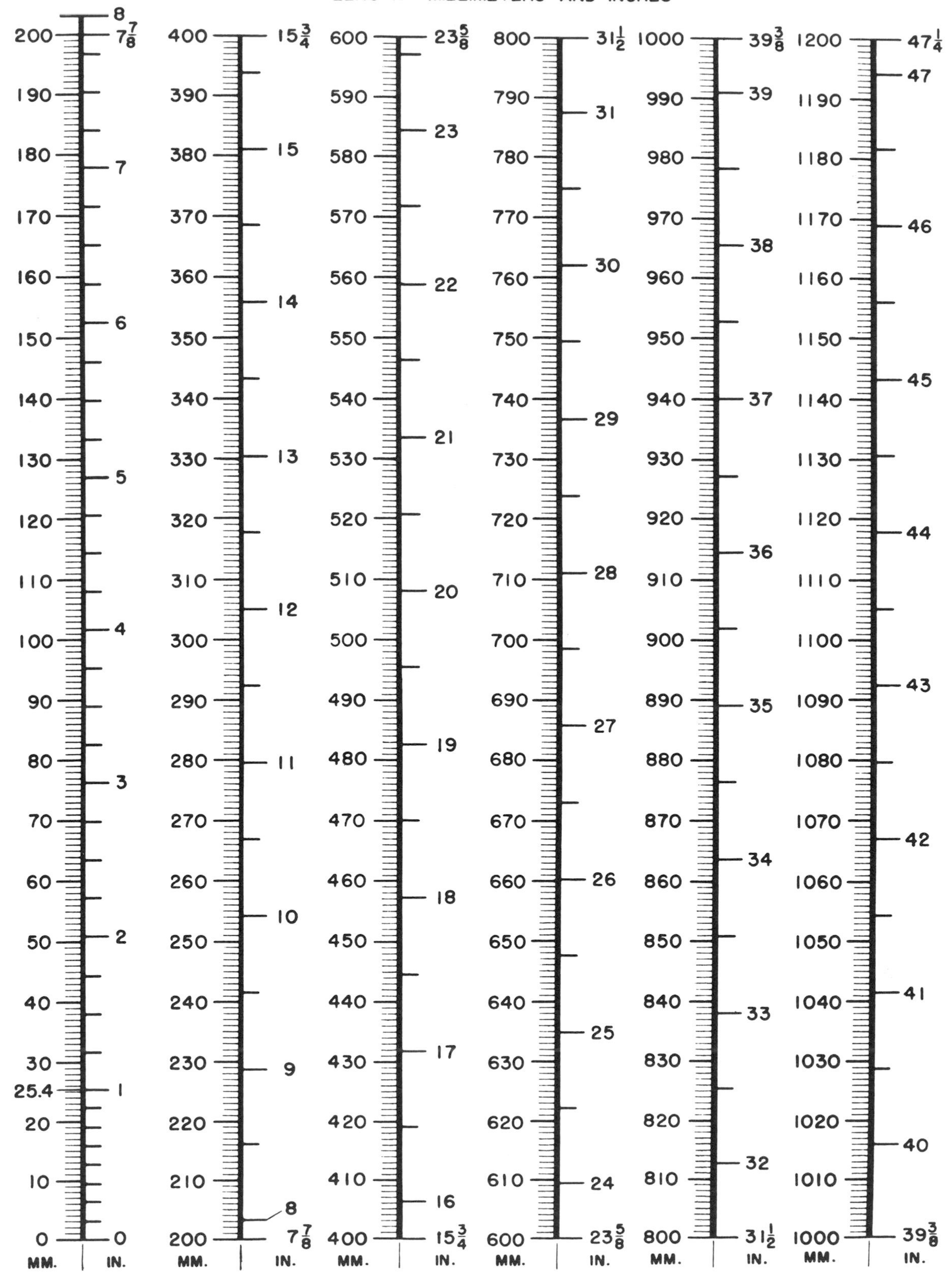

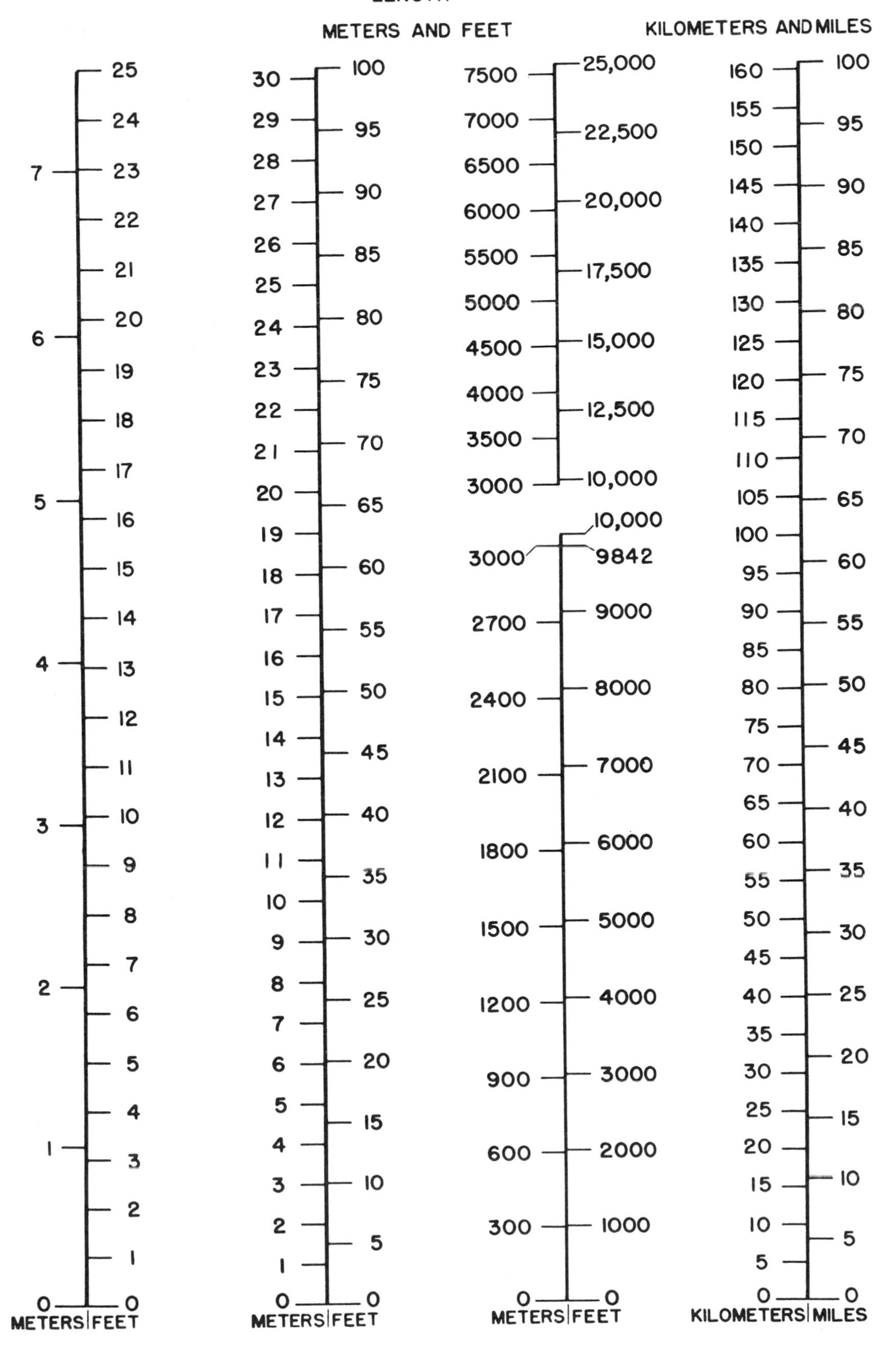
LENGTH
METERS AND FEET
KILOMETERS AND MILES
METERS
FEET
METERS
FEET
METERS
FEET
3000
9842
KILOMETERS
MILES

Supplementary Readings

Allen, G. M. 1939. *Bats.* Harvard University Press, Cambridge (Dover Reprint Available).

Allin, J. T. and E. M. Banks. 1968. Behavioural Biology of the Collared Lemming (*Dicrostonyx groenlandicus*) (Traill): I. Agonistic Behavior. *Animal Behaviour* 16(2–3): 245–262.

Alpers, A. 1961. *A Book of Dolphins.* John Murray, London.

Alpers, A. 1961. *Dolphins, The Myth and the Mammal.* Houghton Mifflin, Boston.

Andersen, H. T. A. 1969. *The Biology of Marine Mammals.* Academic Press, New York.

Andrews, R. C. 1934. *Ends of the Earth.* Garden City Publishing Co., Garden City, New York. (Whales)

Ashlee, T. 1971. *Skana.* Holt, Rinehart & Winston, Toronto. (Cetaceans)

Balbour, R. W. and W. H. Davis. 1969. *Bats of America.* University of Kentucky Press, Lexington.

Barash, D. P. 1973. The Social Biology of the Olympic Marmot. *Animal Behaviour Monographs* 6(3): 173–245.

Barnard, K. H. 1954. *A Guidebook to South African Whales and Dolphins.* South African Museum, Cape Town.

Barnett, S. A. 1963. *A Study in Behaviour; Principles of Ethology and Behavioural Physiology, Displayed Mainly in the Rat.* Methuen, London.

Barnouw, Victor. 1963. *Culture and Personality.* Dorsey Press, Homewood, Illinois.

Bennett, A. G. 1931. *Whaling in the Antarctic.* Wm. Blackwood and Sons, Ltd., Edinburgh and London.

Berry, R. J. 1970. The Natural History of the House Mouse. *Field Studies* 3(2): 219–262.

Bidney, David. 1953. *Theoretical Anthropology.* Columbia University Press, New York.

Bingham, H. 1927. Parental Play of Chimpanzees. *Journal of Mammalogy.* 8: 77–89.

Blond, G. 1955. *The Great Story of Whales.* Hanover House, New York.

Bramson, Leon and George W. Goethals. (Eds.) 1964. *War. Studies from Psychology, Sociology, Anthropology.* Basic Books, New York.

Bourne, G. H. (Ed.) 1970. *The Chimpanzee,* 3 Vols. University Park Press. Baltimore, Md.

Budker, P. 1959. *Whales and Whaling.* Macmillan, New York.

Calhoun, J .B. 1963. *The Ecology and Sociology of the Norway Rat.* U.S. Public Health Service Publication, No. 1008. U.S. Government Printing Office, Washington, D.C.

Campbell, B. 1966. *Human Evolution.* Aldine Publishing Company, Chicago.

Cavalli Sforza, L. L. and W. F. Bodmer. *The Genetics of Human Populations.* Freeman, San Francisco.

Chapple, E. D. *Culture and Biological Man: Explorations in Behavioral Anthropology.* Holt, Rinehart & Winston, New York.

Chapskii, K. K. and V. E. Sokolov, eds. 1973. *Morphology and Ecology of Marine Mammals; Seals, Dolphins, Porpoises.* J. Wiley, New York.

Cherry, Colin. 1957. *On Human Communication.* Technology Press, Cambridge, Massachusetts.

Chiarelli, A. B. 1973. *Evolution of the Primates. An Introduction to the Biology of Man.* Academic Press, New York.

Clark, W. E. Le. G. 1962. *The Antecedents of Man.* Edinburgh University Press, Edinburgh.

Clymer, E. 1968. *The Case of the Missing Link.* Basic Books, New York.

Cockrum, E. L. 1969. *Migration of the Guano Bat, Tadarida brasiliensis.* University of Kansas Museum of Natural History Publication No. 51: 303–336.

Collias, N. E. 1944. Aggressive Behavior among Vertebrate Animals. *Physiological Zoology.* 17: 83–123.

Costello, D. F. 1966. *The World of the Porcupine.* Lippincott, Philadelphia.

Count, Earl W. 1973. *Being and Becoming Human.* Van Nostrand Reinhold Co., New York.

Craig, W. 1918. Appetites and Aversions as Constituents of Instincts. *Biological Bulletin.* 3: 91–107.

Crowcroft, P. 1966. *Mice All Over.* G. T. Foulis & Co. Ltd., London.

Cunningham, Michael. 1972. *Intelligence: Its Organization and Development.* Academic Press, New York.

Dakin, Wm. J. 1938. *Whalemen Adventures.* Angus & Robertson, Ltd., Sydney.

Dampier, Sir William, 1932. *A History of Science and Its Relations with Philosophy and Religion.* The Macmillan Company, New York.

Dart, R. A. & D. Craig. 1959. *Adventures with the Missing Link.* Harper & Row, New York.

Darwin, Charles. 1965. *The Expression of the Emotions in Man and the Animals.* Reissue, University of Chicago Press.

Derennes, C. 1925. *The Life of the Bat.* Thornton Butterworth, London.

Devine, E. and M. Clark (Ed.) 1967. *The Dolphin Smile: Twenty-Nine Centuries of Dolphin Lore.* Macmillan Co., New York.

Dobzhansky, T. 1963. *Evolution, Genetics, and Man.* John Wiley and Sons, New York.

Douglas, C. L. 1969. *Comparative Ecology of Pinyon Mice and Deer Mice in Mesa Verde National Park, Colorado.* University of Kansas Publications, Museum of Natural History 18(5): 421–504.

Eibl-Eibesfeldt, Irenäus. 1970. *Ethology: The Biology of Behavior* (Erich Klinghammer, transl.). Holt, Rinehart & Winston, New York.

Eibl-Eibesfeldt, I. 1971. *Love and Hate.* Holt, Rinehart & Winston, New York.

Eibl-Eibesfeldt, I. 1972. *Die !Ko Buschmann Gesellschaft.* R. Piper & Co., München.

Eibl-Eibesfeldt, I. 1973. *Der Vorprogrammierte Mensch.* Fritz Molden, Vienna.

Eisenberg, J. F. 1963. *The Behavior of Heteromyid Rodents.* University of California Publications in Zoology, No. 69. University of California Press, Berkeley.

Eisenberg, J. F. 1967. A Comparative Study in Rodent Ethology with Emphasis on Evolution of Social Behavior, I. *Proceedings of U.S. National Museum* 122: 1–51.

Eisenberg, J. F. and W. S. Dillon (Ed.) 1971. *Man and Beast: Comparative Social Behavior.* Smithsonian Institution Press, Washington, D.C.

Ekman, Paul. 1971. Universal and Cultural Differences in Facial Expressions of Emotion. *Nebraska Symposium on Motivation.* pp. 207–283.

Ekman, Paul (Ed.). 1972. *Darwin and Facial Expression.* Academic Press, Inc., New York & London.

Ekman, Paul (Ed.). 1973. *Darwin and Facial Expression.* Academic Press, New York.

Ekman, P., Friesen, W. V., Ellsworth, P. 1972. *Emotion in the Human Face.* Pergamon Press, Inc., New York.

Ellerman, J. R. 1941–49. *The Families and Genera of Living Rodents.* British Museum (Natural History), London.

Elton, C. S. 1942. *Voles, Mice and Lemmings: Problems in Population Dynamics.* The Clarendon Press, Oxford.

Errington, P. L. 1962. *Muskrat Populations.* Iowa State University Press, Ames.

Evans, J. 1970 *About Nutria and Their Control.* Bureau of Sport Fisheries & Wildlife Resource Publication No. 86.

Ewer, R. F. 1947. Whales. *New Biology* 2: 53–73.

Fisler, G. F. 1965. *Adaptations and Speciation in Harvest Mice of the Marshes of San Fransisco Bay.* University of California Press, Berkeley and Los Angeles.

Fletcher, Ronald, 1957. *Instinct in Man.* International Universities Press, New York.

Freedman, L. Z. and A. Roe. 1958. *Evolution and Human Behavior.* In: A. Roe & G. G. Simpson, (Eds.) Behavior and Evolution pp. 455–479. Yale University Press, New Haven.

Garn, S. M. 1961. *Human Races.* Charles C. Thomas, Springfield, Illinois.

Gaskin, D. E. 1968. *The New Zealand Cetacea.* Fisheries Research Bulletin, New Series (Wellington) 1: 1–92.

Gaskin, D. E. 1973. *Whales, Dolphins and Seals. With Special Reference to the New Zealand Region.* St. Martin's Press, New York.

Gavan, J. A. (Ed.) 1955. *The Non-Human Primates in Human Evolution.* Wayne State University, Detroit.
Goffman, Erving. 1963. *Behavior in Public Places.* The Free Press, New York.
Goldsby, R. A. 1971. *Race and Races.* Macmillan, New York.
Goldstein, Kurt. 1939. *The Organism.* American Book Co., New York. Beacon Press, Boston (1963).
Goodall, J. 1963. My Life Among Wild Chimpanzees. *National Geographic Magazine,* 124(2): 272–308.
Goodall, J. 1965. *Chimpanzees of the Gombe Stream Reserve.* In: I. DeVore, Ed. Primate Behavior, pp. 524–473. Holt, Rinehart and Winston, New York.
Greenhall, A. M. & J. L. Paradiso. 1968. *Bats and Bat Banding.* U.S. Fish and Wildlife Service Resource Publication 72. Superintendent of Documents, Washington, D.C.
Griffin, D. R. 1959. *Echoes of Bats and Men.* Anchor Books. Doubleday & Company, New York.
Griffin, D. R. 1959. *Listening in the Dark: The Acoustic Orientation of Bats and Men.* Yale University Press, New Haven, Conn.

Hall, Edward T. 1966. *The Hidden Dimension,* Doubleday, New York.
Handler, P. 1971. *Biology and the Future of Man.* Oxford University Press, New York.
Harlow, Harry F. and Woolsey, Clinton N., (eds.) 1958. *Biological and Biochemical Bases of Behavior.* University of Wisconsin Press, Madison.
Harris, Marvin. 1968. *The Rise of Anthropological Theory.* Thomas Y. Crowell, New York.
Harrison, R. J. 1972. *Functional Anatomy of Marine Mammals.* Academic Press, New York.
Hass, Hans. 1970. *The Human Animal.* G. P. Putman's Sons, New York.
Hayes, C. 1951. *The Ape in our House.* Harper & Row. New York.
Herrick, C. Judson. 1956. *The Evolution of Human Nature.* University of Texas Press, Austin.
Hershkovitz, P. 1961. *On the Nomenclature of Certain Whales.* Chicago Natural History Museum, Chicago.
Hershkovitz, P. 1966. *Catalog of Living Whales.* Bulletin 246. Smithsonian Institution. U.S. National Museum, Washington.
Hill, W. C. O. 1972. *Evolutionary Biology of the Primates.* Academic Press. New York.
Hinde, Robert A. 1974. *Biological Bases of Human Social Behavior.* McGraw-Hill Book Co., New York.
Hoffmeister. D. F. 1951. *A Taxonomic and Evolutionary Study of the Pinon Mouse, Peromyscus truei.* Illinois Biological Monographs, V. 21, No. 4. University of Illinois Press, Urbana.
Holloway, Ralph L. (Ed.) 1974. *Primate Aggression, Territoriality, and Xenophobia.* Academic Press, New York.
Horner, B. E. 1954. *Arboreal adaptations of Peromyscus with special reference to the use of the tail.* Contributions, Laboratory of Vertebrate Biology. University of Michigan. 61: 1–85.
Horwich, R. H. 1972. *The Ontogeny of Social Behavior in the Gray Squirrel (Sciurus carolinensis).* Paul Parey, Berlin and Hamburg.
Howell. A. B. 1930. *Aquatic Mammals. Their Adaptations to Life in the Water.* Charles C. Thomas, London (Reprinted in paperback by Dover).
Husson, A. M. 1962. *The Bats of Suriname.* Brill, Leiden.
Huxley, J. 1948. *Man in the Modern World.* The New American Library, New York.

Jackson, John Hughlings: Selected Writings of. 1958. James Taylor, Gordon Holmer, F. M. R.Walske, (Eds.) 2 Vols. Basic Books, New York.
Jarrard, Leonard E., (Ed.) 1971. *Cognitive Processes of Nonhuman Primates.* Academic Press, New York.
Jay, Phyllis C. (Ed.) 1968. *Primates: Studies in Adaptation and Variability.* Holt, Rinehart, and Winston, New York.
Jerison, Harry J. 1973. *Evolution of the Brain and Intelligence.* Academic Press, New York.
Johnson, Roger N. 1972. *Aggression in Man and Animals.* W. B. Saunders Co., Philadelphia.
Jolly, A. 1966. *Lemur behavior; a Madagascar field study.* University of Chicago Press, Chicago.
Jolly, A. 1972. *The Evolution of Primate Behavior.* Macmillan, New York.
Jones, J. K., Jr., J. D. Smith & R. W. Turner. 1971. *Noteworthy Records of Bats from Nicaragua; with a Checklist of the Chiropteran Fauna of the Country.* University of Kansas Museum of Natural History. Occasional Papers, No. 2.
Jones, Richard M. 1970. *The New Psychology of Dreaming.* Grune & Stratton, New York.

Kardiner, Abram. 1945–56. *The Psychological Frontiers of Society.* Columbia University Press, New York.
Kellogg, R. 1940. Whales, Giants of the Sea. *National Geographic Magazine.* 77(1): 35–90.
Kellogg, W. M. 1961. *Porpoises and Sonar.* Phoenix Science Series. University of Chicago Press, Chicago.
King, J. A. (Ed.) 1968. *Biology of Peromyscus (Rodentia).* Spec. pub. No. 2. The American Society of Mammalogists.
Kluckhohln, Clyde and Henry A. Murray. 1956. *Personality in Nature, Society, and Culture.* Alfred A. Knopf, New York.
Koestler, Arthur and J. R. Smythies (Eds.). *Beyond Reductionism.* Hutchinson Pub. Co., 1969. Beacon Press ed. Boston, 1971.
Kohler, W. 1927. *The Mentality of Apes.* Routledge and Kegan Paul, London.
Koopman, K. F. and E. L. Cockrum. 1967. *Bats.* pp. 109–150 in S. Anderson and J. K. Jones (Eds.), Recent Mammals of the World: A Synopsis of Families. Ronald Press, New York.
Kortlandt, A. 1972. *New Perspectives on Ape and Human Evolution.* Stichting voor Psychobiologie, Amsterdam.
Kortlandt, A. and M. Kooij. 1963. Protohominid Behavior in Primates. *Symposia of the Zoological Society of London.* 10: 61–88.
Kubie, Lawrence S. 1956. Influence of Symbolic Processes on the Role of Instincts in Human Behavior. *Psychosomatic Medicine.* 18: 189–208.
Kummer, Hans. 1971. *Primate Societies: Group Techniques of Ecological Adaptation.* Aldine-Atherton, Chicago.

LaBarre, W. 1954. *The Human Animal.* University of Chicago Press, Chicago.
Lawick-Goodall, J. van. 1967. *My Friends the Wild Chimpanzees.* National Geographic Society, Washington.
Lawick-Goodall, J. van. 1968. The Behavior of Free-living Chimpanzees in the Gombe Stream Reserve. *Animal Behavior Monographs.* 1: 165–311.
Lawick-Goodall, J. van. 1969. Mother-offspring Relationship in Wild Chimpanzees. In *Primate Ethology.* Desmond Morris. (Ed.) Doubleday and Co., Garden City, New York.
Lawick-Goodall, J. van. 1971. *In the Shadow of Man.* Houghton Mifflin Company, Boston.
Leakey, L. S. B. 1934 *Adam's Ancestors: The Evolution of Man and his Culture.* Methuen and Co. Ltd., London. (Reprinted in Paperback by Harper & Row, New York).
Lehrman, R. L. 1961. *The Long Road to Man.* Basic Books, Inc., New York.
Lenneberg, Eric H. 1967. *Biological Foundations of Language.* John Wiley and Son, New York.
Lilly, J. C. 1961. *Man and Dolphin.* Doubleday, Garden City, New York.
Lindzey, G. and D. D. Thiessen. 1970. *Contributions to Behavior-genetic Analysis; the Mouse as a Prototype.* Appleton-Century-Crofts, New York.
Lorenz, Konrad. 1965. *Evolution and Modification of Behavior.* University of Chicago Press, Chicago.
Lorenz, Konrad, 1966. *On Aggression.* Harcourt, Brace & World., N.Y.
Lorenz, K. and P. Leyhausen. 1973. *Motivation of Human and Animal Behavior.* Van Nostrand Reinhold Co., New York.

Mackintosh, N. A. 1946. *The Natural History of the Whale-Bone Whales.* In Smithsonian Institution Annual Report, 1946; Washington, D.C.
Mackintosh, N. A. 1965. *The Stocks of Whales.* Fishing News Books, Ltd., London.
MacMillan, R. E. 1964. *Population Ecology, Water Relations, and Social Behavior of a Southern California Semi-Desert Rodent Fauna.* University of California Publications in Zoology. V. 71, Berkeley.
Martinez, D. R. and Klinghammer, E. 1970. The Behavior of the Whale *Orcinus orca*; a Review of the Literature. *Zeitschr. f. Tierpsychol.* 27: 828–839.
Martinez, D. R. 1973. *The Behavior of the Killer Whale (Orcinus orca L.); A Synthesis of Field and Captivity Studies.* M.S. Thesis, Purdue University, Lafayette, Ind.

Matthews, L. H. (Ed.) 1968. *The Whale*. Simon and Schuster, New York.
McCabe, T. T. and B. D. Blanchard. 1950. *Three Species of Peromyscus*. Rood Associates, Santa Barbara, California.
Mead, Margaret. 1935. *Sex and Temperament in Three Primitive Societies*. Routledge, London.
Menzel, Emil W., Jr. (Ed.) 1973. *Precultural Primate Behavior*. Symposia of the IV International Congress, Portland, Oregon, 1972. Vol. I. S. Kaiger, Basel.
Michael, R. P. and J. H. Crook. (Eds.) 1973. *Comparative Ecology and Behavior of Primates*. Academic Press, New York.
Miller, George A. Galanser, Eugene, Pribraus, Karl H. 1960. *Plans and the Structure of Behavior*. Henry Holt & Co., New York.
Miller, G. S. Jr. 1907. *The Families and Genera of Bats*. Bulletin of the U.S. National Museum, 57.
Morgan, L. H. 1970. *The American Beaver and his Works*. B. Franklin, New York.
Mowat, F. *A Whale for the Killing*. McClelland and Stewart, Toronto.
Muul, I. 1968. *Behavioral and Physiological Influences on the Distribution of the Flying Squirrel, Glaucomys volans*. University of Michigan Museum of Zoology, Miscellaneous Publications, No. 734. N.N. (22 authors) 1956. *L'instinct dans le comportement des animaux et de l'homme*. Masson & Cie, Paris.

Napier, J. 1973. *Bigfoot. The Yeti and Sasquatch in Myth and Reality*. Dutton, New York.
Newman, M. A. 1968. *Narwhals: Seagoing Unicorns of the North*. Pacific Search, Seattle.
Nishiwaki, M. 1967. *Distribution and Migration of Marine Mammals in the North Pacific Area*. Bulletin of Ocean Research of the University of Tokyo (Japan). 1.
Nissen, H.W. 1931. *A Field Study of the Chimpanzee*. Comparative Psychology Monographs. 1.
Norris, K. S. (Ed.) 1966. *Whales, Dolphins & Porpoises*. University of California Press, Berkeley and Los Angeles.
Norris, K. S. & J. H. Prescott. 1961. *Observations on Pacific Cetaceans of Californian and Mexican Waters*. University of California Publications in Zoology 63(4): 291–402.
Novick, A. 1969. *The World of Bats*. Photos by N. Leen. Holt, Rinehart and Winston, New York.

Ombredane, André. 1950. *L'aphasie et l'élaboration de la pensée explicite*. Presses Universitaires de France. Paris.
Ommanney, F. D. *Lost Leviathan: Whales and Whaling*. Dodd, Mead, N.Y.
Omura, H. 1950. Whales in the Adjacent Waters of Japan. *Scientific Reports of the Whale Research Institute* 4: 27–113.
Orr, R. T. 1972. *Marine Mammals of California*. University of California Press, Berkeley.

Payne, R. and D. Webb, 1971. Orientation by Means of Long Range Acoustic Signalling in Baleen Whales. *Annals of the New York Academy of Sciences* 188: 110–141.
Peterson, R. 1964. *Silently, By Night*. McGraw-Hill Book Co., New York.
Pfeiffer, J. E. 1972. *The Emergence of Man*. Harper & Row, New York.
Pike, G. C. 1956. *Guide to the Whales, Porpoises and Dolphins of the North-east Pacific and Arctic Waters of Canada and Alaska*. Fisheries Research Board of Canada. Biological Station, Nanimo, B.C. Circular No. 32 (Revised) Aug. 1956.
Pike, G. C. & I. B. Macaskie. 1969. *Marine Mammals of British Columbia*. Fisheries Research Board of Canada. Bulletin 171.
Pilbeam, D. 1972. *The Ascent of Man: An Introduction to Human Evolution*. Macmillan, New York.
Platt, J. R. 1966. *The Step to Man*. John Wiley & Sons, Inc., New York.
Pribram, Karl H. 1971. *Languages of the Brain*. Prentice-Hall, Englewood Cliffs, N.J.
Pringle, J. W. S. 1972. *Biology and the Human Species*. The Herbert Spencer Lectures 1970. Oxford University Press, New York.

Ratcliffe, F. N. 1948. *Flying foxes and drifting sand*. Anugs and Robertson, Sydney.
Reynolds, V. 1964. *Budongo, an African Forest and its Chimpanzees*. Natural History Press, Garden City, New York.
Rice, D. W. and V. B. Scheffer. 1968. *A List of the Marine Mammals of the World*. U.S. Fish & Wildlife Service, Special Scientific Report—Fisheries No. 579.
Rice, D. W. and A. A. Wolman. 1971. *The Life History and Ecology of the Gray Whale (Eschrichtus robustus)*. American Society of Mammalogists Special Publication No. 3.
Robertson, Dougal. 1973. *Survive the Savage Sea*. Praeger Publishers, New York, Washington.
Robertson, R. B. 1954. *Of Whales and Men*. Knopf, New York.
Robinson, J. T. 1973. *Early Hominid Posture and Locomotion*. University of Chicago Press, Chicago.
Roe, Ann and George Gaylord Simpson. (Eds.) 1958. *Behavior and Evolution*. Yale U. Press, New Haven, Conn.
Roeder, K. D. and A. E. Treat. 1961. *The Detection and Evasion of Bats by Moths*. In Smithsonian Institution, Annual Report, 1961, Washington, D.C.
Rood, J. P. 1970. Ecology and Social Behavior of the Desert Cavy (*Microcavia australis*). *American Midland Naturalist* 83(2): 415–454.
Rosenberg, Sheldon. 1965. *Directions in Psycholinguistics*. Macmillan, New York.
Rosevear, J. R. 1965. *The Bats of West Africa*. British Museum (Natural History), London.
Ross, A. 1967. Ecological Aspects of the Food Habits of Insectivorous Bats. *Proceedings of the Western Foundation for Vertebrate Zoology* 1(4): 205–264.
Rue, L. L. 1964. *The World of the Beaver*. Lippincott, Philadelphia.
Rugh, R. 1968. *The Mouse, Its Reproduction and Development*. Burgess Publishing Company, Minneapolis.

Sanborn, C. C. 1931. *Bats from Polynesia, Melanesia and Malaysia*. Field Museum of Natural History, Chicago. Publication No. 286. Zoological Series Vol. 28, No. 2.
Sanborn, C. C. 1933. *Descriptions and Records of African Bats*. Field Museum of Natural History, Chicago. Zoological Series, Vol. 20, No. 14.
Sanborn, C. C. 1954. *Bats From Chimanta—Tepui, Venezuela, with Remarks on Choeroniscus*. Chicago Natural History Museum, Chicago. Publication No. 734.
Sanborn, C. C. and A. J. Nicholson. 1950. *Bats from New Caledonia, The Solomon Islands, and New Hebrides*. Chicago Natural History Museum, Publication No. 646. Chicago
Sanderson, T. 1956. *Follow the Whale*. Cassell and Co., London.
Scheffer, V. B. 1969. *The Year of the Whale*. Scribner, New York.
Scheffer, V. B. & J. W. Slipp. 1948. The Whales and Dolphins of Washington State. *American Midland Naturalist* 39(2): 257–337.
Schiller, Claire M. (Ed.) 1957. *Instinctive Behavior*. International Universities Press, New York.
Scientific American 1949–1967. *Human Variation and Origins*. W. H. Freeman & Co., San Francisco & London.
Scott, John Paul. (Ed.) 1974. *Primate Aggression, Territoriality, and Xenophobia*. Academic Press, New York.
Seed, A. (Ed.) 1971. *Toothed Whales in Eastern North Pacific and Antarctic Waters*. Pacific Search, Seattle.
Sherrington, Sir Charles. 1951. *Man on His Nature*. Cambridge University Press, Second Edition.
Silverstein, A. and V. B. Silverstein. 1968. *Rats and Mice; Friends and Foes of Man*. Lothrop, Lee and Shepard, New York.
Simons, E. L. 1972. *Primate Evolution, An Introduction to Man's Place in Nature*. Macmillan, New York.
Slaughter, B. H. and D. W. Walton. 1970. *About Bats: A Chiropteran Biology Symposium*. Southern Methodist University Press, Dallas.
Slijper, E. J. 1962. *Whales*. Hutchinson and Company, London.
Small, G. L. 1971. *The Blue Whale*. Columbia University Press, New York.
Smith, C. C. 1970. The Coevolution of Pine Squirrels (*Tamiascirus*) and Conifers. *Ecology Monographs*. 40(3): 349–371.
Smith, R. L. 1972. *The Ecology of Man; An Ecosystems Approach*. Harper and Row, New York.
Sommer, Robert. 1969 (Paperback). *Personal Space*. Prentice Hall, Inc.

Tanner, J. M. (Ed.) 1952. *Prospects in Psychiatric Research*. Basil Blackwell, Oxford.
Tavolga, W. N. (Ed.) 1964. *Marine Bio-acoustics*. Pergamon Press, Long Island City.

Teilhard De Chardin, P. 1965. *The Appearance of Man*. Harper & Row, New York.

Theiler, K. 1972. *The House Mouse. Development and Normal Stages from Fertilization to Four Weeks of Age*. Springer-Verlag, New York.

Thompson, Laura. 1961. *Toward a Science of Mankind*. McGraw-Hill, New York.

Thorpe, W. H. 1974. *Animal Nature and Human Nature*. Anchor Press—Doubleday, Garden City, New York.

Thorpe, W. H. 1962. *Biology and the Nature of Man*. Oxford University Press, New York.

Tinbergen, E. A. and N. Tinbergen. 1972. *Early Childhood Autism—An Ethological Approach*. Paul Parey, Berlin and Hamburg.

Tinbergen, Niko. 1951. *The Study of Instinct*. Clarendon Press, Oxford.

Verheyen, W. N. & E. Van Der Straeten. 1969. *An Annotated Bibliography of the African Muridae (Rodentia)*. Museum Royale L'Afrique Centrale Documentation Zoologique, No. 16: 1–237.

Vernberg, W. B. and F. J. Vernberg. 1972. *Environmental Physiology of Marine Animals*. Springer-Verlag, New York.

Vesey-FitzGerald, B. S. 1947. *The Senses of Bats*. Smithsonian Institution, Annual Report. 1947. Washington, D.C.

Vesey-FitzGerald, B. 1949. *British Bats*. Methuen, London.

Vinogradov, B. S. & A. I. Argiropulo. 1968. *Fauna of the U.S.S.R.; Mammals, Key to Rodents*. U.S. Department of Commerce. Clearinghouse for Federal Scientific and Technical Information. Springfield, Virginia.

Waddington, C. H. (Ed.) 1969. *Towards a Theoretical Biology*. 3 volumes. Aldine, Chicago.

Washburn, S. L. (Ed.) 1961. *The Social Life of Early Man*. Viking Fund Publications in Anthropology. No. 31. Aldine Publishing Co., Chicago.

Washburn, S. L. (Ed.) 1963. *Classification and Human Evolution*. Viking Fund Publications in Anthropology No. 37. Aldine Publishing Co., Chicago.

Washburn, S. L. and P. J. Dolhinow. 1968. *Perspectives on Human Evolution*. 2 Vols. Holt, Rinehart & Winston, New York.

Webster, F. A. and O. G. Brazier. 1968. *Experimental Studies on Echolocation Mechanisms in Bats*. Aerospace Medical Research Lab. Clearinghouse For Federal Scientific and Technical Information. Springfield, Virginia.

Wilsson, L. 1968. *My Beaver Colony*. Doubleday & Co., Inc., New York.

Wimsatt, W. A. (Ed.) 1970. *Biology of Bats*. Academic Press, New York.

Wood, F. G. 1973. *Marine Mammals and Man. The Navy's Porpoises and Sea Lions*. Luce, Washington, D.C.

Yerkes, R. M. 1943. *Chimpanzees, A Laboratory Colony*. Yale University Press, New Haven, Conn.

Young, J. Z. 1971. *An Introduction to the Study of Man*. Oxford University Press, New York.

Young, L. B. (Ed.) 1971. *Evolution of Man*. Oxford University Press, New York.

Yovits, M. C. and S. Cameron. (Eds.) 1960. *Self-organizing Systems*. Pergamon Press, Oxford.

Zuckerman, Sally. 1932. *The Social Life of Monkeys and Apes*. Harcourt Brace, New York.

Picture Credits

Artists: P. Barruel (p. 400, 409, 426, 435, 436, 437, 438). E. Bierly (p. 189, 190). W. Eigener (p. 34, 35, 151/152). K. Grossmann (p. 83, 86, 87, 123, 124, 125, 126, 127, 128, 133, 134, 135, 136, 137, 138). E. Hudecek-Neubauer (p. 176, 214/215, 466). F. Reimann (p. 105/106, 257/258, 321/322). W. Weber (p. 78, 199, 229, 230, 239, 240, 284, 297, 300, 371, 372, 382, 399). R. Zieger (p. 463, 464, 465, 474, 479, 480). Scientific advisors to the artists: Prof. Dr. H. Dathe (Reimann, Zieger), Dr. F. Dieterlen (Weber with the exception of p. 78 and 199), Prof. Dr. M. Eisentraut (Grossmann with the exception of p. 83, 86, 123, 124, 127, 133, 138), Dr. H. Felten (Grossmann p. 86, 123, 124, 127, 138), Dr. Ch. O. Handley, Jr. (Bierly), Dr. D. Heinemann (Barruel, Eigener, Grossmann p. 83 and 133), Dr. E. Mohr (Weber p. 199), Prof. Dr. E. Thenius (Hudecek-Neubauer, Weber p. 78).

Color photographs: Baglin/ZFA (p. 77, pictures 46, 47, 48, 49, 50). Bitsch/Institut für Auslandsbeziehungen (p. 77, pictures 26, 28, 29). Blöth/ZFA (p. 76, picture 10). Burton/Photo Researchers (p. 298). Des Bartlett/Photo Researchers (p. 200 below). Doehler (p. 76, picture 7). Dossenbach (p. 88 below). Glinn/Magnum (p. 77, pictures 41, 42). Gourley/Photo Researchers (p. 473 below). Grathwohl/ZFA (p. 77, picture 37). Grzimek/Okapia (p. 25, 29, 30/31, 32, 33 36, 173 above, 174, 200 above, 283, 410, 423, 424, 425). Grundlach/Frickinger (p. 73). Haberland (p. 77, picture 45). Hemmer (p. 76, pictures 1, 2, 3). Hendrickson (p. 84). Hoffmann-Buchardi/ZFA (p. 76, pictures 11, 19). Interfoto/Len Sirman (p. 76, picture 20; p. 77, pictures 30, 31, 32, 33, 34, 38). Kantorowizc/ZFA (p. 77, picture 36). Kanus (p. 76, pictures 15, 16, 17, 18, 21, 22, 23, 24, 25). Kinne/Photo Researchers (p. 484, 497). Klages (p. 213 lower right). Klingele/Institut für Auslandsbeziehungen (p. 76, picture 8). Kohama (p. 77, pictures 43, 44). Kunitsch/Bavaria (p. 76, picture 9). Lane/Collingnon (p. 499). van Lawick (p. 26, 27). Lober/Institut für Auslandsbeziehungen (p. 77, picture 39). Löhr (p. 88 above). Miami Seaquarium (p. 483). Mondadori (p. 75). Müller (p. 28). Müller-Seeberg/ZFA (p. 76, picture 12; p. 77, pictures 35, 40). Ott/Photo Researchers (p. 216 lower left). Panjabi/ZFA (p. 76, picture 14). PIP (p. 85). Riboud/Magnum (p. 74). Root/Grzimek/Okapia (p. 173 middle and lower right, 175, 213 lower left). Schlenker/Institut für Auslandsbeziehungen (p. 76, picture 13). Schmidt/PIP/ZFA (p. 213 above). Schneiders/ZFA (p. 76, picture 6). Schrempp (p. 216 above and lower right, 299, 381). Tessore/Bavaria (p. 76, picture 5). Thau/ZFA (p. 173 lower left). Trenkwalder/ZFA (p. 76, picture 4). Veronese Verlag/Institut für Auslandsbeziehungen (p. 77, picture 27). ZFA (p. 473 above, 498). The photographs on p. 76 and 77 were compiled by Dr. H. Hemmer.

Line drawings: J. Kühn (distribution maps). Drawings based on material supplied by the authors (Fig. 8–3, p. 186; Fig. 10–50, p. 288). From Bibikow, Die Murmeltiere, with the permission of Ziemsen-Verlags, Wittenberg Lutherstadt (Fig. 10-5, p. 210; Fig. 10–7, p. 220). From Eibl-Eibesfeldt, Grundriss der vergleichenden Verhaltensforschung, with the permission of the author (Fig. 1–13, p. 40). From Slijper, Riesen des Meeres, with the permission of Springer-Verlags, Berlin/Göttingen/Heidelberg (Fig. 14–1, p. 458; Fig. 14–2, p. 460; Fig. 14–3, p. 461; Fig. 15–1, p. 477; Fig. 15–2, p. 487; Fig. 15–3, p. 489; Fig. 15–4, p. 491; Fig. 16–1, p. 493; Fig. 16–10, p. 510; Fig. 16–11, p. 511). W. Weber (according to Gaisler Fig. 4–1, p. 69; according to Kortlandt Fig. 1–14, p. 45; according to Krieg Fig. 7–3, p. 160; Fig. 7–17, p. 179; Fig. 7–18, p. 180; according to Kummer Fig. 2–2, p. 52; according to Oakley and Wilson [British Museum] Fig. 2–1, p. 51; Fig. 2–3, p. 53; according to Rahm Fig. 8–4, p. 187; according to Reynolds Fig. 1–2, 1–3, p. 20; Fig. 1–4, 1-5, 1–6, p. 21; Fig. 1–7, 1–8, 1–9, p. 22; Fig. 1–10, 1–11, p. 23; Fig. 1–12, p. 38; according to Walker Fig. 7–11, p. 167). E. Diller (all others, among them according to the authors themselves Fig. 3–2, p. 66; Fig. 4–1, p. 69; Fig. 4–5, p. 80; Fig. 4–6, p. 91; Fig. 5–2, p. 95; Fig. 5–6, p. 101; Fig. 6–1, p. 112; Fig. 6–2, p. 115; Fig. 6–3, p. 120; Fig. 6–8, p. 131; Fig. 6–10, p. 139; Fig. 6–42, p. 148; Fig. 11–14, 11–15, p. 311; Fig. 11–16, p. 311; Fig. 11–21, p. 317; according to Bouliere Fig. 10–27, p. 261; according to Eibl-Eibesfeldt in Handbuch der Zoologie, Volume 8, Fig. 11–17, p. 312; Fig. 11–18, p. 314; Fig. 11–19, 11–20, p. 315; Fig. 11–48, p. 345; Fig. 11–57, p. 357; Fig. 11–58, p. 358; Fig. 11–59, p. 359; Fig. 11–60, p. 360; Fig. 11–62, p. 362; Fig. 11–66, p. 365; Fig. 11–87, p. 396; Fig. 11–95, 11–96, p. 405; Fig. 13–16, p. 433; according to Piechocki Fig. 11–74, p. 368; according to Richard Fig. 10–47, 10–48, p. 280; according to Shelford Fig. 12–6, p. 415).

Index

Abbott, W. L., 430
Abert's squirrel (*Sciurus aberti*), 255m
Abrocoma bennetti, (Bennett's rat chinchilla), 427m
—*cinerea*, (Rat chinchilla), **422**, 427m, 435★
Abrocomidae (Rat chinchillas), 214/215★, 420, **422**
Acanthion, 408
—*brachyura* (Malayan porcupine), 400★, **413**, 414m
—*hodgsoni* (Himalayan crestless porcupine), 414m
—*javanicum* (Javan porcupine), **413**, 414m
—*klossi* (Kloss' porcupine), 414m
—*suberistatum* (Chinese porcupine), 414m
Acomys (Spiny mice), 369f, 374m
—*cahirinus* (Egyptian spiny mouse), **369**
—*dimidiatus* (Sinai spiny mouse) **369**, 371★
—*minous* (Cretan spiny mouse), **369**
Aconaemys fuscus (South American rock rat), **421**, 421m, 435★
Acouchys (*Myoprocta*), **450**
Aegyptopithecus, 51
Aeolopithecus, 50
Aetiocetidae, 476
Aetiocetus, 476
Afghan Marmot (*Marmot caudata dichrous*), **206**
Afghan mole lemming (*Ellobius fuscocapillus*), 300★, **342**, 342m
African banana bat (*Pipistrellus nanus*), **145**
African bush squirrels (*Paraxerus*), **261**
African cane rats (Thryonomyidae), 214/215★, **419**
African climbing mouse (*Dendromus insignis*), 372★, **384**
African dormice (Graphilurinae), 389, **397f**
African ground squirrel (*Xerus rutilus*), 213★
African long-tongued fruit bat (*Megaloglossus woermanni*), 87★, 94, **109**
African mole rat (*Cryptomys hottentotus*), **417f**
African mole rats (*Tachyoryctes*), **347f**, 347m
African native mice (*Leggada*), **36f**, 376m
African palm squirrel, see ebien
African pygmy squirrel (*Myosciurus pumilio*), 230★, **262**
African small flying squirrels (*Idiurus*), 289, **291**
African striped squirrels (*Funisciurus*), **261**
African tree mice (Dendromurinae), 351, **383**
African tree squirrels (Funambulini), 261
African yellow-winged bat (*Lavia frons*), 124★, **131**
Agnotogaster, 286
Agorophiidae, 476
Agorophius, 476
Agoutis (*Dasyprocta*), 449
Aharoni, I., 315
Ainuid race human, 57, 76/77★
Aitken, 101
Alactagulus pygmaeus (Little earth jerboa), **404**, **406m**
Allactaga (Five-toed jerboas), 464
—*bullata* (Gobi jerboa), **406**
—*jaculus* (Northern five-toed jerboa), 399★, **404**, 406
—*sibirica* (Siberian jerboa), **406**, 406m
Allison, 91
Allocricetulus curtatus (Short dwarf hamster), 305m, **306f**
—*eversmanni* (Eversmann's dwarf hamster), 305m, **306**
Alpine chipmunk (*Tamias alpinus*), **242**, 242m
Alpine marmot (*Marmota marmota*), **205ff**, 205m, 210★, 211★, 216★, 239★
Alpine race human, 56, 76/77★
Altai marmot (*Marmota bobak baibacina*), 222
Altai zokor (*Myospalax myospalax*), **316f**, 317★
Alticola (Western voles), **330**
—*roylei* (Royle's high-mountain vole), **330**
—*strelzowi* (High-mountain vole), **330**
Amazon or Amazonian dolphin (*Inia geoffrensis*), 479★, **502f**, 503m
Amazon wild cavy (*Cavia fulgida*), 442m
Amblyrhiza inundata, 451
American harvest mice (*Reithrodontomys*), **302**
American leaf-nosed bats (Phyllonycterinae), 141
American pine vole (*Pitymys pinetorum*), **335**, 335m
American vole, see Richardson's vole
Amir, 397
Amur lemming (*Lemmus amurensis*), 323m
Amur souslik (*Citellus eversmanni menzbieri*), **232**
Andersen, Knud, 100
Angermann, Renate, 396f
Angolan fruit bat (*Rousettus angolensis*), **98f**
Anomaluridae (Scaly-tailed squirrels), 214/215★, **289**
Anomalurinae (Scaly-tailed squirrels), 201, **289**
Anomaluroidea (Scaly-tailed squirrel-like rodents), 201, **289**
Anomalurus, 66★, 289f
—*beecrofti* (Beecroft's scaly-tailed squirrel), 291f, 293m
—*beldeni* (Belden's scaly-tailed squirrel), **291**, 292m
—*chrysophoenus* (Gabon scaly-tailed squirrel), **291**, 292m
—*erythronotus*, **291f**
—*fraseri* (Fraser's scaly-tailed squirrel), 240★, **291**, 293m
—*jacksoni* (Jackson's scaly-tailed squirrel), 288★, 290, **291ff**, 292m
—*neavi* (Neave's scaly-tailed squirrel,) **291**, 292m
—*peli* (Pel's scaly-tailed squirrel), **290**, 292, 292m
—*pusillus* (Pygmy scaly-tailed squirrel), **290**, 292, 292m
Anteaters (Myrmecophagidae), 149, 172ff, 176★
Antelope squirrel, see Harris' antelope squirrel
Anthops ornatus (Flower-faced bat), 131★, **139**
Anthropoid apes (Pongidae), 51
Antrozous pallidus (Pallid bat), 136★, **146**, 146m
Apache squirrel (*Sciurus apache*), 255m
Aperea (*Cavia aperea*), **441**, 442m
Apidium, 51
Aplodontia rufa (Mountain beaver or sewellel), **201ff**, 203m, 229★
Aplodontidae (Mountain beavers or sewellels), **201**, 214/215★
Aplodontoidea, **201**
Apodemus (Wood and field mice), **365**, 366m
—*agrarius* (Striped field mouse), 365f, 366m, 371★
—*flavicollis* (Yellow-necked field mouse), 321/322★, **366f**, 367m, 371★
—*geisha* (Giesha wood mouse), 365
—*microps*, **366**, 366m
—*mystacinus* (Broad-toothed field mouse), **365**, 366m
—*sylvaticus* (Long-tailed field mouse), 299★, 365★, **366f**, 367m, 371★
Aquatic rats (*Ichthyomys*), 296
Aral yellow souslik, see large-toothed souslik
Arboreal soft-furred spiny rat (*Diplomys caniceps*), 429m
Archaeoceti, 458, 466★, **475**
Archaeopteropus transiens **71**, 94
Archaic mongoloid race human, 57, 76/77★
Arctic ground squirrel (*Citellus undulatus*), 150/151★, 227m, **232**, 236
Arctic lemming (*Dicrostonyx torquatus*), **323**, 323m
Ardrey, R., 59
Aristotle, 457
Arizona gray squirrel (*Sciurus arizonensis*), 254m
Armadillos (Dasypodidae), 149, **154f**, 176★
Armenian race human, 56, 76/77★
Armored edentates (Cingulata), 154
Armored spiny rat (*Echimys armatus*), **427**, 429m, 435★
Armorless edentates (Pilosa), 166
Arnoux's whale (*Berardius arnouxii*), **501**
Artibeus, 117, 127★
—*jamaicensis*, 140m, **142**
—*lituratus*, 140m, **142**
—*nanus*, 140m, **142**
Arvicanthis (Grass mice or kusus), 374, 375m
Arvicola (Water voles), 331f
—*sapidus* (Iberian water vole), **331**, 331m
—*terrestris* (European water vole), 105/106★, 257/258★, **331f**, 331m
Asellia tridens (Trident leaf-nosed bat), 80, 125★, **139**
Asian squirrels (Callosciurini), **260**
Asiatic brush-tailed porcupine (*Atherurus macrourus*), **412**
Asiatic dormouse, see mouse-like dormouse
Asiatic striped squirrel, see Swinhoe's striped squirrel
Astraloid type humans, 57, 76/77★
Atherurus (Brush-tailed porcupines), 408, **412**, 412m
—*africanus* (West African brush-tailed porcupine), 409★, **412**, 413m
—*centralis* (Central African brush-tailed porcupine), **413**
—*macrourus* (Asiatic brush-tailed porcupine), **412**
—*turneri* (East African brush-tailed porcupine), 413m
Atlantic blackfish, ee Northern pilot whale
Atlantic right whale (*Eubalaena glacialis*), 463★, 481, **485f**
Atlantoxerus getulus (North African ground squirrel), **243**
Australian giant false vampire bat (*Macroderma gigas*), 86★, 112, 115, 124★, **131**
Australian hopping mice (*Notomys*), **353**, 353m
Australian native mouse (*Leggadina hermannsburgensis*), **353**
Australopithecinae (Protohominids) 52f
Australopithecus africanus, **53**, **78★**
—*transvaalensis*, **53★**
—*robustus*, **54**, **78★**
Azara, 455
Azara's agouti (*Dasyprocta azarae*), **449**, 449m
Bactrian house mouse (*Mus musculus bactrianus*), **363**
Bahama hutia (*Geocapromys ingrahami*), **429**, 430m
Bailey, 91
Bailey's pocket mouse (*Perognathus baileyi*), 272m, **273**
Baiomys taylori (Pygmy mouse), 297★, **302**, 303m
Baird's beaked whale (*Berardius bairdi*), 480★, **501**
Balaena mysticetus (Bowhead whale), 463★, 481, **482f**
Balaenidae (Right whales), 458, 466★, 477, **481**
Balaenoptera (Rorquals), **488**
—*acutorostrata* (Lesser rorqual), 464★, **489f**
—*borealis* (Sei whale), 464★, **489ff**
—*edeni* (Byrde's whale), 465★, **489ff**
—*musculus* (Blue whale), 465★, 481, **488ff**
—*brevicauda*, **488**, 490
—*physalus* (Common rorqual), 460★, 464★, 481, **488ff**
Balaenopteridae (Finback whales), 458, 466★, 477, **488**, 489★
Balantiopteryx, 123★
Baleen whales (Mystacoceti), 458, 466★, **477ff**, 487★
Ballow, 332
Balsac, Heim de, 340
Bamboo rats (Rhizomyidae), 214/215★, 296, **346f**
Bambutic race human, 56, 76/77★
Banana bat (*Musonycteris harrisoni*), 86★, 118, 127★, **141**
Bandicoot rats (*Bandicota*), 353m, **354**
Bandicota (Bandicoot rats), 353m, **354**
—*bengalensis* (Lesser or Bengali bandicoot rat), **354**
—*indica* (Large bandicoot rat), **354f**
—*nemorivaga* (Eastern large bandicoot rat), 371★
Bank vole (*Clethrionomys glareolus*), 300★, **328f**, 329m
Barbastella barbastellus (Barbastelle), 112★, 135★, **145**, 145m
Barbastelle (*Barbastella barbastellus*) 112★, 135★, **145**, 145m
Bartels, 109
Bartholinus, 494
Basilosauridae, 475
Basilosaurus, aka *Zeuglodon*, 476

Bate, 98
Bates, Henry Walter, 292f, 503
Bathyergidae (Mole rats), 214/215★, **417**
Bathyergoidea (Mole rat-like rodents), 407, **416**
Bathyergus (Mole rats,) **417**
—*suillus* (Cape sand mole), **417**, 417m
Bats (Chiroptera), 64, **67ff**
Baumann, 211
Beaked whales (Ziphiidae), 458 466★, 493, **496**
Beamys (Long-tailed pouched rats), 378
Beaver (*Castor fiber*), 257/258★, **276ff**, 276m, 276★, 277m, 279★, 280★, 281★, 283★, 284★
Beaver-like rodents (Castoroidea), 201, **275ff**
Bechstein's bat (*Myotis bechsteini*), 134★, 141m, **144**
Becker, 360
Beebe, William, 452
Beecroft's scaly-tailed squirrel (*Anomalurus beecrofti*), **291**f, 293m
Belden's scaly-tailed squirrel (*Anomalurus beldeni*), **291**, 292m
Belding's ground squirrel (*Citellus beldingi*), 233m
Beluga or belukha, see white whale
Bengali bandicoot rat, see lesser bandicoot rat
Bennett's rat chinchilla (*Abrocoma bennetti*), 427m
Berardius (Giant bottle-nosed whales), **501**
—*arnouxii* (Arnoux's whale), 501
—*bairdi* (Baird's beaked whale), 480★, 501
Berger, 411
Berndt, Rudolf, 281
Betpakdala dormouse (*Selevinia betpakdalaensis*), 382★, 397m, **401**
Big brown bat (*Eptesicus fuscus*), 144m, **145**
Big-eared kangaroo rat (*Dipodomys elephantinus*), **273**, 274m
Big-eared leaf-nosed bats (Phyllostominae), **140**
Birch mice (Sicistinae), **402**, 404m
Black finless porpoise (*Neophocaena phocaenoides*), 480★, **507**
Black rat or house rat (*Rattus rattus*), 351, **361**f, 361m, 362★, 371★
Black-and-white dolphins, see Commerson's dolphins
Blackfish, see pilot whales
Black-footed lemming (*Lemmus nigripes*), 324m
Black-tailed hutia (*Capromys melanurus*), **428**, 430m
Black-tailed prairie dog (*Cynomys ludovicianus*), **225ff**, 225m, 239★
Blasius horseshoe bat (*Rhinolophus blasii*), 130m
Blesmols (*Cryptomys*), **417**f, 417m
Bloedel, 115
"Bloodsuckers" or vampire bats (Desmodontidae), 115, **142**f
Blow hole, 458
Blue whale (*Balaenoptera musculus*), 465★, 481, **488ff**
Blue-white dolphin (*Stenella caeruleoalba*), 480★, **515**
Bobac marmot (*Marmota bobak*), **205**, 206m, 218ff, 220★, 239★
Boehm's African bush squirrel (*Paraxerus boehmi*), **262**
Bog lemmings (*Synaptomys*), **323**
Bohlken, H., 441
Bolau, 281
Bonobo, see pygmy chimpanzee
Bornean black-striped squirrel (*Lariscus hosei*), **261**
Bornean long-tailed porcupine(*Trichys lipura*), 409★, **412**
Bornean thick-spined porcupine (*Thecurus crassispinus*), 414m, 415★
Bottle-nosed dolphin (*Tursiops truncatus*), 458★, 480★, 483★, 499★ 500★, 510, 510★, 511★, **515f**
Bottle-nosed whale (*Hyperoodon ampullatus*), 480★, 495m, **496f**
Boule, M., 50
Bowhead whale (*Balaena mysticetus*), 463★, 481, **482f**
Brachytarsomys albicauda, **318**
Brachyuromys, **318**
Bradypodidae (Sloths), 149, **166ff**, 167★, 176★
Bradypus (Three-fingered sloths), 169, **171f**
—*cuculliger*, 171m, **172**
—*torquatus* (Necklace sloth), 171m, **172**
—*tridactylus* (South American three-fingered sloth), 171m, **172**, **175**★, 190★
Brandt's vole (*Microtus brandti*), **336** 339
Brazilian squirrel (*Sciurus aestuans*), **259**
Brehm, A. E., 156, 396, 454f, 509
Bridgwater, D. D., 188
Bristly ground squirrels (Xerini), 205, **243**
British red squirrel (*Sciurus vulgaris leucourus*), 247
Broad-toothed field mouse (*Apodemus mystacinus*), **365**, 366m
Bronson, 224
Broom, Robert, 53
Brown dwarf squirrel (*Nannosciurus melanotis*), **260**
Brown lemmings (*Lemmus*), **324f**
Brown rat or Norway rat (*Rattus norvegicus*), 356m, **359ff**, 357★, 358★, 359★, 360★, 371★
Brush-tailed porcupines (*Atherurus*), 408, **412**, 412m
Bryde's whale (*Balaenoptera edeni*), 464★, **489ff**
Bucher, 429
Buffeo negro (*Sotalia fluviatilis*), **507**, 507m
Bujakovi, 334
Bulldog bats (Noctilionidae), **129**
Bürger, 252
Burmeister, 166
Burmeisteria retusa (Greater pichiciago), **165**f, 189★
Burrell, 353
Burunduk (*Eutamias sibiricus*), **238f**, 242m
Bushy-tailed cloud rat (*Crateromys schadenbergi*), **385f**, 385m
Bushy-tailed wood rat (*Neotoma cinerea*), 297★, **304**, 304m
Büttikofer, Johannes, 95, 292f
Büttikofer's epauletted fruit bat (*Epomops buettikoferi*), **102**
Byelorussian beaver, see Polish beaver

Cabassous hispidus (Spiny armadillo), **161**, 161m
—*lugubris*, **161**
—*oricatus*, **161**
—*unicinctus* (Eleven-banded armadillo), 161m, **161**, 189★
Cadenat, 460
California ground squirrel (*Citellus beecheyi*), 227m, **232**, 242
California pocket mouse (*Perognathus californicus*), **273**, 273m, 297★
California vole (*Microtus californicus*), 337m
Callosciurini (Asian squirrels), **260**
Callosciurus, 260
—*notatus* (Spotted squirrel), **260**
—*swinhoei* (Swinhoe's or Asiatic striped squirrel), 230★, **260**
Calomys musculinus (Vesper mouse), **303**
Calomyscus bailwardi (Mouse-like hamster), **305f**, 305m
Cameroons dormouse (*Graphiurus murinus haedulus*), **398**
Canadian beaver (*Castor fiber canadensis*), **276**
Cane and rock rats (Petromuroidea), 407, **418f**
Cannomys badius (Lesser bamboo rat), **347**
Cape mole rat (*Georhychus capensis*), **417**, 417m, 426★
Cape pangolin (*Manis temmincki*), 182m, **183**, 185f, 199★, 200★
Cape pouched mouse (*Saccostomus campestris*), **380**
Cape sand mole (*Bathyergus suillus*), **417**, 417m
Cape springhaas (*Pedetes cafer*), **295**, 295m
Capromyidae (Hutias), 214/215★, 420, **428**
Capromys, **428**
—*melanurus* (Black-tailed hutia), **428**, 430m
—*nana* (Dwarf hutia), **428**, 430m
—*pilorides* (Cuban hutia), **428f**, 431m
—*prehensilis* (Prehensile-tailed hutia), **428**, 438★
Capybara or carpincho (*Hydrochoerus hydrochaeris*), 424★, 437★, **446f**
Cardiocraniinae (Dwarf jerboas), 404
Cardiocranius paradoxus (Five-toed dwarf jerboa), 399★, 404
Cardioderma, 131
Carollia perspicillata (Seba's short-tailed bat), 128★, 139★, **141**
Carolliinae (Short-tailed leaf-nosed bats), 117, **141**
Carpincho, see capybara
Carr, 304
Carterodon sulcidens, 428m
Castor fiber (Beaver), 257/258★, **276ff**, 276★, 276m, 277m, 279★, 280★, 281★, 283★, 284★
—*albicus* (Elbe beaver), **276**, 278m, 285f
—*birulai* (Mongolian beaver), **276**, 278m
—*caecator* (Newfoundland beaver) **276**
—*canadensis* (Canadian beaver), **276**
—*fiber* (Scandanavian beaver), **276**, 278m
—*frondator* (Rio Grande beaver), **276**
—*galliae* (Rhone beaver), **276**, 278m, 280★
—*michiganensis* (Michigan beaver) **276**
—*pohlei* (Ural beaver), **276**, 278m
—*subauratus* (Golden-bellied beaver), **276**
—*vistulanus* (Polish or Byelorussian beaver), 276, 278m
Castoridae (Beavers), 201, 214/215★, **275f**
Castoroidea (Beaver-like rodents), 201, **275ff**
Castoroides, 286
Caucasian squirrel (*Sciurus anomalus*), 242m, **255**, 260
Cavia, 441
—*aperea* (Aperea), **441**, 442m
—*porcellus* (Guinea pig), 436★, **441**f
—*tschudii* (Wild cavy), 436★, **441**
—*fulgida* (Amazon wild cavy), 442m
—*stolida*, 442m
Cavies (Caviidae), 214/215★, 420, **441**
Caviidae (Cavies), 214/215★, 420, **441**
Caviinae, 441
Cavioidea, 420, **441**
Caviomorpha (Cavies), 193, 198 **420ff**
Cayenne spiny rat (*Proëchimys guyannensis*), **427**, 427m, 435★
Central African brush-tailed porcupine (*Atherurus centralis*), **41 3**
Central European red squirrel (*Sciurus vulgaris fuscoater*), 229★, 247
Centurio senex (Wrinkled-faced bat), 86★, 128★, 140m, **142**
Cephalorhynchinae, 508, **517**
Cephalorhynchus (Commerson's or black-and-white **dolphins,** aka skunk dolphins), **517**, 517m
—*albifrons* (White-headed dolphin), **517**
—*commersonii* (Commerson's dolphin), 480★, **517**
—*heavisidii* (White-lipped dolphin), **517**
—*hectori* (Hector dolphin), **517**
Cercomys cunicularis (Punare), 428m
Cetacea (Whales), **457ff**
Cetotheriidae, 466★, 476
Chaetomyinae, **453**
Chaetomys subspinosus (Thin-spined porcupine), 409★, **453**, 453m
Chance, M. R. A., 62
Chapman, M. F., 440
Chappel, E. D., 60
Charlesworth, W. R., 61
Cheiromeles parvidens, **147**
—*torquatus* (Naked bat), 112, 137★, **147**
Chickaree (*Tamiasciurus douglasii*), **244**, 244m
Chilonycterinae, **139**
Chilonycteris (Mustache bats), **139**
—*personata*, 126★
Chimpanzee (*Pan troglodytes*), **19ff**, 20m, 20★, 21★, 22★, 23★, 25★, 26★, 27★, 28★, 29★, 30/31★, 32/33★, 34★, 36★, 38★, 40★, 45★
Chinchilla, 433f
—*chinchilla* (Short-tailed chinchilla), 433m, 439f
—*boliviana*, **439**
—*chinchilla*, **439**
—*laniger* (Chinchilla), 433m, 438★, **439ff**
Chinchillidae (Chinchillas), 214/215★, 420, **433**
Chinchilloidea, 420, **433**
Chinese hamster (*Cricetulus griseus*), **306**
Chinese jumping mouse (*Eozapus setchuanus*), **402**
Chinese pangolin (*Manis pentadactyla*), **183**, 183m, 185, 199★
Chinese porcupine (*Acanthion subcristatum*), 414m
Chinese pygmy dormouse (*Typhlomys cinereus*), 382★, 388, **398**, 401

Chinese river dolphin or white flag dolphin (*Lipotes vexillifer*), 479★, **502**, 502m

Chinese white dolphin (*Sousa sinensis*), **507**

Chipmunks or eastern chipmunks (*Tamias*), **241**

Chiropodomys, 383m

—*gliroides* (Pencil-tailed tree mouse), **386**

Chiroptera (Bats), 64, **67ff**

Chlamyphorina (Pichiciagos), 162, **165**

Chlamyphorus truncatus (Lesser pichiciago), 164m, **165f**, 189★

Choeronycteris mexicana (Mexican long-nosed bat), 118, 127★, 140m, **141**

Choloepus (Two-fingered sloths), **169**

—*didactylus* (Common or South American two-fingered sloth), **169ff**, 170m, 174★, 190★

—*hoffmanni* (Hoffmann's two-fingered sloth), **169f**, 170m

Cingulata (Armored edentates), 154

Citellus (Sousliks and ground squirrels), **231f**

—*annulatus* (Ring-tailed ground squirrel), 232m

—*armatus* (Uinta ground squirrel), 233m

—*beecheyi* (California ground squirrel), 227m, **232**, 242

—*beldingi* (Belding's ground squirrel), 233m

—*citellus* (European souslik), 226m, **231**, 235, 239★

—*columbianus* (Columbia ground squirrel), 233m

—*eversmanni* (Long-tailed souslik), 226m, **232**

—*menzbieri* (Amur souslik), **232**

—*franklinii* (Franklin's ground squirrel), 227m, **232**

—*fulvus* (Large-toothed or Aral yellow souslik), **232**, 237

—*harrisi* (Harris' antelope ground squirrel, or antelope squirrel), 234m

—*lateralis* (Golden-mantled ground squirrel), 232m, **233**

—*leucurus* (White-tailed antelope squirrel), 227m, **233**

—*mexicanus* (Mexican ground squirrel), 233m

—*mohavensis* (Mohave ground squirrel), 232m

—*nelsoni* (Nelson's antelope squirrel), 227m, 234

—*primigenius*, 238

—*pygmaeus* (Little souslik), 226m, **232**, 235

—*richardsonii* (Richardson's ground squirrel), 234m

—*rufescens* (Red souslik), 238

—*spilosoma* (Spotted ground squirrel), 234m

—*suslicus* (Spotted souslik), 226m, **232**, 236, 239★

—*tereticaudus* (Round-tailed ground squirrel), 237m, **233**

—*townsendi* (Townsend's ground squirrel), 232m

—*tridecemlineatus* (Thirteenlined ground squirrel), 227m, **232**, 242

undulatus (Arctic ground squirrel), 151/152★, 227m, **232**, 236

—*variegatus* (Rock squirrel), 232m

—*washingtoni* (Washington ground squirrel), 232m

Clarke, R., 495

Clawed jird (*Meriones unguiculatus*), 345

Clethrionomys gapperi (Gapper's red-backed vole), 329m

—*glareolus* (Bank vole), 300★, **328f**, 329m

—*occidentalis* (Western red-backed vole), 329m

—*rufocanus* (Large-toothed red-backed vole), 329m, 330

—*rutilus* (Northern red-backed vole), 329m, **330**

Cloud rats (Phloeomyinae) 351, **385**

Clyomys, 428m

Coëndou, 454m

—*insiduosus* (Woolly prehensile-tailed porcupine), **454**

—*prehensilis* (Prehensile-tailed porcupine), 409★, 423★, **454f**

Coleura, 123★

Collared anteater, see Tamandua

Collared fruit bats (*Myonycteris*), **99**

Collared lemming (*Dicrostonyx hudsonius*), 323m

Colloredo-Mannsfeld, Prince, 333

Colorado chipmunk (*Tamias quadrivittatus*), **242**, 243m

Colugos or flying lemurs (Cynocephalidae), **64ff**

Columbia ground squirrel (*Citellus colombianus*), 233m

Comb-toed jerboa (*Paradipus ctenodactylus*), **404f**, 406m

Commerson's dolphin (*Cephalorhynchus commersonii*), 480★, **517**

Commerson's leaf-nosed bat (*Hipposideros commersonii*), 125★, **132**

Commerson's or black-and-white dolphins, aka skunk dolphins (*Cephalorhynchus*), **517**, 517m

Common African dormouse (*Graphiurus murinus*), 382★, **398**

Common dolphin (*Delphinus delphis*), 480★, **515**

Common dormouse (*Muscardinus avellanarius*), 382★, 389, **395**, 395m

Common giant flying squirrel (*Petaurista petaurista*), **264**

Common hamster (*Cricetus cricetus*), 300★, **309ff**, 310m, 311★, 312★, 314★, 321/322★

Common mountain chinchilla (*Lagidium viscacia*), **434**, 438★, 440m

Common noctule (*Nyctalus noctula*), 69, 69★, 121, 136★, 143m, **145**

Common or South American two-fingered sloth (*Choloepus didactylus*), **169ff**, 170m, 174★, 190★

Common pipistrelle (*Pipistrellus pipistrellus*), 69, 112★, 135★, 142m, **144**

Common porpoise (*Phocaena phocaena*), 480★, **506**

Common rorqual (*Balaenoptera physalus*), 460★, 464★, 481, **488ff**

Common stick-nest rat (*Leporillus conditor*), **352**, 371★

Common vole (*Microtus arvalis*), 300★, 321/322★, **336ff**, 336m, 338★

Congo forest mouse (*Deomys ferrugineus*), 372★, **384f**, 384m

Congo or Emin's rat (*Cricetomys emini*), **378**

Congo striped squirrel (*Funisciurus congicus*), 230★

Cook, James, 471

Coon, C. S., 61

Cordier, Charles, 294

Corn rat (*Rattus rattus frugivorus*), **361**

Coro-coros (Dactylomyinae), 427

Cotton rats (*Sigmodon*), **304f**

Count, E., 60

Coypu or nutria (*Myocastor coypus*), **430ff**, 431m, 432★, 438★

Crandall, Lee S., 429, 450

Crateromys schadenbergi (Bush-tailed cloud rat), **385f**, 385m

Cratogeomys, 269

—*castanops* (Yellow pocket gopher), 271m, **272**

Crawford, 37

Creeping vole (*Microtus oregoni*), 338m

Crested porcupine (*Hystrix cristata*), 400★, 412, **413ff**, 415m

Crested spiny rats (*Echimys*), **427**

Cretan spiny mouse (*Acomys minous*), **369**

Cricetidae (Cricetid rodents), 214/215★, **296**

Cricetinae, 301

Cricetini (Hamsters), 301, **305**

Cricetomyinae (Pouched rats), 351, **378**

Cricetomys Giant pouched rats), **378ff**, 378m

—*emini* (Congo, or Emin's rat), **378**

—*gambianus* (Gambian pouched rat), 371★, **378**

Cricetulus (Ratlike hamsters), **306**

—*barabensis* (Striped hamster), **306**, 306m

—*griseus* (Chinese hamster), **306**

—*lama* (Tibetan hamster), 306m, **307**

—*longicaudatus* (Long-tailed hamster), **306f**, 306m

—*migratorius* (Migratory hamster), **306**, 306m

Cricetus cricetus (Common hamster), 300★, **309ff**, 310m, 311★, 312★, 314★, 321/322★

—*nehringi*, 313

Cro-magnon man (*Homo sapiens "fossilis"*), 54

Crossomys moncktoni (Monckton's water rat), 385m, **388**

Crowcroft, 364

Cryptomys (Blesmols), **417f**, 417m

—*hottentotus* (African mole rat), **417f**

—*zechi* (Togo mole rat), **417**, 426★

Ctenodactylidae (Gundis), 201, 214/215★, **286f**

Ctenodactyloidea (Gundis), 201, **286**

Ctenodactylus, 287

—*gundi* (Gundi), **287**, 426★

Ctenomyidae (Tuco-tucos), 214/215★, 420, **422**

Ctenomys knighti (Knight's tuco-tuco), **422**, 422m, 435★

Cuban hutia (*Capromys pilorides*), **428f**, 431m

Cuniculinae (Pacas), 447

Cuniculus paca (Spotted paca), 437★, **447ff**, 448m

Cururo (*Spalacopus cyanus*), **421**, 421m, 435★

Cuvier, Frédéric, 149, 430

—Georges, 79, 430

Cuvier's beaked whale, aka goose-beak whale (*Ziphius cavirostris*), 479★, **501**

Cuvier's hutia (*Plagiodontia aedium aedium*), **429f**

Cyclopes didactylus (Two-toed or silky anteater), **180f**, 181m, 190★

Cynocephalidae (Colugos or flying lemurs), **64ff**

Cynocephalus temminckii (Temminck flying lemur), **64f**, 65m, 66★

—*volans* (Philippine flying lemur), **64ff**, 65m, 83★, 84★

Cynomys (Prairie dogs), **225ff**

—*gunnisoni* (White-tailed prairie dog), **225ff**, 225m

—*parvidens*

—*ludovicianus* (Black-tailed prairie dog), **225ff**, 225m, 239★

Cynopterinae (Short-nosed fruit bats), 93, **108**

Cynopterus sphinx (Indian short-nosed fruit bat), 85★, 87★, 95★, 96, **108**

Dactylomyinae (Coro-coros), 427

Dactylomys, 429m

Dall's harbor porpoise (*Phocaenoides dalli*), 480★

Dao Van Thien, 355

Dark kangaroo mouse (*Microdipodops megacephalus*), 273m

Dart, R. A., 54

Darwin, Charles, 49, 58, 164, 433

Dasypodidae (*Armadillos*), 149, **154f**, 176★

Dasypodini, 157

Dasyprocta (Agoutis), 449

—*aguti* (Orange-rumped agouti), 437★, **449**, 449m

—*azarae* (Azara's agouti), **449**, 449m

—*fuliginosa* (Sooty agouti), 437★, **449**, 449m

Dasyproctidae (Agoutis), 214/215★, **447**

Dasyproctinae (Agoutis), 420, 441, **449**

Dasypus kappleri (Kappler's armadillo), **157**, 157m

—*novemcinctus* (Nine-banded armadillo), 156, **157ff**, 157m, 173★, 189★

—*pilosus* (aka "hairy armadillo"), **157f**, 157m

—*septemcinctus* (Seven-banded armadillo), **157**, 157m, 159

Dawaa, 339

Dawbin, W. H., 514

Deer mouse (*Peromyscus maniculatus*), 297★, **301**, 302m

Degu (*Octodon degus*), **421**, 421m, 435★

Dekeyser, 293

Delphinapterus leucas (White whale or belukha or beluga), 480★, 484★, **504f**, 505m

Delphinidae (Dolphins), 458, 466★, 493, 506, **508**

Delphininae (True dolphins), 508, **514f**

Delphinoidea, 493, **506**

Delphinus delphis (Common dolphin, 480★, **515**

Delphinus gladiator, see *Orcinus orca*

Dendromurinae (African tree mice), 351, **383**

Dendromus, 384m

—*insignis* (African climbing mouse), 372★, **384**

Dent's horseshoe bat (*Rhinolophus denti*), **132**

Deomys ferrugineus (Congo forest mouse), 372★, **384f**, 384m

Dermoptera, 64

Desert dormice (Selviniidae), 214/215★, 296, 388, **401**

Desert jerboa (*Jaculus jaculus*), 399★, **404f**, 405★

Desert kangaroo rat (*Dipodomys deserti*), **273**, 274m, 297★

Desmodontidae (Vampire bats or "bloodsuckers"), 115, **142f**

Desmodus rotundus (Vampire bat), 86★, 115f, 128★, **143**

Diaemus youngi, **143**

Diclidurus albus (Ghost bat), 113

Dicrostonyx (Collared lemmings), **323**
—*groenlandicus* (Greenland collared lemming), 151/152★, **323**, 323m
—*hudsonius* (Collared lemming), 323m
—*torquatus* (Arctic lemming), **323**, 323m
Dijkgraaf, 79
Dinaric race human, 56, 76/77★
Dinomyidae (Pacaranas), 214/215★, 420, **451**
Dinomyoidea (Pacaranalike rodents), 420, **450**
Dinomys branickii (Pacarana), 437★, **451ff**, 451m
Diphylla ecaudata (Hairy-legged vampire bat), 128★, **143**
Diplomys caniceps (Arboreal soft-furred spiny rat), 429m
Dipodidae (Jerboas), 296, 402, **403**
Dipodinae, 404, 404m
Dipodoidea, 296, **401f**
Dipodomys (Kangaroo rats), 272ff
—*agilis* (Pacific kangaroo rat), **273**, 274m, 297★
—*deserti* (Desert kangaroo rat), **273**, 274m, 297★
—*elator* (Texas kangaroo rat), **273**, 274m
—*elephantinus* (Big-eared kangaroo rat), **273**, 274m
—*merriami* (Merriam kangaroo rat), 274m
—*ordii* (Ord's kangaroo rat), **273f**, 274m
—*venustus* (Narrow-faced kangaroo rat), 274m
Dipus sagitta (Northern three-toed jerboa), 399★, 404, 405★, **405f**, 406m
Disk-winged bats (Thyropteridae), **143**
Displaying, 38f
Dobsonia, 94
Dobson's long-tongued dawn bat (*Eonycteris spelaea*), **109**
Dolichotinae (Patagonian cavies), 441, **444**
Dolichotis patagonum (Patagonian cavy, aka mara), 425★, 437★, **444f**, 444m
Dolphins (Delphinidae), 458, 466★, 493, 506, **508**
"Domestic giant sloth" (*Mylodon domesticum*), 154
Dominican hutia (*Plagiodontia aedium hylaeum*), **429f**, 438★
Dormice (Gliroidea), 296, **388ff**
Dremomys pernyi (Perny's long-nosed squirrel), 230★, **261**
Dryomys nitedula (Forest dormouse), 382★, **396f**, 397m
Dryopithecus, 51
—*punjabicus*, 52
Dubois, Eugène, 54
Dubrowskij, 323
Dunn, 115
Dusky-footed wood rat (*Neotoma fuscipes*), 303★, 304m
Dwarf epauletted fruit bat (*Micropteropus pusillus*), 87★, **103**
Dwarf hamsters (*Phodopus*), **306**
Dwarf hutia (*Capromys nana*), **428** 430m
Dwarf jerboas (*Cardiocraniiae*), 404
Dwarf long-tongued fruit bat (*Macroglossus minimus*), **109**
Dwarf tree squirrels (*Microsciurus*), **260**

East African brush-tailed porcupine (*Atherurus turneri*), 413m
East African gerbil (*Tatera vicina*), 300★, **343**
East African porcupine (*Hystrix galeata*), **413**, 415m
East African springhaas (*Pedetes surdaster*), 284★, **295**, 295m
Eastern chipmunk (*Tamias striatus*), **242**, 242m
Eastern fox squirrel (*Sciurus niger*), 254m, 259
Eastern house mouse (*Mus musculus spicilegus*), **362**
Eastern large bandicoot rat (*Bandicota indica nemorivaga*), 371★
Eastern meadow mouse (*Microtus pennsylvanicus*), 300★, **336**, 337m
Eastern or golden-bellied water rat (*Hydromys chrysogaster*), 372★, **387**
Eastern pocket gophers (*Geomys*), 269
Eastern wood rat (*Neotoma floridana*), 304m
Ebien, or African palm squirrel (*Epixerus ebii*), 230★, **263**
Echimyidae (Spiny rats), 214/215★, 420, **427**
Echimyinae, 427
Echimys (Crested spiny rats), **427**
—*armatus* (Armored spiny rat), **427**, 429m, 435★
Echinoprocta rufescens (Upper Amazonian porcupine), 409★, 453m, **454**
Echo orientation, or echolocation, 71ff
Ectophylla alba, 113
Edentata, 149ff
Egyptian fruit bat (*Rousettus aegyptiacus*), 87★, **98f**
Egyptian spiny mouse (*Acomys cahirinus*), **369**
Ehrenberg's mole rat (*Spalax ehrenbergi*, **348ff**
Eibel-Eibesfeldt, Irenäus, 59, 194, 204, 234, 246, 250ff, 255, 303, 310ff, 345, 358, 364, 393, 405
Eidolon, 97, 97m
—*helvum* (Straw-colored bat), 87★, 91, 95, **97f**
—*dubreanum*, 97
—*helvum*, 97
—*sabaeum*, 97
Eisentraut, Martin, 293, 311, 376, 392, 398
Ekman, P., 60
Elbe beaver (*Castor fiber albicus*), **276**, 278m, 285f
Eleven-banded armadillo (*Cabassous unicinctus*), **161**, 161m, 189★
Eliomys quercinus (Garden dormouse), 381★, 382★, **395f**, 396m, 396★
Eliurus minor, 318
Ellermann, 218
Ellobiini (Mole lemmings), 319, **342**
Ellobius fuscocapillus (Afghan mole lemming), 300★, **342**, 342m
—*talpinus* (Mole lemming), **342f**, 342m
Elton, C., 194
Emballonura, 123★
Emballonuridae (Sac-winged bats), **122**
Emin's rat, see Congo rat
Eocetus, 472
Eonycteris, 94
—*spelaea* (Dobson's long-tongued dawn bat), **109**
Eothenomys melanogaster (Père David's vole), **330**
Eozapus setchuanus (Chinese jumping mouse), **402**
Epauletted fruit bats (Epomophorinae), 93, **102f**
Epigaulus, 203
Epixerus ebii (Ebien, or African palm squirrel), 230★, **263**
—*wilsoni* (Wilson's palm squirrel), **262**
Epomophorinae (Epauletted fruit bats), 93, **102f**
Epomophorus wahlbergi (Wahlberg's epauletted fruit bat), **103**
—*haldenani*, 87★, **103**
Epomops buettikoferi (Büttikofer's epauletted fruit bat), **102**
—*franqueti* (Franquet's epauletted fruit bat), 87★, **102**
Eptesicus fuscus (Big brown bat), 144 m, **145**
—*nilssoni* (Northern bat), 144m, **145**
—*serotinus* (Serotine bat), 135★, 144m, **145**
—*tenuipinnis*, 113, **145**
Eregden-Dagwa, Dordshiin, 220f
Eremodipus lichtensteini (Lichtenstein's jerboa), **405**, 406m
Erethizon dorsatum (North American porcupine), 409★, **454ff**, 455m
Erethizontidae (New World porcupines), 214/215★, 420, **453**
Erethizontinae, 454
Erethizontoidea, 420, **453**
Eschrichtiidae (Gray whales), 458, 466★, 477, **486**
Eschrichtius gibbosus (Gray whale), 464★, **486f**
Eskimos, 57, 76/77★
Ethiopian harsh-furred mouse (*Lophuromys flavopunctatus*), **378**
Ethiopian race human, 56, 76/77★
Eubalaena australis (Southern right whale or ice baleen whale), **485**
—*glacialis* (Atlantic right whale), 463★, 481, **485f**
—*japonica* (Pacific right whale), 481, **485f**
Euchoreutes naso, 404
Euchoreutinae (Long-eared jerboas), 404
Eumegamys, 451
Eumops perotis californicus, 138★, 147m
Eupetaurus cinereus (Woolly flying squirrel), **265**
Euphractina, 162f
Euphractini, 162
Euphractus nationi, **163**
—*pichiy* (Pygmy armadillo), **163f**, 164m, 189★
—*sexcinctus* (Six-banded armadillo), **163f**, 163m, 189★
—*vellerosus*, **163**
—*villosus* (Hairy armadillo), **163ff**, 164m, 173★, 189★
Eurasian dormice (Glirinae), **389**
European flying squirrel (*Pteromys volans*), 240★, **264**, 264m
European or true Alpine marmot (*Marmota marmota marmota*), 210
European pine vole (*Pitymys subterraneus*), **335f**, 335m
European souslik (*Citellus citellus*), 226m, **231**, 235, 239★
European water vole (*Arvicola terrestris*), 105/106★, 257/258★, **331f**, 331m
Europoid type human, 56, 56m
Euryzygomatomys spinosus (Suira), 428m
Eutamias (Old World and western chipmunks), 238
—*sibericus* (Burunduk), **238f**, 242m
Eversmann's dwarf hamster (*Allocricetulus eversmanni*), 305m, **306**
False killer whale (*Pseudorca crassidens*), 480★, 511, **518**
False vampires (*Megaderma*), 131
Fan Shou, 317
Fantz, R., 61
Fat dormouse (*Glis glis*), 382★, **389ff**, 390m
Fat mouse (*Steatomys pratensis*), 372★, **384**
Fat-tailed jerboa (*Pygeretmus*), 404
Fat-tailed mouse (*Pachyuromys duprasi*), **343**, 346
Felovia, 287
Felten, 114, 129, 139ff
Feresa attenuata (Pygmy killer whale), 480★, **518**
Field gerbil (*Gerbillus nanus garamantis*), **343f**
Field vole (*Microtus agrestis*), **336**, 336m, 339
Finback whales (*Balaenopteridae*), 458, 466★, 477, **488**, **489★**
Fischel, 414
Fischer, Wolfgang, 255
Five-toed dwarf jerboa (*Cardiocranius paradoxus*), 399★, 404
Five-toed jerboas (*Allactaga*), 404
Flat-headed bat (*Tylonycteris pachypus*), **144**
Flightless scaly-tailed squirrel (*Zenkerella insignis*), **290f**, 294m
Flower-faced bat (*Anthops ornatus*), 131★, **139**
Flying foxes (*Pteropus*), 97, 99m, **100**
Flying lemurs, see colugos
Flying squirrels (Pteromyinae), 204, **263**
Forest dormouse (*Dryomys nitedula*), 382★, **396f**, 397m
Formidabilis balaenarum hostis, see *Orcinus orca*
Formosan giant flying squirrel (*Petaurista grandis*), 65, 240★, **264**
Formosow, A. N., 219, 252, 401
Foyn, Svend, 470
Frank, F., 196, 325f, 328, 331, 337ff, 368
Franklin's ground squirrel (*Citellus franklinii*), 227m, **232**
Franquet's epauletted fruit bat (*Epomops franqueti*), 87★, **102**
Fraser's scaly-tailed squirrel (*Anomalurus fraseri*), 240★, **291**, 293m
Freedman, D. G., 61
Free-tailed bats (*Molossidae*), **146**
Freye, H.-A., 275
Fruit bats, i.n.s. (*Pteropidae*), 93, **97f**
Funambulini (African tree squirrels), 261
Funambulus (Palm squirrels), **261**
—*palmarum* (Indian palm squirrel), **261**
Funisciurus (African striped squirrels), **261**
—*congicus* (Congo striped squirrel), 230★
—*lemniscatus* (Western African striped squirrel), **261**
Funnel-eared bats (*Natalidae*), **143**
Furipteridae (Smokey bats), **143**
Furipterus horreus, **143**

Gabon scaly-tailed squirrel (*Anomalurus chrysophoenus*), **291**, 292m
Galea, 441
—*musteloides*, 436★, 442m
Gambian pouched rat (*Cricetomys gambianus*), 371★, **378**
Ganges river or gangetic dolphin (*Platanista gangetica*), 479★, **502**, 502m

Gangetic dolphins (*Platanistidae, Platanista*), 458, 466★, 493, **502**
Gapper's red-backed vole (*Clethrionomys gapperi*), 329m
Garden dormouse (*Eliomys quercinus*), 381★, 382★, **395**f, 396m, 396★
Geisha wood mouse (*Apodemus geisha*), 365
Genest-Villard, 378
Geocapromys brownii (Jamaican hutia), **429**f, 430m, 438★
—*columbianus*, **429**, 430m
—*ingrahami* (*Bahama hutia*), **429**, 430m
Geomyidae (Pocket gophers), 201, 214/215★, **268**f
Geomyoidea (Pocket gophers), 201, **268**ff
Geomys (Eastern pocket gophers), 269
—*bursarius* (Plains or Eastern pocket gopher), **270**f, 270m, 297★
—*personatus* (Texas pocket gopher), 270m
—*pinetis* (Southeastern pocket gopher), 270m
Georhychus capensis (Cape mole rat), **417**, 417m, 426★
Gerbillinae (Gerbils), 301, **343**, 343m
Gerbillus (Gerbils), **343**
—*campestris* (Large North African gerbil), 300★, **343**
—*nanus garamantis* (Field gerbil), **343**f
Gerbils (Gerbillinae), 301, **343**, 343m
Gesner, Konrad, 357, 442
Gewalt, Wolfgang, 246, 248ff, 504f
Ghost bat (*Diclidurus albus*), 113
Giant anteater (*Myrmecophaga tridactyla*), 173★, **177**ff, 178m, 179★, 190★
Giant anthropoid ape (*Gigantopithecus*), 51
Giant armadillo (*Priodontes giganteus*), 160m, **161**f, 189★
Giant bottle-nosed whales (*Berardius*), **501**
Giant naked-tailed rat (*Uromys anak*), **353**
Giant noctule (*Nyctalus lasiopterus*), **145**
Giant pangolin (*Manis gigantea*), 182m, **183**, 185, 199★
Giant pouched rats (*Cricetomys*), **378**ff, 378m
Giant sloths (Gravigarda), 153
Giant velvety free-tailed bat (*Molossus major*), 137★
Gigantopithecus (Giant anthropoid ape), 51
Gijzen, Agatha, 162
Girtanner, 217
Glaucomys sabrinus (Northern flying squirrel), 65, **265**f
—*volans* (Southern flying squirrel), 240★, 261★, **265**f, 265m
Glauconycteris, 113
Gliridae (True dormice), 214/215★, 296, **388**f
Glirinae (Eurasian dormice), **389**
Gliroidea (Dormice), 296, **388**ff
Glirulus japonicus (Japanese dormouse), **397**
Glis glis (Fat dormouse), 382★, **389**ff, 390m
Globicephala (Pilot whales or blackfish), 498★, **517**
—*macrorhyncha* (Indian pilot whale or short-finned blackfish), **517**
—*melaena* (Northern pilot whale or Atlantic blackfish), 480★, **517**
—*sieboldii* (Pacific pilot whale), 511, **517**
Glossophaga soricina (Long-tongued bat), 117, 127★, **141**
Glossophaginae (Long-tongued bats), 117, **141**
Glyptodon, 154
Glyptodontidae, 154
Gobi jerboa (*Allactaga bullata*), **406**
Goeldi, Emil, 446, 452
Goertz, 304
Goffman, E., 60
Golden bat (*Myzopoda aurita*), 144
Golden hamster (*Mesocricetus auratus*), 298★, 310m, **315**, 315★
Golden mouse (*Peromyscus nuttalli*), **301**f, 302m
Golden-bellied beaver (*Castor fiber subauratus*), **276**
Golden-bellied water rat, see eastern water rat
Golden-mantled ground squirrel (*Citellus lateralis*), 232m, **233**
Goosebeak whale, see Cuvier's beaked whale
Göring, Anton, 161
Gottwaldt, C., 282
Grampus griseus (Risso's dolphin), 480★, **515**
Graphiurinae (African dormice), 389, **397**f
Graphiurus (African dormice), **398**
—*murinus* (Common African dormouse), 382★, **398**
—*haedulus* (Cameroons dormouse), **398**
—*nanus* (Pygmy dormouse), **398**
Grass mice or kusus (*Arvicanthis*), 374, 375m
Grasshopper mice (*Onychomys*), 296, **302**
Gravigarda (Giant sloths), 153
Gray squirrel (*Sciurus carolinensis*), 229★, 255m, **255**
Gray whale (*Eschrichtius gibbosus*), 464★, **486**f
Gray-headed flying fox (*Pteropus poliocephalus*), 87★, **100**
Great Basin pocket mouse (*Perognathus parvus*), 273m
Great cane rat (*Thryonomys swinderianus*), **419**, 426★
Great gerbil (*Rhombomys opimus*), 300★, **343**, 345
Greater horseshoe bat (*Rhinolophus ferrumequinum*), 69★, 105/106★, 121m, 125★, **132**
Greater mole rat (*Spalax microphthalmus*), 348ff
Greater pichiciago (*Burmeisteria retusa*), **165**f, 189★
Greenland collared lemming (*Dicrostonyx groenlandicus*), 151/152★, **323**, 323m
Griffin, 79
Griffin, Ted, 522
Grizell, 224
Groove-toothed squirrel (*Rheithrosciurus macrotis*), **260**
Grummt, 335
Grundova, 308
Grünthal, E., 510
Grzimek, Bernhard, 19f, 43
—Michael, 20
Guerlinquetus, 259
Guianan river dolphin (*Sotalia guianensis*), **507**, 507m
Guinea pig (*Cavia aperea porcellus*), 436★, **441**f
Guyana acouchy (*Myoprocta acouchi*), 437★, **450**, 450m
Gundi (*Ctenodactylus gundi*), **287**, 426★
Gundis (*Ctenodactylidae*), 201, 214/215★, **286**f
Gunther's vole (*Microtus guentheri*), 337m
Haake, Wilhelm, 454
Hairy armadillo (*Euphractus villosus*), **163**ff, 164m, 173★, 189★
"Hairy armadillo," aka, see *Dasypus pilosus*
Hairy-legged vampire bat (*Diphylla ecaudata*), 128★, **143**
Haitian hutia (*Plagiodontia aedium*), **430**, 430m
Hall, E. T., 60, 341
Hammer-headed fruit bat (*Hypsignathus monstrosus*), 87★, **107**
Hamsters (Cricetini), 301, **305**
Hanney, 378, 384
Harmen, 182
Harpiocephalus, 108
Harpy fruit bats (Harpyionycteridae), 94, **110**
Harpyionycteridae (Harpy fruit bats), 94, **110**
Harpyionycteris whiteheadi (Whitehead's harpy fruit bat), **110**, 110m
Harris' antelope ground squirrel or antelope squirrel (*Citellus harrisi*), 234m
Harrison, J. L., 66, 386
Harsh-furred mice (*Lophuromys*), **377**, 379m
Harvest mouse (*Micromys minutus*), 257/258★, 367m, **368**f, 368★
Hass, Hans, 59, 496
Hayes, 37
Heart-nosed false vampire (*Megaderma cor*), **131**
Heberer, Gerhard, 49
Heck, Ludwig, 243, 310, 346, 444
Hector dolphin (*Cephalorhynchus hectori*), **517**
Hediger, Heini, 60, 415
Heide, 109
Heinrich, G., 366, 397
Heinroth, O., 59
Heliophobius (Sand rats), **417**, 417m
—*argenteocinereus* (Silvery mole rat), **417**
Heliosciurus (Sun squirrels), **262**
—*gambianus* (West African sun squirrel), **263**
—*undulatus*, 230★, **263**
Heliscomys, 275
Hemianthropus peii, 53
Hensel, 159
Heptner, W. G., 305, 341, 395, 505
Herre, Wolf, 162, 166
Hertwig, P., 198
Hesperomyini (New World mice), **301**
Hess, E. H., 62
Heterocephalus glaber (Naked sand rat), **417**f, 417m, 426★
Heteromyidae (Kangaroo mice), 201, 214/215★, 268, **272**f
Heteromys (Spiny pocket mice), 272
—*anomalous* (South American forest mouse), **275**, 275m
—*goldmani* (Goldman's spiny pocket mouse), **274**f, 275m
High-mountain vole (*Alticola strelzowi*), **330**
Hildebrant's horseshoe bat (*Rhinolophus hildebrandii*), 124★
Himalayan crestless porcupine (*Acanthion hodgsoni*), 414m
Himalayan marmot (*Marmota himalayana*), 206m
Hipposideridae (Old World leaf-nosed bats), **131**f
Hipposideros armiger, **139**
—*caffer* (South African lesser leaf-nosed bat), 125★, **132**f
—*commersoni* (Commerson's leaf-nosed bat), 125★, **132**
—*gigas*, **132**
—*jonesi*, 125★
Hispaniolan hutias (*Plagiodontia*), **429**f
Hispid cotton rat (*Sigmodon hispidus*), **304**, 304m
Hispid pocket mouse (*Perognathus hispidus*), 273m
Hoang Trong Chu, 355
Hoary bat (*Lasiurus cinereus*), **145**, 145m
Hoary marmot (*Marmota caligata*), **206**, 206m, 216★
Hofer, 204
Hoffmann's two-fingered sloth (*Choloepus hoffmanni*), **169**f, 170m
Holst, F., 410
Holzmaier, 250
Hominidae (Humans), 49
Hominoidea (Human-like forms), 50
Homo (Man), 54
—*erectus erectus* (Java man), 54, 78★
—*heidelbergensis*, 54
—*mauritanicus*, 54
—*pekinensis* (Peking man), 54
—*habilis*, 54
—*sapiens* (Human or man), **56**, 73★, 74★, 75★, 76★, 77★
—*fossilis* (Cro-Magnon man), 54
—*neanderthalensis* (Neanderthal man) 53f, 78★
—*sapiens* (Modern man), **55**, **78★**
Honduran disk-winged bat (*Thyroptera discifera*), **144**
Hoplomys gymnurus (Armored rat), 428m
Hornung, 247
Horseshoe bats (*Rhinolophidae*), **131**
House mouse (*Mus musculus*), 351, **362**f, 363m, 371★
House rat, see black rat
Human (*Homo sapiens*), 56, 73★, 74★, 75★, 76★, 77★
Humpback whale (*Megaptera novaeangliae*), 465★, 468★, 481, **491**f, 491★, 492m
Hürzeler, J., 51
Hutias (Capromyidae), 214/215★, 420, **428**
Hydrochoeridae (Capybaras), 214/215★, 420, 441, **446**
Hydrochoerus hydrochaeris (Capybara or carpincho), 424★, 437★, **446**f
—*isthmius* (Panama capybara), **446**f
Hydromyinae (Water rats), 351, **387**
Hydromys (Water rats), 385m, **387**
—*chrysogaster* (Eastern or golden-bellied water rat), 372★, **387**
—*habbema* (New Guinea water rat), **387**
—*neobritannicus* (New Britain water rat), **387**
Hylobatidae (Gibbons), 50
Hylopetes lepidus (Indo-Malayan flying squirrel), **264**
Hyosciurus heinrichi (Long-snouted squirrel), **261**
Hyperoodon ampullatus (Bottle-nosed whale), 480★, 495m, **496**f
—*planifrons* (Southern bottle-nosed, whale), 495m, **501**
Hypogeomys antimena (Votsotsa), **318**
Hypsignathus monstrosus (Hammer-headed fruit bat), 87★, **107**
Hystricidae (Old World porcupines), 214/215★, **407**
Hystricoidea, **407**
Hystricomorphia (Porcupines), 193, 198, **407**

Hystrix (Large procupines), 407, **413**
—*africaeaustralis* (South African porcupine), **413**, 415m
—*cristata* (Crested porcupine), 400★, 412, **413**ff, 415m
—*galeata* (East African porcupine), **413**, 415m
—*leucura* (White-tailed porcupine), 410★, **413**, 415m

Iberian water vole (*Arvicola sapidus*) **331**, 331m
Ice baleen whale, see Southern right whale
Ichthyomys (Aquatic rats), 296
Idiurus (African small flying squirrels), 289, **291**
—*langi* (Lang's small flying squirrel), 240★, **291**m, 293
—*macrotis* (Large-eared small flying squirrel), 288★, **291**, 294m
—*zenkeri* (Zenker's small flying squirrel), **291**, 293, 294m
Ihering, von, 305
Indian brown spiny mouse (*Mus platythrix*), **362**
Indian false vampire (*Megaderma lyra*), 115, 124★, **131**
Indian flying fox (*Pteropus giganteus*), 94f, **100**f
Indian giant squirrel (*Ratufa indica*), **262**
Indian or large Indian gerbil (*Tatera indica*), **343**, 346
Indian palm squirrel (*Funambulus palmarum*), **261**
Indian pangolin (*Manis crassicaudata*), **183**, 183m, 186, 199★
Indian pilot whale or short-finned blackfish (*Globicephala macrorhyncha*), **517**
Indian race human, 56, 76/77★
Indian short-nosed fruit bat (*Cynopterus sphinx*), 85★, 87★, 95★, 96, **108**
Indo-Malayan flying squirrel (*Hylopetes lepidus*), **264**
Indo-Malayan giant squirrels (Ratufini), 205, **262**
Inia geoffrensis (Amazon or Amazonian dolphin), 479★, **502f**, 503m
Iniidae (Amazon dolphins), 458, 466★, 493, **502**
Insectivorous bats (Microchiroptera), 68, **111**ff
Irrawaddy dolphin (*Orcaella brevirostris*), 480★, 517m, **518**
Isothrix bistrata, 429m

Jäckel, 336f
Jackson's scaly-tailed squirrel (*Anomalurus jacksoni*), 288★, 290, **291**ff, 292m
Jaculus jaculus (Desert jerboa), 399★, **404**f, 405★
Jamaican hutia (*Geocapromys brownii*), **429**f, 430m, 438★
Japanese dormouse (*Glirulus japonicus*), **397**
Java man (*Homo erectus erectus*), 54, 78★
Javan porcupine (*Acanthion javanicum*), **413**, 414m
Jelski, Constantin, 451
Jerboas (Dipodidae), 296, 407, **403**
Jirds (*Meriones*), **343**
Jones, Blurton, 62
Jones' stick-nest rat (*Leporillus jonesi*), **352**
Jumping hares, see springhaas
Jumping mice (*Zapodidae*), 214/215★, 296, **402**

Kafroid race human, 76/77★
Kahmann, 396
Kaibab squirrel (*Sciurus kaibabensis*), 255m, 259
Kalabuchov, 236
Kalela, 324, 327f
Kälin, J., 50
Kamchatkan marmot (*Marmota bobak camtschatica*), **206**, 206m
Kangaroo mice (Heteromyidae), 201, 214/215★, 268, **272f**
Kangaroo rats (*Dipodomys*), 272ff
Kannabateomys amblyonyx (Rato de Taquara), **427f**, 429m, 435★
Kant, 49
Kappler's armadillo (*Dasypus kappleri*), **157**, 157m
Karroo rat (*Otomys denti*), **383**
Keith, Arthur, 52
Keller, 206, 246
Kemper, 360
Kenyapithecus wickeri, 52
Kerodon (Rock cavies), 441
—*rupestris* (Rock cavy), 436★, 443m, **443**
Khoisanic race human, 56, 56m
Killer whale, or orca (*Orcinus orca*), 480★, **519**ff
King, 226, 228
Kirchshofer, Rosl, 47, 179, 313, 344, 442f
Kittel, R., 315
Klein, 328
Klemm, 238, 241, 243
Kloss' porcupine (*Acanthion klossi*), 414m
Knight's tuco-tuco (*Ctenomys knighti*), **422**, 422m, 435★
Kobb, 114
Kock, 288f
Koenig, Lilli, 390, 393f
—, Otto, 217f
Koford, 225
Kogia breviceps (Pygmy sperm whale), 474★, **494**
Köhler, Wolfgang, 23
Kramer, M. O., 310, 459
Kratochvil, 218, 335
Krieg, Hans, 160ff, 167ff, 172, 178ff, 303, 445
Krumbiegel, Ingo, 235, 237, 263, 270
Kühlhorn, Friedrich, 166, 303
Kull, Albert, 319
Kulzer, 90, 101, 103, 147
Küsthardt, 340
Kusus, see grass mice

Lagenorhynchus, 516m
—*acutus* (White-sided dolphin), 480★, 497★, **516**
—*albirostris* (White-beaked dolphin), **516**
—*obliquidens* (Pacific white-sided dolphin), 511, **517**
Lagidium (Mountain chinchillas), 433f
—*peruanum* (Peruvian mountain chinchilla), 440m
—*viscacia* (Common mountain chinchilla), **434**, 438★, 440m
—*wolffsoni* (Wolffson's mountain chinchilla), 440m
Lagostomus, 433
—*maximus* (Viscacha), **433f**, 433m, 438★
Lagurus curtatus (Sagebrush or rabbittailed meadow vole), **341**
—*lagurus* (Steppe lemming), **341**, 341m
Lamarck, 49
Lamberton's Malagasy rat (*Nesomys lambertoni*), 300★, **318**

Langenstein-Issel, 335
Lang's small flying squirrel (*Idiurus langi*), 240★, **291**, 293
La Plata dolphin (*Stenodelphis blainvillei*), 479★, **502**, 503m
La Plata three-banded armadillo (*Tolypeutes matacus*), **160f**, 160m
Large bandicoot rat (*Bandicota indica*), **354f**
Large gerbils (*Tatera*), **343**
Large Indian gerbil, see Indian gerbil
Large mouse-eared bat (*Myotis myotis*), 68, 86★, 88★, 112★, 114, 134★, 140m, **144**
Large North African gerbil (*Gerbillus campestris*), 300★, **343**
Large pocket gophers or tuzas (*Orthogeomys*), 269
Large porcupines (*Hystrix*), 407, **413**
Large tube-nosed fruit bat (*Nyctimene major*), 87★
Large-eared small flying squirrel (*Idiurus macrotis*), 288★, **291**, 294m
Large-toothed, or Aral yellow souslik (*Citellus fulvus*), **232**, 237
Large-toothed red-backed vole (*Clethrionomys rufocanus*), 329m, **330**
Large-winged bats (*Megadermatidae*), 115, **130**
Larina, 366
Lariscus hosei (Bornean black-striped squirrel), **261**
—*insignis* (Malayan black-striped squirrel), 230★, **261**
Lasiurus borealis (Red bat), 113, 119, 136★, **145**, 145m
—*cinereus* (Hoary bat), **145**, 145m
Lavia frons (African yellow-winged bat), 124★, **131**
Lawick-Goodall, Jane van, 20, 22f, 39ff, 46
Layard's or strap-toothed whale (*Mesoplodon layardii*), 480★, **501**
Leaf-nosed bats (Phyllostomidae), 112, **139**
Leakey, L. S. B., 52f
Least chipmunk (*Tamias minimus*), **242f**, 243m
Least wood rat (*Sigmodon minimus*), 304m
Leche, W., 64
Leggada (African native mice), **369**, 376m
—*minutoides* (Small African native mouse), **369**
Leggadina (Australian native mice), 352m
—*hermannsburgensis* (Australian native mouse), **353**
Lemmini (Lemmings), 319
Lemmus (Brown lemmings), **324f**
—*amurensis* (Amur lemming), 323m
—*lemmus* (Norway lemming), 300★, **324**ff, 324m
—*nigripes* (Black-footed lemming), 324m
—*sibiricus* (Siberian lemming), 323m
—*trimucrontus* (Brown lemming), 324m
Lemniscomys (Striped grass mice) **373**, 375m
—*barbarus* (North African striped grass mouse), **373**
—*griselda* (Single-striped grass mouse), **373**
—*striatus* (Striped grass mouse), 371★, **374**
Leporillus (Stick-nest rats), **351f**, 352m
—*conditor* (Common stick-nest rat), **352**, 371★

—*jonesi* (Jones' stick-nest rat), **352**
Le Sovef, 353
Lesser bamboo rat (*Cannomys badius*), **347**
Lesser cane rat (*Thryonomys gregorianus*), **419**
Lesser horseshoe bat (*Rhinolophus hipposideros*), 69, 69★, 120★, 120m, 125★, **132**
Lesser mole rat (*Spalax leucodon*), 300★, **348**ff
Lesser mouse-eared bat (*Myotis oxygnathus*), 140m
Lesser noctule (*Nyctalus leisleri*), 144m, **145**
Lesser or Bengali bandicoot rat (*Bandicota bengalensis*), **354**
Lesser pichiciago (*Chlamyphorus truncatus*), 164m, **165f**, 189★
Lesser rorqual (*Balaenoptera acutorostrata*), 464★, **489f**
Lewis, G. E., 52
Leyhausen, P., 62
Lichtenstein's jerboa (*Eremodipus lichtensteini*), **405**, 406m
Lilly, John C., 513
Limnopithecus, 51
Linné, 49, 457
Linné's false vampire bat (Vampyrum spectrum), 112, 116f, **140**
Liomys, 272
—*irroratus* (Mexican spiny pocket mouse), **275**, 275m, 297★
—*pictus* (Painted spiny pocket mouse), **275**, 275m
Liponycteris, 129
Lipotes vexillifer (Chinese river dolphin or white flag dolphin), 479★, **502**, 502m
Lipp, 246
Lissodelphinae (Right whale dolphins), 508, **514**
Lissodelphis (Right whale dolphins), **514**
—*borealis* (Northern or Pacific right whale dolphin), 480★, **514**
—*peronii* (Peron's dolphin), **514**
Little earth jerboa (*Alactagulus pygmaeus*), **404**, 406m
Little Indian field mouse (*Mus booduga*), **362**
Little rat (*Rattus exulans*), **356**
Little souslik (*Citellus pygmaeus*), 226m, **232**, 235
Löhrl, 392, 396
Lonchorrhina aurita (Tome's long-eared bat), 126★, **140**
Long-clawed ground squirrel (*Spermophilopsis leptodactylus*), **234**f
Long-clawed mole vole (*Prometheomys schaposchnikowi*), **341f**, 341m
Long-eared bat (*Plecotus auritus*), 70★, 112★, 114, 133★, 135★, **145**
Long-eared jerboas (Euchoreutinae), 404
Long-legged bat (*Macrophyllum macrophyllum*), **140**
Long-nosed squirrel (*Rhinosciurus laticaudatus*), **261**
Long-snouted dolphins (Stenidae), 458, 466★, 493, 506, **507**
Long-snouted squirrel (*Hyosciurus heinrichi*), **261**
Long-tailed climbing mouse (*Vandeleuria oleracea*), **355**
Long-tailed field mouse (*Apodemus sylvaticus*), 299★, 365★, **366f**, 367m, 371★
Long-tailed fruit bat (*Notopteris*), 93f
Long-tailed hamster (*Cricetulus longicaudatus*), **306f**, 306m

Long-tailed marmot (*Marmota caudata*), **205**, 206m, 222
Long-tailed porcupines (*Trichys*), 407, **412**
Long-tailed pouched rats (*Beamys*), 378
Long-tailed rat (*Mystromys longicaudatus*), **316**
Long-tailed souslik (*Citellus eversmanni*), 226m, **232**
Long-tailed tree pangolin (*Manis tetradactyla*), 182m, **183**ff, 199★
Long-tailed vole (*Microtus longicaudus*), 338m
Long-tongued bat (*Glossophaga soricina*), 117, 127★, **141**
Long-tongued fruit bats (Macroglossidae), 93, **108**
Long-winged bat (*Miniopterus schreibersi*), 69, **146**, 146m
Löns, 310
Lophiomyinae (Maned rats), 301, **318**
Lophiomys imhausi (Maned rat), 300★, **318f**
Lophuromys (Harsh-furred mice), **377**, 379m
—*flavopunctatus* (Ethiopian harsh-furred mouse), **378**
—*woosnami* (Uganda harsh-furred mouse), 377
Lorenz, Konrad, 59
Loveridge, 132
Lyroderma, 131

Maclaud's horseshoe bat (*Rhinolophus* maclaudi), **132**
Macroderma gigas (Australian giant false vampire bat), 86★, 112, 115, 124★, **131**
Macroglossidae (Long-tongued fruit bats), 93, **108**
Macroglossus lagochilus, **109**
—*minimus* (Dwarf long-tongued fruit bat), **109**
Macrophyllum macrophyllum (Long-legged bat), **140**
Macrotarsomys bastardi, **318**
Malagasy rats (Nesomyinae), 301, **318**
Malayan black-striped squirrel (*Lariscus insignis*), 230★, **261**
Malayan false vampire (*Megaderma spasma*), 115, **131**
Malayan giant squirrel (*Ratufa bicolor*), 230★, **262**
Malayan long-tailed porcupine (*Trichys fasciculata*), **412**
Malayan pangolin (*Manis javanica*), **183**, 183m, 199★
Malayan porcupine (*Acanthion brachyura*), 400★, **413**, 414m
Man (Hominidae), 49
Manchurian zokor (*Myospalax psilurus*), **316**
Maned rat (*Lophiomys imhausi*), 300★ **318f**
Manidae (Pangolins), 182
Manis (Pangolins), **183**ff, 186★, 200★
—*crassicaudata* (Indian pangolin), **183**, 183m, 186, 199★
—*gigantea* (Giant pangolin), 182m, **183**, 185, 199★
—*javanica* (Malayan pangolin), **183**, 183m, 199★
—*pentadactyla* (Chinese pangolin), **183**, 183m, 185, 199★
—*pusilla*, **183**
—*temmincki* (Cape pangolin), 182m, **183**, 185f, 199★, 200★
—*tetradactyla* (Long-tailed tree pangolin), 182m, **183**ff, 199★
—*tricuspis* (White-bellied tree pangolin), **182**, 182m, 185, 187★, 199★
Mantled African bush squirrel (*Paraxerus palliatus*), **261**
Mara, see Patagonian cavy
Marco Polo, 221
Marmota (Marmots), **205**
—*bobak* (Bobac marmot), **205**, 206m, 218ff, 220★, 239★
—*baibacina* (Altai marmot), 222
—*sibirica* (Mongolian marmot or tarbagan), 219f
—*caligata* (Hoary marmot), **206**, 206m, 216★
—*camtschatica* (Kamchatkan marmot), **206**, 206m
—*vancouverensis*, 206
—*caudata* (Long-tailed marmot), **205**, 206m, 222
—*dichrous* (Afghan marmot), **206**
—*flaviventris* (Yellow-bellied marmot), **206**, 206m
—*himalayana* (Himalayan marmot), 206m
—*marmota* (Alpine marmot), **205**ff, 205m, 210★, 211★, 216★, 239★
—*latirostris*, 218
—*marmota* (European, or true Alpine marmot), 210
—*primigenia*, 218
—*menzbieri* (Tien Shan marmot), 206m
—*monax* (Woodchuck), **206**, 206m, 222ff, 239★
—*bunkeri*, 222
—*canadensis*, 222
—*ignava*, 222
—*johnsoni*, 222
—*ochracea*, 222
—*petrensis*, 222
—*preblorum*, 222
—*rufescens*, 222
Marmotini (Northern ground squirrels), **205**
Marmots (*Marmota*), **205**
Marshall, G. A. K., 398
Massoutiera, 287
Mastomys (Multimammate rats), **376**, 376m
—*coucha*, 371★, **376f**
Masui, M., 188
Meadow jumping mouse (*Zapus hudsonius*), 399★, **402**, 403m
Mediterranean horseshoe bat (*Rhinolophus euryale*), 130m
Mediterranean pine vole (*Pitymys duodecimcostatus*), 335m
Mediterranean race human, 56, 76/77★
Medway, 386
Megachiroptera (Old World fruit bats), 68, **93**ff, 93m
Megaderma (False vampires), 131
—*cor* (Heart-nosed false vampire), **131**
—*lyra* (Indian false vampire), 115, 124★, **131**
—*spasma* (Malayan false vampire), 115, **131**
Megadermatidae (Large-winged bats), 115, **130**
Megadermatoidea, 130
Megaloglossus woermanni (African long-tongued fruit bat), 87★, 94, **109**
Megalonychidae, 153
Megalonyx, 153
Meganthropus palaeojavanicus, 53
Megapedetes, 294
Megaptera novaeangliae (Humpback whale), 465★, 468★, 481, **491**ff, 491★, 492m
Megatheriidae, 153
Megatherium, 153
Mehely's horseshoe bat (*Rhinolophus mehelyi*), 130m
Melanesian race human, 57, 75/77★
Melanid race human, 57
Melomys (Mosaic-tailed rats), **352**
Melville, Herman, 496
Meriones (Jirds), **343**
—*meridianus psammophilus* (Southern Mongolian jird), 345
—*persicus* (Persian jird), 345
—*shawi* (Shaw's jird), 345★
—*tristrami* (Tristram's jird), 300★, 345
—*unguiculatus* (Clawed jird), 345
Merriam kangaroo rat (*Dipodomys merriami*), 274m
Mesocricetus auratus (Golden hamster), 298★, 310m, **315**, 315★
Mesomys hispidus, 428m
Mesoplodon (Beaked whales or cowfish), 501
—*bidens* (Sowerby's whale), 480★, **501**
—*europaeus*, **501**
—*layardi* (Layard's or strap-toothed whale), 480★, **501**
—*mirus* (True's beaked whale), 480★
Mexican bulldog bat (*Noctilio leporinus*), 113, 115, 115★, 124★, **129f**
Mexican free-tailed bat (*Tadarida brasiliensis mexicana*), 91, **147**, 147m
Mexican ground squirrel (*Citellus mexicanus*), 233m
Mexican long-nosed bat (*Choeronycteris mexicana*), 118, 127★, 140m, **141**
Mexican spiny pocket mouse (*Liomys irroratus*), **275**, 275m, 297★
Mexican vole (*Microtus mexicanus*), 337m
Michigan beaver (*Castor fiber michiganensis*), **276**
Microcavia (Mountain cavies), 441, 443m
—*australis* (Southern mountain cavy), 436★, 443m
Microchiroptera (Insectivorous bats), 68, **111**ff
Microdipodops megacephalus (Dark kangaroo mouse), 273m
—*pallidus* (Pale kangaroo mouse), **273**, 273m
Micromys minutus (Harvest mouse), 257/258★, 367m, **368f**, 368★
Micropteropus pusillus (Dwarf epauletted fruit bat), 87★, **103**
Microsciurus (Dwarf tree squirrels), **260**
—*alfari*, **260**
Microtinae (Microtine rodents), 301, **319**
Microtini (Voles), 319, **328**
Microtus (Voles), **336**
—*agrestis* (Field vole), **336**, 336m, 339
—*arvalis* (Common vole), 300★, 321/322★, **336**ff, 336m, 338★
—*brandti* (Brandt's vole), **336**, 339
—*californicus* (California vole), 337m
—*chrotorrhinus* (Rock vole), 338m
—*guentheri* (Gunther's vole), 337m
—*longicaudus* (Long-tailed vole), 338m
—*mexicanus* (Mexican vole), 337m
—*miurus* (Singing vole), 338m
—*nivalis* (Snow vole), 300★, **336**, 337m, 340f
—*ochrogaster* (Prairie vole), 338m
—*oeconomus* (Tundra vole), **336**, 337m, 339
—*oregoni* (Creeping vole), 338m
—*pennsylvanicus* (Eastern meadow mouse), 300★, **336**, 337m
—*richardsoni* (Richardson's or American vole), 338m
—*townsendi* (Townsend's vole), 338m
—*xanthognathus* (Yellow-cheeked vole), 338m
Migratory hamster (*Cricetulus migratorius*), **306**, 306m
Miller, Gerrit S., 430, 450
Miniopterus schreibersi (Long-winged bat), 69, **146**, 146m
Modern man (*Homo sapiens sapiens*), **55**, **78★**
Mohave ground squirrel (*Citellus mohavensis*), 232m
Mohr, Erna, 234, 247, 408, 414f, 429, 444f, 505
Möhres, 79f, 89f
Mole lemming (*Ellobius talpinus*), **342f**, 342m
Mole mice (Myospalacini), 301, **316**
Mole ratlike rodents (Bathyergoidea), 407, **416**
Mole rats (Bathyergidae), 214/215★, **417**
Molina, Juan Ignazio, 440
Molossidae (Free-tailed bats), **146**
Molossops malagai, 137★
Molossus major (Giant velvety free-tailed bat), 137★
—*rufus* (Red velvety free-tailed bat), 137★
Monckton, Oldfield Thomas, 388
Monckton's water rat (*Crossomys moncktoni*), 385m, **388**
Mongolian beaver (*Castor fiber birulai*), **276**, 278m
Mongolian marmot or tarbagan (*Marmota bobak sibirica*), 219f
Mongoloid type humans, 56, 56m
Monodon monoceros (Narwhal), 480★, 493★, **505f**, 506m
Monodontidae (White whales and narwhals), 458, 466★, 493, **503**
Monodontoidea (Narwhals), 493, **503**
Moore, Joseph Curtis, 205
Morris, Desmond, 59
Morrison-Scott, 218
Mosaic-tailed rats (*Melomys*), **352**
Mossmann, 270
Mountain beaver or sewellel (*Aplodontia rufa*), **201**ff, 203m, 229★
Mountain cavies (*Microcavia*), 441, 443m
Mountain chinchillas (*Lagidium*), **433f**
Mountain paca (*Stictomys taczanowskii*), 437★, **447**, 448m, 449
Mouse-eared bats (*Myotis*), **144f**
Mouse-like dormouse or Asiatic dormouse (*Myomimus personatus*), **397**
Mouse-like hamster (*Calomyscus bailwardi*), **305f**, 305m
Mouse-like rodents (Myomorpha), 193, **296**
Müller, R., 193
—, S., 108
Müller-Using, 208ff, 282
Multimammate rats (*Mastomys*), **376**, 376m
Münch, H., 207, 209f
Murid rodents (Muroidea), 296
Muridae (Murid rodents), 214/215★ 296, **350**
Murie, 122
Murina, 108
Murinae, 350
Muroidea (Murid rodents), 296
Mus (House mice, i.n.s.), **362**
—*booduga* (Little Indian field mouse), **362**
—*musculus* (House mouse), 351, **362f**, 363m, 371★

—*bactrianus* (Bactrian house mouse), **363**
—*domesticus* (Western house mouse), **362**, 363m
—*musculus* (Northern house mouse), **362**, 363m
—*spicilegus* (Eastern house mouse), **362**
—*platythrix* (Indian brown spiny mouse), **362**
Muscardinus avellanarius (Common dormouse), 382★, 389, **395**, 395m
Muschketov, 236
Muskrat (*Ondatra zibethica*), 257/258★, **332**, 332m, 334★
Musonycteris harrisoni (Banana bat), 86★, 118, 127★, **141**
Mustache bats (*Chilonycteris*), **139**
Mylagaulidae, 203
Mylodon, 153
—*domesticum* ("Domestic giant sloth"), 154
Mylodontidae, 153
Myocastor coypus (Coypu or nutria), **430**ff, 431m, 432★, 438★
Myocastoridae (Coypus), 214/215★, 420, **431**
Myomimus personatus (Mouse-like dormouse, or Asiatic dormouse), **397**
Myomorpha (Mouse-like rodents), 193, **296**
Myonycteris (Collared fruit bats), **99**
—*torquata*, **99**
Myoprocta (Acouchys), **450**
—*acouchi* (Guyana acouchy), 437★, **450**, 450m
—*exilis*, **450**
Myopus schisticolor (Wood lemming), **324**, 324m
Myosciurus pumilio (African pygmy squirrel), 230★, **262**
Myospalacini (Mole mice), 301, **316**
Myospalax (Zokors), **316**
—*aspalax* (Transbaikal zokor), 300★, **316**
—*myospalax* (Altai zokor), **316**, 317★
—*psilurus* (Manchurian zokor), **316**
Myotis (Mouse-eared bats), **144**f
—*bechsteini* (Bechstein's bat), 134★, 141m, **144**
—*capaccinii*, 141m
—*dasycneme* (Pond bat), 141m, **144**
—*daubentoni* (Water bat), 105/106★, 121, 134★, 141m, **144**
—*emarginatus*, 82, 142m
—*myotis* (Large mouse-eared bat), 68, 86★, 88★, 112★, 114, 134★, 140m, **144**
—*mystacinus* (Whiskered bat), 142m, **144**
—*nattereri* (Natterer's bat), 80★, 86★, 134★, 142m, **144**
—*oxygnathus* (Lesser mouse-eared bat), 140m
—*welwitschii* (Welwitsch's bat), 113
Myrmecophaga tridactyla (Giant anteater), 173★, **177**ff, 178m, 179★, 190★
Myrmecophagidae (Anteaters), 149, **172**ff, 176★
Mystacina tuberculata (New Zealand short-tailed bat), 137★, **146**
Mystacinidae (New Zealand short-tailed bats), 118, **146**
Mystacoceti (Baleen whales), 458, 466★, **477**ff, 487★
Mystromyini (White-tailed rats), 301, **316**
Mystromys, 310m, **315**
—*albicaudatus* (White-tailed rat), **316**
—*longicaudatus* (Long-tailed rat), **316**
Myzopoda aurita (Golden bat), **144**
Myzopodidae (Sucker-footed bats), 118, **144**

McCann, 96

Naked bat (*Cheiromeles torquatus*), 112, 137★, **147**
Naked sand rat (*Heterocephalus glaber*), **417**f, 417m, 426★
Naked-backed bat (*Pteronotus davyi*), 131m, **139**
Nannosciurus (Oriental pygmy squirrels), 230★, **260**
—*melanotis* (Brown dwarf squirrel), **260**
—*whiteheadi* (Whitehead's dwarf squirrel), **260**
Napaeozapus insignis (Woodland jumping mouse), 399★, **402**, 403m
Narrow-faced kangaroo rat (*Dipodomys venustus*), 274m
Narwhal (*Monodon monoceris*), 480★, 493★, **505**f, 506m
Natalidae (Funnel-eared bats), **143**
Natalus stramineus, 134★, **143**
Natterer's bat (*Myotis nattereri*), 80★, 86★, 134★, 142m, **144**
Naundorf, Elisabeth, 314
Neanderthal man (*Homo sapiens neanderthalensis*), 53f, 78★
Neave's scaly-tailed squirrel (*Anomalurus neavei*), **291**, 292m
Necklace sloth (*Bradypus torquatus*), 171m, **172**
Neck-pouched hopping mouse (*Notomys cervinus*), **353**f
Negroid type human, 56, 56m
Nelson's antelope squirrel (*Citellus nelsoni*), 227m, 234
Neobalaena marginata (Pygmy right whale), 463★, **486**
Neofiber alleni (Round-tailed muskrat), 332m, **333**
Neophocaena phocaenoides (Black finless porpoise), 480★, **507**
Neotamias, 242
Neotoma (Wood rats), **304**
—*albigula* (White-throated wood rat), **304**, 304m
—*cinerea* (Bushy-tailed wood rat), 297★, **304**, 304m
—*floridana* (Eastern wood rat), 304m
—*fuscipes* (Dusky-footed wood rat), 303★, 304m
Nero, 397
Nesokia indica (Pest rat), **355**, 355m
Nesomyinae (Malagasy rats), 301, **318**
Nesomys lambertoni (Lamberton's Malagasy rat), 300★, **318**
Nesonycteris, 94
New Britain water rat (*Hydromys neobritannicus*), **387**
New Guinea water rat (*Hydromys habbema*), **387**
New World mice (Hesperomyini), **301**
New World porcupines (Erithizontidae), 214/215★, 420, **453**
New Zealand short-tailed bat (*Mystacina tuberculata*), 137★, **146**
Newfoundland beaver (*Castor fiber caecator*), **276**
Newfoundland muskrat (*Ondatra obscura*), **332**, 332m
Niethammer, Jochen, 206, 416
Nikitin, 309
Nilotic race human, 76/77★
Nine-banded armadillo (*Dasypus novemcinctus*), 156, **157**ff, 157m, 173★, 189★
Nissen, Henry W., 20
Noctilio labialis (Southern bulldog bat), **130**
—*leporinus* (Mexican bulldog bat), 113, 115, 115★, 124★, **129**f
Noctilionidae (Bulldog bats), **129**
Noctule bats (*Nyctalus*), **145**
Nordic race human, 56, 76/77★
North African ground squirrel (*Atlantoxerus getulus*), **243**
North African striped grass mouse (*Lemniscomys barbarus*), **373**
North American porcupine (*Erethizon dorsatum*), 409★, **454**ff, 455m
North American red squirrel (*Tamiasciurus hudsonicus*), 229★, **244**, 244m
North Italian pine vole (*Pitymys multiplex*), 335m
Northern bat (*Eptesicus nilssoni*), 144m, **145**
Northern birch mouse (*Sicista betulina*), 399★, **402**ff, 404m
Northern bog lemming (*Synaptomys borealis*), **323**, 323m
Northern five-toed jerboa (*Allactaga jaculus*), 399★, **404**, 406
Northern flying squirrel (*Glaucomys sabrinus*), 265, **265**f
Northern grasshopper mouse (*Onychomys leucogaster*), **302**, 302m
Northern ground squirrels (Marmotini), **205**
Northern house mouse (*Mus musculus musculus*), **362**, 363m
Northern pilot whale or Atlantic blackfish (*Globicephala melaena*), 480★, **517**
Northern pocket gopher (*Thomomys talpoides*), 271m
Northern red-backed vole (*Clethrionomys rutilus*), 329m **330**
Northern or Pacific right whale dolphin (*Lissodelphis borealis*), 480★, **514**
Northern three-toed jerboa (*Dipus sagitta*), 399★, 404, **405**f, 406m
Norway lemming (*Lemmus lemmus*), 300★, **324**ff, 324m
Norway rat, see brown rat
Notomys (Australian hopping mice), **353**, 353m
—*cervinus* (Neck-pouched hopping mouse), **353**f
Notopteris (Long-tailed fruit bat), 93f
Nutria, see coypu
Nux, de la, 95
Nyctalus (Noctule bats), **145**
—*lasiopterus* (Giant noctule), **145**
—*leisleri* (Lesser noctule), **145**
—*noctula* (Common noctule bat), 69, 69★, 121, 136★, 143m, **145**
Nycteridae (Slit-faced bats), **130**
Nycteris grandis, **130**
—*hispida*, **130**
—*javanica*, **130**
—*thebaica*, 114, 124★, **130**
Nyctimene, 108
—*major* (Large tube-nosed fruit bat), 87★
—*robinsoni*, **108**
Nyctimeninae (Tube-nosed fruit bat), 93, **108**

Obolenskij, S. J., 233
Octodon degus (Degu), **421**, 421m 435★
Octodont rodents (Octodontidae), 214/215★, 418, **420**f
Octodontidae (Octodont rodents), 214/215★, 418, **420**f
Octodontoidea (Octodont rodents), **420**f
Octodontomys gliroides (Bori), 421m
Octomys mimax (Viscacha rat), **421**, 421m
Odontoceti (Toothed whales), 458, 466★, **493**ff
Oenomys hypoxanthus (Rufous-nosed rat), **374**f, 375m
Ogilvie, P. W., 188
Ognew, 219, 235, 238, 249, 333
Oil palm squirrel (*Protoxerus stangeri*), 230★, **262**
Old World fruit bats (Megachiroptera), 68, **93**ff, 93m
Old World and western chipmunks (*Eutamias*), 238
Old World leaf-nosed bats (Hipposideridae), **131**f
Old World porcupines (Hystricidae) 214/215★, **407**
Oliver, W. R. B., 502
Olsen, 490
Ondatra (Muskrats), 332f
—*obscura* (Newfoundland muskrat), **332**, 332m
—*zibethica* (Muskrat), 257/258★, **332**, 332m, 334★
Onychomys (Grasshopper mice), 296, **302**
—*leucogaster* (Northern grasshopper mice), **302**, 302m
—*torridus* (Southern grasshopper mice), **302**, 302m
Oppenheimer, 227
Orange-rumped agouti (*Dasyprocta aguti*), 437★, **449**, 449m
Orca destructor, see *Orcinus orca*
Orca, see killer whale
Orcaella brevirostris (Irrawaddy dolphin), 480★, 517m, **518**
Orcininae, 508, **517**
Orcinus orca (Killer whale or orca), 480★, **519**ff
Ord's kangaroo rat (*Dipodomys ordi*), **273**f, 274m
Oregon pocket gopher (*Thomomys mazama*), 271m
Oreopithecidae, 51
Oreopithecus bambolii, 51
Oriental pygmy squirrels (*Nannosciurus*), 230★, **260**
Oriental race human, 56, 76/77★
Orr, Robert T., 146
Orshoven, van, 44, 46
Orthogeomys (Large pocket gophers or tuzas), 269
—*grandis* (Tuza), **271**, 271m
Oryzomys, **302**f, 303m
—*palustris* (Rice rat), 297★, **303**, 303m
Ostrjakov, 342
Otomyinae (Vlei and karroo rat), 351, **380**
Otomys, 380, 384m
—*denti* (Karroo rat), **383**
—*irroratus* (Vlei rat), 372★, **383**
Oviedo y Valdes, Gonzalo Fernandez de, 450
Owen, Richard, 518

Pacarana (*Dinomys branickii*), 437★, **451**ff, 451m
Pacaranalike rodents (Dinomyoidea), 420, **450**
Pacas (Cuniculinae), **447**
Pachyuromys duprasi (Fat-tailed mouse), **343**, 346
Pacific jumping mouse (*Zapus trinotatus*), 403m
Pacific kangaroo rat (*Dipodomys agilis*), 273, 274m, 297★
Pacific pilot whale (*Globicephala sieboldii*), 511, **517**
Pacific right whale (*Eubalaena japonica*), 481, **485**f

Pacific right whale dolphin, see Northern right whale dolphin
Pacific white-sided dolphin (*Lagenorhyncus obliquidens*), 511, **517**
Painted spiny pocket mouse (*Liomys pictus*), **275**, 275m
Palaeanodonta, 153
Palaeopeltidae, 154
Pale kangaroo mouse (*Microdipodops pallidus*), **273**, 273m
Palearctic mole rats (Spalacidae), 214/215★, 296, **348**
Pallid bat (*Antrozous pallidus*), 136★, **146**, 146m
Palm squirrels (Funambulus), **261**
Pan, **19ff**
—*paniscus* (Pygmy chimpanzee or bonobo), 19f, 20m, 22, 30/31★, 35★, **47**
—*troglodytes* (Chimpanzee), **19ff**, 20m, 20★, 21★, 22★, 23★, 25★, 26★, 27★, 28★, 29★, 30/31★, 32/33★, 34★, 36★, 38★, 40★, 45★
Panama capybara (*Hydrochoerus hydrochaeris isthmius*), **446f**
Pangolins (*Manis*), **183ff**, 186★, 200★
Pappocetus 472
Paradipus ctenodactylus (Comb-toed jerboa), **404f**, 406m
Paramanis, 183
Paramys, 198
Paranyctimene, **108**
Parapedetes, 294
Paraxerus (African bush squirrels),**261**
—*boehmi* (Boehm's African bush squirrel), **262**
—*flavivittis*, 230★, **262**
—*palliatus* (Mantled African bush squirrel), **261**
Particolored bat (*Vespertilio murinus*) 121, 145m
Patagonian cavy, aka mara (*Dolichotis patagonum*), 425★, 437★, **444f**, 444m
Patterson, 198
Pechvel-Loesche, 515
Pectinator, 287
—*spekei* (Speke's pectinator), **289**, 426★
Pedetes, 294
—*cafer* (Cape springhaas), **295**, 295m
—*surdaster* (East African springhaas), 284★, **295**, 295m
Pedetidae (Springhaas, spring hares or jumping hares), 201, **294**
Pedetoidea (Springhaas), 201, **294**
Pediolagus salinicola (Salt-desert cavy), 437★, 444m, **445**
Peking man (*Homo erectus pekinensis*), 54
Pel's scaly-tailed squirrel (*Anomalurus peli*), **290**, 292, 292m
Pencil-tailed tree mouse (*Chiropodomys gliroides*), **386**
Père David's rock squirrel (*Sciurotamias davidianus*), **244**
Père David's vole (*Eothenomys melanogaster*), **330**
Perny's long-nosed squirrel (*Dremomys pernyi*), 230★, **261**
Perognathus (Pocket mice), **272ff**
—*baileyi* (Bailey's pocket mouse), 272m, **273**
—*californicus* (California pocket mouse), **273**, 273m, 297★
—*fasciatus* (Wyoming pocket mouse), 272m
—*flavescens* (Plains pocket mouse), 272m
—*flavus* (Silky pocket mouse), 272m, **273**
—*hispidus* (Hispid pocket mouse), 273m
—*parvus* (Great Basin pocket mouse), 273m
Peromyscus, **301**
—*leucopus* (White-footed mouse), **301f**, 302m
—*tornillo*, 302
—*maniculatus* (Deer mouse), 297★, **301**, 302m
—*nuttalli* (Golden mouse), **301f**, 302m
—*truei* (Piñon mouse), 297★, **301**, 302m
Peron's dolphin (*Lissodelphis peronii*), **514**
Persian jird (*Meriones persicus*), 345
Persian leaf-nosed bat (*Triaenops persicus*), **139**
Peruvian mountain chinchilla (*Lagidium peruanum*), 440m
Pest rat (*Nesokia indica*), **355**, 355m
Petaurista grandis (Formosan giant flying squirrel), 65, 240★, **264**
—*petaurista* (Common giant flying squirrel), **264**
Peters, 451
Petromuridae (Rock rats), 214/215★, **419**
Petromuroidea (Cane and rock rats), 407, **418f**
Petromus typicus (Rock rat), **419**, 426★
Petter, Francis, 374
Petzsch, Hans, 310, 312ff, 411
Peus, 234
Pfeyffers, 44, 46
Phatages, 183
Phataginus, 182
Philippine flying lemur (*Cynocephalus volans*), **64ff**, 65m, 83★, 84★
Philippine thick-spined porcupine (*Thecurus pumilis*), 414m
Phloeomyinae (Cloud rats), 351, **385**
Phloeomys cumingi (Slender-tailed cloud rat), 372★, **385**, 385m
Phocaena dioptrica), (Spectacled porpoise), 480★, **507**
—*phocaena* (Common porpoise), 480★, **506**
Phocaenidae (Porpoises), 458, 466★, 493, **506**
Phocaenoides dalli (Dall's harbor porpoise), 480★
Phodopus (Dwarf hamsters), **306**
—*roborovskii* (Roborovsky's dwarf hamster), 305m, **306f**
—*sungorus* (Striped hairy-footed hamster), 305m, **306**, 308
Pholidota (Pangolins), **182ff**
Phyllonycterinae (American leaf-nosed bats), 141
Phyllostomidae (Leaf-nosed bats), 112, **139**
Phyllostominae (Big-eared leaf-nosed bats), **140**
Phyllostomus discolor, **140**
—*hastatus* (Spear-nosed bat), 115, 126★, **140**
Physeter catodon (Sperm whale), 474★, **494ff**
Physeteridae (Sperm whales), 458, 466★, 493, **494f**
Physeteroidea, 493
Pichiagos (Chlamyphorina), 162, **165**
Piechocki, 241, 368
Pijl, 97
Pilleri, G., 212, 280, 510
Pilosa (Armorless edentates), 166
Pilot whales or blackfish (*Globicephala*), 498★, **517**
Pine voles (*Pitymys*), **335**
Piñon mouse (*Peromyscus truei*), 297★, **301**, 302m
Pipistrellus, 112, 115, **144**
—*kuhli*, 143m
—*nanulus*, 112
—*nanus* (African banana bat), **145**
—*nathusii*, 143m
—*pipistrellus* (Common pipistrelle), 69, 112★, 135★, 142m, **144**
—*savii*, 134m
"*Pithecanthropus*", see *Homo erectus erectus*
Pitymys (Pine voles), **335**
—*duodecimcostatus* (Mediterranean pine vole), 335m
—*multiplex* (North Italian pine vole), 335m
—*pinetorum* (American pine vole), **335**, 335m
—*savii* (Southern European pine vole), 335m
—*subterraneus* (European pine vole), **335f**, 335m
Piveteau, J., 50
Pizonyx vivesi, 115
Plagiodontia (Hispaniolan hutias), **429f**
—*aedium* (Haitian hutia), 430, 430m
—*aedium* (Cuvier's hutia), **429f**
—*hylaeum* (Dominican hutia), **429f**, 438★
—*ipnaeum*, **429**
—*spelaeum*, **429**
Plains pocket mouse (*Perognathus flavescens*), 272m
Planetherium, 64
Platacanthomyidae (Spiny dormice), 296, 388, **398**
Platacanthomys lasiurus (Spiny dormouse), **398**, 401
Platalina genovensium, 141
Platanista gangetica (Ganges river or Gangetic dolphin), 479★, **502**, 502m
Platanistidae (Gangetic dolphins), 458, 466★, 493, **502**
Platanistoidea (River dolphins), 493, **502**
Platypittamys, 420, 456
Plecotus, 146m
—*auritus* (Long-eared bat), 70★, 112★, 114, 133★, 135★, **145**
—*austriacus* (Southern long-eared bat), **146**
Pleske, 326
Pliopeithecus, 51
Pliopithecinae (Short arm gibbons), 50
Plors-Bartels, 41
Pocket gophers (Geomyidae), 201, 214/215★, **268f**
Pocket mice (*Perognathus*), **272ff**
Polish or Byelorussian beaver (*Castor fiber vistulanus*), **276**, 278m
Polynesian race human, 57, 76/77★
Pond bat (*Myotis dasycneme*), 141m, **144**
Pongidae (Anthropoid apes), 51
Poole, Jonas, 470
Pöppig, 421
Porcupines (Hystricomorphs), 193, 198, **407**
Porpoises (Phocaenidae), 458, 466★, 493, **506**
Porsch, 96
Portmann, Adolf, 208
Pouched rats (Cricetomyinae), 351, **378**
Pournelle, 47
Prairie dogs (Cynomys), **225ff**
Prairie vole (*Microtus ochrogaster*), 338m
Praomys (Soft-furred rats), **375**, 376m
—*morio*, **375f**
Prehensile-tailed hutia (*Capromys prehensilis*), **428**, 438★
Prehensile-tailed porcupine (*Coendou prehensilis*), 409★, 423★, **454f**
Priodontes, 154
—*giganteus* (Giant armadillo), 160m, **161f**, 189★
Priodontini, 161
Proboscis bat (*Rhynchonycteris naso*), **122**
Proconsul africanus, 51, 51★
Proechimys guyannensis (Cayenne spiny rat), **427**, 427m, 435★
Prometheomys schaposchnikowi (Long-clawed mole vole), **341f**, 341m
Propliopithecus, 51
Protocatarrhine stage, 50
Protocetus atavus, 475
Protohominids (Australopithecinae), 52f
Protosteiromys, 456
Protoxerini (Oil palm squirrels), 205, **262**
Protoxerus stangeri (Oil palm squirrel), 230★, **262**
Psenner, Hans, 212, 217
Pseudorca crassidens (False killer whale), 480★, 511, **518**
Pteromyinae (Flying squirrels), 204, **263**
Pteromys, 66★
—*volans* (European flying squirrel), 240★, **264**, 264m
Pteronotus davyi (Naked-backed bat), 131m, **139**
—*suapurensis* (Suapure naked-backed bat), 126★, 131m, **139**
Pteropidae (Fruit bats, i.n.s.), 93, **97f**
Pteropinae (Flying foxes), 93, **97**
Pteropus (Flying foxes), 97, 99m, **100**
—*alecto gouldi*, 96
—*capistratus*, 101★
—*giganteus* (Indian flying fox), 94f, **100f**
—*gigantus*, **100**
—*leucocephalus*, **100**
—*niger*, 95
—*poliocephalus* (Gray-headed flying fox), 87★, **100**
—*rufus* (Rufous flying fox), **100**
—*subniger*, **100**
—*tonganus* (Tonga fruit bat), 86★
—*vampyrus* (Red-necked fruit bat), 87★, **100**
—*vampyrus*, **100**
Pygeretmus (Fat-tailed jerboa), 404
Pygmies, 56
Pygmy armadillo (Euphractus pichiy, **163f**, 164m, 189★
Pygmy chimpanzee (*Pan paniscus*), 19f, 20m, 22, 30/31★, 35★, **47**
Pygmy dormouse (*Graphiurus nanus*), **398**
Pygmy jerboa (*Salpingatus crassicaudatus*), 399★, **404**
Pygmy killer whale (*Feresa attenuata*), 480★, **518**
Pygmy mouse (*Baiomys taylori*), 297★, **302**, 303m
Pygmy right whale (*Neobalaena marginata*), 463★, **486**
Pygmy scaly-tailed squirrel (*Anomalurus pusillus*), **290**, 292, 292m
Pygmy sperm whale (*Kogia breviceps*), 474★, **494**

Quemisia gravis (Quemi), 451

Rabbittailed meadow vole, see sagebrush vole
Rahm, Urs, 179
Ramapithecus brevirostris, 52
—*punjabicus*, 52
—*thorpei*, 52

Rat chinchilla (*Abrocoma cinerea*), **422**, 427m, 435★
Rat chinchillas (Abrocomidae), 214/215★, 420, **422**
Ratlike hamster (*Tscherskia triton*), 306m, **307**
Rat-like hamsters (*Cricetulus*), **306**
Rato de Taquara (*Kannabateomys amblyonyx*), **427**f, 429m, 435★
Rats (*Rattus*), **356**, 356m
Rat-tailed bats (Rhinopomatidae), **121**
Rattus (Rats), **356**, 356m
—*exulans* (Little rat), **356**
—*norvegicus* (Brown rat or Norway rat), 356m, **357**ff, 357★, 358★, 359★, 360★, 371★
—*rattus* (Black rat or house rat), 351, **361**f, 361m, 362★, 371★
—*alexandrinus* (Roof rat), **361**
—*frugivorus* (Corn rat), **361**
—*rattus* (House rat), **361**
Ratufa bicolor (Malayan giant squirrel), 230★, **262**
—*indica* (Indian giant squirrel), **262**
Ratufini (Indo-Malayan giant squirrels), 205, **262**
Ray, John, 457
Red bat (*Lasiurus borealis*), 113, 119, 136★, **145**, 145m
Red fruit-eating bats (Stenoderminae), 117, **141**
Red Sea bottle-nosed dolphin (*Tursiops aduncus*), 480★, 516
Red souslik (*Citellus rufescens*), 238
Red squirrel (*Sciurus vulgaris*), 213★, 229★, **245**ff, 252★, 253★
Red velvety free-tailed bat (*Molossus rufus*), 137★
Red-necked fruit bat (*Pteropus vampyrus*), 87★, **100**
Reed, 304
Reithrodontomys (American harvest mice), **302**
—*megalotis* (Western harvest mouse), **302**, 303m
Rengger, 163, 432
Reynolds, Frankie, 20
—, Vernon, 20ff, 40
Rhabdomys pumilio (Striped field mouse), 371★, **374**, 375m
Rheithrosciurus macrotis (Groove-toothed squirrel), **260**
Rhinolophidae (Horseshoe bats), **131**
Rhinolophoidea, 131
Rhinolophus alcyone, **132**
—*blasii* (Blasius horseshoe bat), 130m
—*denti* (Dent's horseshoe bat), **132**
—*euryale* (Mediterranean horseshoe bat), 130m
—*ferrumequinum* (Greater horseshoe bat), 69★, 105/106★, 121m, 125★, **132**
—*hildebrandtii* (Hildebrandt's horseshoe bat), 124★
—*hipposideros* (Lesser horseshoe bat), 69, 69★, 120★, 120m, 125★, **132**
—*landeri*, **132**
—*maclaudi* (Maclaud's horseshoe bat), **132**
—*mehelyi* (Mehely's horseshoe bat), 130m
Rhinopoma hardwickei, **122**
—*microphyllum*, 121, 123★
Rhinopomatidae (Rat-tailed bats), **121**
Rhinosciurus laticaudatus (Long-nosed squirrel), **261**
Rhizomyidae (Bamboo rats), 214/215★, 296, **346**f
Rhizomys, 300★, 347m
—*sumatrensis* (Sumatran bamboo rat), **347**
Rhodes, Cecil, 259
Rhombomys opimus (Great gerbil), 300★, **343**, 345
Rhone beaver (*Castor fiber galliae*), **276**, 278m, 280★
Rhynchomyinae (Shrew rats), 351, **386**
Rhynchomys soricoides (Shrew rat), 372★, 385m, **386**
Rhynchonycteris naso (Proboscis bat), **122**
Rice rat (*Oryzomys palustris*), 297★, **303**, 303m
Richardson's ground squirrel (*Citellus richardsonii*), 234m
Richardson's or American vole (*Microtus richardsoni*), 338m
Right whale dolphins (*Lissodelphis*), **514**
Right whales (Balaenidae), 458, 466★, 477, **481**
Ring-tailed ground squirrels (*Citellus annulatus*), 232m
Rio Grande beaver (*Castor fiber frondator*), **276**
Risso's dolphin (*Grampus griseus*), 480★, **515**
River dolphins (Platanistoidea), 493, **502**
Robertson, Dougal, 522
Roborovsky's dwarf hamster (*Phodopus roborovskii*), 305m, **306**f
Rock cavy (*Kerodon rupestris*), 436★, 443m, **443**
Rock rat (*Petromus typicus*), **419**, 426★
Rock squirrel (*Citellus variegatus*), 232m
Rock squirrels (*Sciurotamias*), **244**
Rock vole (*Microtus chrotorrhinus*), 338m
Rodentia (Rodents), **191**ff, 192★, 214/215★
Rokitansky, von, 403
Roof rat (*Rattus rattus alexandrinus*), **361**
Rorquals (Balaenoptera), **488**
Ross, Sir James C., 471
Roth, H., 289
Rough-toothed dolphin (*Steno bredanensis*), 480★, **507**
Round-tailed ground squirrel (*Citellus terreticaudus*), 227m, **233**
Round-tailed muskrat (*Neofiber alleni*), 332m, **333**
Rousette bats (*Rousettus*), 80, 97, **98**, 98m
Rousettus (Rousette bats), 80, 97, **98**, 98m
—*aegyptiacus* (Egyptian fruit bat), **87★**, **98**f
—*aegyptiacus*, **98**
—*leachi*, **98**
—*occidentalis*, **98**
—*angolensis* (Angolan fruit bat), **98**f
—*angolensis*, **98**
—*ruwenzorii*, **98**
—*smithi*, **98**
Rowe, 364
Royle's high-mountain vole (*Alticola roylei*), **330**
Rubruk, Wilhelm von, 221
Rufous flying fox (*Pteropus rufus*), **100**
Rufous-nosed rat (*Oenomys hypoxanthus*), **374**f, 375m
Rümmler, 351
Ruud, 472
Rwanda mole rat (*Tachyoryctes*), **347**f
Saccolaimus, 129
Saccopteryx (Sheath-tailed bats), **122**
—*bilineata* (Two-lined sheath-tailed bat), 113, **122**, 123★
Saccostomus, 378, 378m
—*campestris* (Cape pouched mouse), **380**
Sac-winged bats (Emballonuridae), **122**
Sagebrush or rabbittailed meadow vole (*Lagurus curtatus*), **341**
Salpingotus crassicaudatus (Pygmy jerboa), 399★, **404**
Salt-desert cavy (*Pediolagus salinicola*), 437★, 444m, **445**
Sand rats (*Heliophobius*), **417**, 417m
Sanden, Walter von, 403
Sanderson, Ivan T., 107, 269f, 401, 418
Scaly-tailed squirrel-like rodents (Anomaluroidea), 201, **289**
Scaly-tailed squirrels (Anomaluridae), 214/215★, **289**
Scammon, 487, 491
Scandanavian beaver (*Castor fiber fiber*), **276**, 278m
Schaub, S., 198
Scheflen, A., 61
Schleidt, 329
Schneider, Karl Max, 414
Schomber, 288f
Schopenhauer, 49
Schöps, P., 247
Schwidetzky, 57
Sciuridae (Typical squirrels, tree squirrels, or sciurids), 201, **203**ff, 214/215★
Sciurinae (Ground an tree squirrels), **204**f
Sciurini (Tree squirrels), 205, **244**ff
Sciuroidea, 201, **203**
Sciuromorpha (Sciurid or squirrel-like rodents), 193, **201**ff
Sciurotamias (Rock squirrels), **244**
—*davidianus* (Père David's rock squirrel), **244**
Sciurus aberti (Abert's squirrel), 255m
—*aestuans* (Brazilian squirrel), **259**
—*anomalus* (Caucasian squirrel), 242m, **255**, 260
—*apache* (Apache squirrel), 255m
—*arizonensis* (Arizona gray squirrel), 254m
—*carolinensis* (Gray squirrel), 229★, 255m, **255**
—*griseus* (Western gray squirrel), 245m
—*kaibabensis* (Kaibab squirrel), 255m, 259
niger (Eastern fox squirrel), 254m, 259
—*vulgaris* (Red squirrel), 213★, 229★, **245**ff, 245m, 252★, 253★
—*exalbidus* (Siberian red squirrel), 247
—*fuscoater* (Central European red squirrel), 229★, 247
—*leucourus* (British red squirrel), 247
Scoresby, William, 482, 506
Scotonycteris ophiodon (Snake-toothed bat), **107**
—*zenkeri* (Zenker's fruit bat), 87★, **107**
Seba's short-tailed bat (*Carollia perspicillata*), 128★, 139★, **141**
Sei whale (*Balaenoptera borealis*), 464★, **489**ff
Selevin, W. A., 401
Selevinia betpakdalaensis (Betpakdala dormouse), 382★, 397m, **401**
Seleviniidae (Desert dormice), 214/215★, 296, 388, **401**
Serebrennikov, 342
Serotine bat (*Eptesicus serontinus*), 135★, 144m, **145**
Seven-banded armadillo (*Dasypus septemcinctus*), **157**, 157m, 159
Sewellel, see mountain beaver
Shaw, 236
Shaw's jird (*Meriones shawi*), 345★
Sheath-tailed bats (*Saccopteryx*), **122**
Shepherd's beaked whale (*Tasmacetus shepherdi*), 494, **502**
Short dwarf hamster (*Allocricetulus curtatus*), 305m, **306**f
Short-arm gibbons (Pliopithecinae), 50
Short-finned blackfish, see Indian pilot whale
Short-nosed fruit bats (Cynopterinae), 93, **108**
Short-tailed chinchilla (*Chinchilla chinchilla*), 433m, **439**f
Short-tailed leaf-nosed bats (Carolliinae), 117, **141**
Shrew rat (*Rhynchomys soricoides*), 372★, 385m, **386**
Sibbaldus, see blue whale
Siberian jerboa (*Allactaga sibirica*), **406**, 406m
Siberian lemming (*Lemmus sibiricus*), 323m
Siberian red squirrel (*Sciurus vulgaris exalbidus*), 247
Sibirid race human, 56
Sicista betulina (Northern birch mouse), 399★, **402**f, 404m
—*subtilis* (Southern birch mouse), **402**, 404m
Sicistinae (Birch mice), **402**, 404m
Sierra pocket gopher (*Thomomys monticola*), 271m
Siewert, Horst, 391
Sigmodon (Cotton rats), **304**f
—*hispidus* (Hispid cotton rat), **304**, 304m
—*minimus* (Least wood rat), 304m
—*ochrognathus* (Yellow-nosed wood rat), 304m
Silky anteater, see two-toed anteater
Silky pocket mouse (*Perognathus flavus*), 272m, **273**
Silvery mole rat (*Heliophobius argenteocinereus*), **417**
Simons, E. L., 52
Sinai spiny mouse (*Acomys dimidiatus*), **369**, 371★
Singing vole (*Microtus miurus*), 338m
Single-striped grass mouse (*Lemniscomys griselda*), **373**
Sinoid race human, 57, 76/77★
Six-banded armadillo (*Euphractus sexcinctus*), **163**f, 163m, 189★
Skunk dolphins, see Commerson's dolphins
Slender-tailed cloud rat (*Phloeomys cumingi*), 372★, **385**, 385m
Sleptsov, 368
Slijper, 472, 475f
Slit-faced bats (*Nycteridae*), **130**
Sloths (Bradypodidae), 149, **166**ff, 167★, 176★
Small African native mouse (*Leggada minutoides*), **369**
Small naked-soled gerbil (*Taterillus emini*), **343**
Smith, 227
Smokey bats (Furipteridae), **143**
Smutsia, 183
Snake-toothed fruit bat (*Scotonycteris ophiodon*), **107**
Snethlage, Margarete, 427, 452

Snow vole (*Microtus nivalis*), 300★, **336**, 337m, 340f
Soft-furred rats (*Praomys*), **375**, 376m
Sommer, R., 60
Sooty agouti (*Dasyprocta fuliginosa*), 437★, **449**, 449m
Sotalia (White dolphins), **507**
—*fluviatilis* (Buffeo negro), **507**, 507m
—*guianensis* (Guianan river or guiana dolphin), **507**, 507m
Sousa sinensis (Chinese white dolphin), **507**
—*teuszi* (West African white dolphin), 480★, **507f**
Sousliks and ground squirrels (*Citellus*), **231f**
South African ground squirrel (*Xerus inauris*), 230★
South African lesser leaf-nosed bat (*Hipposideros caffer*), 125★, **132f**
South African porcupine (*Hystrix africaeaustralis*), **413**, 415m
South American forest mouse (*Heteromys anomalus*), **275**, 275m
South American rock rat (*Aconaemys fuscus*), **421**, 421m, 435★
South American three-fingered sloth (*Bradypus tridactylus*), 172m, **172**, 175★, 190★
South American two-fingered sloth, see common two-fingered sloth
Southeastern pocket gopher (*Geomys pinetis*), 270m
Southern birch mouse (*Sicista subtilis*), **402**, 404m
Southern bog lemming (*Synaptomys cooperi*), **323f**, 323m
Southern bottle-nosed whale (*Hyperoodon planifrons*), 495m, **501**
Southern bulldog bat (*Noctilio labialis*), **130**
Southern European pine vole (*Pitymys savii*), 335m
Southern flying squirrel (*Glaucomys volans*), 240★, 261★, **265f**, 265m
Southern grasshopper mouse (*Onychomys torridus*), **302**, 302m
Southern long-eared bat (*Plecotus austriacus*), **146**
Southern Mongolian jird (*Meriones meridianus*), 345
Southern mountain cavy (*Microcavia australis*), 436★, 443m
Southern pocket gopher (*Thomomys umbrinus*), 271m
Southern right whale or ice baleen whale (*Eubalaena australis*), **485**
Sowerby's whale (*Mesoplodon bidens*), 480★, **501**
Spalacidae (Palearctic mole rats), 214/215★, 296, **348**
Spalacopus cyanus (Cururo), **421**, 421m, 435★
Spalax, 348m, 348f
—*ehrenbergi* (Ehrenberg's mole rat), **348f**
—*leucodon* (Lesser mole rat), 300★, **348ff**
—*microphthalmus* (Greater mole rat), **348ff**
Spallanzani, Lazarro, 79
Spear-nosed bat (*Phyllostomus hastatus*), 115, 126, **140**
Spectacled porpoise (*Phocaena dioptrica*), 480★, **507**
Speke's pectinator (*Pectinator spekei*), **289**, 426★
Sperm whale (*Physeter catodon*), 474★, **494ff**
Spermophilopsis leptodactylus (Long-clawed ground squirrel), **243f**
Spiny armadillo (*Cabassous hispidus*), **161**, 161m
Spiny dormouse (*Platacanthomys lasiurus*), **398**, 401
Spiny mice (*Acomys*), **369f**, 374m
Spiny pocket mice (*Heteromys*), 272
Spiny rats (Echimyidae), 214/215★, 420, **427**
Spix's disk-winged bat (*Thyroptera tricolor*), **144**
Spotted dolphins (*Stenella*), **514**
Spotted ground squirrel (*Citellus spilosoma*), 234m
Spotted paca (*Cuniculus paca*), 437★, **447ff**, 448m
Spotted souslik (*Citellus suslicus*), 226m, **232**, 236, 239★
Spotted squirrel (*Callosciurus notatus*), **260**
Spring hares or jumping hares, see springhaas
Springhaas (Pedetoidea), 201, **294**
Squalodontidae, 476
Starck, Dietrich, 129, 418
Steatomys, 384m
—*pratensis* (Fat mouse), 372★, **384**
Steen, 278
Stein, Georg H. W., 338, 353
Steiner, H. 341
Steiniger, 358ff
Stenella (Spotted dolphins), **514**
—*caeruleoalba* (Blue-white dolphin), 480★, **515**
Stenidae (Long-snouted dolphins), 458, 466★, 493, 506, **507**
Steno bredanensis (Rough-toothed dolphin), 480★, **507**
Stenodelphidae (La Plata dolphins), 493, **502**
Stenodelphis blainvillei (La Plata dolphin), 479★, **502**, 503m
Stenoderminae (Red fruit-eating bats), 117, **141**
Stenofiber, 286
Stephan, 57
Steppe lemming (*Lagurus lagurus*), **341**, 341m
Stick-nest rats (*Leporillus*), **351f**, 352m
Stictomys taczanowskii (Mountain paca), 437★, **447**, 448m, 449
Strap-toothed whale, see Layard's whale
Straus, W. J., 50
Straw-colored bat (*Eidolon helvum*), 87★, 91, 95, **97f**
Striped field mouse (*Apodemus agrarius*), **365f**, 366m, 371★
Striped field mouse (*Rhabdomys pumilio*), 371★, **374**, 375m
Striped grass mouse (*Lemnicomys striatus*), 371★, **374**
Striped hairy-footed hamster (*Phodopus sungorus*), 305m, **306**, 308
Striped hamster (*Cricetulus barabensis*), **306**, 306m
Stubbe, M., 330, 345, 406
Sturnira lilium, **142**
Sturnirinae (Yellow-shouldered bats), **141**
Stylodipus (Thick-tailed three-toed jerboa), 404
Suapure naked-backed bat (*Pteronotus suapurensis*), 126★, 131m, **139**
Subko, 342
Sucker-footed bats (Myzopodidae), 118, **144**
Sudanese race human, 56, 76/77★
Sulzer, Friedrich Gabriel, 310
Sumatran bamboo rat (*Rhizomys sumatrensis*), **347**
Sumatran thick-spined porcupine (*Thecurus sumatrae*), 400★, **413**, 414m
Sun squirrels (*Heliosciurus*), 262
Svolba, 207
Swinhoe's or Asiatic striped squirrel (*Callosciurus swinhoei*), 230★, **260**
Synaptomys (Bog lemmings), **323**
—*borealis* (Northern bog lemming), **323**, 323m
—*cooperi* (Southern bog lemming), **323f**, 323m
Tachyoryctes (African mole rats), **347f**, 347m
—*daemon* (Tanzanian mole rat), 300★, **347**
—*ruandae* (Rwanda mole rat), **347f**
Tadarida brasiliensis, 147m
—*mexicana* (Mexican free-tailed bat), 91, **147**, 147m
—*condylura*, 138★
—*limbata*, **147**
—*teniotis*, 138★, **147**, 147m
Tamandua or collared anteater (*Tamandua tetradactyla*), 173★, **179f**, 180★, 180m, 190★
Tamandua tetradactyla (Tamandua or collared anteater), 173★, **179f**, 180★, 180m, 190★
Tamias (Chipmunks, or eastern chipmunks), **241**
—*alpinus* (Alpine chipmunk), **242**, 242m
—*amoenus* (Yellow pine chipmunk), **242**, 242m
—*minimus* (Least chipmunk), **242f**, 243m
—*quadrivittatus* (Colorado chipmunk), **242**, 243m
—*striatus* (Eastern chipmunk), **242**, 242m
—*townsendi* (Townsend's chipmunk), **242**, 243m
Tamiasciurini (Red squirrels or chickarees), 205, **244**
Tamiasciurus douglasii (Chickaree), **244**, 244m
—*hudsonicus* (North American red squirrel), 229★, **244**, 244m
Tanzanian mole rat (*Tachyoryctes daemon*), 300★, **347**
Taphozous (Tomb bats), 123★, **129**
—*mauritianus*, **129**
—*nudiventris*, **129**
—*perforatus*, **129**
Tarbagan, see Mongolian marmot
Tasmacetus shepherdi (Shepherd's beaked whale), 494, **502**
Tate, 351
Tatera (Large gerbils), **343**
—*indica* (Indian or large Indian gerbil), **343**, 346
—*vicina* (East African gerbil), 300★, **343**
Taterillus emini (Small naked-sole gerbil), **343**
Teichert, 241
Telle, 359
Temminck flying lemur (*Cynocephalus temminckii*), 64f, 65m, 66★
Tent-building bat (*Uroderma bilbobatum*), 113, 127★, **142**
Texas kangaroo rat (*Dipodomys elator*), **273**, 274m
Texas pocket gopher (*Geomys bursarius*), **270f**, 270m, 297★
Thecurus (Thick-spined porcupines), 408, **412**
—*crassispinis* (Bornean thick-spined porcupine), 414m, 415★
—*pumilis* (Philippine thick-spined porcupine), 414m
—*sumatrae* (Sumatran thick-spined porcupine), 400★, **413**, 414m
Thenius, 204
Theridomyidae, 198
Thick-spined porcupines (*Thecurus*), 408, **412**
Thick-tailed three-toed jerboa (*Stylodipus*), 404
Thin-spined porcupine (*Chaetomys subspinosus*), 409★, **453**, 453m
Thirteenlined ground squirrel (*Citellus tridecemlineatus*), 227m, **232**, 242
Thoma, 56
Thomomys, 269
—*bottae* (Western pocket gopher), **271**
—*mazama* (Oregon pocket gopher), 271m
—*monticola* (Sierra pocket gopher), 271m
—*talpoides* (Northern pocket gopher), 271m
—*umbrinus* (Southern pocket gopher), 271m
Three-banded armadillo (*Tolypeutes matacus*), 159, **160**, 160★, 160m, 189★
Three-fingered sloths (*Bradypus*), 169, **171f**
Thrinacodus albicauda, 429m
Thryonomyidae (African cane rats), 214/215★, **419**
Thryonomys, 419
—*gregorianus* (Lesser cane rat), **419**
—*swinderianus* (Great cane rta), **419**, 426★
Thyroptera discifera (Honduran disk-winged bat), **144**
—*tricolor* (Spix's disk-winged bat), **144**
Thyropteridae (Disk-winged bats), **143**
Tibetan hamster (*Cricetulus lama*), 306m, **307**
Tien Shan marmot (*Marmota menzbieri*), 206m
Tiflow-Potapow, 235
Tinbergen, N., 62
Tirler, Hermann, 168, 172
Togo mole rat (*Cryptomys zechi*), **417**, 426★
Tolypeutes matacus (La Plata three-banded armadillo), **160f**, 160m
—*muriei*, **160**
—*tricinctus* (Three-banded armadillo), 159, **160**, 160★, 160m, 189★
Tolypeutini, 159
Tomb bats (*Taphozous*), 123★, **129**
Tome's long-eared bat (*Lonchorrhina aurita*), 126★, **140**
Tonga fruit bat (*Pteropus tonganus*), 86★
Toothed whales (Odontoceti), 458, 466★, **493ff**
Townsend's chipmunk (*Tamias townsendi*), **242**, 243m
Townsend's ground squirrel (*Citellus townsendi*), 232m
Townsend's vole (*Microtus townsendi*), 338m
Trachops airrhosus, 126★
Transbaikal zokor (*Myospalax aspalax*), 300★, **316**
Tratz, 206
Tree squirrels (Sciurini), 205, **244f**
Triaenops persicus (Persian leaf-nosed bat), **139**
Trichys (Long-tailed porcupines), 407, **412**
—*fasciculata* (Malayan long-tailed porcupines), **412**
—*lipura* (Bornean long-tailed porcupine), 409★, **412**

Trident leaf-nosed bat (*Asellia tridens*), 80, 125★, **139**
Tristram's jird (*Meriones tristrami*), 300★, 345
Trogontherium, 286
True Alpine marmot, see European marmot
True dolphins (Delphiniae), 508, **514**f
True dormice (Gliridae), 214/215★, 296, **388**f
True jumping mice (Zapodinae), **402**
True's beaked whale (*Mesophlodon mirus*), 480★
Tscherskia triton (Ratlike hamster), 306m, **307**
Tschudi, J. J. von, 441
Tube-nosed fruit bats (Nyctimeninae), 93, **108**
Tuco-tucos (Ctenomyidae), 214/215★, 420, **422**
Tundra vole (*Microtus oeconomus*), **336**, 337m, 339
Tungic race human, 56, 76/77★
Tursiops aduncus (Red Sea bottle-nosed dolphin,), 480★, 515
—*truncatus* (Bottle-nosed dolphin), 458★, 480★, 483★, 499★, 500★, 510, 510★, 511★, **515**f
Tuza (*Orthogeomys grandis*), **271**, 271m
Two-fingered sloths (*Choloepus*) ,**169**
Two-lined sheath-tailed bat (*Saccopteryx bilineata*), 113, **122**, 123★
Two-toed or silky anteater (*Cyclops didactylus*), **180**f, 181m, 190★
Tylonycteris pachypus (Flat-headed bat), **144**
Tympanoctomys barrerae, 421m
Types of humans, 56
—europoid, 56
—mongoloid, 56
—negroid, 56
Typhlomys cinereus (Chinese pygmy dormouse), 382★, 388, **398**, 401
Typical squirrels (Sciuridae), 201, **203**ff, 214/215★
Tyrannus balaenarum, see *Orcinus orca*

Uganda harsh-furred mouse (*Lophuromys woosnami*), **377**
Uhlenhaut, 406
Uinta ground squirrel (*Citellus armatus*), 233m
Ulmer, 227
Upper Amazonian porcupine (*Echinoprocta rufescens*), 409★, 453m, **454**
Ural beaver (*Castor fiber pohlei*), **276**, 278
Uroderma bilobatum (Tent-building bat), 113, 127★, **142**
Uromanis, 183
Uromys, **352**, 352m
—*anak* (Giant naked-tailed rat), **353**
—*caudimaculatus*, 371★

Vampire bat (*Desmodus rotundus*), 86★, 115f, 128★, **143**
Vampyrum spectrum (Linné's false vampire bat), 112, 116f, **140**
Vandeleuria, 355m
—*oleracea* (Long-tailed climbing mouse), **355**
Veddid race human, 57
Verschuren, 102
Veselovsky, Zdenek, 170, 308
Vesper mouse (*Calomys musculinus*), **303**
Vespertilio murinus (Particolored bat), 121, 145m
Vespertilionidae, **144**
Vespertilionoidea, **143**
Vianden, 249
Vietinghoff-Reisch, von, 390
Villa, de, 227
Vinogradov, B. S., 233, 405
Viscacha (*Lagostomus maximus*), **433**f, 433m, 438★
Viscacha rat (*Octomys mimax*), **421**, 421m
Vlasov, 345f
Vlei rat (*Otomys irroratus*), 372★, **383**
Voles (*Microtus*), **336**
Voronzow, 316
Vosseler, 414, 416
Votsotsa (*Hypogeomys antimena*), **318**

Wagner, Helmuth O., 271, 275
Wahlberg's epauletted fruit bat (*Epomophorus wahlbergi*), 87★, **103**
Walker, Ernest P., 266f, 346, 429, 431, 449f
Washington ground squirrel (*Citellus washingtoni*), 232m
Water bat (*Myotis daubentoni*), 105/106★, 121, 134★, 141m, **144**
Water rats (*Hydromys*), 385m, **387**
Water voles (*Arvicola*), 331f
Waterhouse, 315, 385
Webb, John, 222
Weddell, James, 471
Weihe, 197
Weinert, 51
Welwitsch's bat (*Myotis welwitschii*), 113
Wendt, Herbert, 447
West African brush-tailed porcupine (*Atherurus africanus*), 409★, **412**, 413m
West African sun squirrel (*Heliosciurus gambianus*), **263**
West African white dolphin (*Sousa teuszi*), 480★, **507**f
Western African striped squirrel (*Funisciurus lemniscatus*), **261**
Western gray squirrel (*Sciurus griseus*), 254m
Western harvest mouse (*Reithrodontomys megalotis*), **302**, 303m
Western house mouse (*Mus musculus domesticus*), **362**, 363m
Western jumping mouse (*Zapus princeps*), 403m
Western pocket gopher (*Thomomys bottae*), **271**
Western red-backed vole (*Clethrionomys occidentalis*), 329m
Western voles (*Alticola*), **330**
Wettstein, 403
Whales (Cetacea), **457**ff
Whaling, 469ff, 473★
Wharton, H., 65f
Whiskered bat (*Myotis mystacinus*), 142m, 144
White dolphins (*Sotalia*), **507**
White flag dolphin, see Chinese river dolphin
White whale or belukha or beluga (*Delphinapterus leucas*), 480★, 484★, **504**f, 505m
White whales and narwhals (Monodontidae), 458, 466★, 493, **503**
White-beaked dolphin (*Lagenorhynchus albirostris*), **516**
White-bellied tree pangolin (*Manis tricuspis*), **182**, 182m, 185, 187★, 199★
White-footed mouse (*Peromyscus leucopus*), **301**f, 302m
White-headed dolphin (*Cephalorhynchus albifrons*), **517**
Whitehead's dwarf squirrel (*Nannosciurus whiteheadi*), **260**
Whitehead's harpy fruit bat (*Harpyioncteris whiteheadii*), **110**, 110m
White-lipped dolphin (*Cephalorhynchus heavisidii*), **517**
White-sided dolphin (*Lagenorhynchus acutus*), 480★, 497★, **516**
White-tailed antelope squirrel (*Citellus leucurus*), 227m, **233**
White-tailed porcupine (*Hystrix leucura*), 410★, **413**, 415m
White-tailed prairie dog (*Cynomys gunnisoni*), **225**f, 225m
White-tailed rat (*Mystomys albicaudatus*), **316**
White-throated wood rat (*Neotoma albigula*), **304**, 304m
Whitman, C. O., 59
Wickler, W., 62
Wild cavy (*Cavia aperea tschudii*), 436★, **441**
Wilson's palm squirrel (*Epixerus wilsoni*), **262**
Wolfe, John B., 37
Wolffson's mountain chinchilla (*Lagidium wolffsoni*), 440m
Wood, A. E., 198
Wood and field mice (*Apodemus*), **365**, 366m
Wood lemming (*Myopus schisticolor*), **324**, 324m
Wood rats (*Neotoma*), **304**
Woodchuck (*Marmota monax*), **206**, 206m, 222ff, 239★
Wood-Jones, F., 49, 352, 354
Woodland jumping mouse (*Napaeozapus insignis*), 399★, **402**, 403m
Woolly flying squirrel (*Eupetaurus cinereus*), **265**
Woolly prehensile-tailed porcupine (*Coendou insiduosus*), **454**
Wrinkled-faced bat (*Centurio senex*), 86★, 128★, 140m, **142**
Wroughton, 96
Wyoming pocket mouse (*Perognathus fasciatus*), 272m

Xenarthra, 149
Xerini (Bristly ground squirrels), 205, **243**
Xerus, 243
—*inauris* (South African ground squirrel), 230★
—*rutilis* (African ground squirrel), 213★

Yellow pine chipmunk (*Tamias amoenus*), **242**, 242m
Yellow pocket gopher (*Cratogeomys castanops*), 271m, **272**
Yellow-bellied marmot (*Marmota flaviventris*), **206**, 206m
Yellow-cheeked vole (*Microtus xanthognathus*), 338m
Yellow-necked field mouse (*Apodemus flavicollis*), 321/322★, **366**f. 367m, 371★
Yellow-nosed wood rat (*Sigmodon ochrognathus*), 304m
Yellow-shouldered bats (*Sturnirinae*), **141**
Young, 226

Zapodidae (Jumping mice), 214/215★, 296, **402**
Zapodinae (True jumping mice), **402**
Zapus, 402
—*hudsonius* (Meadow jumping mouse), 399★, **402**, 403m
—*princeps* (Western jumping mouse), 403m
—*trinotatus* (Pacific jumping mouse), 403m
Zenkerella, 289
—*insignis* (Flightless scaly-tailed squirrel), **290**f, 294m
Zenkerellinae (Flightless scaly-tailed squirrels), 289
Zenker's fruit bat (*Scotonycteris zenkeri*), 87★, **107**
Zenker's small flying squirrel (*Idiurus zenkeri*), **291**, 293, 294m
Zeuglodon, see *Basilosaurus*
Zimmermann, K., 194, 196, 339f, 396
"*Zinjanthropus boisei*," see *Australopithecus robostus*
Ziphiidae (Beaked whales), 458, 466★, 493, **496**
Ziphius cavirostris (Cuvier's beaked whale, aka goosebeak whale), 479★, **501**
Zokors (*Myospalax*), **316**
Zon, van, 45f

Abbreviations and Symbols

C, °C .... Celsius, degrees centigrade

C.S.I.R.O. .. Commonwealth Scientific and Industrial Res. Org. (Australia)

f .... following (page)

ff. .... following (pages)

L. .... total length (from tip of nose [bill] to end of tail)

I.R.S.A.C. .. Institute for Scientific Res. in Central Africa, Congo

I.U.C.N. ..... Intern. Union for Conserv. of Nature and Natural Resources

BH .... body height

HRL .... head-rump length (from nose to base of tail or end of body)

N, N- ...... North, Northern, North-

NE, NE- .... Northeast, Northeastern, Northeast-

E, E- .... East, Eastern, East-

S, S- .... South, Southern, South-

TL .... Tail length

SE, SE- ..... Southeast, Southeastern, Southeast-

SW, SW- .. Southwest, Southwestern, Southwest-

W, W- ...... West, Western, West-

♂ .... male

♂♂ .... males

♀ .... female

♀♀ .... females

♂♀ .... pair

+ .... extinct

$\frac{2 \cdot 1 \cdot 2 \cdot 3}{2 \cdot 1 \cdot 2 \cdot 3}$ ...... tooth formula, explanation in Volume 10

▻ .... following (opposite page) color plate

▻▻ .... Color plate or double color plate on the page following the next

▻▻▻ ...... Third color plate or double color plate (etc.)

◊ ◊ .... Endangered species and subspecies